T0181535

Lecture Notes in Computer Science 13421

More information about this series at https://link.springer.com/bookseries/558

Bohan Li · Lin Yue · Chuanqi Tao · Xuming Han · Diego Calvanese · Toshiyuki Amagasa (Eds.)

Web and Big Data

6th International Joint Conference, APWeb-WAIM 2022 Nanjing, China, November 25–27, 2022 Proceedings, Part I

Editors
Bohan Li
Nanjing University of Aeronautics
and Astronautics
Nanjing, China

Chuanqi Tao
Nanjing University of Aeronautics
and Astronautics
Nanjing, China

Diego Calvanese
Free University of Bozen-Bolzano
Bolzano, Italy

Lin Yue
Newcastle University
Callaghan, NSW, Australia

Xuming Han
Jinan University
Guangzhou, China

Toshiyuki Amagasa
University of Tsukuba
Tsukuba, Japan

ISSN 0302-9743 ISSN 1611-3349 (electronic)
Lecture Notes in Computer Science
ISBN 978-3-031-25157-3 ISBN 978-3-031-25158-0 (eBook)
https://doi.org/10.1007/978-3-031-25158-0

This Springer imprint is published by the registered company Springer Nature Switzerland AG
The registered company address is: Gewerbestrasse 11, 6330 Cham, Switzerland

Preface

These volumes (LNCS 13421–13423) contain the proceedings of the 6th Asia-Pacific Web (APWeb) and Web-Age Information Management (WAIM) Joint Conference on Web and Big Data (APWeb-WAIM). Researchers and practitioners from around the world came together at this leading international forum to share innovative ideas, original research findings, case study results, and experienced insights in the areas of the World Wide Web and big data, thus covering web technologies, database systems, information management, software engineering, knowledge graphs, recommender system and big data.

The 6th APWeb-WAIM conference was held in Nanjing during 25–27 November 2022. As an Asia-Pacific flagship conference focusing on research, development, and applications in relation to Web information management, APWeb-WAIM builds on the successes of APWeb and WAIM. Previous APWeb events were held in Beijing (1998), Hong Kong (1999), Xi'an (2000), Changsha (2001), Xi'an (2003), Hangzhou (2004), Shanghai (2005), Harbin (2006), Huangshan (2007), Shenyang (2008), Suzhou (2009), Busan (2010), Beijing (2011), Kunming (2012), Sydney (2013), Changsha (2014), Guangzhou (2015), and Suzhou (2016). And previous WAIM events were held in Shanghai (2000), Xi'an (2001), Beijing (2002), Chengdu (2003), Dalian (2004), Hangzhou (2005), Hong Kong (2006), Huangshan (2007), Zhangjiajie (2008), Suzhou (2009), Jiuzhaigou (2010), Wuhan (2011), Harbin (2012), Beidaihe (2013), Macau (2014), Qingdao (2015), and Nanchang (2016). The combined APWeb-WAIM conferences have been held in Beijing (2017), Macau (2018), Chengdu (2019), Tianjin (02020), and Guangzhou (2021). With the ever-growing importance of appropriate methods in these data-rich times and the fast development of web-related technologies, we believe APWeb-WAIM will become a flagship conference in this field.

The high-quality program documented in these proceedings would not have been possible without the authors who chose APWeb-WAIM for disseminating their findings. APWeb-WAIM 2022 received a total of 297 submissions and, after the double-blind review process (each paper received at least three review reports), the conference accepted 75 regular papers (including research and industry track) (acceptance rate 25.25%), 45 short research papers, and 5 demonstrations. The contributed papers address a wide range of topics, such as big data analytics, advanced database and web applications, data mining and applications, graph data and social networks, information extraction and retrieval, knowledge graphs, machine learning, recommender systems, security, privacy and trust, and spatial and multimedia data. The technical program also included keynotes by Ihab F. Ilyas Kaldas, Aamir Cheema, Chengzhong Xu, Lei Chen, and Haofen Wang. We are grateful to these distinguished scientists for their invaluable contributions to the conference program.

We would like to express our gratitude to all individuals, institutions, and sponsors that supported APWeb-WAIM 2022. We are deeply thankful to the Program Committee members for lending their time and expertise to the conference. We also would like to acknowledge the support of the other members of the organizing committee. All of them helped to make APWeb-WAIM 2022 a success. We are grateful for the guidance of the Honorary Chairs (Zhiqiu Huang), Steering Committee representative (Yanchun Zhang) and the General Co-chairs (Aoying Zhou, Wojciech Cellary and Bing Chen) for their guidance and support. Thanks also go to the Workshop Co-chairs (Shiyu Yang and Saiful Islam), Tutorial Co-chairs (Xiang Zhao, Wenqi Fan and Ji Zhang), Demo Co-chairs (Jianqiu Xu and Travers Nicolas), Industry Co-chairs (Chen Zhang Hosung Park), Publication Co-chairs (Chuanqi Tao, Lin Yue and Xuming Han), and Publicity Co-chairs (Yi Cai, Siqiang Luo and Weitong Chen).

We hope the attendees enjoyed the exciting program of APWeb-WAIM 2022 as documented in these proceedings.

November 2022

Toshiyuki Amagasa
Diego Calvanese
Xuming Han
Bohan Li
Chuanqi Tao
Lin Yue

Organization

General Chairs

Aoying Zhou	East China Normal University, China
Bing Chen	Nanjing University of Aeronautics and Astronautics, China
Wojciech Cellary	WSB University, Poland

Steering Committee

Aoying Zhou	East China Normal University, China
Divesh Srivastava	AT&T Research Institute, USA
Jian Pei	Simon Fraser University, Canada
Lei Chen	Hong Kong University of Science and Technology, China
Lizhu Zhou	Tsinghua University, China
Masaru Kitsuregawa	University of Tokyo, Japan
Mingjun Huang	National University of Singapore, Singapore
Tamor Özsu	University of Waterloo, Canada
Xiaofang Zhou	University of Queensland, Australia
Yanchun Zhang	Victoria University, Australia

Program Committee Chairs

Diego Calvanese	Free University of Bozen-Bolzano, Italy
Toshiyuki Amagasa	University of Tsukuba, Japan
Bohan Li	Nanjing University of Aeronautics and Astronautics, China

Publication Chairs

Chuanqi Tao	Nanjing University of Aeronautics and Astronautics, China
Lin Yue	University of Newcastle, Australia
Xuming Han	Jinan University, China

Program Committee

Alex Delis	University of Athens, Greece
An Liu	Soochow University, China
Aviv Segev	University of South Alabama, USA
Bangbang Ren	National University of Defense Technology, China
Baokang Zhao	NUDTCS, China
Baoning Niu	Taiyuan University of Technology, China
Bin Guo	Northwestern Polytechnical University, China
Bin Cui	Peking University, China
Bin Xia	Nanjing University of Posts and Telecommunications, China
Bin Zhao	Nanjing Normal University, China
Bolong Zheng	Huazhong University of Science and Technology, China
Byung Suk Lee	University of Vermont, USA
Carson Leung	University of Manitoba, Canada
Cheng Long	Nanyang Technological University, Singapore
Chengliang Chai	Tsinghua University, China
Cheqing Jin	East China Normal University, China
Chuanqi Tao	Nanjing University of Aeronautics and Astronautics, China
Cuiping Li	Renmin University of China, China
Dechang Pi	Nanjing University of Aeronautics and Astronautics, China
Dejun Teng	Shandong University, China
Demetrios Zeinalipour-Yazti	University of Cyprus, Cyprus
Derong Shen	Northeastern University, China
Dong Li	Liaoning University, China
Donghai Guan	Nanjing University of Aeronautics and Astronautics, China
Fabio Valdés	FernUniversität in Hagen, Germany
Fei Chen	Shenzhen University, China
Genoveva Vargas-Solar	CNRS, France
Giovanna Guerrini	University of Genoa, Italy
Guanfeng Liu	Macquarie University, Australia
Guodong Long	University of Technology Sydney, Australia
Guoqiong Liao	Jiangxi University of Finance & Economics, China
Haibo Hu	Hong Kong Polytechnic University, China
Hailong Liu	Northwestern Polytechnical University, China
Haipeng Dai	Nanjing University, China

Haiwei Pan	Harbin Engineering University, China
Haiwei Zhang	Nankai University, China
Hancheng Wang	Nanjing University, China
Hantao Zhao	Southeast University, China
Harry Kai-Ho Chan	University of Sheffield, UK
Hiroaki Ohshima	University of Hyogo, Japan
Hong Chen	Renmin University, China
Hongzhi Wang	Harbin Institute of Technology, China
Hongzhi Yin	University of Queensland, Australia
Hua Dai	Nanjing University of Posts and Telecommunications, China
Hua Wang	Victoria University, Australia
Hui Li	Xiamen University, China
Javier A. Espinosa-Oviedo	University of Lyon, France
Ji Zhang	University of Southern Queensland, Australia
Jia Yu	Washington State University, USA
Jiajie Xu	Soochow University, China
Jiali Mao	East China Normal University, China
Jian Yin	Sun Yat-Sen University, China
Jiangtao Cui	Xidian University, China
Jianqiu Xu	Nanjing University of Aeronautics and Astronautics, China
Jianxin Li	Deakin University, Australia
Jilin Hu	Aalborg University, Denmark
Jing Jiang	University of Technology Sydney, Australia
Jizhou Luo	Harbin Institute of Technology, China
Jun Gao	Peking University, China
Junhu Wang	Griffith University, Australia
Junjie Yao	East China Normal University, China
K. Selçuk Candan	Arizona State University, USA
Kai Zeng	Microsoft, USA
Kai Zheng	University of Electronic Science and Technology of China, China
Kazutoshi Umemoto	University of Tokyo, Japan
Krishna Reddy P.	International Institute of Information Technology, India
Ladjel Bellatreche	ISAE-ENSMA, France
Le Sun	Nanjing University of Information Science and Technology, China
Lei Duan	Sichuan University, China
Lei Zou	Peking University, China
Leong Hou U.	University of Macau, China

Liang Hong	Wuhan University, China
Lin Li	Wuhan University of Technology, China
Lin Yue	University of Queensland, Australia
Liyan Zhang	Yanshan University, China
Lizhen Cui	Shandong University, China
Lu Chen	Zhejiang University, China
Luyi Bai	Northeastern University, China
Makoto Onizuka	Osaka University, Japan
Maria Damiani	University of Milan, Italy
Maria Luisa Damiani	University of Milan, Italy
Markus Endres	University of Augsburg, Germany
Meng Wang	Southeast University, China
Miao Xu	University of Queensland, Australia
Mingzhe Zhang	University of Queensland, Australia
Min-Ling Zhang	Southeast University, China
Mirco Nanni	ISTI-CNR Pisa, Italy
Mizuho Iwaihara	Waseda University, Japan
My T. Thai	University of Florida, USA
Nicolas Travers	Léonard de Vinci Pôle Universitaire, France
Peer Kroger	Christian-Albrechts-Universität Kiel, Germany
Peiquan Jin	University of Science and Technology of China, China
Peisen Yuan	Nanjing Agricultural University, China
Peng Peng	Hunan University, China
Peng Wang	Fudan University, China
Qian Zhou	Nanjing University of Posts and Communications, China
Qilong Han	Harbin Engineering University, China
Qing Meng	Southeast University, China
Qing Xie	Wuhan University of Technology, China
Qiuyan Yan	China University of Mining and Technology, China
Qun Chen	Northwestern Polytechnical University, China
Quoc Viet Hung Nguyen	Griffith University, Australia
Reynold Cheng	University of Hong Kong, China
Rong-Hua Li	Beijing Institute of Technology, China
Saiful Islam	Griffith University, Australia
Sanghyun Park	Yonsei University, Korea
Sanjay Madria	Missouri University of Science & Technology, USA
Sara Comai	Politecnico di Milano
Sebastian Link	University of Auckland, New Zealand

Sen Wang	University of Queensland, Australia
Senzhang Wang	Central South University, China
Shanshan Li	Yanshan University, China
Shaofei Shen	University of Queensland, Australia
Shaojie Qiao	Chengdu University of Information Technology, China
Shaoxu Song	Tsinghua University, China
Sheng Wang	Wuhan University, China
Shengli Wu	Jiangsu University, China
Shiyu Yang	Guangzhou University, China
Shuai Xu	Nanjing University of Aeronautics and Astronautics, China
Shuigeng Zhou	Fudan University, China
Siqiang Luo	Nanyang Technological University, Singapore
Tianrui Li	Southwest Jiaotong University, China
Tingjian Ge	University of Massachusetts, USA
Tung Kieu	Aalborg University, Denmark
Victor Sheng	Texas Tech University, USA
Wang Lizhen	Yunnan University, China
Wei Hu	Nanjing University, China
Wei Shen	Nankai University, China
Wei Song	Wuhan University, China
Weiguo Zheng	Fudan University, China
Weijun Wang	University of Goettingen, Germany
Weitong Chen	Adelaide University, Australia
Wcn Zhang	Wuhan University, China
Wenqi Fan	Hong Kong Polytechnic University, China
Wolf-Tilo Balke	TU Braunschweig, Germany
Xiang Lian	Kent State University, USA
Xiang Zhao	National University of Defense Technology, China
Xiangfu Meng	Liaoning Technical University, China
Xiangguo Sun	Chinese University of Hong Kong, China
Xiangliang Zhang	University of Notre Dame, USA
Xiangmin Zhou	RMIT University, Australia
Xiao Pan	Shijiazhuang Tiedao University, China
Xiao Zhang	Shandong University, China
Xiaohui Yu	Shandong University, China
Xiaohui Tao	University of Southern Queensland, Australia
Xiaoli Wang	Xiamen University, China
Xiaowang Zhang	Tianjin University, China
Xie Xiaojun	Nanjing Agricultural University, China

Xin Cao	University of New South Wales, Australia
Xin Wang	Tianjin University, China
Xingquan Zhu	Florida Atlantic University, USA
Xuelin Zhu	Southeast University, China
Xujian Zhao	Southwest University of Science and Technology, China
Xuming Han	Jinan University, China
Xuyun Zhang	Macquarie University, Australia
Yaokai Feng	Kyushu University, Japan
Yajun Yang	Tianjin University, China
Yali Yuan	University of Göttingen, Germany
Yanda Wang	University of Bristol, UK
Yanfeng Zhang	Northeastern University, China
Yanghua Xiao	Fudan University, China
Yang-Sae Moon	Kangwon National University, Korea
Yanhui Gu	Nanjing Normal University, China
Yanjun Zhang	Deakin University, Australia
Yasuhiko Morimoto	Hiroshima University, Japan
Ye Liu	Nanjing Agricultural University, China
Yi Cai	South China University of Technology, China
Yingxia Shao	BUPT, China
Yong Tang	South China Normal University, China
Yong Zhang	Tsinghua University, China
Yongpan Sheng	Chongqing University, China
Yongqing Zhang	Chengdu University of Information Technology, China
Youwen Zhu	Nanjing University of Aeronautics and Astronautics, China
Yu Gu	Northeastern University, China
Yu Liu	Huazhong University of Science and Technology, China
Yu Hsuan Kuo	Amazon, USA
Yuanbo Xu	Jilin University, China
Yue Tan	University of Technology Sydney, Australia
Yunjun Gao	Zhejiang University, China
Yuwei Peng	Wuhan University, China
Yuxiang Zhang	Civil Aviation University of China, China
Zakaria Maamar	Zayed University, UAE
Zhaokang Wang	Nanjing University of Aeronautics and Astronautics, China
Zheng Zhang	Harbin Institute of Technology, China
Zhi Cai	Beijing University of Technology, China

Zhiqiang Zhang	Zhejiang University of Finance and Economics, China
Zhixu Li	Soochow University, China
Zhuoming Xu	Hohai University, China
Zhuowei Wang	University of Technology Sydney, Australia
Ziqiang Yu	Yantai University, China
Zouhaier Brahmia	University of Sfax, Tunisia

Additional Reviewers

Bo Tang	Southern University of Science and Technology, China
Fang Wang	Hong Kong Polytechnic University, China
Genggeng Liu	Fuzhou University, China
Guan Yuan	China University of Mining and Technology, China
Jiahao Zhang	Hong Kong Polytechnic University, China
Jinguo You	Kunming University of Science and Technology, China
Long Yuan	Nanjing University of Science and Technology, China
Paul Bao	Nagoya University, Japan
Philippe Fournier-Viger	Shenzhen University, China
Ruiyuan Li	Chongqing University, China
Shanshan Yao	Taiyuan University of Technology, China
Xiaofeng Ding	Huazhong University of Science and Technology, China
Yaoshu Wang	Shenzhen University, China
Yunpeng Chai	Renmin University of China, China
Zhiwei Zhang	Beijing Institute of Technology, China

Contents – Part I

Big Data Analytic and Management

An Improved Yin-Yang-Pair Optimization Algorithm Based on Elite Strategy and Adaptive Mutation Method for Big Data Analytics 3
Hui Xu, Mingchao Ding, Yanping Lu, and Zhiwei Ye

MCSketch: An Accurate Sketch for Heavy Flow Detection and Heavy Flow Frequency Estimation 20
Jie Lu, Hongchang Chen, and Zhen Zhang

The Influence of the Student's Online Learning Behaviors on the Learning Performance 28
Chunying Li, Junjie Yao, Zhikang Tang, Yong Tang, and Yanchun Zhang

A Context Model for Personal Data Streams 37
Fausto Giunchiglia, Xiaoyue Li, Matteo Busso, and Marcelo Rodas-Britez

ACF2: Accelerating Checkpoint-Free Failure Recovery for Distributed Graph Processing 45
Chen Xu, Yi Yang, Qingfeng Pan, and Hongfu Zhou

Death Comes But Why: An Interpretable Illness Severity Predictions in ICU 60
Shaofei Shen, Miao Xu, Lin Yue, Robert Boots, and Weitong Chen

Specific Emitter Identification Based on ACO-XGBoost Feature Selection 76
Jianjun Cao, Chumei Gu, Baowei Wang, Yuxin Xu, and Mengda Wang

NED-GNN: Detecting and Dropping Noisy Edges in Graph Neural Networks 91
Ming Xu, Baoming Zhang, Jinliang Yuan, Meng Cao, and Chongjun Wang

A Temporal-Context-Aware Approach for Individual Human Mobility Inference Based on Sparse Trajectory Data 106
Shuai Xu, Donghai Guan, Zhuo Ma, and Qing Meng

Unsupervised Online Concept Drift Detection Based on Divergence and EWMA 121
Qilin Fan, Chunyan Liu, Yunlong Zhao, and Yang Li

A Knowledge-Enabled Customized Data Modeling Platform Towards Intelligent Police Applications ... 135
Tiexin Wang, Hong Jiang, Huihui Zhang, and Xinhua Yan

Integrated Bigdata Analysis Model for Industrial Anomaly Detection via Temporal Convolutional Network and Attention Mechanism ... 150
Chenze Yang, Bing Chen, and Hai Deng

Advanced Database and Web Applications

WSNet: A Wrapper-Based Stacking Network for Multi-scenes Classification of DApps ... 163
Yu Wang, Gang Xiong, Zhen Li, Mingxin Cui, Gaopeng Gou, and Chengshang Hou

HaCache: A Hybrid Adaptive Cache for Persistent Memory Based Key-Value Systems ... 180
Lixiao Cui, Gang Wang, Yusen Li, and Xiaoguang Liu

An Energy-efficient Routing Protocol Based on Two-Layer Clustering in WSNs ... 190
Feng Xu, Qian Ni, and PanFei Liu

An Energy-efficient Routing Protocol Based on DPC-MND Clustering in WSNs ... 199
Feng Xu, Jing Liu, and PanFei Liu

Cloud Computing and Crowdsourcing

Lightweight Model Inference on Resource-Constrained Computing Nodes in Intelligent Surveillance Systems ... 209
Zhuohang Wang, Yunfeng Zhao, Yong Wang, Li Yan, Zhicheng Liu, Chao Qiu, Xiaofei Wang, and Qinghua Hu

Trajectory Optimization for Propulsion Energy Minimization of UAV Data Collection ... 224
Juan Xu, Di Wu, Jiabin Yuan, Hu Liu, Xiangping Zhai, and Kun Liu

Robust Clustered Federated Learning with Bootstrap Median-of-Means ... 237
Ming Xie, Jie MA, Guodong Long, and Chengqi Zhang

SAPMS: A Semantic-Aware Privacy-Preserving Multi-keyword Search Scheme in Cloud ... 251
Qian Zhou, Hua Dai, Zheng Hu, Yuanlong Liu, and Geng Yang

Task Assignment with Spatio-temporal Recommendation in Spatial Crowdsourcing 264
Chen Zhu, Yue Cui, Yan Zhao, and Kai Zheng

DE-DQN: A Dual-Embedding Based Deep Q-Network for Task Assignment Problem in Spatial Crowdsourcing 280
Yucen Gao, Dejun Kong, Haipeng Dai, Xiaofeng Gao, Jiaqi Zheng, Fan Wu, and Guihai Chen

Dynamic Vehicle-Cargo Matching Based on Adaptive Time Windows 296
Chong Feng, Jiajun Liao, Jiali Mao, Jiaye Liu, Ye Guo, and Weining Qian

Microservice Workflow Modeling for Affinity Scheduling to Improve the QoS 313
Yingying Wen, Guanjie Cheng, ShuiGuang Deng, and Jianwei Yin

Data Mining

SynBERT: Chinese Synonym Discovery on Privacy-Constrain Medical Terms with Pre-trained BERT 331
Lingze Zeng, Chang Yao, Meihui Zhang, and Zhongle Xie

Improving Motor Imagery Intention Recognition via Local Relation Networks 345
Lin Yue, Yuxuan Zhang, Xiaowei Zhao, Zhe Zhang, and Weitong Chen

EAS-GCN: Enhanced Attribute-Aware and Structure-Constrained Graph Convolutional Network 357
Jijie Zhang, Yan Yang, Shaowei Yin, and Zhengqi Wang

Mining Frequent Patterns with Counting Quantifiers 372
Yanxiao He, Xin Wang, Yuji Sha, Xueyan Zhong, and Yu Fang

MST-GNN: A Multi-scale Temporal-Enhanced Graph Neural Network for Anomaly Detection in Multivariate Time Series 382
Zefei Ning, Zhuolun Jiang, Hao Miao, and Li Wang

Category Constraint Spatial Keyword Preference Query Based Spatial Pattern Matching 391
Yi Li and Shaopeng Wang

Many-to-Many Pair Trading 399
Yingying Wang, Xiaodong Li, Pangjing Wu, and Haoran Xie

Proximity Preserving Graph Convolutional Networks 408
Zhenglin Yu, Hui Yan, and Ling Guo

Discovering Prevalent Weighted Co-Location Patterns on Spatial Data Without Candidates 417
Vanha Tran, Lizhen Wang, Muquan Zou, and Hongmei Chen

A Deep Looking at the Code Changes in OpenHarmony 426
Yuqing Niu, Lu Zhou, and Zhe Liu

GADAL: An Active Learning Framework for Graph Anomaly Detection 435
Wenjing Chang, Jianjun Yu, and Xiaojun Zhou

Fake News Detection Based on the Correlation Extension of Multimodal Information 443
Yanqiang Li, Ke Ji, Kun Ma, Zhenxiang Chen, Jin Zhou, and Jun Wu

SPSTN: Sequential Precoding Spatial-Temporal Networks for Railway Delay Prediction 451
Junfeng Fu, Limin Zhong, Changyu Li, Hao Li, Chuiyun Kong, and Jie Shao

Graph Data and Social Networks

Mining Periodic *k*-Clique from Real-World Sparse Temporal Networks 461
Zebin Ren, Hongchao Qin, Rong-Hua Li, Yongheng Dai, Guoren Wang, and Yanhui Li

LSM-Subgraph: Log-Structured Merge-Subgraph for Temporal Graph Processing 477
Jingyuan Ma, Zhan Shi, Shang Liu, Wang Zhang, Yutong Wu, Fang Wang, and Dan Feng

ForGen: Autoregressive Generation of Sparse Graphs with Preferential Forest 495
Yao Shi, Yu Liu, and Lei Zou

Iterative Deep Graph Learning with Local Feature Augmentation for Network Alignment 511
Jiuyang Tang, Zhen Tan, Hao Guo, Xuqian Huang, Weixin Zeng, and Huang Peng

OntoCA: Ontology-Aware Caching for Distributed Subgraph Matching 527
Yuzhou Qin, Xin Wang, Wenqi Hao, Pengkai Liu, Yanyan Song, and Qingpeng Zhang

A Social-Aware Deep Learning Approach for Hate-Speech Detection 536
George C. Apostolopoulos, Panagiotis Liakos, and Alex Delis

Community Detection Based on Deep Dual Graph Autoencoder 545
Zhiyuan Jiang, Kai Xu, Zhixiang Wu, Zhenyu Wang, and Hui Zhu

SgIndex: An Index Structure Supporting Multiple Graph Queries 553
Shibiao Zhu, Yuzhou Huang, Zirui Zhang, and Xiaolin Qin

Correction to: Lightweight Model Inference on Resource-Constrained Computing Nodes in Intelligent Surveillance Systems C1
Zhuohang Wang, Yunfeng Zhao, Yong Wang, Li Yan, Zhicheng Liu, Chao Qiu, Xiaofei Wang, and Qinghua Hu

Author Index 563

Contents – Part II

Graph Data and Social Networks

Accelerating Hypergraph Motif Counting Based on Hyperedge Relations 3
Yuhang Su, Yu Gu, Yang Song, and Ge Yu

Next POI Recommendation Method Based on Category Preference and Attention Mechanism in LBSNs 12
Xueying Wang, Yanheng Liu, Xu Zhou, Zhaoqi Leng, and Xican Wang

Information Extraction and Retrieval

ToSA: A Top-Down Tree Structure Awareness Model for Hierarchical Text Classification 23
Deji Zhao, Bo Ning, Shuangyong Song, Chao Wang, Xiangyan Chen, Xiaoguang Yu, and Bo Zou

Clause Fusion-Based Emotion Embedding Model for Emotion-Cause Pair Extraction 38
Zhiwei Li, Guozheng Rao, Li Zhang, Xin Wang, Qing Cong, and Zhiyong Feng

Accelerated Algorithms for α-Happiness Query 53
Min Xie

SummScore: A Comprehensive Evaluation Metric for Summary Quality Based on Cross-Encoder 69
Wuhang Lin, Shasha Li, Chen Zhang, Bin Ji, Jie Yu, Jun Ma, and Zibo Yi

DASH: Data Aware Locality Sensitive Hashing 85
Zongyuan Tan, Hongya Wang, Ming Du, and Jie Zhang

AOPSS: A Joint Learning Framework for Aspect-Opinion Pair Extraction as Semantic Segmentation 101
Chengwei Wang, Tao Peng, Yue Zhang, Lin Yue, and Lu Liu

Dual Graph Convolutional Networks for Document-Level Event Causality Identification 114
Yang Liu, Xiaoxia Jiang, Wenzheng Zhao, Weiyi Ge, and Wei Hu

Self-supervised Label-Visual Correlation Hashing for Multi-label Image Retrieval 129
Yu Liu, Yanzhao Xie, Jingkuan Song, Rukai Wei, and Ke Zhou

Shallow Diffusion Motion Model for Talking Face Generation from Speech 144
Xulong Zhang, Jianzong Wang, Ning Cheng, Edward Xiao, and Jing Xiao

Hierarchical Clustering and Measure for Tourism Profiling 158
Sonia Djebali, Quentin Gabot, and Guillaume Guerard

Non-stationary Dueling Bandits for Online Learning to Rank 166
Shiyin Lu, Yuan Miao, Ping Yang, Yao Hu, and Lijun Zhang

High-Order Correlation Embedding for Large-Scale Multi-modal Hashing 175
Junfeng An, Yingjian Li, Zheng Zhang, Yongyong Chen, and Guangming Lu

A Self-training Approach for Few-Shot Named Entity Recognition 183
Yudong Qian and Weiguo Zheng

Knowledge Graph

Reasoning Path Generation for Answering Multi-hop Questions Over Knowledge Graph 195
Yuxuan Xiang, Jiajun Wu, Tiexin Wang, Meng Wang, Tianlun Dai, Gaoxu Wang, Shidong Xu, and Jing Li

E^3L: Experience Enhanced Entity Linking for Question Answering Over Knowledge Graphs 210
Zhirong Hou, Meiling Wang, Min Li, and Ying Li

M2R: From Mathematical Models to Resource Description Framework 225
Chenxin Zou, Xiaodong Li, Pangjing Wu, and Haoran Xie

KEP-Rec: A Knowledge Enhanced User-Item Relation Prediction Model for Personalized Recommendation 239
Lisha Wu, Daling Wang, Shi Feng, Yifei Zhang, and Ge Yu

Probing the Impacts of Visual Context in Multimodal Entity Alignment 255
Yinghui Shi, Meng Wang, Ziheng Zhang, Zhenxi Lin, and Yefeng Zheng

Incorporating Prior Type Information for Few-Shot Knowledge Graph Completion 271
Siyu Yao, Tianzhe Zhao, Fangzhi Xu, and Jun Liu

Answering Why-Not Questions on GeoSPARQL Queries ... 286
Yin Li and Bixin Li

Multi-Information-Enhanced Knowledge Embedding in Hyperbolic Space ... 301
Jiajun Wu, Qian Zhou, Yuxuan Xiang, Tianlun Dai, Hua Dai, Hao Wen, and Qun Yang

Knowledge Graph Entity Alignment Powered by Active Learning ... 315
Jiayi Pan and Weiguo Zheng

POSE: A Positional Embedding Model for Knowledge Hypergraph Link Prediction ... 323
Zirui Chen, Xin Wang, Chenxu Wang, and Zhao Li

Machine Learning

TraVL: Transferring Pre-trained Visual-Linguistic Models for Cross-Lingual Image Captioning ... 341
Zhebin Zhang, Peng Lu, Dawei Jiang, and Gang Chen

From Less to More: Common-Sense Semantic Perception Benefits Image Captioning ... 356
Feng Chen, Xinyi Li, Jintao Tang, Shasha Li, and Ting Wang

MACNet: Multi-Attention and Context Network for Polyp Segmentation ... 369
Xiuzhen Hao, Haiwei Pan, Kejia Zhang, Chunling Chen, Xiaofei Bian, and Shuning He

Transformer-Based Representation Learning on Temporal Heterogeneous Graphs ... 385
Longhai Li, Lei Duan, Junchen Wang, Guicai Xie, Chengxin He, Zihao Chen, and Song Deng

SCBERT: Single Channel BERT for Chinese Spelling Correction ... 401
Hong Gao, Xuezhen Tu, and Donghai Guan

Improving Robustness of Medical Image Diagnosis System by Using Multi-loss Hybrid Adversarial Function with Heuristic Projection ... 415
Chufan Cheng and Fang Chen

SSCG: Spatial Subcluster Clustering Method by Grid-Connection ... 430
Yihang Zhang, Xuming Han, Limin Wang, Weitong Chen, and Linliang Guo

Shap-PreBiNT: A Sentiment Analysis Model Based on Optimized Transformer 444
Kejun Zhang, Liwen Feng, and Xinying Yu

Universum-Inspired Supervised Contrastive Learning 459
Aiyang Han and Songcan Chen

A Task-Aware Attention-Based Method for Improved Meta-Learning 474
Yue Zhang, Xinxing Yang, Feng Zhu, Yalin Zhang, Meng Li, Qitao Shi, Longfei Li, and Jun Zhou

One-Stage Deep Channels Attention Network for Remote Sensing Images Object Detection 483
Jinyun Tang, Wenzhen Zhang, Guixian Zhang, Rongjiao Liang, and Guangquan Lu

Multi-objective Global Path Planning for UAV-assisted Sensor Data Collection Using DRL and Transformer 492
Rongtao Zhang, Jie Hao, Ran Wang, Hai Deng, and Hui Wang

SAC-PER: A Navigation Method Based on Deep Reinforcement Learning Under Uncertain Environments 501
Xinmeng Wang, Lisong Wang, Shifan Shen, and Lingling Hu

MAFT: An Image Super-Resolution Method Based on Mixed Attention and Feature Transfer 511
Xin Liu, Jing Li, Yuanning Cui, Wei Zhu, and Luhong Qian

Automatic Report Generation Method based on Multiscale Feature Extraction and Word Attention Network 520
Xin Du, Haiwei Pan, Kejia Zhang, Shuning He, Xiaofei Bian, and Weipeng Chen

Dictionary-Induced Manifold Learning for Incomplete Multi-modal Fusion 529
Bingliang Xu, Haizhou Ye, Zheng Zhang, Daoqiang Zhang, and Qi Zhu

BSAM: A BERT-Based Model with Statistical Information for Personality Prediction 538
Bin Xu, Tongqing Wang, Kening Gao, and Zhaowu Zhang

A Combined Model Based on GRU with Mahalanobis Distance for Oil Price Prediction 546
Shichen Zhai and Zongmin Ma

Author Index 555

Contents – Part III

Query Processing and Optimization

MSP: Learned Query Performance Prediction Using MetaInfo and Structure of Plans 3
Honghao Liu, Zhiyong Peng, Zhe Zhang, Huan Jiang, and Yuwei Peng

Computing Online Average Happiness Maximization Sets over Data Streams 19
Zhiyang Hao and Jiping Zheng

A Penetration Path Planning Algorithm for UAV Based on Laguerre Diagram 34
Dan Li, Xu Chen, Panpan Ding, and Jiani Huang

Scheduling Strategy for Specialized Vehicles Based on Digital Twin in Airport 42
Hongying Zhang, Minglong Liu, Chang Liu, Qian Luo, and Zhaoxin Chen

Recommender Systems

Graph-Based Sequential Interpolation Recommender for Cold-Start Users 57
Aoran Li, Jiajun Wu, Shuai Xu, Yalei Zang, Yi Liu, Jiayi Lu, Yanchao Zhao, Gaoxu Wang, Qing Meng, and Xiaoming Fu

Self-guided Contrastive Learning for Sequential Recommendation 72
Hui Shi, Hanwen Du, Yongjing Hao, Victor S. Sheng, Zhiming Cui, and Pengpeng Zhao

Time Interval Aware Collaborative Sequential Recommendation with Self-supervised Learning 87
Chenrui Ma, Li Li, Rui Chen, Xi Li, and Yichen Wang

A2TN: Aesthetic-Based Adversarial Transfer Network for Cross-Domain Recommendation 102
Chenghua Wang and Yu Sang

MORO: A Multi-behavior Graph Contrast Network for Recommendation 117
Weipeng Jiang, Lei Duan, Xuefeng Ding, and Xiaocong Chen

Eir-Ripp: Enriching Item Representation for Recommendation with Knowledge Graph 132
Kaiwen Li, Chunyang Ye, and Jinghui Wang

User Multi-behavior Enhanced POI Recommendation with Efficient and Informative Negative Sampling 149
Hanzhe Li, Jingjing Gu, Haochao Ying, Xinjiang Lu, and Jingyuan Yang

Neighborhood Constraints Based Bayesian Personalized Ranking for Explainable Recommendation 166
Tingxuan Zhang, Li Zhu, and Jie Wang

Hierarchical Aggregation Based Knowledge Graph Embedding for Multi-task Recommendation 174
Yani Wang, Ji Zhang, Xiangmin Zhou, and Yang Zhang

Mixed-Order Heterogeneous Graph Pre-training for Cold-Start Recommendation 182
Wenzheng Sui, Xiaoxia Jiang, Weiyi Ge, and Wei Hu

MISRec: Multi-Intention Sequential Recommendation 191
Rui Chen, Dongxue Chen, Riwei Lai, Hongtao Song, and Yichen Wang

MARS: A Multi-task Ranking Model for Recommending Micro-videos 199
Jiageng Song, Beihong Jin, Yisong Yu, Beibei Li, Xinzhou Dong, Wei Zhuo, and Shuo Zhou

Security, Privacy, and Trust and Blockchain Data Management and Applications

How to Share Medical Data Belonging to Multiple Owners in a Secure Manner 217
Changsheng Zhao, Wei Song, and Zhiyong Peng

LAP-BFT: Lightweight Asynchronous Provable Byzantine Fault-Tolerant Consensus Mechanism for UAV Network Trusted Systems 232
Lingjun Kong, Bing Chen, and Feng Hu

A Secure Order-Preserving Encryption Scheme Based on Encrypted Index 247
Haobin Chen, Ji Liang, and Xiaolin Qin

A Deep Reinforcement Learning-Based Approach for Android GUI Testing 262
Yuemeng Gao, Chuanqi Tao, Hongjing Guo, and Jerry Gao

RoFL: A Robust Federated Learning Scheme Against Malicious Attacks 277
Ming Wei, Xiaofan Liu, and Wei Ren

FDP-LDA: Inherent Privacy Amplification of Collapsed Gibbs Sampling via Group Subsampling 292
Tao Huang, Hong Chen, and Suyun Zhao

Multi-modal Fake News Detection Use Event-Categorizing Neural Networks 301
Buze Zhao, Hai Deng, and Jie Hao

Unified Proof of Work: Delegating and Solving Customized Computationally Bounded Problems in a Privacy-Preserving Way 309
Yue Fu, Qingqing Ye, Rong Du, and Haibo Hu

TSD3: A Novel Time-Series-Based Solution for DDoS Attack Detection 318
Yifan Han, Yang Du, Shiping Chen, He Huang, and Yu-E Sun

Executing Efficient Retrieval Over Blockchain Medical Data Based on Exponential Skip Bloom Filter 334
Weiliang Ke, Chengyue Ge, and Wei Song

IMPGA: An Effective and Imperceptible Black-Box Attack Against Automatic Speech Recognition Systems 349
Luopu Liang, Bowen Guo, Zhichao Lian, Qianmu Li, and Huiyun Jing

FD-Leaks: Membership Inference Attacks Against Federated Distillation Learning 364
Zilu Yang, Yanchao Zhao, and Jiale Zhang

Spatial and Multi-media Data

When Self-attention and Topological Structure Make a Difference: Trajectory Modeling in Road Networks 381
Guoying Zhu, Yu Sang, Wei Chen, and Lei Zhao

Block-Join: A Partition-Based Method for Processing Spatio-Temporal Joins 397
Ting Li and Jianqiu Xu

RN-Cluster: A Novel Density-Based Clustering Approach for Road Network Partition 412
Yingying Ding and Jianqiu Xu

Fine-Grained Urban Flow Inferring via Conditional Generative Adversarial Networks ... 420
Xv Zhang, Yuanbo Xu, Ying Li, and Yongjian Yang

TDCT: Transport Destination Calibration Based on Waybill Trajectories of Trucks ... 435
KaiXuan Zhu, Tao Wu, Wenyi Shen, Jiali Mao, and Yue Shi

OESCPM: An Online Extended Spatial Co-location Pattern Mining System ... 441
Jinpeng Zhang, Lizhen Wang, Wenlu Lou, and Vanha Tran

gTop: An Efficient SPARQL Query Engine ... 446
Yuqi Zhou, Lei Zou, and Gang Cao

Multi-SQL: An Automatic Multi-model Data Management System ... 451
Yu Yan, Hongzhi Wang, Yutong Wang, Zhixin Qi, Jian Ma, Chang Liu, Meng Gao, Hao Yan, Haoran Zhang, and Ziming Shen

Demonstration on Unblocking Checkpoint for Fault-Tolerance in Pregel-Like Systems ... 456
Zhenhua Yang, Yi Yang, and Chen Xu

Author Index ... 461

Big Data Analytic and Management

An Improved Yin-Yang-Pair Optimization Algorithm Based on Elite Strategy and Adaptive Mutation Method for Big Data Analytics

Hui Xu(✉), Mingchao Ding, Yanping Lu, and Zhiwei Ye

School of Computer Science, Hubei University of Technology, Wuhan 430068, China
xuhui@hbut.edu.cn

Abstract. In order to analyze and explore the potential value of big data more effectively, intelligent optimization algorithms are applied to this field increasingly. However, data often has the characteristics of high dimension for big data analytics. As the dimension of data increases, the performance of the optimization algorithm degrades dramatically. The Yin-Yang-Pair Optimization (YYPO) is a lightweight single-objective optimization algorithm, which has stronger competitive performance compared with other algorithms and has significantly lower computational time complexity. Nevertheless, it also suffers from the drawbacks of easily falling into local optimum and elitism deficiency, resulting in unsatisfactory performance on high-dimensional problems. To further improve the performance of YYPO in solving high-dimensional problems for big data analytics, this paper proposes an improved Yin-Yang-Pair Optimization based on elite strategy and adaptive mutation method, namely CM-YYPO. First, the crossover operator using an elite strategy is introduced to record the individual optimal generated by point P_1 as elite. After the splitting stage, the elite to cross-disturb point P_1 is utilized to improve the global search performance of YYPO. Subsequently, the mutation operator with an improved adaptive mutation method is used to mutate point P_1 to improve the local search performance of YYPO. The proposed CM-YYPO is evaluated by 28 test functions used in the Single-Objective Real Parameter Algorithm competition of the Congress on Evolutionary Computation 2013. The performance of CM-YYPO is compared with YYPO, YYPO-SA1, YYPO-SA2, A-YYPO, and four other single-objective optimization algorithms with superior performance, which are Salp Swarm Algorithm, Sine Cosine Algorithm, Grey Wolf Optimizer and Whale Optimization Algorithm. The experimental results show that, the proposed CM-YYPO can achieve more stable optimization capability and higher computational accuracy on high-dimensional problems, which prospects a promising idea to solve high-dimensional problems in the field of big data analytics.

Keywords: Big data analytics · Yin-Yang-pair optimization · Elite strategy · Adaptive mutation method

B. Li et al. (Eds.): APWeb-WAIM 2022, LNCS 13421, pp. 3–19, 2023.
https://doi.org/10.1007/978-3-031-25158-0_1

1 Introduction

In the era of big data, more and more information is being recorded, collected and stored. Through big data analytics, people can discover and summarize the laws hidden behind the data, which can improve the efficiency of the system, as well as predict and judge future trends [1]. To fully exploit the value of big data and solve a series of technical problems, a growing number of intelligent optimization algorithms are applied to this field [2–4].

Yin-Yang-Pair Optimization (YYPO) is proposed by Punnathanam et al. [5] in 2016, which is a lightweight single-objective optimization algorithm. The advantages of YYPO are few setup parameters and low time complexity. However, the original YYPO has the following disadvantages. (1) The tendency to premature convergence when solving complex and difficult optimization problems [6]; (2) The quality of candidate solutions is poor in the exploration process due to the lack of elites [7].

Some scholars have proposed improved algorithms for the basic YYPO algorithm. Punnathanam et al. [8] proposed Adaptive Yin-Yang-Pair Optimization (A-YYPO), which modified the probability of one-way splitting and D-way splitting in YYPO into a function related to the dimension of the problem. Xu et al. [6] introduced chaos search in YYPO to improve global exploration ability and backward learning strategy to improve local exploitation ability. Li et al. [9] proposed Yin-Yang-Pair Optimization-Simulated Annealing (YYPO-SA) based on simulated annealing strategy, which is further divided into YYPO-SA1 and YYPO-SA2 by switching strategy.

In the field of big data analytics, the data often has high dimensional characteristics [10]. As the dimension increases, the search space becomes dramatically larger. However, none of these improvement methods consider the problem of elite missing. The lack of information interaction with the global optimal solution makes the search ability of YYPO significantly less efficient. Therefore, this paper proposes an improved Yin-Yang-Pair Optimization algorithm based on elite strategy and adaptive mutation method (CM-YYPO).

Proposed CM-YYPO aims to integrate crossover operator and mutation operator based on YYPO, with details as follows. (1) In the crossover stage, an elitist strategy is adopted to enhance the global optimization ability of the algorithm. (2) In the mutation stage, an improved adaptive mutation probability is introduced to improve the ability of the algorithm to jump out of local optimum. (3) Use only D-way splitting to improve the performance of YYPO for solving high-dimensional problems.

The rest of this paper is organized as follows. Section 2 describes the YYPO algorithm. Section 3 presents the CM-YYPO algorithm proposed in this paper. Section 4 gives the experimental analysis of CM-YYPO by benchmark function test results. Section 5 concludes the work.

2 Yin-Yang-Pair Optimization (YYPO)

The algorithmic idea of YYPO comes from Chinese philosophy, which uses two points (P_1 and P_2) representing Yin and Yang to balance local exploitation and global exploration. The points P_1 and P_2 are the centers of the hypersphere volumes in the variable space explored with radii δ_1 and δ_2. δ_1 tends to decrease periodically, while δ_2 is the opposite.

YYPO normalizes all decision variables (between 0 and 1) and scales them appropriately according to the variable boundaries when performing fitness evaluation. I_{min}, I_{max}, α are three user-defined parameters, where I_{min} and I_{max} are the minimum value and maximum value of the archive count I, and α is the scaling factor of the radius.

2.1 Splitting Stage

Points P_1 and P_2 will go through the splitting stage in turn, while entering their corresponding radii. The splitting mode is decided equally by the following two methods.

One-way splitting:

$$\begin{cases} S_j^j = S^j + r \times \delta \\ S_{D+j}^j = S^j - r \times \delta \end{cases}, j = 1, 2, 3, \cdots, D \tag{1}$$

D-way splitting:

$$\begin{cases} S_k^j = S^j + r \times \left(\delta/\sqrt{2}\right), B_k^j = 1 \\ S_k^j = S^j - r \times \left(\delta/\sqrt{2}\right), B_k^j = 0 \end{cases}, k = 1, 2, 3, \cdots, 2D, j = 1, 2, 3, \cdots, D \tag{2}$$

In Formula (1) and Formula (2), S is a matrix consisting of $2D$ identical copies of the point P, which has a size of $2D \times D$; B is a matrix of $2D$ random binary strings of length D (each binary string in B is unique); k denotes the point number and j denotes the decision variable number that will be modified; r is a random number between 0 and 1.

In both splitting methods, random values in the interval [0, 1] are used to correct for out-of-bounds variables. Then, the generated $2D$ new points are evaluated for fitness separately, and the point that undergoes the splitting stage is replaced by the one with the best fitness.

2.2 Archive Stage

The archive stage begins after the required number of archiving updates are completed. In this phase, the archive contains $2I$ points, which correspond to the two points (P_1 and P_2) added in each update before the splitting stage. If the best point in the archive is fitter than point P_1, then swap it with point P_1. After that, if the best point in the archive is fitter than point P_2, then P_2 is replaced by it. Next, update the search radii (δ_1 and δ_2) using the following equations.

$$\begin{cases} \delta_1 = \delta_1 - (\delta_1/\alpha) \\ \delta_2 = \delta_2 + (\delta_2/\alpha) \end{cases} \tag{3}$$

At the end of the archive stage, the archive is cleared, a new value of I for the number of archive updates is randomly generated in the specified range $[I_{min}, I_{max}]$, and the archive counter is set to 0.

3 Improved Yin-Yang Pair Optimization (CM-YYPO)

3.1 Crossover Operator with Elite Strategy

The splitting method of YYPO determines the quality of new solutions generated by the algorithm. Before entering the archive stage, point P_1 is restricted only by radius and lacks information interaction with the global optimal solution. When solving high-dimensional problems, the optimization effect is not satisfactory due to the lack of elites.

Therefore, in this paper, the individual optimal position of P_1 is taken as the elite and recorded with P_{best}. P_{best} is initialized to the best point of P_1 and P_2. At the beginning of the iteration, points P_2 and P_1 are exchanged frequently, and although there is a case that P_2 is better than P_1, there is a high probability that the exchanged points are only locally optimal. Update P_{best} only in the crossover and mutation stages in order to delay the tendency to fall into local optima.

During each iteration, the point P_1 is cross-perturbed by P_{best} to try to produce a higher quality solution. There are various methods of crossover, of which the formula for uniform arithmetic crossover is:

$$Y = Y_a + \lambda \times (Y_b - Y_a) \tag{4}$$

In Formula (4), Y_{a} and Y_{b} represent the two parent chromosomes; Y is the offspring chromosome generated by crossover; λ is a constant belonging to [0, 1].

This crossover method limits the search range and prevents the offspring from exploring more around the parents. In order to enable the offspring to jump out of the solution region enclosed by P_{best} and P_1, an extended arithmetic crossover method [11] is used in this paper to expand the range of λ. At the same time, the influence of the fitness value is added to the crossover stage to improve the chance of generating high-quality offspring. In this work, the crossover operation is performed according to the following equation to produce the children ${P_1}^*$.

$$P_1^* = P_1 + \lambda \times F\left(f_{P_1}, f_{P_{best}}\right) \times (P_{best} - P_1) \tag{5}$$

In Formula (5), f_{P1} and f_{Pbest} are the fitness values of point P_1 and point P_{best}, respectively. $F(f_{\text{P1}}, f_{\text{Pbest}})$ is calculated according to Formula (6).

$$F\left(f_{P_1}, f_{P_{best}}\right) = \begin{cases} 1, f_{P_1} - f_{P_{best}} > 0 \\ -1, f_{P_1} - f_{P_{best}} < 0 \\ rand(-1, 1), f_1 - f_b = 0 \end{cases} \tag{6}$$

At the end of the crossover stage, the parents are updated sequentially according to the fitness values of the offspring. For example, if the point ${P_1}^*$ is fitter than P_1, replace P_1 with ${P_1}^*$, otherwise do not update. In addition, n is used to record the number of times P_{best} has not been updated. If P_{best} is updated, n needs to be reset to 0; otherwise $n = n + 1$.

3.2 Mutation Operator with Adaptive Mutation Probability

If P_{best} is updated in the crossover stage, the algorithm proceeds directly to the next phase. Otherwise, the point P_1 mutates with a certain probability. In this paper, the adaptive mutation probability formula shown below is used.

$$p_m = c \times \frac{1}{1 + n^{-(1-t/T)}} \tag{7}$$

In Formula (7), c is the control parameter, which determines the range of variation probability; t is the number of iterations; n indicates the number of times P_{best} has not been updated. As is indicated in Formula (7), the mutation probability p_m increases as n increases. This is because, as the number of times P_{best} un-updated increases, the likelihood of the algorithm falling into a local optimum also becomes larger. Further, as t increases, the p_m gradually decreases to ensure the convergence of the algorithm.

When the mutation probability is satisfied, the point P_1 is mutated according to the following equation.

$$P_1^{**j} = P_1^j + \delta_1 \times rand(-1, 1), j = 1, 2, 3, \cdots, D \tag{8}$$

In Formula (8), $P_1{}^{**}$ denotes the point after the variation and j denotes the decision variable number that will be modified.

Update P_1 and P_{best} using the update method in the crossover stage. The difference is that if P_{best} is un-updated, n remains unchanged; otherwise, let $n = 0$.

3.3 Splitting Method

In YYPO, one-way splitting and D-way splitting are equally utilized with a probability of 0.5. However, it was found that multidirectional search is more effective than one-way search in high-dimensional problems [8]. Therefore, only D-way splitting is used in this paper.

3.4 Computational Complexity

In this section, the symbol O is used to analyze the computational complexity in a single iteration. CM-YYPO performs crossover stage in each iteration. However, mutation occurs only under conditions that satisfy the probability of mutation. Here, the time complexity analysis is performed in the worst case, with crossover and mutation in each iteration.

In crossover and mutation, there are four identical operations: (a) calculate to obtain a new point; (b) scale the new point for fitness evaluation; (c) update P_1; (d) update P_{best}. In addition, the probability of mutation needs to be calculated at the beginning of the mutation. The computational complexities are shown as follows.

$$O(crossover) = O(D + D + 1 + 1) = O(2D+2)=O(D) \tag{9}$$

$$O(mutation)=O(1 + D + D + 1+1)=O(2D+3)=O(D) \tag{10}$$

Since the computational complexity of original YYPO is $O(D^2)$, the computational complexity of CM-YYPO is as follow.

$$\begin{aligned} O(CM-YYPO) &= O(O(YYPO) + O(crossover) + O(mutation)) \\ &= O(D^2 + D + D) \\ &= O(D^2) \end{aligned} \tag{11}$$

Thus, CM-YYPO is a polynomial-time algorithm with the same time complexity as YYPO for a single iteration. So, proposed CM-YYPO is also a lightweight optimization algorithm.

4 Validation and Discussion

4.1 Experimental Results

To test the performance of CM-YYPO, 28 test functions used in the single-objective real parameter algorithm competition of IEEE Congresson Evolutionary Computation (CEC) 2013 [12] are adopted in this paper. There are 5 unimodal functions (f_1-f_5), 15 basic multimodal functions (f_6-f_{20}) and 8 composition functions (f_{21}-f_{28}). The parameter range of the test functions is uniformly set to [-100, 100].

In this paper, CM-YYPO, YYPO, YYPO-SA1, YYPO-SA2, A-YYPO and four representative single-objective optimization algorithms: Salp Swarm Algorithm (SSA), Sine Cosine Algorithm (SCA), Grey Wolf Optimizer (GWO), and Whale Optimization Algorithm (WOA) are compared for performance in the experiments. The three user-defined parameters ($I_{\min}$, $I_{\max}$, α) are set according to literature [5]. The parameters for CM-YYPO were not rigorously tuned, and were intuitively set as: $\lambda = 0.4$; $c = 0.01$. In addition, the parameters in the simulated annealing mechanism of YYPO-SA1 and YYPO-SA2 are set according to literature [9]. SSA, SCA, GWO and WOA all adopted the parameter values set in the original papers, and the population number was uniformly set to 100.

For fairness, all algorithms are run independently 51 times. The number of iterations for YYPO and its improved algorithms is set to 2500. Iteration times (*iter*) for the remaining algorithms is related to dimension(D) and population number (*pop*):

$$iter = 10^4 * D / pop \tag{12}$$

4.2 Performance Analysis of CM-YYPO

Tables 1–3 give the test results of all algorithms for 28 test functions in 10, 30, and 50 dimensions, respectively. The terms "mean" and "std. Dev." are used to refer to the mean and standard deviation of the error obtained over the 51 runs. A smaller mean value indicates a better average performance of the algorithm, and a smaller standard deviation means that the algorithm is more stable.

As can be seen from Table 1, CM-YYPO ranked first for eleven times on the 10-dimensional test functions, including two unimodal functions, seven multimodal functions, and two composition functions. The performance in the multimodal functions is significantly better than other comparison algorithms, and slightly worse than YYPO in the composition function. CM-YYPO ranked first the most times. YYPO-SA1 was next, but only ranked first for five times. It shows that the improved strategy proposed in this work, solving the low-dimensional for problem is greatly improved.

As can be shown from Table 2, CM-YYPO ranked first for seventeen times on the 30-dimensional test functions, with three unimodal functions, eight multimodal functions, and six composition functions. It has the best optimization performance on all three types of test functions, especially the solution capability of the composite functions is significantly improved. It is shown that CM-YYPO has obvious superiority in solving high-dimensional problems.

As can be demonstrated from Table 3, CM-YYPO ranked first for seventeen times on the 50-dimensional test functions, including four unimodal functions, eight multimodal functions, and five composition functions. For further increases in dimensionality, CM-YYPO still maintains a superb search capability compared to other algorithms.

The crossover stage of CM-YYPO leads to unsatisfactory performance for f_5 and f_{11}. However, the solving ability on f_5 is significantly improved at dimension 50. It can be concluded that the performance of CM-YYPO is stable and significant for high-dimensional problems.

To verify the convergence performance of CM-YYPO, eleven representative test functions are selected. Figure 1 shows the convergence curves of all algorithms in 50 dimensions. The horizontal coordinates of the graph indicate the percentage of the maximum number of iterations, and the vertical coordinates indicate the average fitness value over the current number of iterations.

It can be seen from Fig. 1 that CM-YYPO can converge faster on the unimodal functions (f_2, f_3 and f_4); on the multimodal functions (f_7 f_9 f_{12} f_{18} and f_{20}), the tendency to fall into local optimum is slowed down. For the more difficult composition functions (f_{23} f_{25} and f_{27}), the convergence speed and solution accuracy of CM-YYPO are also better than other comparative algorithms.

In summary, the candidate solutions can effectively improve the quality under the perturbation of elites during the search process. And the mutation operation on the candidate solutions also increases the probability of jumping out of the local optimum.

Table 1. Results of CM-YYPO on the 10-dimensional test functions

F		CM-YYPO	YYPO	YYPO-SA1	YYPO-SA2	A-YYPO	SSA	SCA	GWO	WOA
f_1	Mean	3.13E-09	1.44E-09	1.10E-09	1.28E-09	9.65E-10	**4.68E-10**	5.11E + 02	7.87E + 00	2.63E-02
	Std. Dev	1.15E-09	3.61E-09	1.09E-09	2.21E-09	1.79E-09	**1.52E-10**	1.57E + 02	3.18E + 01	2.30E-02

(continued)

Table 1. (*continued*)

F		CM-YYPO	YYPO	YYPO-SA1	YYPO-SA2	A-YYPO	SSA	SCA	GWO	WOA
f_2	Mean	**4.11E + 04**	6.30E + 04	5.26E + 04	7.17E + 04	6.87E + 04	2.05E + 05	4.21E + 06	1.47E + 06	3.26E + 06
	Std. Dev	**4.26E + 04**	4.86E + 04	4.79E + 04	5.88E + 04	5.58E + 04	1.98E + 05	2.05E + 06	2.12E + 06	2.18E + 06
f_3	Mean	**4.83E + 04**	6.54E + 05	1.96E + 06	1.18E + 06	7.98E + 05	8.56E + 06	6.39E + 08	3.11E + 07	1.54E + 09
	Std. Dev	**1.21E + 05**	1.49E + 06	7.34E + 06	3.98E + 06	1.68E + 06	3.58E + 07	2.87E + 08	4.89E + 07	2.05E + 09
f_4	Mean	8.25E + 01	8.36E + 01	**5.17E + 01**	8.83E + 01	1.03E + 02	1.08E + 03	4.68E + 03	6.86E + 03	2.79E + 04
	Std. Dev	1.21E + 02	1.49E + 02	**6.22E + 01**	1.15E + 02	1.42E + 02	8.59E + 02	1.82E + 03	3.29E + 03	1.59E + 04
f_5	Mean	3.74E-04	9.03E-06	**8.93E-06**	1.03E-05	1.27E-05	1.11E-03	1.11E + 02	2.40E + 01	1.25E + 01
	Std. Dev	9.81E-05	8.88E-06	9.44E-06	**6.23E-06**	3.00E-05	2.53E-04	3.90E + 01	1.54E + 01	1.14E + 01
f_6	Mean	**4.30E + 00**	7.06E + 00	5.96E + 00	8.00E + 00	6.26E + 00	9.57E + 00	3.92E + 01	1.77E + 01	3.41E + 01
	Std. Dev	4.86E + 00	**4.33E + 00**	4.72E + 00	1.10E + 01	4.76E + 00	1.71E + 01	2.50E + 01	1.83E + 01	3.57E + 01
f_7	Mean	**8.27E-01**	2.92E + 00	3.55E + 00	3.72E + 00	4.44E + 00	2.42E + 01	4.06E + 01	8.66E + 00	8.92E + 01
	Std. Dev	**2.63E + 00**	3.94E + 00	5.27E + 00	6.17E + 00	8.55E + 00	1.99E + 01	1.02E + 01	7.87E + 00	4.28E + 01
f_8	Mean	2.03E + 01	2.03E + 01	2.04E + 01	2.04E + 01	2.03E + 01	2.03E + 01	2.03E + 01	2.03E + 01	**2.03E + 01**
	Std. Dev	**6.57E-02**	7.00E-02	6.71E-02	7.25E-02	7.66E-02	7.86E-02	7.39E-02	8.01E-02	8.12E-02
f_9	Mean	**1.61E + 00**	3.19E + 00	2.83E + 00	3.25E + 00	2.97E + 00	4.37E + 00	7.88E + 00	3.56E + 00	7.57E + 00
	Std. Dev	1.03E + 00	1.17E + 00	1.18E + 00	1.30E + 00	1.55E + 00	1.78E + 00	**9.01E-01**	1.19E + 00	1.16E + 00
f_{10}	Mean	**1.07E-01**	2.31E-01	2.19E-01	2.33E-01	2.47E-01	3.00E-01	7.18E + 01	1.12E + 01	8.24E + 00
	Std. Dev	**6.17E-02**	1.23E-01	1.01E-01	1.25E-01	1.48E-01	1.64E-01	2.62E + 01	1.34E + 01	4.16E + 00
f_{11}	Mean	1.16E + 01	4.62E-02	8.71E-02	**3.36E-02**	3.95E-02	2.47E + 01	5.20E + 01	1.11E + 01	6.18E + 01
	Std. Dev	5.48E + 00	2.34E-01	2.71E-01	**1.70E-01**	1.95E-01	1.17E + 01	8.31E + 00	6.94E + 00	2.16E + 01
f_{12}	Mean	**1.29E + 01**	2.08E + 01	2.06E + 01	2.02E + 01	2.14E + 01	2.27E + 01	5.44E + 01	1.67E + 01	7.66E + 01
	Std. Dev	**5.64E + 00**	8.52E + 00	9.07E + 00	9.74E + 00	8.78E + 00	9.16E + 00	8.59E + 00	8.87E + 00	3.18E + 01
f_{13}	Mean	2.32E + 01	2.36E + 01	2.73E + 01	2.47E + 01	2.76E + 01	3.18E + 01	5.30E + 01	**1.96E + 01**	7.06E + 01
	Std. Dev	1.08E + 01	8.55E + 00	1.10E + 01	8.99E + 00	1.16E + 01	1.36E + 01	**8.46E + 00**	9.48E + 00	2.26E + 01

(*continued*)

Table 1. (*continued*)

F		CM-YYPO	YYPO	YYPO-SA1	YYPO-SA2	A-YYPO	SSA	SCA	GWO	WOA
f_{14}	Mean	7.40E + 02	1.62E + 00	1.64E + 00	1.51E + 00	**9.49E-01**	8.15E + 02	1.36E + 03	4.17E + 02	1.04E + 03
	Std. Dev	4.44E + 02	2.24E + 00	2.26E + 00	2.25E + 00	**1.69E + 00**	2.32E + 02	2.25E + 02	1.87E + 02	3.10E + 02
f_{15}	Mean	**4.95E + 02**	6.47E + 02	6.90E + 02	6.51E + 02	6.37E + 02	7.33E + 02	1.44E + 03	5.89E + 02	1.15E + 03
	Std. Dev	3.39E + 02	2.64E + 02	2.36E + 02	2.61E + 02	2.86E + 02	3.11E + 02	**1.72E + 02**	3.63E + 02	3.00E + 02
f_{16}	Mean	1.14E + 00	8.51E-01	6.17E-01	8.06E-01	8.25E-01	**3.14E-01**	1.11E + 00	1.19E + 00	7.55E-01
	Std. Dev	2.06E-01	2.36E-01	2.89E-01	2.30E-01	2.36E-01	2.22E-01	**1.70E-01**	2.00E-01	2.57E-01
f_{17}	Mean	2.41E + 01	**9.14E + 00**	9.76E + 00	9.92E + 00	9.91E + 00	3.56E + 01	5.99E + 01	2.61E + 01	7.53E + 01
	Std. Dev	6.68E + 00	3.50E + 00	2.67E + 00	2.43E + 00	**2.40E + 00**	1.07E + 01	1.02E + 01	5.58E + 00	2.53E + 01
f_{18}	Mean	**2.36E + 01**	2.68E + 01	2.63E + 01	2.44E + 01	2.48E + 01	3.22E + 01	6.64E + 01	3.66E + 01	7.72E + 01
	Std. Dev	**6.78E + 00**	7.96E + 00	9.07E + 00	7.54E + 00	9.57E + 00	8.90E + 00	1.16E + 01	7.37E + 00	2.24E + 01
f_{19}	Mean	9.19E-01	4.53E-01	**4.27E-01**	4.37E-01	4.60E-01	1.39E + 00	8.36E + 00	1.62E + 00	6.10E + 00
	Std. Dev	3.84E-01	1.72E-01	1.76E-01	1.70E-01	**1.47E-01**	7.05E-01	1.99E + 00	7.14E-01	3.57E + 00
f_{20}	Mean	2.62E + 00	2.74E + 00	**2.55E + 00**	2.73E + 00	2.65E + 00	3.16E + 00	3.45E + 00	2.59E + 00	3.73E + 00
	Std. Dev	6.13E-01	5.43E-01	6.80E-01	4.87E-01	5.84E-01	4.66E-01	**3.19E-01**	5.01E-01	5.09E-01
f_{21}	Mean	4.00E + 02	3.51E + 02	3.55E + 02	**3.48E + 02**	3.51E + 02	4.00E + 02	4.07E + 02	3.91E + 02	3.89E + 02
	Std. Dev	6.55E-11	1.07E + 02	8.93E + 01	8.93E + 01	9.25E + 01	**9.16E-12**	1.47E + 01	5.00E + 01	4.72E + 01
f_{22}	Mean	6.19E + 02	**7.43E + 01**	9.83E + 01	9.01E + 01	8.47E + 01	9.51E + 02	1.56E + 03	5.04E + 02	1.30E + 03
	Std. Dev	2.98E + 02	**7.46E + 01**	8.47E + 01	8.14E + 01	7.92E + 01	3.22E + 02	2.27E + 02	2.84E + 02	3.84E + 02
f_{23}	Mean	**6.19E + 02**	8.77E + 02	8.56E + 02	8.56E + 02	8.43E + 02	1.02E + 03	1.68E + 03	7.49E + 02	1.56E + 03
	Std. Dev	2.97E + 02	2.36E + 02	2.72E + 02	3.06E + 02	3.01E + 02	3.29E + 02	**1.80E + 02**	3.62E + 02	2.88E + 02
f_{24}	Mean	1.93E + 02	1.87E + 02	**1.82E + 02**	1.86E + 02	1.90E + 02	2.12E + 02	2.24E + 02	2.10E + 02	2.25E + 02
	Std. Dev	2.74E + 01	3.52E + 01	3.87E + 01	3.75E + 01	3.56E + 01	2.01E + 01	8.73E + 00	4.84E + 00	**4.04E + 00**
f_{25}	Mean	2.00E + 02	1.95E + 02	1.94E + 02	**1.87E + 02**	1.90E + 02	2.13E + 02	2.24E + 02	2.10E + 02	2.24E + 02
	Std. Dev	1.32E + 01	2.72E + 01	2.74E + 01	3.35E + 01	3.17E + 01	1.50E + 01	**2.55E + 00**	6.11E + 00	3.67E + 00

(*continued*)

Table 1. (*continued*)

F		CM-YYPO	YYPO	YYPO-SA1	YYPO-SA2	A-YYPO	SSA	SCA	GWO	WOA
f_{26}	Mean	1.43E + 02	**1.29E + 02**	1.31E + 02	1.32E + 02	1.30E + 02	1.93E + 02	1.97E + 02	1.90E + 02	1.99E + 02
	Std. Dev	3.91E + 01	2.20E + 01	2.41E + 01	2.80E + 01	2.07E + 01	2.29E + 01	**1.15E + 01**	6.79E + 01	5.17E + 01
f_{27}	Mean	**3.06E + 02**	3.34E + 02	3.22E + 02	3.31E + 02	3.44E + 02	5.03E + 02	6.00E + 02	4.06E + 02	5.92E + 02
	Std. Dev	**2.37E + 01**	6.68E + 01	4.68E + 01	5.31E + 01	6.30E + 01	6.60E + 01	3.39E + 01	9.70E + 01	7.97E + 01
f_{28}	Mean	2.92E + 02	**2.72E + 02**	2.89E + 02	2.87E + 02	2.80E + 02	4.08E + 02	6.86E + 02	3.65E + 02	7.29E + 02
	Std. Dev	**3.92E + 01**	9.39E + 01	8.19E + 01	7.91E + 01	6.01E + 01	1.70E + 02	1.04E + 02	1.30E + 02	2.14E + 02

Table 2. Results of CM-YYPO on the 30-dimensional test functions

F		CM-YYPO	YYPO	YYPO-SA1	YYPO-SA2	A-YYPO	SSA	SCA	GWO	WOA
f_1	Mean	2.38E-08	6.16E-09	7.02E-09	9.96E-09	8.62E-08	**4.53E-09**	1.09E + 04	7.41E + 02	3.67E-01
	Std. Dev	9.23E-09	6.53E-09	7.15E-09	1.69E-08	4.02E-08	**8.50E-10**	2.18E + 03	5.68E + 02	1.46E-01
f_2	Mean	**6.37E + 05**	1.45E + 06	1.37E + 06	1.75E + 06	1.17E + 06	2.46E + 06	1.40E + 08	1.78E + 07	3.30E + 07
	Std. Dev	**3.34E + 05**	7.52E + 05	8.22E + 05	9.34E + 05	5.84E + 05	1.39E + 06	3.52E + 07	9.54E + 06	1.24E + 07
f_3	Mean	**7.19E + 06**	7.12E + 07	1.02E + 08	6.45E + 07	3.28E + 07	2.03E + 08	3.34E + 10	2.22E + 09	9.97E + 09
	Std. Dev	**1.26E + 07**	1.64E + 08	1.59E + 08	7.45E + 07	6.94E + 07	3.09E + 08	1.13E + 10	2.63E + 09	5.37E + 09
f_4	Mean	**3.46E + 01**	2.07E + 03	2.48E + 03	2.29E + 03	1.29E + 03	5.62E + 02	3.49E + 04	2.64E + 04	4.75E + 04
	Std. Dev	**4.99E + 01**	1.34E + 03	1.63E + 03	1.32E + 03	7.49E + 02	3.77E + 02	5.16E + 03	6.99E + 03	1.64E + 04
f_5	Mean	1.49E-03	8.17E-05	**3.28E-05**	3.83E-05	1.52E-03	2.95E-03	2.07E + 03	5.98E + 02	8.42E + 01
	Std. Dev	1.52E-04	1.57E-04	1.02E-04	**7.76E-05**	1.89E-04	3.27E-04	4.98E + 02	2.61E + 02	1.69E + 01
f_6	Mean	**3.49E + 01**	4.95E + 01	5.56E + 01	5.38E + 01	4.09E + 01	4.35E + 01	6.46E + 02	1.22E + 02	1.02E + 02
	Std. Dev	2.72E + 01	2.82E + 01	2.85E + 01	3.01E + 01	2.80E + 01	2.78E + 01	1.99E + 02	**2.23E + 01**	3.87E + 01
f_7	Mean	**1.11E + 01**	6.52E + 01	6.49E + 01	6.08E + 01	4.20E + 01	1.02E + 02	1.68E + 02	4.55E + 01	1.46E + 03
	Std. Dev	**9.16E + 00**	2.91E + 01	2.60E + 01	2.70E + 01	2.62E + 01	2.83E + 01	3.89E + 01	1.50E + 01	7.66E + 03
f_8	Mean	2.09E + 01	2.09E + 01	2.09E + 01	2.09E + 01	2.09E + 01	2.10E + 01	2.09E + 01	2.09E + 01	**2.09E + 01**
	Std. Dev	5.11E-02	4.95E-02	5.07E-02	4.89E-02	5.49E-02	**4.08E-02**	5.63E-02	4.66E-02	6.02E-02

(*continued*)

Table 2. (*continued*)

F		CM-YYPO	YYPO	YYPO-SA1	YYPO-SA2	A-YYPO	SSA	SCA	GWO	WOA
f_9	Mean	**1.20E + 01**	2.43E + 01	2.00E + 01	2.44E + 01	1.80E + 01	2.31E + 01	3.93E + 01	1.86E + 01	3.69E + 01
	Std. Dev	2.94E + 00	4.66E + 00	4.75E + 00	5.26E + 00	3.95E + 00	4.14E + 00	**1.15E + 00**	3.25E + 00	2.44E + 00
f_{10}	Mean	**1.02E-02**	5.18E-02	4.05E-02	4.32E-02	1.70E-02	8.39E-02	1.48E + 03	2.19E + 02	5.41E + 01
	Std. Dev	**9.36E-03**	5.34E-02	3.57E-02	3.32E-02	1.33E-02	4.52E-02	2.32E + 02	1.17E + 02	2.44E + 01
f_{11}	Mean	6.76E + 01	3.66E-01	3.20E-01	**2.58E-01**	7.96E + 00	1.34E + 02	3.63E + 02	8.23E + 01	4.96E + 02
	Std. Dev	1.77E + 01	6.84E-01	**5.02E-01**	6.07E-01	2.74E + 00	4.11E + 01	3.11E + 01	2.33E + 01	1.11E + 02
f_{12}	Mean	**6.24E + 01**	1.65E + 02	1.62E + 02	1.65E + 02	1.09E + 02	1.32E + 02	3.80E + 02	1.26E + 02	5.22E + 02
	Std. Dev	**2.04E + 01**	5.16E + 01	5.42E + 01	5.47E + 01	3.66E + 01	5.07E + 01	2.77E + 01	5.92E + 01	1.04E + 02
f_{13}	Mean	**1.35E + 02**	1.89E + 02	1.88E + 02	2.09E + 02	1.69E + 02	2.18E + 02	3.73E + 02	1.71E + 02	4.98E + 02
	Std. Dev	**3.03E + 01**	3.88E + 01	4.09E + 01	3.63E + 01	3.83E + 01	6.10E + 01	3.52E + 01	4.21E + 01	7.67E + 01
f_{14}	Mean	4.09E + 03	5.53E + 00	5.35E + 00	**4.77E + 00**	1.91E + 01	3.66E + 03	6.95E + 03	3.07E + 03	5.00E + 03
	Std. Dev	2.10E + 03	2.84E + 00	**2.37E + 00**	2.60E + 00	8.94E + 00	5.21E + 02	3.52E + 02	1.04E + 03	7.61E + 02
f_{15}	Mean	3.74E + 03	4.04E + 03	4.02E + 03	3.89E + 03	4.15E + 03	3.78E + 03	7.38E + 03	**3.51E + 03**	5.44E + 03
	Std. Dev	2.03E + 03	5.83E + 02	7.56E + 02	7.42E + 02	7.77E + 02	7.14E + 02	**3.83E + 02**	1.38E + 03	7.34E + 02
f_{16}	Mean	2.44E + 00	1.38E + 00	5.71E-01	1.56E + 00	1.63E + 00	**4.40E-01**	2.47E + 00	2.39E + 00	1.74E + 00
	Std. Dev	3.16E-01	1.61E 01	4.03E-01	3.15E-01	3.16E-01	**2.23E-01**	3.20E-01	2.86E-01	3.84E-01
f_{17}	Mean	1.03E + 02	3.22E + 01	3.22E + 01	**3.15E + 01**	4.32E + 01	1.68E + 02	4.83E + 02	1.56E + 02	5.91E + 02
	Std. Dev	1.85E + 01	**8.17E-01**	9.29E-01	4.14E + 00	5.34E + 00	4.62E + 01	5.05E + 01	4.03E + 01	1.20E + 02
f_{18}	Mean	**1.05E + 02**	1.47E + 02	1.47E + 02	1.52E + 02	1.25E + 02	1.73E + 02	4.86E + 02	2.36E + 02	5.88E + 02
	Std. Dev	**2.04E + 01**	3.45E + 01	3.28E + 01	3.48E + 01	2.58E + 01	4.25E + 01	4.68E + 01	3.70E + 01	1.38E + 02
f_{19}	Mean	3.89E + 00	1.59E + 00	1.54E + 00	**1.54E + 00**	2.51E + 00	7.52E + 00	2.68E + 03	2.83E + 01	6.11E + 01
	Std. Dev	9.96E-01	**3.06E-01**	3.49E-01	3.85E-01	5.69E-01	2.98E + 00	1.76E + 03	4.77E + 01	2.04E + 01
f_{20}	Mean	**1.09E + 01**	1.16E + 01	1.17E + 01	1.22E + 01	1.13E + 01	1.28E + 01	1.39E + 01	1.19E + 01	1.47E + 01
	Std. Dev	8.96E-01	7.86E-01	8.01E-01	6.99E-01	8.72E-01	1.20E + 00	**3.48E-01**	1.59E + 00	3.61E-01

(*continued*)

Table 2. (*continued*)

F		CM-YYPO	YYPO	YYPO-SA1	YYPO-SA2	A-YYPO	SSA	SCA	GWO	WOA
f_{21}	Mean Std. Dev	**2.99E + 02** **7.55E + 01**	3.04E + 02 8.56E + 01	3.05E + 02 8.54E + 01	3.18E + 02 7.88E + 01	3.20E + 02 7.70E + 01	3.39E + 02 8.65E + 01	1.83E + 03 1.97E + 02	7.11E + 02 2.43E + 02	3.48E + 02 7.66E + 01
f_{22}	Mean Std. Dev	3.26E + 03 7.86E + 02	1.24E + 02 5.67E + 01	**1.23E + 02** **5.29E + 01**	1.25E + 02 6.22E + 01	1.47E + 02 5.44E + 01	4.30E + 03 6.54E + 02	7.58E + 03 4.31E + 02	2.97E + 03 6.44E + 02	5.87E + 03 9.96E + 02
f_{23}	Mean Std. Dev	**3.26E + 03** 9.41E + 02	4.60E + 03 7.36E + 02	4.73E + 03 7.36E + 02	4.59E + 03 8.06E + 02	4.68E + 03 6.62E + 02	4.11E + 03 7.10E + 02	7.82E + 03 **3.46E + 02**	3.77E + 03 1.48E + 03	6.17E + 03 8.81E + 02
f_{24}	Mean Std. Dev	**2.15E + 02** 8.80E + 00	2.66E + 02 1.79E + 01	2.56E + 02 1.52E + 01	2.66E + 02 1.89E + 01	2.45E + 02 1.64E + 01	2.69E + 02 9.03E + 00	3.18E + 02 **5.72E + 00**	2.50E + 02 1.08E + 01	3.11E + 02 8.96E + 00
f_{25}	Mean Std. Dev	**2.48E + 02** 2.35E + 01	2.90E + 02 1.41E + 01	2.91E + 02 1.42E + 01	2.93E + 02 1.42E + 01	2.82E + 02 1.54E + 01	2.79E + 02 1.05E + 01	3.28E + 02 **4.64E + 00**	2.70E + 02 7.52E + 00	3.17E + 02 1.06E + 01
f_{26}	Mean Std. Dev	**2.00E + 02** **1.79E-02**	2.00E + 02 2.69E-02	2.00E + 02 3.69E-02	2.00E + 02 2.99E-02	2.00E + 02 3.90E-02	2.45E + 02 7.32E + 01	2.13E + 02 4.55E + 00	2.99E + 02 6.74E + 01	3.27E + 02 9.43E + 01
f_{27}	Mean Std. Dev	**5.45E + 02** 1.11E + 02	9.14E + 02 1.14E + 02	8.88E + 02 1.24E + 02	9.23E + 02 1.31E + 02	7.72E + 02 1.37E + 02	9.44E + 02 9.11E + 01	1.36E + 03 **3.64E + 01**	7.90E + 02 8.12E + 01	1.29E + 03 8.01E + 01
f_{28}	Mean Std. Dev	3.17E + 02 1.49E + 02	3.55E + 02 2.79E + 02	3.08E + 02 1.83E + 02	**3.00E + 02** **1.48E-03**	3.42E + 02 2.10E + 02	3.61E + 02 3.01E + 02	2.56E + 03 1.64E + 02	1.02E + 03 3.00E + 02	4.06E + 03 6.89E + 02

Table 3. Results of CM-YYPO on the 50-dimensional test functions

F		CM-YYPO	YYPO	YYPO-SA1	YYPO-SA2	A-YYPO	SSA	SCA	GWO	WOA
f_1	Mean Std. Dev	6.18E-08 2.95E-08	1.56E-07 1.35E-07	1.35E-07 1.16E-07	1.57E-07 1.25E-07	2.41E-07 9.87E-08	**1.27E-08** **1.90E-09**	2.61E + 04 3.80E + 03	3.18E + 03 1.63E + 03	1.27E + 00 6.03E-01
f_2	Mean Std. Dev	**1.56E + 06** **5.41E + 05**	3.54E + 06 1.59E + 06	3.43E + 06 1.06E + 06	3.24E + 06 1.43E + 06	2.13E + 06 6.63E + 05	2.92E + 06 1.26E + 06	4.02E + 08 9.18E + 07	3.75E + 07 1.97E + 07	3.92E + 07 1.17E + 07

(*continued*)

Table 3. (*continued*)

F		CM-YYPO	YYPO	YYPO-SA1	YYPO-SA2	A-YYPO	SSA	SCA	GWO	WOA
f_3	Mean	**4.42E + 07**	4.24E + 08	3.31E + 08	4.57E + 08	1.36E + 08	5.88E + 08	8.24E + 10	1.36E + 10	2.02E + 10
	Std. Dev	**4.28E + 07**	4.22E + 08	3.65E + 08	4.05E + 08	2.09E + 08	5.10E + 08	1.75E + 10	4.20E + 09	9.96E + 09
f_4	Mean	**7.58E + 00**	5.70E + 03	5.91E + 03	5.47E + 03	1.28E + 03	2.78E + 02	5.67E + 04	4.08E + 04	3.40E + 04
	Std. Dev	**1.18E + 01**	2.46E + 03	2.48E + 03	2.41E + 03	7.41E + 02	1.87E + 02	4.33E + 03	7.85E + 03	6.99E + 03
f_5	Mean	**2.48E-03**	3.58E-03	3.60E-03	3.49E-03	2.53E-03	4.30E-03	2.82E + 03	7.98E + 02	1.05E + 02
	Std. Dev	2.25E-04	4.64E-04	5.88E-04	6.08E-04	**2.12E-04**	2.77E-04	6.86E + 02	2.49E + 02	2.29E + 01
f_6	Mean	**4.41E + 01**	6.30E + 01	5.87E + 01	5.40E + 01	4.57E + 01	5.67E + 01	1.74E + 03	2.46E + 02	1.49E + 02
	Std. Dev	**1.63E + 00**	3.22E + 01	2.73E + 01	2.26E + 01	7.98E + 00	2.28E + 01	2.96E + 02	6.91E + 01	5.14E + 01
f_7	Mean	**2.70E + 01**	9.63E + 01	9.86E + 01	9.61E + 01	6.68E + 01	1.14E + 02	1.84E + 02	6.18E + 01	5.66E + 02
	Std. Dev	**1.17E + 01**	2.66E + 01	2.51E + 01	2.05E + 01	2.10E + 01	2.32E + 01	2.30E + 01	1.73E + 01	7.43E + 02
f_8	Mean	2.11E + 01	2.11E + 01	2.11E + 01	2.11E + 01	2.11E + 01	**2.11E + 01**	2.11E + 01	2.11E + 01	2.11E + 01
	Std. Dev	**2.97E-02**	3.69E-02	3.41E-02	3.04E-02	4.14E-02	4.46E-02	3.73E-02	4.66E-02	3.27E-02
f_9	Mean	**2.32E + 01**	4.55E + 01	4.42E + 01	4.57E + 01	3.44E + 01	4.74E + 01	7.25E + 01	3.77E + 01	6.87E + 01
	Std. Dev	4.09E + 00	8.30E + 00	6.84E + 00	8.12E + 00	7.18E + 00	5.75E + 00	**1.41E + 00**	3.88E + 00	3.72E + 00
f_{10}	Mean	1.28E-02	3.00E-02	3.53E-02	3.81E-02	**1.10E-02**	1.01E-01	3.35E + 03	6.28E + 02	1.23E + 02
	Std. Dev	9.16E-03	2.09E-02	3.16E-02	2.37E-02	**7.17E-03**	5.25E-02	5.30E + 02	2.16E + 02	3.38E + 01
f_{11}	Mean	1.34E + 02	1.68E + 00	1.97E + 00	**1.60E + 00**	2.75E + 01	2.93E + 02	6.80E + 02	2.10E + 02	7.42E + 02
	Std. Dev	3.19E + 01	1.47E + 00	1.39E + 00	**1.30E + 00**	7.20E + 00	7.29E + 01	3.94E + 01	4.90E + 01	9.18E + 01

(*continued*)

Table 3. (*continued*)

F		CM-YYPO	YYPO	YYPO-SA1	YYPO-SA2	A-YYPO	SSA	SCA	GWO	WOA
f_{12}	Mean Std. Dev	**1.41E + 02** **3.23E + 01**	3.82E + 02 1.21E + 02	3.79E + 02 9.71E + 01	3.82E + 02 1.30E + 02	2.27E + 02 7.99E + 01	2.83E + 02 7.76E + 01	7.10E + 02 4.99E + 01	2.57E + 02 9.47E + 01	9.04E + 02 1.31E + 02
f_{13}	Mean Std. Dev	**2.69E + 02** 5.63E + 01	4.01E + 02 6.47E + 01	4.13E + 02 7.77E + 01	4.29E + 02 7.59E + 01	3.41E + 02 6.14E + 01	4.76E + 02 1.02E + 02	7.27E + 02 **5.41E + 01**	3.81E + 02 5.88E + 01	9.82E + 02 1.26E + 02
f_{14}	Mean Std. Dev	6.64E + 03 3.28E + 03	1.04E + 01 4.19E + 00	**1.01E + 01** **3.70E + 00**	1.11E + 01 3.95E + 00	1.36E + 02 1.10E + 02	7.10E + 03 9.06E + 02	1.33E + 04 3.85E + 02	5.72E + 03 1.36E + 03	8.66E + 03 1.28E + 03
f_{15}	Mean Std. Dev	**6.89E + 03** 2.81E + 03	7.81E + 03 9.38E + 02	8.19E + 03 1.21E + 03	7.77E + 03 1.06E + 03	8.14E + 03 1.55E + 03	7.46E + 03 8.44E + 02	1.43E + 04 **3.31E + 02**	7.30E + 03 2.50E + 03	1.08E + 04 1.32E + 03
f_{16}	Mean Std. Dev	3.37E + 00 **2.83E-01**	1.86E + 00 4.31E-01	7.97E-01 4.42E-01	1.97E + 00 4.65E-01	2.33E + 00 4.23E-01	**6.91E-01** 3.25E-01	3.30E + 00 2.96E-01	3.32E + 00 2.95E-01	2.55E + 00 4.91E-01
f_{17}	Mean Std. Dev	2.03E + 02 2.92E + 01	**5.79E + 01** 2.50E + 00	5.83E + 01 **2.38E + 00**	5.80E + 01 2.96E + 00	9.80E + 01 1.30E + 01	3.48E + 02 8.06E + 01	9.37E + 02 7.71E + 01	3.13E + 02 5.46E + 01	1.08E + 03 1.06E + 02
f_{18}	Mean Std. Dev	**1.95E + 02** **3.06E + 01**	3.33E + 02 7.67E + 01	3.30E + 02 7.50E + 01	3.15E + 02 6.99E + 01	2.56E + 02 4.61E + 01	3.60E + 02 8.61E + 01	9.42E + 02 8.12E + 01	5.04E + 02 5.56E + 01	1.08E + 03 1.19E + 02
f_{19}	Mean Std. Dev	7.71E + 00 1.67E + 00	**3.19E + 00** 6.42E-01	3.25E + 00 7.46E-01	3.43E + 00 **6.06E-01**	5.65E + 00 1.16E + 00	1.51E + 01 3.48E + 00	2.29E + 04 1.87E + 04	5.25E + 02 7.34E + 02	1.34E + 02 4.03E + 01
f_{20}	Mean Std. Dev	**1.98E + 01** 8.14E-01	2.13E + 01 8.88E-01	2.15E + 01 1.08E + 00	2.18E + 01 7.22E-01	2.05E + 01 1.12E + 00	2.33E + 01 1.26E + 00	2.36E + 01 4.88E-01	2.05E + 01 1.02E + 00	2.45E + 01 **2.27E-01**
f_{21}	Mean Std. Dev	8.62E + 02 3.69E + 02	**7.48E + 02** 4.07E + 02	9.10E + 02 3.31E + 02	8.49E + 02 3.80E + 02	7.66E + 02 4.03E + 02	8.72E + 02 2.88E + 02	3.80E + 03 **1.43E + 02**	2.14E + 03 5.64E + 02	9.62E + 02 2.35E + 02
f_{22}	Mean Std. Dev	6.40E + 03 1.07E + 03	**8.59E + 01** **7.95E + 01**	1.19E + 02 9.63E + 01	9.51E + 01 8.82E + 01	1.80E + 02 1.00E + 02	7.94E + 03 1.02E + 03	1.43E + 04 4.54E + 02	6.50E + 03 1.39E + 03	1.14E + 04 1.54E + 03

(*continued*)

Table 3. *(continued)*

F		CM-YYPO	YYPO	YYPO-SA1	YYPO-SA2	A-YYPO	SSA	SCA	GWO	WOA
f_{23}	Mean Std. Dev	**6.30E + 03** 9.22E + 02	9.11E + 03 1.37E + 03	9.31E + 03 1.18E + 03	9.46E + 03 1.33E + 03	8.73E + 03 1.49E + 03	8.01E + 03 1.03E + 03	1.48E + 04 **5.34E + 02**	7.62E + 03 1.48E + 03	1.21E + 04 1.49E + 03
f_{24}	Mean Std. Dev	**2.47E + 02** 1.57E + 01	3.32E + 02 2.39E + 01	3.30E + 02 1.93E + 01	3.31E + 02 2.39E + 01	2.79E + 02 2.17E + 01	3.39E + 02 1.51E + 01	4.19E + 02 **7.47E + 00**	2.99E + 02 1.44E + 01	4.03E + 02 1.47E + 01
f_{25}	Mean Std. Dev	**3.23E + 02** 1.13E + 01	3.92E + 02 2.45E + 01	3.90E + 02 2.07E + 01	3.92E + 02 2.55E + 01	3.48E + 02 1.82E + 01	3.65E + 02 1.43E + 01	4.43E + 02 **5.93E + 00**	3.38E + 02 1.30E + 01	4.23E + 02 1.55E + 01
f_{26}	Mean Std. Dev	2.20E + 02 5.15E + 01	2.04E + 02 2.60E + 01	**2.01E + 02** **1.62E + 00**	2.01E + 02 2.40E + 00	2.04E + 02 2.23E + 01	4.17E + 02 4.60E + 01	3.32E + 02 1.19E + 02	3.85E + 02 4.70E + 01	4.52E + 02 8.38E + 01
f_{27}	Mean Std. Dev	**9.18E + 02** 1.21E + 02	1.50E + 03 2.05E + 02	1.59E + 03 2.45E + 02	1.55E + 03 1.85E + 02	1.24E + 03 1.75E + 02	1.59E + 03 1.46E + 02	2.31E + 03 **5.84E + 01**	1.28E + 03 9.57E + 01	2.23E + 03 1.26E + 02
f_{28}	Mean Std. Dev	**4.58E + 02** **4.14E + 02**	7.69E + 02 1.02E + 03	7.07E + 02 9.40E + 02	7.67E + 02 1.01E + 03	5.20E + 02 6.02E + 02	1.13E + 03 1.40E + 03	4.67E + 03 6.79E + 02	1.80E + 03 1.32E + 03	7.27E + 03 1.49E + 03

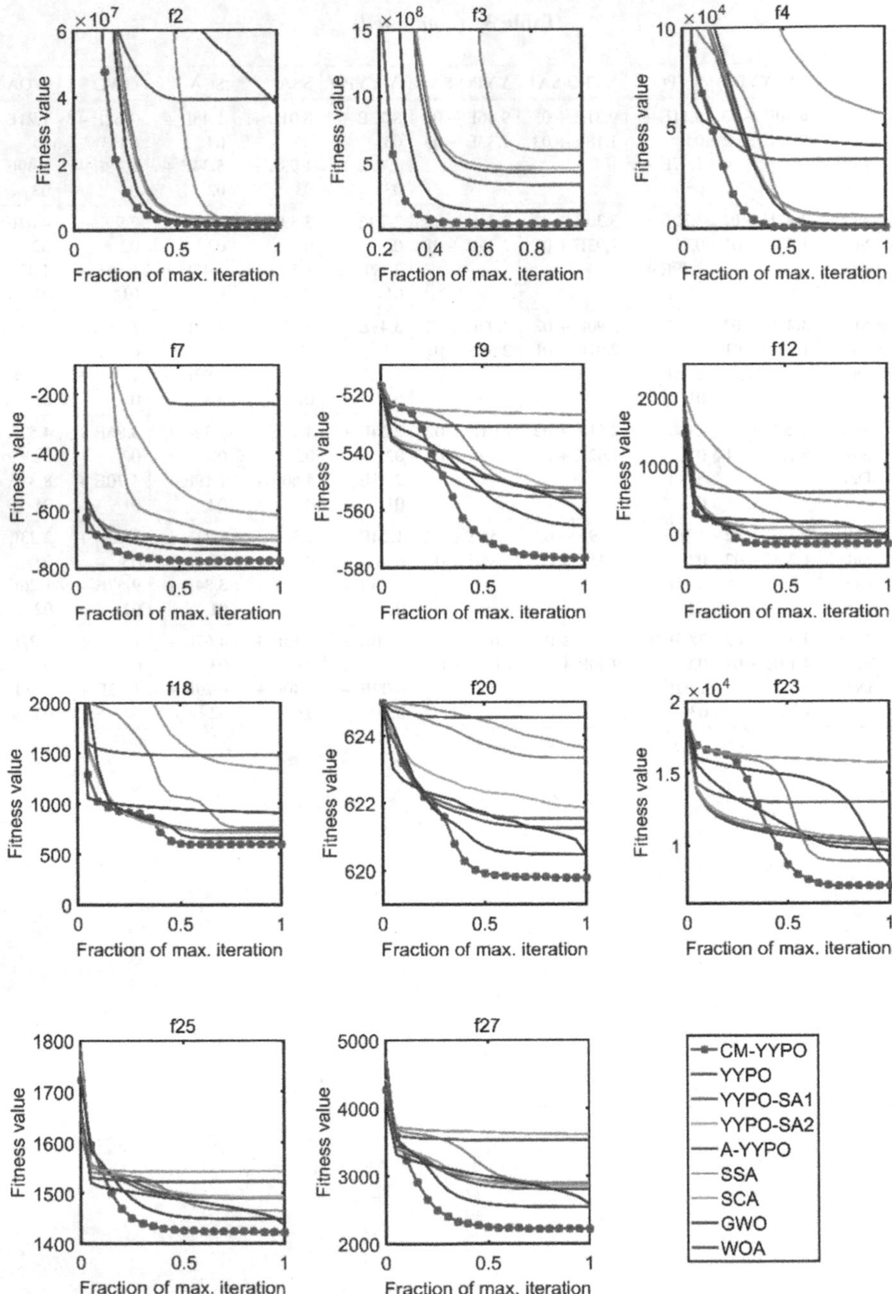

Fig. 1. Convergence curves for the 50-dimensional case

5 Conclusions

In this paper, the crossover and mutation operators are added to the search mechanism of YYPO for the optimization problems in high dimensions. The splitting method is also improved. According to the comparison results of test functions, proposed CM-YYPO not only overcomes the drawback of early convergence, but also greatly improves the solving ability in each dimension. Especially in solving high-dimensional problems, its search performance is obviously better than other algorithms. However, optimization problems in big data analytics involve more optimization objectives than just high-dimensional features. It is still worthwhile to continue to explore the research on how to extend CM-YYPO effectively to multi-objective optimization problems.

Acknowledgments. This work has been supported by the National Natural Science Foundation of China (No. 61602162).

References

1. Li, G., Cheng, X.: Research status and scientific thinking of big data. Bulletin of Chinese Academy Sci. **6**, 647–657 (2012)
2. Jaume, B., Xavier, L.: Large-scale data mining using genetics-based machine learning. Wiley Interdisciplinary Reviews: Data Mining Knowledge Discovery **3**(1), 37–61 (2013)
3. Chen, Y., Miao, D., Wang, R.: A rough set approach to feature selection based on ant colony optimization. Pattern Recogn. Lett. **31**(3), 226–233 (2010)
4. Samariya, D., Ma, J.: A new dimensionality-unbiased score for efficient and effective outlying aspect mining. Data Science Eng. **7**(2), 120–135 (2022)
5. Varun, P., Prakash, K.: Yin-Yang-pair optimization: a novel lightweight optimization algorithm. Eng. Appl. Artif. Intell. **54**, 62–79 (2016)
6. Xu, Q., Ma, L., Liu, Y.: Yin-Yang-pair optimization algorithm based on chaos search and intricate operator. J. Computer Appl. **40**(08), 2305–2312 (2020)
7. Wang, W., et al.: An orthogonal opposition-based-learning Yin–Yang-pair optimization algorithm for engineering optimization. Engineering with Computers (4), (2021)
8. Varun, P., Prakash, K.: Adaptive Yin-Yang-Pair Optimization on CEC 2016 functions. In: 2016 IEEE Region 10 Conference (TENCON), pp. 2296–2299 (2016)
9. Li, D., Liu, Q., Ai, Z.: YYPO-SA: novel hybrid single-object optimization algorithm based on Yin-Yang-pair optimization and simulated annealing. Application Research of Computers **38**(07), 2018–2024 (2021)
10. Guo, P., et al.: Computational intelligence for big data analysis: current status and future prospect. Journal of Software **26**(11), 3010–3025 (2015)
11. Chen, X., Yu, S.: Improvement on crossover strategy of real-valued genetic algorithm. Acta Electron. Sin. **31**(1), 71–74 (2003)
12. Liang, J., et al.: Problem Definitions and Evaluation Criteria for the CEC 2013 Special Session on Real-Parameter Optimization, pp. 281295 (2013)

MCSketch: An Accurate Sketch for Heavy Flow Detection and Heavy Flow Frequency Estimation

Jie Lu[1(✉)], Hongchang Chen[1,2], and Zhen Zhang[1,3]

[1] PLA Information Engineering University, Zhengzhou 450008, China
lujie_cs@zju.edu.cn
[2] Songshang Laboratory, Zhengzhou 450007, China
[3] Network Communication and Security Purple Mountain Laboratory, Nanjing 211100, China

Abstract. Accurately finding heavy flows in data streams is challenging owing to limited memory availability. Prior algorithms have focused on accuracy in heavy flow detection but cannot provide the frequency of a heavy flow exactly. In this paper, we designed a two-mode counter, called Matthew Counter, for the efficient use of memory and an accurate record flow frequency. The key ideas in Matthew Counter are the use of idle high-bits in the counter and the adoption of a power-weakening method. Matthew Counter allows sufficient competition during the early stages of identifying heavy flows and amplifying the relative advantage when the counter is sufficiently large to ensure the level of accuracy. We also present an invertible sketch, called MCSketch, for supporting heavy-flow detection with small and static memory based on Matthew Counter. The experiment results show that MCSketch achieves a higher accuracy than existing algorithms for heavy flow detection. Moreover, MCSketch reduces the average relative error by approximately 1 to 3 orders of magnitude in comparison to other state-of-art approaches.

Keywords: Data stream mining · Heavy flow detection · Sketch

1 Introduction

Data stream processing is a significant issue in many applications, including natural language processing [1] and network security [2, 3]. Heavy flow detection is one of the most fundamental tasks in data stream processing. A heavy flow means that its frequency is over a predefined threshold. Finding heavy flows in a data stream with limited memory is a challenging task. It is impossible to accurately track all flows because recording massive data streams wastes a significant amount of memory.

Many researchers have applied sketch as compact structure to heavy flow detection and achieved remarkable results with limited memory, including MVSketch [4], HeavyKeeper [5], and WavingSketch [6]. Nevertheless, the detection accuracy of these existing algorithms is insufficient, and the estimation error of the frequency of a heavy flow is large. The error of the flow frequency given under certain scenarios is twice the real frequency of a heavy flow.

B. Li et al. (Eds.): APWeb-WAIM 2022, LNCS 13421, pp. 20–27, 2023.
https://doi.org/10.1007/978-3-031-25158-0_2

To understand the difficulty of estimating the flow frequency, suppose we have one counter that can record the candidate key and count the frequencies of the candidate and other flows. Intuitively, when an incoming item belongs to the candidate key, the count of candidate flows increases by 1; otherwise, the count of the other flows increases by 1. When the count of the other flows is larger than that of the candidate flow, the key of the incoming item is used as a new candidate flow. However, if the frequency of the heavy flow is not significantly greater than that of the other flows, it will be difficult to detect a heavy flow based on the previous strategy. Supposing we have a data stream < A, A, A, B, C, B, C >, the record is then (A,1,0), (A,2,0), (A,3,0), (A,3,1), (A,3,2), (A,3,3), (C,1,0) according to the above strategy. When the record is (A,3,3) and then the last item C is processed, we increase the count of other flows to 4 which is greater than the count of candidate flow A of 3, so the candidate flow is changed to C. As a result, the final recorded heavy flow is C, which is not the real heavy flow.

To solve this problem, we are inspired by the Matthew Effect, which can be summarized by the adage "the rich get richer and the poor get poorer." Specifically, we adopt a probabilistic method called a power-weakening increment for items that do not belong to the current candidate flow. For example, when the incoming item belongs to the candidate key, the count of the candidate flows increases by 1; otherwise, the count of the other flows increases by 1 with the probability of the inverse of the current candidate flow count. For data stream < A, A, A, B, C, B, C >, the record is (A,3,0) after three items are inserted. For the fourth item B, the count of other flows increases by 1 with a probability of $1/3$, and thus the probability of candidate flow A being replaced is significantly reduced. Using the power-weakening increment strategy, the heavy flow with relative advantages can be amplified, which is consistent with the Matthew effect.

In addition, the floating-point number inspired us to allow different parts of one counter to take on different functions. We found that when the count is small, the high bit of the counter is in a vacant state. Thus, we can divide the counter into two parts: a high-bit part to count items that do not belong to the candidate flow, and a low-bit part of counting items belonging to the candidate flow. Based on these two ideas, we present Matthew Counter with two modes. In competitive mode, the nvote-part in Matthew Counter counts the number of items with different fingerprints from the candidate flow, whereas the pvote-part counts the number of items with the same fingerprint as the candidate flow. In exclusive mode, the nvote-part is swallowed by the pvote-part. The nvote-part adopts a power-weakening increment method, which increases the nvote-part with a probability.

To summarize, this study makes the following contributions.

1) We describe the design of Matthew Counter with two modes that can accurately track a heavy flow. Then, based on Matthew Counter, we present MCSketch, which can achieve a high accuracy for a heavy flow detection, and provides an estimated frequency extremely close to the true frequency of a heavy flow.
2) We present a mathematical analysis of the error bounds on the heavy flow frequency of MCSketch and theoretically prove its high precision
3) We implemented MCSketch and other algorithms. Trace-driven evaluations show that MCSketch achieves a high accuracy in heavy flow detection compared to state-of-art methods. The average relative error of a heavy flow frequency estimation is

only 0.1 of the current algorithms while maintaining an almost optimal F1 score. We released the source code of MCSketch and the related algorithms on Github [7].

2 Background

The data stream in an observation window is viewed as a set of M flows $F = \{f_1, f_2, ..., f_M\}$, where each flow f_i is composed of items sharing the same flow fingerprint, for example, the source IP in network data stream. The flow frequency refers to the number of items belonging to this flow. Let n_i and $\hat{n}_i$ represent the real and estimated frequencies of flow f_i, respectively. Given a threshold Φ, heavy flow f_i is defined as $n_i \geq \Phi$.

There are two traditional methods used to find a heavy flow: the *record-all* method and the *record-some* method. The *record-all* method uses a sketch to track all flows (including the Count-Min [8] sketch, CU Sketch [9], or Count Sketch [10]). Because of the hash collisions, the flow frequency estimation is inaccurate, although there are numerous strategies to improve the accuracy, such as a flow separation [11, 12] and noise correction [13–15]. As a more severe problem, they are non-invertible because we must check every flow in the entire flow key space to recover all heavy flows. Therefore, this method is slow and inaccurate.

Many algorithms use the *record-some* method, including, LDSketch [16], HeavyKeeper [5], WavingSketch [6], and ActiveKeeper [17]. The *record-some* method uses a key-value counter to record the partial flow key and flow value. The space-saving algorithm captures each incoming flow and increases its value by 1 if the flow has already been recorded. The smallest flow in a summary with size will be replaced by a new flow with size, which leads to a significant lack of precision. HeavyKeeper uses the exponential decay probability to decay or be replaced. The probability of a decay decreases as the count increases, and thus any flow whose size exceeds a certain threshold will be difficult to replace. WavingSketch uses a hash function to hash incoming items to $+1$ or -1, and then increases or decreases the waving counter by 1. A waving counter is then used to obtain the estimated frequency.

However, these methods only consider the accuracy of heavy flow detection, but cannot precisely determine the frequency of a heavy flow.

3 Design of MCSketch

3.1 Matthew Counter

Matthew Counter works in two different modes: competitive and exclusive. Matthew Counter is composed of the following three parts: 1) a flag part indicating the current mode of the counter, where a flag of 0 means competitive mode, and a flag of 1 means exclusive mode. 2) The nvote-part counts the number of items with different fingerprints from the candidate flow in the bucket when in competitive mode. In exclusive mode, the nvote-part is swallowed by the pvote-part. 3) The pvote-part counts the number of items with the same fingerprint as the candidate flow.

A two-mode Matthew counter is shown in Fig. 1. Let the number of bits in the counter be r and apply 1 bit for the flag-part. For competitive mode, Matthew Counter has s bits for the nvote-part and $r - s - 1$ bits for the pvote-part. For exclusive mode, the pvote-part is $r - 1$ bits. Figure 1 (a) shows that the incoming item belongs to the candidate flow in competitive mode, and the pvote-part is incremented by 1. If the value of the pvote-part is larger than $2^{r-s-1} - 2$, the pvote-part and nvote-part are merged into the pvote-part where the count range of the pvote-part is $[0, 2^{r-1} - 1]$. The incoming item does not belong to the candidate flow in competitive mode, as shown in Fig. 1 (b). The incremental probability in the power-weakening method for the nvote-part is defined as follows:

$$P_{nvote-inc} = \begin{cases} 1 & \text{if } C < \beta \\ (\beta/C)^{\alpha} & \text{if } C \geq \beta \end{cases} \tag{1}$$

where C is the value of the current pvote-part, and α and β are predefined parameters. Therefore, the larger the flow, the harder it is for the nvote-part size to be incremented, and this flow is more likely to be a candidate flow.

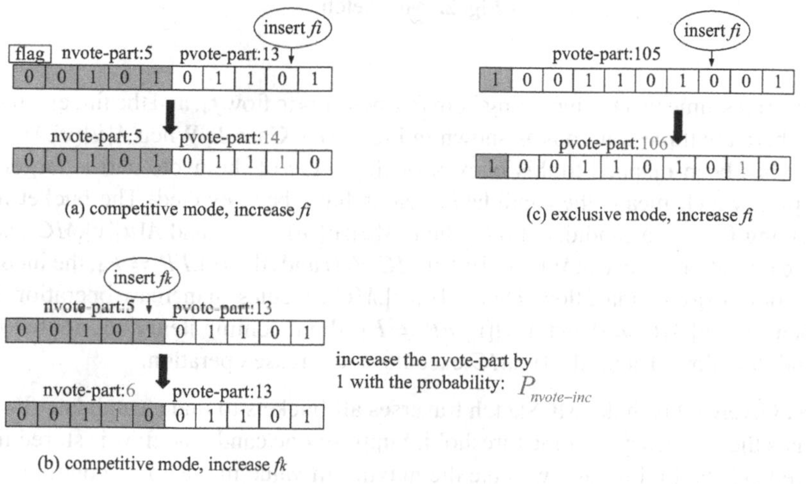

Fig. 1. Matthew counter

Assuming $\beta = 1$ and $\alpha = 1$, we incremented the nvote-part by 1 with a probability of 13^{-1}. If the value of the nvote-part is larger than the value of pvote-part, we set these two parts to 0, which means that the current candidate flow has been evicted and the next incoming flow is accepted as the candidate flow. Otherwise, if the value of the nvote-part is larger than $2^r - 2$, the pvote-part and nvote-part are merged into the pvote-part. In exclusive mode, only the incoming item belonging to the candidate flow is counted. Figure 1 (c) shows that the incoming item belongs to the candidate flow, and the pvote-part is incremented by 1. At this time, the value of pvote-part is sufficiently large, and thus we will no longer replace the candidate flow.

3.2 Data Structure of MCSketch

As shown in Fig. 2 (a), MCSketch comprises d arrays, and each array has w buckets. Each bucket has a fingerprint field and Matthew Counter. The fingerprint field stores the key of the candidate flow, and Matthew Counter is devoted to choosing and counting this flow. We use $A[u][v]$ to denote the v-th bucket in the u-th array. We then use $A[u][v].MC$ and $A[u][v].FP$ to represent its Matthew Counter and fingerprint, respectively. Each incoming item is inserted into d arrays with d pairwise independent hash functions.

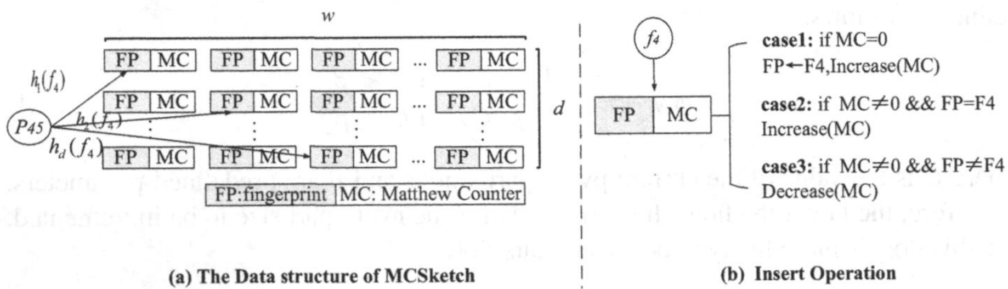

Fig. 2. MCSketch

Insertion: Assume that the incoming item P_{45} belongs to flow f_4, and the fingerprint of f_4 is F_4. There are three solutions as shown in Fig. 2 (b). **Case 1:** When $A[u][v].MC = 0$, no flow has been mapped to this bucket, or the value of the nvote-part is larger than pvote-part, which means the candidate streams have been evicted. The bucket tracks the arriving flow as a candidate flow. Then, $A[u][v].FP = F_i$ and $A[u][v].MC$ execute increase operations. **Case 2:** When $A[u][v].MC \neq 0$ and $A[u][v].FP = F_4$, the incoming item belongs to a candidate flow. Then, $A[u][v].MC$ executes an increase operation. **Case 3:** When $A[u][v].MC \neq 0$ and $A[u][v].FP \neq F_4$, the incoming item does not belong to the candidate flow. Then, $A[u][v].MC$ executes a decrease operation.

Query: Given a threshold, MCSketch traverses all buckets to find a candidate flow that estimates the frequency over the threshold. Suppose one candidate flow is stored in two or more buckets. In this case, we take the maximum value in the corresponding bucket as the heavy flow frequency because MCSketch has no overestimation error.

4 Experiment Results

4.1 Experiment Setup

Platform: Our experiment is run on a server with four 16-core Intel Xeon Gold 5218 @2.3 GHz CPUs and 128 GB of DDR4 memory. **Dataset:** The IP trace dataset contains anonymized IP traces collected in 2016 by CAIDA [18].

Implementation: We compare MCSketch (MC) with state-of-the-art heavy hitter algorithms, including LDSketch (LD), MVSkech (MV), HeavyKeeper (HK), and WavingSketch (WS). For all algorithms, the fingerprint field was 32 bits. For MVSketch, the sum field and counter field are both also 32 bits. For LDSketch, because it dynamically expands the associative array of its buckets, we first determine the most available array based on the given memory, and to ensure balance, we limit the most associative arrays of each bucket. According to the original study, the decay probability in HeavyKeeper is 1.08, and the slot number in the heavy part is 8. In the following experiment, we set $\alpha = 0.6$ and $\beta = 1$ for MCSketch. We also set $d = 2$ for all algorithms, if necessary.

Metrics: We used the throughput to measure the insert speed. The F1 score were used to measure the accuracy of the heavy flow detection. The AAE were used to measure the error of the flow frequency estimation. The throughput is defined as N/T, where T is the total processing time. F1-score is the harmonic mean of the precision and recall, where the precision is the fraction of true heavy flows reported over all reported flows and the recall is the fraction of true heavy flows reported over all true heavy flows. The average absolute error (AAE) is defined as $\left(\sum_{f_i \in \Psi} |n_i - \hat{n}_i|\right) / \Psi$, where Ψ is the true heavy flow reported.

4.2 Experiment on Memory

In this section, we conducted experiments with various memory sizes to study the impact of memory size on the accuracy, error, and throughput. We set the threshold of a heavy flow to 100 and varied the memory size from 5 KB to 30 KB on the CAIDA datasets.

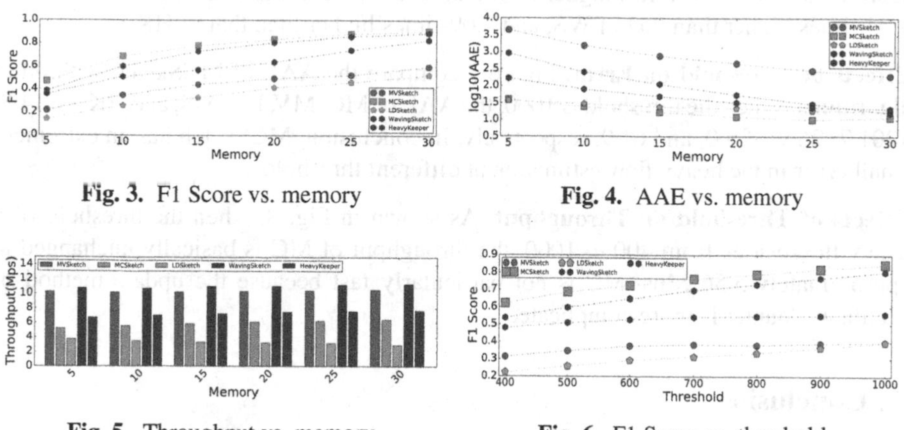

Fig. 3. F1 Score vs. memory

Fig. 4. AAE vs. memory

Fig. 5. Throughput vs. memory

Fig. 6. F1 Score vs. threshold

Effects of Memory on the Accuracy. As shown in Fig. 3 the F1 score of MC is always the largest, and as the memory increases, it continues to increase. When the memory is 10 KB, the F1 scores of MC, MV, LD, HK, and WS are 0.69, 0.34, 0.26, 0.51, and 0.60, respectively.

Effects of Memory on Error. Figure 4 compares the error in the heavy flow frequency estimation of MCSketch with those of the other algorithms. For convenience, we consider the logarithms of AAE. The estimated error of MCSketch is much smaller than that of the other algorithms.

Effects of Memory on Throughput. Figure 5 compares the throughput of MC with that of the other algorithms. When the memory size is 30 KB, the throughput of MCSketch is 6.5 Mps while those of MV, LD, WS, and HK are 10.5, 2.9, 11.3, and 7.7 Mps, respectively.

4.3 Experiments on Threshold

In this section, we describe experiments conducted with various thresholds to study the impact of the threshold. We set the memory to 10 KB and varied the threshold from 500 to 1000.

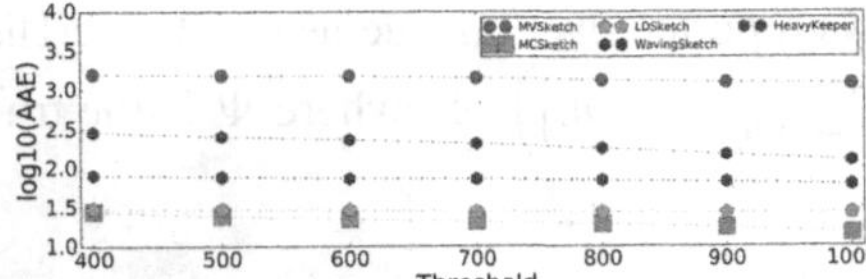

Fig. 7. AAE vs. Threshold

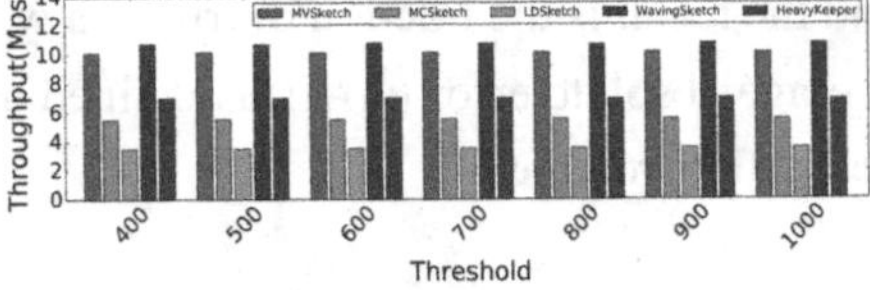

Fig. 8. Throughput vs. Threshold

Effects of Threshold on Accuracy. As shown in Fig. 6 When the threshold is 1000, the F1 scores of these algorithms are 0.73, 0.40, 0.25, 0.70, and 0.67, respectively. The F1 score of MC was 1.8-times higher than that of MV, 2.9-times higher than that of LD, 1.04-times higher than that of WS, and 1.09-times higher than that of HK.

Effects of Threshold on Error. Figure 7 compare the AAE of MCSketch with other algorithms. When the threshold is 1000, the AAE of MC, MV, LD, MS, and HK are 14.8, 1201.7, 26.9, 125.0, and 61.9, respectively. In conclusion, MCSketch has an extremely small error in the heavy flow estimation at different thresholds.

Effects of Threshold on Throughput. As shown in Fig. 8, when the threshold of a heavy flow varies from 400 to 1000, the throughput of MC is basically unchanged at approximately 5.56 Mps. MC is not particularly fast because the update method of Matthew Counter is more complicated.

5 Conclusion

This paper proposes a novel algorithm called MCSketch for heavy flow detection and heavy flow frequency estimation. The key ideas are using idle high-bits and a power-weakening method to allow sufficient competition in the early stages of identifying heavy flows and amplifying the relative advantage when the counter is sufficiently large. The experiment results show that MCSketch achieves a higher accuracy in heavy flow detection and heavy flow frequency estimation.

References

1. Goyal, A., Daume, H., Cormode, G., et al.: Sketch algorithms for estimating point queries in NLP. In: Proceedings of the 2012 joint conference on empirical methods in natural language processing and computational natural language learning, pp. 1093–1103 (2012)
2. Wang, J., Wen, R., Li, J., et al.: Detecting and mitigating target link-flooding attacks using SDN. IEEE Trans Dependable Secur Comput **16**, 944–956 (2018). https://doi.org/10.1109/TDSC.2018.2822275
3. Zheng, J., Li, Q., Gu, G., et al.: Realtime DDoS defense using COTS SDN switches via adaptive correlation analysis. IEEE Trans. Inf. Forensics Secur. **13**, 1838–1853 (2018). https://doi.org/10.1109/TIFS.2018.2805600
4. Tang, L., Huang, Q., Lee, P.P.C.: MV-Sketch: a fast and compact invertible sketch for heavy flow detection in network data streams. In: Proceedings - IEEE INFOCOM 2019-April, pp. 2026–2034 (2019). https://doi.org/10.1109/INFOCOM.2019.8737499
5. Gong, J., Yang, T., Zhang, H., et al.: HeavyKeeper: An accurate algorithm for finding top-k elephant flows. In: Proc 2018 USENIX Annual Tech Conference USENIX ATC 2018, pp. 909–921 (2018). https://doi.org/10.1109/tnet.2019.2933868
6. Li, J., Li, Z., Xu, Y., et al.: WavingSketch: An Unbiased and Generic Sketch for Finding Top-k Items in Data Streams. In: Proceedings ACM SIGKDD International Conference Knowledge Discovery Data Mining, pp. 1574–1584 (2020). https://doi.org/10.1145/3394486.3403208
7. Source code related to MCSketch. https://github.com/Paper-commits/MCSketch
8. Cormode, G., Muthukrishnan, S.: An improved data stream summary: the count-min sketch and its applications. J. Algorithms **55**, 58–75 (2005). https://doi.org/10.1016/j.jalgor.2003.12.001
9. Estan, C., Varghese, G.: New directions in traffic measurement and accounting. Comput. Commun. Rev. **32**, 75 (2002). https://doi.org/10.1145/510726.510749
10. Cormode, G., Hadjieleftheriou, M.: Finding frequent items in data streams. Proc. VLDB Endow. 1(2), 1530–1541 (2008). https://doi.org/10.14778/1454159.1454225
11. Yang, T., Huang, Q., Miao, R., et al.: Elastic sketch: adaptive and fast network-wide measurements. In: SIGCOMM 2018 - Proceedings of the 2018 Conference of the ACM Special Interest Group on Data Communication. Association for Computing Machinery, Inc, pp. 561–575 (2018)
12. Zhou, Y., Yang, T., Jiang, J., et al.: Cold filter: a meta-framework for faster and more accurate stream processing. In: Proceedings of the ACM SIGMOD Int. Conf. Manage. Data, pp 741–756 (2018)
13. Ting, D.: Count-min: optimal estimation and tight error bounds using empirical error distributions. In: Proc ACM SIGKDD Int. Conf. Knowl. Discov. Data Min., pp. 2319–2328 (2018). https://doi.org/10.1145/3219819.3219975
14. Huang, Q., Lee, P.P.C., Bao, Y.: SketChlearn: Relieving user burdens in approximate measurement with automated statistical inference. In: SIGCOMM 2018 - Proceedings of the 2018 Conference of the ACM Special Interest Group on Data Communication. Association for Computing Machinery, Inc, pp. 576–590 (2018)
15. Jie, L., Hongchang, C., Penghao, S., et al.: OrderSketch: an unbiased and fast sketch for frequency estimation of data streams. Comput. Networks **201**, 108563 (2021)
16. Huang, Q., Lee, P.P.C.: A hybrid local and distributed sketching design for accurate and scalable heavy key detection in network data streams. Comput. Networks **91**, 298–315 (2015). https://doi.org/10.1016/j.comnet.2015.08.025
17. Wu, M., Huang, H., Sun, Y., et al.: ActiveKeeper : an accurate and efficient algorithm for finding top- k elephant flows. IEEE Commun. Lett. **7798**, 1–5 (2021). https://doi.org/10.1109/LCOMM.2021.3077902
18. The caida anonymized internet traces 2016. http://www.caida.org/data/overview/

The Influence of the Student's Online Learning Behaviors on the Learning Performance

Chunying Li[1], Junjie Yao[1], Zhikang Tang[1](✉), Yong Tang[2], and Yanchun Zhang[3,4]

[1] School of Computer Science, Guangdong Polytechnic Normal University, Guangzhou 510665, China
fzutang@126.com

[2] School of Computer, South China Normal University, Guangzhou 510631, China

[3] Cyberspace Institute of Advanced Technology, Guangzhou University, Cuangzhou 510006, China

[4] New Cyber Research Department, Peng Cheng Laboratory, Shenzhen 518055, China

Abstract. The emergence of online learning platforms means learners have a variety of learning behavior patterns. Many studies have found that there is a certain correlation between online learning behavior and learning performance. To better optimize the function of an online learning platform in hybrid teaching mode and further improve the quality of teaching and learning, this paper takes the 5y online learning platform as the target scene, and uses the online learning behavior data of 2205 learners and final exam score data as the breakthrough point of learning analytics. Through factor analysis on the behavior data of 13 measurement indicators of learners, this paper uses multiple linear regression model to analyze the correlation between learners' online learning behavior and their final exam scores. The research found that the final examination results of learners are obviously positively correlated with the basic question factors and comprehensive question factors. Therefore, teachers and students who use 5y platform should focus on the use of knowledge point tests and unit tests to improve the quality of teaching and learning within the limited class time.

Keywords: Online learning · Learning analytics · Linear regression

1 Research Background

The Online and Offline hybrid teaching model can stimulate the learning initiative and motivation of the learners. The 7*24 h online learning service in the hybrid teaching model can satisfy the seamless online learning demands of diverse levels and types of learners. Therefore, in-depth analysis of the impact of learner online learning behavior on learner performance can further improve the quality of teaching and learning in the mixed teaching model, as well as provide specific and operable opinions and suggestions for online learning platform developers to improve platform functions. 5y Platform[1] is an online learning platform independently developed by the Teaching and Examination

[1] http://5ystudy.gdoa.net/.

B. Li et al. (Eds.): APWeb-WAIM 2022, LNCS 13421, pp. 28–36, 2023.
https://doi.org/10.1007/978-3-031-25158-0_3

Management Center of Guangdong Provincial Institutions of Higher Education. It provides online learning services in various teaching modalities to over 100 institutions in Guangdong, HeBei, Guangxi, and Fujian, and has had a positive social influence. Since 2017, the course of Computer Application Basics as a public fundamental course for non-computer specialty in Guangdong Polytechnic Normal University, has cooperated with the 5y Platform and has achieved good teaching results for mixed teaching, and separation of teaching and examination. However, it has been discovered that a considerable proportion of the course learners did not achieve ideal results in the final examination. Therefore, this paper takes the online learning behavior data and the final exam score data of the course of "Computer Application Foundation" as the starting point for learning analytics, analyzes the impact of test behavior on the end-of-term learning performance, and aims to provide practical opinions and suggestions on the improvement of teaching quality and the system optimization of platform functions in the mixed teaching mode of the Computer Application Foundation course for our school and other universities using the 5y platform.

2 Related Research

2.1 Introduction to Learning Analytics

Learning analytics has gotten a lot of interest in recent years as a study subject for extracting useful information from educational data. The concept of learning analytics was officially introduced in 2011 with the help of "measuring, collecting, analyzing and reporting data on learners and their learning environments to understand and optimize learning and the environment in which it occurs" [1]. Muldner believes that learning analytics is conducive to the self-monitoring of learners' learning status and learning activities, improving their motivation and enabling them to have a positive emotional experience [2]. Han et al. proposed a review framework of learning analytics, including concepts and overview, composition and model, technical system, organization and evaluation [3]. G. Siemens et al. analyzed the value of learning analytics based on large data sets for education, including guiding the reform of higher education and promoting teaching, etc. [4].

2.2 Research on the Relationship Between Online Learning Behavior and Learning Performance

With the rapid development of online education, the factors that affect learners' learning performance and the relationship between learning behavior and their learning performance are scientific issues worth investigating. Song Jia et al. developed a multiple linear regression equation to better understand the impact mechanism of online learning institutions, communication frequency, communication time, and communication mode on in-depth learning [5]. Shen et al. constructed a performance evaluation model of online learning behavior and online learning through stepwise regression analysis of learning behavior data on the school online platform [6], Research by Liu et al. shows that learning analytics and personalized learning resource recommendation set up a personalized learning path for learners, which helps to increase learners' enthusiasm for participating in learning activities and improve their academic performance [7]. Liu et al. have shown the cognitive input in online learning input. There is a significant positive correlation between emotional input and social input and learning performance [8].

2.3 Other Related Learning Analytics Methods

In recent years, some new methods have also been widely applied in learning analytics in addition to the traditional methods of educational research. For example, multimodal learning analytics is a new direction formed by the intersection of multimodal interaction, learning science, machine learning and other fields, which uses multimodal data to analyze learning behavior in complex environments to optimize the learning experience [9]. Kent et al. used social network analysis to assess the balance between the interactive benefits and the cost of coordination of the learner community [10]. Shen used a variety of intelligent algorithms such as artificial neural network and ant colony probability recommendation to develop a personalized learning path recommendation for users [11]. Karthikeyan et al. used basic information and behavioral performance data of learners' learning to predict academic performance and assess learner performance through the Naive Bayesian and J48 classifiers [12].

3 Platform Functions and Research Samples

The 5y platform provides the supporting video, exercises, tests and exams for the course, as well as the functions of learning notes, learning statistics, learning groups and discussion areas. Besides having the functions that most MOOCs have, the 5y platform's greatest feature is the ability to score subjective and objective topics in the course test, such as Word typesetting, Excel statistical analysis, etc. It greatly frees up the time for teachers to check/correct their homework and reduces the rate of misjudgment in examination corrections. 5y Platform roles mainly include four types: faculty, administrator, teacher and learner. The faculty and administrators are used internally by platform data maintenance and function maintenance personnel. The teacher performs basic classroom teaching functions such as customizing test papers, job publishing, notification publishing, and interactive communication. The learner side includes functions such as video learning, knowledge point testing, unit testing, comprehensive testing and interactive classroom communication. At any time, the learner can check his own learning results and data information such as the rankings of the class score, and then adjust learning strategies in real time according to his own situation. With the help of 5y platform, this paper collects 133297 online learning behavior data and their final exam results data from 2236 students of the Computer Application Basics course in the first semester of Guangdong Polytechnic Normal University from 2020 to 2021. The data set is analyzed by statistical analysis. Predict the final exam results of the learners according to the analysis results.

4 Research Results

4.1 Data Preprocessing

Python 3.7 was used in this study for data preprocessing, which include desensitization, deletion of duplicate and abnormal records, missing value processing, and filling in part of the dimension outliers with the mean or median of the dimension. 2205 learners and 131917 valid data were obtained after pretreatment. The data will be analyzed for learning analytics.

4.2 Learning Analytics

The research mainly includes descriptive statistics of end-term performance, and the impact of online behaviors on learning performance. The impact of learners' online behavior on learning performance is analyzed by factor analysis and multiple linear regression by selecting relevant online behavior indicators. The analysis software is SPSS 23.

4.2.1 Descriptive Statistics

Collect the final exam scores of 2205 learners in this semester. The lowest score is 7, the highest score is 97, the average score is 74.9, and the median score is 80. 318 of them failed, and the failure rate on the exam is 14.4%. There 325 (14.7%) students scored 60 to 69, 449 (20.3%) students scored 70 to 79, 766 (34.7%) students scored 80 to 89, and 347 (15.7%) students scored 90 or more. Most of the students' final examination results are focused on more than 70 points, which indicates that the mixed teaching mode has achieved good teaching quality in general.

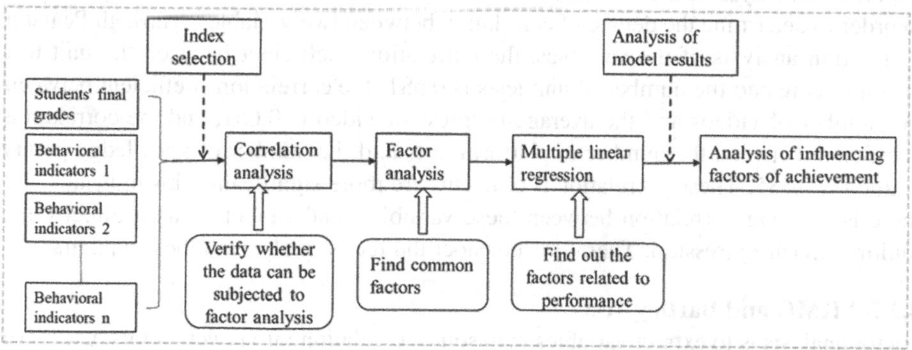

Fig. 1. The process of influencing factors analysis model for achievement

4.2.2 The Effect of Learners' Online Behavior on Learning Performance

The process of influencing factors of learners' online behavior on learning performance is shown in Fig. 1. Firstly, the appropriate online behavior indicators are selected. Then, the correlation analysis, factor analysis and multiple linear regression are carried out between the selected behavior indicators and the learners' end-of-term performance in turn. Finally, the results of the multiple linear regression model are analyzed to find the main factors affecting the performance.

13 representative online behaviors are selected and named based on the characteristics of the 5y platform and the valid data generated by the platform's learners. Among them, the number of platform tests represents the number of tests performed on the platform, as detailed in Table 1.

Table 1. Variables and their meanings

Variable	Variable meaning	Variable	Variable meaning
A1	Number of unit tests	A8	Average score on comprehensive test
A2	Unit test average score	A9	Average video progress
A3	Number of intensive trainings	A10	Number of video learning
A4	Average score of intensive training	A11	Number of platform tests
A5	Number of knowledge point tests	A12	Number of learning notes
A6	Average score of knowledge point test	A13	Number of comment statements
A7	Number of comprehensive tests		

4.2.2.1 Correlation Test

Correlation analysis is the examination of two or more variable elements for correlation in order to determine the degree of correlation between two variables. Through Pearson correlation analysis of the variables, the correlation coefficient between the unit test average score and the number of unit tests is 0.681, the correlation coefficient between the number of videos and the average progress of video is 0.651, and the correlation coefficient between the number of platform tests and the number of knowledge points learned is 0.885. These correlation coefficients are more significant. This indicates that there is a strong correlation between these variables, and cannot be used directly for multiple linear regression. Therefore, consider the factor analysis of these data first.

4.2.2.2 KMO and Bartley Test

Factor analysis is to extract variables with some correlation into fewer factors, use these factors to represent the original variables, and also classify the variables according to the factors. Its greatest advantage is that the new factors can be named and interpreted so that they can be interpreted. Before factor analysis, *KMO* and *Bartlett* tests are performed on the selected variables to determine whether the selected independent variables are suitable for factor analysis. The calculation formulas for *KMO* statistics are as follows:

$$KMO = \frac{\sum\sum_{i \neq j} r_{ij}^2}{\sum\sum_{i \neq j} r_{ij}^2 + \sum\sum_{i \neq j} \beta_{ij}^2} \quad (1)$$

In the Eq. (1), *R* is the correlation coefficient. β For the partial correlation coefficient. The *KMO* is between 0 and 1, the closer to 1, the stronger the correlation between variables, the weaker the partial correlation, and the better the effect of factor analysis. As shown in Table 2, *KMO* statistic 0.667, *KMO* above 0.6 can be used for factor analysis [13], The *Bartlett* test significance level is less than 0.01, indicating that the selected sample data meet the requirements of factor analysis.

Table 2. KMO and Bartlett test

Number of *KMO* Sampling Appropriateness Quantities		.668
Bartlett sphericity test	Approximate Chi Square	10841.457
	Freedom	78
	Significance	.000

4.2.2.3 Calculate Eigenvalue and Variance Contribution Ratio

The characteristic values of each principal component factor obtained from the online learning behavior indicators selected in this paper are 4, and the cumulative contribution rate of variance of the four factors has reached 64.775%. This shows that the extracted four common factors can better explain most of the 14 selected learning behavior indicators. Therefore, the number of common factors is determined to be 4, and they are named *F1*, *F2*, *F3*, and *F4*. The explanatory rate of factor *F1* is 21.633%, which is higher than other factors. It is the first factor that learners' online behavior affects their performance.

4.2.2.4 Refining Analysis Results

Factor rotation using the maximum variance orthogonal rotation (Varimax) method improves the interpretability of the common factor. After five iterations, the matrix converges after 5 iterations.

The factor load factor of 12 variables in the rotated factor load matrix is greater than 0.5, which makes the analysis better. Horizontally, the number of intensive training sessions A3 does not belong to any dimension, so it is an invalid variable and is deleted. The first common factor has a large load on the number of platform tests, the number of knowledge point tests, the average score of intensive training, and the average score of knowledge point tests, which can be named as the basic question factor. The second factor has a large load on the average score of comprehensive tests, the number of unit tests, the number of comprehensive tests, and the average score of unit tests. It can be named as the comprehensive question factor. The third factor has a large load on the average video progress and the number of video learning, which can be named the video viewing factor. The fourth factor has a large load on the number of comments and learning notes, which can be named learning activity factor.

4.2.2.5 Calculating Factor Score

The factor score and the final reflection of the factor analysis. By calculating the factor score, we can know the scores of the 13 selected learning behavior variables in the four extracted common factors, and analyze the end-term performance level of each variable in the common factor according to the results, as shown in Eq. (2):

$$\begin{cases} F1 = 0.003A1 + 0.120A2 + 0.119A3 + 0.285A4 + 0.290A5 \\ \quad +0.303A6 - 0.099A7 - 0.040A8 - 0.079A9 \\ \quad +0.002A10 + 0.304A11 - 0.003A12 + 0.017A13 \\ F2 = 0.322A1 + 0.215A2 + 0.064A3 - 0.059A4 + 0.021A5 \\ \quad -0.189A6 + 0.328A7 + 0.351A8 + 0.030A9 \\ \quad +0.050A10 + 0.049A11 + 0.022A12 + 0.007A13 \\ F3 = -0.095A1 - 0.200A2 + 0.049A3 - 0.102A4 + 0.042A5 \\ \quad -0.175A6 + 0.212A7 + 0.071A8 + 0.521A9 \\ \quad +0.481A10 - 0.003A11 - 0.022A13 \\ F4 = -0.014A1 - 0.055A2 + 0.068A3 + 0.027A4 - 0.003A5 \\ \quad -0.030A6 + 0.056A7 + 0.020A8 - 0.035A9 \\ \quad +0.001A10 - 0.037A11 + 0.649A12 + 0.649A13 \end{cases} \tag{2}$$

4.2.2.6 Multiple Linear Regression

Multivariate linear regression is a method of studying the relationship between a dependent variable and multiple independent variables, and it is used to explain the linear relationship between the dependent variable and other independent variables. This section performs multivariate linear regression with the four principal component factors as independent variables and the results as dependent variables to get the following regression models, as shown in Eq. (3):

$$score = 74.974 + 5.695F1 + 5.946F2 \tag{3}$$

The *R-Square* of the model is 0.279, which indicates that the model independent variable can explain 27.9% of the dependent variable change, and the *VIF* value is less than 5. This indicates that there is no multiple collinearity among independent variables, and the data residuals follow the normal distribution, indicating that the model is essentially valid. From this model, we can see that the basic factor *F1* and the comprehensive factor *F2* in the principal component factor have a positive influence on the results.

5 Research Conclusions and Recommendations

This study analyzed the online behavior of four different types of tests as well as 13 representative online behaviors. Based on the analysis of the learners' online learning behavior data on this platform and the construction of a multiple linear regression model, it is found that the basic and comprehensive problem factors have a positive impact on performance, while the video factors and learning activity factors have no direct impact on performance. In view of this conclusion, from the point of view of improving the final examination results, it is suggested that the learners should spend more time and energy on the test questions, and try to ensure the correct rate of the test, rather than pursuing the number of questions. Teachers should guide learners to complete more knowledge point tests and unit tests based on the learners' actual situation in order to improve the learning effect under the premise of limited hours and time for learners. For 5y platform, there is no significant improvement in learning performance for video viewing factor and learning activity factor. One reason is that the data of video viewing factor and learning activity factor are too sparse and not representative. On the other hand, the course developer should improve the video in the course to attract the learning interest of the learners. Although this paper only performed an in-depth analysis on the data of the students enrolled in the course of Computer Application Foundation of Guangdong Normal University for one semester, the selected data are representative in Guangdong Polytechnic Normal University and other applied for undergraduate colleges and universities. Therefore, the results of the study analysis have sufficient reference and practical significance. It provides a relevant reference for the next stage of 5y platform function improvement and the improvement of teaching quality of "Computer Application Foundation" course under the mixed teaching mode.

References

1. LAK.: Shaping the future of the field. https://lak20.solaresearch.org/ (2020)
2. Muldner, K., Wixon, M., Rai, D., et al.: Exploring the impact of a learning dashboard on student affect. In: Int. Conf. Artificial Intelligence in Education, pp. 307–317. Springer International Publishing (2015). https://doi.org/10.1007/978-3-319-19773-9_31
3. Han, X.B., Huang, Y., Ma, J., et al.: A systematic review of learning analysis: review, identification and prospect. Research On Education Tsinghua University **38**(03), 41–51+124 (2017)
4. Siemens, G., Long, P.: Penetrating the fog: analytics in learning and education. EDUCAUSE Review **46**(5), 30 (2011)
5. Song, J., Feng, J.B., Qu, K.C.: Research on the influence of teacher-student interaction on deep learning in online teaching. China Educ. Technol. **11**, 60–66 (2020)
6. Shen, X.Y., Liu, M.C., Wu, J.W., et al.: Research on MOOC learners' online learning behavior and learning performance evaluation model. Distance Education in China (10), 1–8+76 (2020)
7. Liu, M., Zheng, M.Y.: Learning analysis and personalized resource recommendation in the view of intelligent education. China Educ. Technol. **09**, 38–47 (2019)
8. Liu, F.H., Yi, X.T.: Analysis model construction and application research of online learning input. E-education Research **42**(09), 69–75 (2021)
9. Mou, Z.J.: Multimodal learning analysis: learning analysis and analysis of new growth points. E-education Res. **41**(05), 27–32+51 (2020)
10. Kent, C.,Cukurova, M.: Investigating collaboration as a process with theory- driven learning analytics. **7**(1), 59–71 (2020)
11. Shen, Y.F.: A personalized learning path recommendation model based on multiple intelligent algorithms. China Educ. Technol. **11**, 66–72 (2019)

12. Karthikeyan, V.G., Thangaraj, P., Karthik, S.: Towards developing hybrid educational data mining model (HEDM) for efficient and accurate student performance evaluation. Soft. Comput. **24**(24), 18477–18487 (2020)
13. Wu, M.L.: SPSS statistics application practice: Questionnaire analysis and application statistics. Science Press, Beijing (2003)

A Context Model for Personal Data Streams

Fausto Giunchiglia, Xiaoyue Li(✉), Matteo Busso, and Marcelo Rodas-Britez

Department of Information Engineering and Computer Science, University of Trento, Trento, Italy
{fausto.giunchiglia,xiaoyue.li,matteo.busso, marcelo.rodasbritez}@unitn.it

Abstract. We propose a model of the *situational context* of a person and show how it can be used to organize and, consequently, reason about massive streams of sensor data and annotations, as they can be collected from mobile devices, e.g. smartphones, smartwatches or fitness trackers. The proposed model is validated on a very large dataset about the everyday life of one hundred and fifty-eight people over four weeks, twenty-four hours a day.

Keywords: Personal situational context · Data streams

1 Introduction

A lot of prior work has focused on collecting and exploiting massive streams of data, e.g., sensor data and annotations. A first line of work has concentrated on using the streams of personal data for learning daily human behavior, including physical activity, see, e.g., [11], assessment personality states, see, e.g., [12], and visiting points of interest [3]. The *Reality Mining* project [4] collected smartphone sensors, including call records, cellular tower IDs, and Bluetooth proximity logs to study students' social networks and daily activities. In the same vein, the *StudentLife* project [9,15] employed smartphone sensors and questionnaires as the means for inferring the mental health, academic performance, and other behavioral trends of university students, under different workloads and term progress. Slightly different in focus, but still based on the collection of streams of data, is the work on the Experience Sampling Method (ESM). The ESM is an intensive longitudinal social and psychological research methodology, where participants are asked to report their thoughts and behaviours [14]. Here the focus is not so much on learning from the sensor data but, rather, on collecting the user provided answers. In all this work, little attention has been posed on how to represent and manage these data streams. The most common solution has

Xiaoyue receives funding from the China Scholarships Council (No.202107820014). Marcelo, Fausto and Matteo receive funding from the project "DELPhi - DiscovEring Life Patterns" funded by the MIUR (PRIN) 2017.

B. Li et al. (Eds.): APWeb-WAIM 2022, LNCS 13421, pp. 37–44, 2023.
https://doi.org/10.1007/978-3-031-25158-0_4

been that of collecting these data, *as is*, into (multiple) files in some common format, e.g. CSV. Which was good enough, given that data were exploited *a posteriori*, once the data collection was finished, by doing the proper off-line data analysis.

Our focus is on the exploitation of data, *at run-time*, while being collected, as the basis for supporting person-centric services, e.g., predicting human habits or better human-machine interaction. This type of services are in fact core for the development of *human-in-the-loop* Artificial Intelligence systems [2]. Towards this end, our proposed solution is to represent the input streams, no matter whether coming from sensors or from the user feedback, as sequences of *personal situational contexts* [7]. Here by context we mean *"a theory of the world which encodes an individual's subjective perspective about it"* [5,6]. Many challenges still need to be solved towards this goal. For instance, these data are highly heterogeneous, e.g., categorical, numerical, in natural language, and unstructured, usually collected with different time frequencies. Furthermore, different data may be at different levels of abstraction, for instance the current location can be described as, e.g., GPS coordinates, my office, the University, or the city of Trento.

The main goal of this paper is to provide a representation of data streams at the *knowledge level* [10], rather than only at the *sensor or data level*, fully understandable by the user, in the user terms, thus enabling the kind of Human-Machine interactions which we need. We realize this requirement by representing streams as sequences of situational contexts, and by modeling them as Knowledge Graphs (KGs) [1]. In this context, by KG we mean a graph where the nodes are the entities involved in the current user context, e.g., friends, the current location, the current event, for instance a meeting, while links are the relations occurring among entities, e.g., the fact that two people are classmates or that a person is on a car or talking to another person. Notice how various notions of context model have been proposed in the past. Some work focused on representing the current situation with reference to the location, see, e.g., [13]. Other approaches have used hierarchical context models [16]. However, these proposals did not deal with the problem of how to provide an abstract user-level representation of ever growing streams of data.

The proposed design the knowledge level representation of the personal situational context is articulated in three steps, as follows:

1. An abstract conceptualization of the notion of context in terms of the person space and time localisation plus the people and objects populating the context itself;
2. A schema of the KG, what we call an ETG (Entity Type Graph), which defines the data structure used the current situational context as it occurs in a certain period of time;
3. The actual data streams, memorised as sequences of context KGs each with the same ETG, differently populated.

The paper is organized as follows. Section 2 formalizes the notion of situational context. Section 3 described the details of the situational context KG. Section 4 presents a large scale case study. Finally, in Sect. 5, we present our conclusions.

2 The Situational Context

A situational context represents a real world scenario from the perspective of a specific person, whom we call `me`, e.g., Mary. A *Life sequence* is a set of situational contexts during a certain period of time. We define the life sequence of `me`, $S(me)$, as follows:

$$S\left(me\right) = \langle C_1\left(me\right), C_i\left(me\right), \ldots, C_n\left(me\right)\rangle; \quad 1 \leq i \leq n \tag{1}$$

where C_i is the i_{th} situational context of `me`. We assume that `me` can be in only one context at any given time, based on the fact that a person can be in only one location at any time. Hence, S is a sequence of `me`'s contexts, occurring one after the other, strictly sequentially, with no time in between. In turn, we model the *Situational context* of `me` $C(me)$ as follows:

$$C(me) = \langle L(C(me)), E(L(C(me)))\rangle. \tag{2}$$

In the following, we drop the argument `me` to simplify the notation. $L(C)$ is the (current) *Location* of `me`. $L(C)$ defines the boundaries inside which the current scenario evolves. The location is an endurant, which is wholly present whenever it is present, and it persists in time while keeping its identity [6]. $E(L(C))$ is an *Event* within which `me` is involved. The event is a perdurant, which is composed of temporal parts [6]. $L(C)$ and $E(L(C))$, as the priors of experience, define the scenario being modeled and the space-time volume within which the current scenario evolves. This is a consequence of the foundational modeling decision that contexts are the space-time prior to experience. In other words, the situational context of `me` is univocally defined by `me`'s *spatial position* and *temporal position.* In practice, any electronic device can easily provide us with the spatial position (via GPS, annotations, etc.) and temporal position (via timestamp) of a person.

In a certain context, `me` can be inside one or multiple locations as follows:

$$L(C) = \langle L_1(C), L_i(C), \ldots, L_n(C)\rangle; \quad 1 \leq i \leq n \tag{3}$$

where $L_i(C)$ is a spatial part of $L(C)$, we call $L_i(C)$ is a *sub-location* of $L(C)$. If `me` is inside one location, we have $L(C) = L_1(C) = \cdots = L_n(C)$, and the context is static, e.g., Mary is at the university library, or Mary is at home. Otherwise, the context is dynamic, e.g., Mary travels around Trento ($L(C)$), going from the university ($L_1(C)$), to the central station ($L_2(C)$), and then to her home ($L_3(C)$). Inside contexts, multiple events will occur:

$$E(L(C)) = \langle E_1(L(C)), E_i(L(C)), \ldots, E_n(L(C))\rangle; \quad 1 \leq i \leq n \tag{4}$$

where $E_i(L(C))$ is a part of $E(L(C))$. We call an $E_i(L(C))$ a *sub-event* of $E(L(C))$. Different sub-events may occur in parallel or be sequential or mixed, but a sub-event can not be part of another sub-event. A simple event is the event where $E(L(C)) = E_1(L(C)) = \cdots = E_n(L(C))$. A complex event is the event where there are multiple distinct sub-events.

Finally, the context contains various types of things interacting with one another. We define a *Parts of a Context* as follows:

$$P(C) = \langle me, \{P\}, \{O\}, \{F\}, \{A\} \rangle \tag{5}$$

where $\{P\}$ and $\{O\}$ are, respectively, a set of *persons* (e.g., Bob) and *objects* (e.g., Mary's smartphone) populating the current context. $\{F\}$ and $\{A\}$ are, respectively, a set of *functions* and *actions* involving `me`, persons and objects. We define a *Generic object* G, consisting of `me`, $\{P\}$, and $\{O\}$, i.e., $G = \mathtt{me} \cup \{P\} \cup \{O\}$. Functions define the roles that different generic objects have towards one another [8]. Thus a person can be a *friend* with another person, a horse can be a *transportation means* for person, while a phone can be a *communication medium* among people. Functions are endurants. Actions model how generic objects G change in time [8], e.g., Mary touches her smartphone in a certain moment, while she walks or eats at some other times. Actions are perdurants. Functions are characterized by the set of actions which enable them [8]. Thus for instance, the function *friend* might be associated with the actions *talking to*, *helping*, or *listening to*. Similarly, a smartphone (i.e., G_a) can be recognized as an entertainment tool for Mary (i.e., G_b), because the smartphone allows certain actions related to the entertainment of Mary, e.g., playing videos, playing music, etc. Hence, for two generic objects G_a and G_b, in the context, we have the following:

$$F(G_a, G_b) = \langle A_1(G_a, G_b), \ldots, A_n(G_a, G_b) \rangle; \tag{6}$$

where a function F relates G_a with G_b, namely, it is associated with the set of actions $(A_1, \ldots, A_n)$ involving G_b that G_a can do or allow.

Table 1. Properties of the principal Entity types of a situational context.

Property types	Entity types			
	Location; Sub-location	Event; Sub-event	Person; `me`	Object
Spatial property: relating to or occupying space	Coordinates Volume	None	Coordinates	Coordinates
Temporal property: relating to time	None	Start-EndTime	None	None
Function property: indicating attributes of functions	Location functions	None	Person functions	Object functions
Action property: indicating attributes of actions	None	None	Person actions	Object actions
External property: relating to outward features	Name ID	Name ID	Name ID Gender	Name ID Color
Internal property: relating to persons' internal states	None	None	InPain InMood InStress	None

3 The Entity Type Graph

We define Location, Sub-location, Event, Sub-event and Generic object as `Entity types (etypes)`, where an entity is anything which has a name and can be distinctly identified via its properties and where, in turn, an etype is a set of entities. Functions and Actions are modeled as `Object properties` representing the relations among Generic objects. In Table 1, we define and provide examples of `Spatial`, `Temporal`, `External`, and `Internal data property types` as well as of `Function` and `Action object property types`.

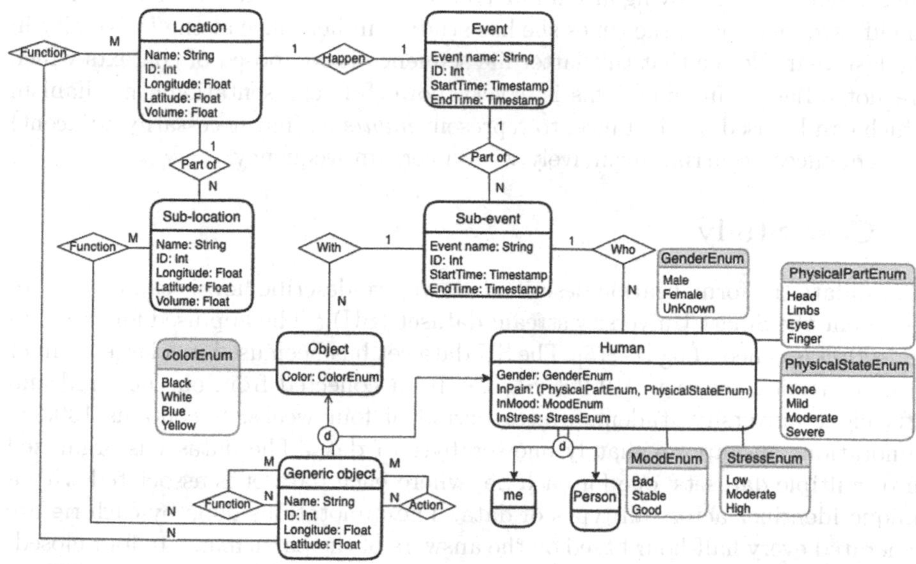

Fig. 1. An example of ETG modeling the situational context.

We represent the schema of the situational context of `me` as an eType Graph (ETG), i.e., an Enhanced Entity-Relationship (EER) model. See Fig. 1 for a simplified version of an ETG representing a personal situational context. An ETG is a knowledge graph where nodes are etypes, decorated by data properties, which in turn are linked by object properties. Each etype (represented as a box) is decorated with its data properties. For example, the etype Human has the data property Gender, and the data type (the green box) of Gender is GenderEnum. Also, etypes are connected with object properties showing their relations (represented as rhombuses). One such example, is the relation *With* which in turn is associated its own cardinality. Finally, as from EER models, it is possible to have inheritance relations among he etypes, e.g., a Generic Object is specialized into Object and Human.

Given that a context is represented by a single ETG, we represent the evolution in time of the life of a person as a sequence of ETGs, each representing the state of affairs at a certain time and for a certain time interval. In turn, this sequence of ETGs is populated by the input data streams, where each element of the stream will populate the ETG for that time slot. Of course for each input stream there will be a dedicated suitable property for the proper etype. Thus, for instance, the GPS will populate the data property *GPSLocation* of the etypes *person* and/ or *phone*, while the label of a location, e.g., *Trento* will be used to create an object property link between the etype *person* and the etype *location*. Given the above, a life sequence, as defined in Sect. 2, is just a sequence of contexts satisfying a certain property, namely, a subset of the overall sequence of ETGs, populated by the input data streams. So for instance we may have Mary's life sequence of her moving around in Trento, see example above, or we can have the life sequence of all the times she has studied in her office at the University in the last year. Notice that this latter life sequence is composed of contexts which are not adjacent in time. This is a very powerful representational mechanism which can be used, for instance, to represent *habits* as (not necessarily adjacent) life sequences occurring recursively with a certain frequency.

4 Case Study

To validate the formalization described above, we describe how it can be used to represent the Smart University stream dataset (SU).[1] The app used for the data collection is called `iLog` [17,18]. The SU data set has been used in a large number of case studies, see, e.g., [19,20]. SU has been collected from one hundred and fifty-eight university students over a period of four weeks. It contains 139.239 annotations and approximately one terabyte of data. The dataset is organized into multiple datasets, one for each `me`, where each dataset is associated with a unique identifier across all types of data. The annotations done by each `me` are generated every half-hour based on the answers of the participants to four closed-ended questions. Based on this, the best choice is to build a sequence of ETGs, one for every half an hour for each `me`. The four questions are "Where are you?", "What are you doing?", "With whom are you?", and "What is your mood?" and are based on the *HETUS* (*Harmonized European Time Use Surveys*) standard.[2]

Figures 2 and 3 provide a small, clean and anonymized subset of SU. In both figures, the first part (in white) provides the timestamps when this data were collected. In the first figure, the location of me (in green) is represented together with some of her attributes (in orange). The second figure reports the current event (in yellow) in which me is involved, her function towards the person she is with (in red) and her phone with some of its attributes (in blue). It is easy to compare the contents of Figs. 2 and 3 with the notions defined in the previous sections. Let us consider some examples:

- `Human Entities`: They are `me` and `Person`, both associated with External and Internal properties.

[1] See https://livepeople.datascientia.eu/dataset/smartunitn2 for a detailed description of the dataset plus the possibility of downloading it.

[2] https://ec.europa.eu/eurostat/web/time-use-survey.

Identifier of *me* {14}		me					Location {L}			
id	timestamp	gender	faculty	perceived_stress	extraversion	mood	where	latitude	longitude	accuracy
14	2021-05-01 10:56:00 UTC	Female	Computer science	23	25	5	Bar, Pub, etc.	46.0670° N	11.15° E	151
14	2021-05-01 10:57:00 UTC	Female	Computer science	23	25	5	Bar, Pub, etc.	46.0675° N	11.15° E	151
14	2021-05-01 10:58:00 UTC	Female	Computer science	23	25	5	Bar, Pub, etc.	46.083° N	11.1632° E	151
14	2021-05-01 10:59:00 UTC	Female	Computer science	23	25	5	Bar, Pub, etc.	46.06° N	11.15° E	151
14	2021-05-01 11:00:00 UTC	Female	Computer science	23	25	4	University Library	46.1° N	11.2° E	1230
14	2021-05-01 11:01:00 UTC	Female	Computer science	23	25	4	University Library	46.0675° N	11.15° E	-
14	2021-05-01 11:02:00 UTC	Female	Computer science	23	25	4	University Library	46.0675° N	11.15° E	50
14	2021-05-01 11:03:00 UTC	Female	Computer science	23	25	4	University Library	46.0668° N	11.1497° E	50
14	2021-05-01 11:04:00 UTC	Female	Computer science	23	25	4	University Library	46.0668° N	11.1497° E	50
14	2021-05-01 11:05:00 UTC	Female	Computer science	23	25	4	University Library	46.067° N	11.164° E	-
14	2021-05-01 11:06:00 UTC	Female	Computer science	23	25	4	University Library	46.0670° N	11.15° E	160

Fig. 2. Me and the current location.

Identifier of *me* {14}		Event {E}	Person {P(Other)}	Object {O}				
id	timestamp	what	with	notification	app_name	touch_event	wifi_connection	wifi_network_available
14	2021-05-01 10:56:00 UTC	Coffee Break	Classmate(s)	-	mail, sms	44	uni_wifi	Bob's Hotspot, uni_wifi
14	2021-05-01 10:57:00 UTC	Coffee Break	Classmate(s)	-	-	-	uni_wifi	Bob's Hotspot, uni_wifi
14	2021-05-01 10:58:00 UTC	Coffee Break	Classmate(s)	-	-	-	uni_wifi	Bob's Hotspot, uni_wifi
14	2021-05-01 10:59:00 UTC	Coffee Break	Classmate(s)	-	-	-	uni_wifi	Bob's Hotspot, uni_wifi
14	2021-05-01 11:00:00 UTC	Studying	Classmate(s)	-	-	-	uni_wifi	Bob's Hotspot, uni_wifi
14	2021-05-01 11:01:00 UTC	Studying	Classmate(s)	1	mail	32	uni_wifi	Bob's Hotspot, uni_wifi
14	2021-05-01 11:02:00 UTC	Studying	Classmate(s)	-	mail	-	uni_wifi	Bob's Hotspot, uni_wifi
14	2021-05-01 11:03:00 UTC	Studying	Classmate(s)	-	mail	-	uni_wifi	Bob's Hotspot, uni_wifi
14	2021-05-01 11:04:00 UTC	Studying	Classmate(s)	-	mail	-	uni_wifi	Bob's Hotspot, uni_wifi
14	2021-05-01 11:05:00 UTC	Studying	Classmate(s)	2	mail	-	uni_wifi	Bob's Hotspot, uni_wifi
14	2021-05-01 11:06:00 UTC	Studying	Classmate(s)	-	mail, sms	8	uni_wifi	Bob's Hotspot, uni_wifi

Fig. 3. The current event, the people and the object with Me.

- `Human's External Properties`: They are mainly collected synchronously and are represented by the variables "gender" and "faculty".
- `Human's Internal Properties`: They are both synchronic, i.e., "extraversion", and diachronic, i.e.,"mood".
- `Location Entity`: It is defined by "where" and it is annotated by the data properties "latitude" and "longitude" (with their respective "accuracy").

According to the research purpose, many additional data points may be used as proxies for characterizing the main notions of the context. For instance, concerning the `Location`, the WiFi router can be used as a proxy of a facility; a question posed in the online questionnaire about a daily routine can be a proxy of a travel path. By imputation on the GPS, it is possible to derive the Point Of Interest (POI), which can be understood as the set of `Objects` surrounding a given spatial coordinate. And so on.

5 Conclusion

This paper proposes a model of the situational context of a person and it shows how it can be used to provide a knowledge level representation over the data collected in time, both sensor data and user provided label, from mobile devices.

References

1. Bonatti, P.A., Decker, S., Polleres, A., Presutti, V.: Knowledge graphs: new directions for knowledge representation on the semantic web Dagstuhl seminar 18371. Schloss Dagstuhl-Leibniz-Zentrum fuer Informatik **8** (9), 29–111 (2019)
2. Bontempelli, A., et al.: Lifelong personal context recognition (2022). https://doi.org/10.48550/ARXIV.2205.10123, https://arxiv.org/abs/2205.10123
3. Do, T.M.T., Gatica-Perez, D.: The places of our lives: Visiting patterns and automatic labeling from longitudinal smartphone data. IEEE Trans. Mob. Comput. **13**(3), 638–648 (2013)
4. Eagle, N., Pentland, A.S.: Reality mining: sensing complex social systems. Pers. ubiquit. comput. **10**(4), 255–268 (2006)
5. Giunchiglia, F.: Contextual reasoning. Epistemologia spec. issue Linguaggiele Mach. **16**, 345–364 (1993)
6. Giunchiglia, F., Bignotti, E., Zeni, M.: Personal context modelling and annotation. In: 2017 IEEE Int. PerCom Workshops, 117–122. IEEE (2017)
7. Giunchiglia, F., Britez, M.R., Bontempelli, A., Li, X.: Streaming and learning the personal context. In: Twelfth International Workshop Modelling and Reasoning in Context, p. 19 (2021)
8. Giunchiglia, F., Fumagalli, M.: Teleologies: Objects, Actions and *Functions*. In: Mayr, H., Guizzardi, G., Ma, H., Pastor, O. (eds.) ER 2017. LNCS, vol. 10650, pp. 520–534. Springer, Cham (2017). https://doi.org/10.1007/978-3-319-69904-2_39
9. Harari, G.M., et al.: Sensing sociability: Individual differences in young adults' conversation, calling, texting, and app use behaviors in daily life. J. pers. soc. psychol. **119**(1), 204 (2020)
10. Newell, A.: The knowledge level. Artif. Intell. **18**(1), 87–127 (1982)
11. Patterson, K., Davey, R., Keegan, R., Freene, N.: Smartphone applications for physical activity and sedentary behaviour change in people with cardiovascular disease: A systematic review and meta-analysis. PloS one **16**(10), e0258460 (2021)
12. Rüegger, D., Stieger, M., Nißen, M., Allemand, M., Fleisch, E., Kowatsch, T.: How are personality states associated with smartphone data. Eur. Pers. **34**(5), 687–713 (2020)
13. Schilit, B.N., Theimer, M.M.: Disseminating active map information to mobile hosts. IEEE network **8**(5), 22–32 (1994)
14. Van Berkel, N., Ferreira, D., Kostakos, V.: The experience sampling method on mobile devices. ACM Comput. Surv. (CSUR) **50**(6), 1–40 (2017)
15. Wang, R., et al.: StudentLife: assessing mental health, academic performance and behavioral trends of college students using smartphones. In: Proceedings of ACM - UBICOMP, 3–14 (2014)
16. Wang, X.H., Da Qing Zhang, T.G., Pung, H.K.: Ontology based context modeling and reasoning using owl. In: PERCOMW'04. Citeseer (2004)
17. Zeni, M., Bison, I., Gauckler, B., Reis, F., Giunchiglia, F.: Improving time use measurement with personal big collection - the experience of the european big data hackathon 2019. J. Official Stat. (2020)
18. Zeni, M., Zaihrayeu, I., Giunchiglia, F.: Multi-device activity logging. In: Proceedings of ACM - UBICOMP: Adjunct Publication, 299–302 (2014)
19. Zeni, M., Zhang, W., Bignotti, E., Passerini, A., Giunchiglia, F.: Fixing mislabeling by human annotators leveraging conflict resolution and prior knowledge. Proc. ACM - IMWUT **3**(1), 1–23 (2019)
20. Zhang, W., et al.: Putting human behavior predictability in context. EPJ Data Sci. **10**(1), 1–22 (2021). https://doi.org/10.1140/epjds/s13688-021-00299-2

ACF2: Accelerating Checkpoint-Free Failure Recovery for Distributed Graph Processing

Chen Xu[1,2](✉), Yi Yang[1,2], Qingfeng Pan[1,2], and Hongfu Zhou[3]

[1] East China Normal University, Shanghai, China
cxu@dase.ecnu.edu.cn, {yiyang,qfpan}@stu.ecnu.edu.cn
[2] Shanghai Engineering Research Center of Big Data Management, Shanghai, China
[3] Shanghai Ruanzhong Information Technology Company Limited, Shanghai, China
hfzhou@softline.sh.cn

Abstract. Iterative computation in distributed graph processing systems typically incurs a long runtime. Hence, it is crucial for graph processing to tolerate and quick recover from intermittent failures. Existing solutions can be categorized into checkpoint-based and checkpoint-free solution. The former writes checkpoints periodically during execution, which leads to significant overhead. Differently, the latter requires no checkpoint. Once failure happens, it reloads input data and resets the value of lost vertices directly. However, reloading input data involves repartitioning, which incurs additional overhead. Moreover, we observe that checkpoint-free solution cannot effectively handle failures for graph algorithms with topological mutations. To address these issues, we propose ACF2 with a *partition-aware backup strategy* and an *incremental protocol*. In particular, the partition-aware backup strategy backs up the sub-graphs of all nodes after initial partitioning. Once failure happens, the partition-aware backup strategy recovers the lost sub-graphs from the backups, and then resumes computation like checkpoint-free solution. To effectively handle failures involving topological mutations, the incremental protocol logs topological mutations during normal execution which would be exploited for recovery. We implement ACF2 based on Apache Giraph and our experiments show that ACF2 significantly outperforms existing solutions.

Keywords: Failure recovery · Graph processing · Checkpoint-free

1 Introduction

Graph processing is widely employed in various application (e.g., social network analysis and spatial data processing). In big data era, to efficiently process big graph data, a set of large-scale distributed graph processing systems such as Pregel/Giraph [9], GraphLab [7] and PowerGraph [3] has emerged. Distributed graph processing usually involves iterative computation, where each iteration is regarded as a *superstep* [9]. Typically, the computation with a serial of supersteps leads to a long execution time. During this prolonged time span, certain nodes

B. Li et al. (Eds.): APWeb-WAIM 2022, LNCS 13421, pp. 45–59, 2023.
https://doi.org/10.1007/978-3-031-25158-0_5

of a distributed graph processing system may encounter failures due to network disconnection, hard-disk crashes, etc. Hence, it is vital that distributed graph processing systems tolerate and recover from failures automatically.

A common solution in systems like GraphLab [7] and Pregel/Giraph [9] to handle failures is to periodically write checkpoints, i.e., checkpoint-based solution. However, it consumes runtime costs to write checkpoints even though no failure happens. In contrast, some studies propose the checkpoint-free solutions, e.g., Phoenix [2] and Zorro [11], to tolerate failure. Once failure happens, it reloads input data to recover the lost sub-graphs on the failed nodes, and recovers the values of vertices on these sub-graphs by applying a user-defined compensation function. Hence, checkpoint-free solution outperforms checkpoint-based solution by saving the overhead cost of checkpointing.

Nonetheless, checkpoint-free solution still requires further improvement. First, it has to rebuild the lost sub-graphs from input data once failure happens, which involves repartitioning. Particularly, the repartitioning incurs additional recovery overhead, since it shuffles graph data during recovery. Moreover, checkpoint-free solution fails to handle the failures involving topological mutations (e.g., deletions of edges). Once failures occur, checkpoint-free solution re-initializes the topology on the failed nodes from input data, but keeps the topology on the normal nodes unchanged. This may lead to result inaccuracy, since checkpoint-free solution discards topological mutations of vertices on failed nodes.

In this work, we propose a prototype system, namely ACF2, with a *partition-aware backup strategy* to reduce the recovery overhead, and an *incremental protocol* to handle failures involving topological mutations. The partition-aware backup strategy backs up the sub-graphs of all nodes into reliable storage after partitioning. Once failure happens, the partition-aware backup strategy recovers the lost sub-graphs from the backup and then restarts computation like checkpoint-free solution. Our experimental results show that, in case of failures, the partition-aware backup strategy is able to reduce the overall execution time by 29.4% in comparison to Phoenix [2], a state-of-the-art checkpoint-free solution. The incremental protocol logs the topological mutations during execution. Upon failure, our protocol utilizes these logs to recover the topology on all nodes to a certain superstep before failure. In our experiments, we find the incremental protocol improves the result accuracy by up to 50% compared to Phoenix.

In the rest of this paper, we introduce the background of the checkpoint-free solution and our motivation in Sect. 2, and make the following contributions.

- We propose a *partition-aware backup strategy* in Sect. 3 that decreases the overhead imposed by checkpoint-free solution on the lost sub-graphs.
- We devise an *incremental protocol* in Sect. 4, so as to support fault tolerance with graph topological mutations.
- We describe the implementation of ACF2 based on Giraph in Sect. 5 and demonstrate the ACF2 outperforms state-of-the-art solutions via experimental evaluation in Sect. 6.

We discuss the related work in Sect. 7 and conclude this paper in Sect. 8.

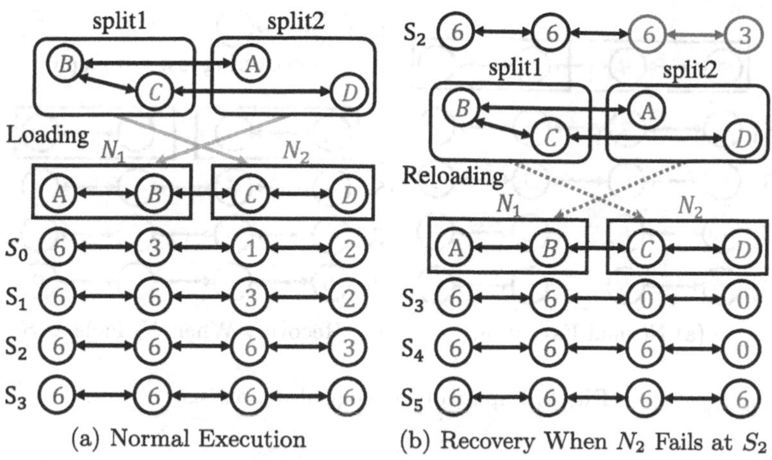

Fig. 1. The working process of checkpoint-free solution

2 Background and Motivation

Many graph processing systems such as GraphX [4] employ a checkpoint-based solution to tolerate failures. During normal execution, the checkpoint-based solution writes checkpoints periodically. Once failure happens, the systems read the latest checkpoint for rollback and start recomputation. Clearly, this checkpoint-based solution incurs a high overhead even though no failure happens.

To avoid the checkpointing overhead, some studies propose checkpoint-free solution, such as Phoenix [2], optimistic recovery [12] and Zorro [11]. Differently, the solution does not write any checkpoint during normal execution. Once failure happens, it loads input data to recover the lost sub-graphs on failed nodes via partitioning. After that, the checkpoint-free solution employs a user-defined compensation function to recover the value of vertices on lost sub-graphs and continues computation. In particular, it does not rollback the value of vertices and starts recomputation, but utilizes semantic properties of graph algorithms to set the value of vertices, so as to decrease the overhead of recomputation [2].

Figure 1 takes maximum value calculation as an example to illustrate the checkpoint-free solution. Here, the job executed on two nodes N_1 and N_2 consists of four suspersteps from superstep S_0 to S_3. As shown in Fig. 1(a), before executing S_0, N_1 and N_2 load sub-graphs from input data splits split1 and split2, respectively. However, according to partition function, the vertex C on the sub-graph in split1 does not belong to N_1. Likewise, A does not belong to N_2. Hence, N_1 and N_2 exchange vertices with each other during loading, called *shuffle*. After that, the job executes computation from S_0 to S_3. During computation, the checkpoint-free solution achieves zero overhead, since it does not write checkpoints. Once N_2 fails at S_2, as shown in Fig. 1(b), the sub-graph consisting of C and D on N_2 is lost. To recover the lost sub-graph, the checkpoint-free solution requires N_1 and N_2 to reload the sub-graphs from split1 and split2.

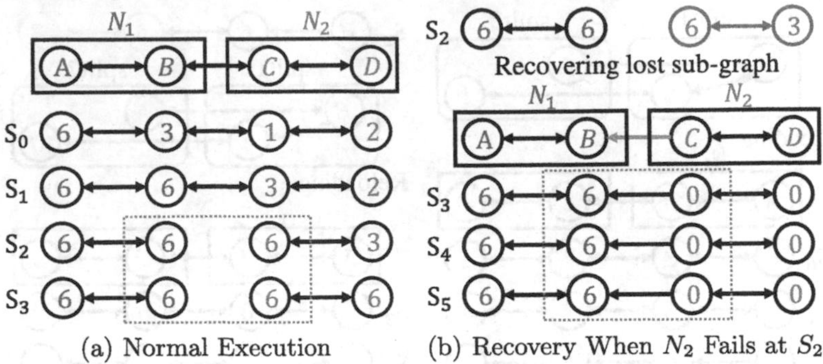

(a) Normal Execution (b) Recovery When N_2 Fails at S_2

Fig. 2. Impact of topological mutations

Then, the checkpoint-free solution compensates the value of vertices C and D to minimum value, i.e., 0. This avoids the recomputation overhead on updating the value of D to 3. After that, the job continues the computation and terminates the iteration at S_5.

Recovery Overhead. Once failure occurs, the checkpoint-free solution has to reload the splits of input data and executes partitioning. However, the partitioning shuffles data via the network during recovery, which introduces additional overhead. We take Fig. 1(b) as an example to illustrate this. Once N_2 fails at S_2, the checkpoint-free solution asks N_1 and N_2 to reload input data. Then, N_1 recovers the vertex C by shuffle, i.e., N_1 sends C via the network to N_2 after reading C from split1. Clearly, the checkpoint-free solution avoids the additional overhead incurred by shuffle if N_2 can read the vertex C directly.

Inaccuracy. Besides recovery overhead, the checkpoint-free solution leads to a decrease in the result accuracy to recover graph algorithms with topological mutations. To elaborate this, Fig. 2 shows the normal execution and failure recovery of maximum value graph job with topological mutations. As shown in Fig. 2(a), the vertex B sends the maximum value 6 as a messages to C by edge $\langle B, C\rangle$ during normal execution. Then, the vertex C updates its value to 6 at S_2. Meanwhile, the job removes the edges $\langle B, C\rangle$ and $\langle C, B\rangle$. Subsequently, the job updates the value of vertex D to 6 and finishes the computation at S_3. Once N_2 fails at S_2, as shown in Fig. 2(b), the checkpoint-free solution employs the same way as Fig. 1(b), i.e., reloading input data, to recover the topology of N_2 to initial topology at S_0. Moreover, for normal node N_1, the checkpoint-free solution keeps its topology unchanged. Hence, at superstep S_3, there is an edge $\langle C, B\rangle$ but no edge $\langle B, C\rangle$ in the graph. Without the edge $\langle B, C\rangle$, vertex B is unable to send the maximum value 6 as a messages to C again. Hence, C and D cannot update their values to the maximum value 6, so that the job eventually outputs inaccurate results.

Further, combining the above example, we summarize the reasons for the inaccuracy introduced by the checkpoint-free solution. There two reasons for the inaccuracy: (a) computation on the failed nodes requires messages from

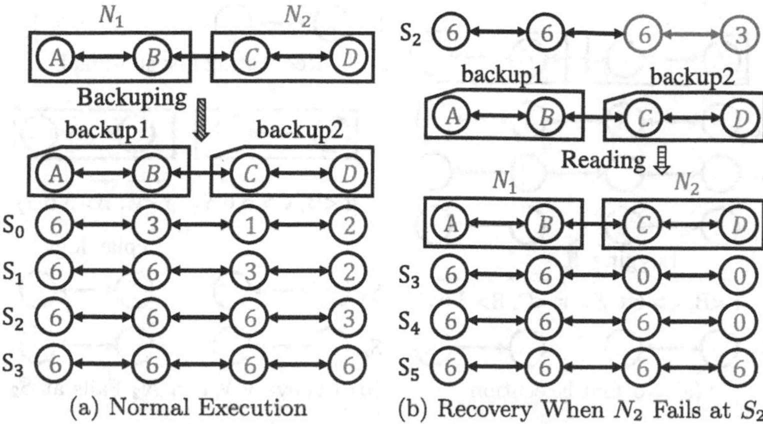

Fig. 3. Partition-aware backup strategy

the normal nodes, while topological mutations break the link used to transfer message from the vertices on normal nodes to the vertices on failed nodes; (b) the checkpoint-free solution does not fix these links, since it keeps the topology of the normal nodes unchanged during recovery.

3 Partition-aware Backup Strategy

To avoid the additional overhead caused by the checkpoint-free solution during recovery, we propose partition-aware backup strategy which saves the sub-graphs of all nodes to distributed file system (DFS) during normal execution. Once failures occur, our strategy recovers the lost sub-graphs directly from the backups instead of from the input data. Then, we are able to apply a user-defined function following the checkpoint-free solution.

During normal execution, the design of the partition-aware backup strategy involves two questions: *how many* times to backup; and *when* to backup. For the first question, the times of backuping depend on whether the system repartitions the sub-graphs on the nodes during computation. However, repartitioning sub-graphs does not offer significant performance improvements except under particular conditions [6,10]. Based on this observation, we assume that the system does not repartition the sub-graphs, and therefore back up the sub-graphs of all nodes only once. Moreover, we employ an unblocking manner to minimize the influence of backup on computation. This manner backs up the sub-graphs in parallel with the computation. For the second question, our strategy immediately backs up sub-graphs after partitioning, i.e., at superstep S_0, so as to make the backup available as soon as possible. As an example in Fig. 3(a), at superstep S_0, our strategy asks the sub-graph writers on N_1 and N_2 to back up their sub-graphs. Here, the backup is in parallel with the iterative computation, which involves a low overhead cost.

Once failure happens, our strategy checks the availability of backup. Then, our strategy allows the sub-graph readers on the failed nodes to recover the

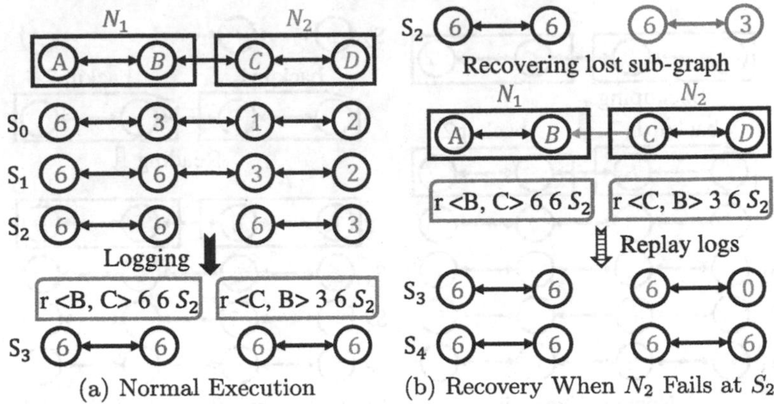

Fig. 4. Incremental protocol

lost sub-graphs from the backup when the backup is available. Otherwise, it lets these readers recover the lost sub-graphs from the input data. Then, our strategy employs the same way as checkpoint-free solution to recover the values of the vertices. As illustrated by the example in Fig. 3(b), the backup is available when N_2 fails at superstep S_2. Hence, the sub-graph reader on N_2 recovers the lost sub-graph from the backup rather than the input data, and resets the values of vertices following checkpoint-free solution.

4 Incremental Protocol

In this section, we propose an incremental protocol to deal with failure involving topological mutations, so as to ensure the result accuracy. To improve the result accuracy, a naïve solution is to reset the topology of normal nodes to their initial topology at S_0, which ensures that normal nodes can resend messages to failed nodes. However, this solution forces the job to recompute from the initial graph topology, wasting the computation.

To reduce the overhead on recomputation, we propose an incremental protocol, which logs topological mutations of all nodes during execution. Once failure occurs, our protocol utilizes these logs to redo mutations on failed nodes, so as to avoid the overhead of recomputation involving message transmission and processing. In detail, our protocol consists of the following two phases, i.e., normal execution and failure recovery.

During normal execution, once the node N_i modifies the topology of graph at superstep S_j, our protocol lets the logger on N_i log these modifications to DFS. In particular, the logger also stores the type of modifications, e.g., adding or removing edges, as well as the superstep where the mutations occur. Moreover, when the modifications of all nodes at superstep S_j is fully stored, our protocol creates a log flag file of S_j on DFS. This log flag file indicates that it is viable to redo the mutations from S_0 to S_j on failed nodes once failure happens.

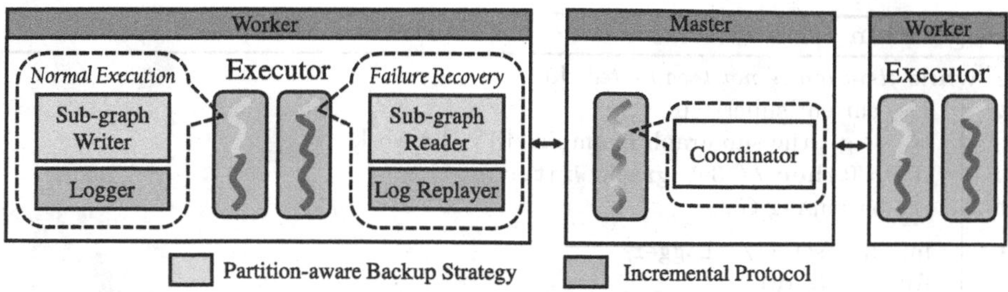

Fig. 5. System architecture

Figure 4(a) provides an example on how the incremental protocol works during normal execution. At superstep S_2, the vertex C on N_2 updates the value to 6 and removes the edge $\langle C, B\rangle$ after receiving the message from B on N_1. Meanwhile, the logger on N_2 logs the information related to the modification "r $\langle C, B\rangle$ 3 6 S_2" to DFS. These information contains the type of modification r, the removed edges $\langle C, B\rangle$, the change in the value of vertex from 3 to 6 as well as the superstep S_2. Likewise, the logger on N_1 logs the information related to the modification. Once these information at superstep S_2 is fully stored, our protocol creates a log flag file of S_2 on DFS.

Once failure occurs at superstep S_f, our protocol loads all log flag files and generates a list L consisting of the supersteps in these flag files. Then, our protocol requires all nodes to load their respective log from DFS. For failed nodes, our protocol obtains the superstep S_t of each modification in log and redoes the modification when S_t belongs to L. In contrast, our protocol does not redo the modification when S_t does not belong to L, since our protocol may not have wrote fully the mutations on failed nodes to DFS during execution. To redo these mutations, the failed nodes still requires messages from the normal nodes. Hence, our protocol undoes the modification when S_t does not belong to L for normal nodes, so as to fix the link used to transfer messages.

Figure 1(b) depicts how the incremental protocol works during recovery. As shown in Fig. 4, once N_2 fails at superstep S_2, our protocol generates a list L including S_2 and lets the log replayers on N_1 and N_2 read logs from the DFS. Then, the log replayer on N_2 redoes the modification "r $\langle C, B\rangle$ 3 6 S_2", since L contains S_2. Meanwhile, the log replayer on N_1 does not redo the modification involving B, since it is normal node and L contains S_2. It is worth noting that the log replayer undoes the modification involving B if L does not contain S_2, so as to fix the link from B to C. Then, the vertex D receives the message from C and performs the subsequent computation. Eventually, the computation will produce accurate results.

Algorithm 1: Normal Execution of the Worker n

```
1  while iteration is not terminated do
2      i ← current superstep
3      G_n ← get the sub-graph maintained by the worker n
4      if i = 0 then // Sub-graph Writer
5          backuping G_n
6      initialize set L // Logger
7      for v ∈ G_n do
8          if v modifies topology then
9              R_n^i ← topological mutations
10             add R_n^i into L
11     write L to DFS
12     i ← i + 1
```

5 System Implementation

In this section, we present the implementation of ACF2 which is based on Apache Giraph. Figure 5 shows the architecture of our proposed ACF2, which inherits the master/worker architecture of the existing systems. Specifically, the master is responsible for coordinating worker activity, whereas the worker executes computation via a series of executor threads. Moreover, the worker integrates the failure recovery mechanism implemented through an executor thread. To achieve the partition-aware backup strategy, we introduce a new component *Sub-graph Writer* for the worker and integrate it into the thread responsible for the computation. Also, we modify the execution logic of the thread responsible for failure recovery on the worker, so as to integrate the new component *Sub-graph Reader*. Likewise, to achieve the incremental protocol, we add two new components to worker, i.e., *Logger* and *Log Replayer*. Moreover, we modify the execution logic of the *Coordinator* component on the master to record the failed supersteps, so as to coordinate these new components in the worker.

Algorithm 1 illustrates the implementation details of one worker (say worker n) during normal execution. Our partition-aware backup strategy employs sub-graph writer obtains and backs up the sub-graph of each node after partitioning, i.e., superstep S_0 (line 3–5). The logger in incremental protocol monitors the behavior of each vertex v at each superstep. When v modifies the graph topology, the logger logs the topological mutations R_n^i in superstep S_i of worker n, and adds it to a set L that stores all topological mutations (line 6–10). Then, it uploads logs to DFS (line 11).

Algorithm 2 describes the implementation details of worker n upon failure. The partition-aware strategy adopts sub-graph reader to load lost sub-graphs from the backup if it exists (line 2). Otherwise, it reloads the input data (line 4). The log repalyer in incremental protocol get a list F of failed supersteps from master and read logs from DFS into L (line 5–6). For each record R_n^i in logs,

Algorithm 2: Failure Recovery of the Worker n

1 **if** *backup exists* **then** // sub-graph reader
2   load lost sub-graphs from the backup
3 **else**
4   recover lost sub-graphs from input data
5 list $F \leftarrow$ get failed supersteps from Master
6 $L \leftarrow$ read logs for worker n
7 **for** *each* R_n^i *in* L **do** // log replayer
8   **if** $i \in F$ **then**
9   undo R_n^i
10   **else**
11   redo R_n^i

the protocol obtains the superstep S_i saved in R_n^i and undoes it if S_i belongs to F (line 9). Otherwise, our protocol redoes R_n^i (line 11).

6 Experiments

This section introduces our experimental setting and demonstrates the efficiency of the partition-aware backup strategy and the incremental protocol.

6.1 Experimental Setting

Cluster Setup. We conduct all experiments on a cluster with 13 compute nodes. Here, each compute node has a eight-core Intel Xeon CPUs, a 32GB RAM, a 300GB SSD and 1Gbps Ethernet. We deploy Hadoop 2.5.1 on this cluster, since the Giraph program is executed via a MapReduce job on the top of Hadoop. By default, we issue 13 map tasks and each task owns 24GB memory. Here, we follow previous studies [1,8,13] to set the size of task memory.

Algorithms. We choose two algorithms to evaluate the effectiveness of partition-aware backup strategy and incremental protocol. They are the single source *shortest path* without topological mutations and the *maximum weight matching* with topological mutations. These algorithms are popular in graph analysis applications [5]. In the rest of this paper, we refer to above algorithms as *SP* and *MWM*, respectively. We set the number of iterations of *MWM* to 50. Moreover, we conduct the experiments over three real-life graphs, Orkut[1], WebCC12[2] and Friendster[3], with million-scale or billion-scale edges.

[1] http://networkrepository.com/orkut.php.
[2] http://networkrepository.com/web-cc12-PayLevelDomain.php.
[3] https://snap.stanford.edu/data/com-Friendster.html.

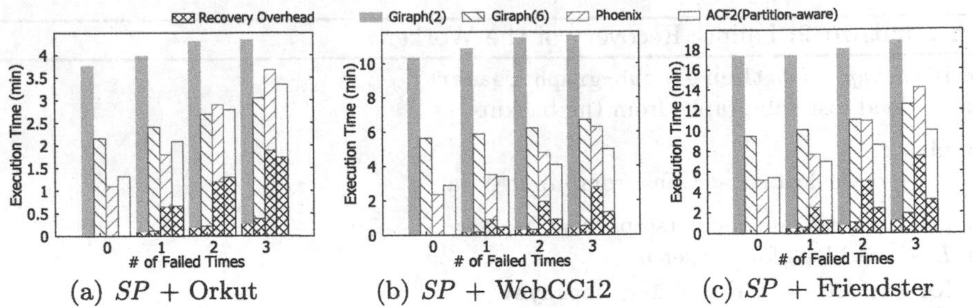

(a) SP + Orkut (b) SP + WebCC12 (c) SP + Friendster

Fig. 6. Impact of the number of failures

Baselines. We take the original checkpoint-based recovery solution in Giraph as a baseline. The checkpoint interval for this solution imposes an impact on performance. For fairness, based on previous studies [12,15,17], we set the checkpointing interval to two and six, respectively. Also, we take the state-of-the-art checkpoint-free solution, i.e., Phoenix, as a baseline and implement it in Giraph. Compared to these baselines, we evaluate our proposed ACF2 consisting of a partition-aware backup strategy and an incremental protocol.

6.2 Efficiency of Partition-aware Backup Strategy

In this section, we evaluate performance of the partition-aware backup strategy which means that ACF2 does not enable the incremental protocol, denoted as ACF2(Partition-aware) in Fig. 6, 7, 8, 9 and 10. In particular, we consider the impact of two factors on performance: (i) the number of failures and (ii) the time at which they occur, i.e., failed superstep.

Number of Failures. We focus on the case where the number of failures is less than three, since checkpoint-free solution assumes that modern clusters has large mean time between failures of a machine [11]. Figure 6 depicts the execution time of failure recovery with different number of failures n ranging from zero to three. Particularly, the failure does not happen when $n = 0$. When $n = 1$, we issue one failure at superstep S_6. When $n = 2$, we issue two failures at S_6 and S_8. When $n = 3$, we issue three failures at S_6, S_8 and S_{10}.

As shown in Fig. 6, Giraph with a checkpointing interval of six, i.e., Giraph(6) in Fig. 6, always outperforms Giraph with a checkpointing interval of two denoted as Giraph(2). The reason is Giraph with a checkpointing interval of six writes fewer checkpoints. Based on this observation, in the following discussion, we only discuss the differences between Giraph with a checkpoint interval of 6 and other recovery solutions.

In general, Phoenix performs better than the Giraph. However, Phoenix performs worse than the Giraph in some cases because of its additional overhead involving repartitioning. As an example in Fig. 6(a), in comparison to Giraph, Phoenix decreases the overhead by up to 58.6% when $n = 0$, while they exceed the execution time of the Giraph when $n = 2$.

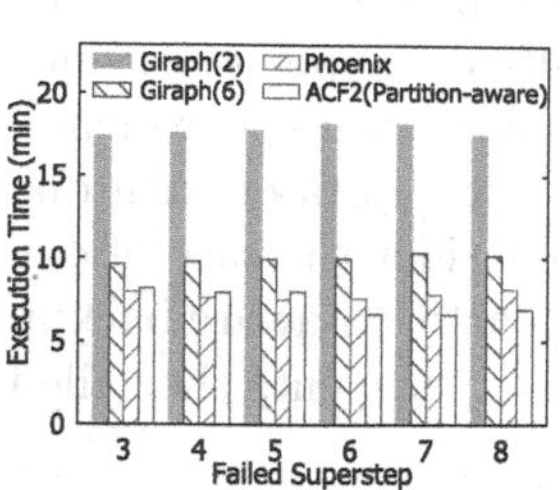

Fig. 7. Different failed supersteps

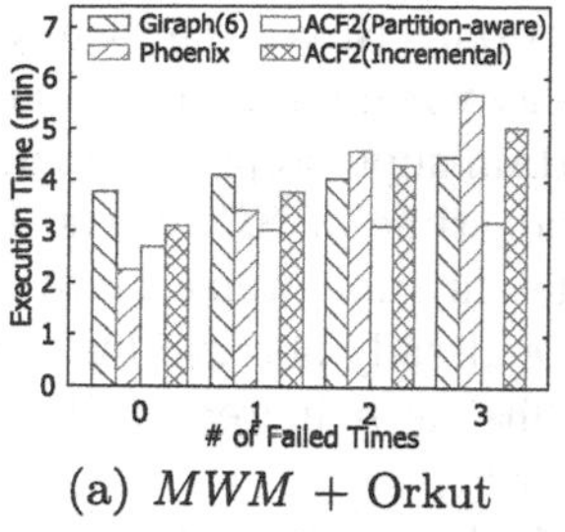

(a) *MWM* + Orkut

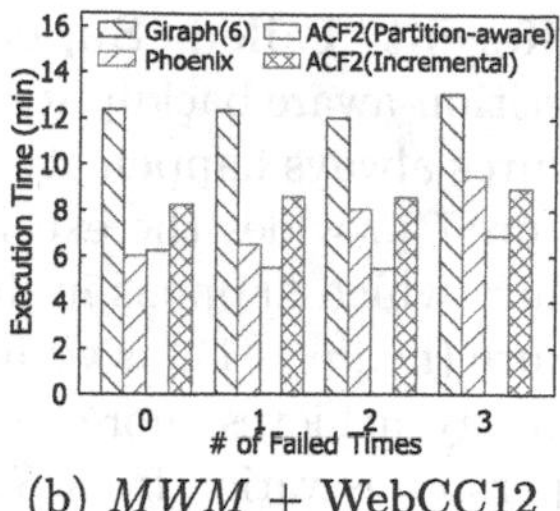

(b) *MWM* + WebCC12

Fig. 8. Impact of the number of failures

The partition-aware backup strategy keeps low overhead under failure-free cases, so as to achieve similar performance to Phoenix when $n = 0$. As shown in Fig. 6(c), the overhead cost of the partition-aware backup strategy on the Orkut dataset only increases 4.6% as against Phoenix. In case of failure, the partition-aware backup strategy outperforms other recovery solutions, and achieves the best performance. For example, in Fig. 6(c), the partition-aware backup strategy decreases the overhead by 31% compared to the Giraph, and by up to 9% compared to Phoenix when $n = 1$. Notably, for individual cases depicted in Fig. 6(a), the partition-aware backup strategy performs worse than Giraph. The reason is that the backup is not completed in time when the *SP* algorithm is executed on the small dataset Orkut. This leads to the partition-aware backup strategy to recover lost partition data using a similar way to Phoenix.

As the number of failures n increases, the partition-aware backup strategy benefits more than Phoenix on the execution time. As shown in Fig. 6(c), the percentage of overhead cost reduced by the partition-aware backup strategy increases from 9% to 29.4% as against Phoenix when n increases from 1 to 3.

Overhead of Recovering Lost Sub-graphs. The change of n has an impact on the overhead of recovering lost sub-graphs, which decides the performance of different recovery solutions. Next, we further evaluate the overhead of recovering lost sub-graphs caused by these solutions.

In Fig. 6, the partition-aware backup strategy achieves a lower recovery overhead than Phoenix, since it recovers lost sub-graphs from backups which avoids the repartitioning during recovery. As depicted in Fig. 6(b), the Giraph decreases the overhead of recovering lost sub-graphs by up to 81.2%, compared to Phoenix when $n = 1$. In particular, Giraph obtains a similar recovery overhead as the partition-aware backup strategy, since it recovers lost sub-graphs from the checkpoint which also avoids the repartitioning.

Along with n increases, advantages of the partition-aware backup strategy during recovery phase becomes significant. As shown in Fig. 6(c), our strategy decreases the overhead by 67.2% as against Phoenix when $n = 1$. Moreover, as n increases, the overhead of recovering lost sub-graphs for Phoenix grows faster than that of the partition-aware backup strategy, since the cost on reloading the input data is higher than the cost on loading backup.

Different Failed Supersteps. Next, we study the performance of the partition-aware backup strategy when failure occurs in different supersteps, since failures always happen at certain supersteps in the aforementioned experiments. Figure 7 provides the execution time of *SP* on the Friendster dataset with one failure which happens in different supersteps. Here, we consider the case of one failure (i.e., $n = 1$), as we have already observed that the partition-aware backup strategy achieves more benefits as n increases. In this experiment, the failed superstep is varied from S_3 to S_8.

Clearly, as shown in Fig. 7, the partition-aware backup strategy outperforms other recovery solutions as long as the backup of sub-graphs is available, i.e., the failure occurs after superstep S_5. Otherwise, the partition-aware backup strategy behaves similarly to Phoenix. In other words, the partition-aware backup strategy degenerates to Phoenix when the backup is not available.

In summary, the partition-aware backup strategy outperforms Giraph and Phoenix in most cases. Moreover, with the increasing of the number of failures, the partition-aware backup strategy obtains more benefits as against Phoenix.

6.3 Efficiency of Incremental Protocol

In this section, based on the partition-aware backup strategy, we compare the performance of the incremental protocol against other recovery solutions. This means that ACF2 enables both the partition-aware backup strategy and the incremental protocol, denoted as ACF2(Incremental). In particular, we exclude Giraph with a checkpointing interval of two in this group experiments, since previous experiments have demonstrated that Giraph with a checkpointing interval of six outperforms it.

Number of Failures. Figure 8 reports the total execution time of different recovery solutions with the number of failures n ranging from zero to three. Here, we follow the previous failures setting and issue failures at the same supersteps. Moreover, due to space limitation, we only take *MWM* on Orkut and WebCC12 datasets as examples. However, similar results hold on Friendster.

As shown in Fig. 8, the incremental protocol outperforms Giraph in most cases. As an example depicted in Fig. 8(b), compared to Giraph, the incremental protocol reduces the overhead by 28.7% on the execution time when $n = 2$. Due to the logging overhead, our incremental protocol performs worse than the partition-aware backup strategy and Phoenix. For example, in Fig. 8(b), the execution time of incremental protocol is always longer than that of the partition-aware backup strategy, no matter how many times the failure occurs. Moreover, compared to Phoenix, the overhead of incremental protocol increases 32% when $n = 1$. However, incremental protocol saves the execution time up to 5% when $n = 3$.

Result Accuracy. Next, we focus on the result accuracy, since the incremental protocol affects it. To evaluate the accuracy of *MWM*, we define the *correct match* metric, which means the fraction of matched vertices with the same value as the original result, i.e., the result under the failure-free case. This metric is

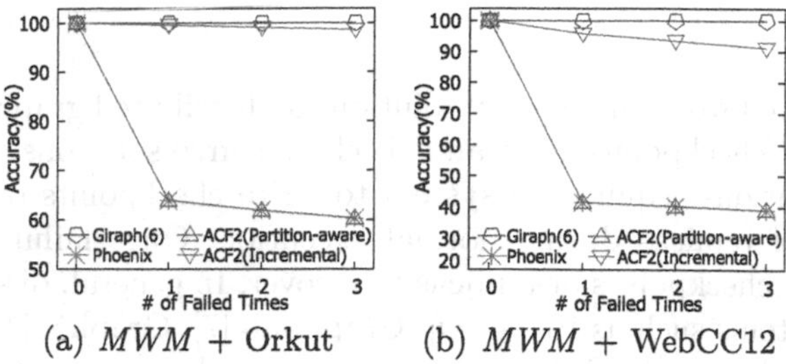

(a) MWM + Orkut (b) MWM + WebCC12

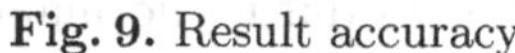

Fig. 9. Result accuracy

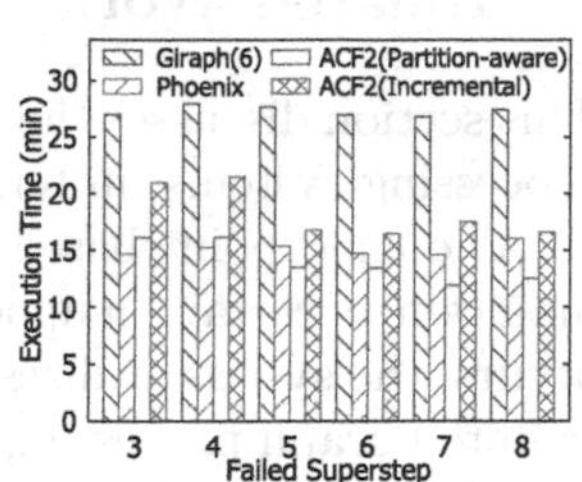

Fig. 10. Different failed supersteps on MWM

computed via $m_c(v)/m_a(v) * 100\%$, where $m_c(v)$ denotes the matched vertices with the same value as the original result and $m_a(v)$ denotes all matched vertices in the original result.

As shown in Fig. 9, Phoenix and the partition-aware backup strategy do not effectively handle failures when running graph algorithms with topological mutations. For example, in Fig. 9(b), Giraph achieves no accuracy loss in all cases. However, Phoenix and the partition-aware backup strategy achieve the maximum accuracy of around 40% when $n = 1$. Moreover, the accuracy of Phoenix and the partition-aware backup strategy continue to decline as n increases from 1 to 3.

The incremental protocol efficiently handles failures when running graph algorithms with topological mutations. As shown in Fig. 9(a), compared to Phoenix and the partition-aware backup strategy, the incremental protocol leads to an inaccuracy of 0.4% when $n = 1$. Further, when n increases from 1 to 3, the incremental protocol still ensures an accuracy over 98.6%, even though its accuracy decreases. The incremental protocol still sacrifices the accuracy of the results, which is because we limit the number of iterations. However, our protocol achieves 100% accurate results if we do no limit the number of iterations [2].

Different Failed Supersteps. Similarly, we continue to investigate the performance trend of the incremental protocol when failure occurs in different supersteps. Also, we follow the previous failure setting. Figure 10 provides the results on the Friendster dataset.

As shown in Fig. 10, no matter in which superstep the failure happens, the incremental protocol always outperforms Giraph. Moreover, the incremental protocol performs worse than Phoenix and the partition-aware backup strategy, when the failure occurs before S_5. This is because our protocol introduces logging overhead and employs a similar way as Phoenix to recover lost sub-graphs. However, the performance of the incremental protocol is close to that of Phoenix when the failure occurs after S_5, due to the backup of sub-graphs.

In summary, our incremental protocol outperforms Phoenix, since we effectively handle failures involving topological mutations. Also, our protocol achieves similar accuracy as Giraph and faster execution than Giraph.

7 Related Work

This section discusses the related work on recovery solutions of distributed graph processing systems, in term of checkpoint-based and checkpoint-free solutions.

The checkpoint-based solutions require the system to write checkpoints or maintain a certain number of replicas during normal execution. Once failure occurs, the system utilizes the checkpoints or replicas to recover. In general, distributed graph processing systems such as Pregel [9], GraphLab [7], GraphX [4] and PowerGraph [3] employ the checkpoint-based solution to tolerate failure. These systems periodically write the checkpoints during normal execution and reload the latest checkpoint upon failure. Rather than writing all the edges in each checkpoint, the lightweight checkpointing [17] reduces the data volume of checkpoint by saving the vertices and incrementally storing edges. Xu et al. [16] explore unblocking checkpoint to decrease the overhead of blocking checkpoint for graph processing on dataflow systems. CoRAL [14] applies unblocking checkpointing to asynchronous graph processing systems. Shen et al. [13] propose a repartition strategy to reduce the communication cost during recovery, so as to accelerate recovery. However, these works focus on checkpoint-based failure recovery, whereas our work targets checkpoint-free failure recovery. Also, Spark [18] employs checkpoint-based solutions to tolerate failure. In particular, Spark utilizes the lineage of RDD to accelerate recovery. During recovery, it replays the transformation to recompute the lost RDD sub-graphs. However, our work directly loads the lost sub-graphs from the backups and resets the vertex value by semantic properties of graph algorithms, which avoids the recomputation.

In contrast to the checkpoint-based solutions, the checkpoint-free solutions achieve failure recovery without any checkpoint or replica. Schelter et al. [12] propose the optimistic recovery which reloads the lost partitions from input data and applies the algorithmic compensations in the lost partitions once failure happens. As a variant of optimistic recovery, Zorro [11] utilizes the implicit replicas of vertices in graph systems to recover the value of lost vertices, so as to accelerate recovery. The optimistic recovery is applicable only to partial graph algorithms. Different from optimistic recovery, Phoenix [2] classifies existing graph algorithms into four classes and provides APIs for these different classes of algorithms to achieve failure recovery. However, these works do not consider the additional overhead caused by recovering lost sub-graphs, and the failure involving topological mutations.

8 Conclusions

This paper proposes ACF2 to mitigate the shortcoming of checkpoint-free solution. In specific, ACF2 includes a partition-aware backup strategy and an incremental protocol. Instead of reloading input data, the partition-aware backup strategy recovers the lost sub-graphs from the backups on DFS, so as to reduce the recovery overhead incurred by checkpoint-free solution. The incremental protocol which logs topological mutations during normal execution and employs these logs for failure recovery. The experimental studies show that ACF2 outperforms existing checkpoint-free solutions in general. Presently, we implement ACF2 based on Giraph. Nevertheless, it is possible to integrate ACF2 in other distributed graph processing systems such as GraphLab.

Acknowledgments. This work has been supported by the National Natural Science Foundation of China (No. 61902128).

References

1. Ammar, K., et al.: Experimental analysis of distributed graph systems. Proc. VLDB Endow. **11**(10), 1151–1164 (2018)
2. Dathathri, R., et al.: Phoenix: a substrate for resilient distributed graph analytics. In: ASPLOS, pp. 615–630 (2019)
3. Gonzalez, J.E., et al.: Powergraph: distributed graph-parallel computation on natural graphs. In: OSDI, pp. 17–30 (2012)
4. Gonzalez, J.E., et al.: Graphx: graph processing in a distributed dataflow framework. In: OSDI, pp. 599–613 (2014)
5. Kalavri, V., et al.: High-level programming abstractions for distributed graph processing. IEEE Trans. Knowl. Data Eng. **30**(2), 305–324 (2018)
6. Li, B., et al.: : A trusted parallel route planning model on dynamic road networks. TITS (2022)
7. Low, Y., et al.: Distributed graphLab: a framework for machine learning in the cloud. Proc. VLDB Endow. **5**(8), 716–727 (2012)
8. Lu, Y., et al.: Large-scale distributed graph computing systems: an experimental evaluation. Proc. VLDB Endow. **8**(3), 281–292 (2014)
9. Malewicz, G., et al.: Pregel: a system for large-scale graph processing. In: SIGMOD, pp. 135–146 (2010)
10. McCune, R.R., Weninger, T., Madey, G.: Thinking like a vertex: a survey of vertex-centric frameworks for large-scale distributed graph processing. ACM Comput. Surv. **48**(2), 1–39 (2015)
11. Pundir, M., et al.: Zorro: zero-cost reactive failure recovery in distributed graph processing. In: SoCC, pp. 195–208 (2015)
12. Schelter, S., et al.: "All roads lead to rome": optimistic recovery for distributed iterative data processing. In: CIKM, pp. 1919–1928 (2013)
13. Shen, Y., et al.: Fast failure recovery in distributed graph processing systems. Proc. VLDB Endow. **8**(4), 437–448 (2014)
14. Vora, K., et al.: Coral: confined recovery in distributed asynchronous graph processing. In: ASPLOS, pp. 223–236 (2017)
15. Wang, P., et al.: Replication-based fault-tolerance for large-scale graph processing. In: DSN, pp. 562–573 (2014)
16. Xu, C., et al.: Efficient fault-tolerance for iterative graph processing on distributed dataflow systems. In: ICDE, pp. 613–624 (2016)
17. Yan, D., et al.: Lightweight fault tolerance in Pregel-like systems. In: ICPP, pp. 1–10 (2019)
18. Zaharia, M., et al.: Resilient distributed datasets: a fault-tolerant abstraction for in-memory cluster computing. In: NSDI, pp. 15–28 (2012)

Death Comes But Why: An Interpretable Illness Severity Predictions in ICU

Shaofei Shen[1], Miao Xu[1], Lin Yue[2], Robert Boots[3], and Weitong Chen[4(✉)]

[1] The University of Queensland, St Lucia, Australia
{shaofei.shen,miao}@uq.net.au
[2] University of Newcastle, Newcastle, Australia
Lin.Yue@newcastle.edu.au
[3] Royal Brisbane and Women's Hospital, Herston, Australia
robert.boots@health.qld.gov.au
[4] University of Adelaide, Adelaide, Australia
t.chen@adelaide.edu.au

Abstract. Predicting the severity of an illness is crucial in intensive care units (ICUs) if a patient's life is to be saved. However, most methods do not consider the correlations between features in the patient's condition over time. Moreover, the existing prediction methods fail to provide sufficient evidence for the time-critical decisions required in such dynamic and changing environments. For ICU caregivers, the facts and reasoning behind a prediction are the most important criteria when deciding what medical actions to take. However, the current methods lack organ level predictions and reliable interpreted prediction results. They focus on either the overall physiological severity without insight for medical staff or the illness severity that lacks generalizability. The existing interpretable models only provide the feature importance but the importance cannot be used as reliable guidance of treatment. In this paper, we propose an interpretable organ failure prediction method as a benchmark on the MIMIC-III dataset. We build some state-of-the-art (SOTA) models to implement the predictions of different organs that are used in the Sequential Organ Failure Assessment (SOFA)scores. Then we interpret prediction results by introducing the counterfactual explanation techniques. The experiment results show the high performances of the predictions on coagulation, liver, nervous, and respiration failures with F1 scores of more than 0.9. Furthermore, the counterfactual explanations can also capture the key features that can prevent the organ failure trend by comparing with true normal records in two case studies.

Keywords: Counterfactual explanation · Deep learning · Illness severity prediction

1 Introduction

The accumulation of over 23.3 million Electronic Health Records (EHR) from over 207 thousand people available in the *My Health Record System*[1], has

[1] https://myhealthrecord.gov.au.

B. Li et al. (Eds.): APWeb-WAIM 2022, LNCS 13421, pp. 60–75, 2023.
https://doi.org/10.1007/978-3-031-25158-0_6

attracted great attention from the machine learning and data mining learning communities. Learning such a large volume of data could provide a stronger evidence base to provide effective and robust decision support, which could benefit clinical practice. In general, clinical decision-making in the ICU is fundamentally driven by the probability of clinical outcome [11]. As a result, the outcome prediction of ICU patients has become a crucial task in the medical domain and the numerous scoring systems have been developed and progressively refined to assist with rapid patient assessment. Examples include the sequential organ failure assessment score (SOFA) [28], APACHE II [14], and SAPS II [16]. The produced scores reflect the current clinical condition of a patient based on a set of basic physiological indicators.

These scoring systems cannot provide time-critical information as they are simply hand-crafted evaluation of a patient's vital signs calculated at various times throughout the day. Moreover, many medical evaluations, such as pathology tests, may only happen every day or two. The longer the time between updated information, the less time there is to respond to a patient in critical condition, which is one of the reasons why the most important indicators, such as heart rate, are monitored continuously. Over the years, recurrent neural networks (RNNs) and their variants have been explored as deep models for handling time series data, and many have achieved significant results with clinical prediction tasks like mortality risk. Given a sequence of multivariate features, the typical outlook of mortality risk with the prediction techniques of today is about 24 h -barely enough time for clinicians to intervene. More importantly, short-term mortality risk predictions may have ethical implications. For example, the mortality risk to a patient over the next week may be, say, 80% but the prediction for the next 24 h may only be 5%. If faced with an unaffordable treatment, many patients and caregivers may choose not to continue with clinical services unbeknownst to the consequences of that decision beyond tomorrow. Thus, continuously predicting the medical trajectory not only offers more detailed information at a finer time granularity but could also help caregivers concentrate on planning effective treatments with better consideration of an illness's true severity.

Despite their solid results to date, deep learning models have some deficiencies. For instance, they normally treat all multivariate time-series variables as an entire input stream without considering the correlations between the physiological variables. For instance, [26] compares the performance of several probabilistic models, like the logistic model and BayesNet on the mortality prediction. [2] constructs a CNN-LSTM model to realise the mortality risk prediction. These two works both concern the overall mortality risk of patients in ICU. However, human organs are highly correlated to each other and to a patient's deterioration. When one or two organs start to malfunction, others tend to follow over a short period. For example, systolic blood pressure is positively correlated with diastolic blood pressure and pulse pressure, whereas diastolic blood pressure is inversely correlated with pulse pressure. Also, a deterioration in the fraction of inspired oxygen can asynchronously affect cerebral blood flow. Thus, exploiting correlations between medical time-series variables can further improve classification performance for ICU prediction tasks. [5] firstly separates the overall severity score prediction into a multi-task learning task. Apart from the works on the prediction of overall physiological severity, some works also focus on some particular illness.

However, some urgent issues still exist in current severity prediction methods. Firstly, the works only aim at the prediction of overall mortality, without considering the specific organs or illness. Such prediction results only give a rough judgement of the future physiological status but cannot provide enough information for doctors about the exacerbation of organs or illnesses. Secondly, some existing algorithms predict the overall physiological status by Sequential Organ Failure Assessment (SOFA) score [28] or Acute Physiology and Chronic Health Evaluation (APACHE) score [13]. However, the level of overall score does not mean the specific trend of organ failures or mortality. For instance, the two patients in Fig. 1 have similar SOFA scores in the first 14 h and the SOFA score green line grows rapidly in the last 10 h. However, the patient denoted by the green line is finally alive and another patient who has a lower SOFA score is finally dead because the later patient has a severer respiration failure than the first patient. Thus, the lower overall score does not always mean the healthier status and we still need to focus on the specific organ status. Lastly, the current interpretation methods in severity prediction mostly use the SHAP algorithm [19] to gain feature importance of the prediction results as explanations. [2,7,25]. Although the first method is commonly used to interpret severity prediction results, it still has two drawbacks in the ICU situation. Firstly, the explanation may give a high importance factor on the unchangeable features, for instance, age, weight, or gender. Secondly, based on the feature importance, the medical care staff still do not know how to adjust the corresponding indicators can prevent the trend of organ failures. Therefore, the feature importance explanation cannot provide much guidance about further treatment.

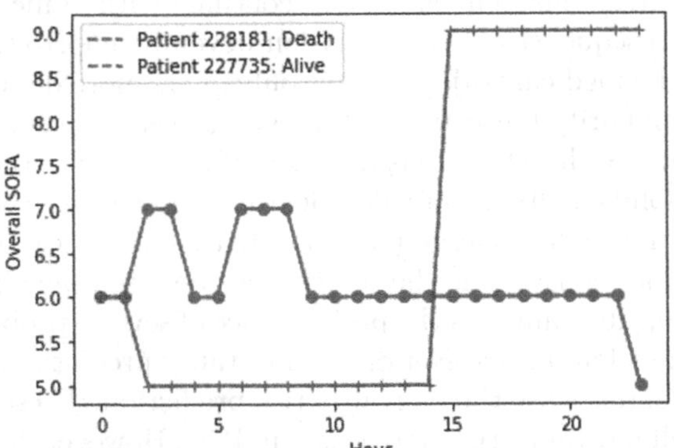

Fig. 1. SOFA score during the last 24 ICU-staying hours of patient 228181 and 227735.

To overcome the aforementioned challenges, we choose to build prediction models for multiple organ failures and employ the counterfactual explanation method for the current prediction results. The different organ predictions can work together on one patient and realise systematic organ failure predictions, which can provide more precise predictions on the physiological status of patients. As for the explanation methods, compared with feature importance explanations, the counterfactual explanation has two significant strengths. Firstly, the counterfactual explanation can give an intuitional explanation about why some organ is predicted to fail and how it can be predicted to be normal. Secondly, it is more reliable because it is generated from true patients data and consistent with the learned data distribution.

Specifically, We first measure the severity of six different organs by the SOFA algorithm. Then, we construct different models including the gated recurrent units (GRU) model, tree-based models, boosting models, and other deep neural networks to realise a regression prediction of different organ failure scores. After finishing the prediction task, we introduce one causal inference technique (i.e. counterfactual inference [21]) to realise the counterfactual explanation of our prediction models as medical evidence of the prediction results and guidance of further treatment. The code will be available on Github[2].

In summary, our contributions are in three folds:

- We evaluate the performance of different types of prediction models on the systematic severity prediction of organ failures instead of overall patient severity.
- We firstly introduce the counterfactual explanation for the organ failure predictions which can provide evidence of prediction results and possible treatment effectiveness analysis that can alleviate the organ failure.
- We conducted experiments and compare different models. The result demonstrated that our prediction models and explanation method can work as a benchmark for future work on organ failure predictions.

2 Related Works

In this section, we review previous works related to the severity prediction of ICU patients and some widely used explanation algorithms, especially the algorithms that have been introduced into the severity prediction area. Then, we summarize the prior works and briefly present our research methodology as a benchmark.

2.1 Severity Prediction on ICU Patients

Through the years, the research on the severity prediction of patients has developed in different directions: the severity prediction which focuses on the overall future physiological status [1–3,5,8,12]; and the severity prediction which is related to the specific illness [10,15,20,24].

The existing works on overall severity prediction have employed different machine learning models. [8,26] compare some probabilistic machine learning models such as logistic regression (LR) with different regularizations, and gradient boosting decision trees (GBDT). [3] tried more models like Random Forest (RF), the predictive Decision Trees (DT), the probabilistic Naive Bayes (NB),

[2] https://github.com/ShaofeiShen768/Organ_failure.

and the rule-based Projective Adaptive Resonance Theory (PART). As for deep neural networks, [2] construct a CNN-LSTM model for mortality risk prediction while [5] introduce an attention-based RNN model to implement the classification of different severity status according to SOFA score. Another type of work focus on one specific illness, especially Sepsis. [15] compares the performance of different probabilistic models to predict Sepsis while [10,24] both use RNN models. In addition, [20] summarize a systematic review about Sepsis prediction.

Among these works, the common approach in pre-processing is to extract a group of physiological features ranging from individual information like gender and age to some physiological indicators like heart rate and body temperature from the raw dataset. Furthermore, the works above handle the extracted features in different ways based on their prediction models. [5,10,24] utilize the time series data to implement the following RNN models and [2] process the data in each time point into one dimension as input of 1D-CNN layers. In our organ failure prediction work, we will extract the time-series physiological indicators.

2.2 Interpretable Deep Learning Models

The explainable deep learning models also attract more attention with the development of neural networks. However, compared with the traditional models like DT, and LR, the deep models are usually hard to be interpreted because of the complex model structure and tens of thousands of parameters. Currently, the popular explanation methods in ICU patient prediction can be divided into three categories: Feature Importance [2,7,25], Attention Based [23], and Counterfactuals [30].

The first type of work mostly utilizes the SHAP algorithm [19] which has been widely used in deep model explanation to gain insight into how different features influence the prediction results. In particular, considering that the features used in severity prediction of ICU patients are all physiological indicators, the important factors can show the abnormal indicators that lead to mortality. Another type of explanation is based on the parameter values of the attention layer in neural networks. However, the relationship between the attention value and the interpretability of model prediction is still not reliable because the attention layers cannot represent the whole model layers [4]. The last category makes use of counterfactuals, which means a series of external facts that lead to an opposite prediction [21]. Different from the first two types of methods that focus on the feature impacts on the current prediction, the counterfactual explanations focus on the changes in features that can lead to a different prediction result. The counterfactual method has also been widely used in machine learning and [27] gives a counterfactual algorithms review. This method is more meaningful in clinical medicine because the counterfactuals can give advice on how to rescue critical patients by analysing the feature differences between counterfactuals and prototypes. However, in the field of severity prediction of ICU patients, the counterfactual explanation is still a novel technique and only one recent work tries to introduce it into the medical treatment recommendation [30]. In this paper, we will introduce this technique into the severity prediction.

3 Benchmark Dataset and Methods

In this section, we introduce the dataset for benchmarking and how data is pre-processed for the experiment including data filtering, feature extraction, and data segmentation. Then, we present the comparison of state-of-the-art methods for predictions. Finally, we demonstrate the interpretablility using the counter-factual algorithm. The GitHub link of the code of the pre-processing, prediction, and explanation has already been attached in the introduction.

3.1 MIMIC-III Dataset

We use the Medical Information Mart for Intensive Care (MIMIC-III) dataset [9] to construct the benchmark dataset for the organ failure prediction. We choose this public dataset because MIMIC-III dataset contains detailed clinical information including admission records, laboratory events, note events, and so on from 53423 adult patients. In addition, this dataset also contains relatively complete laboratory and chart event data which contains a great number of physiological indicators which can represent the current individual organ status directly and reflect the physiological status of patients, such as heart rate, or partial pressure of oxygen (PO2). Therefore, this dataset can provide the adequate physiological indicators for extraction and does not require heavy imputation work. In the following pre-processing section, we will show the details of the construction of the benchmark dataset for the organ failure prediction tasks.

3.2 Pre-processing

In the pre-processing procedure, we firstly selected the same 41 physiological indicator features as shown in [5]. Then we need to decide the length of the time series. The data distribution of ICU staying lengths is shown in Fig. 2. From this figure, we can find the lengths of most of the patient data are within the interval of 10 h to 120 h. On the one hand, we do not want to discard many patients. On the other hand, we want to use a slightly longer time length to preserve more information about patients. Therefore, we decide to set the length of the time series as 24 h. So, we filter the records which are no more than 24 h which can account 20.2% of all records.

Next, we extracted the 41 common indicators as [5]. Considering that some laboratory indicator data are not recorded every hour, we impute the raw data of 41 features by the forward-fill imputation strategy which is also used in [5,18]. In addition, inspired by [5], we divide all the features into 6 organ-specific groups related to respiration, coagulation, liver, cardiovascular, nervous, and renal and 1 shared group to delete the redundant physiological indicator when implementing prediction on one specific organ. The organ-specific groups contain some indicators that are mainly related to one specific organ. For instance, bilirubin is mainly used in liver prediction and O2 flow is related to the respiratory. The shared group contains some indicators that are related to the whole physiological status like body temperature. In the training procedure, the data from the organ-specific group and shared group will be combined for different prediction tasks.

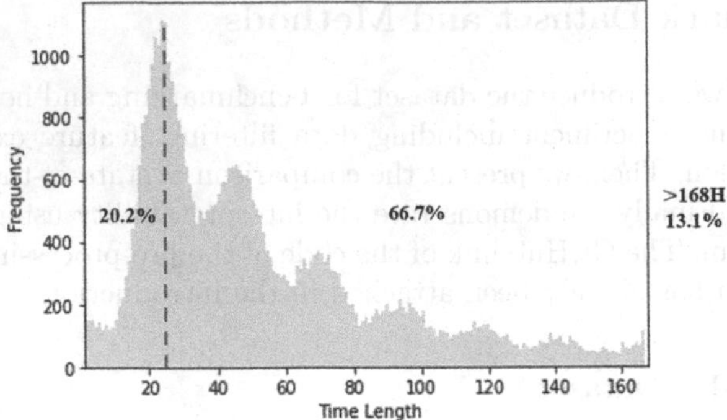

Fig. 2. ICU staying lengths of patients in MIMIC-III

Apart from the feature data used in prediction, we also construct the supervised information for organ failure predictions. Specifically, we extract the indicators related to the SOFA score [28] and implement the same imputation strategy on the laboratory indicator data. Then we use calculate six different organ failure scores for each hour using these data. The organ failure scores include respiration, coagulation, liver, cardiovascular, nervous, and renal scores [28] and each score has 5 different levels from 0 to 4, where the larger level means the severer physiological status. These organ failure scores can be used as the supervised information for the further prediction tasks.

After finishing the feature extraction and the construction of supervised information, we implement the normalization on the whole dataset to scale different magnitudes of features. Then, we divide the selected patients whose ICU staying time longer than 24 h as 0.8 training set, 0.1 validation set, and 0.1 test set. For patients in each dataset, we implement segmentation as Fig. 3 shown. We extract data in the first 24 h as the first feature point and the SOFA scores in the 25th hour as the first target. Then the segmentation moves forward one hour to choose the data from 2nd to 25th and the SOFA scores in the 26th hour as the second feature point and target and so on. Considering different patients do not have the same ICU stay hours, the real proportions of training, validation, and test dataset are not exactly the same but close to 0.8, 0.1, and 0.1.

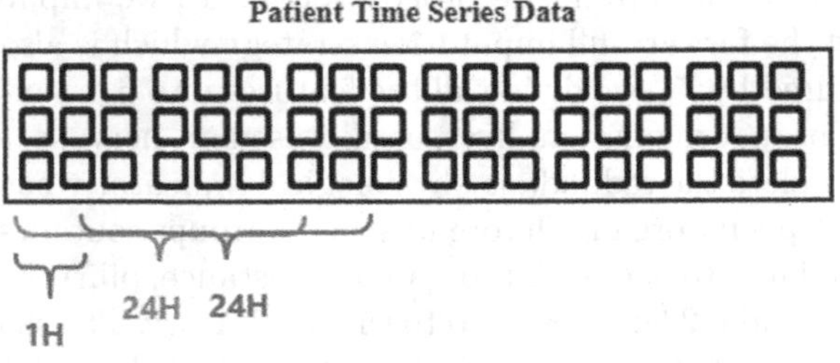

Fig. 3. The segmentation method in pre-process

In addition, after the construction of the whole benchmark dataset, we conduct a basic statistic on the data distribution, which is shown in Table 1. The statistic results shows an high imbalance among different classes in our constructed dataset and the class proportions of different organs also have huge differences. For instance, over 70% of the data in liver and coagulation tasks are labelled as class 0 while only 8.5% of respiration data belong to the class 0.

Table 1. Class distributions in benchmark dataset

Task	Class 0	Class 1	Class 2	Class 3	Class 4
Nervous	0.356	0.113	0.259	0.195	0.078
Coagulation	0.702	0.157	0.103	0.032	0.005
Liver	0.797	0.061	0.086	0.027	0.031
Cardiovascular	0.683	0.247	0.014	0.034	0.022
Respiration	0.085	0.158	0.411	0.278	0.067
Renal	0.542	0.164	0.075	0.071	0.147

3.3 Prediction Models

Model Details. For the prediction models, we will mainly use the aforementioned attention-based GRU (AGRU) model which is a variant of the RNN model [6]. The GRU model has similar performance as the long short-term memory (LSTM) model but has fewer parameters than LSTM and less risk of overfitting.

We also choose the tree-based models because the physiological indicators usually have normal and abnormal intervals. Such interval bounds can be learned through the tree-based models to split trees. We choose DT, RF, and XGBoosting algorithms to build regression models for comparison which are shown in Table 1. In addition, we also compare the performance with the classification methods. we build logistic regression (LR), Multi-Layer Perceptron (MLP), and Convolutional Neural Network (CNN) which are shown in Table 2.

Evaluation Metric. Considering that we use regression and classification models here, we transform the numerical prediction results into 5 classes as the organ failure score in SOFA to make the performance of regression models more intuitive to be compared with multi-classification models. Then we can use the accuracy to evaluate the model performance. In addition, the datasets are imbalanced (Table 3), so we also choose the Macro F1 score, which is the average F1 of all classes, to make further evaluations.

3.4 Counterfactual Explanation

Some models we used in this paper can be easily interpreted. For instance, the decision path of the DT models can be used as an explanation, and the parameters of the linear function in LR models can also be used to interpret. However, for the RF, boosting model, and other deep neural network models, interpreting the results are not easy [4]. Although the current SHAP [19] and LIME [22] have been widely used in the explanation of boosting and deep models, the explanation, i.e. feature importance, cannot guarantee the validity of the changes on the points with high importance in the real medical situation. However, the counterfactual explanation can be introduced for the treatment guidance in this situation because it is usually generated from the truly existing samples and the counterfactual explanation can be applied to most models, including deep models.

The counterfactual explanation of model f usually requires a piece of sample data X to be explained and a desired target value y_0. Then the counterfactual $\tilde{X}$ can be generated based on two rules: (a)the prediction result of counterfactual $f(\tilde{X})$ should be close to the desired target y_0 (b) the feature difference between sample data X and counterfactual $\tilde{X}$ should be minimized. The two rules can be realised by the following loss function [29], where L_1 measures the rule (a) and L_2 is the distance function for the rule (b):

$$Loss = \alpha L_1(f(\tilde{X}) - y_0) + \beta L_2(X, \tilde{X}) \tag{1}$$

To enhance the reliability of the counterfactuals, the distractors, which are real data with the desired target from the training set, can be used as the initial data to optimal. Figure 4 shows an intuitive procedure of how counterfactual samples are generated from the true sample data.

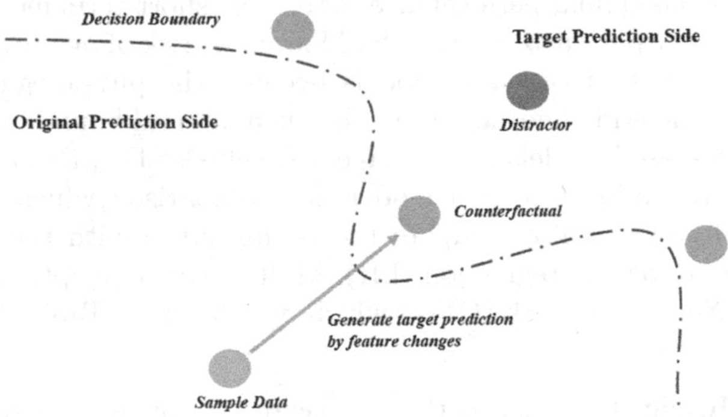

Fig. 4. Counterfactual explanation generation

However, the counterfactual explanation generated directly from (1) may lead to unreliable counterfactual samples sometimes. On the one hand, for ICU patients, some physiological indicators are hard to be changed, for example, weight. On the other hand, the changes of different physiological indicators may have different influence on the predictions. Considering the two issues in organ failure prediction, we seperate the signal hyper-parameter β into a vector of feature weights in loss function L_2, which is shown in (2) where $\mathbf{E}$ denotes the feature sets:

$$Loss = \alpha L_1(f(\tilde{X}) - y_0) + \sum_{i \in \mathbf{E}} \beta_i L_2(X_i, \tilde{X}_i) \tag{2}$$

The algorithm based on this improved loss function (2) is summarized in the following. This counterfactual explanation algorithm is improved from the basic loss function in counterfactual explanation [17]. In the experiment, we actually use the MSE as L_1, and use Euclidean distance as L_2. For the distractor X_d, we use the nearest-neighbour algorithm to select the data points with the desired prediction y_0 from different patients. We then choose $\alpha = 10$ and β_i are selected from 1,0.1, 0.01, where most of features use $\beta_i = 1$ and some features are not easy to changed in medical choose smaller β_i. Then we set a max optimization iteration of 200 times.

Algorithm 1: Counterfactual Explanation via Optimization

Data: Test data X, distractor X_d, selected feature set $\mathbf{E}$, prediction model f, desired prediction y_0, hyper-parameters α, $\{\beta_i\}$, learning rate γ, and max epoch K, two loss functions: L_1, L_2

Result: Counterfactual explanation $\tilde{X}$

Initialize $\tilde{X} = X_d$, and a zero mask matrix A **while** *epoch ¡ K* **do**

  Update $\tilde{X} = A\tilde{X} + (I - A)X$ and predict $\tilde{y} = f(\tilde{X})$;

  ; Calculate prediction $l_1 = L_1(\tilde{y}, y)$;

  Calculate weighted distance $l_2 = \sum_{i \in \mathbf{E}} \beta_i L_2(X_i, \tilde{X}_i)$;

  Calculate total loss $l = l_1 + l_2$;

  Apply backpropagation to update mask matrix A;

end

return $\tilde{X}$ as the counterfactual explanation;

4 Experiment Result and Analysis

In this section, we show our experiment results on the organ failure prediction including performances of different regression and classification models. Then we select 2 patient cases to demonstrate the counterfactual explanations.

4.1 Model Performance

Table 2. Model accuracy ± standard deviations

Model	Nervous	Coagula	Liver	Respira	Renal	Cardio
GRU	**0.955** ± 0.006	0.987 ± 0.001	**0.996** ± 0.001	**0.945** ± 0.005	0.876 ± 0.017	0.739 ± 0.020
MLP	0.954 ± 0.111	0.977 ± 0.006	0.994 ± 0.001	0.927 ± 0.002	0.869 ± 0.001	**0.783** ± 0.024
CNN	0.785 ± 0.068	0.787 ± 0.127	0.851 ± 0.092	0.767 ± 0.098	0.670 ± 0.097	0.605 ± 0.058
XGB	0.945 ± 0.004	0.990 ± 0.008	0.996 ± 0.009	0.936 ± 0.013	0.847 ± 0.035	0.696 ± 0.042
RF	0.946 ± 0.001	**0.991** ± 0.000	0.996 ± 0.000	0.930 ± 0.001	**0.884** ± 0.000	0.726 ± 0.001
DT	0.938 ± 0.012	0.990 ± 0.002	0.995 ± 0.001	0.929 ± 0.008	0.881 ± 0.012	0.726 ± 0.008
LR	0.952 ± 0.002	0.972 ± 0.003	0.980 ± 0.006	0.909 ± 0.009	0.775 ± 0.047	0.781 ± 0.001

Table 3. Model F1 ± Standard deviations

Model	Nervous	Coagula	Liver	Respira	Renal	Cardio
GRU	**0.948** ± 0.001	0.948 ± 0.003	0.985 ± 0.001	**0.939** ± 0.001	0.771 ± 0.005	**0.312** ± 0.002
MLP	0.946 ± 0.004	0.749 ± 0.007	0.981 ± 0.001	0.917 ± 0.002	0.712 ± 0.012	0.307 ± 0.002
CNN	0.763 ± 0.056	0.639 ± 0.233	0.764 ± 0.083	0.664 ± 0.057	0.480 ± 0.012	0.307 ± 0.011
XGB	0.936 ± 0.002	0.959 ± 0.009	0.988 ± 0.001	0.925 ± 0.002	0.756 ± 0.002	0.281 ± 0.013
RF	0.938 ± 0.012	**0.969** ± 0.008	**0.988** ± 0.004	0.917 ± 0.004	**0.793** ± 0.007	0.291 ± 0.009
DT	0.930 ± 0.011	0.960 ± 0.003	0.988 ± 0.002	0.915 ± 0.003	0.788 ± 0.004	0.292 ± 0.004
LR	0.944 ± 0.001	0.735 ± 0.004	0.948 ± 0.003	0.896 ± 0.002	0.524 ± 0.003	0.303 ± 0.001

Tables 2 and 3 show the average performances of different models using 3 different random seeds. From the two tables, we can find the AGRU model has the best performance under both the two metrics in nervous, and respiration failure prediction tasks, while the tree-based model works well in the liver failure prediction. In the coagulation task, the AGRU model, two tree-based models and the boosting model all have high performance under the accuracy and F1 score. However, all the three multi-classification models have significantly lower F1 scores than the regression-based models. Similarly, four regression models outperform the other three multi-classification models in the renal prediction tasks. This is mainly because some class has much less training data than other classes. which is shown in Table 1. For the cardiovascular failure predictions, all of our selected models are influenced by the imbalanced distribution and have rather low F1 scores.

When comparing the performances of different prediction tasks, we can find the predictions on nervous, coagulation, respiration, and liver failures have greatly high performances. In particular, the predictions on the liver have the highest accuracy and F1 score which are both more than 98%. However, the predictions on renal failure do not have the same high level of performance as the other four tasks, and the performance on cardiovascular failure is even worse. Two possible causes may lead to these results. Firstly, the physiological indicators selected from renal and cardiovascular groups are mainly extracted from laboratory events, which usually have a long time interval to be recorded. Therefore, the raw data will have a great number of missing values and the forward-fill imputation may fail to reflect the correct trend of organ failures. On the other hand, the high proportions of the first two classes may also be one cause.

4.2 Explanation Cases

To demonstrate the counterfactual explanation of our deep models, we choose our AGRU models as examples and select two ICU patient cases: the first patient with the id of 207251 stayed in ICU for more than 15 d during which the patient experienced a coagulation failure problem and then become normal in the last two days; the second patient with the id of 221950 stayed in ICU around 3 d and recovered from a heavier liver failure. We choose these two patients for the following three reasons: firstly, the prediction models of coagulation and liver failures have high ac curacies, which can lead to the reliable predictions and the explanation results are also more reliable. Secondly, we choose one survival case and another death case for comparisons.

In the generating procedure, for the first patient, we choose the data on the eleventh day in ICU and extract the coagulation features as sample data with a coagulation failure score of 1. As for the second patient, we choose the data on the first day which has a ground-truth coagulation failure score of 1. The two records are correctly predicted. We aim to find the counterfactual explanations with the target failure score of 0 for both patients. We then select three records that are predicted as normal as distractors to generate counterfactual explanations and compare the feature differences between the prototype sample data and the counterfactual sample. In addition, in the explanation procedure, we control the max iterations to prevent the loss in (2) being globally optimized. This is because we want to preserve some data characteristics of distractors which can increase also the reliablity and variance of counterfactual explanations. The result of two case studies is shown in Fig. 5 and Fig. 6.

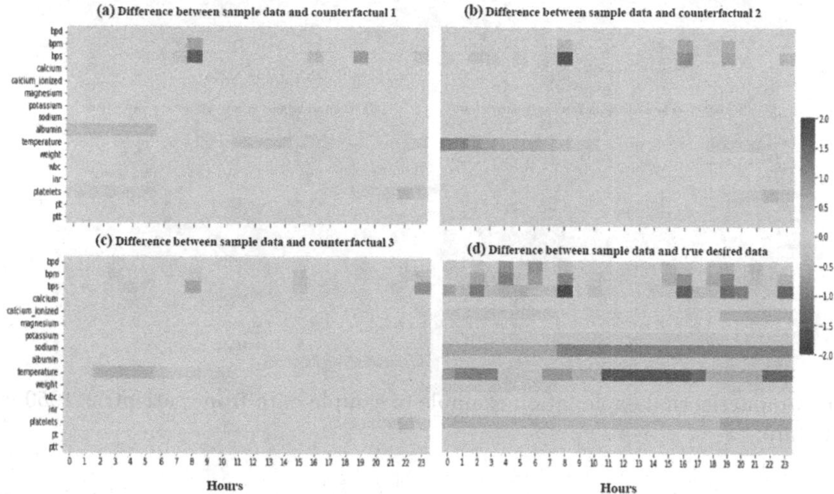

Fig. 5. Counterfactual explanation example of sample data from patient: 207251. The true data with target label is also from this patient when his Coagulation becomes normal. The rows of each figure are physiological indicators and the columns are the hours. The red point means the value in sample data is increased compared with the counterfactual explanations while the blue point means the value is decreased in sample data. Moreover, the shade of colour denotes the size of the changed value. (Color figure online)

In the first case in Fig. 5, we can find that all the three counterfactual samples can successfully capture the significantly different points in bps, and platelets, which are also shown in the comparison of true data 5(d). In addition, 5(b) and 5(c) also show changes in temperature in the early hours while 5(a) only captures the differences in the middle hours. The differences in three counterfactual samples can provide three different explanations for why this patient is predicted to have a slight coagulation failure. Combining the three explanations, we can provide some directions for further treatments:

- Control the patient's blood pressure to the normal level.
- Take treatment to lower body temperature of patients.
- Increase the platelet value of patients.

In addition, we can also notice some differences in sodium and temperature of Fig. 5(d) do not happen in the counterfactual explanations. This is because this patient was finally died because of the failure of other organs and his temperature, and sodium suddenly become abnormal from the 8-th hour. However, our counterfactual samples are generated based on distractors from alive patients and they will not capture these invalid changes when being generated.

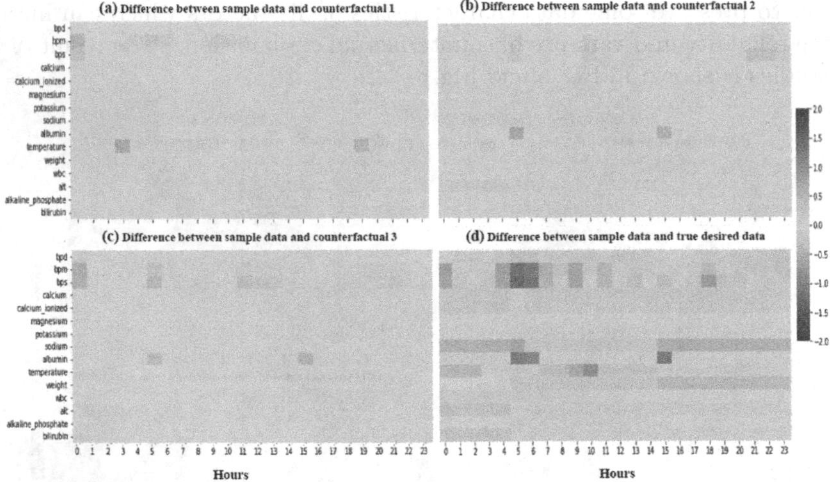

Fig. 6. Counterfactual explanation example of sample data from patient: 221950 w.r.t liver failure.

As for the second case, the three counterfactual explanations are similar. From Fig. 6, the changes in bpm, bps, and albumin are captured in all three counterfactual samples to explain why this patient is predicted to have a trend of liver failure. The changes in features in the liver group, i.e. alt and bilirubin, are not significant in the counterfactual samples because the alt and bilirubin in the prototype sample are quite close to the normal values. Even the true bilirubin value is less than the global mean value. Therefore, we can also summarize some possible treatment directions for this patient based on the counterfactual explanations:

- Control the patient's blood pressure to the normal level.
- Balance the value of blood sodium, blood albumin and body temperature of patients.
- Slightly reduce the alt and bilirubin value.

5 Conclusion and Future Work

In conclusion, we apply different models to the organ failure prediction tasks and introduce counterfactual explanation techniques in the fields of patient severity prediction. As we have mentioned before, organ failure prediction can provide more specific and precise information about the future status of ICU patients compared with the mortality prediction based on the overall physiological status. As a benchmark for this research task, we conduct experiments on different models including the tree-based models, boosting models, and deep neural networks to compare the performances of different prediction tasks. We find the prediction results on nervous, coagulation, respiration, and liver have high performances. In addition, as a further exploration of the organ failure predictions, we also employ the counterfactual explanation algorithm and improve it based on the situation of ICU patients. From the two cases in our experiment result, the counterfactual explanations are reliable as a guide for the treatment direction.

References

1. Aczon, M., et al.: Dynamic mortality risk predictions in pediatric critical care using recurrent neural networks. arXiv preprint arXiv:1701.06675 (2017)
2. Alves, T., Laender, A., Veloso, A., Ziviani, N.: Dynamic prediction of ICU mortality risk using domain adaptation. In: 2018 IEEE International Conference on Big Data Big Data, pp. 1328–1336. IEEE (2018)
3. Awad, A., Bader-El-Den, M., McNicholas, J., Briggs, J.: Early hospital mortality prediction of intensive care unit patients using an ensemble learning approach. Int. J. Med. Inform. **108**, 185–195 (2017)
4. Bodria, F., Giannotti, F., Guidotti, R., Naretto, F., Pedreschi, D., Rinzivillo, S.: Benchmarking and survey of explanation methods for black box models. arXiv preprint arXiv:2102.13076 (2021)
5. Chen, W., Long, G., Yao, L., Sheng, Q.Z.: AMRNN: attended multi-task recurrent neural networks for dynamic illness severity prediction. World Wide Web **23**(5), 2753–2770 (2020)
6. Chung, J., Gulcehre, C., Cho, K., Bengio, Y.: Empirical evaluation of gated recurrent neural networks on sequence modeling. arXiv preprint arXiv:1412.3555 (2014)
7. Ichikawa, K., Tamano, H.: Unsupervised qualitative scoring for binary item features. Data Sci. Eng. **5**(3), 317–330 (2020)
8. Johnson, A.E., Mark, R.G.: Real-time mortality prediction in the intensive care unit. In: AMIA Annual Symposium Proceedings. Am. Med. Inf. Assoc. vol. 2017, p. 994 (2017)
9. Johnson, A.E., et al.: Mimic-iii, a freely accessible critical care database. Scientific data **3** (1), 1–9 (2016)
10. Kam, H.J., Kim, H.Y.: Learning representations for the early detection of sepsis with deep neural networks. Comput. Biol. Med. **89**, 248–255 (2017)
11. Keegan, M.T., Gajic, O., Afessa, B.: Severity of illness scoring systems in the intensive care unit. Crit. Care Med. **39**(1), 163–169 (2011)

12. Khope, S.R., Elias, S.: Critical correlation of predictors for an efficient risk prediction framework of ICU patient using correlation and transformation of mimic-iii dataset. Data Sci. Eng. **7**(1), 71–86 (2022)
13. Knaus, W.A., Draper, E.A., Wagner, D.P., Zimmerman, J.E.: Apache ii: a severity of disease classification system. Crit. Med. **13**(10), 818–829 (1985)
14. Knaus, W.A., Wagner, D.P., Draper, E.A., Zimmerman, J.E., Bergner, M., Bastos, P.G., Sirio, C.A., Murphy, D.J., Lotring, T., Damiano, A., et al.: The apache iii prognostic system: risk prediction of hospital mortality for critically iii hospitalized adults. Chest **100**(6), 1619–1636 (1991)
15. Kong, G., Lin, K., Hu, Y.: Using machine learning methods to predict in-hospital mortality of sepsis patients in the ICU. BMC med. Inf. Decis. making **20**(1), 1–10 (2020)
16. Le Gall, J.R., Lemeshow, S., Saulnier, F.: A new simplified acute physiology score (saps ii) based on a european north american multicenter study. JAMA **270**(24), 2957–2963 (1993)
17. Leung, V.J., Ates, E., Aksar, B., Coskun, A.K.: Comte: counterfactual explanations for supervised machine learning frameworks on multivariate time series data. Tech. rep., Sandia National Lab. (SNL-NM), Albuquerque, NM (United States) (2021)
18. Lipton, Z.C., Kale, D.C., Wetzel, R., et al.: Modeling missing data in clinical time series with RNNs. Mach. Learn. Healthc. **56**, pp. 253–270 (2016)
19. Lundberg, S.M., Lee, S.I.: A unified approach to interpreting model predictions. Adv. Neural Inf. proc. syst. **30**, 4765–4774 (2017)
20. Moor, M., Rieck, B., Horn, M., Jutzeler, C.R., Borgwardt, K.: Early prediction of sepsis in the ICU using machine learning: a systematic review. Front. Med. **8**, 348 (2021)
21. Pearl, J.: Causality. Cambridge University Press (2009)
22. Ribeiro, M.T., Singh, S., Guestrin, C.: why should i trust you explaining the predictions of any classifier. In: Proceedings of the 22nd ACM SIGKDD international conference on knowledge discovery and data mining, 1135–1144 (2016)
23. Rosnati, M., Fortuin, V.: MGP-AttTCN: an interpretable machine learning model for the prediction of sepsis. PLoS ONE **16**(5), e0251248 (2021)
24. Scherpf, M., Gräßer, F., Malberg, H., Zaunseder, S.: Predicting sepsis with a recurrent neural network using the mimic iii database. Comput. Biol. Med. **113**, 103395 (2019)
25. Subudhi, S., Verma, A., Patel, A.B., Hardin, C.C., Khandekar, M.J., Lee, H., McEvoy, D., Stylianopoulos, T., Munn, L.L., Dutta, S., et al.: Comparing machine learning algorithms for predicting ICU admission and mortality in COVID-19. NPJ Digital Med. **4**(1), 1–7 (2021)
26. Veith, N., Steele, R.: Machine learning-based prediction of ICU patient mortality at time of admission. In: Proceedings of the 2nd International Conference on Information System and Data Mining, 34–38 (2018)
27. Verma, S., Dickerson, J., Hines, K.: Counterfactual explanations for machine learning: a review. arXiv preprint arXiv:2010.10596 (2020)
28. Vincent, J., et al.: The sofa score to describe organ dysfunction/failure. on behalf of the working group on sepsis-related problems of the european society of intensive care medicine. Intensive Care Med 22(7), 707–710 (1996) https://doi.org/10.1007/BF01709751

29. Wachter, S., Mittelstadt, B., Russell, C.: Counterfactual explanations without opening the black box: Automated decisions and the GDPR. Harv. JL Tech. **31**, 841 (2017)
30. Wang, Z., Samsten, I., Papapetrou, P.: Counterfactual explanations for survival prediction of cardiovascular ICU patients. In: Tucker, A., Henriques Abreu, P., Cardoso, J., Pereira Rodrigues, P., Riaño, D. (eds.) AIME 2021. LNCS (LNAI), vol. 12721, pp. 338–348. Springer, Cham (2021). https://doi.org/10.1007/978-3-030-77211-6_38

Specific Emitter Identification Based on ACO-XGBoost Feature Selection

Jianjun Cao[1], Chumei Gu[1,2](✉), Baowei Wang[2], Yuxin Xu[1,2], and Mengda Wang[1,2]

[1] The Sixty-Third Research Institute, National University of Defense Technology, Nanjing, China
caojj@nudt.edu.cn, m15261820030@163.com

[2] School of Computer Science, Nanjing University of Information Science and Technology, Nanjing, China

Abstract. In order to improve the performance of Specific Emitter Identification (SEI) in complex electromagnetic environment, a SEI model based on Ant Colony Optimization-eXtreme Gradient Boosting (ACO-XGBoost) feature selection is proposed. Firstly, lifting wavelet package transformation is introduced to extract features of original emitter signals and data normalization is carried out on the data. Secondly, ACO is used to optimize the main parameters for XGBoost and the optimized XGBoost algorithm is applied to score the importance of the features. Thirdly, the importance of each feature is used as threshold to determine the best feature subset. Finally, combined with XGBoost classifier, specific emitter has been correctly identified. The proposed model is investigated on three datasets and the experimental result shows that the optimal feature subset can be effectively acquired and reasonably classified by the proposed model. Comparing with Decision Tree (DT), Random Forest (RF), Gradient Boosting Decision Tree (GBDT) and XGBoost feature selection methods, the identification performance of the proposed model is effectively improved.

Keywords: Lifting wavelet package transformation · Specific emitter identification · Ant colony optimization · XGBoost · Feature selection

1 Introduction

With the rapid development of communication engineering and the population of mobile communication device, wireless communication has more and more indispensable of modern society. The obtained signal from specific emitter not only carries signal information, but also contains the hardware information of the emitter, which is usually called fingerprint feature of emitter [1]. SEI method identifies different emitters based on fingerprint features which is detectable, immutable and tamper-resistant. SEI has great application value in military and civil fields. In the military field, it can fundamentally improve the investigation and confrontation ability against enemy emitter by extracting subtle feature of signal and identifying the emitter. In the civil field, this method can locate the emitter lacking detection and eliminate the fault timely. SEI is mainly divided

B. Li et al. (Eds.): APWeb-WAIM 2022, LNCS 13421, pp. 76–90, 2023.
https://doi.org/10.1007/978-3-031-25158-0_7

into three steps: signal preprocessing, feature extraction and classification [2]. However, the feature vector of emitter after using some preprocessing and feature extraction methods is high dimensional and low sample size. Data of these characteristics pose a challenge to effective classification and further analysis. One feasible and powerful method to solve the above problems is feature selection.

Feature selection (FS) assesses the importance score of every feature and constructs an optimal feature subset efficient for pattern recognition tasks [3]. The purpose of feature selection is to obtain the best predictive accuracy by selecting relevant features and removing irrelevant or redundant features [4]. Feature selection techniques usually be divided into three categories: filter, wrapper and embedded methods [5]. The filter methods evaluate the importance of data according to its inherent characteristics. Filter methods first use an evaluation function to score every feature and then all features are sorted. Finally, some top-*k* features are selected to constitute a feature subset. This method is simple, fast and independent of learning algorithm but ignores the dependence between features. Typical filter feature selection methods are Relief, Pearson Correlation Coefficient and Mutual Information (MI), etc. [6]. Wrapper methods wrap feature selection around learning algorithm, and achieve the feature subset with the most discriminative ability by minimizing the prediction error of classifier. This method considers the correlation between features, but indeed increases computation complexity and overfitting risk. Recursive Feature Elimination (RFE) is typical wrapper feature selection method [7]. Embedded methods embed feature selection into specified learning algorithm like Linear SVM, DT, RF, GBDT and XGBoost, etc. [8]. After training, the importance of each feature is achieved and then the optimal feature subset can be obtained. This method avoids the repeated execution of classifier.

Because embedded methods can be regarded as a balance between filter and wrapper methods, scholars have done lots of researches on feature selection based on embedded methods. Chen et al. [9] proposed a network intrusion detection model based on RF and XGBoost to improve the accuracy and real-time performance in complex network environment. The feature importance was calculated based on RF and a hybrid feature selection method combining filtering and embedded was used for feature selection. Then XGBoost method was used to detect and recognize the optimal feature subset. The model greatly reduced processing time with high detection accuracy. Zhou et al. [10] proposed a model to solve the multi-classification problem of unbalanced data in network intrusion detection. GBDT was used to calculate the importance of features and rank them, and Recursive Feature Elimination was used for feature selection. Then RF was used for feature conversion. The model had significant advantages in solving the multi-classification problem of unbalanced data. Based on the negative impact of imbalanced data on multi-classification problem, Feng et al. [11] proposed a stacked model based on XGBoost and RF feature selection. The new feature importance was the harmonic mean of the feature importance of RF and XGBoost. The threshold value was set of 0.01 and features less importance than this threshold is eliminate. This method effectively improved the classification performance. Gong et al. [12] proposed an interpretable Traditional Chinese Medicine treatment model based on XGBoost to overcome the problem of ignoring the trust mechanism of decision-making process. XGBoost was used to calculate feature importance and filter non representative features according to feature importance. Besides, the model training combined base feature selection, parameter optimization and model integration. This model was considered interpretable in the feature selection and classification. To solve the problem of low neuropsychiatric disorder classification accuracy caused by single data type, Liu et al. [13] proposed an ensemble hybrid feature selection method. 3D DenseNet was used to select image features from magnetic resonance imagines and XGBoost was used to select phenotypic features from feature importance and then image features and phenotypic features were spliced in the form of vectors. The hybrid features improved the performance of classification algorithms. Qiao et al. [14] proposed an intrusion detection model based on the combination of XGBoost and RF to address the problem of inaccurate classification. XGBoost was used to score the importance of features and improved RF was use to judge whether the network traffic is normal or abnormal. The optimal features could be effectively selected and classified by this model.

According to the advantages and disadvantages of the above ideas, we propose a new method for SEI based on ACO-XGBoost feature selection. This method mainly combines lifting wavelet package transformation, ACO and XGBoost. The main contributions of this paper are summarized as follows:

- We use lifting wavelet package transformation to extract features. Then the feature parameter system is established.
- ACO algorithm is used to optimize the parameters of XGBoost. The paper relates the maximum depth of tree (χ), the minimum sum of instance weight needed in a child (δ), the number of decision trees (ε), the L2 regularization term on weights (γ), the minimum loss reduction required to make a further partition on a leaf node (λ) and learning rate (ω). The best parameters will beneficial for SEI.
- The optimized XBGoost algorithm is used to obtain the importance of every feature and every importance is used as threshold to obtain the final best feature subset. XGBoost will also serve as a classifier to determine the classification result.
- Experiments on radio datasets in different states show the efficiency and effectiveness of the proposed model by comparing the performance of different feature selection methods, including DT, RF, GBDT and XGBoost.

The remainder of this paper is organized as follows. Section 2 introduces the model of SEI based on ACO-XGBoost FS. In Sect. 3, three specific emitter datasets are employed to prove the effectiveness and efficiency of the proposed method. Finally, we summarize this paper and propose the future work in Sect. 4.

2 Mathematical Model of Parameter Optimization and FS of XGBoost

In order to identify different specific emitters, the features should be extracted of collected original signals. The extracted features can be used to constructed a feature set $set = \{v_i | v_i = v_1, v_2, \ldots, v_n\}$, $i = 1, 2, \ldots, n$. The feature vector corresponding to the features in the set is recorded as V, and the existing Y feature vector sample of class

W is recorded as V_{wyi}, $w = 1, 2, ..., W$, $y = 1, 2, ..., Y$, $i = 1, 2, ..., n$, V_{wyi} is the ith feature value of the yth sample vector in class w. Some of the extracted features are highly relevant to the prediction result, but there are irrelevant and redundant features. The input of these features will greatly reduce the identification performance of specific emitters. Effective feature selection can reduce the feature dimension and improve the performance of SEI.

The essence of SEI is classification and embedded feature selection methods embed feature selection into specified learning algorithm. The selection of classifier parameters and feature subsets directly affect the final classification performance. In this paper, the classification accuracy of classifier and the number of features in the selected subset are directly used as the objective function. The mathematical model of parameter optimization and FS of XGBoost can be described as: selecting the value of six parameters of XGBoost and a subset $subset^q$ with cardinality q from the original feature set, where the classification accuracy A is maximal and the number of features q is minimal. The specific mathematical model are as follows:

$$\max A(\chi;\ \delta;\ \varepsilon;\ \gamma;\ \lambda;\ \omega;\ subset^q) \tag{1}$$

$$\min q \tag{2}$$

$$\text{s.t.}|subset^q| = q,\ 1 \le q \le n \tag{3}$$

where A is the classification accuracy and its equation is as follows:

$$A = \frac{TP + TN}{TP + TN + FP + FN} \tag{4}$$

TP: the real category of the sample is positive, and the result of model recognition is also positive. FN: the real category of the sample is positive, but the result of model recognition is negative. FP: the real category of the sample is negative, but the result of model recognition is positive. TN: the real category of the sample is negative, and the result of model recognition is also negative. Accuracy represents the ratio of the number of correctly classified samples to the total number of samples.

3 SEI Based on ACO-XGBoost FS

The basic framework of SEI based on ACO-XGBoost FS is shown in Fig. 1. Firstly, the features are extracted based on lifting wavelet package transformation, then the original feature data set is formed. Secondly, the feature preprocessing is carried out, and the training set and testing set are divided according to a certain proportion. Thirdly, XGBoost is optimized by using ACO, then the optimized XGBoost algorithm is used to evaluate the importance of each feature. The obtained importance value will be used as threshold to obtain the optimal feature subset. Finally, the obtained feature subsetwill be detected and recognize by XGBoost classifier. Main steps of the proposed model are described in detail in the rest of Sect. 3.

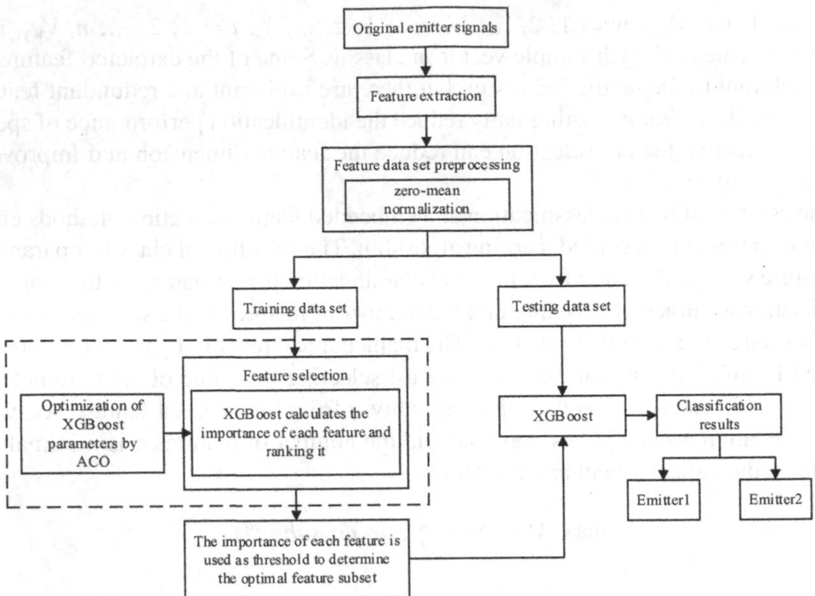

Fig. 1. The model of specific emitter identification based on ACO-XGBoost feature selection

3.1 Feature Extraction Based on Lifting Wavelet Package Transformation

Feature extraction is a key step of SEI. Feature extraction extracts the effective classification features in the original emitter signals through data conversion or data mapping, and obtains a new feature space. This step can reduce signal flow, directly affects the performance of classifiers and is an effective data compression method [15].

In view of the excellent time-frequency resolution and efficient operation ability of lifting wavelet package transformation, it is used as a tool for feature extraction in this paper. This method is helpful to expand feature set and even obtain more feature information through secondary change, so as to provide more choices for the final selection of relevant features.

Twelve statistical characteristic parameters, namely mean value, mean amplitude, square root amplitude, standard deviation, effective value, peak-to-peak value, shape factor, impulse factor, crest factor, skewness, kurtosis and clearance factor, are used. In addition, standardized relative energy is used, which is defined in reference [16].

Combined with lifting wavelet package decomposition and reconstruction, the characteristic parameter system is established as follows [17]:

Firstly, for a group of signals, the number of decomposition layers *CS* is limited to 5, and lifting wavelet packet tree is constructed. The optimal wavelet packet tree is obtained by using first order decomposition and then order search algorithm, and then the optimal wavelet packet tree is adjusted to a two-layer full binary tree.

Secondly, the information is concentrated on node (2, 0), (2, 1), (2, 2) and (2, 3) of the full binary tree. Twelve statistical characteristic parameters of four node's coefficient are extracted respectively, and standardized relative energy which were mentioned above of four nodes are extracted.

Thirdly, single branch reconstruction is carried out for each node, and twelve statistical characteristic parameters are extracted from four single branch reconstruction signals respectively.

Finally, the second layer nodes are used to reconstruct the original signal and twelve statistical characteristic parameters of the reconstructed original signal are extracted.

According to the twelve statistical characteristic parameters of reconstructed original signal (label 1–12), each twelve statistical characteristic parameters of four node's coefficient of the second layer of wavelet packet decomposition (13–60), each twelve statistical characteristic parameters of four single branch reconstructed signals (61–108) and standardized relative energy of four nodes (109–112), the feature set $set = \{v_i | v_i = v_1, v_2, ..., v_n\}$, $i = 1, 2, ..., n = 112$ is constructed.

3.2 Feature Importance Scoring Based on XGBoost

XGBoost [18] is a kind of boosting algorithm. Based on GBDT algorithm, XGBoost algorithm carries out second-order Taylor expansion of loss function and adds a regular term, which avoids overfitting and effectively accelerates the convergence speed. By continuously adding new decision trees to fit the residual of previous prediction, XGBoost algorithm reduces the residual between the predicted value and the real value. Finally, the prediction accuracy is improved and the feature importance score is obtained. The prediction of XGBoost is described as follows:

$$\hat{y}_i = \sum_{k=1}^{K} f_k(x_i), f_k \in F \tag{5}$$

where K is the number of decision trees, f_k is the kth sub model, x_i is the ith input sample, F is the set of all decision trees.

The objective function of XGBoost consists of loss function and regular term:

$$L^{(t)} = \sum_{i=1}^{n} l(y_i, \hat{y}_i^{(t-1)} + f_t(x_i)) + \Omega(f_t) \tag{6}$$

where t is the number of iterations, l is a differentiable convex loss function that measures the difference between the prediction $\hat{y}_i$ and the target y_i, $\hat{y}_i^{(t-1)}$ is the prediction of the previous $t-1$ iteration, $\Omega(f_t)$ is the regular term of the kth iteration and its equation is as follows:

$$\Omega(f) = \gamma T + \frac{1}{2}\lambda \|w\|^2 \tag{7}$$

where γ and λ are regular term coefficients which can prevent the decision tree from being too complex, T is the number of leaf nodes, w is the leaf weight.

The feature importance score can be calculated as follows:

$$IS_i = \{x | x = w_i v_i\} \tag{8}$$

where v_i is the feature set $set = \{v_i | v_i = v_1, v_2, ..., v_n\}$, $i = 1, 2, ..., n$ and w_i is the weight of corresponding feature.

XGBoost algorithm carries out second-order Taylor expansion of loss function [19], then finds the minimum value of the objective function and calculates the corresponding optimal value by

$$\tilde{L}^{(t)}(q) = -\frac{1}{2}\sum_{j=1}^{T}\frac{(\sum_{i\in I_j} g_i)^2}{\sum_{i\in I_j} h_j + \lambda} + \gamma T \tag{9}$$

where q represents the structure of each tree that maps an example to the corresponding leaf index, $I_j = \{i | q(x_i) = j\}$ is the instance set of leaf j, g_i is the first derivative of sample x_i, h_i is the second derivative of sample x_i. Equation (9) can be used to evaluate a tree structure. The smaller the value, the better the model.

The loss reduction which is also known as gain after split is shown as follows:

$$L_{split} = \frac{1}{2}\left[\frac{(\sum_{i\in I_L} g_i)^2}{\sum_{i\in I_L} h_j + \lambda} + \frac{(\sum_{i\in I_R} g_i)^2}{\sum_{i\in I_R} h_j + \lambda} - \frac{(\sum_{i\in I} g_i)^2}{\sum_{i\in I} h_j + \lambda}\right] - \gamma \tag{10}$$

where the four terms on the right side of the equation represent the left subtree and right subtree score after splitting, the score before splitting, and the complexity regularization penalty coefficient respectively. In Eq. (10), $I = I_L \cup I_R$, when the reaches the depth limit or $L_{split} < 0$, the tree will stop splitting.

3.3 XGBoost Optimization Based on ACO

XGBoost has many parameters, and the selection of parameters affects the performance of the model. Reasonable parameters setting can significantly improve the accuracy of XGBoost. Some parameters are χ, δ, and ε which are discrete, and γ, λ and ω which are continuous. In this paper, ACO is used to optimize these parameters.

ACO is a swarm intelligence algorithm developed by natural genetics and evolution. It has great global search ability and can get positive feedback results. The basic idea of ACO is to use walking path of ants to represent the feasible solution, and all paths of the all ant colony constitute the solution space of the optimal problem. Ant with shorter paths releases more pheromone. With the advance of time, the pheromone on the shorter path gradually increase, and the number of ants choosing the path is also increased. Finally, the all ants will focus on the best path under the action of positive feedback and the best path is the optimal solution. The main steps of ACO are path selection and pheromone updating [20]. Path selection formula is shown as follows:

$$P_{ij}^{k}(t) = \frac{\{\max(\tau_i) - [\tau_{ij}(t-1)]^{\alpha}\}\eta_i^{\beta}}{\max(\tau_i)\sum_{e_{wj}\notin tabu_k}\eta_w^{\beta}}, \; e_{ij} \notin tabu_k \tag{11}$$

where $P_{ij}^{k}(t)$ denotes the probability that ant k choose its path from node s_i to s_j at time t, $\tau_{ij}(t)$ is pheromone amounts, η_i is the heuristic function and the specific expression depends on the specific problem, α is the importance of pheromone amounts, reflecting the impact extent of information for an ant to select a new path or not, β is the importance of heuristic function, e_{ij} is the edge from node s_i to s_j, taboo list $tabu_k$ records sides ant k have walked. In Eq. (11), $\tau_{ij}(t)$ is calculated as follow:

$$\tau_{ij}(t) = (1 - \rho)\tau_{ij}(t - 1) + Q\varphi'(tabu^t) \tag{12}$$

where $\rho(0 < \rho < 1)$ is the pheromone evaporation coefficient, Q is a constant, which is determined according to ρ and adjusts the size of pheromone increment. $\varphi'(tabu^t)$ is the objective function value of the path $tabu^t$ at time t [21].

The ACO-XGBoost model is shown in algorithm 1.

Algorithm 1. ACO-XGBoost

Input: the parameters of ACO and the parameters to be optimized of XGBoost
k : the number of ants; $\tau_{ij}(0)$: pheromone amounts before executing ACO;
α : importance of pheromone amounts; β : importance of heuristic function;
$\rho(0 < \rho < 1)$: pheromone evaporation coefficient; Q : a constant;
K : the whole number of ants; N : iterations;
the range of discrete parameters and continuous parameters namely C
Output: optimal combination of parameters namely C'
initialize continuous parameters and discrete parameters;
for i =1: N **do**
  for j =1: K **do**
    ant k calculates the transition probability according to equation (11)
    if (selected edges do not satisfy constraints)
      {discard this selection; **break**}
  if (the iteration optimal solution is better than the current global optimal solution)
    {update the current global optimal solution; update the pheromone in this iteration according to equation (12)}
return C';

In the training phase of ACO-XGBoost, the accuracy of XGBoost is used to evaluate the optimization method. Firstly, we initialize the parameters of ACO, set the value range of parameters which should be optimized in XGBoost. Then three discrete parameters are firstly optimized according to ACO and the results are assigned to XGBoost. In this condition, three continuous parameters are also optimized by ACO. Secondly, the test set is used to evaluate the model after training and determines whether the current parameter value is the optimal value. Finally, ACO-XGBoost follows these steps to find the best parameters after the specified number of iterations.

4 Experiment Procedure and Result Analysis

4.1 Datasets

The CPU is Inter Xeon E5–2630, memory size is 192GB, operating system is Ubuntu 16.04, programming environment is Python 3.8 of the experiment.

The experimental data comes from two emitters. The acquisition environment is basically a clean environment without noise. The signal data are obtained under 10 different acquisition states. The 10 specific signal parameters are shown in Table 1.

Table 1. Signal parameters

	Carrier wave/MHz	Modulation mode	Signal bandwidth/Hz	Sampling frequency/M	Interval time/ms
1	55	QPSK	25 k	1	20
2	75	QPSK	25 k	1	20
3	420	QPSK	2 M	20	2
4	420	QPSK	5 M	50	1
5	420	QPSK	10 M	80	1
6	420	QPSK	20 M	100	2
7	2000	QPSK	2 M	20	2
8	2000	QPSK	5 M	50	1
9	2000	QPSK	10 M	80	1
10	2000	QPSK	20 M	100	1

For each radio station, 200 groups (4096 data of a group) data are selected in each acquisition state (4000 $\times$ 4096 data in total), and selecting 75% (3000) for training, selecting 25% (1000) for testing. According to Sect. 3.1, after feature extraction, a feature $set = \{v_i | v_i = v_1, v_2, ..., v_{112}\}, i = 1, 2, ..., 112$ can be constructed. In this paper, the features are extracted from amplitude, channel I and channel Q respectively. And then a dataset named $dataset_{\text{original}} = \{v_i | v_i = v_1, v_2, ..., v_{336}\}, i = 1, 2, ..., 336$ is achieved. To verify the efficiency and effectiveness of the proposed method, Gaussian white noise is added to the original signal to make the signal-to-noise ratio be 10 dB and 5 dB respectively. The other two datasets are named $dataset_{10\text{dB}}$ and $dataset_{5\text{dB}}$.

4.2 Data Preprocessing

In order to unify the order of magnitude, increase comparability and speed up convergence of data samples, zero-mean normalization is used and the formula is shown as follows [14]:

$$v'_{tb} = \frac{v_{tb} - \overline{X}_{vt}}{\delta_{vt}} \tag{13}$$

where v_{tb} is the bth feature value of feature t, $\overline{X}_{vt}$ is the mean value of feature t, δ_{vt} is the standard deviation of feature t.

4.3 Feature Selection by Using Important Value of XGBoost

Based on Eq. (8), the score map of feature importance in the original radio dataset (the top 20 largest is shown) is got as Fig. 2. In Fig. 2, f46 represents the 47th feature.

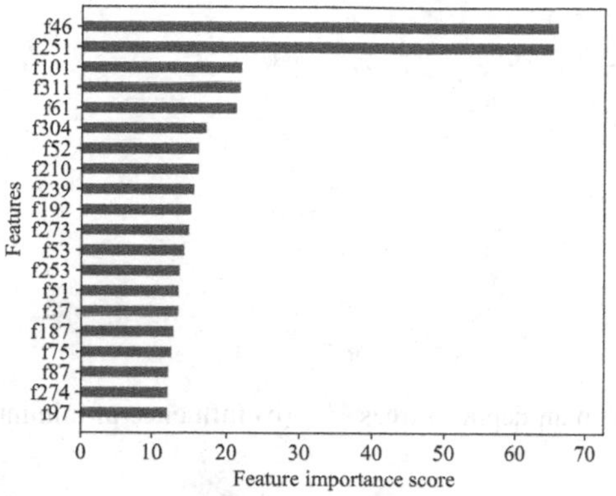

Fig. 2. Ranking of feature importance

To obtain the optimal feature subset, the importance of each feature is used as threshold to select features. For $dataset_{\text{original}}$, The threshold and the number of features is shown in Table 2.

Table 2. Feature selection accuracy of thresholds

Threshold	Numbers of features	*A* (%)
0.000	336	98
0.001	129	97.8
0.002	**84**	**98.7**
0.003	73	98
0.005	45	98
...	...	...
0.091	1	51

In Table 2, by setting each importance as the threshold, the features are divided into some subsets and the accuracy of the current subset is calculated. In the experiment, the feature importance is sorted from small to large, and then the sorted importance is used as every threshold to select the best feature subset.

It can be seen from Table 2 that the subset composed of the first 84 features with the highest importance has the highest accuracy. In some cases, adding features cannot improve the accuracy or even reduce the accuracy, so feature selection is rather necessary.

4.4 The Effect of Parameters on XGBoost

The parameters in XGBoost will affect the final prediction result of the model. In this section, the effect of different parameters on XGBoost are shown Fig. 3. The classification accuracy without parameter optimization in $dataset_{\text{original}}$ was 0.98.

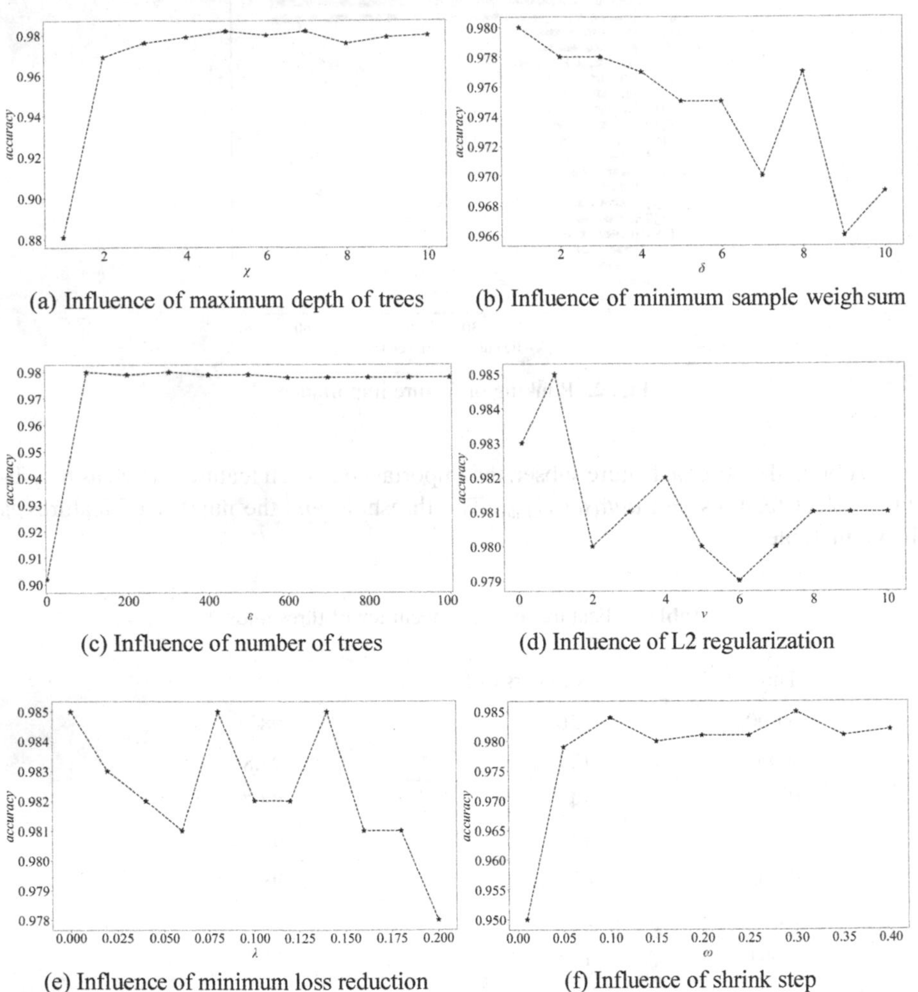

(a) Influence of maximum depth of trees (b) Influence of minimum sample weigh sum

(c) Influence of number of trees (d) Influence of L2 regularization

(e) Influence of minimum loss reduction (f) Influence of shrink step

Fig. 3. Influence of XGBoost parameters on accuracy

Figure 3(a)(b)(c) are the change rule of prediction accuracy in the process to test the effect of three discrete parameters. As shown in Fig. 3(a), when $\chi = 5$, a higher accuracy can be obtained, and with the continuous increase of the tree depth, accuracy will not be further improved. The larger tree depth makes the model more complex and lead to overfitting easily, which will in turn reduce the prediction accuracy of test set. Figure 3(b) shows that the accuracy of the model basically decreases with the increase of δ, which indicates that the larger the parameter is, the less easy it is to overfit, but it may result in underfit. It can be seen from Fig. 3(c) that when the value of ε is between 100 and 300, a higher accuracy can be obtained. However, with the continuous increase of the number of trees, the change range of accuracy is small and the model will be easy to overfit.

For three continuous parameters, several values within the corresponding value range are selected and the prediction accuracy of each value is calculated. Then line charts are drawn. Figure 3(d)(e)(f) show the influence and change law of three continuous parameters on the accuracy of SEI. In Fig. 3(d) and Fig. 3(e), when the values of γ and λ are relatively small, the accuracy has been improved. With the increase of parameter value, although the accuracy has been fluctuating, a better accuracy value has not obtained. Figure 3(f) shows that a relatively highest accuracy can be obtained when ω is around 0.3. The smaller the learning rate is, the smaller the calculation speed is. If the learning rate is too large, it may not converge.

By changing the parameters of XGBoost, the accuracy of SEI is greatly affected, which shows that the most appropriate value of parameters will maximize the use of all characteristics of signal data and obtain the optimal model. In this section, the influence of various parameters on XGBoost has been explored. In order to obtain the optimal value of these parameters, ACO is used to futher optimize the parameters.

4.5 XGBOOSt's Parameter Optimization Based on ACO

This section aims to use ACO to optimize XGBoost and the variation curve of the predicted value with the iterations of ACO is shown in Fig. 4.

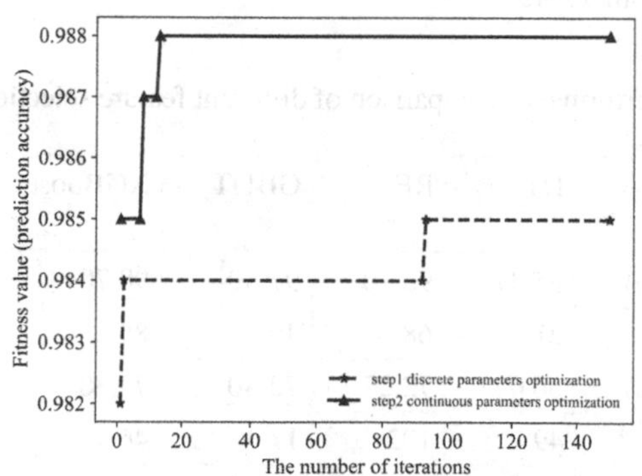

Fig. 4. Iteration curves of XGBoost optimized by ACO

In Fig. 4, the range we set and the default value of six parameters which have mentioned above are all shown in Table 3. Initializing $k = 45$, $\tau_{ij}(0) = 0$, $\alpha = 1$, $\beta = 0$, $\rho = 0.8$, $Q = 1$, $N = 150$. In Fig. 4, firstly, ACO is used to optimize three discrete parameters. Secondly, we optimize the three continuous parameters when we have determined and brought the discrete parameters into XGBoost model. Compared with default parameters of XGBoost, the accuracy is improved by 0.80% after two steps of optimization. Under the given termination condition, the parameters optimized by ACO are shown in Table 3.

Table 3. Model parameters

Model parameters	Parameter description	Range we set	Default	ACO-XGBoost
χ	Maximum depth of trees	[1, 10]	6	5
δ	Minimum sample weigh sum in leaf node	[1, 10]	1	1
ε	Number of trees	[1,1000]	100	202
γ	L2 regularization penalty coefficient	[0.1,10]	1	5.91199579
λ	Minimum loss reduction	[0,0.2]	0	0
ω	Shrink step	[0.01,0.4]	0.3	0.24792934

4.6 Performance Comparison

Based on the above introduction, datasets and evaluation criteria, the classification prediction experiment is designed, and the method proposed in this paper is compared with DT, RF, GBDT and XGBoost with default parameters. Due to certain randomness, we conduct three experiments of DT, RF and GBDT respectively and take the average value. The results are shown in Table 4.

Table 4. Performance comparison of different feature selection models

Dataset and evaluation criteria		DT	RF	GBDT	XGBoost	ACO_XGBoost
$dataset_{\text{original}}$	A (%)	95.37	98.10	97.80	98.70	**98.90**
	q	**21**	68	49	84	39
$dataset_{10\text{dB}}$	A (%)	66.07	72.27	72.40	72.30	**73.00**
	q	49	192	**11**	46	35
$dataset_{5\text{dB}}$	A (%)	55.90	59.10	58.40	59.20	**59.30**
	q	92	141	179	40	**35**

In Table 4, the bold value is the optimal value and the underlined value is the sub-optimal value. It can be seen that ACO-XGBoost method proposed in this paper can effectively improve the prediction ability and obtain a relatively small feature subset of most datasets. In $dataset_{\text{original}}$, accuracy of ACO_XGBoost has been improved by 0.20%-3.53% compared with the other four comparison algorithms and q is the second smallest. In $dataset_{10\text{dB}}$, accuracy index has been improved by 0.60%-6.93%. Also, q is the second smallest. In $dataset_{5\text{dB}}$, accuracy increased by 0.10%-3.40%, and the number of selected features q is smallest. Experiment shows that the proposed method in this paper not only obtains the optimal classification results but also acquired the feature subsets with relatively small feature number. A balance is got between the classification accuracy and the number of selected numbers.

5 Conclusion

In this paper, we present a new model of SEI based on ACO-XGBoost feature selection to improve the identification ability of specific emitter. Our method consists of the following three steps: 1) Lifting wavelet packet decomposition and deconstruction are used for feature extraction and then a characteristic parameter system is constructed. 2) ACO is used to optimize the parameters of XGBoost. The optimization of discrete parameters is carried out firstly and the results are brought into the process of continuous parameter optimization. 3) The optimized XGBoost model is used for feature selection. The importance of each feature is calculated, and all the importance values are used as thresholds to select the optimal features to form the optimal feature subset. Experimental results on three datasets show that our proposed method can keep a balance between accuracy and selected features. In our future work, we will further control the number of selected features without reducing the prediction accuracy.

Acknowledgement. This paper is supported by the National Natural Science Foundation of China (71901215, 61371196), China Postdoctoral Science Foundation Funded Project (20090461425, 201003797), and the Nation Important Scientific and Technological Special Project (2015ZX01040201-003).

References

1. Satija, U., Trivedi, N., Biswal, G., Ramkumar, B.: Specific emitter identification based on variational mode decomposition and spectral features in single hop and relaying scenarios. In: IEEE Transactions on Information Forensics and Security, pp. 581–591. (2019)
2. Han, J., Zhang, T., Wang, H., Ren, D.: Communication emitter individual identification based on 3D-Hibert energy spectrum and multi-scale fractal features. In: Journal on Communications **38**(4), 99–109 (2017, in Chinese). https://doi.org/10.11959/j.issn.1000-436x.2017080
3. Xiu, Y., Zhao, S., Chen, H., Li, C.: I-mRMR: incremental max-relevance, and min-redundancy feature selection. In: Shao, J., Yiu, M.L., Toyoda, M., Zhang, D., Wang, W., Cui, B. (eds.) APWeb-WAIM 2019. LNCS, vol. 11642, pp. 103–110. Springer, Cham (2019). https://doi.org/10.1007/978-3-030-26075-0_8
4. Xue, H., Jiang, J., Shao, Y., Cui, B.: FeatureBand: a feature selection method by combining early stopping and genetic local search. In: Shao, J., Yiu, M.L., Toyoda, M., Zhang, D., Wang, W., Cui, B. (eds.) APWeb-WAIM 2019. LNCS, vol. 11642, pp. 27–41. Springer, Cham (2019). https://doi.org/10.1007/978-3-030-26075-0_3

5. Zhang, W., Chen, C., Jiang, L., Bai, X.: A new feature selection algorithm based on category difference for text categorization. In: Shao, J., Yiu, M.L., Toyoda, M., Zhang, D., Wang, W., Cui, B. (eds.) APWeb-WAIM 2019. LNCS, vol. 11642, pp. 322–336. Springer, Cham (2019). https://doi.org/10.1007/978-3-030-26075-0_25
6. Wang, J., Wei, J., Yang, Z.: Supervised feature selection by preserving class correlation. In: CIKM 2016, pp. 1613–1622. ACM, (2016). https://dx.doi.org/https://doi.org/10.1145/2983323.2983762
7. Rostami, M., Berahmand, K., Nasiri, E., Forouzandeh, S.: Review of swarm intelligence-based feature selection methods. Engineering Appl. Artificial Intelligence **100**, 104210 (2021)
8. Wang, T., Wang, X., Liu, J.: Prediction of moisture content of Aquilaria Sinensis leaves based on RFE_RF algorithm model. In: Journal of Nanjing Forestry University (Natural Sciences Edition) [J/OL] (2022, in Chinese)
9. Chen, Z., Lu, N.: Network intrusion detection model based on Random Forest and XGBoost. In: Journal of Signal Processing **36**(7), 1055–1064 (2020, in Chinese)
10. Zhou, J., He, P., Qiu, R., Chen, G., Wu, W.: Research on intrusion detection based on random forest and gradient boosting tree. Journal of Software **32**(10), 3254–3265 (2021, in Chinese)
11. Feng, J., Liang, J., Qiang, Z., Li, X., Chen, Q., Liu, G., et al.: Effective techniques for intelligent cardiotocography interpretation using XGB-RF feature selection and stacking fusion. In: International Conference on Bioinformatics and Biomedicine, pp. 2667–2673 (2021)
12. Gong, H., Zhang, H., Zhou, L., Liu, Y.: An interpretable artificial intelligence model of Chinese medicine treatment based on XGBoost algorithm. In: Proceedings of the 2020 IEEE International Conference on Bioinformatics and Biomedicine, pp. 1550–1554 (2020)
13. Liu, L., Tang, S., Wu, F., Wang, Y., Wang, J.: An ensemble hybrid feature selection method for neuropsychiatric disorder classification. IEEE/ACM Transactions on Computational Biology and Bioinformatics 19(3), 14591471 (2021). https://doi.org/10.1109/TCBB.3021.3053181
14. Qiao, N., Li, Z., Zhao, G.: Intrusion detection model of internet of things based on XGBoost-RF. In: Journal of Chinese Computer System **43**(1), 152–158 (2022, in Chinese)
15. Liu, M., Yang, Z., Zhang, J.: Specific emitter identification method for aerial target. In: Systems Engineering and Electronics **41**(11), 2408–2415 (2019, in Chinese). https://doi.org/10.16798/j.issn.1001-506X(2019)11-2408-08
16. Cao, J., Zhang, P., Zhang, Y.: Feature extraction of an engine cylinder head vibration signal based on lifting wavelet package transformation. Journal of Vibration and Shook **27**(2), 34–37 (2008, in Chinese)
17. Cao, J., Zhang, P., Ren, G.: Feature selection of vibration signal based on ant colony optimization. J. Vibration and Shock **27**(5), 24–31 (2008, in Chinese)
18. Chen, T., Guestrin, C.: XGBoost: A scalable tree boosting system. In: Proceedings of the 22nd ACM SIGKDD International Conference on Knowledge Discovery and Data Mining, pp. 785–794. San Francisco (2016)
19. Li, L., et al.: Spectroscopy-based food internal quality evaluation with XGBoost algorithm. In: U, L.H., Xie, H. (eds.) APWeb-WAIM 2018. LNCS, vol. 11268, pp. 56–64. Springer, Cham (2018). https://doi.org/10.1007/978-3-030-01298-4_6
20. Yang, L., Li, K., Zhang, W., Ke, Z., Xiao, K., Du, Z.: An improved chaotic ACO clustering algorithm. In: Proceedings of the 2018 IEEE 20th International Conference on High Performance Computing and Communications, pp. 1642–1649 (2018)
21. Mu, N., Xu, X., Zhang, X.: Ant colony optimization based salient object detection for weak light images. In: Proceedings of the 2018 IEEE SmartWorld, Ubiquitous Intelligence, Internet of People and Smart City Innovations, pp. 1432–1437 (2018)

NED-GNN: Detecting and Dropping Noisy Edges in Graph Neural Networks

Ming Xu[1,2], Baoming Zhang[1,2], Jinliang Yuan[1,2], Meng Cao[1,2], and Chongjun Wang[1,2(✉)]

[1] State Key Laboratory for Novel Software Technology, Nanjing, China
{zhangbm,yuanjl19,caomeng}@smail.nju.edu.cn, chjwang@nju.edu.cn
[2] Nanjing University, Nanjing, China

Abstract. Graph neural networks have become the standard learning architectures in graph-based learning and achieve great progress in real-world tasks. Existing graph neural network methods are mostly based on message passing neural network(MPNN), which aggregates messages from neighbor nodes to update representations of target nodes. The framework follows the assumption of homophily that nodes linked by edges are similar and share the same labels. In the real world, the graphs can mostly follow the assumption. However, for nodes in the graph, the connections between nodes are not always connecting two similar nodes. We regard the edges as noisy edges. Such edges will introduce noise to message passing in the training process and hurt the performance of graph neural networks. To figure out the noisy edges and alleviate their influence, we propose the framework called *Noisy Edge Dropping Graph Neural Network*, short as NED-GNN. By evaluating the weights between sampled negative edges and existing edges for each node, NED-GNN detects and removes noisy edges. Extensive experiments are conducted on benchmark datasets and the promising performance compared with baseline methods indicates the effectiveness of our model.

Keywords: Graph neural networks · Noisy edges · Graph learning · Data mining

1 Introduction

Deep Learning has achieved great success in enhancing machine learning tasks with data in Euclidean space, like computer vision [18], natural language processing [17], etc. In recent years, research of the non-Euclidean graph-structured data is increasingly popular as graphs are widely used in presenting objects and their complex interactions. Graph learning is promoted to capture node features and topology information in graphs by representing nodes to low-dimensional embeddings. In this way, the learned embeddings can be well applied to machine learning methods, thus solving downstream tasks. Graph Neural Networks (GNNs) introduce deep learning to graph learning and show promising capacity, thus

B. Li et al. (Eds.): APWeb-WAIM 2022, LNCS 13421, pp. 91–105, 2023.
https://doi.org/10.1007/978-3-031-25158-0_8

being broadly used in tackling problems such as node classification [13,22], link prediction [8,28] and graph classification [9,24]. In the real world, GNN also becomes a promising solution for recommendation [16,21], social networks [5], text extraction [26], knowledge graphs [1], etc.

Existing GNN methods mostly follow the manner of message passing neural network (MPNN) [10]. The main idea of MPNN is to aggregate features from neighbor nodes. Then the representation of nodes is updated in each iteration and the final embeddings can be used for downstream tasks. The effectiveness of MPNN benefits from the assumption of homophily. When the two nodes linked by one edge are similar and share the same labels, nodes can aggregate similar and useful information from neighbor nodes. After that, the final representations can be robust and generalized. In real-world graphs, nodes with similar features tend to gather, which can well satisfy the assumption. However, the particular node in the graphs sometimes connects to dissimilar neighbors. For instance, nodes in the boundaries between classes connect to different-labeled nodes in the connected graph. Besides, the adversarial attacks in graphs often happens through connecting nodes of different classes. In particular, we call the edges constructed by nodes of different classes **Noisy Edges**. To validate our observations, we calculate the number of edges with nodes of different classes in widely-used citation datasets. The results are represented in Table 1.

Table 1. Noisy Edges Counting in Datasets

	Overall Edges	Noisy Edges	Propotion(%)
Cora	5278	1003	19.00
Citeseer	4614	1220	26.45
PubMed	44325	8759	19.76

From Table 1, we can find noisy edges exists commonly even in widely-used benchmark datasets. Such noisy edges may hurt the performance of GNNs as they bring the noise to target nodes by aggregating unnecessary information and mislead the training of the model. Here we show an example in the sample graph:

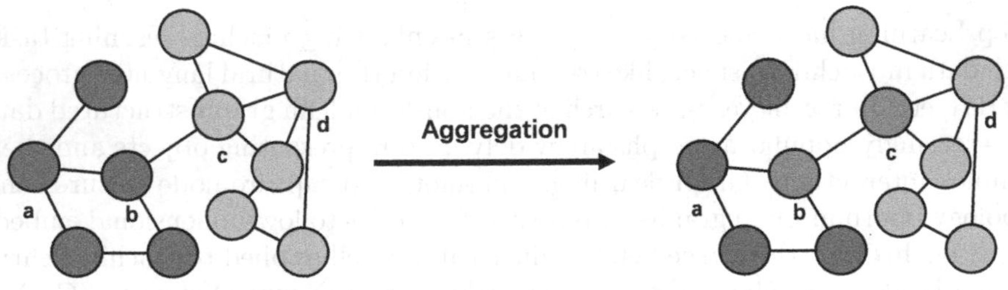

Fig. 1. An example of aggregation in graph neural networks.

In Fig. 1, we conducted 1-step aggregation in the sample graph, and different colors indicate different features. According to the figure, as the node a and node d share neighbors with same colors, respectively, their colors remain the same after aggregation. However, node b and node c change their colors as different-colored neighbors' information propagated, and their final representations are unreliable. What's worse, the noisy information will influence other nodes with the procedure of iterative aggregation.

To alleviate the undesirable effects of noisy edges, an intuitive approach is to remove such edges. Some works are proposed to update the structure with different strategies. DropEdge [20] tries to drop edges randomly to improve the generalization capacity of GNNs. However, the edge dropping without supervision sometimes breaks the topology structure and discards the essential relations among nodes in graphs. Besides, some methods update topology structure by the prediction of the model or co-training both structure learning and GNN models. The methods achieve success, but somehow they get stuck in the confidence of predictions with the setting of prediction threshold or the interpretability of the structural modifications in the graph.

In this paper, we focus on modifying the graph structure to obtain better node embeddings. In particular, we proposed an effective method, **Noisy Edges Dropping Graph Neural Networks(NED-GNN)**, to evaluate the edges and remove the noisy ones. We first sample negative edges for nodes, which connect the node with negative nodes of different classes in the graph. After that, we evaluate the weights of edges and the negative edges after sampling by calculating the similarity between linked nodes. If the weights of edges connecting neighbors are smaller than negative edges, we tend to regard the edges as noisy edges and remove them in the graph. As nodes in the graphs are mostly unlabeled, we propose that high-order neighbor nodes are more likely to be negative nodes that are different-labeled. So we sample negative edges by connecting nodes to corresponding high-order neighbor nodes. By evaluating similarity weights, we can drop the noisy edges to improve the node representations learned by graph neural networks. From the extensive experiments on benchmark datasets, promising performance indicates the effectiveness of our method.

We summarize the main contributions of this paper as follows:

- We proposed an effective framework to detect and drop noisy edges in the graph.
- We designed a method called NED-GNN by dropping detected noisy edges to update the topology structure and obtain better node embeddings.
- We conducted extensive experiments on semi-supervised node classification and prove the effectiveness of our method.

The remaining part of the paper is organized as follows. Section 2 reviews the related works. We show an observation of real-world datasets in Sect. 3. In Sect. 4 we introduce some preliminaries and our method in detail. Extensive experiments are conducted in Sect. 5 to evaluate the performance of our model. At last, Sect. 6 concludes the paper with discussions and future works.

2 Related Works

In this section, we briefly review the related works, including graph neural networks and modifications to graph structure. For more details in graph neural networks, we refer readers to some surveys [11,30].

2.1 Graph Neural Networks

Graph Neural Networks achieve great success in tackling graph learning tasks in recent years. GNN models encode nodes in the graph into low-dimensional dense embeddings and preserve both node features and structure topology at the same time. Graph convolution is first proposed in [2] from the respective of graph signal processing, and many variants are proposed to simplify the framework in both spectral and spatial domain [11,14,30]. Kipf e.t. [13] simplify the model by considering the direct neighbors of nodes in GCN. The simpleness and conciseness of GCN make the model popular and become the baseline in graph learning. Existing graph neural network models are mainly based on message passing neural network (MPNN) [10], which aggregates messages from neighbors nodes and then update representations of target nodes [22,24].

The models based on MPNN mostly follow the assumption of homogeneity, which states that nodes connected by edges are similar and beneficial information can be propagated in the graph. However, the goal is hard to achieve as there always unintentional or intentional exists noise in real world graphs. We focus on the structural noise, mainly noisy edges, and give a brief introduction to graph structure modification in the next subsection.

2.2 Graph Structure Modification

Graph neural networks utilize both node features and graph structure to encode nodes. However, the commonly existing structural noise, like noisy edges, may hurt the performance of GNNs as the noise may mislead information spread in graphs. To alleviate the influence of structural noise, training with a better topology structure is crucial for better node embeddings. For instance, DropEdge [20] randomly removes a certain number of edges at each epoch to improve the generalization capacity and alleviate over-smoothing. NeuralSparse [31] learns k-neighbor subgraphs for robust graph representation learning by selecting at most k edges for each nodes. Methods like self-enhanced GNN [25], EGAI [15], and AdaInf [6] add or remove edges based on the predicted labels by the model. Bayesian GCN [29], LDS [7] and IDGL [4] adopt different strategies to learn the graph structure and node embeddings simultaneously to make graph structure more suitable for model learning. There are contrasting models try to construct multi views by modifying the structure [3,27]. The methods achieve great progress in acquiring better node embeddings with modified structure, however they may get stuck in randomicity or lack interpretability of the modified structure.

In this paper, we try to update the structure of the graph to help the training of GNNs. In particular, we proposed a simple but effective method to detect end drop noisy edges in the graph.

3 Case Study

GNNs utilize message passing framework to propagate messages from neighbor nodes by assuming that nodes connected in graphs are similar. In this section, we did some simple analysis on the widely used benchmark citation dataset Cora from *DGL* [23], and the findings prove the necessity and rationality of our method.

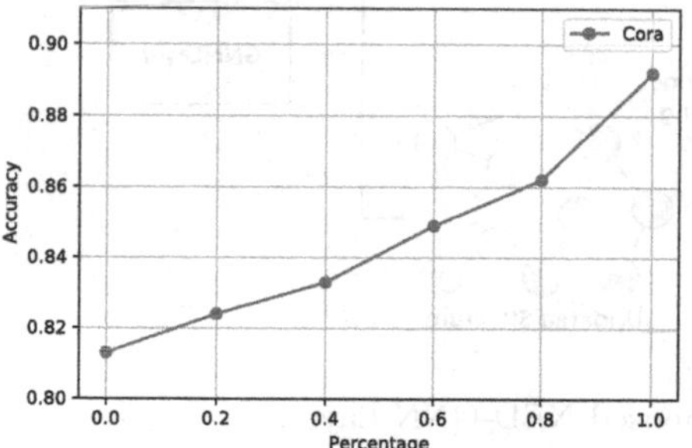

Fig. 2. Accuracy with Different Percentage of Noisy Edges Dropping

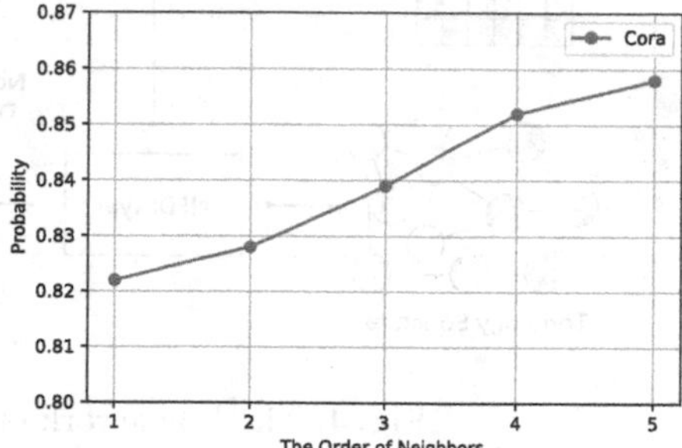

Fig. 3. Probability of Connecting Different-labeled Nodes beyond k-Order Neighbors

Model Performance Without Noisy Edges. We design an empirical experiment to figure out the influence of noisy edges. In particular, we randomly remove noisy edges in Cora with given percentages. After that, we train GCN in the updated graph and use trained embeddings for node classification. The results of classification accuracy are shown in Fig. 2. The results indicate that the representation capacity of GCN would be better with more noisy edges dropping.

The statistical results show that noisy edges are common in real-world graphs, and they will harm the performance of GNNs. Naturally, the conclusion prompts us to consider dropping noisy edges in the training process.

Nodes and the Corresponding High-Order Neighbors. According to the assumption of homophily, nodes with similar features and labels tend to gather, while different-labeled nodes keep away from each other. We calculate the probabilities of nodes sharing different labels with the corresponding high-order neighbors in Cora, and the results are presented in Fig. 3. From the figure, we can figure out that nodes are becoming more dissimilar with the distance between nodes growing. The observation can help find out the noisy edges in the graph.

4 Our Approach

The case study shows the motivation of our work. To better explain and prove our idea, we propose the method called NED-GNN. The main framework is shown in Fig. 4. Our key insight is to drop noisy edges in the graph and update the topology structure to improve the performance of GNNs.

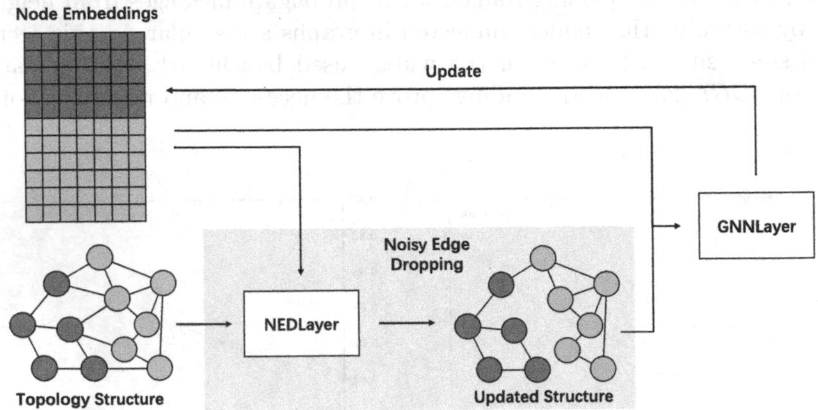

Fig. 4. The Framework of proposed NED-GNN Layer.

4.1 Notations and Preliminaries

This paper mainly focuses on undirected graphs, but the method can also be used in directed graphs. We present $\mathcal{G} = (\mathcal{V}, \mathcal{E}, \mathcal{X})$ as a graph, where $\mathcal{V}$ consists of the set of nodes in $\mathcal{G}$, with $|\mathcal{V}| = N$. $\mathcal{E}$ is the collection of edges. $\mathcal{X} \in \mathbb{R}^{N \times F}$ denotes node feature matrix, where $\mathbf{x_i} \in \mathbb{R}^F$ represents the attributes of node v_i, and F is the dimension of node features. Adjacency matrix $\mathbf{A} \in \{0,1\}^{n \times n}$ is the topological structure of graph $\mathcal{G}$, where $\mathbf{A}_{ij} > 0$ indicates that there is an edge between nodes i and j. Otherwise, $\mathbf{A}_{ij} = 0$.

Given topological structure $\mathbf{A}$ and feature matrix $\mathcal{X}$ as input, the goal of GNN framework is to learn low-dimensional dense node embedding matrix $\mathbf{Z} \in \mathbb{R}^{N \times d}$ with $d \ll F$. The learned node embeddings can well preserve topology and feature information so as to be applied to downstream tasks.

The training procedure of GNNs is as follows. First, the framework learns node representations by aggregating the features of neighbor nodes. The output of the k-th layer of the framework can be generally expressed as,

$$\mathbf{h}_i^{(k)} = \sigma(\mathbf{h}_i^{(k-1)}, AGG(\mathbf{h}_j^{(k-1)})), \quad j \in \mathcal{N}(i) \tag{1}$$

where $h_i^{(l)}$ is the node representation of node v_i at the k-th layer with $h_i^{(0)} = \mathbf{x}_i$ and $\mathcal{N}(i)$ is the direct neighbors of node v_i. $AGG(\cdot)$ is the aggregation function and $\sigma(\cdot)$ is the non-linear function.

After that, the framework calculates the loss of prediction on labeled nodes, $\mathcal{V}_L$, to update the parameters in the framework. The loss function can be expressed as,

$$\mathcal{L} = -\frac{1}{|\mathcal{V}_L|} \sum_{i \in \mathcal{V}_L} \sum_{l=1}^{k} \mathbf{y}_{il} \log \hat{\mathbf{y}}_{il}, \tag{2}$$

where $\hat{\mathbf{y}}_i$ is the prediction of v_i, $\mathbf{y}_i$ indicates the original label for node v_i in one-hot embedding, k is the length of label embedding.

4.2 Noisy Edges Dropping

Before the introduction of our method, we regard the nodes which are labeled differently to target nodes as **negative nodes** and the connected edges as **negative edges**.

The main idea of our method is to figure out noisy edges and update the topology structure by removing the edges. Intuitively, if we find the neighbor node is less similar to the target node than a negative node, we think the edge connecting the neighbor is the noisy edge. Figure 5 is an example. We sample a negative node j for the target node t and constructed a negative edge (t, j). Then we evaluate the weight of edge (t, i) and (t, j), namely the similarities between nodes. If $sim(t, i) < sim(t, j)$, the edge between t and i is possible noisy. As a result, the edge will be removed.

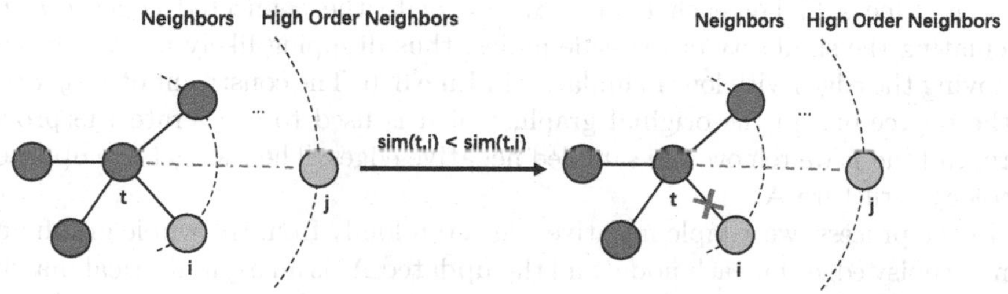

Fig. 5. An example of noisy edge dropping

To calculate the similarity between nodes, we set,

$$sim(i, j) = similarity(\mathbf{h}_i, \mathbf{h}_j) \tag{3}$$

$\mathbf{h}_i$ is the representation of node v_i, the choice of $similarity(\cdot)$ can be any function that measures the similarity of two embeddings, like Cosine similarity and Euclidean Distance. For the simplicity, we choose Cosine similarity as our similarity function.

However, the nodes in the graph are mostly unlabeled, which is unable for us to directly sample negative nodes and negative edges. To solve the problem, we proposed the idea of sampling negative edges according to the property of graphs. The similarity of nodes becomes weaker as the distance increasing in graphs. The case study in Sect. 3 also validates our assumption. In particular, for each node, no more than r-hop neighbors can be regarded as the positive nodes which are more likely to be similar, while the other nodes are seen as the negative nodes. So we sample neighbor nodes beyond r-order as the negative nodes and then constructing negative edges.

Algorithm 1: NEDLayer Algorithm

Input: Adjacency matrix $\mathbf{A}$, Representation Matrix $\mathbf{Z}$
Output: Updated Adjcency Matrix $\mathbf{A}'$

1 $\mathbf{A}_{neg} = Sample(A^{\neg r}) \leftarrow$ Sample k negative edges beyond r-order neighbors
2 $\hat{\mathbf{A}} = \mathbf{A} + \mathbf{A}_{neg} \leftarrow$ Concatenate the negative edges with original graph
3 **for** *node i in* $\hat{\mathbf{A}}$ **do**
4   **for** *node j where* $\hat{\mathbf{A}}_{ij} > 0$ **do**
5         $sim_{ij} = similarity(\mathbf{Z}_i, \mathbf{Z}_j) \leftarrow$ Evaluate the weight of edge (i, j) in $\hat{\mathbf{A}}$
6     $indices_i = Topk(sim_{i,}, k = Degree(i)) \leftarrow$ Remove the less similar edges constrained by the degree in original graph
7     $\mathbf{A}'_{i,} = \{\mathbf{A}'_{ij} = 1 | \mathbf{A}_{ij} > 0 \ and \ j \in indices_i\} \leftarrow$ Remove sampled negative edges
8 return $\mathbf{A}'$

Algorithm 1 summarizes the overall training of the NEDLayer. In each epoch, we sample k negative edges connecting nodes that are beyond r-order neighborhood in Line 1–2. For each node i, we evaluate the connected edges(i, j) by calculating the similarity of two-side nodes, thus dropping likely noisy edges by removing the edges with lower similarity in Line 3–6. The constraint of $Degree(i)$ is the degree of i in the original graph, which is used to accelerate the procedure. In Line 7, we remove the sampled negative edges. Then we get the updated topology structure $\mathbf{A}'$.

In the process, we sample negative edges randomly from the whole graph and remove noisy edges for each node, and the updated $\mathbf{A}'$ is an asymmetrical matrix. With the NEDLayer, we update the topology structure in a self-supervised manner by considering the embeddings of nodes and the property of graphs.

4.3 NED-GNN

After the procedure of NEDLayer, we detect and drop the likely noisy edges in the graph and generate the modified graph structure. With the generated structure, we can update the aggregation process by setting,

$$\mathbf{A}'^{(k)} = normalize(NEDLayer(\mathbf{A}, \mathbf{H}^{(\mathbf{k-1})})) \tag{4}$$

$$\mathbf{H}^{(\mathbf{k})} = \sigma(\mathbf{A}'^{(k)} \mathbf{H}^{(\mathbf{k-1})} \mathbf{W} + \mathbf{B}) \tag{5}$$

where $normalize(\cdot)$ performs L_1 normalization by row to the adjacency matrix. $\sigma(\cdot)$ is the non-linear method, $\mathbf{W}$ is the parameters of GNNLayer, and $\mathbf{B}$ is the bias. As the representations of nodes are unstable at early epochs and may result in mistakenly removing edges connecting two similar nodes, we update the topology from the original graph in each epoch to avoid the situation.

4.4 The Variant of NED-GNN

Though NED-GNN can alleviate unnecessary information aggregation in graph neural networks, the modification of graph structure sometimes brings side effects. As we focus on homophily graphs in the paper, where nodes mostly follow the assumption of homophily, there is no need for the negative edges sampling for all nodes. Besides, most of the graphs are sparse in the real world, so the process of edge dropping may break the substructure and result in poor performance when excessive edges are dropped.

According to the training process of graph neural networks, we can find that the parameters of the framework mainly depend on the labeled nodes by calculating the prediction loss. Inspired by the observation, we promote the variant of NED-GNN, **Noisy Edges Dropping Graph Neural Networks on Training Nodes(NED-GNN-t)**, which focuses on the training nodes, namely sampling negative nodes and dropping noisy edges only for labeled nodes. Then, we can decrease the number of negative edges and the calculation of similarity between nodes. In this way, we can accelerate the process of NEDLayer and improve training efficiency. What's more, as NED-GNN-t only removes the edges of the training nodes, the modification of the structure is mild with less edges dropped, and the model can preserve more topology information.

5 Experiments

The goal of our method is to detect and drop noisy edges in the graph which may harm the aggregation in the training of GNNs. To prove the effectiveness of our method in improving GNN representation capacity, we designed the semi-supervised node classification experiments on benchmark datasets. The procedure is introduced as follows.

5.1 Experimental Settings

Datasets. Following previous works [13], we utilize three benchmark datasets of citation network, Cora, Citeseer and Pubmed. In these datasets, nodes and edges represent documents and citation relations between documents, respectively. Each node is represented by the bag-of-words features extracted from the content of the document. Each node corresponds a label with one-hot encoding of the document category. We employ the same data split in *DGL* [23] module and the data distribution is shown in Table 2.

Table 2. Data distribution of Three Datasets

	Cora	Citeseer	Pubmed
Nodes	2078	3327	19717
Edges	5278	4614	44325
Features	1433	3703	500
Classes	7	6	3
Training Nodes	140	120	60
Validation Nodes	500	500	500
Test Nodes	1000	1000	1000

Baseline Methods. To evaluate the effectiveness of our method, we compare with the following *state-of-the-art* methods.

- **DeepWalk** [19]. The typical shallow network embedding model.
- **GCN** [13]. The baseline of graph nerual networks, which considers aggregating messages from direct neighbors.
- **GraphSage** [12]. Extending the mean aggregator of GCN to perform multi aggregation and performing a sampling strategy before aggregation.
- **GAT** [22]. Considering introducing attention mechanism to GCN and assigns different weights to neighbor nodes according to attention scores.
- **GAT-single**. The method which only utilizes a single attention head for GAT. As our method can somehow be regarded as an further step for single-head GAT by removing low weights.
- **DropEdge** [20]. Randomly removing a certain number of edges at each epoch to improve the generalization capacity of GCN.
- **NED-GNN**. Our method, we choose the GCN and GAT as our baseline method.
- **NED-GNN-t**. The variant of NED-GNN, which samples negative edges only for labeled nodes.

Parameter Settings. In parameter settings, We designed 2-layer neural networks with the same hidden layer dimension and the same output dimension simultaneously for every method. For baseline methods like DeepWalk, GCN, GAT, and DropEdge, we follow the instruction of original codes in Github published by the authors. For GraphSage, we only consider the situation with mean aggregator, and the model is implemented the same as the authors' guidance. With our methods, we follow most settings of GCN except that we use an early stopping strategy the same as GAT with a patience of 100 epochs. For all models, we conducted 10 times and got the average results to show the performance.

5.2 Experiments Comparison

Table 3 shows the results of the compared methods on semi-supervised node classification. From the table, we can find the following observations. Our method achieves the best or competitive performance compared with baseline methods in all datasets. The promising results show the effectiveness of our method which removes the noisy edges in the graph by comparing with negative edges. NED-GAT performs better than NED-GCN as GAT can still assign different weights to noisy edges which are not detected. DropEdge randomly removes a certain percentage of edges in the graph to achieve better performance. However, the randomicity is hard to control and sometimes discards the important interactions between nodes, resulting in unsatisfying results. GraphSage also performs a sampling strategy before aggregation and would get stuck in the randomicity with a performance drop.

Table 3. The results of node classification accuracy(%) on the three datasets

Data	Cora	Citeseer	Pubmed
DeepWalk	67.2	43.2	65.3
GraphSage	77.4	67.0	76.6
GIN	77.6	66.1	77.0
DropEdge	79.6	67.6	73.4
GCN	81.6	70.5	78.7
NED-GCN	**82.9**	70.9	**79.0**
NED-GCN-t	82.6	**73.4**	78.8
GAT	82.6	70.3	77.5
GAT-single	81.6	70.3	77.4
NED-GAT	**83.3**	71.0	**78.1**
NED-GAT-t	**83.3**	**73.1**	**78.1**

NED-GNN-t shows competitive performance in Cora compared with NED-GNN, which indicates that the representative embeddings of labeled nodes are essential for graph neural networks. In the sparse graph like Citeseer, NED-GNN-t outperforms NED-GNN as the side-effects of NED-GNN break the important topology structure, which result in sub-optimal performance. While NED-GNN-t focuses on labeled nodes and modifies a small part in the graph to preserve more topology information.

Besides, GAT outperforms other baselines in Cora as they can prevent the aggregation of nodes from different classes by assigning the noisy edges small weights. However, as GAT uses $softmax(\cdot)$ functions to calculate the weights of attention, the weights of edges will always be positive. Our method tries to remove a part of noisy edges, namely setting the weights of the edges to 0, thus our method performs competitive results. In fact, our method can somehow be regarded as a single-head attention mechanism that calculates the similarity of neighbors directly, but we apply the weights to filter out the dissimilar nodes with a further step. And the results show that our method can learn better node embeddings than GAT with a single head.

Among the baseline methods, the GCN model performs better than DeepWalk as graph convolution is powerful by combining node features and topology information. GAT outperforms GCN in some cases as GAT introduces attention to graph convolution to decide the more correlative neighborhoods. The results are consistent with those in previous works.

5.3 Parameter Study

We conduct the ablation study to verify the effect of the number of negative edges k and the order of sampling neighbors r. The results of node classification on Cora dataset are shown in Fig. 6 and Fig. 7. We traverse k from 200 to 2000 and r from 2 to 6 to show the influence of the parameters.

The Number of Sampled Edges. From the results of Fig. 6, we can figure out that when we sample more edges, the performance of the model increase at first as more negative edges are sampled in the graph which can help detect noisy edges. However, more is not always better. When we sample more than 800 edges in the graph, the performance decreases. As the discussion in Sect. 3, noisy edges occupy a small proportion in the graph. The capacity gain brought by negative edges is then offset by the side-effects of NED-GNN. In conclusion, there is no need for a large number of sampled negative edges.

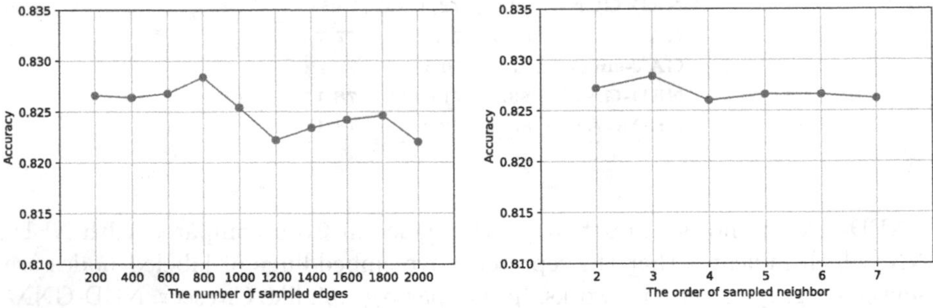

Fig. 6. Accuracy with Different Number of Sampled Edges

Fig. 7. Accuracy with Different Neighbor Order

The Order of Sampled Neighbors. In Fig. 7, classification accuracy is robust with the increasing order of sampled neighbors. The results are consistent with our statistical results. When we sample nodes without nodes in 3-order neighbors, the probability of noisy edges is larger than 83%. So in the practice of our method, the choice with low order neighbors can well satisfy the model training.

5.4 Training Visualization

In this section, we conduct experiments on Cora and Citeseer to verify the noisy edges dropping in the training of NED-GNN. The proportion of noisy edges dropped by our methods in the training process is shown in Fig. 8 and 9. The blue lines in the figures are the probability of randomly dropping noisy edges.

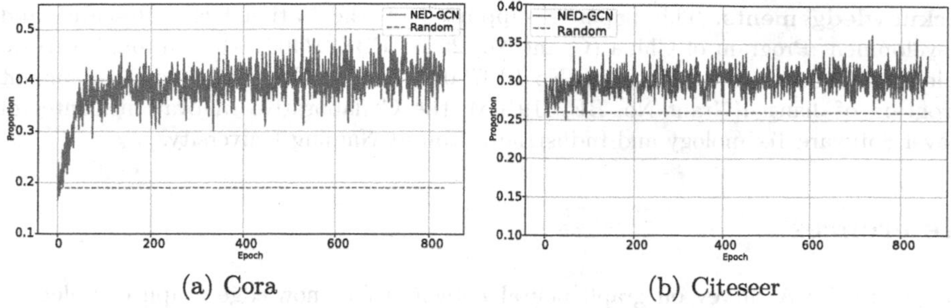

(a) Cora (b) Citeseer

Fig. 8. The proportion of noisy edges in dropped edges when training NED-GCN.

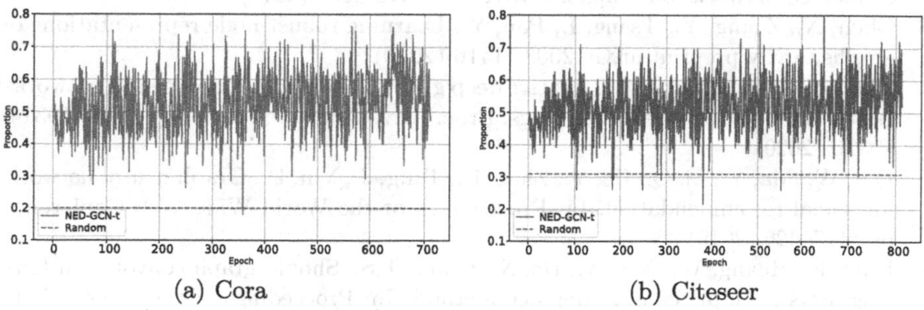

(a) Cora (b) Citeseer

Fig. 9. The proportion of noisy edges in dropped edges when training NED-GCN-t.

From the figures, we can conclude that our method can effectively detect and drop the noisy edges in the graph compared with randomly dropping the edges. Considering the performance in Citeseer, which is too sparse with the average degree of node is 2.8, we can also figure out that NED-GCN-t can better detect and drop noisy edges than NED-GCN. The reason is that NED-GCN-t performs a wilder edge dropping strategy and can preserve more topology structure information in the graph.

6 Conclusion

Existing graph neural network methods mostly follow the assumption of homogeneity, which states that nodes connected by edges are similar and share the same labels. However, particular nodes in the graph sometimes break the assumption. In this paper, we try to discuss noisy edges in the graph and remove the edges to obtain better embeddings. In particular, we proposed a simple and effective method called *Noisy Edge Dropping Graph Neural Network*, short as NED-GNN, to detect and remove noisy edges in the graph by sampling negative edges. Extensive experiments are conducted on benchmark datasets and promising performance shows the effectiveness of our method. In the future, we aim to improve the capacity of our method in detecting noisy edges by more strategies.

Acknowledgements. This paper is supported by the National Key Research and Development Program of China (Grant No. 2018YFB1403400), the National Natural Science Foundation of China (Grant No. 61876080), the Key Research and Development Program of Jiangsu(Grant No. BE2019105), the Collaborative Innovation Center of Novel Software Technology and Industrialization at Nanjing University.

References

1. Arora, S.: A survey on graph neural networks for knowledge graph completion. arXiv preprint arXiv:2007.12374 (2020)
2. Bruna, J., Zaremba, W., Szlam, A.D., Lecun, Y.: Spectral networks and locally connected networks on graphs. CoRR abs/1312.6203 (2014)
3. Chen, X., Zhang, Y., Tsang, I., Pan, Y.: Learning robust node representations on graphs. arXiv preprint arXiv:2008.11416 (2020)
4. Chen, Y., Wu, L., Zaki, M.: Iterative deep graph learning for graph neural networks: Better and robust node embeddings. Proc. Adv. Neural Inf. Proc. Syst. **33**, 19314–19326 (2020)
5. Fan, W., Ma, Y., Li, Q., He, Y., Zhao, E., Tang, J., Yin, D.: Graph neural networks for social recommendation. In: Proceedings of the World Wide Web Conference, pp. 417–426 (2019)
6. Feng, F., Huang, W., Xin, X., He, X., Chua, T.S.: Should graph convolution trust neighbors a simple causal inference method. In: Proceedings of the International ACM SIGIR Conference on Research and Development in Information Retrieval, pp. 1208–1218 (2021)
7. Franceschi, L., Niepert, M., Pontil, M., He, X.: Learning discrete structures for graph neural networks. In: Proceedings of the International Conference on Machine Learning, pp. 1972–1982 (2019)
8. Gao, H., et al.: CSIP: enhanced link prediction with context of social influence propagation. Big Data Res. **24**, 100217 (2021)
9. Gao, H., Ji, S.: Graph u-nets. In: Proceedings of the International Conference on Machine Learning, pp. 2083–2092 (2019)
10. Gilmer, J., Schoenholz, S.S., Riley, P.F., Vinyals, O., Dahl, G.E.: Neural message passing for quantum chemistry. In: Proceedings of the International Conference on Machine Learning, pp. 1263–1272 (2017)
11. Hamilton, W.L.: Graph representation learning. Synth. Lect. Artif. Intell. Mach. Learn. **14**(3), 1–159 (2020)
12. Hamilton, W.L., Ying, R., Leskovec, J.: Inductive representation learning on large graphs. In: Proceedings of the International Conference on Neural Information Processing Systems, pp. 1025–1035 (2017)
13. Kipf, T.N., Welling, M.: Semi-supervised classification with graph convolutional networks. In: Proceedings of the International Conference on Learning Representations (2017)
14. Kipf, T.N., et al.: Deep learning with graph-structured representations (2020)
15. Liu, C., Wu, J., Liu, W., Hu, W.: Enhancing graph neural networks by a high-quality aggregation of beneficial information. Neural Netw. **142**, 20–33 (2021)
16. Liu, Y., Li, B., Zang, Y., Li, A., Yin, H.: A knowledge-aware recommender with attention-enhanced dynamic convolutional network. In: Proceedings of the 30th ACM International Conference on Information Knowledge Management, pp. 1079–1088 (2021)

17. Mikolov, T., Sutskever, I., Chen, K., Corrado, G.S., Dean, J.: Distributed representations of words and phrases and their compositionality. In: Proceedings of Advances in neural information processing systems, pp. 3111–3119 (2013)
18. O'Mahony, N., et al.: Deep learning vs. traditional computer vision. In: Proceedings of Science and Information Conference, pp. 128–144 (2019)
19. Perozzi, B., Al-Rfou, R., Skiena, S.: Deepwalk: Online learning of social representations. In: Proceedings of the ACM SIGKDD International Conference on Knowledge Discovery and Data Mining, pp. 701–710 (2014)
20. Rong, Y., Huang, W., Xu, T., Huang, J.: Dropedge: Towards deep graph convolutional networks on node classification. In: Proceedings of the International Conference on Learning Representations (2020)
21. Sun, J., et al.: Neighbor interaction aware graph convolution networks for recommendation. In: Proceedings of the International ACM SIGIR Conference on Research and Development in Information Retrieval, pp. 1289–1298 (2020)
22. Veličković, P., Cucurull, G., Casanova, A., Romero, A., Liò, P., Bengio, Y.: Graph Attention Networks. In: Proceedings of the International Conference on Learning Representations (2018)
23. Wang, M., et al.: Deep graph library: A graph-centric, highly-performant package for graph neural networks. arXiv preprint arXiv:1909.01315 (2019)
24. Xu, K., Hu, W., Leskovec, J., Jegelka, S.: How powerful are graph neural networks In: Proceedings of the International Conference on Learning Representations (2019)
25. Yang, H., Yan, X., Dai, X., Chen, Y., Cheng, J.: Self-enhanced GNN: Improving graph neural networks using model outputs. arXiv preprint arXiv:2002.07518 (2020)
26. Yao, L., Mao, C., Luo, Y.: Graph convolutional networks for text classification. In: Proceedings of the AAAI Conference on Artificial Intelligence, pp. 7370–7377 (2019)
27. You, Y., Chen, T., Sui, Y., Chen, T., Wang, Z., Shen, Y.: Graph contrastive learning with augmentations. Adv. Neural Inf. Proc. Syst. **33**, 5812–5823 (2020)
28. Zhang, M., Li, P., Xia, Y., Wang, K., Jin, L.: Revisiting graph neural networks for link prediction. arXiv preprint arXiv:2010.16103 (2020)
29. Zhang, Y., Pal, S., Coates, M., Ustebay, D.: Bayesian graph convolutional neural networks for semi-supervised classification. In: Proceedings of the AAAI Conference on Artificial Intelligence, pp. 5829–5836 (2019)
30. Zhang, Z., Cui, P., Zhu, W.: Deep learning on graphs: a survey. IEEE Trans. Knowl. Data Eng. **34**(1), 249–270 (2020)
31. Zheng, C., et al.: Robust graph representation learning via neural sparsification. In: Proceedings of the International Conference on Machine Learning, pp. 11458–11468 (2020)

A Temporal-Context-Aware Approach for Individual Human Mobility Inference Based on Sparse Trajectory Data

Shuai Xu[1,2](✉), Donghai Guan[1], Zhuo Ma[3], and Qing Meng[4]

[1] Nanjing University of Aeronautics and Astronautics, Nanjing, China
xushuai7@nuaa.edu.cn

[2] Key Laboratory of Computer Network and Information Integration (Southeast University), Ministry of Education, Nanjing, China

[3] Jiangsu Police Institute, Nanjing, China

[4] Singapore University of Technology and Design, Singapore, Singapore

Abstract. Inferring individual human mobility at a given time is not only beneficial for personalized location-based services, but also crucial for trajectory tracking of the confirmed cases in the context of the COVID-19 pandemic. However, individual generated trajectory data using mobile Apps is characterized by implicit feedback, which means only a few individual-location interactions can be observed. Existing studies based on such sparse trajectory data are not sufficient to infer individual's missing mobility in his/her historical trajectory and further predict individual's future mobility given a specific time. To address this concern, in this paper, we propose a temporal-context-aware approach that incorporates multiple factors to model the *time sensitive individual-location interactions* in a bottom-up way. Based on the idea of feature fusion, the driving effect of heterogeneous information such as time, space, category and sentiment on individual's mobile behavior is gradually strengthened, so that the temporal context when a check-in occurs can be accurately depicted. We leverage Bayesian Personalized Ranking (BPR) to optimize the model, where a novel negative sampling method is employed to alleviate data sparseness. Based on three real-world datasets, we evaluate the proposed approach with regard to two different tasks, namely, missing mobility inference and future mobility prediction at a given time. The empirical results encouragingly demonstrate that our approach outperforms multiple baselines in terms of two evaluation metrics, i.e., accuracy and average percentile rank.

Keywords: Mobility inference · Preference modeling · Feature fusion · Trajectory mining · Context awareness

B. Li et al. (Eds.): APWeb-WAIM 2022, LNCS 13421, pp. 106–120, 2023.
https://doi.org/10.1007/978-3-031-25158-0_9

1 Introduction

With the popularity of 4G/5G wireless communication technology, mobile Apps like Yelp, Foursquare, and Dianping that provide location-based services have brought significant convenience to people's daily life. These applications enable users to check in and publish reviews (sometimes with numerical ratings and related images attached) after they have visited certain places, therefore such user generated data can be regarded as trajectory data that reflects users' mobile behavior in the physical world. An example concerning a Point-of-Interest (POI) and a user's generated trajectory on Dianping[1] is presented in Fig. 1. By systematically analyzing and extracting temporal-spatial information from such trajectory data, various individual mobility patterns and preferences can be mined, so that not only an individual's future mobility can be effectively predicted, but also the missing mobility record can be accurately inferred. The former has proved to bring benefits to precise product marketing [17], while the latter can assist to track the historical trajectory of the confirmed cases, which is of practical significance for confronting the COVID-19 pandemic [26].

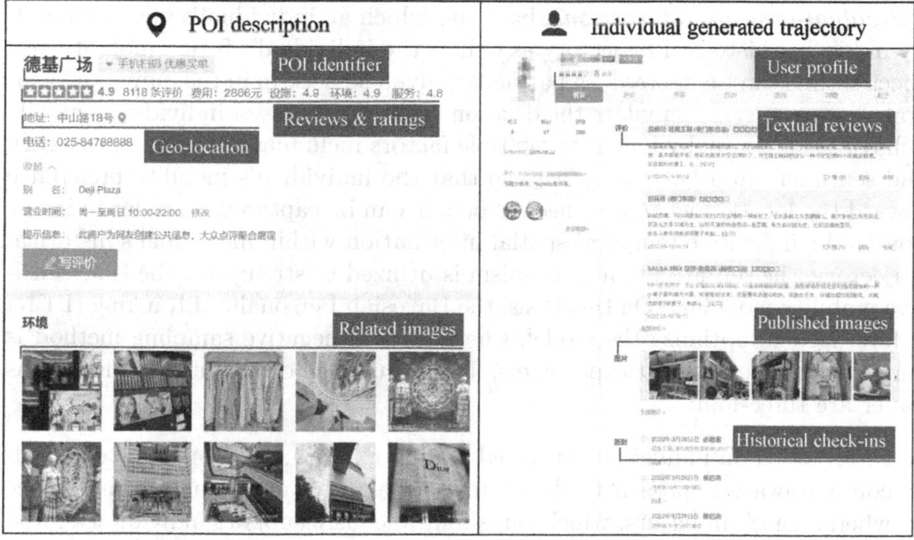

Fig. 1. Overview of an example concerning a POI and a user on Dianping. The left part illustrates the available description of this POI, while the right part shows this user's generated trajectory. For the sake of preserving user privacy, we blur the user profile photos and nicknames.

However, individual generated trajectory data is characterized by implicit feedback, which means only a few individual-location interactions can be

[1] https://www.dianping.com/.

observed [25]. The unobserved mobility may be either because the individual has a negative opinion on some venues, or because the individual does not visit the relevant venue. In real life, due to the needs of personal privacy protection, even if some users have visited some venues, they would not check in and share their location in the Apps [23]. For the above reasons, the interactions between individuals and locations within such trajectory data are extremely sparse. Although human mobility modeling and prediction have been widely studied in the past decades, existing studies are mainly conducted based on cellular networking data [10] and GPS positioning data [11], which are quite different from the individual generated trajectory data using mobile Apps. Due to the sampling mechanism, GPS positioning data and cellular networking data do not have the characteristics of implicit feedback, and they are less in semantics compared with individual generated trajectory data (see Fig. 1). Moreover, existing studies either try to recover the missing mobility records in individual's historical trajectory [7,14], or focus on predicting an individual's next mobility [18,25], few of them can support the inference of individual's missing mobility and further predict individual's future mobility at a given time.

In this paper, taking into account the heterogeneous semantic information, we propose a temporal-context-aware approach to model the *time sensitive individual-location interactions*, based on which an individual's missing mobility in his/her historical trajectory as well as the individual's future mobility at a specific time can be inferred. In particular, given a check-in in the individual generated trajectory, to simulate the decision-making process of individual's mobile behavior, we gradually integrate multiple factors including time, space, category and sentiment in a bottom-up way, so that the individual's mobility preference toward the location when the check-in occurs can be captured. In order to incorporate the indicative temporal-spatial information within individual's historical trajectory, the self-attention mechanism is utilized to strengthen the time sensitive mobility modeling. On this basis, the Bayesian Personalized Ranking (BPR) is leveraged to optimize the model, where a novel negative sampling method is employed to alleviate data sparseness. To sum up, the core contributions of this paper are three-fold:

- First, based on individual generated trajectory data, we propose a temporal-context-aware approach to depict individual's dynamic mobility preference when a check-in occurs, which can simultaneously support individual's missing mobility inference and future mobility prediction at a given time. To our knowledge, this is the first attempt to infer the individual's time-sensitive mobility both in the past and in the future.
- Second, we regard the temporal-context when a check-in occurs as the combination of time, space, category and sentiment, and exploit the heterogeneous information to portray the individual-location interaction based on the idea of deep feature fusion.
- Third, we conduct a comprehensive evaluation based on three real-world datasets. Specifically, we evaluate the approach with regard to two different tasks, notably, missing mobility inference and future mobility prediction

at a given time, and further compare it with multiple baselines. The results using *Acc@k* and *APR* metrics reveal that our approach can match the performance with state-of-the-art methods in terms of the missing mobility inference task, and it outperforms other state-of-the-art methods in terms of the future mobility prediction at a given time.

The rest of this paper is structured as follows: Sect. 2 revisits related works. In Sect. 3, we describe the user mobility inference problems. Section 4 introduces the proposed approach. Section 5 elaborates the experiments. Finally, Sect. 6 concludes the paper.

2 Related Works

Zhang et al. [24] point out that human's demand for visiting different places is generally driven by a variety of factors, and the predictability of individual's mobile behavior can be strengthened by effectively integrating multi-dimensional structural features such as temporal cyclic effect, geographical influence, transition patterns, semantic information and so on. Whether it is the individual missing mobility inference problem or the individual future mobility prediction problem, the key point is to model individual human mobility preference, the essence of which is to perceive individual's visit interest under different temporal-spatial contexts [4]. In the following, we revisit the representative literature regarding human trajectory recovery and mobility prediction, respectively.

Individual Mobility Recovery. To prevent and control the COVID-19 pandemic, the historical trajectory of confirmed cases should be traced, so it has become a pressing need to recover individual trajectories from existing trajectory data. Earlier studies like [5] deem human mobility recovery as time series data recovery problem. However, as the individual generated trajectory data is extremely sparse, these methods' performance is not acceptable because they are unable to capture individual human mobility patterns. Recent studies either tend to leverage the information before the missing check-in location and model the spatial-temporal dependency between the missing check-in location and the check-in locations afterwards [13,14], or attempt to capture the transition patterns as well as the shifting nature of human mobility periodicity [12]. However, these studies basically assume that there exist observable missing locations (for some check-ins, the timestamp are known, but the visit locations are unobserved) in individual's historical trajectory data. This is not the case in the scenario of tracing trajectory of the confirmed cases in the context of COVID-19 pandemic, where their missing mobility (if any) in the reported trajectories is not clear.

Individual Future Mobility Prediction. In real life, human's visit preference toward different types of locations will change with their temporal-spatial context. Recent works can be divided into two lines. The first line often applies the variants of RNNs to model sequential influence and temporal dynamics in individual trajectories, so that individual's preference for the next mobility can be perceived, based on which the individual's next visit location can be predicted. In this process, the difficulty lies in how to capture the correlation of successive check-ins and the dynamic changing rule of individual's mobility preference with trajectory extension [16]. Yang et al. [18] and Xu et al. [15] introduce time gates to control the influence of the hidden state of a previous RNN unit based on the time interval between successive check-ins, which encodes the temporal correlation for context-aware user location prediction. Feng et al. [2] and Yu et al. [22] divide individual's whole trajectory into several sub-trajectories and input them into the reformed RNN unit, with the matching ability of the self-attention mechanism, the informative check-in behavior in the past can be fused to improve the learning of individual's up-to-date mobility preference. The second line designs embedding models to obtain the embeddings of various kinds of nodes such as user (individual), time, location, category and so on, based on which the prediction can be made by calculating the vector similarity [19,20]. Note that as the embedding of time slots is obtained in the same latent space, the second line is often used to predict individual's visit location at a given time.

To sum up, existing studies generally regard the individual trajectory recovery problem and future mobility prediction problem as independent problems, so far few of them can offer the approach to infer individual's missing mobility in the past and further predict individual's mobility in the future at a given time. Different from previous studies, we aim to solve these two dependent tasks by a temporal-context-aware approach.

3 Preliminaries

Suppose we have a dataset $\mathcal{D}$ containing user trajectories collected from a mobile App (like Foursquare). The user set is represented as $\mathcal{U} = \{u_1, u_2, ..., u_M\}$, the POI set is represented as $\mathcal{V} = \{v_1, v_2, ..., v_N\}$ and the time slot set is represented as $\mathcal{T} = \{t_1, t_2, ..., t_T\}$, where M, N and T are the number of users, number of POIs, and number of time slots, respectively. A user's trajectory is defined as this user's chronologically ordered check-in sequence. For user u, let $\mathcal{L}^u = \{(u, v_i, t_i, r_i) | u \in \mathcal{U}, v_i \in \mathcal{V}, t_i \in \mathcal{T}, 1 \leq i \leq n\}$ denote his/her historical trajectory, where n is the number of check-ins, and the quadruple (u, v_i, t_i, r_i) means that user u visited POI v_i at time t_i with numerical rating r_i. In this paper, the time slot t_i is accurate to hour-level (24 h in total), and the numerical rating $r_i \in \{1, 2, 3, 4, 5\}$.

Based on the notations above, without loss of generality, we formally define the individual mobility inference problem as follows:

Problem Statement. Given user u's historical trajectory $\mathcal{L}^u$, we aim to infer where user u visited at a past time t_p, and further predict where user u will visit at a future time t_q. Note that for both tasks, we aim to produce a top-K location list, so that the correct location user u would visit at time t_p and t_q can be ranked as high as possible. To make it clear, we present a visual description of the problem definition in Fig. 2.

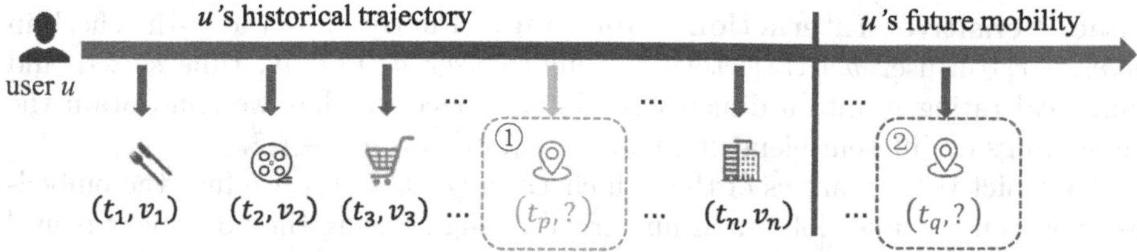

Fig. 2. A visual description the individual mobility inference problems in this paper, where ① corresponds to *individual missing mobility inference* at a given time in the past, and ② corresponds to *individual future mobility prediction* at a given time in the future.

4 Methodologies

We propose a temporal-context-aware approach to model the time sensitive individual-location interaction, where individual's mobility preference is gradually depicted by integrating multiple factors including time, space, category and sentiment. The architecture of the modeling process is shown in Fig. 3. We can see that the architecture consists of three parts, i.e., time sensitive interaction modeling, POI attribute modeling, as well as geo-influence modeling. For each check-in in user u's historical trajectory, these three parts gradually integrate the heterogeneous information from bottom to top, in this way the probability of user u checking in at POI v at time t can be finally approximated. The detailed modeling process is elaborated as follows.

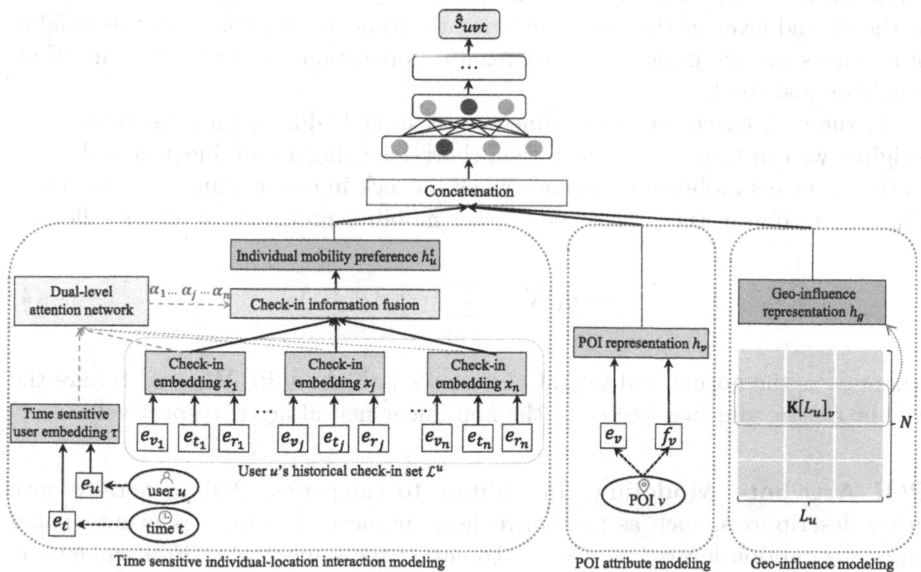

Fig. 3. Architecture of the proposed temporal-context-aware approach to model individual mobility preference.

Time Sensitive Interaction Modeling. Firstly, for the i-th check-in (u, v_i, t_i, r_i) in user u's trajectory, we embed user u, POI v_i, time slot t_i and numerical rating r_i into d-dimensional latent space, so that we can obtain the embeddings of different elements, namely, e_u, e_{v_i}, e_{t_i}, $e_{r_i} \in \mathbb{R}^d$.

To depict the semantics of the i-th check-in (u, v_i, t_i, r_i), we fuse the embeddings of POI v_i, time slot t_i and numerical rating r_i using the non-linear neural layer:

$$x_i = g(e_{v_i} \oplus e_{t_i} \oplus e_{r_i}) \tag{1}$$

where $\oplus$ represents concatenation operation, and $g(\cdot)$ is the multi-layer perceptron function with non-linear activation. In this way, vector x_i not only contains temporal-spatial information, but also contains user u's sentiment toward POI v_i.

Secondly, in order to depict user u's preference toward time slot t_i, we also fuse the embeddings of user u and time slot t_i using the non-linear neural layer:

$$\tau_i = g(e_u \oplus e_{t_i}) \tag{2}$$

Third, we apply the self-attention mechanism to discover informative check-ins in user u's trajectory, so that the indicative temporal-spatial information within user u's historical trajectory can be used to strengthen the time sensitive interaction modeling. Specifically, the dual-level attention network is adopted to compute the weight of user u's j-th check-in:

$$\alpha_j^* = \mathbf{w}_2^T \cdot \tanh(\mathbf{W}_1 \cdot (x_j \oplus \tau_i) + \mathbf{b}_1) + b_2 \tag{3}$$

where $\mathbf{W}_1$ and $\mathbf{b}_1$ are the weight matrix and bias vector in the first layer of the attention network, respectively, while $\mathbf{w}_2$ and b_2 are the weight vector and bias in the second layer of the attention network, respectively. Based on the weights of all check-ins, we perform a normalization operation to ensure the sum of all weights equals to 1.

In the end, based on the obtained check-in embeddings and the normalized weights, we can fuse user u's historical check-ins using a non-linear neural layer, so that user u's mobility preference when a check-in occurs can be represented. The way to obtain the time sensitive user mobility preference h_u^t is as follows:

$$h_u^t = g(\mathbf{W}_u \cdot (\sum_{j=1}^{n} \alpha_j x_j) + \mathbf{b}_u) \tag{4}$$

where α_j is the normalized weight of user u's j-th check-in, $\mathbf{W}_u$ and $\mathbf{b}_u$ are the weight matrix and bias vector in the non-linear neural layer, respectively.

POI Attribute Modeling. In addition to categories, POIs generally have other descriptions, such as textual reviews, numerical ratings, relevant images and consumption levels (see Fig. 1). To enrich the semantics of POI embedding learned in the section above, we further incorporate numerical ratings associated with textual reviews toward a POI. Specifically, we average all rating embeddings associated with a POI v to obtain the attribute embedding f_v. Based on the POI embedding e_v and the attribute embedding f_v, we fuse them using the non-linear neural layer to obtain the enriched POI attribute embedding h_v:

$$h_v = g(e_v \oplus f_v) \tag{5}$$

Note that the POI attribute modeling method is scalable. Actually, not only the numerical ratings, but also the textual reviews and the images (if provided in the dataset) can be used to attain the attribute embedding f_v, after which the enriched POI attribute embedding h_v can be obtained.

Geo-influence Modeling. According to Tobler's first law [8], there exists mutual influence between any two locations, and the closer the geographical distance is, the greater the degree of influence. To model such geo-influence, we adopt kernel density estimation (KDE) to quantify the correlation between any two POIs, in this way we can obtain the geo-factor matrix $\mathbf{K} \in \mathbb{R}^{N \times N}$, where each element $\mathbf{K}(l_i, l_j)$ is calculated as follows:

$$\mathbf{K}(l_i, l_j) = exp(-\gamma ||l_i - l_j||^2) \tag{6}$$

where l_i, l_j are the geo-coordinates of POI v_i and POI v_j, respectively, and γ is the bandwidth parameter, which is set according to [6].

As the locations visited by user u can be counted, we slice the matrix $\mathbf{K}$ to attain user u's geo-matrix $\mathbf{K}[L_u] \in \mathbb{R}^{|L_u| \times |L_u|}$, where L_u is the set of POIs that user u has visited. On this basis, we get the geo-vector $\mathbf{K}[L_u]_v \in \mathbb{R}^{|L_u|}$, which represents the geo-influence that POI set L_u exerts on POI v.

In the end, we perform nonlinear mapping on geo-vector $\mathbf{K}[L_u]_v$ to obtain the implicit representation of geo-influence h_g:

$$h_g = g(\mathbf{W_g} \cdot \mathbf{K}[L_u]_v + \mathbf{b_g}) \tag{7}$$

where $\mathbf{W_g}$ and $\mathbf{b_g}$ are the weight matrix and bias vector in the non-linear mapping layer.

Check-in Probability Approximation. As shown in Fig. 3, based on the three modeling parts above, we concatenate the time sensitive user mobility preference h_u^t, the enriched POI attribute embedding h_v and the implicit representation of geo-influence h_g, and feed it into a multi-layer neural network architecture, which finally outputs the check-in probability $\hat{s}_{uvt}$ of user u checking in at POI v at time t. This process is shown as follows:

$$\begin{aligned} l_1 &= h_u^t \oplus h_v \oplus h_g \\ l_2 &= \varphi(l_1) \\ &\dots \\ l_m &= \varphi(l_{m-1}) \\ \hat{s}_{uvt} &= \mathbf{w}_m^T \cdot l_m + b_m \end{aligned} \tag{8}$$

where m is number of fully connected layers, $\mathbf{w}_m$ and b_m are the weight vector and bias in the last linear layer.

Model Training. Similar to [17], we train the model with a learning-to-rank method. To be concrete, we employ Bayesian Personalized Ranking (BPR) [9] to optimize the parameters that involved in the model. An observed check-in triple $\langle u, v^+, t\rangle$ is indeed a positive sample, and an unobserved check-in triple $\langle u, v^-, t\rangle$ is a negative sample because user u does not visit location v^- at time t. BPR can make use of the unobserved time sensitive individual-location interactions by learning a pair-wise ranking loss in the training process. The BPR loss function with a regularization term is described below:

$$Loss_{BPR} = -\sum_{u\in\mathcal{U}}\sum_{<v^+,v^->\in R_u} \ln\delta(\hat{s}(u,v^+,t) - \hat{s}(u,v^-,t)) + \varepsilon||\Theta||^2 \quad (9)$$

where R_u is the training set containing positive samples and negative samples collected from user u's trajectory, $\hat{s}(u,v,t)$ is the check-in probability (i.e., $\hat{s}_{uvt}$) derived from Eq. 8, $\delta(\cdot)$ is the sigmoid function, Θ represents the parameter set in the non-linear neural network layers, and ε is the regularization coefficient.

Note that parameters to be learned include the set of weight matrices, the set of weight vectors as well as the set of biases in the neural network layers. As the whole model is indeed a feed-forward neural network, where all parameters in the loss function are differentiable, we can apply Adam algorithm [3] with mini-batch strategy to minimize the BPR loss automatically. Considering that individual trajectory data is extremely sparse, in order to efficiently optimize the above objective function, we employ a novel negative sampling method. For each positive sample $\langle u, v^+, t\rangle$, we randomly sample 10 negative samples $\langle u, v^-, t\rangle$. For each epoch during model training, we would re-sample negative samples for each positive sample, so that each negative sample only gives very weak negative signal in the training process.

Mobility Inference. Once the model is well trained, all parameters are determined. Whether it is the task to infer user u_o's missing mobility at a specific time in the past or the task to predict user u_o's mobility at a specific time in the future, we can produce the result list based on the model in an intuitive manner. Suppose the given time is t', we can calculate user u_o's time sensitive check-in probability $\hat{s}_{u_o v' t'}$ at a candidate POI v' ($v' \in \mathcal{V}$). After the probabilities toward all possible POIs are calculated, we select the POIs with top-K probabilities as the result list. Algorithm 1 below illustrates the process to infer the target individual's visit location at a given time in the past or in the future.

5 Experiments

The experiments are conducted on a Dell workstation with dual processors (2 × Intel Xeon E5 @ 2.10GHz), four graphic processing units (Nvidia Titan Xp, 12GB), and 188GB RAM. The operating system of the workstation is 64-bit Ubuntu 16.04. All the codes are written in Python 3.7 and the model architecture is built using Pytorch 1.1.0.

Algorithm 1: Individual mobility inference algorithm

Input : u, t', $\mathcal{L}^u$, $\mathcal{V}$, Θ, K.
Output: the predicted top-K POI list L.
obtain the time sensitive user preference $h_u^{t'}$;
for *each candidate POI* $v' \in \mathcal{V}$ **do**
    obtain the POI attribute $h_{v'}$ and the geo-influence h_g;
    compute $\hat{s}(u, v', t')$ according to Eq. 8;
end
sort POIs based on the computed probabilities in the descending order;
produce the prediction list L by selecting the top-K POIs from $\mathcal{V}$;
return L

5.1 Datasets

We evaluate the approach on three real-world datasets, i.e., Foursquare NYC, Foursquare TKY, and Yelp LAS, the former two datasets are provided in [21], and the third dataset is provided in [17]. In Foursquare NYC dataset and Foursquare TKY dataset, each user has at least 10 check-ins, and each POI has been visited by at least 10 users. In Yelp LAS dataset, each user has at least 20 check-ins and each POI has been visited at least 20 times. The statistics of the selected datasets are shown in Table 1.

The experiment is based on train-validation-test mode. For each user u in each dataset, we use the former 80% check-ins for training, the middle 10% check-ins for validation and the last 10% check-ins for testing. Note that to evaluate the individual missing mobility inference task, we randomly mask one check-in in the training data.

Table 1. Statistics of the selected datasets.

Dataset	City	# Users	# POIs	# Check-ins	Density
Foursquare NYC	New York	950	1,167	45,632	4.12%
Foursquare TKY	Tokyo	2,274	2,865	333,184	5.11%
Yelp LAS	Las Vegas	13,542	7,392	641,652	0.60%

5.2 Baselines

To demonstrate the performance of the proposed approach, we compare it with the following four baselines:

- **TOP**: This is a simple counting-based model, which directly selects the most popular POI in the training set as the result.

- **MfM** [1]: This is a linear scoring model, which computes the weighted sum of multiple features like temporal cyclic effect, geographical influence, categorical information and social relationship. The weighted sum is regarded as the final score measuring a user's preference toward a POI.
- **LBSN2Vec** [19]: This is a graph embedding model, which learns node embeddings from hyper-edges by preserving the cosine proximity between nodes in LBSNs. It computes two similarities, one between the user embedding and the POI embedding, and one between the time slot embedding and the POI embedding, then the two similarities are added up to measure a user's preference toward a POI.
- **Bi-STDDP** [12]: This is the state-of-the-art model for user missing mobility inference. Assume the t-th check-in of a user is missing, this model can infer which POI the user visited when the t-th check-in occurs based on the previous check-in sequence and the backward check-in sequence.

5.3 Experimental Setup

Evaluation Metrics. We select two metrics from different aspects to verify the performance of different methods [1,17]. The first metric is $Acc@K$, which measures the ratio of successfully predicted POIs in the top-K list. The second metric is Average Percentile Rank (APR), which considers the rank of correctly predicted POI in the top-K list. The higher the rank of ground-truth POI in the produced list, the larger the APR metric value, and vice versa.

Parameter Setting. The parameter set is initialized according to the uniform distribution $U(-0.01, 0.01)$, the batch size is set to 100, and the dimension of the embeddings is fixed to 100. Depth of the multi-layer neural network architecture that maps l_1 to $\hat{s}_{uvt}$ (refer to Eq. 8) is set to 2. The initial learning rate of Adam is 0.001. For the regularization parameter ε, we set it to 0.001.

5.4 Empirical Analysis

In this paper, we propose a **T**emporal-context-aware approach for **I**ndividual **M**obility **I**nference (TIMI for short), which not only enables individual's missing mobility inference in the past, but also supports individual's mobility prediction in the future. Therefore, we need to verify its performance regarding these two different tasks, and further compare the results with representative baselines.

Individual Missing Mobility Inference. As aforementioned in Sect. 5.1, since we randomly mask ONE check-in in the training data of each user in each dataset, when evaluating the performance for this task, we should fix $K = 1$. The experimental results of different methods are reported in Table 2, where the best method w.r.t. each metric is highlighted, and the second-best is underlined, so that the improvement percentage (the last column) can be calculated intuitively.

Table 2. Performance comparisons of all methods on individual missing mobility inference task.

Dataset	Metric	Top	MfM	LBSN2Vec	Bi-STDDP	TIMI	Imp.
Foursquare NYC	*Acc*@1	0.138	0.152	0.162	**0.170**	<u>0.166</u>	−2.4%
	APR	0.694	0.702	<u>0.706</u>	**0.715**	0.702	−1.8%
Foursquare TKY	*Acc*@1	0.142	0.164	0.166	**0.178**	<u>0.173</u>	−2.8%
	APR	0.743	0.755	0.762	**0.766**	<u>0.764</u>	−0.26%
Yelp LAS	*Acc*@1	0.015	0.018	0.024	**0.028**	<u>0.026</u>	−7.1%
	APR	0.427	0.432	0.435	**0.440**	<u>0.438</u>	−0.45%

From Table 2, we can observe that no matter on which dataset, Bi-STDDP can consistently achieve the best performance inferring where a user has visited at a specific time in the past. Compared to other methods, Bi-STDDP is meticulously designed for user missing mobility inference task, which combines the bi-directional global spatial and local temporal information together to capture complex dependence within user trajectory. As a contrast, the proposed method TIMI and other methods only consider the single-directional temporal-spatial information, which can not dig into the transition patterns within the trajectory before the missing check-in and the trajectory after the missing check-in. Except for Bi-STDDP, the proposed method TIMI can achieve better performance (especially on *Acc*@1 metric) than baseline methods like LBSN2Vec. Basically, for the task of individual missing mobility inference at a given time, the performance of the proposed method TIMI can match the state-of-the-art method, even though sequential dependencies in individual trajectory are not considered.

Individual Future Mobility Prediction. For this task, we tune the value of K ($K \in \{1, 5, 10\}$) to comprehensively evaluate the predictive power of various methods. The evaluation is based on the testing data, i.e., the last 10% check-ins of each user. For each check-in in the testing data, the corresponding check-in time t' is taken as input, and we can produce a POI list according to Algorithm 1 . Then we observe whether the correct POI v' is in the top-K position. Thereafter, *Acc*@K metric is obtained by averaging all user check-ins in the testing data. The experimental results of different methods are reported in Table 3.

As we can see from Table 3, no matter on which dataset, our proposed method TIMI consistently outperforms other baselines. Surprisingly, Bi-STDDP, which performs the best in individual missing mobility inference task, is no longer dominant. The reason can be roughly analyzed. In individual future mobility prediction task, given a specific time in the future, we can only use the single-direction temporal-spatial information before this time for Bi-STDDP. Note that Bi-STDDP are designed to capture local temporal information and global spatial information, this means Bi-STDDP may lose all information within user trajectory after the given time, which severely decrease the model's ability to perceive bi-directional sequential dependency from user trajectory. Additionally, the interval between the given time and the time of the last check-in in user training set may be very long, which further weakens the model's performance. In comparison with LBSN2Vec, although this method can learn structure-preserving embeddings based on sparse interactions among different types of nodes, the weakness is that it can not distinguish the importance among check-ins, which means all POIs in the sampled sequence are assigned with the same weight. With the self-attention mechanism, the proposed method TIMI performs better than LBSN2Vec because it can find similar temporal-spatial context from user's historical trajectory and further make use of it to update user's time sensitive mobility preference. In addition, for each dataset, individual-location interactions are extremely sparse, it is difficult to effectively capture user dynamic preference only by preserving the topological properties of the static location-based social network (LBSN). However, in terms of TIMI, we incorporate heterogeneous information including time, space, category and sentiment, which not only alleviates the problem of data sparsity, but also strengthen the modeling of time sensitive user mobility preference.

Table 3. Performance comparisons of all methods on individual future mobility prediction task.

Dataset	Metric	K	Top	MfM	LBSN2Vec	Bi-STDDP	TIMI	Imp.
Foursquare NYC	$Acc@K$	1	0.106	0.137	0.164	0.160	**0.168**	+1.2%
		5	0.129	0.186	0.201	0.194	**0.207**	+2.9%
		10	0.152	0.203	0.234	0.235	**0.252**	+7.2%
	APR	/	0.671	0.679	0.718	0.707	**0.731**	+1.8%
Foursquare TKY	$Acc@K$	1	0.088	0.114	0.168	0.166	**0.173**	+3.0%
		5	0.110	0.138	0.208	0.202	**0.216**	+3.8%
		10	0.136	0.172	0.250	0.244	**0.255**	+2.0%
	APR	/	0.774	0.792	0.825	0.820	**0.828**	+0.4%
Yelp LAS	$Acc@K$	1	0.010	0.018	0.022	0.020	**0.025**	+13.6%
		5	0.014	0.023	0.026	0.024	**0.030**	+15.4%
		10	0.022	0.028	0.033	0.029	**0.039**	+18.2%
	APR	/	0.312	0.433	0.451	0.432	**0.456**	+1.1%

We note that the proposed TIMI method is able to gain practical value in real applications. For one thing, this temporal-context-aware approach can be used for individual human mobility recovery and future mobility prediction simultaneously. For another thing, for both tasks, TIMI delivers considerable performance through empirical evaluations based on real-world datasets. Taking $Acc@1$ as an example, as we can rank a location a target individual will visit at a given time at the top-1 position with nearly 20 percent confidence, it is indeed a remarkable performance because there are thousands candidate locations to be considered in a city.

6 Conclusion

In this paper, we study the problem of individual human mobility inference, which covers individual missing mobility inference in the past and individual future mobility prediction in the future. To this end, we propose a temporal-context-aware approach, which incorporates heterogeneous information to model the time sensitive individual-location interactions in a bottom-up way. Experimental results reveal that the proposed method can achieve comparable performance with state-of-the-art methods with regard to individual missing mobility inference task, and it outperforms other methods with regard to individual future mobility prediction task.

For future work, we consider to study an extended version of the individual mobility inference problem. We intend to make a joint prediction toward the overall individual mobility event of who, when, where and what.

Acknowledgements. This work is supported by the Natural Science Foundation of Jiangsu Province (No. BK20210280), the Fundamental Research Funds for the Central Universities (NO. NS2022089), the Jiangsu Provincial Innovation and Entrepreneurship Doctor Program under Grants No. JSSCBS20210185.

References

1. Cao, J., Xu, S., Zhu, X., Lv, R., Liu, B.: Effective fine-grained location prediction based on user check-in pattern in LBSNs. J. Netw. Comput. Appl. **108**, 64–75 (2018)
2. Feng, J., et al.: Deepmove: Predicting human mobility with attentional recurrent networks. In: Proceedings of the 2018 World Wide Web Conference, pp. 1459–1468 (2018)
3. Kingma, D.P., Ba, J.: Adam: a method for stochastic optimization. In: Proceedings of the 3rd International Conference on Learning Representation, pp. 1–15 (2015)
4. Luca, M., Barlacchi, G., Lepri, B., Pappalardo, L.: A survey on deep learning for human mobility. ACM Comput. Surv. (CSUR) **55**(1), 1–44 (2021)
5. Luo, Y., Cai, X., Zhang, Y., Xu, J., Yuan, X.: Multivariate time series imputation with generative adversarial networks. In: Proceedings of the 32nd International Conference on Neural Information Processing Systems, pp. 1603–1614 (2018)
6. Ma, C., Zhang, Y., Wang, Q., Liu, X.: Point-of-interest recommendation: exploiting self-attentive autoencoders with neighbor-aware influence. In: Proceedings of the 27th ACM International Conference on Information and Knowledge Management, pp. 697–706 (2018)
7. Mahajan, R., Mansotra, V.: Predicting geolocation of tweets: using combination of CNN and BiLSTM. Data Sci. Eng. **6**(4), 402–410 (2021)
8. Miller, H.J.: Tobler's first law and spatial analysis. Ann. Assoc. Am. Geogr. **94**(2), 284–289 (2004)
9. Rendle, S., Freudenthaler, C., Gantner, Z., Schmidt-Thieme, L.: BPR: bayesian personalized ranking from implicit feedback. In: Proceedings of the 25th Conference on Uncertainty in Artificial Intelligence, pp. 452–461 (2009)
10. Teixeira, D.D.C., Viana, A.C., Almeida, J.M., Alvim, M.S.: The impact of stationarity, regularity, and context on the predictability of individual human mobility. ACM Trans. Spat. Algorithms Syst. **7**(4), 1–24 (2021)

11. Wang, P., Yang, L.T., Peng, Y., Li, J., Xie, X.: M^2T^2: the multivariate multistep transition tensor for user mobility pattern prediction. IEEE Trans. Netw. Sci. Eng. **7**(2), 907–917 (2020)
12. Xi, D., Zhuang, F., Liu, Y., Gu, J., Xiong, H., He, Q.: Modelling of Bi-directional spatio-temporal dependence and users' dynamic preferences for missing poi check-in identification. In: Proceedings of the AAAI Conference on Artificial Intelligence. vol. 33, pp. 5458–5465 (2019)
13. Xia, T., et al.: Attnmove: History enhanced trajectory recovery via attentional network. arXiv preprint arXiv:2101.00646 (2021)
14. Xu, F., Tu, Z., Li, Y., Zhang, P., Fu, X., Jin, D.: Trajectory recovery from Ash: user privacy is NOT preserved in aggregated mobility data. In: Proceedings of the 26th International Conference on World Wide Web, p. 1241–1250 (2017)
15. Xu, J., Zhao, J., Zhou, R., Liu, C., Zhao, P., Zhao, L.: Predicting destinations by a deep learning based approach. IEEE Trans. Knowl. Data Eng. **33**(2), 651–666 (2021)
16. Xu, S., Fu, X., Cao, J., Liu, B., Wang, Z.: Survey on user location prediction based on geo-social networking data. World Wide Web **23**(3), 1621–1664 (2020). https://doi.org/10.1007/s11280-019-00777-8
17. Xu, S., Pi, D., Cao, J., Fu, X.: Hierarchical temporal-spatial preference modeling for user consumption location prediction in geo-social networks. Inf. Process. Manage. **58**(6), 102715 (2021)
18. Yang, D., Fankhauser, B., Rosso, P., Cudre-Mauroux, P.: Location prediction over sparse user mobility traces using RNNs: flashback in hidden states! In: Proceedings of the 29th International Joint Conference on Artificial Intelligence, pp. 2184–2190 (2020)
19. Yang, D., Qu, B., Yang, J., Cudre-Mauroux, P.: Revisiting user mobility and social relationships in LBSNs: a hypergraph embedding approach. In: Proceedings of the 2019 World Wide Web Conference, pp. 2147–2157 (2019)
20. Yang, D., Qu, B., Yang, J., Cudré-Mauroux, P.: Lbsn2vec++: heterogeneous hypergraph embedding for location-based social networks. IEEE Trans. Knowl. Data Eng. **34**(4), 1843–1855 (2022)
21. Yang, D., Zhang, D., Zheng, V.W., Yu, Z.: Modeling user activity preference by leveraging user spatial temporal characteristics in LBSNs. IEEE Trans. Syst. Man Cybern. Syst. **45**(1), 129–142 (2015)
22. Yu, F., Cui, L., Guo, W., Lu, X., Li, Q., Lu, H.: A category-aware deep model for successive poi recommendation on sparse check-in data. In: Proceedings of the 2020 World Wide Web Conference, pp. 1264–1274 (2020)
23. Zhan, Y., Kyllo, A., Mashhadi, A., Haddadi, H.: Privacy-aware human mobility prediction via adversarial networks. arXiv preprint arXiv:2201.07519 (2022)
24. Zhang, C., Zhao, K., Chen, M.: Beyond the limits of predictability in human mobility prediction: Context-transition predictability. In: IEEE Transactions on Knowledge and Data Engineering, pp. 1–14 (2022)
25. Zhang, M., Li, B., Wang, K.: HGTPU-Tree: an improved index supporting similarity query of uncertain moving objects for frequent updates. In: Li, J., Wang, S., Qin, S., Li, X., Wang, S. (eds.) ADMA 2019. LNCS (LNAI), vol. 11888, pp. 135–150. Springer, Cham (2019). https://doi.org/10.1007/978-3-030-35231-8_10
26. Zhang, Q., Gao, J., Wu, J.T., Cao, Z., Dajun Zeng, D.: Data science approaches to confronting the covid-19 pandemic: a narrative review. Philos. Trans. Roy. Soc. A **380**(2214), 20210127 (2022)

Unsupervised Online Concept Drift Detection Based on Divergence and EWMA

Qilin Fan, Chunyan Liu, Yunlong Zhao(✉), and Yang Li

Nanjing University of Aeronautics and Astronautics, Nanjing, China
zhaoyunlong@nuaa.edu.cn

Abstract. Concept drift problem is a common challenge for data stream mining, while the underlying distribution of incoming data unpredictably changes over time. The classifier model in data stream mining must be self-adjustable to the concept drift, otherwise it will get terrible classification results. To detect concept drift timely and accurately, this paper proposes an unsupervised online **C**oncept **D**rift **D**etection algorithm based on Jensen-Shannon **D**ivergence and **E**WMA(CDDDE), which detects concept drift through measuring the difference of data distribution within sliding windows and calculating the drift threshold dynamically by Exponentially Weighted Moving Average (EWMA), during the detection without the use of labels. Once concept drift is detected, a new classifier would be trained using the current and subsequent data. Experiments on artificial and real-world datasets show that CDDDE algorithm can efficiently detect the concept drift, and the retrained classifier effectively improves the classification accuracy for the subsequent data. Compared with some supervised algorithms, the detection accuracy and classification accuracy are higher for most datasets.

Keywords: Concept drift · Unsupervised learning · Jensen-shannon divergence · EWMA · Data distribution

1 Introduction

Traditionally, the data studied in machine learning tends to be static data, which can be stored in memory and processed for the entire dataset. But in recent years, there has been a tremendous increase of interest in algorithms that can learn from data streams. Data streams are different from traditional data mining methods because of their large volume of data, real-time arrival, and the fact that once the data is processed, it cannot be taken out again for processing, unless it is deliberately saved. Data in the real-world environment may have dynamic behavior, and the concept can change, which is known as the concept drift problem [1]. The concept drift problem was first proposed in [2], where this author modeled a supervised learning task that concept drift occurs due to the environment changes. The definition of concept drift is described as follows. Given a time period $[0, t]$, the data stream in that time period is represented as $S_{0,t} = \{d_0, \cdots, d_t\}$ where

B. Li et al. (Eds.): APWeb-WAIM 2022, LNCS 13421, pp. 121–134, 2023.
https://doi.org/10.1007/978-3-031-25158-0_10

$d_i = (X_i\ , y_i)$ denotes a data instance in the data stream, X_i is the feature vector, y_i is the label, and the data stream $S_{0,t}$ follows some distribution $F_{0,t}\,(X, y)$. If it appears that $F_{0,t}\,(X, y) \neq F_{t+1,\infty}\,(X, y)$, it means that a concept drift occurs at moment $t+1$, denoted as $\exists t : P_t\,(X, y) \neq P_{t+1}\,(X, y)$ [3]. This means that the probability of the same feature vector classification result changes before and after moment t.

Concept drift occurs when the concept about which data are being collected shifts from time to time after a minimum stability period. Such changes are reflected in incoming instances and decreases the accuracy of classifiers learned from past training instances. Examples of real life concept drifts including monitoring systems, financial fraud detection, spam categorization, weather predictions, and customer preferences [4].

Changes of target concepts are categorized into abrupt, gradual, incremental and so on, sometimes with noisy data interspersed in the data stream. Different detection algorithms can handle different types of concept drift, some algorithms can handle only a specific type of drift, while others can accommodate multiple drift types.

This paper proposes an unsupervised online concept drift detection algorithm based on Jensen-Shannon divergence and EWMA using knowledge related to information theory. The algorithm firstly divides the data stream into sliding windows and detects the change of Jensen-Shannon divergence of the feature attributes within the windows, and then dynamically calculate the threshold of the change of data distribution between the sliding windows by EWMA, if concept drift is detected there would incrementally train a new classifier to deal with the decrease of classification accuracy. The algorithm detects concept drift without true labels and can be used in an online environment. The experiments show that the algorithm has a high accuracy improvement in dealing with various concept drifts.

This paper is organized as follows. Section 2 we review some outstanding research work dealing with the concept drift in data streams. Section 3 details the concept drift detection algorithm CDDDE. Section 4 explains our experimental setup and analyzes our experimental results. Finally, Sect. 5 presents our conclusions and directions for future work.

2 Related Work

In dynamically changing and non-stationary environments, the data distribution changes over time, giving rise to the phenomenon of concept drift, which proposed by Schlimmer and Granger in 1986 [5]. Since its introduction, researchers have proposed many relevant algorithms [6] for the concept drift problem and have achieved many results.

2.1 Detection Algorithms Based on Error Rate

According to the literatures over the years it can be seen that error rate based drift detection algorithms are the largest class of algorithms, which focus on tracking the change of online error rate of the classifier in real time. In PAC learning models, if the sample data are stably distributed, the error rate of the learning algorithm decreases with the input of the data, and when the probability distribution changes, the error rate of the model increases. The DDM algorithm is the first algorithm based on error rate, which sets two thresholds for the error rate, and when the error rate reaches the warning threshold, it indicates the precursor of a change in the probability distribution, and when the error rate reaches the drift threshold, it indicates a change in the probability distribution, and the model would learn with the data after the drift point [2]. The basic idea of the EDDM algorithm is slightly different from the DDM in that it considers the distance between error rates in addition to the error rate variation, which not only detects the abrupt type drift as effectively as the DDM algorithm but also compensates for the deficiency of the DDM in the gradual type drift [7]. The HDDM algorithm proposes a new method to monitor the measurement metrics during the learning process, and it applies some probabilistic inequalities to obtain theoretical guarantees for detecting changes in the distribution [8]. Most of these algorithms are based on supervised learning, which assumes that the labels are available and it is undoubtedly time and resource intensive upfront.

2.2 Detection Algorithms Using a Small Number of Labels

The majority of the concept drift detection algorithms rely on the instantaneous availability of true labels for all already classified instances. This is a strong assumption that is rarely fulfilled in practical applications. Kolmogorov-Smirnov test is a hypothesis test to check whether two samples have the same distribution and the test depends on the p-value and significance value of the samples [9]. In [10] it used the Kolmogorov-Smirnov hypothesis test for two samples that vary over time, using a random tree to perform insertion and deletion operations on the data, with no true labels used in the detection and only a limited number of labels used in updating the classification model. However, the method is mostly used for one-dimensional data and cannot be easily extended to multidimensional data [11], and in practical scenarios data streams are not limited to univariate data but may also arrive as multivariate streams. Clustering is an unsupervised machine learning method and in [12] the algorithm uses a sliding window to cluster the data in the window, divides the data into individual clusters and outliers, compares the proportion of clustered instances within adjacent windows and tolerates a certain change in the proportion of clustered instances and a certain number of outlier points, and gives a drift signal when a specified threshold is reached. Confidence voting is also the concept drift detection method based on unsupervised learning, which maintain multiple drift detection trajectories during detection and determine whether concept drift is generated based on changes in confidence voting [13]. Margin density-based methods, which rely on the margin of the classifier to detect concept drift, it calculates the proportion of data instances in the margin and when the margin density exceeds a density threshold would alarm a drift [14]. The use of Chernoff Bound to define the number of instances in data streams that deviate from the mean [15], the key step of this approach is to determine the total amount of instances needed to indicate that the learning algorithm has expired and that a new one should be learned from data [16].

2.3 Detection Algorithms Using Divergence

There are some explorations of concept drift detection algorithms based on divergence. Borchani [17] used Kullback-Leibler divergence to calculate the variability of different subsets of the data stream, which suffers from distance asymmetry. Wang [18] used Kullback-Leibler divergence to measure distribution differences, and then, used their own proposed multi-scale drift detection test to check whether the current data concepts are different from the historical concepts. Sun [19] used Jensen-Shannon divergence to measure the distribution difference, but their algorithm uses a fixed threshold to measure the difference resulted in a poor applicability.

3 The Proposed Method

In this section, We describe in detail the algorithm proposed in this paper. Firstly describing how to construct the data distribution within the sliding window, then measuring the distribution differences using divergence, and finally calculating the drift threshold using EWMA.

3.1 Constructing Data Distribution Functions Based on Sliding Windows

Common concept drift detection algorithms for data streams are per-data-instance-based and block-based, and the algorithm in this paper would follow the second form. Let $x_1, x_2, \cdots$ denote the data stream, where each x_i denotes a data instance and $w = \{x_1, x_2, \cdots, x_n\}$ denotes the data window of n data. We use a double window mechanism [20], where the data in one window is used to construct the initial distribution, which remains relatively fixed and updates it when the concept drift is detected. The other window is used to follow the data stream for sliding, so as to indicate the latest distribution of data in the data stream.

The next to be considered is how to map the multidimensional data within the window to the distribution. we denote the relative proportion $P_w(x)$ of each vector x in w

$$P_w(x) = \frac{N\left(x\,|w\right)}{n} \tag{1}$$

where $N\left(x\,|w\right)$ denotes the number of vectors x in the window and n denotes the number of data in the window. We use data frequency to calculate each attribute within the window, then the combination of the frequencies of each attribute constitutes the empirical distribution function P_w for the current window, and the empirical distribution can be understood as the maximum likelihood estimate of the true distribution. Although it is often infeasible to accurately estimate the probability distribution of concept drift, it helps to design drift detection algorithms [21].

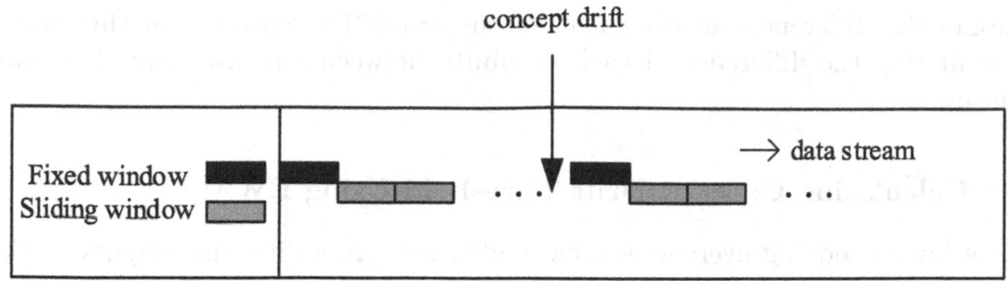

Fig. 1. Sliding window model

3.2 Measuring Differences in Data Distribution Between Windows Using Jensen-Shannon Divergence

We firstly introduce the Kullback-Leibler divergence and then extend to the Jensen-Shannon divergence through its shortcomings in this algorithm. Kullback-Leibler divergence, also called relative entropy, is widely used in the field of information theory. It is a metric often used to quantify the variability between two probability distributions. Denote by X some discrete random variable, and the two probability distributions on the random variable are $P(x)$ and $Q(x)$, respectively, the Kullback-Leibler divergence between them is defined as

$$KL\left(P||Q\right) = \sum_{i=1}^{n} P(x)log\frac{P(x)}{Q(x)} \tag{2}$$

The smaller the difference between the data distributions, the smaller the value of Kullback-Leibler divergence, which is 0 when the two distributions are identical. The formula for Kullback-Leibler divergence shows that it is not symmetric.

$$KL\left(P||Q\right) \neq KL\left(Q||P\right) \tag{3}$$

Therefore, when calculating the data distribution within the window, there may be an abnormal result, so Jensen-Shannon divergence is used in this algorithm. Jensen-Shannon divergence is actually a correction on the asymmetry problem of Kullback-Leibler divergence, and the formula after Jensen-Shannon divergence is expanded is

$$JS\left(P||Q\right) = \sum_{i=1}^{n} P(x)log\frac{2P(x)}{P(x)+Q(x)} + \sum_{i=1}^{n} Q(x)log\frac{2Q(x)}{P(x)+Q(x)} \tag{4}$$

It solves the asymmetry problem of Kullback-Leibler divergence and provides a more accurate measure of similarity. With the data distribution which constructed in Sect. 3.1 the differences in data distribution between windows can be measured. One of the methods to calculate the differences in data distribution between windows is using a certain way to divide the feature subspace, and then combine the differences in each feature subspace. The strategy of this paper is calculating the difference of each attribute between windows and then sum them up.

3.3 Calculating Concept Drift Threshold Using EWMA

The weighted moving average is a method which gives different weights to the observations separately, calculates the moving average by different weights, and uses the moving average as a basis to determine the forecast value. EWMA (exponentially weighted moving average), is a method in which the weighting coefficient of each value decreases exponentially with time, the closer the value to the current moment the greater the weighting coefficient [22]. Why choose the exponentially weighted moving average is that the recent observations have a greater influence on the forecast value and it can reflect the trend of recent changes, which is a powerful indicator in the concept drift detection.

After the first two steps of the algorithm we are able to obtain the value of Jensen-Shannon divergence between sliding windows, which we use as a statistical indicator. Then the EWMA statistic for the current sliding window is expressed as

$$z_i = \lambda j_i + (1 - \lambda) z_{i-1} \tag{5}$$

where z_i denotes the EWMA value of the i-th sliding window in which no concept drift occurred, λ denotes the weight coefficient of EWMA on the historical data, whose value is closer to 1, indicating a lower weight on the historical data, and j_i denotes the Jensen-Shannon divergence between the current window and the fixed window. It is also necessary to recalculate the variance σ_z of the EWMA value at each sliding window. σ_{zi} denotes the variance of the i-th sliding window, which is calculated as [22]

$$\sigma_{zi}^2 = \sigma^2 \left(\frac{\lambda}{2-\lambda}\right) \left[1 - (1-\lambda)^{2i}\right] \tag{6}$$

where σ denotes the overall variance of the EWMA computed before the current window when no concept drift has occurred. When i gradually increases, $(1-\lambda)^{2i}$ will soon converge to zero, but when i is small, retaining this part is beneficial to improve the effects of EWMA. With the constant arrival of the data stream, we can then set a variable upper and lower threshold by the calculated value of EWMA and the mean variance. We use UCL and LCL to denote the upper and lower thresholds, respectively, which are calculated as

$$UCL = \mu + L\sigma \sqrt{\left(\frac{\lambda}{2-\lambda}\right) \left[1 - (1-\lambda)^{2i}\right]} \tag{7}$$

$$LCL = \mu - L\sigma \sqrt{\left(\frac{\lambda}{2-\lambda}\right) \left[1 - (1-\lambda)^{2i}\right]} \tag{8}$$

where μ denotes the average value of the EWMA calculated before the current window when no concept drift occurs, and L as a control limit width factor that can be dynamically adjusted according to the variation of the Jensen-Shannon divergence detected. The adjustment of L can make the algorithm adapt to more drift types and drift datasets which provide higher robustness and applicability.

Algorithm 1. CDDDE concept drift detection algorithm

Input: data stream S, window size n, EWMA history weighting factor λ, limit width L

Output: Concept drift detection result F

for each instance $x_i \in S$ **do**
    while initial window w_1 size < n **do**
        add x_i to w_1;
    end while
    Calculate probability distribution of w_1 by (1)
    while sliding window w_2 size < n **do**
        add x_i to w_2;
    end while
    Calculate probability distribution of w_2 by (1)
    Calculate $JS(w_1||w_2)$ by (4)
    Calculate z_i by (5)
    Calculate threshold UCL and LCL by EWMA
    if $z_i >$ UCL or $z_i <$ LCL **then**
        alarm concept drift
        clear w_1,clear w_2
    else
        clear w_2
    end if
end for

4 Experiments

4.1 Datasets

Massive Online Analysis (MOA) is a framework for data stream analysis. It includes many machine learning algorithms and tools for evaluation. This algorithm is developed and implemented based on the MOA framework, and the experiments use artificial datasets and real-world datasets. The basic information of the datasets is shown in Table 1.

Artificial Datasets. The artificial dataset is generated based on the data generation function in MOA.

Table 1. Basic information of the datasets

Name	Number of instances	Number of features	Number of class	Number of drifts	Type of drifts
$Agrawal_a$	100K	9	2	3	Abrupt
$Agrawal_g$	100K	9	2	3	Gradual
SEA	100K	3	2	3	Abrupt
Hyperplane	100K	10	2	-	Incremental
Airlines	539K	7	2	Unknown	Unknown
Covertype	581K	54	7	Unknown	Unknown
Spambase	4K	57	2	Unknown	Unknown

Agrawal Dataset. It is a dataset that determines whether or not to loan based on information about an individual, and contains both loanable and non-loanable categories. It contains 6 numerical attributes and 3 categorical attributes, and uses ten different predefined loan functions to generate the data. The dataset contains 100K instances, with drift occurring every 25K, and is divided into two types of abrupt and gradual drift, with three drift points set for both types.

SEA Dataset. It contains 3 numerical attributes and uses 4 different functions defined to generate the data. The dataset contains 100K instances, with abrupt drift occurring every 25K, and a total of three drift points set.

Hyperplane Dataset. It contains 10 numerical attributes, and the data is incrementally drifted by constant small changes in the decision boundary. The dataset contains 100K instances, and the probability of change for each generated instance is 0.001, and its drift type is incremental.

Real-world Datasets. In addition to artificial datasets, we also chose some common real-world datasets for concept drift detection to conduct experiments.

Airlines Dataset. It contains 3 numerical attributes and 4 categorical attributes. This dataset contains 539383 data and it is a binary classification dataset that determines whether the plane will be delayed based on the condition.

Covertype Dataset. It contains 10 numeric attributes and 44 categorical attributes with only 0 and 1 values. The dataset contains 581,012 data and it aims to predict the type of cover of a forest in an area with 7 different class labels.

Spambase Dataset. It contains 57 attributes and 4601 instances. The dataset is mainly used for spam identification filtering, where spam resources are obtained from mail administrators and individuals who submit spam, and the dataset is also often used to construct spam filters.

4.2 Experimental Settings

The concept drift detection algorithm in this paper is based on unsupervised learning, and the detection algorithm does not use the labels of the dataset, but for the sake of relevant statistical metrics and comparison experiments, we assume that the labels are immediately available after the detection of concept drift and can be used for the calculation of classifier accuracy. The classifier accuracy will serve as an important evaluation metric for our detection algorithm, since the consequence of concept drift is a dramatic decrease in the classification accuracy of the classifier [23]. In order to evaluate only the impact of the concept drift detection algorithm on the classification accuracy, our experiments are computed using Naïve Bayes for classification, which does not have an automatic adaptation strategy for concept drift, making the drift detection completely dependent on our detection algorithm.

In addition to the classification accuracy we care about the number of detected drift points in the dataset and whether the drift points are incorrectly located, so we use true positive TP to indicate the number of detected drift points as correct drift points, false positive FP to indicate the number of detected drift points as incorrect drift points, and false negative FN to indicate the number of undetected correct drift points, since most algorithms are have a certain delay in detection, so we allow a certain delay in counting TP and include it in the number of TP. Due to the difficulty in estimating these performance measures on incremental datasets and real-world datasets, so we will only count the number of drift points output by the concept drift detection algorithm and the accuracy of the classifier.

The data stream S is read by MOA using the generated and real datasets, the window size n is defaulted to 500. λ denotes the EWMA historical weight factor, a larger λ makes the algorithm increases the weight of the Jensen-Shannon divergence calculated in the current window, thus decreasing the EWMA statistic calculated in the previous window, and it can be adjusted according to the changes in the datasets, the default value of λ is 0.1. L denotes the limit width factor, it is used to adjust the threshold of whether the concept drift is detected or not, a smaller L can be set to detect small changes in the datasets, conversely, a larger L can be set to detect large changes in the datasets and avoid small changes due to noise, the default value of L is 3. The other parameters of the detection algorithm or classifier are adopted as the default values in the MOA framework. The following experimental results are all run under MOA experimental platform.

4.3 Results and Analysis

The experiments are evaluated using the Evaluate Prequential strategy in MOA, which means that each instance is used as test data and then used as training data to incrementally train the classifier, thus maximizing the use of each instance and ensuring smooth accuracy.

The focus is on the concept drift detection and whether classifier can adapt the new data after changes, so we evaluate the performance of the proposed algorithm by comparing with the algorithm which are integrated in MOA and has high detection efficiency. NoChangeDetection detector is used as a benchmark to demonstrate whether the classification accuracy is affected when the datasets appear concept drift. Besides, we compare it with the concept drift detection algorithm to evaluate the accuracy gain of the detection algorithm on the classifier. Most of these comparative algorithms have been described in related work.

Table 2, Table 3 and Table 4 show the experimental results of comparing our proposed concept drift detection algorithm with common concept drift detection algorithms on artificial datasets where the exact drift points are known. The types of data drift used in these three datasets are abrupt type and gradual type, and it can be seen that our algorithm has good results in determining the detection of concept drift points, and basically there is no leakage and wrong detection, while other algorithms either have leakage or wrong detection values are higher. In terms of the accuracy of the classifier, we can see that the algorithm CDDDE either reaches the highest on the three datasets or has a slight difference compared to the highest accuracy in the comparison algorithms. But the advantage of CDDDE is that it does not require labels and it can be used in an online environment, while other algorithms require a large number of labels during the detection of concept drift.

Table 2. The experimental results in $Agrawal_a$ dataset

Detector	TP	FP	FN	Accuracy
NoChangeDetection	0	0	3	73.93
DDM	1	0	2	76.84
EDDM	3	15	0	79.52
STEPD	3	18	0	80.34
HDDM-A-Test	3	0	0	**80.89**
CDDDE	**3**	**0**	**0**	80.49

Table 3. The experimental results in $Agrawal_g$ dataset

Detector	TP	FP	FN	Accuracy
NoChangeDetection	0	0	3	73.81
DDM	1	1	2	75.17
EDDM	1	6	2	75.2
STEPD	3	30	0	76.34
HDDM-A-Test	3	4	0	77.25
CDDDE	**3**	**1**	**0**	**77.61**

Table 4. The experimental results in SEA dataset

Detector	TP	FP	FN	Accuracy
NoChangeDetection	0	0	3	86.47
DDM	1	2	2	87.35
EDDM	0	18	3	86.55
STEPD	3	9	0	87.7
HDDM-A-Test	3	0	0	**87.87**
CDDDE	**3**	**0**	**0**	87.84

In order to show the comparison results of each algorithm more visually, we export the real-time accuracy changes calculated in $Agrawal_a$ dataset in MOA and visualize them. As shown in Fig. 2, since we set drift points at 25K, 50K, and 75K of the dataset, we can see that the accuracy of the classifier drops sharply without concept drift detection, but with the concept drift detection algorithm we can see that our algorithm can detect concept drift accurately and adapt immediately as most concept drift detection algorithms. Similar to the $Agrawal_a$ dataset, the other datasets used for experiment are also able to identify the drift points well and improve the adaptation of the classifier.

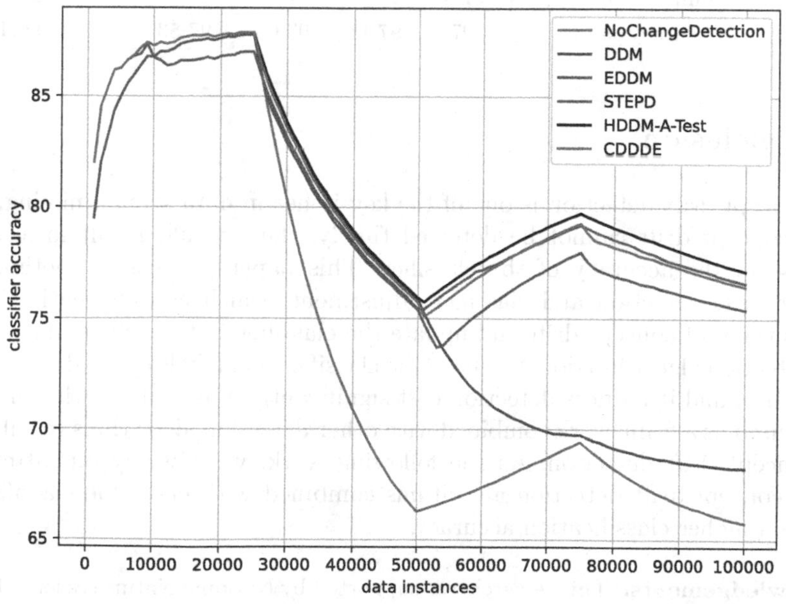

Fig. 2. Comparison of real-time classification accuracy in the $Agrawal_a$ dataset

Table 5 show the results on the incremental drift and the real-world datasets, both of them are commonly used for concept drift detection. Due to the difficulty in estimating those performance measures which on artificial datasets, we only count the number of drift points output by the concept drift detection algorithm and the accuracy of the classifier. It can be seen that our algorithm is effective in detecting concept drift, and the output of the number of drift points is much smaller than other algorithms. It may allow our algorithm to detect critical variation points and make adjustments, so that other parts related to the detection algorithm will not have to change too often in order to reduce the impact of concept drift.

Table 5. The experimental results in Hyperplane, Airlines, Covertype and Spambase

Detector		NoChange-Detection	DDM	EDDM	STEPD	HDDM-A-Test	CDDDE
Hyperplane	Num	0	10	44	24	10	12
	Accuracy	77.13	81.5	83.91	85.42	84.64	85.06
Airlines	Num	0	13	23	644	72	30
	Accuracy	66.99	67.71	67.17	66.75	68.23	67.9
Covertype	Num	0	4634	2416	3731	3284	166
	Accuracy	66.98	86.62	85.03	86.88	86.6	80.75
Spambase	Num	0	1	2	47	1	2
	Accuracy	90.15	97.42	97.18	97.68	97.53	96.46

5 Conclusion

The concept drift detection is one of the key issues in data stream mining, and if the concept drift cannot be detected timely, there would result in a sharp decrease in the accuracy of the classifier. This paper proposes a method for concept drift detection and classifier adjustment, which can efficiently detect various types of concept drift and update the classifier at the drift point timely. The detection algorithm does not need the classification labels of the data stream in advance, and it reduces detection cost significantly. Particularly, this method can be used as a framework combined with other detection algorithms to enhance the concept drift detection. As the following work, we will pay our attention to the concept drift detection algorithms combined with ensemble classifier to achieve a higher classification accuracy.

Acknowledgements. This research was supported by National Natural Science Foundation of China under Grant No. 62072236.

References

1. Iwashita, A.S., Papa, J.P.: An Overview on Concept Drift Learning. IEEE Access **7**, 1532–1547 (2019)
2. Gama, J., Medas, P., Castillo, G., Rodrigues, P.: Learning with Drift Detection. In: Bazzan, A.L.C., Labidi, S. (eds.) SBIA 2004. LNCS (LNAI), vol. 3171, pp. 286–295. Springer, Heidelberg (2004). https://doi.org/10.1007/978-3-540-28645-5_29
3. Lu, J., Liu, A. Dong, F., Gu, F., Gama, J., Zhang, G.: Learning under Concept Drift: a Review. IEEE Trans. Knowl. Data Eng. **31**, pp. 2346–2363 (2019)
4. Dongre P.B., Malik L.G.: A review on real time data stream classification and adapting to various concept drift scenarios. IEEE International Advance Computing Conference (IACC), pp. 533–537 (2014)
5. Gama, J., Žliobaitė, I., Bifet, A., Pechenizkiy, M., Bouchachia, A.: A survey on concept drift adaptation. ACM Comput. Surv. **46**(4), 1–37 (2014)
6. Patil, M.M.: Handling Concept Drift in Data Streams by Using Drift Detection Methods. In: Balas, V.E., Sharma, N., Chakrabarti, A. (eds.) Data Management, Analytics and Innovation. AISC, vol. 839, pp. 155–166. Springer, Singapore (2019). https://doi.org/10.1007/978-981-13-1274-8_12
7. Baena-Garcıa M, del Campo-Ávila J, Fidalgo R, et al.: Early drift detection method. In: Fourth International Workshop on Knowledge Discovery from Data Streams, pp. 77–86. (2006)
8. Frias-Blanco, I., del Campo-Avila, J., Ramos-Jimenez, G., Morales-Bueno, R., Ortiz-Diaz, A., Caballero-Mota, Y.: Online and Non-Parametric drift detection methods based on Hoeffding's bounds. IEEE Trans. Knowl. Data Eng. **27**(3), 810–823 (2015)
9. Wang, Z., Wang, W.: Concept Drift Detection Based on Kolmogorov–Smirnov Test. In: Liang, Q., Wang, W., Mu, J., Liu, X., Na, Z., Chen, B. (eds.) Artificial Intelligence in China. LNEE, vol. 572, pp. 273–280. Springer, Singapore (2020). https://doi.org/10.1007/978-981-15-0187-6_31
10. Dos Reis, D. M., Flach, P., Matwin, S., Batista, G.: Fast Unsupervised Online Drift Detection Using Incremental Kolmogorov-Smirnov Test. In: Proceedings of the 22nd ACM SIGKDD International Conference on Knowledge Discovery and Data Mining, pp. 1545–1554. Association for Computing Machinery, San Francisco, California, USA (2016)
11. Lu, N., Zhang, G., Lu, J.: Concept drift detection via competence models. Artif. Intell. **209**, 11–28 (2014)
12. Chen, H.-L., Chen, M.-S., Lin, S.-C.: Catching the Trend: A Framework for Clustering Concept-Drifting Categorical Data. IEEE Trans. Knowl. Data Eng. **21**(5), 652–665 (2009)
13. D'Ettorre, S., Viktor, H.L., Paquet, E.: Context-Based Abrupt Change Detection and Adaptation for Categorical Data Streams. In: Yamamoto, A., Kida, T., Uno, T., Kuboyama, T. (eds.) DS 2017. LNCS (LNAI), vol. 10558, pp. 3–17. Springer, Cham (2017). https://doi.org/10.1007/978-3-319-67786-6_1
14. Sethi, T.S., Kantardzic, M.: Don't Pay for Validation: Detecting Drifts from Unlabeled data Using Margin Density. Procedia Comput. Sci. **53**, 103–112 (2015)
15. Diaz-Rozo, J., Bielza, C., Larrañaga, P.: Clustering of Data Streams With Dynamic Gaussian Mixture Models: an IoT Application in Industrial Processes. In: IEEE Internet of Things Journal **5**(5), pp. 3533–3547 (2018)
16. Ghani, N. L. A., Aziz, I. A., Mehat, M.: Concept Drift Detection on Unlabeled Data Streams: a Systematic Literature Review. In: 2020 IEEE Conference on Big Data and Analytics (ICBDA), pp. 61–65 (2020)

17. Hanen, B., Pedro, L., Concha, B..: Classifying Evolving Data Streams with Partially Labeled Data. Intell. Data Anal. **15**(5), 655–670 (2011)
18. Wang, X., Kang, Q., An, J., Zhou, M.: Drifted Twitter Spam Classification Using Multiscale Detection Test on K-L Divergence. IEEE Access 7, pp. 108384–108394 (2019)
19. Yange, S., et al.: Adaptive ensemble classification algorithm for data streams based on information entropy J. Univ. Sci. Technol. Chin. **47**(7), 575–582 (2017)
20. Guo, H., Li, H., Ren, Q., Wang, W.: Concept drift type identification based on multi-sliding windows: Husheng Guo, Hai Li, Qiaoyan Ren, Wenjian Wang. Inf. Sci. **585**, 1–23 (2022)
21. Webb, G.I., Hyde, R., Cao, H., Nguyen, H.L., Petitjean, F.: Characterizing concept drift. Data Min. Knowl. Disc. **30**(4), 964–994 (2016). https://doi.org/10.1007/s10618-015-0448-4
22. Ross, G.J., Adams, N.M., Tasoulis, D.K., Hand, D.J.: Exponentially weighted moving average charts for detecting concept drift. Pattern Recogn. Lett. **33**(2), 191–198 (2012)
23. Khamassi, I., Sayed-Mouchaweh, M., Hammami, M., Ghédira, K.: Discussion and review on evolving data streams and concept drift adapting. Evol. Syst. **9**(1), 1–23 (2016). https://doi.org/10.1007/s12530-016-9168-2

A Knowledge-Enabled Customized Data Modeling Platform Towards Intelligent Police Applications

Tiexin Wang[1(✉)], Hong Jiang[1], Huihui Zhang[2], and Xinhua Yan[3]

[1] College of Computer Science and Technology, Nanjing University of Aeronautics and Astronautics, 29#, Jiangjun Road, Jiangning District, Nanjing 211106, China
{tiexin.wang,jiang1997}@nuaa.edu.cn
[2] Weifang University, Weifang 261061, China
huihui@wfu.edu.cn
[3] Nanjing DENET System Technology Co. LTD, Nanjing, China

Abstract. With the rapid development of information and communication technologies, massive amounts of data continue to be generated and flood all aspects of society. As one of the key departments of the government, the public security bureau masters all kinds of heterogonous data. Deep analysis of these data will help to detect and prevent public security cases and maintain social stability. Therefore, it is an urgent demand for grassroots police officers to better manage and use these data. To address this demand, in this paper, we present the work of designing and implementing a customized data modeling platform. With the modeling platform, which owns a visual interface, police officers can have a better overview and understanding of collected data and use the drag-and-drop method to build data analysis models. As a core component of this modeling platform, after analyzing 211 tables of practical police data, we built a public security domain knowledge model. Cooperating with the Sucheng branch of Suqian Public Security Bureau, we conducted a set of experiments with police officers on real police data. Experiment results show that the modeling platform has better user-friendliness and outperforms the traditional SQL-based querying method considering the integrity of querying results.

Keywords: Public security bureau · Knowledge model · Customized modeling platform · Query expansion

1 Introduction

In recent years, the sustainable development of information technologies such as big data analysis and artificial intelligence has promoted the construction of public security informatization [1]. Massive amounts of heterogeneous data are quickly generated in all aspects of society. To better use these data, the public security bureau has formed a public security basic information network and built big data centers.

With the advent of the big data era, traditional data processing methods and technologies are no longer able to meet the increasing demands of data, and explosive data

B. Li et al. (Eds.): APWeb-WAIM 2022, LNCS 13421, pp. 135–149, 2023.
https://doi.org/10.1007/978-3-031-25158-0_11

growth has forced people to seek new data processing methods [2–5]. This kind of data processing demand is particularly prominent in the field of public security [6, 7]. The public security bureau masters a large number of data sources, such as public security business systems (e.g., criminal records, sentence records), government management systems (e.g., tax records, bank records), Internet data (e.g., online shopping records, online chat records), social data (e.g., travel records), etc. We regard all these data being collected by the public security bureau as police data.

However, due to the wide range of police data sources and the continuous emergence of new data sources, it hinders a thorough understanding of police data by grassroots police officers. The lack of correlations between data of various sources has formed more and more data islands [8], which severely hampers the use of police data. Currently, there exist several intelligent police data management platforms, such as the Tianhe Big Data Platform[1], which provides a visual operating environment, and completes the configuration and construction of data processing and analysis processes by dragging and dropping pre-defined elements. However, since it lacks unified data specifications and standards of police data, these systems are always developed towards specific police data sets and show poor generality and reusability. In addition, data query methods provided by existing visualization platforms are mainly based on string matching [9], which often leads to inaccurate querying results due to inconsistency in semantics.

In this paper, we propose and implement a customized police data modeling platform (CPDMP). With CPDMP, police officers may have a better understanding of multi-source data in a visualized environment and digitally convert their practical experience as customized models. To better organize multi-source data and eliminate semantic inconsistency issues in CPDMP, we construct a public security domain knowledge model (PSDKM) considering collected police data. Additionally, we have also added query expansion capabilities to CPDMP to enhance its usability.

The three main contributions of this paper are as follows.

- We built eight police thematic databases and a police domain knowledge model to help police officers better understand and use multi-source data.
- We implemented a customized modeling platform to provide a visual method to help police officers convert their own experiences into digital models.
- The domain knowledge model, which brings query expansion capability to CPDMP, can be further extended as a foundation.

The rest of the paper is organized as follows. In Sect. 2, we present the related concepts and a running example. In Sect. 3, we illustrate the architecture and main functions of CPDMP, and detail PSDKM. Experiment settings and results analysis are given in Sect. 4. In Sect. 5, we present the related work. Finally, we conclude in Sect. 6.

[1] https://www.linewell.com/linewell/gw

2 Related Concepts and Running Example

In this section, we provide relevant concepts and a running example to introduce the work of this paper, including ontology (Sect. 2.1), query expansion (Sect. 2.2), and a running example (Sect. 2.3, Fig. 1).

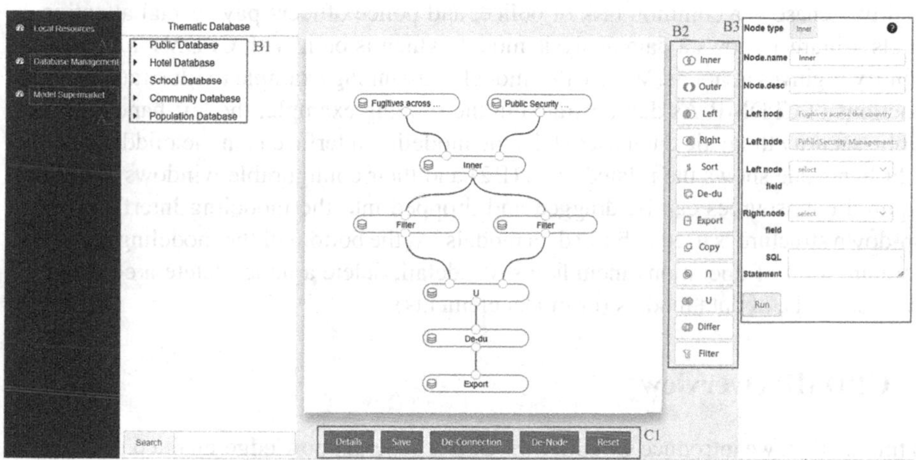

Fig. 1. A running example of the modeling platform. A1 and A2 show the database management and model supermarket interfaces. B1 presents data tables that can be dragged and dropped. B2 and B3 show modeling tools and their configurations. C1 shows a set of other operations.

2.1 Ontology

Ontology is a conceptual model that describes concepts and relationships between concepts in the related fields [10]. The structure of the ontology is a five-tuple [11] O: = {C, R, H, Rel, A}. C and R are two disjoint sets. Among them: the elements in C are called concepts; R is a relationship set between elements in C; H represents the concept level, that is, the taxonomy relation between concepts; Rel represents a non-taxonomy relation between concepts; A stands for ontological axiom.

In this paper, we use a bottom-up method to construct the public security domain knowledge model (PSDKM). Through the analysis of police data, the concepts, terms and their relationships in the field of public security are clarified.

2.2 Query Expansion

The relational database is usually accessed through structured query language SQL, which is a deterministic and precise query, that is, users need to construct accurate SQL query statements. The querying conditions should be accurate, and the querying results are also accurate. However, this kind of methods is all based on string matching and takes no advantage of semantic relationships between data (such as synonym, subordinate relationships, etc.).

Query expansion is one of the primary methods to implement semantic queries in relational databases [12]. It replaces the keywords in original query statements with related words, concepts, etc. to construct a new query statement.

2.3 Running Example

Fugitives chase is a common task of police, and police officers pay special attention at hotels. Figure 1 shows a data analysis model, which is built with CPDMP and aims to identify fugitives at hotels. We use this model as a running example to illustrate the main functions of CPDMP. As demonstrated in the running example, the left-hand side lists public security thematic databases (B1), the modeling interface is in the middle, and the right-hand side shows the related tools (B2) and their configurable windows (B3). The required data sources can be dragged and dropped into the modeling interface, and a top-down structure is used to build data models. At the bottom of the modeling interface, there are a set of operations including save, detail, delete a node, delete a connection, and reset (C1), of data models (or model elements).

3 CPDMP Overview

In this section, we introduce CPDMP, which is a general knowledge-enabled police data customized modeling platform. First, we briefly introduce the implementation architecture of CPDMP in Subsect. 3.1. Then, we introduce PSDKM and CPDMP in detail in Subsect. 3.2 and Subsect. 3.3, respectively. Finally, we illustrate the use of CPDMP in Subsect. 3.4.

3.1 Implementation of CPDMP

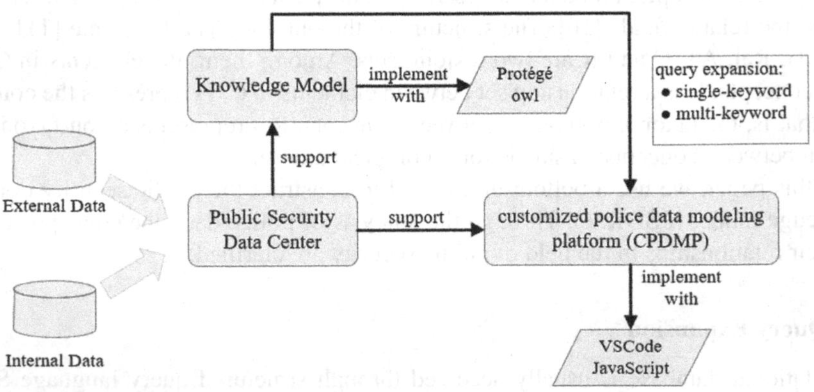

Fig. 2. The implementation architecture of CPDMP.

The implementation architecture of CPDMP is shown in Fig. 2. The public security data center is a police data storage center built on the police security intranet (physically isolated from the Internet). The construction of the knowledge model and the output of the data model in CPDMP all depend on the support of the public security data center. The construction of CPDMP adopts the method of separating the front-end and back-end. The front end displays the platform and provides a user interface, and the back end provides business services support.

3.2 PSDKM

Towards managing large amounts of police data, CPDMP was established. Especially, in CPDMP, we construct different thematic databases and a knowledge model "PSDKM" to express the connections among police data.

According to the real requirements of a specific local Public Security Bureau, we have established eight thematic databases to support police officers' daily work. Table 1 shows the detailed information (i.e., thematic name, tables involved, data size, main contents and data sources) of each thematic database. The total number of tables contained in the eight theme databases is 211.

Table 1. Detailed information of eight thematic databases.

Thematic	Tables	Size	Contents	Data sources
School Management	21	249.2 GB	Schools, teachers, students, security staff, etc	MPPDB, Oracle
Hotel Management	36	897.3 GB	Hotels, hotel guests, hotel staffs, etc	MPPDB, Oracle
Public Opinion Management	12	224.3 GB	Police stations, petition incidents, etc	SQL server, MPPDB, Oracle
Legal Management	28	747.7 GB	Cases, suspects, etc	SQL server, MPPDB, Oracle
Command and Control Management	32	797.6 GB	Duty officer, address, body worn camera, etc	MySQL, MPPDB, Oracle
Community Management	39	997.0 GB	Houses, vehicles, security staff, use water, etc	MPPDB
Population Management	30	760.2 GB	The floating population, residence permit, etc	SQL server, MPPDB, Oracle
Risk Management	13	586.1 GB	Units, practitioner, etc	MPPDB, MySQL

To define and show potential relationships between data, we adopted a semi-automatic approach to build PSDKM [10, 13]. First, we determine that the scope of knowledge is mainly from 211 tables. Then, we look for data and their relationships from databases. Related steps are as follows.

- Analyze the information of tables to sort out the domain scope and important concepts in PSDKM.
- Extract important concepts and construct the conceptual framework of PSDKM.
- Semi-automatically converting the conceptual framework of PSDKM into an owl file (a knowledge model) using mapping rules.
- Use Protégé to refine and evaluate the initial constructed knowledge model.

The relevant mapping rules are as follows.

- Tables in relational databases are mapped to classes.
- Columns of tables are mapped to properties of classes.
- Each row of tables is mapped to an entity.
- The value of each cell in tables is mapped to a property value. When a cell corresponds to a foreign key, we replace it with the entity or property pointed to by the foreign key.

As shown in Fig. 3, we sorted out 178 classes, 1033 data properties and 791 object properties. Notice that, building PSDKM is a continuous iterative process.

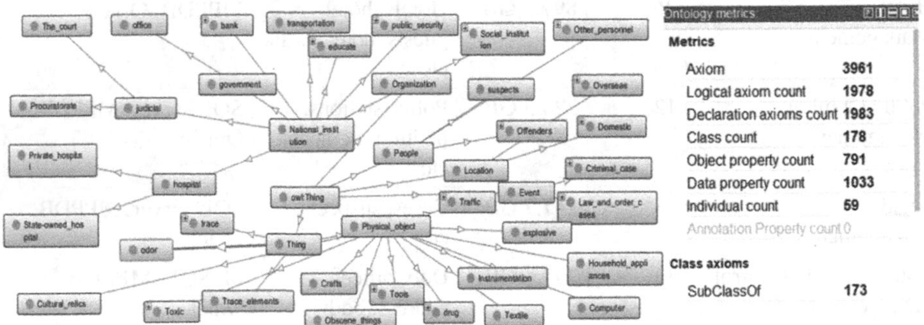

Fig. 3. An illustration of PSDKM.

3.3 CPDMP

CPDMP uses graphics to describe actual combat experience [14]. The data modeling process is visualized and can be modified or extended at any time.

As shown in Fig. 4, CPDMP consists of six modules (five functional modules and one application module) and a knowledge model (PSDKM).

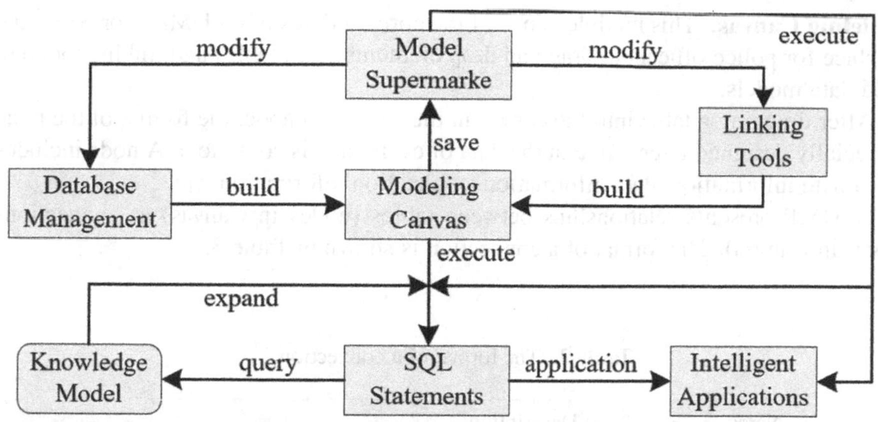

Fig. 4. The connections between modules and the PSDKM in CPDMP.

Database Management. The main function provided by this module is to adapt to different databases usage requirements. It allows police officers to view and operate original databases. To facilitate users' operations, we provide a visual interface to realize operations such as viewing databases, viewing, adding and deleting tables, etc. We categorized eight thematic databases. Take "risk management" as an example, as shown in Table 2, this thematic database includes 13 tables, such as the basic information of face grasping, camera position, etc. These tables come from different original databases and relevant data is synchronized to the "risk management" thematic database in real-time.

Table 2. Risk management thematic database.

Table name	Description	Original databases
hengcheng.xcy_rlzp_jbxx	Basic information of face grasping	MPPDB
hengcheng.wg_camera_jbxx	Camera position	MPPDB
cmf_gxgj_dwlb_ga	Type of unit	MySQL
cmf_gxgj_shxx	Business information	MySQL
cmf_branch	Police station	MySQL
cmf_jxkp_database	Dictionary table	MySQL
cmf_gxgj_wpbb	Item filing	MySQL
cmf_gxgj_jcjl_main	Inspection record	MySQL
cmf_gxgj_zgjl_main	Rectification record	MySQL
cmf_gxgj_cyry	Practitioner information	MySQL
cmf_gxgj_shxx_cccs	Storage location information	MySQL
cmf_gxgj_wpbb_syyz	Usage threshold	MySQL
cmf_users	User table	MySQL

Modeling Canvas. This module is one of the core modules of CPDMP, it provides the interface for police officers to drag and drop elements (e.g., tables and linking tools) to build data models.

After dragging a table into the canvas in the form of a node, the format of the node is specially designed to ensure that the data of each node is consistent. A node includes coordinate information, data information, connection information, etc.

CPDMP presents relationships between tables (nodes in Canvas) as connections (edges in Canvas). The format of a connection is shown in Table 3.

Table 3. The format of a connection.

Name	Description
ID	Connection ID
Start	Node exit where connection starts
End	Node exit where connection ends

Linking Tools. For building data models, it is necessary to connect different tables and generate new tables. In this module, we design 14 kinds of configurable linking tools, including 4 connection tools, 1 copy tool, 1 export tool, 1 sort tool, 1 de-duplication tool, 3 Boolean tools, 1 conditional filter tool, and 2 warning tools.

As shown in Table 4, the 4 connection tools are: inner connection, outer connection, left connection, and right connection.

Table 4. Four connection tools.

Tool name	Function description
Inner connection	When the left and right input have the same value in a fixed column, output the result of this row
Outer connection	Completely return all rows of input, and merge the same rows
Left connection	Completely return all rows of left input, if right input exists and is the same as the value of a fixed column, the same row returns the corresponding right input
Right connection	Completely return all rows of right input, if left input exists and is the same as the value of a fixed column, the same row returns the corresponding left input

The main function of the "copy tool" is to copy a connection node to adjust output fields. The "export tool" is used to execute output results. The function of "sort tool" includes ascending and descending sorting, which can be configured according to practical needs. The "de-duplication tool" can remove the same fields between different tables, and only keep one of them to display. The function of "conditional filter tool" is to set the selection conditions on one or more fields (of a table) and return the filtered data rows.

Table 5 shows the functions of "Boolean tools" and "warning tools". Boolean tools include three types: intersection, union and difference set. Warning tools include conditional warning and timed warning.

Table 5. The functions of Boolean tools and warning tools.

Tool name	Function description
Intersection	Return rows that exist in both left and right input
Union	Return all rows of left and right input and the rows that appear in both left and right input will only be returned once
Difference set	Return the left input rows that are not in the right input
Conditional warning	When the result meets the given conditions, send reminders
Timed warning	Automatically send reminders at a fixed time interval

SQL Statements. The function of this module is to find the dependencies between nodes in a data model and execute data models.

A complete data model has a visually top-down structure. In order to discover the data model structure during execution, we start from the execution node and look up all dependent nodes in turn. In this way, we not only find all nodes in a data model, but also find the dependencies between nodes.

During the execution of data models, the information contained in data models is converted into SQL statements. These modules support two kinds of querying forms: single-keyword and multi-keyword.

In the single-keyword-based querying method, to improve the accuracy of querying results, PSDKM is involved to perform query expansion. The query expansion steps with PSDKM are as follows.

- Extract the generated query statements, identify the keywords (replace with synonym, upper and lower levels), and generate new query statements.
- Execute the newly generated query statements and output new querying results.
- Compare new querying results with original querying results.
- Classify the results of all queries.

For the multi-keyword querying method, it concerns the realization of combining concepts, individuals and relations in PSDKM. The keywords may include concepts and individuals in PSDKM, and individual is an instance of a concept. If the keywords contain an individual, we can directly find the concept corresponding to the individual, and then reason the user's query intention through semantic relationships defined between concepts.

Model Supermarket. To save and reuse data models being built by police officers, we design this module. As shown in Table 6, a complete data model includes model ID, model title, model nodes, model connections, model creator, model description, and model type.

This information can be used to identify data models. If a data model is modified, the corresponding information of this data model will also be automatically updated.

Table 6. All fields of a data model stored in model supermarket.

Field	Description
model_id	Model id
model_title	Model title
model_nodes	Model node information
model_lines	Model connection information
model_creator	The creator of a model
model_desc	Model descriptions
model_type	Model type

Intelligent Applications. The majority of police applications require police data to be real-time [15]. Executing one same data model, different results may be obtained in different time slots. Therefore, CPDMP provides intelligent applications, such as early warning functions that realize notifications to police officers.

3.4 Methodology

In this section, we illustrate the use of CPDMP with a UML activity diagram.

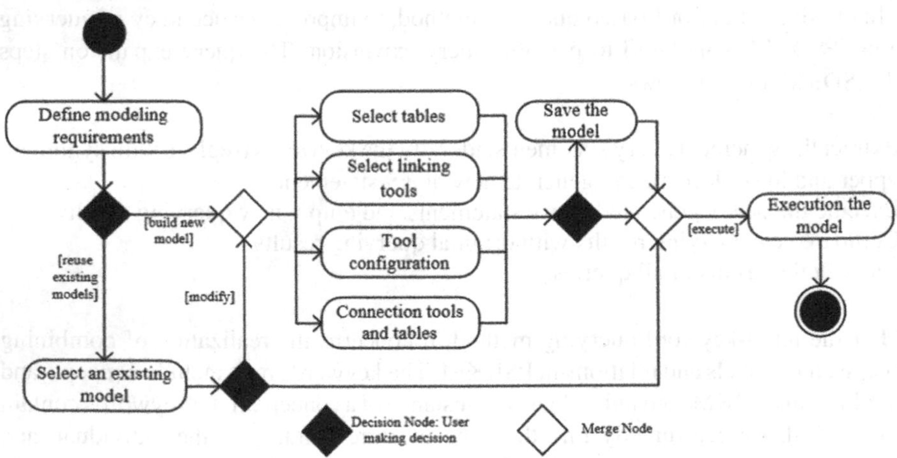

Fig. 5. The methodology of employing CPDMP.

As shown in Fig. 5, first, a user should start to clarify their specific needs based on their practical experience. Then, the user can choose whether to create a new data model or reuse an existing data model. During the process of creating a new data model, the user can choose to select tables, select linking tools, tool configuration and connection tools and tables. In order to ensure the reliability of created data models, the user can execute part of the data model during the modeling process. After connecting tools and tables, the user can choose whether to save the model or execute the query directly.

When choosing to reuse an existing data model, the user can select an existing data model in the model supermarket. Then, the selected data model will be presented in the modeling canvas. The user can modify the data model or use the data model directly. In the process of modifying data models, the user can perform the same operations as creating a new data model.

4 Experiments

In this section, we evaluate CPDMP with a specific use case. Considering the sensitivity of police data, we conducted experiments with our collaboration partner "Sucheng branch of Suqian Public Security Bureau" (S2PSB). Four grassroots police officers are invited to take part in experiments.

4.1 Experiments Setting

The four under-test police officers have rich actual combat experience. They range in age from 28 to 35 and have at least 3-year working experience. Before experiments started, we gave the four police officers a two-hour training lecture, which included data presentation, linking tools description, modeling demonstration, etc. Then, leave them one hour to get acquainted with CPDMP and allow them to ask questions. After that, we distributed a specific modeling requirement to four police officers. They were asked to complete the modeling task and output the results within one hour.

The experimental data model describes the process of finding people with criminal records from hotel occupants. We compared the results obtained by CPDMP and the traditional SQL-based querying method (currently used in S2PSB) to verify the integrity of CPDMP. To verify that CPDMP is easier to use than the traditional SQL-based querying method, we collected subjective opinions with a questionnaire on four police officers after the experiment. When they completed the experiments, the four police officers were asked to answer this predefined questionnaire immediately. There are 15 questions in the questionnaire (e.g., Is the visualized data easy to understand? How visible is the modeling process?). Answers to these questions are used to measure the degree of satisfaction. For each question, police officers can choose a value between 0–100 as the answer. Different value ranges indicate different satisfaction of police officers (<60 means dissatisfied. > 60 & < 80 means neutral. > 80 means satisfaction). Finally, we count the responses of each police officer. Calculate the average of all questions as the final evaluation criterion.

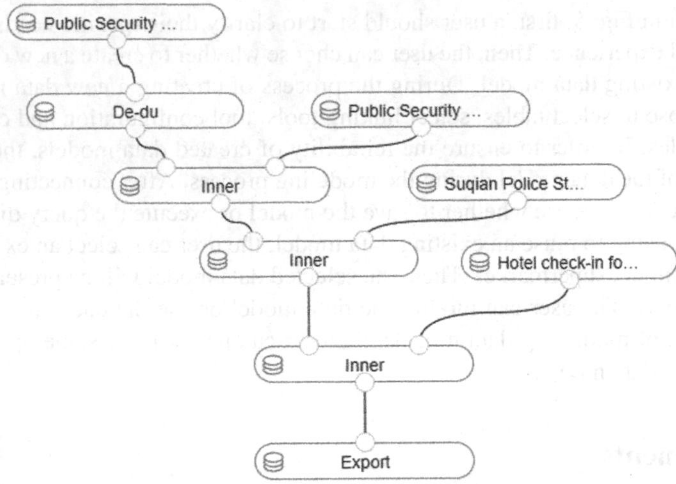

Fig. 6. The experimental data model.

An example of the data model is shown in Fig. 6. Through the investigation and analysis of police officers, we know that the person with higher crime rates at hotels are those involved in pornography and drugs. So, we only use these two types of criminal recorders as the results of the experiment to meet the actual needs.

4.2 Data Set

In the current study, we use the practical data from S2PSB. In the experiment, we use the hotel occupants in Suqian City of February 2022 as the dataset. There is a total of 298 657 records in the dataset. The data includes hotel information, identity information and check-in information of each hotel occupant.

In addition, we also used some other data tables as the input of the data model, including the basic information of personnel, hotel unit information, Sucheng Police Station Information and hotel check-in form in S2PSB. Considering the confidentiality of police data and to prevent data leakage, we do not display detailed data in experiments.

4.3 Experiment Results and Analysis

Table 7 shows the querying results of CPDMP and the traditional SQL-based querying method. For different types of persons with criminal records, the number of results obtained by CPDMP is higher than that of the traditional SQL-based querying method.

Experiment results show that CPDMP outperforms traditional SQL-based querying methods in the integrity of querying results. In practical applications, the traditional SQL-based querying method requires a combination of police officers and technicians to perform the query. On CPDMP, police officers can build data models and obtain querying results by themselves. Table 8 shows the statistics on answers to the questionnaire. Through the analysis of these answers, we found that police officers thought CPDMP was superior to traditional methods in terms of convenience. This also indicates that CPDMP outperforms the traditional SQL-based querying method.

Table 7. Experiment results.

Querying method	Pornography related	Drug related
Traditional SQL-based querying method	449	452
CPDMP	452	457

Table 8. The results of the questionnaire statistics.

Subjects	Results
Police officer 1	84.9
Police officer 2	83.5
Police officer 3	87.3
Police officer 4	85.2
Average value	85.2

5 Related Work

Big data analysis is a hot issue in many application areas. In the context of public security, Yu et al. [1] built a unified police big data analysis platform to provide strong support for the public security bureau in carrying out various police activities. Alic et al. [16] focused on data analysis in the field of public transportation and developed a platform "BIGSEA" that can process data under various constraints. Khorshidi et al. [17] presented how interpretable models can be constructed of police officer risk assessments and discussed issues of fairness that may arise when constructing models for police officer complaints and misconduct. Xi et al. [18] optimized the data visualization design model from the perspective of user experience, and combined human vision and cognitive laws to study the police system data visualization design strategy to guide the design practice. However, these methods are only simple data analyses and do not consider the deep-level characteristics of police data.

Many studies on knowledge model have been conducted with the goal of enhancing use of domain knowledge. Sitar-Taut et al. [19] proposed a recommender pipeline integrating domain-specific modeling, knowledge graphs, and a multi-criteria decision method that bridges the decision-makers' priorities with the customer-facing output. Zheng et al. [20] adopted a combination of top-down and bottom-up methods to build a knowledge model, and applied the knowledge model to the field of hazardous chemical management. Luo et al. [21] constructed a large-scale E-commerce Cognitive Concept Net named "AliCoCo", which is practiced in Alibaba, the largest Chinese e-commerce platform in the world. Donalds et al. [22] presented and illustrated a new cybercrime knowledge model that incorporates multiple perspectives and offered a more holistic viewpoint for cybercrime classification. However, these knowledge models do not provide a solution to the semantic inconsistency problem in police data. Considering the heterogeneity and sensitivity of policing data, a specific domain knowledge model (ontology) is necessary to be built, iteratively improved and customized extended.

Taking query expansion into consideration, Alfred et al. [23] focused on ontology-based query expansion methods in the field of agriculture. The method expands the original query by adopting new words and helps users to search for satisfactory results of queries. Jain et al. [24] proposed an information query method based on fuzzy ontology. The fuzzy ontology is constructed by using domain-specific knowledge. Based on the constructed fuzzy ontology, the most semantically relevant words to the query are identified and the query is expanded. However, there is still a lack of applications in the field of public security.

6 Conclusion

This paper presents the work of developing a customized police data modeling platform. First, after systematically analyzing the police data from 211 tables collected by a public security bureau branch, we propose a police domain knowledge model. Then, we detail the architecture and functions of the customized modeling platform mainly through six core modules. Next, we use a UML activity diagram to illustrate the using mechanism of the customized modeling platform. Finally, we evaluate the performance and prove the feasibility and integrity of CPDMP by carrying out experiments with police officers.

For the future work, we will focus on improving the scalability of CPDMP, continuously optimizing and improving the quality of PSDKM, and building more applicable models with CPDMP.

Acknowledgement. The authors would like to thank the policemen from Sucheng branch of Suqian Public Security Bureau for their cooperation and assistance. This work was partially supported by the Shandong Provincial Natural Science Foundation (No. ZR2021MF026).

References

1. Yu, H., Hu, C.: A police big data analytics platform: framework and implications. In: 2016 IEEE First International Conference on Data Science in Cyberspace (DSC), pp. 323–328. IEEE (2016)
2. Elgendy, N., Elragal, A.: Big data analytics in support of the decision making process. Procedia Computer Science **100**, 1071–1084 (2016)
3. Chong, D., Shi, H.: Big data analytics: a literature review. J. Manage. Analytics **2**(3), 175–201 (2015)
4. Silva, B.N., et al.: Urban planning and smart city decision management empowered by real-time data processing using big data analytics. Sensors **18**(9), 2994 (2018)
5. Che, D., Safran, M., Peng, Z.: From big data to big data mining: challenges, issues, and opportunities. In: International conference on database systems for advanced applications, pp. 1–15. Springer (2013). https://doi.org/10.1007/978-3-642-40270-8_1

6. Lum, C., Koper, C.S., Willis, J.: Understanding the limits of technology's impact on police effectiveness. Police Q. **20**(2), 135–163 (2017)
7. Chan, J.B.: The technological game: how information technology is transforming police practice. Criminal Justice **1**(2), 139–159 (2001)
8. Zhang, Y., Tang, X., Du, B., Liu, W., Pu, J., Chen, Y.: Correlation feature of big data in smart cities. In: International Conference on Database Systems for Advanced Applications. pp. 223–237. Springer (2016). https://doi.org/10.1007/978-3-319-32055-7_19
9. Kim, W.: Xrel: a path-based approach to storage and retrieval of xml documents using relational databases. ACM Trans. Internet Technol. (TOIT) **1**(1), 110–141 (2001)
10. Westerinen, A., Tauber, R.: Ontology development by domain experts (without using the "o" word). Applied Ontology **12**(3–4), 299–311 (2017)
11. Maedche, A., Staab, S.: Ontology learning for the semantic web. IEEE Intell. Syst. **16**(2), 72–79 (2001)
12. Azad, H.K., Deepak, A.: Query expansion techniques for information retrieval: a survey. Inf. Process. Manage. **56**(5), 1698–1735 (2019)
13. Zhao, Y., Dong, J., Peng, T.: Ontology classification for semantic-web-based software engineering. IEEE Trans. Serv. Comput. **2**(4), 303–317 (2009)
14. Kamsu-Foguem, B., Chapurlat, V.: Requirements modelling and formal analysis using graph operations. Int. J. Prod. Res. **44**(17), 3451–3470 (2006)
15. Carnaz, G., Nogueira, V.B., Antunes, M., Ferreira, N.: An automated system for criminal police reports analysis. In: International Conference on Soft Computing and Pattern Recognition, pp. 360–369. Springer (2018). https://doi.org/10.1007/978-3-030-17065-3_36
16. Alic, A.S., et al.: Bigsea: a big data analytics platform for public transportation information. Futur. Gener. Comput. Syst. **96**, 243–269 (2019)
17. Khorshidi, S., Carter, J.G., Mohler, G.: Repurposing recidivism models for forecasting police officer use of force. In: 2020 IEEE International Conference on Big Data (Big Data), pp. 3199–3203 (2020). https://doi.org/10.1109/BigData50022.2020.937817
18. Xi, Z., Chunyu, W.: Research on data visualization design for police system. In: 2021 IEEE International Conference on Artificial Intelligence and Industrial Design (AIID), pp. 463–468 (2021). https://doi.org/10.1109/AIID51893.2021.9456583
19. Sitar-Taut, D.A., Mican, D., Buchmann, R.A.: A knowledge-driven digital nudging approach to recommender systems built on a modified onicescu method. Expert Syst. Appl. **181**, 115170 (2021)
20. Zheng, X., Wang, B., Zhao, Y., Mao, S., Tang, Y.: A knowledge graph method for hazardous chemical management: ontology design and entity identification. Neurocomputing **430**, 104–111 (2021)
21. Luo, X., et al.: Alicoco: Alibaba e-commerce cognitive concept net. In: Proceedings of the 2020 ACM SIGMOD International Conference on Management of Data, pp. 313–327. SIGMOD'20, Association for Computing Machinery, New York, NY, USA (2020)
22. Donalds, C., Osei-Bryson, K.M.: Toward a cybercrime classification ontology: a knowledge-based approach. Comput. Hum. Behav. **92**, 403–418 (2019)
23. Alfred, R., et al.: Ontology-based query expansion for supporting information retrieval in agriculture. In: The 8th International Conference on Knowledge Management in Organizations, pp. 299–311. Springer (2014). https://doi.org/10.1007/978-94-007-7287-8_24
24. Jain, S., Seeja, K., Jindal, R.: A fuzzy ontology framework in information retrieval using semantic query expansion. Int. J. Information Manage. Data Insights **1**(1), 100009 (2021)

Integrated Bigdata Analysis Model for Industrial Anomaly Detection via Temporal Convolutional Network and Attention Mechanism

Chenze Yang, Bing Chen(✉), and Hai Deng

Nanjing University of Aeronautics and Astronautics, Nanjing, China
{ycz,cb_china}@nuaa.edu.cn

Abstract. Bigdata analysis has been the key to the abnormal detection of industrial systems using the Industrial Internet of Things (IIoT). How to effectively detect anomalies using industrial spatial-temporal sensor data is a challenging issue. Deep learning-based anomaly detection methods have been widely used for abnormal detection and fault identification with limited success. Temporal Convolutional Network (TCN) has the advantages of parallel structure, larger receptive field and stable gradient. In this work, we propose a new industrial anomaly detection model based on TCN, called IAD-TCN. In order to highlight the features related to anomalies and improve the detection ability of the model, we also introduce attention mechanism into the model. The experimental results over real industrial datasets show that the IAD-TCN model outperforms the traditional TCN model, the long short-term memory network (LSTM) model, and the bidirectional long short-term memory network model (BiLSTM).

Keywords: Anomaly detection · Big data · Industrial Internet of Things (IIoT) · Temporal convolutional network · Attention mechanism

1 Introduction

With increasingly wide deployment of 5G and the Industrial Internet of Things (IIoT), traditional process-based manufacturing industries, such as chemical, paper, steel, pharmaceutical, food and beverage, have begun to turn to intelligent manufacturing for more efficient mass production [1]. Since the beginning of the IIoT and big data eras, the data-driven optimization closed-loop analysis has been regarded as the key to realizing the value of the IIoT. In order to improve the profitability of factories via smart manufacturing, a large number of distributed monitoring sensors are used to collect multi-characteristic data to provide low-cost monitoring services [1]. Therefore, the amount of industrial data are becoming more diverse and growing exponentially. Furthermore, in the actual production environment, data acquisition devices such as sensors operate in harsh environments or some components could fail and invalid or even misleading data are generated as a result [2]. The erroneous data could negatively

B. Li et al. (Eds.): APWeb-WAIM 2022, LNCS 13421, pp. 150–160, 2023.
https://doi.org/10.1007/978-3-031-25158-0_12

affect the factory's products, but also could be used for fault detection through big data analysis. However, industrial big data generally have the characteristics of being correlated in both spatial and temporal domains, unevenly sampled and distributed, low value density and high complexity. Therefore, how to efficiently and accurately detect anomalies in industrial systems based on temporal-spatial IIoT data to reduce factory trial and error costs and reduce unplanned downtime, becomes particularly important and intriguing.

In recent years, industrial time series data analysis has attracted more and more attention, especially with the availability of the periodic operation of production lines and regular data collection in the industry. We can conduct data analysis to accurately identify whether the manufacturing systems generating the data are abnormal, or extract valuable diagnosis information by integrating and processing relevant data of the corresponding time period [3]. Anomaly detection in time series data generally refers to identifying problems that are inconsistent with expected or normal behavior. For example, the data at the current time point may be related to a previous time point or a time point long ago, known as short- and long-term time dependencies [4]. This dependence indicates that the occurrence of anomalies in sensor data at the current time point may also be related to the data of the previous time point due to system or component failures, so we can make good use of this feature for anomaly detection, by issuing early warnings and diagnosing the corresponding equipment, and preventing potential catastrophes from ever happening [5].

Traditional anomaly detection approaches are not effective due to the randomness and complexity of industrial data. For example, the distance-based anomaly detection algorithms [6], have a shallow architecture, and the detection performance is not satisfactory. Furthermore, several sequential predication models are used for processing sequential data, such as Markov models [7] and Kalman filtering [8], but they still lack the ability to learn long-term hidden correlations in the data. In recent years, the recurrent neural network (RNN) [9] was widely employed to solve the problem of long-term predication tasks, but due to the defects of its own structure, it is easy for gradient explosion or vanishing gradient to happen on time-series tasks that are longer [10]. This makes long-term predication of temporal dependencies difficult. With the introduction of Gated Recurrent Unit networks (GRU), Long Short-Term Memory (LSTM) networks somehow alleviate the vanishing and exploding gradient problems in RNNs [11]. At the same time, Bidirectional LSTM (BiLSTM), as an extension of recurrent neural network, has achieved some good results under certain circumstances [12]. But in a recent study, the work of Bai et al. [13] showed that for modeling long sequences, these methods are still difficult to capture extended sequence information, and proposed a novel general model for sequence modeling termed Temporal Convolutional Network (TCN). It is shown to have very stable gradients and flexible receptive fields, as well as good parallelism in the network.

In this paper, we propose a new detection model based on TCN, called Temporal Convolutional Networks for Industrial Anomaly Detection (IAD-TCN). Specifically, we use a TCN model to capture the correlations in the time series data, and then combine attention mechanisms for optimization. After getting a satisfactory detection model, we will test it on time-series anomalies in real industrial datasets. The main contributions of this article are given as follows:

1. We propose to adopt the TCN model to capture the temporal correlation with its powerful learning ability. Unlike RNN, the TCN model has good parallelism and the TCN convolution kernel is shared and the memory usage is low [13].
2. The model IAD-TCN overcomes some drawbacks of the traditional TCN model. The traditional TCN assigns equal weights to each input feature, and may bring a lot of negative effects and interference in anomaly detection. For effective abnormality detection, some more important features need to be emphasized in the model. Therefore, we embedded the attention mechanism [14] into the model to solve this problem. Through the attention mechanism, we can adjust the weights of different features in the data, according to their importances in the data, thereby improving the performance of the model. We validate the model on real industrial datasets, and the results show that our method outperforms the traditional TCN model, the BiLSTM model, the traditional LSTM model, and the simple RNN model.

The rest of this article is organized as follows. In Sect. 2, the relevant literature on anomaly detection based on industrial time series is reviewed. Section 3 gives the details of the proposed the model. Section 4 discusses the experimental results. Section 5 presents the conclusions of this paper.

2 Related Works

In this part, we will further discuss the related research of time series data anomaly detection in the industrial field, as well as their problems. Then the application of TCN model will be analyzed. Finally, the attention mechanism will be discussed on how to optimize the TCN model.

2.1 Anomaly Detection

Anomaly detection is defined as the process of identifying abnormal events or behaviors from normal time series. In the current research, there are several common anomaly detection methods: including 1) anomaly detection based on statistics; 2) anomaly detection based on similarity measure; and 3) anomaly detection based on deep learning.

Statistics-based anomaly detection is an early method in the field of anomaly detection. The basic process is to find the probability density distribution of the studied data, and then use consistency test to detect outliers according to the probability density distribution [15], but this type of method may have poor detection performance for high-dimensional data.

Anomaly detection methods based on similarity metrics mainly include the following: 1) methods based on distance metrics, such as KNN (K-Nearest Neighbor) algorithm [16]; 2) methods based on density metrics, such as LOF (Local

Outlier Factor) algorithm, The algorithm calculates an outlier factor for each point in the data set, and determine whether it is an outlier factor by judging whether the LOF is close to 1 [17]; 3) KDE kernel density estimation, by using a smooth peak function to fit the observed data points can be used to simulate the real probability distribution curve [18]; 4) clustering-based methods, such as K-means algorithm [19]; 5) tree-based partitioning methods, such as isolation forest, etc. [20].

Time series anomaly detection algorithms based on deep learning are mainly divided into two categories: supervised learning and unsupervised learning. Supervised learning relies on sufficient data labels to train the model parameters, and then classifies unknown samples according to the trained model, thereby judging whether the unknown samples are normal or abnormal [21]. Conversely, for the cases without sufficient training data, semi-supervised and unsupervised learning are often employed, avoiding the data labeling process [21].

2.2 Application of Temporal Convolutional Network

In previous research, recurrent neural network (RNN) was generally used to model time series problems, because the inherent recurrent autoregressive structure of RNN can represent time series well [10]. Traditional convolution neural networks are generally considered unsuitable for modeling time series problems, mainly due to the limitation of the size of their convolution kernels, and incapability to capture long-term dependency information well. However, recent studies have shown that specific convolutional neural network structures, such as the wavenet [22] proposed by Goolgle for speech synthesis can achieve good results as well.

Originally proposed as an extension of CNN, TCN was creatively used by Lea et al. for action segmentation and detection in the field of computer vision [23]. Yan et al. used TCN in weather prediction tasks and achieved good research results, leading to the discussion of TCN in time series prediction tasks [24]. Benefiting from the more flexible and longer receptive field size, TCN was recently demonstrated that it can outperform general recurrent architectures such as LSTM and GRU in different tasks [13]. Compared to the traditional CNN, TCN is able to integrate causal convolutions to preserve sequential information without leaking from the future to the past. TCN is also more flexible in processing sequences of arbitrary length and generating output sequences of the same length.

3 IAD-TCN Framework

3.1 Temporal Convolutional Network to Capture Time Series Dependence

In this part, we will introduce the structure of each part of TCN in detail.

Causal Convolution. The causal convolution is shown in Fig. 1. For the value of the previous layer at time t, it only depends on the value of the next layer at time t and before, so as to ensure that the input and output follow the temporal causal relationship. Unlike traditional convolutional neural networks, causal convolution cannot use future data. It is a one-way structure, and strictly time-constrained.

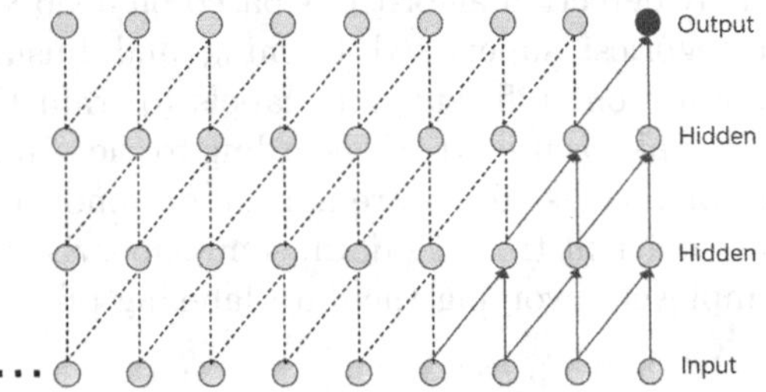

Fig. 1. A causal convolution

Dilated Convolution. Pure causal convolution cannot get rid of the problem of traditional convolutional neural networks, in which the length of time modeling is limited by the size of the convolution kernel. If we want to capture longer dependencies, more layers need to be stacked. The difference is that dilated convolution allows spaced sampling of the input during convolution, and the sampling rate is controlled by the spaced parameter d in Fig. 2.

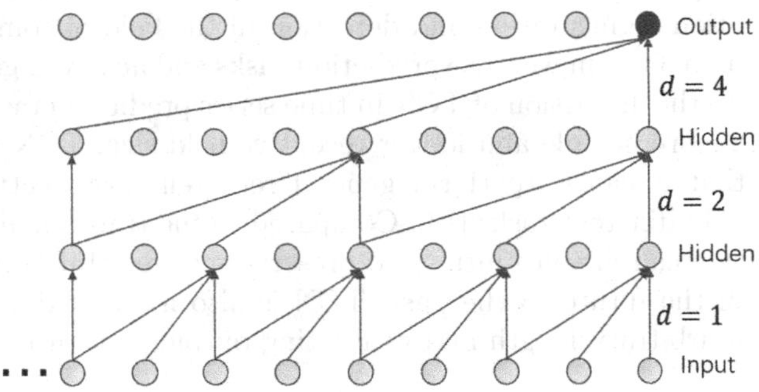

Fig. 2. An expansion causal convolution with an expansion factor of 1, 2 ,4

For example, $d = 1$ in the bottom layer means that every point is sampled, and $d = 2$ in the middle layer means that every 2 points are sampled at input.

In general, the higher the level, the larger the value of d. Increased dilated convolution makes the effective window size to grow exponentially with the number of layers. As a result, the convolutional network can obtain a large receptive field with fewer layers, making TCN have a strong ability to deal with time series problems.

Residual Connections. Residual connections have been proven to be effective in training deep networks, which performs better in dealing with the problem of gradient explosion and disappearance, by letting the entire network use skip connections to speed up the training process. For a residual block, it consists of two layers of convolution and nonlinear mapping, and weightNorm and dropout are added to each layer to regularize the network (Fig. 3).

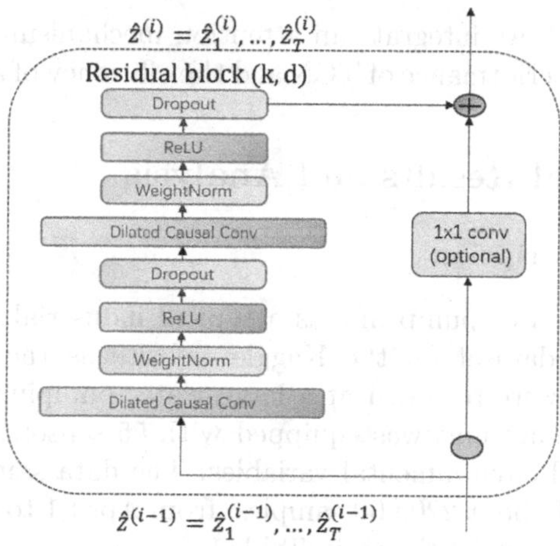

Fig. 3. Residual module of TCN

3.2 Attention Mechanism for Improved Model

In the learning of neural network, the more parameters of the model mean that the network is more powerful in modeling and dealing with data and information. By introducing the attention mechanism, the model can focus on the information that is more critical to the current task and improve the efficiency and accuracy of task processing. At present, the attention mechanism has been successfully applied in the fields of medicine, industry, machine translation and computer vision [25].

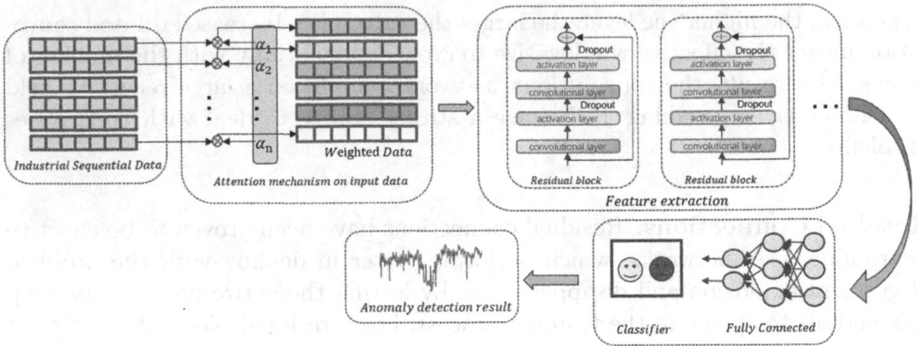

Fig. 4. IAD-TCN framework

As shown in Fig. 4, we integrate an attention mechanism into the TCN network to improve the performance of TCN and the efficiency of anomaly detection.

4 Experimental Results and Analysis

4.1 Datasets Description

The operation of a water pump unit is a typical industrial scenario. Take the water pump sensor dataset on the Kaggle website as the training and test datasets [26], which were recorded at a large water pumping station for urban water supply. The pump unit was equipped with 55 sensors to detect different unit components and environmental variables. The data sampling frequency is 1 min, with a total of about 220,000 samples, from April 1 to August 31, including health and fault data, as shown in Table 1.

Table 1. Original dataset description

Dataset	Features	Normal	Recovery	Broken	Total
Original	55	205836	14477	7	220300

4.2 Evaluation Metrics

In order to verify the anomaly detection performance of our proposed model, we use four metrics in precision, recall, F1 score and accuracy for performance evaluation. They are calculated as follows:

Precision is our most common evaluation metric. As shown in Eq. 1, precision is defined as how many of all the samples the model judges positive are true positives.

$$Pr = \frac{TP}{TP + FP} \tag{1}$$

As shown in Eq. 2, the recall rate represents the ratio of correctly detected positives to all actual positives.

$$Re = \frac{TP}{TP + FN} \tag{2}$$

In order to evaluate the pros and cons of different algorithms, taking into account Precision and Recall, the F1 score is used to evaluate Precision and Recall as a whole, as shown in Eq. 3.

$$F_1 = \frac{2Pr \cdot Re}{Pr + Re} \tag{3}$$

Accuracy is one of the most common evaluation metrics. As shown in Eq. 4, it represents the ratio of correctly classified records to the total records.

$$Acc = \frac{TP + TN}{TP + FN + FP + TN} \tag{4}$$

where TP, FP, TN, and FN represent the true-positive samples, the false-positive samples, the true-negative samples, and the false-negative samples, respectively.

4.3 Anomaly Detection Results

To evaluate the performance of our proposed algorithm model in industrial big data anomaly detection, we compare it with traditional TCN model, BiLSTM model, traditional LSTM model and simple RNN model, and the results are shown in Table 2.

In the Table 2 we can find useful information, IAD-TCN model gets the best precision, recall, F1 score and accuracy with 0.9817, 0.9686, 0.9751 and 0.9804. This shows that our proposed model is effective.

To further describe the evaluation metrics of the proposed algorithm model and illustrate the experimental results, a comparison with other algorithm models is plotted in Fig. 5. The results show that our proposed model, IAD-TCN, can perform better in anomaly detection for industrial time series data.

Table 2. Metrics comparison with different methods

Algorithm	Evaluation metrics (%)			
	Pr	Re	F1	Acc
IAD-TCN	0.9817	0.9686	0.9751	0.9804
TCN	0.9583	0.9365	0.9473	0.9496
BiLSTM	0.9379	0.9267	0.9323	0.9343
LSTM	0.9106	0.8957	0.9031	0.9279
Simple RNN	0.8794	0.8593	0.8692	0.8862

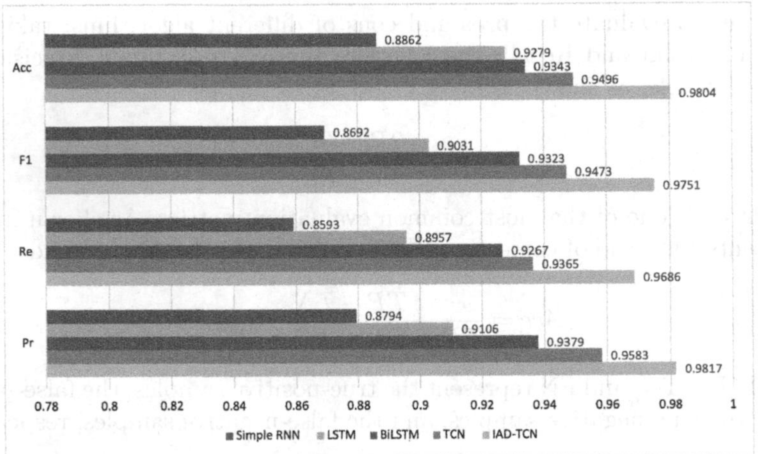

Fig. 5. Performance comparison

5 Conclusions

In this paper, we propose an IAD-TCN deep network model to detect industrial time series anomalies. We exploit the TCN model to capture correlations in time series data, and then integrate attention mechanisms into the model for optimization. The experimental results show that the model achieves better performance than other algorithms in anomaly detection of industrial time series data. For future work, we will further improve the structure of the model for better performance.

Acknowledgements. This work was supported in part by the National Natural Science Foundation of China, under Grant 62176122, 62001217; in part by A3 Foresight Program of NSFC, under Grant No. 62061146002; and in part by the Key Research and Development Program of Jiangsu Province, under Grant BE2019012.

References

1. Zhang, X., Ge, Z.: Local parameter optimization of LSSVM for industrial soft sensing with big data and cloud implementation. IEEE Trans. Industr. Inf. **16**, 2917–2928 (2019)
2. Wang, T., et al.: Big data cleaning based on mobile edge computing in industrial sensor-cloud. IEEE Trans. Ind. Inform. **16**, 1321–1329 (2019)
3. Siegel, B.: Industrial anomaly detection: a comparison of unsupervised neural network architectures. IEEE Sens. Lett. **4**, 1–4 (2020)
4. De Gooijer, J. G., Hyndman, R.J.: 25 Years of IIF time series forecasting: a selective review. Tinbergen Institute Discussion Papers (2005)
5. LSTM Learning With Bayesian and Gaussian Processing for Anomaly Detection in Industrial IoT (2020)
6. Shaikh, S.A., Kitagawa, H.: Efficient distance-based outlier detection on uncertain datasets of Gaussian distribution. World Wide Web **17**(4), 511–538 (2014)
7. Li, X., Gao, T., Chen, W.: Fuzzy dynamic Markov model for time series anomaly detection. In: 2021 IEEE Asia-Pacific Conference on Image Processing, Electronics and Computers (IPEC) IEEE (2021)
8. Xu, M., Han,, M.: Online prediction for multivariate time series by echo state network based on square-root cubature Kalman filter. In: Control Conference IEEE (2014)
9. Miljanovic, M.: Comparative analysis of recurrent and finite impulse response neural networks in time series prediction. Indian J. Comput. Sci. Eng. **3**(1), 180–191 (2012)
10. Pascanu, R., Mikolov, T., Bengio, Y.: On the difficulty of training Recurrent Neural Networks. JMLR.org (2012)
11. Nanduri, A., Sherry, L.: Anomaly detection in aircraft data using Recurrent Neural Networks (RNN). In: Integrated Communications Navigation & Surveillance IEEE (2016)
12. Sunny, M., Maswood, M., Alharbi, A.G.: Deep learning-based stock price prediction using LSTM and bi-directional LSTM model. In: 2020 2nd Novel Intelligent and Leading Emerging Sciences Conference (NILES) (2020)
13. Bai, S., Zico Kolter, J., Koltun, V.: An empirical evaluation of generic convolutional and recurrent networks for sequence modeling. arXiv preprint arXiv:1803.01271 (2018)
14. Bahdanau, D., Cho, K., Bengio, Y.: Neural machine translation by jointly learning to align and translate. Comput. Sci. (2014)
15. Zhao, Z., et al.: Research on time series anomaly detection algorithm and application. In: 2019 IEEE 4th Advanced Information Technology, Electronic and Automation Control Conference (IAEAC) IEEE (2019)
16. Knorr, E.M., Ng, R.T.: Finding intensional knowledge of distance-based outliers. Proc. VLDB **99**, 211–222 (1999)
17. Breunig, M.M., Kriegel, H.-P., Ng, R.T., Sander, J.: LoF: Identifying density-based local outliers. SIGMOD Rec. **29**(2), 93–104 (2000)
18. Rosenblatt, M.: Remarks on some nonparametric estimates of a density function. Ann. Math. Stat. **27**(3), 832–837 (1956)
19. Kumari, R., et al.: Anomaly detection in network traffic using K-mean clustering. In: 2016 3rd International Conference on Recent Advances in Information Technology (RAIT) IEEE (2016)

20. Liu, F.T., Ting, K. M., Zhou, Z.H.: Isolation forest. In: IEEE International Conference on Data Mining IEEE (2008)
21. Ren, H., El-Khamy, M., Lee, J.: Stereo disparity estimation via joint supervised, unsupervised, and weakly supervised learning. In: 2020 IEEE International Conference on Image Processing (ICIP) IEEE (2020)
22. Oord, A., et al.: WaveNet: a generative model for raw audio (2016)
23. Lea, C., Vidal, R., Reiter, A., Hager, G.D.: Temporal convolutional networks: a unified approach to action segmentation. In: Hua, G., Jégou, H. (eds.) ECCV 2016. LNCS, vol. 9915, pp. 47–54. Springer, Cham (2016). https://doi.org/10.1007/978-3-319-49409-8_7
24. Yan, J., et al.: Temporal convolutional networks for the advance prediction of ENSO. Sci. Rep. **10**(1), 8055 (2020)
25. Lasswell, H.D.: The world attention survey. Public Opin. Q. **3**, 456–462 (1941)
26. kaggle.pumpsensor. www.kaggle.com/datasets/nphantawee/pump-sensor-data

Advanced Database and Web Applications

WSNet: A Wrapper-Based Stacking Network for Multi-scenes Classification of DApps

Yu Wang[1,2,3], Gang Xiong[1,2], Zhen Li[1,2], Mingxin Cui[1,2], Gaopeng Gou[1,2], and Chengshang Hou[1,2(✉)]

[1] Institute of Information Engineering, Chinese Academy of Sciences, Beijing, China
{wangyu1996,xionggang,lizhen,cuimingxin,gougaopeng, houchengshang}@iie.ac.cn

[2] School of Cyber Security, University of Chinese Academy of Sciences, Beijing, China

[3] ZhongGuanCun Laboratory, Beijing, China

Abstract. Decentralized applications (DApps) are growing rapidly with the prevalence of blockchain, but security and performance issues plague network managers and developers. Encrypted network traffic classification (ETC) plays a fundamental role in application management, security detection, and QoS improvement and requires different granularity for different scenarios. Existing work focuses on a single scenario, and objects of them are traditional centralized applications (Apps). Since DApps use similar encrypted traffic settings and the same communication interface, the traffic is more complex than Apps. Under the premise of manual-design features, sophisticated architecture, or lots of training data, existing methods have good results, otherwise suffering from low accuracy. In this paper, we propose Wrapper-based Stacking Network (WSNet). According to traffic characteristics of different scenarios, WSNet adaptively selects optimal features for different algorithms to filter out irrelevant and redundant features without human intervention, thereby improving classification efficiency. Combining with stacking technology to integrate advantages of primary learners, hence it has good performance in complex traffic scenarios. Our comprehensive experiments on two real-world datasets show that WSNet adapts to and outperforms the state-of-the-art methods.

Keywords: Decentralized applications · Encrypted traffic classification · Wrapper · Stacking · Feature selection · Ensemble learning

1 Introduction

With the growth in usage of blockchain technology and resistance to censorship, DApps attract a lot of attention from people. Smart contracts act as backend of DApps, which are Turing-complete scripts running on a peer-to-peer (P2P)

B. Li et al. (Eds.): APWeb-WAIM 2022, LNCS 13421, pp. 163–179, 2023.
https://doi.org/10.1007/978-3-031-25158-0_13

network. Until now, a total of 3886 DApps are deployed on blockchain [1], such as Ethereum (75.1%), EOS (8.5%), which can be divided into decentralized gambling, finance, exchanges, etc. Since Ethereum has the largest DApps community with nearly 190 thousand daily active users, existing work takes DApps on Ethereum as the research object [9,10,13,14].

However, DApps are emerging as a new service paradigm and face similar problems as Apps, such as revealing plenty of sensitive information of users, security risks, and bad QoS. ETC is a good solution [7] and plays an important role in ensuring Apps security, stable operation of Apps, and Apps management. According to traffic priority policies, DApps traffic can be identified for better network management, and abnormal traffic can be detected to find behaviors of attack. In accordance with user's requirements, ETC can be divided into multiple scenarios, such as network service classification [6], traditional centralized application classification, including websites [3,11] and mobile application classification [2,7,8,12], user behaviors (UBs) classification [4,5,15], DApps classification [10,14], but these studies only focus on a single scene.

Unlike Apps, most DApps are deployed on the same platform to manage backend and data, and they have similar encrypted traffic settings and the same communication interface, increasing similarity of traffic. Although existing studies have achieved good results in different scenarios, machine learning models rely on fine-tuning features which are selected through expert knowledge, and deep learning models rely on sophisticated network architecture which is rarely used in real scenarios. Figure 1 shows that encrypted traffic has different characteristics in three scenarios. Compared with DApps traffic, user behavior traffic is more fine-grained and difficult to distinguish (i.e., the gap between classes is small, and intra-class samples are more scattered), resulting in different optimal feature subsets. The existence of redundant and irrelevant features will increase time cost of training and testing. Therefore, in multiple classification scenarios of DApps encrypted traffic, how to adaptively extract feature sets for different scenarios and identify samples with high precision are more challenging.

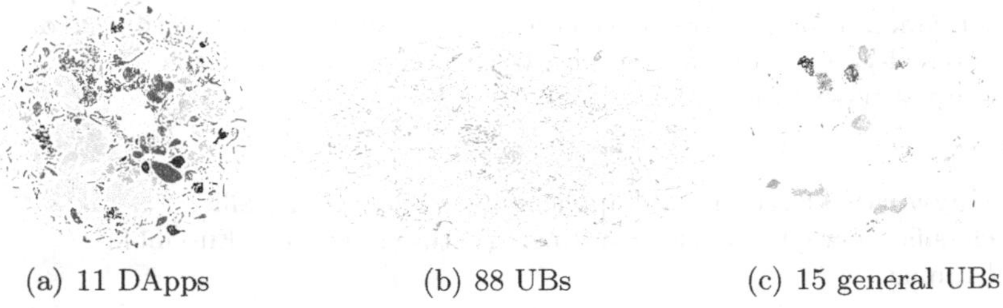

(a) 11 DApps (b) 88 UBs (c) 15 general UBs

Fig. 1. Traffic characteristics of different DApps classification scenarios

In this paper, we propose Wrapper-based Stacking Network, dubbed WSNet, an efficient model adapted to multiple classification scenarios of DApps encrypted traffic. Specifically, we first combine the features that are commonly used for traffic classification to form a base feature set. Then, WSNet selects appropriate features for models through wrapper-based feature selection algorithm according to traffic characteristics of different scenarios and combines stacking technology to integrate the advantages of primary learners, so that WSNet can achieve the best accuracy in multiple classification scenarios. Extensive experiments conducted on three benchmark traffic classification, i.e., DApps identification, UBs classification, and general UBs classification, with two real-world datasets to verify superiority of WSNet. Our contributions are as follows:

1) We present Wrapper-based Stacking Network (WSNet), an effective model for multiple classification scenarios of DApps encrypted traffic to meet user needs for different granularity. The model includes feature extraction and ensemble Learning.
2) We propose a base feature set that contains the features used by most of the existing encrypted traffic methods. It is fed to WSNet, thereby adaptively removing irrelevant and redundant features for different scenarios and reducing expert participation.
3) We conduct extensive experiments on two real-world datasets for three benchmark tasks. WSNet achieves outstanding results and outperforms the state-of-the-art methods.

Roadmap. We first review the related work in Sect. 2. Section 3 summarizes the preliminaries. The detailed design architecture of WSNet is presented in Sect. 4. We give the evaluation results and make a comprehensive comparison with existing methods in Sect. 5. Finally, we conclude this paper in Sect. 6.

2 Related Work

Recent studies on encrypted traffic classification resort to many different techniques. Here, we only focus on studies that are closely related to our work. Hence, prior work falls into three categories as shown in Table 1.

Table 1. Survey of encrypted traffic classification methods

Category	Refs	Scenarios	Features	Classifiers
Application classification	[3]	Website	The edit-distance	SVM
	[3]		Accumulated sum of packet sizes	SVM
	[11]		Packet direction/length sequence	CNN
	[6]	VPN services	Time related features	C4.5, KNN
	[8]	Mobile app	Certificate packet length	Markov model
	[2]		Packet sizes of the first 64 packets	Gaussian/Multinomial naïve bayes
	[12]		Statistical features of packet length	Random forest
	[7]		Flow sequence	Bi-GRU
User Behavior classification	[5]	iMessage	Sizes of exchanged packets	Linear regression
	[4]	Mobile app	Dynamic warping distance and cluster	Random forest
	[15]	WeChat	Packet length, statistics of burst	Random forest
DApps classification	[9]	DApps	Fusing time series, packet length, burst	Random forest
	[14]		4096-length payload	CNN
	[13]	User behavior	Statistical features, flow sequence	Random forest
	This paper	Multi-scenario	Base feature set	Stacking model

Application Classification. The target of this research category is to identify encrypted traffic generated by websites or mobile Apps. For identifying website, Cai et al. [3] leveraged a custom kernel based on an edit-distance that is fed to a Support Vector Machine (SVM). In [3], the author used accumulated sum of packet sizes to present a website fingerprinting method at Internet Scale. The research [11] designed sophisticated architecture of Convolutional Neural Networks (CNN), which used packet direction/length sequence to identify Tor websites. For classifying mobile Apps, Liu et al. [7] takes multi-layer bi-GRU encoder and decoder to learn features for 18 Apps identification. Recent studies extracted certificate packet lenght, statistical features of packet length, packet size to train models such as Markov [8], Random forest [12], and Naive bayse [2].

User Behavior Classification. The target of this research category is to identify fine-grained encrypted traffic generated by users. Condi et al. [4] clustered streams of UBs by clustering methods, and calculated the dynamic warping distance. Zhou et al. [16] used inference techniques to reveal UBs of Twitter. Yan et al. [15] extracted packet length, number of TCP handshakes to identify fund transfers and red packet transactions. In [5], the author utilized packet size to identify iMessage UBs, such as starting to write, stopping to write, sending messages, sending attachments.

DApps Classification. DApps are emerging as a new service paradigm that relies on blockchain [9]. As for DApps classification, Shen et al. [10] fused three sequences, timestamp, packet length, and burst sequence, to obtain high-dimensional features. Although this method increases accuracy, the time cost of training and testing is much longer because of the large input vectors. Wang et al. [14] constructed quadruplets that can obtain more constraints and used CNN to classify 61 DApps. As for multiple classification scenarios, Wang et al. [8] extracted features manually for different scenarios to obtain high accuracy, which requires expert knowledge and labor cost. In this paper, we attempt to use a wrapper-based stacking network structure without needing professional knowledge and fine-tuning features.

3 Preliminaries

Previous work has obtained some insights about single classification scenario of Apps encrypted traffic, but it is still a considerable challenge in multiple classification scenarios of DApps encrypted traffic. In this section, we describe DApps background, the definition of DApps ETC in multiple scenarios, and the limitation of existing methods.

3.1 DApps Background

Different from mobile or website applications deployed on centralized servers, the backend of Dapps, smart contracts, running on a decentralized P2P network, which is completely open-source, autonomous, and not controlled by a single entity, with a high degree of freedom. Most DApps can be visited through plug-ins via browsers (e.g., Firefox, Chrome). The way to access DApps is different from that of Apps, as shown in Fig. 2.

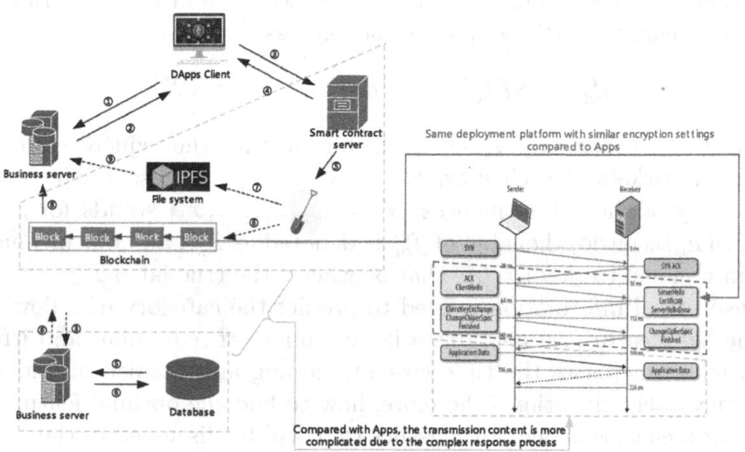

Fig. 2. The difference between the way to access DApps and Apps.

To initialize a visit, users request UI static files through DApps client to present the interface. Then clients send a transaction to the server equipped with the corresponding smart contracts, and its IP address can be got through DNS query. Call the smart contracts, judge the logic, and determine the result of each operation performed by the user in DApps. Further, miners package all data of these operations and results in blocks and store them in the corresponding distributed ledgers. Then, the list of blockchain or file system servers is obtained from the smart contract server, through which users can get the latest information to update the interface. The difference between Apps and DApps is that the business server sends a request to the application server and queries from the database. Finally, the record is transmitted to the business server through the application server to update the interface.

Therefore, compared with Apps, transmission content is more complicated due to complex response process. DApps on Ethereum employ similar traffic encryption settings, data and process run on the same blockchain platform, causing encrypted traffic of DApps more challenging to distinguish than that of Apps.

3.2 Problem Definition

The goal of the paper is to classify encrypted traffic of DApps in multiple scenarios, using the sequence information as the unique identifier of a flow. For different scenarios, flows can be classified into specific categories, such as application-level, fine-grained level of user behavior, and general user behavior level. An original flow can be represented as the same sequence formed by combining multiple types of information in different scenarios, including packet length information, packet arrival time information, statistical information, and specific information. Assume that there are a total of N samples, C scenes, and L_s categories in each scene. In c_1 scene, the m*th* sample can be expressed as:

$$f_m^{c_1} = \left[SEQ^1, \ldots, SEQ^v, S^1, \ldots, S^k, X\right] \tag{1}$$

where $SEQ^v = [seq_1^v, seq_2^v, \ldots, seq_r^v]$, which represents the sequence information of the first r packets of each flow, $S^k = \left[s_1^k, \ldots, s_b^k\right]$ represents the statistical characteristics of the kth sequences, $X = [x_1, x_2, \ldots, x_z]$ stands for z specific features. In c_1 scenario, the label of $f_m^{c_1}$ is denoted as $L_m^{c_1}$, and our ultimate goal is to train a model to predict $\hat{L}_m^{c_1}$ that is exactly the true label $L_m^{c_1}$.

However, if all information is used to predict the category of a flow in each scene, the base feature set contains a large number of redundant and irrelevant features, which increases the time cost of training and prediction and reduces the accuracy of classification. Therefore, how to find the optimal feature subset for different scenarios is one of the issues that will be discussed in the following.

3.3 Limitation of Existing Methods

The existing methods of ETC improve the final accuracy from all levels. At the data level, models can be fully trained by constructing a dataset containing a large number of samples (e.g., nearly one million flows in [7]). If there is not enough data, researchers increase the amount through data augmentation methods. At the feature level, different feature sets are constructed for different scenarios by manual extraction. Although a whole feature (e.g., 100-dimensional sequence of bidirectional packet length) can be constructed, only one or several features in it are helpful for the model's judgment. With expert knowledge, it is impossible to judge which feature is the most effective. In addition, if all feature sets need to be filtered manually for different scenarios, which consumes a lot of resources. Therefore, how to automatically select features is still a big challenge. At the model level, because deep learning can extract effective features, it becomes more popular. With the development of technology and the complex and changeable characteristics of scenarios, deep learning needs to design complex models to meet the needs of scenarios. However, due to the large memory consumption and high time complexity, it cannot be used in actual scenarios.

4 WSNet

In this section, we introduce wrapper-based stacking network (WSNet) to circumvent the problems raised in Sect. 3.3. The overall architecture of WSNet is shown in Fig. 3. We outline the underlying ideas of WSNet, including feature extraction, ensemble learning.

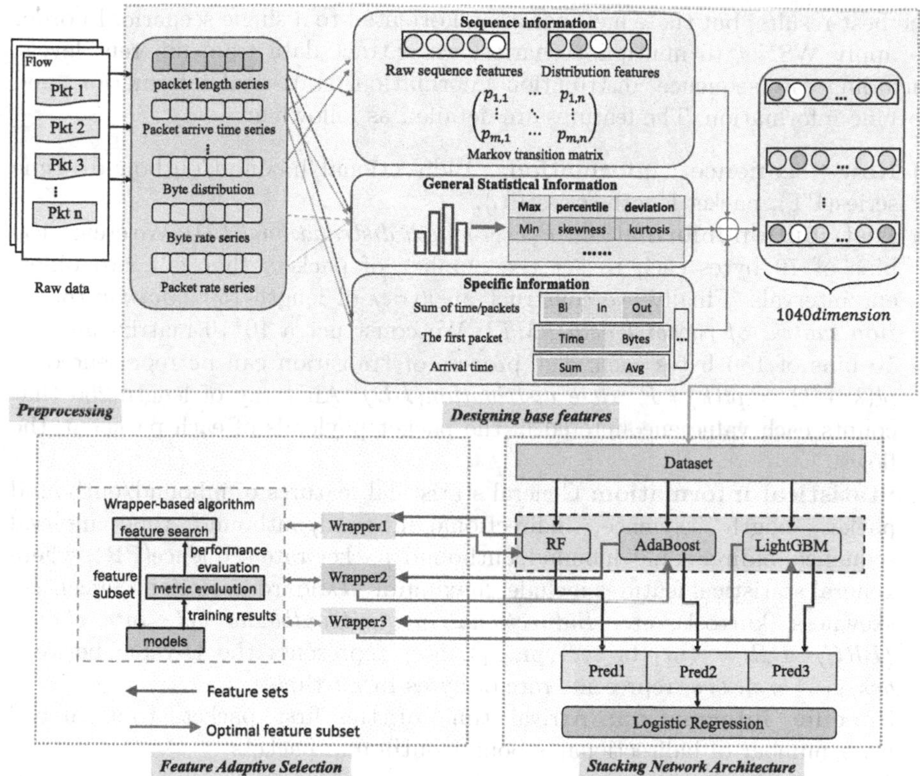

Fig. 3. The architecture overview of WSNet.

4.1 Feature Extraction

Although most researches use deep learning methods, the problems of complex models, large models, and large memory usage have not been completely solved, the methods are rarely used in practical scenarios. Therefore, we go back to the classics and use machine learning models. Whether it is a machine learning model or a deep learning model, features determine the upper limit of the final effect. It is necessary to select appropriate features for different scenarios through expert knowledge or extracting high-level features through feature fusion methods which undoubtedly increase the time cost. When the model needs to be updated due to data drift, features need to be filtered manually again.

By reviewing previous work in Sect. 2, we find that there are common features in different scenarios, and various combinations and fusions can be used to obtain the best results, but these methods are all oriented to a single scenario. In order to apply WSNet to multiple scenarios, we extract data from different levels, including raw sequence, distribution information, statistical information, and specific information. The features are detailed as following:

1) **Raw sequence information:** Bidirectional/inbound/outbound time series(PT), packet length series(PL).
2) **Distribution information:** *Packet length distribution(PLD):* We create 150 bins of 10 bytes each to get the number of packets that fall into different intervals. Finally, we construct an array of length-150. *Markov transition matrix of packet length(PMT):* We construct a 10*10 matrix, and set 10 bins of 150 bytes each. The process of transition can be represented as $pl(k+1) = pl(k) * P$. *Byte distribution(BD):* An array of length-256 that counts each value encountered in the packet payloads of each packet in the flows.
3) **Statistical information:** General statistical features of inbound/outbound packet length sequence, bidirectional/inbound/outbound time interval sequence, bidirectional/inbound/outbound packet rate sequence(PR), where general statistical features include: maximum, standard deviation, percentile, skewness, kurtosis, etc. *Bidirectional/inbound/outbound byte rate change (BRC):* $BR = (\mathrm{br}_1, \mathrm{br}_2, \ldots, \mathrm{br}_{n-1})$, brc_i represents the D-value between $brs_{i+1} - brs_i$, brs_i represents rate of bytes in Δt time.
4) **Specific information:** Arrival time of the first packet, total arrival time/number of bidirectional/inbound/outbound packets

After the combination of the above features, the encrypted flow m of DApps can be uniformly represented as follows:

$$f_m^{c_1} = [PL, PT, PLD, BD, PMT, S^{PL}, S^{\Delta PT}, S^{BRC}, S^{PR}, X] \quad (2)$$

Finally, $f_m^{c_1}$ has a total of 1040-dimensional features, which is the base feature set of m*th* flow. In order to automatically extract effective features for different scenarios to form optimal feature subsets, it is necessary to remove irrelevant and redundant features. It is impossible to manually determine which dimensional features are important features. Therefore, We introduce algorithms of feature selection, including filter-based algorithms, embedded-based algorithms, and wrapper-based algorithms.

1) *Filter-based algorithm:* Enhance the correlation between features and classes or weaken the redundancy between different features by evaluating criteria based on distance, information entropy, correlation, consistency, etc., which is independent of model. However, there is no guarantee that the size of the feature subset is the smallest, especially when there is a large correlation between the feature and the classifier; Secondly, its ability to select combined features is poor. Therefore, the attribute effects obtained by this algorithm cannot meet the needs of subsequent algorithms.

2) *Embedded-based algorithm:* By embedding it into the learning model, the objective function is optimized with the mechanisms of reward and punishment accuracy and the fixed number of features. However, the number of features needs to be determined in advance, which often cannot achieve the optimal feature subset, and the robustness is not high.
3) *Wrapper-based algorithm:* The evaluation of the optimal feature subset depends on the classification model. The selected feature subset is small in scale and has a high accuracy, which is suitable for scenarios with the requirements of high classification accuracy. The classification scenarios of DApps user behavior is complex and the traffic characteristics are different. Therefore, the Wrapper-based algorithm is finally selected, which can rely on different classification models for different scenarios. The algorithm determines the final feature subset as optimal feature subset according to the classification model.

4.2 Ensemble Learning

After feature extraction through wrapper-based algorithm, we choose stacking technology to process the encrypted data of DApps. The primary learner includes random forest, adaboost, and lightGBM algorithms, and the secondary learner is logistic regression.

1) *Random Forest* (RF) randomly selects decision trees to divide feature list. When the dimension of features is high, it still has high performance of training, and finally the minority obeys the majority to determine the predicted label. RF uses random samples, random features, and random subtrees, which has good stickiness, suitable for various application scenarios;
2) *Adaboost* (ADA) assigns weights to samples. The indistinguishable samples will become more important in the training process. The final weighted vote determines the predicted label.
3) *LightGBM* (LGBM) adopts gradient-based one-sided sampling strategy and mutually exclusive feature bundling strategy, which alleviates the complexity caused by large data and high feature dimensions. LGBM greatly improves performance while ensuring accuracy. Since the encrypted dataset of DApps is encoded by one-hot and other methods, many zeros are generated, and the problem of sparse features can be effectively solved by LGBM.

5 Experiments

In this section, we first introduce datasets, experimental settings. Then, we evaluate the effectiveness of WSNet by experimental results, ablation studies.

Table 2. The statistical information of 61 DApps

Categories	DApps (Number of flows)
Exchanges	1inch(2703), SushiSwap(1508), dYdX(1641), Curve(1460), Matcha(6690), Nash(2023), Mirror(2649), Tokenlon(2356)
Development	Enigma(1543), Rocket Pool(3822), Aelf(999), MyContract(1573)
Finance	Tether(1032), MakerDAO(3863), Nexo(1016), AaveP(4513), Paxos(3495), Harvest(4589), Ampleforth(926), Synlev(1041), BarnBridge(3108)
Gambling	Dice2win(1551), FunFair(1062), Edgeless(2059), Kingtiger(1129)
Governance	Kleros(1510), Decentral Games(1536), Iotex(1026), Aragon(3574)
Marketplace	Knownorigin(2009), Ocean Market(6462), OpenSea(2099), Superrare(1306), District0x(1011), Cryptokitties(4722)
Media	AVXChange(46282), Refereum(1074), Megacryptopolis(13924)
Game	Axie Infinity(1049), BFH(3415), Evolution Land(1516), F1deltatime(6119), Nftegg(1006), Pooltogether(2353)
Property	Decentraland(3932), FOAM(6558)
Social	Livepeer(1337), Loom Network(1073), Catallax(1219), 2key(3705)
High risk	DoubleWay(1045), E2X(1054), Gandhiji(1143), Forsage(3501), Minereum(4557)
Security	Chainlink(1173), Quantstamp Security Network(3203)
Storage	Storj(1020), Numerai(2107)
Identity	LikeCoin(1021), SelfKey(1521)

5.1 Dataset

We verify WSNet on two datasets, including ECML-CQNet [14] and Securecomm-88UBs [13], which are captured from a real-world network environment. ECML-CQNet was collected through the routers of a laboratory, and it only have one scene, concerning DApps encrypted traffic. After techniques of packet recombination and flow reduction, the dataset consists of 199+ thousand flows referring to top-61 DApps on Ethereum. Due to space limitations, please refer to the corresponding paper for details. The scale of ECML-CQNet is shown in Table 2.

Table 3. The statistical information of Multi-label dataset. UB represents User behavior. GUBL represents general user behavior labels.

ID	DApps labels	UB labels	Flows	GUBL	ID	DApps label	UB labels	Flows	GUBL
1	**Superrare**	open superrare	2819	1	45	**Ethlance**	open Ethlance	937	1
2		open market	4268	2	46		user homepage	659	8
3		select artwork	3018	3	47		look work	737	2
4		like artwork	2841	5	48		look worker	731	2
5		user homepage	1294	8	49		become employer	2123	6
6		view activities	1498	9	50		learn DApp	976	13
7		search	1855	7	51		different category	1926	7
8	**Editional**	open editional	1419	1	52		search	2142	7
9		learn DApp	1220	13	53		become employee	2118	6
10		select collectible	1238	3	54	**Knownorigin**	open Knownorigin	1578	1
11		artist homepage	1216	8	55		view gallery	1388	2
12		artist create	1207	8	56		select artwork	1714	3
13		artist collect	1072	8	57		like artwork	2504	5
14		view support page	1135	15	58		view activities	1592	9
15		search support	1329	7	59		browse all artists	1859	2
16		like article	1230	5	60		select artist	2761	3
17	**John Orion Young**	add to shopping cart	1428	10	61		search	1804	7
18		open market	3085	2	62	**Crowdholding**	open Crowdholding	2648	1
19		select joy	1350	3	63		select an article	2422	3
20		open shop	1342	2	64		comment	1343	12
21		refresh shopping cart	1585	14	65		agree article	2478	5
22		open john orion young	2719	1	66		user homepage	2118	8
23		view collector	2428	8	67		follow a person	2391	4
24	**Thomas Crown Art**	add to shopping cart	1443	10	68		create a project	2601	6
25		browse all artists	1117	2	69		view project	2749	3
26		browse all artworks	1251	2	70		search	3198	7
27		open the DApp	1264	1	71		learn DApp	2552	13
28		search	1491	7	72	**Latium**	open Latium	3800	1
29		view blog	1408	9	73		select task	3128	3
30		look shopping cart	1392	14	74		user homepage	3749	8
31		select artist	1341	3	75		look transaction web	6761	15
32		select artwork	1398	3	76		look my tasks	3885	15
33	**Cryptoboiler**	open the Cryptoboiler	1329	1	77		homepage	4632	15
34		view questions	1398	2	78		create a task	4290	6
35		comment post	2735	12	79	**Viewly**	open Viewly	6004	1
36		post a problem	1631	6	80		watch video	2882	11
37		like post	1832	5	81		user homepage	2420	8
38		user homepage	1662	8	82		look video transaction	2475	3
39		view blog	1678	9	83		follow person	2294	4
40		view question	1276	3	84		look game rank page	2887	15
41		search	1606	7	85		create new channel	1329	15
42	**Staybit**	open Staybit	3909	1	86		upload video to draft	1172	6
43		create payment	2706	6	87		search	3233	7
44		view contract	4632	3	88	**Staybit**	accept payment	4288	7

Securecomm-88UBs focuses more on fine-grained ETC of DApps, and contains 11 representative DApps, as a multi-label dataset. Different from the data collection method of ECML-CQNet, a total of 88 valid UBs were extracted. In order to achieve specific goals, it is necessary to perform the same sequence of actions as in the real world when collecting encrypted traffic generated by each user behavior. By classifying the user behavior traffic of each DApps into one category, we can obtain 11 categories of DApps traffic with coarser-grained. According to the service type, ignoring the difference between DApps, it can be divided into 15 categories of general UBs. Securecomm-88UBs contains a total of 192+ thousand encrypeted traffic flows, as shown in Table 3.

According to the proportion of each class, we randomly select packets, and divide training, validation, and test according to 7:1:2. We preprocess each flow by removing ACK and retransmission packets.

5.2 Experiments Settings

Methods to Compare. In order to have a comprehensive understanding on the effects of WSNet, we compare it with the following models:

FFP [10] fuses packet length, burst,and time series to the high-dimensional feature through kernel functions for DApps ETC with random forest classifier.

FS-net [7] uses multi-layer bi-GRU encoder and decoder to learn features, aiming to classify 18 applications.

Appscanner(AppS) [12] combines statistical features of packet length, such as standard deviation, percentile of bidirectional, inbound, and outbound flows. It uses random forest to classify encrypted traffic of mobile applications.

DeepFingerprinting(DF+L) [13] uses the information of packet length sequence, and sends it into CNN to classify the encrypted flows of websites.

RF+LT [13] captures packet length and time series. Random forest classifier is utilized to identify 11 DApps in multi-scenarios.

CQNet [14] constructs quadruplets which can obtain more constraints and uses CNN for classifying 61 DApps.

Metrics. Six common metrics are adopted for evaluation, including Accuracy(Acc), Precision(Pre), Recall, F1-score and Receiver Operating Characteristic Curve(rocAUC). The two datasets have the phenomenon of data imbalance, so Balanced accuracy (balAcc) is used, which avoids inflated performance.

5.3 Experimental Results

Evaluation of WSNet. We evaluate WSNet on the two datasets, as shown in Fig. 4. Table 4 shows the number of features is fed to each primary learner, we have the following findings from the results:

Table 4. The number of features selected by each primary

	ECML-CQNet	Securecomm-88UBs 11 DApps	Securecomm-88UBs 15 general UBs	Securecomm-88UBs 88 UBs
RF	150	150	288	500
Adaboost	150	150	250	500
LGBM	150	288	473	658

1) In the two datasets, no matter in the training set or the test set, the various metrics of WSNet are optimal compared to primary learners, indicating that WSNet concentrates the advantages of the primary learner on itself through the ensemble learning, stacking technology, so that it has optimal results.
2) For primary learners, it can be found that the effects of the three learners in multiple scenarios, RF is better than Adaboost and LightGBM. RF, as the representative of the Bagging algorithm, has high robustness due to randomness. So that it can be flexibly applied to various scenarios.
3) WSNet can adapt to traffic characteristics according to different scenarios, and select more effective features for complex traffic scenarios to obtain higher classification accuracy by capturing more complete information.
4) In each scene, we set the minimum number of features, and WSNet adaptively increases the number of features according to the difficulty in different scenes, so as to achieve the highest final accuracy and obtain the optimal feature subset to reduce the time cost of training and prediction.

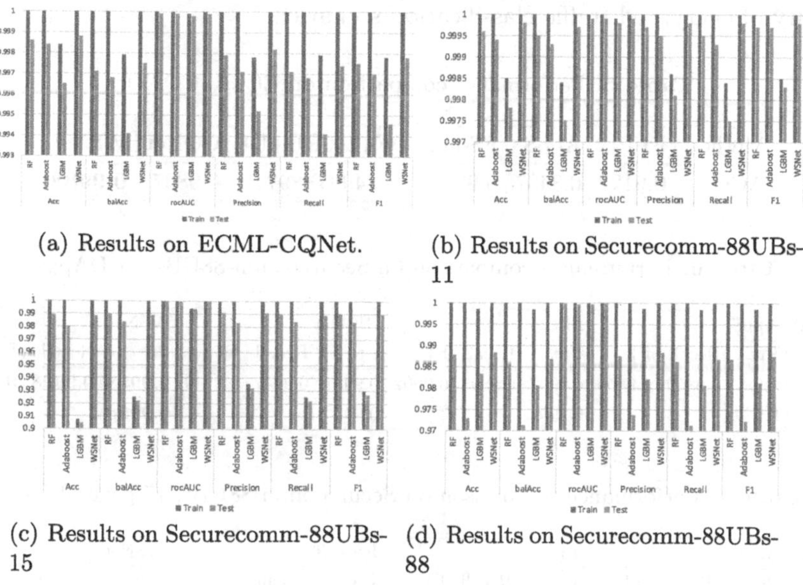

(a) Results on ECML-CQNet.

(b) Results on Securecomm-88UBs-11

(c) Results on Securecomm-88UBs-15

(d) Results on Securecomm-88UBs-88

Fig. 4. Evaluation of WSNet

Performance Comparison. For different scenarios, we choose different comparison methods, the comparison results are shown in Table 5, Table 6, Table 7, Table 8. We can obtain the following conclusions:

1) Overall, WSNet achieves the best performance on Precision, Recall, and F1. For the same scene, encrypted traffic classification of DApps, the final accuracy rate reached 99.9% on the two datasets of ECML-CQNet, Securecomm-UBs 11 DApps. In the other two classification scenarios of fine-grained user

behavior and general user behavior, the accuracy rate also exceeds 98%. The results also show that user behavior classification scenarios are more difficult to detect than DApps classification scenario.

2) In the same scenario, the encrypted traffic classification of DApps, ECML-CQNet dataset contains 6 times as many kinds as Securecomm-UBs 11 DApps, but the final classification accuracy only drops by 0.1%. Compared with experimental results of AppS, FFP, and RF+LT in these two datasets, when the categories are increased, the accuracy rates drop by 14.07%, 7.79%, and 9.73%, respectively. It indicates that our WSNet can better handle more categories, select the optimal feature subset according to the complexity of the traffic, and integrate the advantages of multiple models through stacking technology to achieve the highest classification accuracy.
3) In more complex scenarios of traffic classification, 88 UBs and 15 general UBs of Securecomm-UBs, the performance of AppS and FFP is greatly degraded, AppS is the method proposed for traditional applications, while FFP is proposed for DApps. The two methods cannot be well adapted to more fine-grained classification scenarios, However, WSNet can used to various granularity of encrypted traffic classification scenarios.

Table 5. Performance comparison on ECML-CQNet

Method	FFP	AppS	FS-net	DF+L	RF+LT	CQNet	WSNet
ACC	0.9122	0.8437	0.9611	0.7849	0.8977	0.9837	**0.9988**

Table 6. Performance comparison on Securecomm-88UBs, 11 DApps

Method	AppS			FFP			RF+LT			WSNet		
	Pre	Recall	F1	Pre	Recall	F1	Pre	Recall	F1	Pre	Recall	F1
Value	0.987	0.9852	0.9859	0.9907	0.9885	0.9902	0.9951	0.9947	0.9949	**0.9998**	**0.9998**	**0.9998**
ACC	0.9844			0.9901			0.995			**0.9998**		

Table 7. Performance comparison on Securecomm-88UBs, 15 general UBs

Method	AppS			FFP			RF+LT			WSNet		
	Pre	Recall	F1	Pre	Recall	F1	Pre	Recall	F1	Pre	Recall	F1
Value	0.6539	0.5727	0.5648	0.6292	0.5794	0.5635	0.9875	**0.9867**	0.9871	**0.9883**	**0.9867**	**0.9875**
ACC	0.5485			0.5357			0.9857			**0.9883**		

Table 8. Performance comparison on Securecomm-88UBs, 88 UBs

Method	AppS			RF+LT			WSNet		
	Pre	Recall	F1	Pre	Recall	F1	Pre	Recall	F1
Value	0.7027	0.6541	0.6664	0.9617	0.9565	0.9553	**0.9909**	**0.9887**	**0.9897**
ACC	0.6808			0.9566			**0.9884**		

5.4 Ablation Studies

Scale of Feature Set on ECML-CQNet. To evaluate the contribution of the feature selection component of WSNet, we conduct the following experiments on ECML-CQNet dataset. We force WSNet to select 150-dimensional, 250-dimensional, and 500-dimensional features for each classifier, named as WSNet-150, WSNet-250, and WSNet-500. The results are shown in Fig. 5:

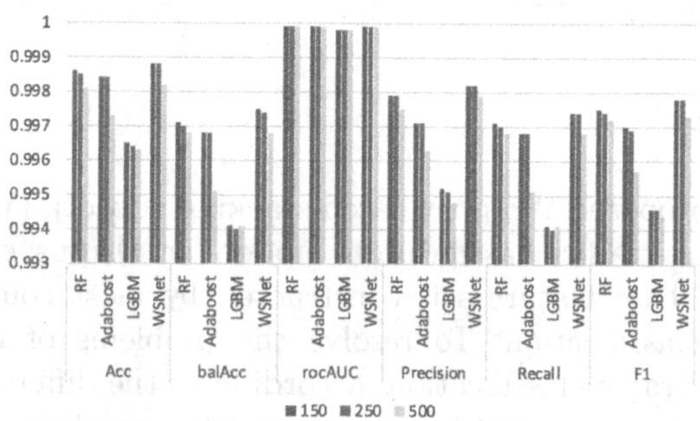

Fig. 5. Performances by different features scale on ECML-CQNet.

1) **Irrelevant features.** The effect of WSNet-150 is the same or slightly better than that of WSNet-250, indicating that the additionally extracted 100-dimensional features are less helpful for the classification of WSNet.
2) **Redundant features.** Comparing WSNet-150 and WSNet-500, the performace of WSNet-150 are better that WSNet-500, indicating that the information containing by these features has played a reverse guiding role in the training process of model, so WSNet can remove redundant features.

Scale of Feature Set on Securecomm-UBs. We also conducted ablation experiments on Securecomm-UBs. The experimental settings are that the features selected by WSNet in each scene and the WSNet using 1040-dimensional features, named as WSNet and WSNet-1040, respectively. Table 9 shows the results. Regardless of primary learners or the secondary learner, WSNet performs the best performance except RF in 11 DApps classification and LGBM in 15 general user behavior classification. It indicates that for different scenarios, the component of feature selection deletes redundant and irrelevant features that are difficult to remove manually, thereby improving the accuracy of classifiers. For RF of 11 DApps classification, WSNet is only 0.01% lower than WSNet-1040. For LGBM of 15 general UBs classification, WSNet is slightly inferior to WSNet-1040 by 1.27%, considering that LGBM needs more features to obtain more comprehensive hidden information.

Table 9. Performances by different features scale on Securecomm-UBs.

ACC	11 DApps		88 Ubs		15 general Ubs	
Methods	WSNet-1040	WSNet	WSNet-1040	WSNet	WSNet-1040	WSNet
Random Forest	**0.9991**	0.999	0.9801	**0.9878**	0.9824	**0.9892**
Adaboost	0.9987	**0.9996**	0.9624	**0.972**	0.9699	**0.9799**
LightGBM	0.9976	**0.9981**	0.9775	**0.9835**	**0.9188**	0.9061
Logistic Regression	0.9992	**0.9998**	0.9825	**0.9884**	0.9814	**0.9883**

6 Conclusion

In this paper, we proposed Wrapper-based stacking network (WSNet), an effective multi-scenarios traffic classification model for DApps encrypted traffic. WSNet includes a base feature set, constructed by most common features of encrypted traffic classification. To resolve the problems of multi-scenario of DApps encryption traffic classification, according to the different traffic characteristics, WSNet utilizes wrapper-based feature selection algorithm to adaptively extract optimal feature subsets and can be easily extended with other classification scenarios. Secondly, WSNet integrates the advantages of multiple models, including random forest, adaboost, and LightGBM, so as to achieve the highest accuracy. We verify the effect of WSNet on two datasets collected in the real-world, showing that WSNet outperforms the state-of-the-art methods in terms of six metircs.

Acknowledgements. This work is supported by The National Key Research and Development Program of China (No.2020YFB1006100, No.2020YFE0200500 and No.2018YFB1800200) and Key research and Development Program for Guangdong Province under grant No. 2019B010137003.

References

1. State of the DApps 2022. https://www.stateofthedapps.com/
2. Alan, H.F., Kaur, J.: Can android applications be identified using only TCP/IP headers of their launch time traffic? In: Proceedings of the 9th ACM Conference on Security and Privacy in Wireless and Mobile Networks, pp. 61–66 (2016)
3. Cai, X., Zhang, X.C., Joshi, B., Johnson, R.: Touching from a distance: website fingerprinting attacks and defenses. In: Proceedings of the ACM Conference on Computer and Communications Security (CCS), pp. 605–616 (2012)
4. Conti, M., Mancini, L.V., Spolaor, R., Verde, N.V.: Analyzing android encrypted network traffic to identify user actions. IEEE Trans. Inf. Forensics Secur. **11**(1), 114–125 (2016)
5. Coull, S.E., Dyer, K.P.: Traffic analysis of encrypted messaging services: apple imessage and beyond. Comput. Commun. Rev. **44**(5), 5–11 (2014)
6. Draper-Gil, G., Lashkari, A.H., Mamun, M.S.I.: Characterization of encrypted and VPN traffic using time-related features. In: Proceedings of the 2nd International Conference on Information Systems Security and Privacy, pp. 407–414 (2016)

7. Liu, C., He, L., Xiong, G.: FS-Net: a flow sequence network for encrypted traffic classification. In: IEEE INFOCOM 2019 - IEEE Conference on Computer Communications (INFOCOM), pp. 1171–1179 (2019)
8. Shen, M., Wei, M., Zhu, L., Wang, M.: Classification of encrypted traffic with second-order markov chains and application attribute bigrams. IEEE Trans. Inf. Forensics Secur. **12**(8), 1830–1843 (2017)
9. Shen, M., Zhang, J., Zhu, L., Xu, K., Du, X.: Accurate decentralized application identification via encrypted traffic analysis using graph neural networks. IEEE Trans. Inf. Forensics Secur. **16**, 2367–2380 (2021)
10. Shen, M., Zhang, J., Zhu, L., Xu, K., Du, X., Liu, Y.: Encrypted traffic classification of decentralized applications on ethereum using feature fusion. In: Proceedings of the International Symposium on Quality of Service. IWQoS, pp. 18:1–18:10 (2019)
11. Sirinam, P., Imani, M., Juárez, M., Wright, M.: Deep fingerprinting: undermining website fingerprinting defenses with deep learning. In: Proceedings of the 2018 ACM SIGSAC Conference on Computer and Communications Security (CCS), pp. 1928–1943 (2018)
12. Taylor, V.F., Spolaor, R., Conti, M., Martinovic, I.: AppScanner: automatic fingerprinting of smartphone apps from encrypted network traffic. In: IEEE European Symposium on Security and Privacy, EuroS&P, pp. 439–454 (2016)
13. Wang, Y., Li, Z., Gou, G.: Identifying DApps and user behaviors on ethereum via encrypted traffic. In: Security and Privacy in Communication Networks (2020)
14. Wang, Y., Xiong, G., Liu, C.: CQNet: a clustering-based quadruplet network for decentralized application classification via encrypted traffic. In: Machine Learning and Knowledge Discovery in Databases, pp. 518–534 (2021)
15. Yan, F., et al.: Identifying wechat red packets and fund transfers via analyzing encrypted network traffic. In: TrustCom/BigDataSE, pp. 1426–1432 (2018)
16. Zhou, X., Demetriou, S., He, D., Naveed, M., Pan, X.: Identity, location, disease and more: inferring your secrets from android public resources. In: ACM SIGSAC Conference on Computer and Communications Security. (CCS), pp. 1017–1028 (2013)

HaCache: A Hybrid Adaptive Cache for Persistent Memory Based Key-Value Systems

Lixiao Cui, Gang Wang, Yusen Li, and Xiaoguang Liu(✉)

TKLNDST, College of Computer Science, Nankai University, Tianjin, China
{cuilx,wgzwp,liyusen,liuxg}@nbjl.nankai.edu.cn

Abstract. Many previous studies have proposed high-performance key-value systems based on Persistent memory (PM). However, these work ignores the fact that the read performance of PM is also lower than DRAM. In this paper, we propose HaCache, a well-designed hybrid cache for PM-based KV systems to improve read performance. HaCache combines key-value (KV) cache, key-pointer (KP) cache, and Block cache to retain the advantages of the three types of cache schemes. It can adaptively adjust the partition of cache space among the three cache schemes to adapt to workload changing. The evaluation results show that HaCache outperforms pure KV cache, pure KP cache, and pure Block cache by 2.7x, 2x, and 18% respectively.

Keywords: Cache · Key-value system · Persistent memory

1 Introduction

The emerging byte-addressable persistent memory (PM) [4] provides opportunities to design high-performance persistent key-value (KV) systems. There are many previous works that aim to design persistent KV systems [6,7,9,10,12] dedicated for PM. These works mainly focus on optimizing write operations. However, according to recent analyses of Intel Optane PM [4], the read performance of PM is only 1/3 of DRAM [13], which implies that optimizing the read performance of PM-based KV systems is also important. Cache technology is suitable for read optimization of PM-based KV systems.

There are usually two types of read operations of PM-based KV systems, point lookup (`GET`) and range query (`SCAN`). To accelerate point lookup, we can cache *Key-Value* pairs (KV cache) or cache *Key* and *Pointers* to associated values (KP cache). Compared with caching key-pointer pairs, caching key-value pairs can improve overall read performance more significantly. However, KP cache is space-efficient because it can save more DRAM space than KV cache.

This research is supported in part by the NSF of China (No. 62141412, 61872201), the Science and Technology Development Plan of Tianjin (20JCZDJC00610, 19YFZCSF00900), the Fundamental Research Funds for the Central Universities, the Graduate Student Scientific Research Innovation Project of Tianjin (2021YJSB082).

B. Li et al. (Eds.): APWeb-WAIM 2022, LNCS 13421, pp. 180–189, 2023.
https://doi.org/10.1007/978-3-031-25158-0_14

For ordered KV systems, range query is also an important operation, we could cache *Blocks* (Block cache) to accelerate it. These three cache schemes have their advantages and disadvantages. We aim to design a cache solution able to combine the virtues of performance-efficient, space-efficient, and support for range query. In this paper, we propose **HaCache**, **H**ybrid **A**daptive **Cache**, for PM-based KV systems. HaCache is a general cache solution to optimize the read performance of PM-based KV systems (with hash index, B+tree index, and so on), which contains point cache (include KV cache and KP cache) and block cache to serve point lookups and range queries respectively.

With limited space, it is difficult to combine three caches to achieve all their advantages. So the cache partition is important. HaCache follows a top-down cache partition strategy. The high-level space partition is between point cache and block cache. It determines which kind of read operation (point lookup and range query) HaCache is more advantageous. In different scenarios, the performance requirements for these two operations may be different. Therefore, we allow users to set preferences for different read operations through certain metrics (e.g. latency) and partition cache to meet the preference. The low-level space partition is between KV cache and KP cache. In this part, we partition cache to achieve higher performance with smaller space. Since the workloads usually change with time [2], the main challenge of HaCache is how to dynamically adjust cache space to achieve the purposes under changing workloads. For the high-level partition, HaCache leverages Proportional-Integral-Derivative (PID) control algorithm to dynamically adjust the cache space partition, so HaCache can be stable around the metrics set by users. For the low-level partition, we proposed KV adaptive caching (KV-AC) algorithm motivated by Adaptive Replacement Cache (ARC) [8]. To achieve the best partition, when adjusting the capacity, we take the size, access frequency, and benefit of cache items into consideration.

We evaluate HaCache using real PM devices. Under pure `GET` workload, compare with `KV cache` and `KP cache`, HaCache improves the throughput by up to 2.7x and 2x respectively. Under mixed `GET-SCAN` workload, HaCache can satisfy user's preference precisely and outperforms `Block cache` by 18%.

2 Background and Motivation

2.1 PM-Based KV Systems

Similar to DRAM, persistent memory device is attached to the memory bus and can be accessed in byte granularity. Meanwhile, it can guarantee data persistence after the power is off. The emergence of PM provides an opportunity to make traditional in-memory KV systems [2] persistent while maintaining high performance. In the past few years, dozens of works are proposed to optimize PM oriented KV systems [6,7,9,10,12]. However, these works mainly optimize the write performance but ignore the fact that the read performance of PM is slow than DRAM. We measure the read latency of two PM-based KV systems (CCEH [9] and FAST&FAIR [7]). For `GET` operation, the latency of KV systems in PM is about 3x of DRAM alternatives. For `SCAN` operation, the performance gap is widened to 5 times. These observations imply the necessity of optimizing the read performance of PM-based KV systems.

2.2 KV, KP and Block Cache

To figure out the impact on KV systems of different cache schemes, we evaluated KV, KP and, block cache in a B+tree (i.e. FAST&FAIR) based KV system. The results of GET are shown in Table 1. With a large cache, the performance of KV cache is better than KP cache because no extra PM accesses are needed. With a small cache, KP cache has a higher hit ratio than KV cache because of the smaller cache item. A very low hit ratio leads to the poor performance of KV cache, even though every KV cache hit has no further overhead. **Motivation 1: Combining KV cache and KP cache to maximum GET performance.**

Table 1. GET throughput and hit ratio under different cache sizes. The key size is 8B and the value size is 100B. We use YCSB [3] workload with 50M key-value pairs.

	Throughput (M ops/s)			Hit ratio		
Cache size	KV	KP	Block	KV	KP	Block
200MB	1.04	2.48	0.6	58.2%	92.7%	84.8%
500MB	1.33	3.75	0.63	69.1%	100%	88.5%
1GB	2.03	3.75	0.65	82.4%	100%	90.6%
2GB	7.48	3.75	0.7	100%	100%	93.1%

Table 2. The performance under mixed GET -SCAN workload. The cache size is 1GB and we allocate the space to KV cache and Block cache in different proportions.

KV/Block cache ratio	0/1	1/7	1/1	7/1	1/0
Throughput (K ops/s)	201	218	213	217	189
GET latency (ns)	1798	1241	889	641	541
SCAN latency (ns)	12210	12270	13448	13811	16640
Latency ratio (SCAN/GET)	6.79	9.88	15.12	21.54	30.75

SCAN is another important read operation. We measured the performance of a simple hybrid cache that contains KV cache and block cache under mixed GET-SCAN workload. The number of SCAN is 30% of total operations. Table 2 shows the results. With block cache, the SCAN latency is significantly reduced compared to using KV cache only. Another observation is the performance gap (latency ratio) between GET and SCAN changes drastically varying different KV/Block cache ratios. Different application scenarios may favor different read operations. For example, users could require the GET to be processed as soon as possible, but relax the requirements for SCAN. Therefore, the respective performance of different read operations should be considered. **Motivation 2: Balancing the performance of GET and SCAN when handling mixed GET-SCAN workloads.**

3 Related Work

Optimizing Read for PM-Based KV Systems. The representative work of using cache is Bullet [6]. It maintains a KV cache in DRAM to accelerate GET. Some studies store partial of indexes into DRAM to reduce the overhead of lookup. FPtree [10] stores the leaf nodes of B+tree into PM while placing the internals into DRAM. HiKV [12] uses a hybrid index composed of hash table and B+tree. The hash table is placed at PM and the B+tree is placed at DRAM.

Hybrid Cache. SwapKV [5] is a hotness aware KV caches that places the metadata and hot data to DRAM and stores cold data to PM. Cassandra [1] is an LSM-tree-based system. It uses KP cache and KV cache to replace the block cache. The users can configure the size of different cache statically.

Adaptive Cache Algorithms. Some previous studies use multiple caching schemes and adaptively adjust the capacity of different caches. The Adaptive Replacement Cache (ARC) [8] is the most representative. ARC is designed for page cache. It partitions the cache into a frequency cache and a recency cache. A page that is first accessed will be placed into the recency cache. If the pages in the recency cache are accessed again, they are promoted into the frequency cache. ARC uses two ghost cache regions to help space partition of the recency cache and the frequency cache adaptively. The ghost recency cache and the ghost frequency cache keep the metadata of recently evicted pages from the two real caches respectively. A ghost cache hit implies that the hit item should not be evicted, therefore, the corresponding real cache should be expanded to accommodate more cache items.

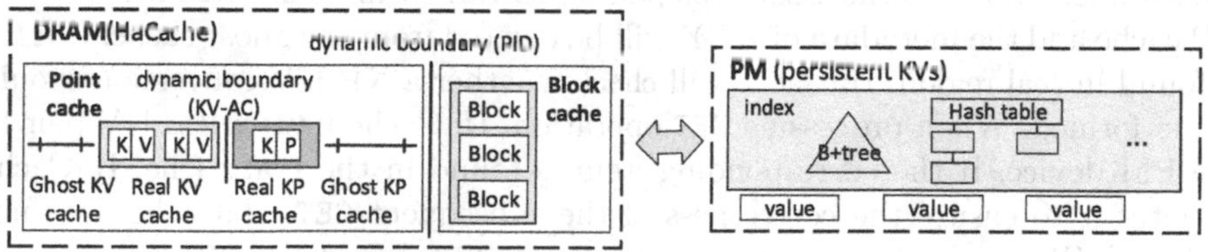

Fig. 1. Overview of HaCache.

4 Design and Implementation

Figure 1 shows the structure of HaCache. Following motivation 1, we design a KV-AC algorithm inspired by ARC, to adjust the cache capacity between KV and KP cache. The two ghost cache is used to guide the movement of the dynamic boundary. Following motivation 2, a dynamic boundary adjusted by PID algorithm is responsible for capacity partition between point and block cache.

4.1 Basic Operations of HaCache

Cache Replacement. HaCache uses CLOCK rather than LRU as the cache replacement algorithm, which uses less space (1-bit metadata for each cache item). To make the hit ratio of CLOCK close to LRU, we develop Frequency-based CLOCK for real cache. It selects several cache items (default 2 items) as the candidate evicted items at a time. The candidate with minimal access frequency is evicted.

Conversion Between KV and KP Caches. The items fetched from PM will be inserted into KP cache first. When an item in the KP cache is accessed and the corresponding CLOCK bit is active (implying that the item has been accessed recently), the KP item is converted into a KV item. Note that for the same key, only one cache item is stored in the KP cache or the KV cache. An item evicted from the KV cache is considered cold and converted to a KP item. To guide the adjustment of cache capacity, HaCache inserts the metadata of evicted items to ghost cache. For an evicted item that has been converted to a KV item, the corresponding metadata will be inserted into the ghost KV cache. Otherwise, the metadata of it will be inserted into the ghost KP cache.

Processing of Commands. When processing `GET` operation, The real KV cache and the real KP cache are searched first. If *KEY* (denote the requested key) does not exist in both of them, HaCache will examine the ghost KV cache and the ghost KP cache. If *KEY* exists in the ghost region, the corresponding real region will be expanded according to the KV-AC algorithm (Sect. 4.2). Then the persistent KV system will search *KEY* and return the associated value directly. Meanwhile, *KEY* and the associated pointer to value will be inserted into the real KP cache and the metadata of *KEY* will be deleted from the ghost cache. If *KEY* is found in real region, HaCache will check whether a KP-KV conversion should be performed. When processing `SET` operation, HaCache first writes KV pair to the PM device. If the corresponding item is found in the real cache, HaCache updates it to ensure the correctness of the subsequent `GET`. HaCache does not lookup *KEY* in ghost region when processing `SET`. When the `SET` operation is applied to an existing KV pair (i.e. update operation), the value is modified but the key is not. The items in the ghost region do not contain values or pointers, so there is no need to update the items in the ghost region when processing `SET`.

4.2 Adaptive Adjustment Within Point Cache

The purpose of point cache is to combine the performance efficiency of KV cache with the space efficiency of KP cache. HaCache comprehensively consider the space and performance of different cache items. HaCache uses the *adjustment factor* F to weigh the performance and space. Every cache item has its factor. The variable k denotes the access count of the cache item, b presents the benefit if it is cached and s is its size.

$$F = (\log(k) + 1)(b/s) \tag{1}$$

Among these parameters, the benefit of caching is difficult to determine. We define the benefit to be reduced latency (i.e. latency difference between DRAM cache and PM for the same item). However, for different sizes of cache items, the saved latency is different. To address these issues, we develop a *sampling method* to estimate benefits for specific KV systems. For KV cache items, we sample the average latency of acquiring KV pairs from PM and HaCache under different value sizes. Then we use piecewise linear regression (PLR) to estimate the relation between value size and latency reduction. In this way, the saved latency can be estimated for each size. We measure the latency reduction in a B+tree-based KV system is roughly linearly related to the value size. For KP cache items, the benefit is independent of value size. We measure the latency of accessing pointers from PM and HaCache respectively.

HaCache uses KV adaptive caching (KV-AC) algorithm to manage the capacity partition of point cache. It uses *adjustment factor* as the granularity of capacity adjustment. C_{point} presents the overall point cache capacity. $[KV_{real}]$ and $[KP_{real}]$ represent The maximum size of the real KV cache and KP cache. $[KV_{ghost}]$ and $[KP_{ghost}]$ represent the maximum size of the ghost KV cache and KP cache, assumed that all of the items were converted to real KV/KP items. KV-AC always maintains Eq. 2. Same with ARC, the ghost cache size in KV-AC is the difference between overall size and corresponding real cache size.

$$[KV_{real}] + [KP_{real}] = [KV_{real}] + [KV_{ghost}] = [KP_{real}] + [KP_{ghost}] = C_{point} \quad (2)$$

If the requests miss in the real cache but hit in the corresponding ghost cache, the cache capacity will be adjusted. For example, if a key is found in the ghost KV cache, the real KV cache will be expanded. Then the new size of the real KV cache should be $[KV_{real}] + nF$, where F is the *adjustment factor* and n is an adjustable coefficient. Correspondingly, the new size of the real KP cache should be $[KP_{real}] - nF$. The value of n determines the magnitude of capacity adjustment. After the new real KV cache size is determined, the subsequent requested items will be brought to KV cache without evicting until $[KV_{real}]$ reaches the new size. Meanwhile, some items will be evicted from the real KP cache until $[KP_{real}]$ is smaller than or equal to its new size. After the real cache size is updated, $[KV_{ghost}]$ and $[KP_{ghost}]$ are also adjusted according to Eq. 2. The processing is similar when a key is found in the ghost KP cache.

4.3 Adaptive Adjustment Between Point Cache and Block Cache

In HaCache, the capacity partition between block cache and point cache will affect the performance of `GET` and `SCAN` directly. To balance the optimization of `GET` and `SCAN`, HaCache allows users to set preferences for different read operations to adapt to different application scenarios. In our implementation, the metric of preference is the average latency ratio of `SCAN` and `GET`, that is, $Ratio = Latency_{\texttt{SCAN}}/Latency_{\texttt{GET}}$. The larger the *Ratio*, the more conducive to `GET`, otherwise it is conducive to `SCAN`. To meet an expected ratio from users, HaCache uses PID to adjust the cache partition. Eq. 3 shows the PID algorithm,

$E(T_k)$ presents the error at time T_k. PID algorithm contains proportional, integral and derivative terms. K_p, K_i, and K_d are tuning factors for the three terms respectively. When using PID, we need to set a target stable state. Then we measure the error ($E(T_k)$) between the current state and the target state. By bringing $E(T_k)$ into the equation, we get the output value $U(T_k)$ to adjust the controlled system to approach and stabilize at the target state.

$$U(T_k) = K_p E(T_k) + K_i \sum_{n=0}^{T_k} E(n) + K_d(E(T_k) - E(T_{k-1})) \tag{3}$$

We use the user's expected latency ratio ($Ratio_{setting}$) as the target stable state. We measure the average latency of GET and SCAN over a period of time (default every 100k operations) and calculate the real *Ratio*. Then E_k is calculated ($E_k = Ratio_{setting} - Ratio_{real}$). We try different values and finally decide to set K_p to 100k, K_i to 5k, and K_d to 10k. Putting the above parameters into Eq. 3, we can get $U(T_k)$. If $U(T_k)$ is positive, the real *Ratio* is usually less than the user's setting. In this case, the latency of GET is too large. Therefore, HaCache allocate more space for point cache and the new size of it should be $C_{point} + U(T_k)$. On the contrary, if $U(T_k)$ is negative, HaCache allocates more space for block cache and shrinks KP and KP cache. By using PID, the *Ratio* will be stable on the user's setting latency ratio.

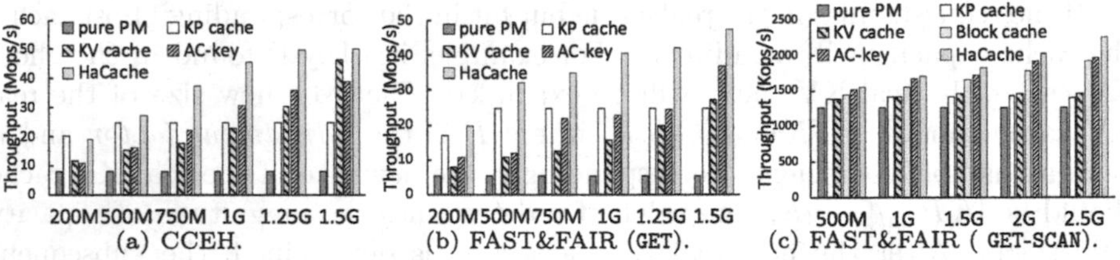

Fig. 2. Throughput under different cache sizes (Byte).

5 Evaluation

The evaluation was conducted on an x86-64 server that has two Intel Xeon Gold 5220 processors, 128 GB of DRAM and 512GB of Intel Optane DC PM. The size of key is 8B and value is 100B. The total number of KV pairs is 50M. We use YCSB [3] to generate the skewed workloads (0.99 skewness). We conduct pure GET workload and mixed GET-SCAN workload (30% SCAN and 70% GET, the range of SCAN is a random number less than 100.). We select CCEH (a persistent hashing) [9] and FAST&FAIR (a persistent B+tree) [7] as the underlying KV systems. The adjustable coefficient n is set to 500. The default latency ratio between SCAN and GET is set to 15:1. We compare HaCache with pure PM, KV cache, KP cache, and Block cache. We also compare AC-key [11], which uses a hierarchical ARC algorithm to adjust capacity among KV, KP, and block cache.

5.1 Impact of Cache Size

Figure 2 (a) and (b) show the `GET` throughput under 8 threads. We first compare HaCache and `KP cache`. When the cache size is small, HaCache tends to allocate most of the cache size to store KP items to guarantee a high hit ratio. When cache size is large, the effectiveness of caching KV items becomes obvious. In CCEH, HaCache outperforms `KP cache` by 1.1x to 2x. We then turn to the comparison between HaCache and `KV cache`. When the cache size is large, HaCache tends to store more KV items because it can bring more benefits. When cache size becomes small, HaCache can maintain a high hit ratio by adjusting the partition between KV items and KP items. In FAST&FAIR, the throughput of HaCache is 2.1x to 3x of `KV cache`. HaCache also outperforms `AC-key` significantly. This mainly because HaCache takes the access frequency into account to decide capacity partition and cache replacement, which can keep frequently accessed items in cache to improve the overall performance. Figure 2(c) shows the throughput of mixed `GET-SCAN` workload. Under different cache sizes, HaCache outperforms `KV cache` by 11% $\sim$ 54% and `KP cache` by 12% $\sim$ 62%. HaCache perform better than `Block cache` because it leverages point cache to serve `GET`. The throughput gap between HaCache and `Block cache` is 7% $\sim$ 18%. Meanwhile, HaCache can control the real latency ratio close to the setting ratio. The measured real latency ratio is 15.13:1.

5.2 Adaptive Adjustment

We conduct two changed workload to verify the effectiveness of HaCache's adaptive adjustment strategy. The first is a two-phase skewness-changed workload on CCEH-based HaCache. We first execute the workload with 0.99 skewness, then change the skewness to 0.88. The total cache size is 1GB. We plot the changes of the KV cache size and the KP cache size in Fig. 3(a). When the skewness is 0.99, most of the accesses are concentrated on a small amount of KV pairs. Therefore, HaCache keeps the frequently accessed KV items to get high performance. When the skewness decreases (100Mops in Fig. 3(a)), HaCache allocates more capacity to keep KP items (KP items are space-efficient) to maintain the high hit ratio. When the capacity partition becomes stable, the throughput of HaCache achieves 28.95 Mops/s, which is almost the same as the throughput of the workload that initially uses 0.88 skewness.

The second evaluation is a `GET-SCAN` ratio changed workload. We conduct it on FAST&FAIR-based HaCache. In the beginning, the number of `SCAN` is 30% of total operations. Then we turn to conduct a pure `GET` workload. As shown in Fig. 3(b), after workload changing, the block cache shrinks quickly. When the cache partition becomes stable (after 150Mops), the cache capacity distribution is almost the same as it that initially conducts pure `GET` workload. The above results confirm that HaCache could adapt to the workload's changes.

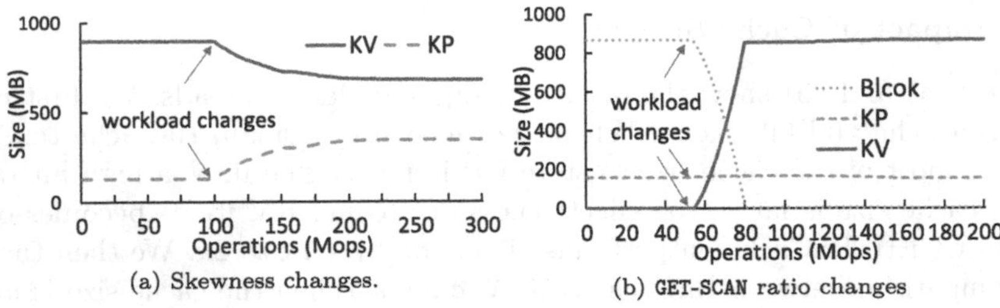

(a) Skewness changes.

(b) GET-SCAN ratio changes

Fig. 3. Adaptive adjustment when workload changes.

6 Conclusion

In this paper, we propose HaCache, which incorporates key-value, key-pointer, and block cache schemes, to optimize the read performance for PM-based KV systems. HaCache employs PID algorithm and KV-AC algorithm to dynamically adjust the capacity allocation among KV, KP, and block parts. Our HaCache outperforms other single cache schemes in different read operations.

References

1. Apache: Cassandra. http://cassandra.apache.org/
2. Chen, J., et al.: HotRing: a hotspot-aware in-memory key-value store. In: USENIX Conference on File and Storage Technologies (FAST 2020), pp. 239–252 (2020)
3. Cooper, B.F., Silberstein, A., Tam, E., Ramakrishnan, R., Sears, R.: Benchmarking cloud serving systems with YCSB. In: Proceedings of the 1st ACM Symposium on Cloud Computing, pp. 143–154. SoCC 2010 (2010)
4. Corporation, I.: Intel(R) Optane(TM) DC Persistent Memory. https://www.intel.com/content/www/us/en/architecture-and-technology/optane-dc-persistent-memory.html
5. Cui, L., et al.: SwapKV: a hotness aware in-memory key-value store for hybrid memory systems. IEEE Trans. Knowl. Data Eng. **35**(1), 1–1 (2021)
6. Huang, Y., Pavlovic, M., Marathe, V.J., Seltzer, M., Harris, T., Byan, S.: Closing the performance gap between volatile and persistent key-value stores using cross-referencing logs. In: USENIX Annual Technical Conference (USENIX ATC 2018), pp. 967–979 (2018)
7. Hwang, D., Kim, W.H., Won, Y., Nam, B.: Endurable transient inconsistency in byte-addressable persistent B+-tree. In: USENIX Conference on File and Storage Technologies (FAST 2018), p. 187 (2018)
8. Megiddo, N., Modha, D.S.: ARC: a self-tuning, low overhead replacement cache. In: USENIX Conference on File and Storage Technologies (FAST 2003), p. 9 (2003)
9. Nam, M., Cha, H., ri Choi, Y., Noh, S.H., Nam, B.: Write-optimized dynamic hashing for persistent memory. In: USENIX Conference on File and Storage Technologies (FAST 2019), pp. 31–44 (2019)
10. Oukid, I., Lasperas, J., Nica, A., Willhalm, T., Lehner, W.: FPTree: a hybrid SCM-DRAM persistent and concurrent B-tree for storage class memory. In: Proceedings of the 2016 International Conference on Management of Data (SIGMOD 2016), p. 371–386 (2016)

11. Wu, F., Yang, M.H., Zhang, B., Du, D.H.: AC-Key: adaptive caching for LSM-based key-value stores. In: USENIX Annual Technical Conference (USENIX ATC 2020), pp. 603–615 (2020)
12. Xia, F., Jiang, D., Xiong, J., Sun, N.: HiKV: a hybrid index key-value store for DRAM-NVM memory systems. In: USENIX Annual Technical Conference (USENIX ATC 2017), pp. 349–362 (2017)
13. Yang, J., Kim, J., Hoseinzadeh, M., Izraelevitz, J., Swanson, S.: An empirical guide to the behavior and use of scalable persistent memory. In: USENIX Conference on File and Storage Technologies (FAST 2020), pp. 169–182 (2020)

An Energy-efficient Routing Protocol Based on Two-Layer Clustering in WSNs

Feng Xu[1,2], Qian Ni[1(✉)], and PanFei Liu[1]

[1] College of Computer Science and Technology/College of Artificial Intelligence Nanjing University of Aeronautics and Astronautics, Nanjing, China
niqian_nuaa@163.com

[2] Collaborative Innovation Center of Novel Software Technology and Industrialization, Nanjing, China

Abstract. The design of energy-efficient routing protocol for large-scale wireless sensor network (WSN) has become a challenge nowadays. In order to balance the energy consumption of nodes and prolong network lifetime, this paper proposes an energy-efficient routing protocol DKB-DCERP based on two-layer clustering. It uses an improved DPC-MND clustering algorithm to cluster nodes at first layer. Then, in each cluster, first-level cluster heads (CHs) are elected according to residual energy of the node and distance to base station (BS). After that, using all the first-level CHs as second layer, K-Means algorithm is used to cluster, and the second-level CHs are elected according to remaining energy of the node and distance between the node and center of mass and BS. Finally, path weight of the second-level CHs is designed according to energy, distance, deflection angle and other factors. On the basis of Dijkstra algorithm, the theory of immune algorithm is introduced to obtain the optimal inter-cluster multi-hop routing. The simulation result shows that DKBDCERP has excellent performance in balancing node energy consumption and prolonging network lifetime.

Keywords: Wsn routing algorithm · Two-level clustering · Immune algorithm · Dpc-mnd

1 Introduction

WSN is a distributed sensor network, which has ushered in a broad stage of historical development. People can use WSN to obtain a large amount of high-density and high-accuracy information at any time and place. In particular, WSN are widely used in many scenarios including intrusion detection, weather monitoring, security and tactical reconnaissance [1].

Routing protocol is an important component of WSN and plays a key role in the collection and transmission of data. Sensor nodes are inherently energy-starved in general, which has led to the formation of innovative techniques to limit any unnecessary energy dissipation. However, factors such as the uneven

B. Li et al. (Eds.): APWeb-WAIM 2022, LNCS 13421, pp. 190–198, 2023.
https://doi.org/10.1007/978-3-031-25158-0_15

distribution of sensor nodes and the continuous expansion of the network scale have led to some gaps and deficiencies in the application scope of most studies [4]. It will bring great challenges to the design of large-scale WSN routing protocols.

2 Related Work

To optimize performance in WSN, researchers have concentrated more on clustering and routing techniques [6]. Heinzelman.WR et al. epoch-making proposed an energy constrained protocol named LEACH [2] , which proposed the idea of clustering in WSN. Li Jianzhou et al. proposed EBCRP [5] based on LEACH, which replaces the random factor with the distance when selecting CHs. Unfortunately, in the case of uneven distribution of nodes, CHs in above protocols may be concentrated around BS, which makes energy hole near BS. Clustering result will determine the performance of routing protocol by applying classical clustering algorithm [7]. Zhang Haiyan et al. proposed a K-Means-based clustering routing protocol KBECRA [8]. Hou Yating et al. proposed a low-energy security-enhanced routing protocol based on DBSCAN [3]. Unfortunately these methods still perform poorly on network performance.

In order to optimize the clustering routing protocol of WSN, this paper proposes the energy-efficient routing protocol DKBDCERP (DPC-MND and K-Means based double clustering energy-efficient routing protocol). The improved DPC-MND clustering algorithm is used to alleviate the energy hole near BS. Dual-CHs structure is improved, that is, multiple first-level CHs correspond to one second-level CHs to disperse the energy consumption of CH. The multi-hop transmission method is used to plan the inter-cluster routing, and path weight of each node is constructed from the distance, deflection angle and residual energy.

3 Description of DKBDCERP

DKBDCERP runs in rounds, and each round mainly includes six stages: first-level cluster establishment, election of the first-level CH, second-level cluster establishment, election of the second-level CH, inter-cluster routing establishment and stable data transmission. The first-level CH is selected from the first-level cluster, which is used to receive and merge the data sent by ordinary nodes. Then send it to the second-level CH, which is used to receive and merge the data from the first-level CH. After inter-cluster routing is established, each second-level CH sends the data packet to BS through multi-hop transmission. The two-layer clustering structure is shown in Fig 1. Clustering in each round will affect the overall efficiency for the reason of large resource consumption. Therefore, in DKBDCERP, the first-level cluster is established once every ten rounds, and the second-level cluster is established once every round.

3.1 First-level Cluster

DPC-MND (Density Peaks Clustering based on Mutual Neighborhood Degrees) calculates the local density of the node distribution while finding the density peak

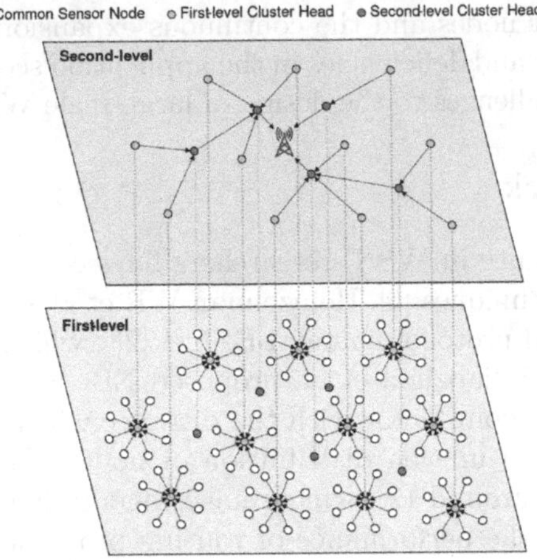

Fig. 1. Double layer clustering structure diagram

point, and then adds k nodes adjacent to the density peak point to the cluster. The nodes in the cluster are extended and recursively clustered according to the concept of mutual proximity, and can be clustered dynamically according to the node distribution density. The idea of recursive clustering is obviously inappropriate in WSN so that DPC-MND has been modified and adapted. The introductions of main definitions in improved DPC-MND are as follows:

Local density ρ**:** The local density of each sensor node x_i is the relative density in its local range, and its local density ρ_i is defined as follows:

$$\rho_i = \frac{\sum_{j=knn(i)} \sum_{v=knn(j)} d_{vj}^2}{2 \cdot k \cdot \sum_{j=knn(i)} d_{ij}^2} \tag{1}$$

where d_{ij} is the Euclidean distance from x_i to x_j, and $knn(i)$ is the set of k neighboring points of x_i.

Relative distance δ**:** The relative distance between a sensor node x_i and other nodes refers to the distance of the node whose local density is greater than itself and closest to itself. Its definition is as follows:

$$\delta_i = \max_{i \neq j} (\delta_j) \tag{2}$$

$$\delta_i = \min_{j:\rho_j > \rho_i} (d_{ij}) \tag{3}$$

Formula (2) indicates that the node with the highest local density does not have a node with higher local density, and it is artificially set as the highest.

Decision value γ**:** The decision value is to judge whether the node can be used as an indicator of the density peak. In DKBDCERP, its definition is as follows:

$$\gamma_i = \frac{\rho_i \cdot \delta_i}{d_{toSink}} \tag{4}$$

Proximity ω: Proximity ω_{ij} is defined by the distance index between sensors. The definition is as follows:

$$\omega_{ij} = \begin{cases} e^{-\frac{d_{ij}^2}{\theta^2}}, & j \in knn(i) \\ 0, & \text{others} \end{cases} \tag{5}$$

where $\sigma^2 = \frac{1}{N \cdot k} \sum_{i=1}^{N} \sum_{j \in knn(i)} d_{ij}^2$ and N is the total number of nodes in the network.

Relative proximity *deg***:** The above-mentioned proximity only considers the global information of the node, but local information needs to be considered to achieve the effect, so the relative proximity is defined as follows:

$$deg_{i \to j} = \frac{1}{k+1} \sum_{v \in [knn(i), i]} \omega_{vj}, \quad i \neq j \tag{6}$$

Mutual proximity A **:** Combining the relative proximity of two nodes, it is defined as follows:

$$A_{i,j} = deg_{i \to j} \cdot deg_{j \to i} \tag{7}$$

The higher the mutual proximity of the two sensor nodes, that is, the higher the proximity of the two nodes, the closer the connection.

In the first-level clustering, the Euclidean distance between each node is first calculated, and the local density of each node and its relative distance to other nodes are obtained according to formulas (1)(2)(3). Then, according to formula (4), the decision value of each node is obtained and arranged in descending order, and the largest M peak nodes are selected as the final cluster center set. Finally, the remaining nodes select the peak node with the largest mutual proximity to join the cluster according to formulas (5)(6)(7).

Considering the energy consumption of each round, if CH is selected every ten rounds as in the cluster establishment phase, it may cause the energy of CH to be exhausted in a certain ten rounds. Therefore, it chooses to select a new first-level CH within the first-level cluster in each round. The first-level CH is responsible for receiving and fusing sensing data from common nodes in the first-layer cluster, so its election algorithm needs to consider the remaining energy of the node and the sum of the distances between the node and other nodes in the cluster. The election factor algorithm is as follow:

$$\lambda_{\text{fir}CH} = \mu \frac{E_{res}}{\overline{E_{res}}} + \nu \left(1 - \frac{d_{toNodes}}{\overline{d_{toNodes}}} \right) \tag{8}$$

where $\mu + \nu = 1$. $d_{toNodes}$ and $\overline{d_{toNodes}}$ are the sum and average sum, respectively, of the distances from the candidate node to other nodes in the cluster.

3.2 Second-level Cluster

The similarity evaluation index of K-Means is distance. It is considered that the smaller the distance between two sample data, the more similar of them.

Given the first-level CH $set\,(x_1, x_2, \ldots, x_n)$, where each CH is a two-dimensional real vector, K-Means must divide these n first-level CHs into k sets ($k \leq n$), making the sum of squares within the group the smallest. That is to find the cluster S_i that satisfies the following formula, the formula is as follows:

$$\underset{S}{\arg min} = \sum_{i=1}^{k} \sum_{x \in S_i} ||x - \mu_i||^2 \tag{9}$$

where $k = 0.3 * N_{first}$ and N_{first} is the number of first-level CHs. μ_i is the mean value of all points in S_i.

The algorithm first determines the cluster centroids of k initial clusters, and then classifies each sample object into the nearest cluster according to the distance. After all nodes are classified, the centroid is recalculated in the new cluster. Iterate in this way again, until the position of the cluster centroid is unchanged or the change is less than a given value, the clustering process completed.

The second-level CH is mainly used to receive and fuse the data of the first-level CH, and then transmits it to BS. Therefore, the residual energy of the candidate node, the distance between the candidate node and BS, and the distance between the candidate node and the centroid need to be considered in the second-level CH election algorithm. Its election factor algorithm is as follows:

$$\lambda_{\text{sec}CH} = \mu \frac{E_{res}}{Eo} + \nu \left(1 - \frac{d_{toCen}}{d_{diagonal}}\right) + \gamma \left(1 - \frac{d_{toSink}}{d_{diagonal}}\right) \tag{10}$$

where $\mu + \nu + \gamma = 1$ and you can adjust the corresponding parameters as needed. E_{res} and E_O represent the current remaining energy and the initial energy of the candidate node respectively. d_{toCen}, $d_{diagonal}$ and d_{toSink} represent the distance from the candidate node to the centroid, the diagonal length of the area and the distance from the candidate node to BS respectively.

3.3 Inter-cluster Routing

Multi-hop transmission routing is used between CHs to transmit data to BS. It will help reduce the number of relay hops and communication collisions.

Firstly, the path weight matrix between second-level CHs and BS is constructed. Define a two-dimensional path weight matrix $D\,[i]\,[j]$, which represents the path weight from point i to point j where $i, j = 1, 2, \ldots, n, n+1$.The n CHs are sorted from near to far according to the distance from BS, and BS is ranked first. Then the weight of the path between each point is expressed as follows:

$$D\,[i]\,[j] = \begin{cases} 0, & i = j \\ 1e^{-6}, & i = 0 \\ ve^{(d-d_0)/100}, & i \neq 0 \wedge j = 0 \wedge d \geq d_0 \\ 1e^{-6}, & i \neq 0 \wedge j = 0 \wedge d < d_0 \\ \mu(1 - \frac{E_j}{E_o}) + ve^{(d-d_0)/100} + \gamma \frac{\theta_{ij}}{\pi}, & \text{others} \end{cases} \tag{11}$$

where $\mu+\nu+\gamma=1$, and the parameters are adjustable. 05 in experiment. E_j is the remaining energy of node j and E_O is the initial energy. d is the Euclidean distance between node i and j and d_0 is the threshold in system model. θ_{ij} is the deflection angle of node j to i.

Secondly, according to the path weight matrix, the initial sub-optimal path is generated by the Dijkstra algorithm, and then the optimal path is calculated according to the immune algorithm. The individual encoding and fitness functions in immune algorithm are as follows:

- The objective function that will solve the optimal path corresponds to the antigen, and the solution to the problem corresponds to the antibody.
- Antibodies are coded in the order of traversing nodes, and each antibody code string is in the form of: $v_1, v_2 \dots\ v_n$, where v_i represents the sequence number of the traversed node. The fitness function takes the reciprocal of the path weight W_d, as follows:

$$Fitness(i)=\frac{1}{W_d(i)} \tag{12}$$

Where, $W_d(i)=\sum_{i=1}^{n-1} d(v_i, v_{i+1})$, represents the weight of the traversal path represented by the $i-th$ antibody.

4 Simulation and Performance Analysis

The DKBDCERP is simulated by using python and compared with LEACH, EBCRP, KBECRA and DBSCAN based on DBSCAN oposed in recent years. The simulation parameters are set as in Table 1:

Table 1. Simulation Parameter

Parameter	Value
Number of sensor nodes	1000
Network size	600*600(m^2)
Base station location	(300,300)
Initial energy of sensor nodeE_0	0.5(J)
Energy dissipation: receiving E_{ele}	50(nJ/bit)
Energy dissipation: free space model E_{fs}	10($pJ/bit/m^2$)
Energy dissipation: multi-path fading model E_{mp}	0.0013($pJ/bit/m^4$)
Energy dissipation: aggregation E_{DA}	5(nJ/bit)
Data fusion ratio ar	0.6
Control message length CPL	200($bits$)
Data message length DPL	4000($bits$)
Max number of rounds r	2000
Neighboring nodes k in DPC-MND	10
Ratio of clusters in DPC-MND	4%

When the number of CHs is not optimal, the energy consumption of WSN must increases, and Fig 2 shows this number for 100 rounds of each model. It can be seen that in LEACH, EBCRP and DBSCAN, the number of CHs varies greatly. The rapid energy reduction leads to a rapid reduction in nodes and CHs in KBECRA. The number of CHs in DKBDCERP is stably distributed at the optimal value, and there isn't no rapid reduction of energy.

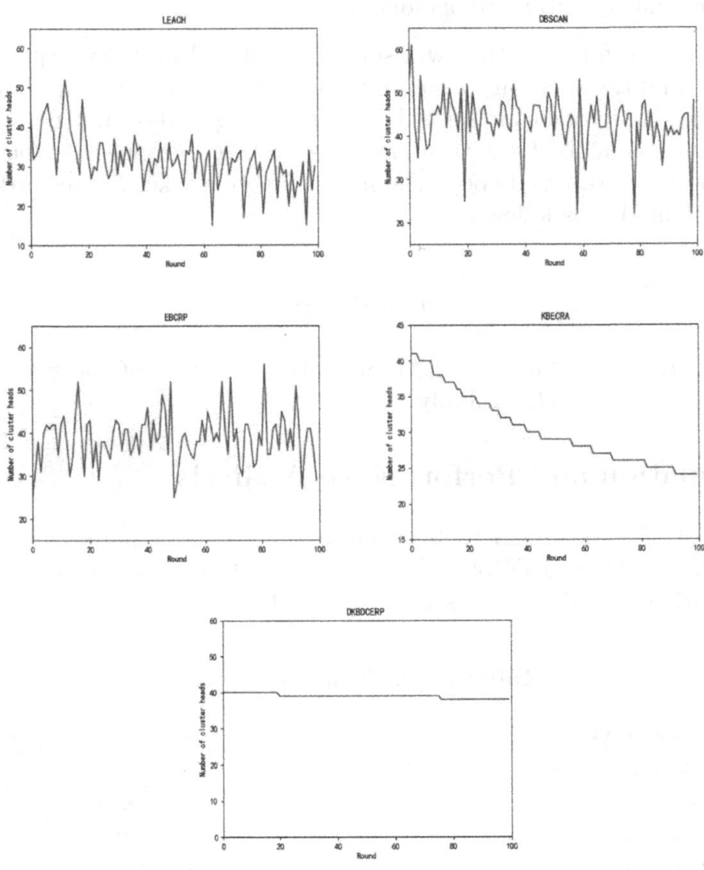

Fig. 2. Clustering performance

Fig 3 and Fig. 4 show the survival of nodes and the energy consumption curve in large-scale WSN respectively. The routing method adopted by LEACH and KBECRA is single-hop transmission, which consumes faster energy, and others are all multi-hop transmission so that the overall survival time is longer. It can be seen that the inflection point of DKBDCERP appears around the 800th round. Before that, there are fewer dead nodes and the network is relatively stable. It embodies the design idea of DKBDCERP, which is to distribute the energy consumption of the entire network evenly among nodes as much as possible.

The inflection points for DBSCAN and EBCRP occurred earlier. At the same time, the energy consumption curve of DKBDCERP is smoother, indicating that the energy consumption distribution of the entire network is more uniform and energy utilization is more efficient.

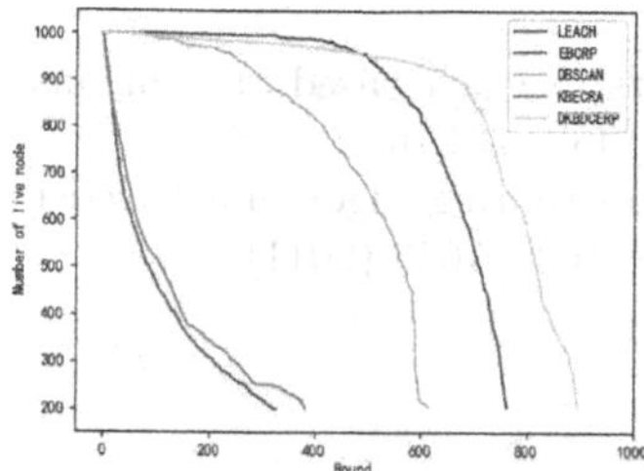

Fig. 3. Nodes lifetime

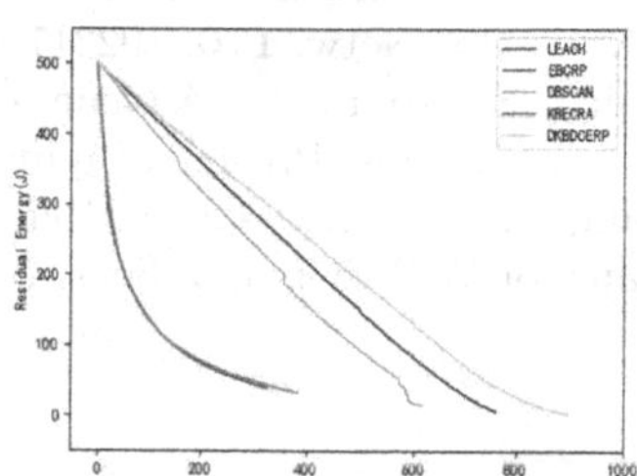

Fig. 4. Network residual energy

Based on the above analysis, DKBDCERP in large-scale WSN greatly extends the lifetime of the entire network. It distributes energy consumption more evenly to each node, resulting in higher energy utilization efficiency, optimizing overall performance compared to several other protocols.

5 Conclusion

This paper proposes an energy-efficient routing protocol DKBDCERP for WSNs based on two-layer clustering. In the first layer, through the improved DPC-MND algorithm, the problem of energy holes near the BS is avoided to a certain extent. In the second layer, the second-level CH is selected by influencing factors such as centroid, residual energy, distance, etc., and then a new path weight is constructed to replace the Euclidean distance. Then, the immune algorithm theory is introduced into the Dijkstra algorithm to obtain the optimal path. After experimental analysis, DKBDCERP effectively prolongs the survival time of each node and the survival time of WSN. In addition, the performance research of WSN includes not only energy consumption, but also network security, QoS, etc., and in-depth research will be carried out in these aspects in the future.

References

1. Al-Karaki, J.N., Kamal, A.E.: Routing techniques in wireless sensor networks: a survey. IEEE wirel. commun. **11**(6), 6–28 (2004)
2. Heinzelman, W.R., Chandrakasan, A., Balakrishnan, H.: Energy-efficient communication protocol for wireless microsensor networks. In: Proceedings of the 33rd annual Hawaii international conference on system sciences, p. 10 IEEE (2000)
3. Hou, Yating, Xu, Feng, Ding, Ruilin: Low-Energy Security-Enhanced Routing Protocol Based on DBSCAN Partition. In: Sun, Xingming, Chao, Han-Chieh., You, Xingang, Bertino, Elisa (eds.) ICCCS 2017. LNCS, vol. 10602, pp. 349–360. Springer, Cham (2017). https://doi.org/10.1007/978-3-319-68505-2_30

4. Li, C., Zhang, H., Hao, B., Li, J.: A survey on routing protocols for large-scale wireless sensor networks. Sensors **11**(4), 3498–3526 (2011)
5. Li, J., Wang, H., Tao, A.: An energy balanced clustering routing protocol for WSN. Chin. J. Sens. Actuators **26**(3), 396–401 (2013)
6. Maheshwari, P., Sharma, A.K., Verma, K.: Energy efficient cluster based routing protocol for WSN using butterfly optimization algorithm and ant colony optimization. Ad Hoc Netw. **110**, 102317 (2021)
7. Rabiaa, E., Noura, B., Adnene, C.: Improvements in leach based on k-means and gauss algorithms. Procedia Comput. Sci. **73**, 460–467 (2015)
8. Zhang, H., Liu, H.: Balanced energy consumption routing algorithm based on k-means for WSN. Chin. J. Sens. Actuators **24**(11), 1639–1643 (2011)

An Energy-efficient Routing Protocol Based on DPC-MND Clustering in WSNs

Feng Xu[1,2], Jing Liu[1(✉)], and PanFei Liu[1]

[1] College of Computer Science and Technology/College of Artificial Intelligence Nanjing University of Aeronautics and Astronautics, Nanjing, China
18751953271@163.com

[2] Collaborative Innovation Center of Novel Software Technology and Industrialization, Nanjing, China

Abstract. Energy Efficiency has become a primary issue in wireless sensor networks (WSNs). Due to the non-uniform distribution and difficulty in supplementing energy of nodes in large-scale WSNs, the design of energy-efficient routing protocols for wireless sensor networks has become a challenge nowadays. It has been already proved that one of the essential approaches to improve WSNs performance is clustering, routing, and data aggregation. However, the main problem dealt with the selection of optimal CH that reduces energy consumption. Till now, more research works have been processing on solving this issue by considering different constraints. Under this scenario, this paper proposes An Energy-efficient Routing Protocol DMBERP(DPC-MND based energy-efficient routing protocol)Based on DPC-MND(Density Peaks Clustering based on Mutual Neighborhood Degrees). This protocol uses an improved DPC-MND clustering algorithm to cluster the nodes in WSNs. Then, in each cluster, two cluster heads are elected according to the residual energy of the node and the distance to the base station. The primary cluster head is responsible for routing and forwarding, and the secondary cluster head is responsible for data collection in the cluster. Finally, a dynamic multi-hop routing between clusters is designed according to the energy, distance and other factors. The experimental results show that the routing protocol proposed in this paper achieves better performance in balancing node energy consumption and prolonging network lifetime.

Keywords: Wireless sensor network · Routing algorithm · multi-hop · Clustering · Density Peaks Clustering based on Mutual Neighborhood Degrees(DPC-MND)

1 Introduction

WSNs is a distributed sensor network [10], composed of a large number of stationary or moving sensors in a self-organizing and multi-hop manner. These sensors cooperatively sense, collect, process and transmit the information of the perceived objects in the geographical area covered by the network, and finally send the information to the owner of the network. Wireless sensor networks are widely

B. Li et al. (Eds.): APWeb-WAIM 2022, LNCS 13421, pp. 199–206, 2023.
https://doi.org/10.1007/978-3-031-25158-0_16

used and have the application scenarios such as intrusion detection, weather monitoring, security and tactical reconnaissance [2]. At present, the main research works are concentrated on the routing algorithms about the energy consumption, security, QOS, etc. [7,12], which derives a series of technologies which includes low-energy routing protocols, data aggregation technologies, energy harvesting technology and secure routing technology, etc. [1].

Due to the continuous expansion of the application, most of the researches [9,13] have a certain degree of inappropriate content. The wireless sensor routing protocol mentioned in this article is the core technology of WSN, which is also the basic guarantee for information collection and transmission, closely related to the performance of the entire network. In large-scale wireless sensor networks, the distribution of sensor nodes is generally uneven. Under the circumstance of huge number of sensor nodes and huge size of network area, uneven energy consumption or even broken network structure may be happened, which definitely brings great challenges to the design of routing protocols for large-scale wireless sensor networks.

2 Related Work

Aiming at solving the problem of energy consumption in WSN, Heinzelman.W.R et al. proposed the classic LEACH [5] protocol. Each node randomly generates a 0-1 decimal. If it is less than the threshold T(n), it will immediately be elected as the cluster head, and other nodes choose a cluster head closest to join the cluster; in recent years, Li Jianzhou and others proposed the EBCRP [6] protocol on the basis of the LEACH, which adds the consideration of distance when electing the cluster head, so that the cluster head probability function changes continuously with distance. In [11] , the authors used K means clustering method as Cluster Head selection algorithm is used over LEACH as base protocol. Initially distance as a parameter is used and later distance plus residual energy of node is chosen. The later proves to be 4 times efficient.

In order to further optimize the clustering routing protocol of WSN, this paper proposes an energy-efficient routing protocol DMBERP based on DPC-MND [14] clustering algorithm. The protocol in this paper is optimized from three aspects:

(1) A dynamic factor related to the distance between the node and the base station is added on the basis of the original clustering algorithm of taking the peak density, the closer to the base station, the more density peak nodes are selected, so that the energy holes near the base station is appropriately avoided.
(2) Since the energy consumption of nodes is mainly in the transmission of data, a large amount of redundant data will shorten the life cycle of wireless sensor networks. Based on the idea of dual cluster heads in the literature [3], this paper distributes the energy consumed by data fusion and data transmission on two nodes, and optimizes the cluster head election to a certain extent.

(3) The multi-hop transmission method is adopted between clusters considering the idea of routing optimization int the literature [4,8], and three factors including distance, direction, and remaining energy are comprehensively considered in the selection of next-hop. This method avoids the influence that the sensor nodes far from the sink node consume more energy due to the long transmission distance in single-hop routing.

3 The Description of DMBERP Protocol

The DMBERP protocol runs in rounds, and each round mainly includes four stages: cluster establishment, cluster head election, inter-cluster routing establishment and stable data transmission. Clustering in each round will affect the overall efficiency for the reason of large resource consumption. Therefore, clusters are established every ten rounds in DMBERP protocol.

3.1 Cluster Establishment Phase

(1)Introduction to Main Definitions

Local density ρ**:** The local density of each sensor node x_i is the relative density in its local range, and its local density ρ_i is defined as follows:

$$\rho_i = \frac{\sum_{j=knn(i)} \sum_{v=knn(j)} d_{vj}^2}{2 \cdot k \cdot \sum_{j=knn(i)} d_{ij}^2} \tag{1}$$

Relative distance δ**:** The relative distance δ_i between a sensor node x_i and other nodes refers to the distance of the node whose local density is greater than itself and closest to itself. Its definition is as follows:

$$\delta_i = \max_{i \neq j} (\delta_j) \tag{2}$$

$$\delta_i = \min_{j:\rho_j > \rho_i} (d_{ij}) \tag{3}$$

Decision value γ**:** The decision value is to judge whether the node can be used as an indicator of the density peak. The closer to the base station, the higher the decision value and the more the clusters. Its definition is as follows:

$$\gamma_i = \frac{\rho_i \cdot \delta_i}{d_{toSink}} \tag{4}$$

Decision value ω**:** Proximity ω_{ij} is defined by the distance index between sensors. The further between the two nodes, the lower the similarity and the smaller the proximity. The definition is as follows:

$$\omega_{ij} = \begin{cases} e^{-\frac{d_{ij}^2}{\theta^2}}, & j \in knn(i) \\ 0, & \text{others} \end{cases} \tag{5}$$

Relative proximity *deg*: The above-mentioned proximity only considers the global information of the node, and local information needs to be considered to achieve the effect, so the relative proximity is defined as follows:

$$deg_{i \to j} = \frac{1}{k+1} \sum_{v \in [knn(i), i]} \omega_{vj}, \quad i \neq j \tag{6}$$

Mutual proximity *A*: Combining the relative proximity of two nodes, it is defined as follows:

$$A_{i,j} = deg_{i \to j} \cdot deg_{j \to i} \tag{7}$$

(2) **Algorithm Flow**

Input: sensor node set $Nodes$, neighboring number k, estimated number of clusters M

Output: clustering results C

a) Calculate the Euclidean distance between nodes, and obtain the local density ρ of each node and the relative distance δ between it and other nodes according to formula (7)(8)(9);
b) Calculate the decision value γ of each node, arrange them in descending order, and select the largest M peak nodes as the final cluster center set C_n;
c) Calculate the mutual proximity from other nodes not included in C_n to all nodes in C_n, select the peak node with the largest mutual proximity for each node, and join the cluster;
d) Repeat step c) until all nodes are clustered, and the cluster establishment is now completed.

3.2 Cluster Head Election Phase

The main cluster head election factor algorithm:

$$\lambda_{\mathrm{maj}CH} = \mu \frac{E_{res}}{\overline{E_{res}}} + \nu \left(1 - \frac{d_{toSink}}{\overline{d_{toSink}}}\right) + \gamma \left(1 - \frac{d_{toOthers}}{\overline{d_{toOthers}}}\right) \tag{8}$$

In this paper, $\mu = 0.5$, $\nu = 0.25$, $\gamma = 0.25$; E_{res} is the current remaining energy of the candidate node, $\overline{E_{res}}$ is the average value of the remaining energy of the candidate nodes; d_{toSink} is the distance from the candidate node to the base station, $\overline{d_{toSink}}$ is the average value of the distances from the candidate nodes to the base station; $d_{toOthers}$ is the sum of the distances between the centers of the point sets of other clusters, $\overline{d_{toOthers}}$ is the average of the sum of the distances between the candidate nodes and the center of other clusters; the cluster center $C = \sum_{i=1} x_i/n$, belongs to the point set of the cluster. Secondary cluster head election factor algorithm:

$$\lambda_{\mathrm{sec}CH} = \mu \frac{E_{res}}{\overline{E_{res}}} + \nu \left(1 - \frac{d_{toNodes}}{\overline{d_{toNodes}}}\right) \tag{9}$$

In this experiment, $\mu = 0.5$, $\nu = 0.5$; $d_{toNodes}$ is the sum of the distances from the candidate nodes to other nodes in the cluster, and $\overline{d_{toNodes}}$ is the average of the sum of the distances from the candidate nodes to other nodes in the cluster.

3.3 Inter-cluster Routing Establishment Phase

In the process of data transmission between clusters, the multi-hop is adopted between cluster heads, and the data is finally transmitted to the base station. Combining the three factors of energy, distance and direction to select the next hop cluster head will help reduce the number of relay hops and communication conflicts. The calculation formula of the relay node metric coefficient α is as follows:

$$\alpha = \frac{E \cdot \cos\theta}{D} \tag{10}$$

Among them, θ is the above-mentioned angle; $E = \frac{E_{res}}{E_{all}}$, E_{all} is the remaining energy of the entire network. When the remaining energy of the entire network is less, the next hop considers the higher proportion of the remaining energy; $D = e^{(d-d_0)/100}$, where d is the distance from the cluster head to the next hop cluster head, and d_0 is a threshold. When the transmission distance is greater than d_0, the energy consumption increases exponentially, that is, when the distance of the next hop cluster head exceeds d_0, the metric coefficient is lower. The following figure is a schematic diagram of the multi-hop transmission path of the cluster heads:

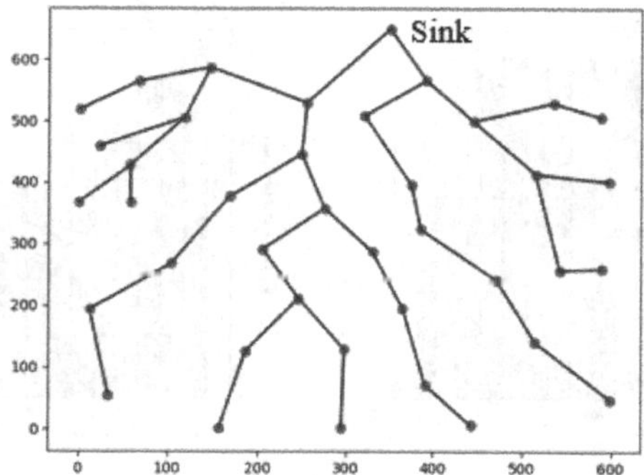

Fig. 1. Schematic diagram of nodes path.

4 Simulation and Performance Analysis

The experimental environment is as follows: the area of network is 600 * 600 and the number of sensor nodes is 1000. The energy consumption performance of DMBERP is analyzed through specific experiments, and the entire wireless sensor network is simulated through Python. The comparison experiments use the classic routing protocol LEACH, the EBCRP routing protocol proposed in recent years, the KBECRA routing protocol and the routing protocol based on DBSCAN clustering.

In this experiment, the maximum number of running rounds is set to 2000. It is determined that when the number of alive nodes is less than 20%, the network structure is severely damaged and cannot operate normally. Simulation experiment of the initialization parameters are shown in Table 1:

Table 1. Simulation Parameter

Parameter	Value
Number of sensor nodes	1000
Network size(m^2)	600*600
Base station location(sink x, sink y)	(300,650)
Initial energy of sensor node$E_0(J)$	0.5
Energy dissipation: receiving $E_{ele}(nJ/bit)$	50
Energy dissipation: free space model $E_{fs}(pJ/bit/m^2)$	10
Energy dissipation: multi-path fading model $E_{mp}(pJ/bit/m^4)$	0.0013
Energy dissipation: aggregation $E_{\mathrm{DA}}(J/bit/message)$	5
Data fusion ratio ar	0.6
Control message length $CPL(bits)$	200
Data message length $DPL(bits)$	4000

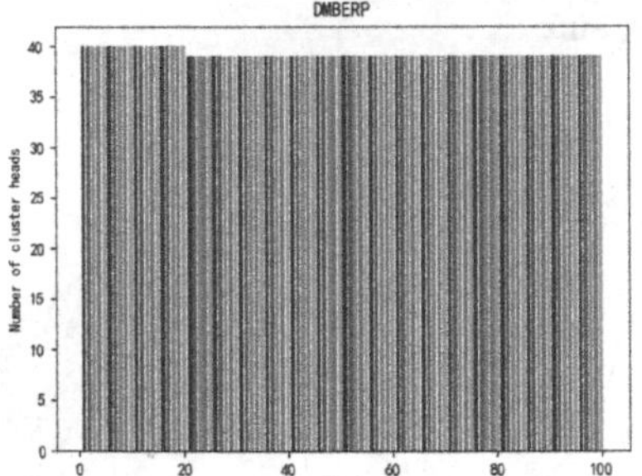

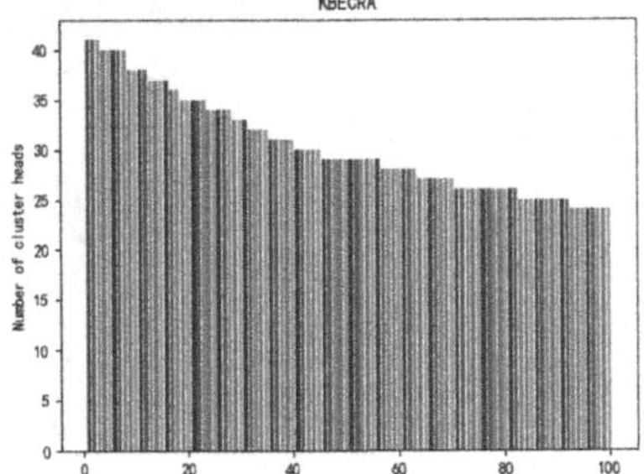

Fig. 2. Cluster adjustment

In this paper, the DPC-MND clustering algorithm requires two parameters to be set manually: the number of neighboring nodes and the number of clusters. Through experiments, it is concluded that when the number of neighboring nodes k is 10 and the number of clusters is 4% of the total number of points, the experiment effect behaves the best.

The number of cluster heads is an important indicator for evaluating the protocol. As can be seen from Fig. 2, the number of the cluster heads in the DMBERP proposed in this paper is stably distributed at the optimal value with good clustering effect and slow energy consumption, so the number of cluster heads does not show a downward trend at the beginning like the KBECRA. Therefore, the DMBERP protocol has high reliability.

We randomly select ten rounds from the experiment and record the average energy consumption of the cluster heads in each protocol and compare them. Figure 3 shows that the energy consumption of cluster heads in the EBCRP and the DMBERP is small and close, the DMBERP consumption is lower than the EBCRP. These two protocols have stable energy consumption and small fluctuation, indicating that the algorithm is stable and effectively balance the energy consumption.

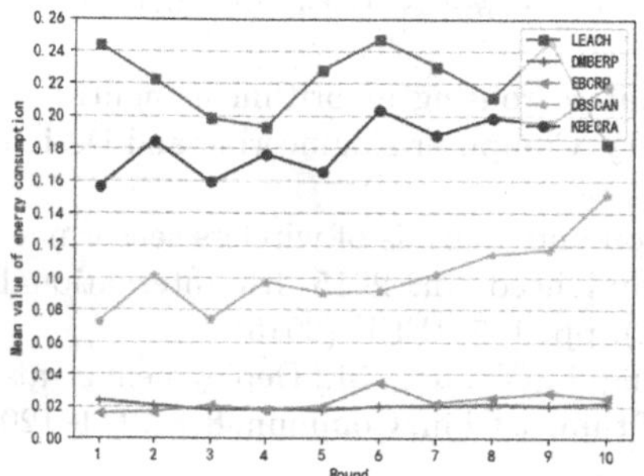

Fig. 3. Mean value of energy cost

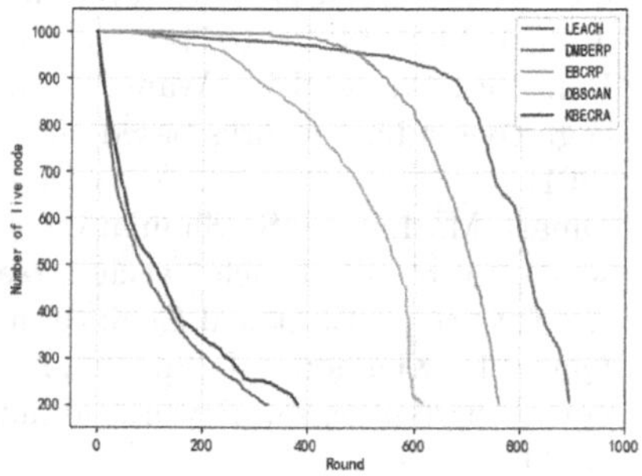

Fig. 4. Nodes lifetime.

Figure 4 shows the survival of sensor nodes in large-scale WSN. It can be seen that the network lifetime of the DMBERP algorithm based on the WSNs node is generally higher than that of the network under other comparison protocols.

5 Conclusion

Aiming at the problem of energy consumption in the process of data transmission, this paper proposed a routing protocol the DMBERP based on density for clustering wireless sensor network nodes improving and optimizing cluster head elections, inter-cluster routing, etc. Experimental results show that DMBERP can improve the energy utilization efficiency of the entire network, and effectively extends the survival period of the entire wireless sensor network. In the future, we will try to figure out the performance of the method in terms of network security.

References

1. Akkaya, K., Younis, M.: A survey on routing protocols for wireless sensor networks. Ad Hoc Netw. **3**(3), 325–349 (2005)
2. Al-Karaki, J.N., Kamal, A.E.: Routing techniques in wireless sensor networks: a survey. IEEE Wirel. Commun. **11**(6), 6–28 (2004)
3. Dai, Z., Yan, C., Wu, Z.: New uneven double cluster head clustering algorithm for WSN-PUDCH algorithm. Chin. J. Sensors Actuators **29**(12), 1912–1918 (2016)

4. Dong, G., Jin, Y., Wang, S., Li, W., Tao, Z., Guo, S.: Db-kmeans: an intrusion detection algorithm based on dbscan and k-means. In: 2019 20th Asia-Pacific Network Operations and Management Symposium (APNOMS), pp. 1–4. IEEE (2019)
5. Heinzelman, W.R., Chandrakasan, A., Balakrishnan, H.: Energy-efficient communication protocol for wireless microsensor networks. In: Proceedings of the 33rd Annual Hawaii International Conference on System Sciences, p. 10. IEEE (2000)
6. Li, J., Wang, H., Tao, A.: An energy balanced clustering routing protocol for WSN. Chin. J. Sensors Actuators **26**(3), 396–401 (2013)
7. Liu, F., Wang, Y., Lin, M., Liu, K., Wu, D.: A distributed routing algorithm for data collection in low-duty-cycle wireless sensor networks. IEEE Internet Things J. **4**(5), 1420–1433 (2017)
8. Liu, J.F., Wang, J.X., Wang, Z.H.: Cooperative routing algorithm in multi-hop cognitive radio networks based on game theory. Computer Engineering and Design (2017)
9. Omari, M., Laroui, S.: Simulation, comparison and analysis of wireless sensor networks protocols: Leach, leach-c, leach-1r, and heed. In: 2015 4th International Conference on Electrical Engineering (ICEE), pp. 1–5. IEEE (2015)
10. Qiang, L., Xiaohong, H., Supeng, L., Longjiang, L., Yuming, M.: Deployment strategy of wireless sensor networks for internet of things. Chin. Commun. **8**(8), 111–120 (2011)
11. Rahim, R., Murugan, S., Priya, S., Magesh, S., Manikandan, R.: Taylor based grey wolf optimization algorithm (tgwoa) for energy aware secure routing protocol. Int. J. Comput. Netw. Appl. (IJCNA) **7**(4), 93–102 (2020)
12. Shokouhifar, M., Jalali, A.: A new evolutionary based application specific routing protocol for clustered wireless sensor networks. AEU-Int. J. Electr. Commun. **69**(1), 432–441 (2015)
13. Yang, C., Gao, W., Liu, N., Song, C.: Low-discrepancy sequence initialized particle swarm optimization algorithm with high-order nonlinear time-varying inertia weight. Appl. Soft Comput. **29**, 386–394 (2015)
14. Zhao, J., Yao, Z., Lü, L., et al.: Density peaks clustering based on mutual neighbor degree. Contr. Dec. **36**(3), 543–552 (2021)

Cloud Computing and Crowdsourcing

Lightweight Model Inference on Resource-Constrained Computing Nodes in Intelligent Surveillance Systems

Zhuohang Wang[1], Yunfeng Zhao[1], Yong Wang[2], Li Yan[2], Zhicheng Liu[1], Chao Qiu[1], Xiaofei Wang[1](✉), and Qinghua Hu[1]

[1] College of Intelligence and Computing, Tianjin University, Tianjin, China
{xiaofeiwang,huqinghua}@tju.edu.cn

[2] Information & Telecommunications Company State grid Shandong electric power Company, Jinan Shandong, China

Abstract. Intelligent Surveillance System (ISS) is an important application combining deep learning with IoT technologies. Meanwhile, multiple targets multiple camera tracking (MTMCT) has been widely recognized as a promising solution for ISS. However, current terminal devices have limited memory, power and computing power. Deep learning models deployed on these resource-constrained devices need to maintain a low-level number of parameters while ensuring inference delay and accuracy as much as possible. In this paper, we propose a lightweight model inference approach for resource-constrained edge computing nodes, which leverages the limited computing capability of edge devices for collaborative processing of model inference tasks. We consider a system model that includes a combination of both horizontal and vertical methods to dynamically partition deep neural networks. Further, we propose an adaptive strategy that dynamically controls the execution part in the terminal devices. In addition, a learning-based algorithm is used to obtain the near-optimal solutions for dynamic resource allocation decisions. Finally, we evaluate the system in terms of two metrics, i.e., computational latency and network throughput, based on a real-world dataset of campus surveillance. The experimental results reveal the superior effectiveness of the proposed scheme.

Keywords: Lightweight model inference · Edge intelligence · Intelligent surveillance system · Multiple targets multiple tracking · Deep reinforcement learning

1 Introduction

Nowadays, city-wide, enterprise park-wide, and campus-wide surveillance systems are increasing to enhance the efficiency of monitoring, instant field response, and instant identification. Particularly, with the breakthroughs of artificial intelligence (AI), recent years have witnessed a booming of intelligence surveillance systems

The original version of this chapter was revised: the text in acknowledgement section has been corrected. The correction to the chapter is available at https://doi.org/10.1007/978-3-031-25158-0_46

B. Li et al. (Eds.): APWeb-WAIM 2022, LNCS 13421, pp. 209–223, 2023.
https://doi.org/10.1007/978-3-031-25158-0_17

(ISSs). Driving by these trends, ISSs have been used in various scenarios, such as self-driving cars [23], traffic control and home monitoring [22]. There are numerous issues remaining to be solved in ISSs, namely low latency, high throughput, good stability, etc. More recently, Multiple Targets Multiple Cameras Tracking (MTMCT) has been widely recognized as a promising solution [26].

The key reasons for the success of MTMCT are the booming of deep learning (DL) technologies and Internet of Things (IoT) [8], such as image recognition [11], object detection [5] and object tracking [1]. Figure 1 shows the workflow of the MTMCT system. Data generated in the MTMCT needs a huge computing capability to fully unlock its potential. Cloud computing is a silver bullet. As the number of surveillance cameras proliferates, it suffers from heavy transmission pressure, high latency, expensive cost and low security [29].

Increasing interest in designing real-time video analysis systems based on edge computing aim to address these problems. Researchers have begun to learn about the solutions. CLONE [17] focuses on solving the MTMCT task with a collaborative learning strategy. However, the Deep Neural Networks (DNN) still run in the cloud, leading to system performance limitations. Zeng et al. [27] propose to adaptively balance the workloads across smart cameras and partition the workloads between cameras and edge clusters to achieve optimized system performance. However, the properties of DNN have not been considered. And AI models deployed on these resource-constrained devices need to **keep the number of parameters low while tolerating high inference latency and low accuracy**. It is meaningful to combine DNN partitioning with cross-camera video processing.

Driving by these trends, many types of research are attaching importance to partition DNN for decreasing latency and resource-saving. There are two basis category partition strategies: (i) horizontal partition. (ii) vertical partition. For the former, the Neurosurgeon [12] designs a fine-grained, layer-level computation partitioning strategy by utilizing the characteristics of DNN. Hu et al. [10] transfer the problem of minimizing latency to an equivalence Min-Cut problem by viewing DNN as a directed acyclic graph (DAG). For the latter, MoDNN [18] partition the feature projection based on the computing ability of a single node for the convolution layer and then merge the outputs of all nodes in the host. DeepThings [28] applies a scalable Fused Tile Partitioning (FTP) strategy of convolution layers to minimize memory footprint and decrease the transmission between nodes. Although the vertical partition enables the parallel computing of convolution layers, it is still an open problem to partition residual structure and attention-based layer.

Utilizing dynamic computation offloading and resource allocation make it possible to achieve real-time IoT-based video analysis. Real-time video streaming analysis architecture based on edge computing validates the possibilities of edge computing in the field of vision. However, the wide range of application scenarios leads to different architectures in edge-cloud clusters, and there is a lack of discussion on the fine-grained DNN models. And it is still an open problem in the field of DNN model partition to consider the cooperation among multi-camera clusters. Therefore, there are still three challenges in edge real-time video analysis systems: (i) **Offloading the complex DNN models to the limited edge device.** (ii) **Cooperating the offload among tiers or nodes in the end-edge-cloud heterogeneous clusters.** (iii) **Relieving the computing pressure of clouds, making full use of systems computing ability and minimizing the computing latency in the scenarios of the multiple video streams.**

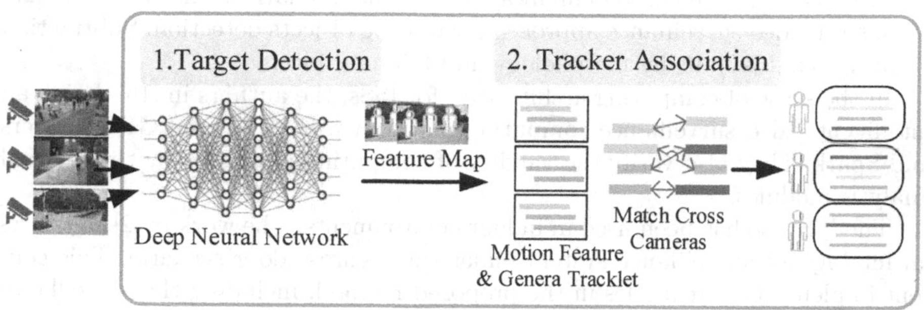

Fig. 1. Workflow of MTMCT System.

In this paper, we present an edge intelligence collaborative lightweight model inference for distributed real-time intelligent surveillance systems. Our approach has the superiority to perform adaptive computing offloading and resource allocation for cooperative multi-camera tracking tasks on the edge-cloud system. Our contributions are summarized as follows:

- We investigate different types of DNN partitioning mechanisms and use a regression model to perform the relationship between the parameters, arithmetic demand, and time cost of the DNN model.
- Our system combines both horizontal and vertical partitioning methods to make full use of computing ability and reduce the time cost of complex models.
- Our system constructs mathematical optimization problems for computational offloading and resource allocation decisions based on characteristic information and a dynamic, realistic environment, i.e., the bandwidth speed between nodes, the available computing power of the device, and the maximum user latency limit setting.
- For such NP-hard problems, we design a deep reinforcement learning-based solution scheme to obtain an approximate optimal solution for dynamic resource allocation, which accelerates the collaborative inference of the edge-cloud system.

The rest of this article is organized as follows. In Sect. 2, some related works about collaborative inference for ISSs are introduced. Section 3 presents the system model, followed by the problem formulation in Sect. 4. Section 5 shows and discusses the experimental results. Finally, the conclusion and future works are shown in Sect. 6.

2 Related Work

2.1 Intelligent Surveillance Systems

The increasing secure requirements from cities, factories, as well as homes inspire further research ideas on intelligent surveillance systems. For example, in terms of security requests from citizens, the authors in [14] study intelligent CCTV monitoring with the help of mobile edge computing and 5G.

ISSs also have been used in highway surveillance scenarios. The authors in [20] consider a video surveillance approach for moving objects detection. Subtraction of background is built by morphology and bilateral filter.

In the sense of computing architecture for ISSs, the authors in [15] design an intelligent video surveillance computing platform with CPU and GPU clusters and Spark. They also verify the validity of computing platforms by three visual analysis modules.

The ISS also has been used in indoor environments. The work in [2] considers an intelligence surveillance system under the secure indoor scenario. This content implements four modes in the proposed method, including the surveillance machine, smart fire detection scheme, man-machine interface, and the server.

To summarize, ISSs have been widely used and studied. In our article, we design an intelligent surveillance system to achieve multiple targets multiple cameras tracking tasks.

2.2 Multiple Targets Multiple Cameras Tracking (MTMCT)

As one of the most promising solutions in ISSs, recent years have witnessed a booming of MTMCT. For example, from the perspective of the tracking algorithm, the authors in [25] learn a multi-hypothesis tracking algorithm. This algorithm utilizes the tracklets feature of spatio-temporal association from multi-camera to distinguish the number of tracks by solving an online weighted clique problem. Meanwhile, an optimal tracks method for every frame is formulated to solve the NP-hard problem. Also, considering the multi-camera tracklet matching problem, the work in [9] learns the assignment problem for tracklet to target matching, especially thinking over the practical limitations. Thus, a non-negative matrix factorization method is discussed to achieve the global trajectory so as to compute the assignment problem.

From the perspective of tracking performance, the authors in [3] study a real-time multi-target multi-camera tracking approach, which is an online multi-layer algorithm. By creating the dynamic real-time gallery of multi-object tracking for a single camera, the time effectiveness of the scheme is greatly improved in the real scenario. The representation of the global model is hard to describe the local features, such as similarity, among targets. Thus, the authors in [26] provide a Pose Association method for MTMCT. Such method includes pose generation, adjustment, and modification. Then the final similarity matrix is generated by the received pose similarity and the spatial similarities. The experiments are also presented to show their effectiveness.

It can be seen that MTMCT is widely researched in the ISSs, ranging from tracking algorithms to tracking performance. In this article, we solve the multi-target multi-camera tracking problem from the angle of the DNN tracking algorithm.

2.3 Collaborative Learning Based ISSs

With the proliferation of edge intelligence, collaborative learning is widely adopted in the MTMCT to pursue improved tracking efficiency and accuracy, especially the required large amount of learning capacity. For example, considering the required large amount of data in ISSs, the authors in [13] present an improved shot multi-box detector with the assistance of edge artificial intelligence. Such a scheme is employed in a smart parking surveillance scenario, which has been confirmed the better target effectiveness and correctness.

The visual surveillance system generally needs a large volume of a resource. In this context, the work in [24] studied an edge-cloud resource allocation of ISSs. Here, a collaborative deep learning method is designed to detect, classify, and track the visual surveillance task. Through the simulation results, the proposed method has a better inference rate and computational efficiency. From the perspective of collaborative information integration, the authors in [7] probe into a fused method, including deep learning and information integration. DL is first used to detect fast objects, and a multi-information integration method is deployed for spatial trajectories of multi-target. The simulation results are shown to evaluate the superior performance.

From the above discussion, it can be found that there is an urgent need to collaborate among edge devices for better surveillance performance. In this article, we discuss collaborative inference in the real world multi-camera intelligent surveillance scenarios.

3 System Model and Problem Formulation

3.1 System Model

Figure 2 shows the workflow of the MTMCT under the end-edge-cloud architecture. Concretely, (i) The whole MTMCT system is divided into two parts by offloading decisions executed at different devices. A configuration such as model structure, parameters, and computation resource requirements are saved in each device. (ii) "Always-on" cameras upload the video streams to a device that executes part of the visual surveillance task. (iii) When the device finishes processing the received tasks, it sends a feature map to the second device according to an offloading decision. (iv) The cloud maintains a central database with dynamic global environmental information covering network conditions, computing node information, executing tasks, etc. The cloud performs central training of intelligent decisions on the model partition and computation offloading.

We consider a system $DES = \{\mathcal{D} \cup \mathcal{E} \cup \mathcal{S}\}$ under the end-edge-cloud architecture. Extensive cameras $\mathcal{D} = \{d_0, d_1, ..., d_{M-1}\}$ and edge devices $\mathcal{E} =$

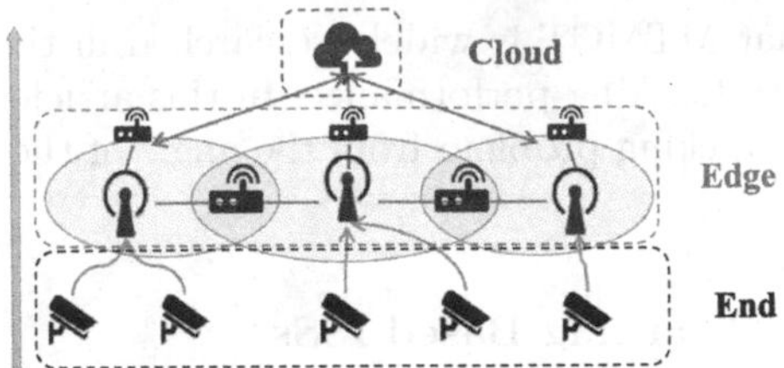

Fig. 2. Workflows of MTMCT system on edge-cloud system.

$\{e_0, e_1, ..., e_{N-1}\}$ are deployed on the end and edge of the networks, respectively. They are managed by a cloud $\mathcal{S} = \{e_N\}$. Cameras produce video streams and request computation resource for tracking tasks. The edge devices consist of resource-limited devices (e.g., base stations, IoT devices, etc.) and provide computation resource. Though the cloud can provide more powerful computing capabilities, it suffers from long transmission distances.

We set the computing capabilities of the node n as C_n, which represents the speed per bit of CPU or GPU disks. Moreover, let $b_{(i,j)}$, $i \in \mathcal{E} \cup \mathcal{S}, j \in \mathcal{D}$ be the bandwidth speed between two nodes and its unit is kb/s. The camera d_m generates a video stream of x_m frames per second. Due to the limited memory of computing nodes, the larger the batch size is, the more memory and computing capacity are consumed when performing image detection. Therefore, to relieve the overall pressure on the system, the camera is considered to discard some frames per second and the frame loss rate is set to F. The effective data frames of camera d_m per second is $f_m = (1 - F) \cdot x_m$ and the actual amount of data for transmitting the video stream is $K_m = k \cdot f_m$ where k is the average unit amount of each video frame. If the computing node n will perform the detection task of d_m, the transmission delay is expressed as $T^{\text{tr}}_{(n,m)} = K_m / b_{(n,m)}$.

A complex DNN model can be viewed as a directed acyclic graph (DAG). Let $\mathcal{G} = (\mathcal{V}, \mathcal{L})$ present the DAG of the DNN model where layer v in the model is a vertex v_v and $\mathcal{V} = \{v_0, v_1, \cdots, v_{V-1}\}$. The edge device $(v_i, v_j) \in \mathcal{L}$ means the input of v_j is the output of v_i in the network architecture of the DNN model. In this paper, we consider the YOLOv4 [21] as the DNN model for the cross-camera tracking task. As Fig. 3 shows, there are three types of base units in the model: the CBM, CBL and Resunit.

3.2 Horizontal Partition Algorithm

Considering the pressure of cross-camera workloads, the limitation of the heterogeneous computing capacities of computing nodes, and the DNN model computing and memory requirements, the DNNs partition strategy is proposed. The horizontal partition algorithm (HPA) splits the model into two parts by DNNs layer granularity, which increases the system's computing ability utilization and reduces higher cloud service costs.

YOLOv4 has a large number of residual structures. To better represent the backbone of each layer, [19] is made to obtain the longest path from input to

Table 1. Modeling parameters and notations.

Symbol	Description
$\mathcal{D}, \mathcal{E}, \mathcal{S}$	The set of cameras, edge devices and the cloud
C_n	The computing capabilities of node n
$b_{(i,j)}$	The bandwidth speed between node i and node j
$T^{\text{tr}}_{(n,m)}$	The transmission delay between m and n
$\mathcal{P}$	The set of horizontal partition point
$H_{p^m,0}$, $H_{p^m,1}$	The two parts of the split network
$\alpha_{p^m,i}$, $\beta_{p^m,i}$, $\gamma_{p^m,i}$	The number of Resunit, CBL, CBM
w_{CBL}, w_{CSP},w_{Resunit}	The number of parameters of Resunit, CBL, CBM
Y^m_i	The execution choices of the network part $H_{p^m,i}$
$T^{\text{tr}}_{Y^m_0}$	The data transmission delay for camera d_m
$T^{\text{wait}}_{Y^m_0}$	The waiting time for the camera d_m
T_{p^m,Y^m_0}	The execution time of d_m's task on computing node Y^m_0
$\xi^0_{p^m,Y^m_0}$	The regression parameters
$K^{\text{out}}_{p^m,Y^m_0}$	The amount of output data from the network part $H_{p^m,0}$
$T^{\text{Total}}_{p^m,Y^m}$	The execution time of task from camera d_m
ρ	The time threshold for tasks

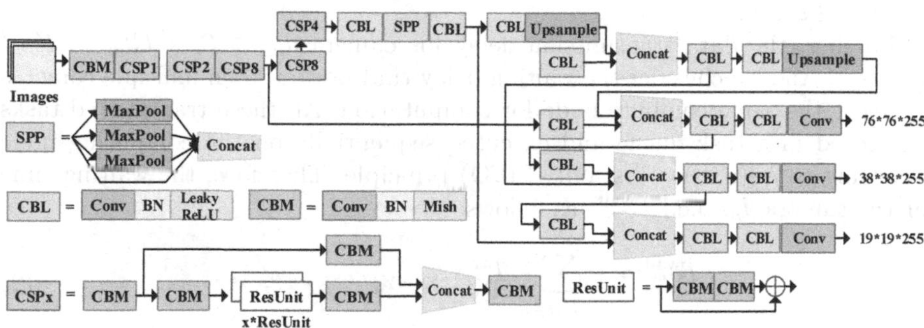

Fig. 3. The structure of YOLOv4.

output by the **Depth-First Search** algorithm [4]. Inspired by this paper, we propose to use a depth-first search algorithm to divide each layer of the neural network horizontally and obtain the horizontal structure of the network. Further, [16], Li *et al.* propose a regression model to predict DNN layer latency with input and output data size. [19] extrapolates the computation formulas of DNN layers to obtain the corresponding floating-point computational formulas. The process of the horizontal model partition is shown in Fig. 4. First, we perform regression and parameter fitting for cameras', edge devices' and cloud's computation resource, respectively. Then, we will make partition point decisions based on the real dynamic networking environment and finally get the segmentation results.

We set the P horizontal partition points in the YOLOv4 network as $\mathcal{P} = \{p_0, ..., p_{P-1}\}$ and $|\mathcal{P}| = P$. If camera d_m selects the partition point $p^m \in \mathcal{P}$,

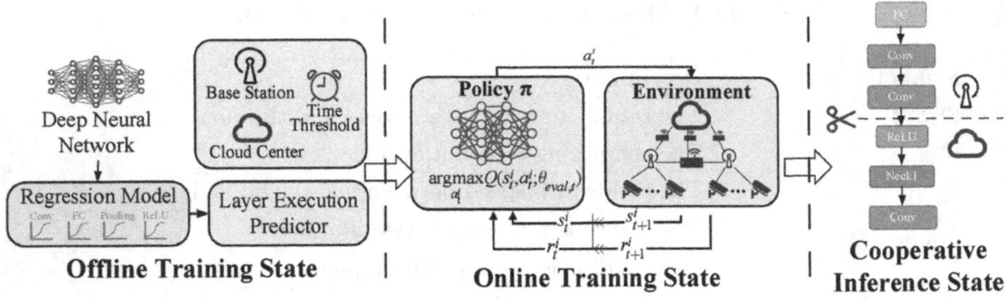

Fig. 4. The process of horizontal model segmentation.

it splits the whole network H into two parts: $H_{p^m,0}$ and $H_{p^m,1}$. Concretely, the structure of the network $H_{p^m,i}$, $i = 0, 1$ includes $\alpha_{p^m,i}$ Resunit, $\beta_{p^m,i}$ CBL (*conv* + *bn* + *leakyRelu*) and $\gamma_{p^m,i}$ CBM (*conv* + *bn* + *mish*). Therefore, the number of parameters of this part is:

$$W(H_{p,i}) = \alpha_{p^m,i} \times w_{\text{Resunit}} + \beta_{p^m,i} \times w_{\text{CBL}} + \gamma_{p^m,i} \times w_{\text{CBM}}, \tag{1}$$

where w_{Resunit}, w_{CBL} and w_{CSP} are the number of parameters of Resunit, CBL and CBM, respectively. The execution choices of the network part $H_{p^m,i}$ is Y_i^m and $Y_i^m \in \mathcal{E} \cup \mathcal{S}$.

Further, the data transmission delay for camera $d_m \in \mathcal{D}$ is $T_{Y_0^m}^{\text{tr}} = K_m / b_{(Y_0^m,d_m)}$. And we consider the waiting delay that occurs when multiple cameras all select the same compute node for computation. All these transmitted tasks are stored in a task queue and executed sequentially on the computing node according to a First-in First-out (FIFO) principle. Therefore, the waiting time for the camera d_m, i.e., $T_{Y_0^m}^{\text{wait}}$, as follows:

$$T_{Y_0^m}^{\text{wait}} = \sum_{j \in \mathcal{D}, j \neq m} T_{Y_0^j}^{\text{tr}} \mathbb{I}_{\{Y_0^j = Y_0^m, T_{Y_0^j}^{\text{tr}} < T_{Y_0^m}^{\text{tr}}\}}, \tag{2}$$

where $\mathbb{I}(\cdot)$ is an indicator function with the value of one when the argument is true and zero otherwise.

We use the regression function to construct the relationship among the number of parameters that need to be executed and the input size, output size, and the amount parameters of $H_{p^m,0}$, i.e., $W(H_{p^m,0})$. Therefore, the execution time of camera d_m's task on computing node Y_0^m, i.e., T_{p^m,Y_0^m}, can be written as

$$T_{p^m,Y_0^m} = \frac{\xi_{p^m,Y_0^m}^0 K_m + \xi_{p^m,Y_0^m}^1 K_{p^m,Y_0^m}^{\text{out}} + \xi_{p,Y_0^m}^2 W(H_{p^m,0}) + \eta_{p^m,Y_0^m}}{C_{Y_0^m}}, \tag{3}$$

where $\xi_{p^m,Y_0^m}^0$, $\xi_{p^m,Y_0^m}^1$, $\xi_{p^m,Y_0^m}^2$ and η_{p^m,Y_0^m} are the regression parameters. $K_{p^m,Y_0^m}^{\text{out}}$ is the amount of output data from the network part $H_{p^m,0}$ and λ_{p^m,Y_0^m} is a constant.

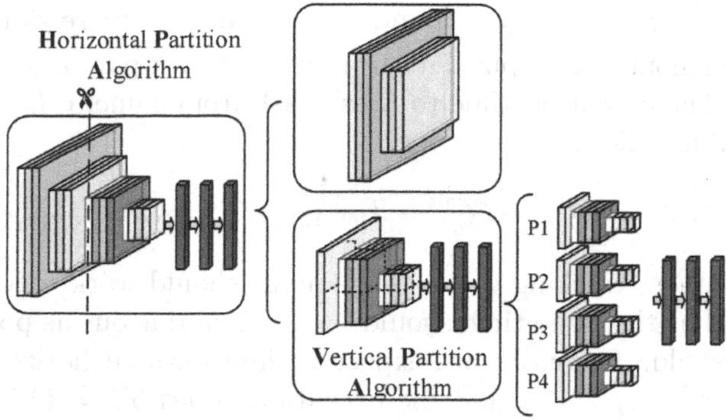

Fig. 5. Two partition strategies: horizontal and vertical

3.3 Vertical Partition Algorithm

For the network part $H_{p^m,1}$, there are finally three output branches of the model structure, where the vertical partition algorithm (VPA) can be performed. In addition to the number of its own parameters $W(H_{p^m,1})$, the output features from $H_{p^m,0}$ are to be received before the computing process. Therefore, when $H_{p^m,1}$ is executed on the computing node Y_1^m, there is a transmission delay between Y_0^m and Y_1^m:

$$T^{\text{tr}}_{p^m,Y^m} = K^{\text{out}}_{p^m,Y_0^m} / b_{(Y_0^m,Y_1^m)} \tag{4}$$

We found that after the YOLOv4 backbone, the network includes three main output branches, which contain a large number of residual connection structures between each other. Using the horizontal partitioning algorithm will increase the transmission of information between nodes. Therefore, we propose using a Vertical Partition Algorithm (VPA) to offload this part of the network.

Vertical partition based on fused tiles: As Fig. 5 shows, The model is divided into two parts by the horizontal partition algorithm, and the first part of the model is assigned to node Y_0^m for execution. The other part is divided vertically by the vertical partition algorithm. We divide the neural network into $l * l$ fused tiles according to the $l * l$ grid method for parallel computation at computing node Y_1^m and merge the final results for the final fully connected operation. Similarly, we use the regression function to construct the relationship among all the amount of parameters for executing network part $H_{p^m,1}$ among input size, output size and the amount parameters of $H_{p^m,1}$, i.e., $W(H_{p^m,1})$. Therefore, the execution time of camera d_m's task on computing node Y_1^m, i.e., T_{p^m,Y_1^m} can be written as

$$T_{p^m,Y_1^m} = \frac{\xi^0_{p^m,Y_1^m} K^{\text{out}}_{p^m,Y_0^m} + \xi^1_{p^m,Y_1^m} K^{\text{out}}_{p^m,Y_1^m} + \xi^2_{p^m,Y_1^m} W(H_{p^m,1}) + \eta_{p^m,Y_1^m}}{C_{Y_1^m}} \tag{5}$$

where $\xi^0_{p^m,Y_1^m}$, $\xi^1_{p^m,Y_1^m}$, $\xi^2_{p^m,Y_1^m}$ and η_{p^m,Y_1^m} are the regression parameters. $K^{\text{out}}_{p^m,Y_1^m}$ is the amount of output data from the network part $H_{p^m,1}$ and λ_{p^m,Y_1^m} is a constant. The execution time to finish task from camera d_m for the whole system is given as follows:

$$T^{\text{Total}}_{p^m,Y^m} = T^{\text{tr}}_{Y_0^m} + T^{\text{wait}}_{Y_0^m} + T_{p^m,Y_0^m} + T^{\text{tr}}_{p^m,Y^m} + T_{p^m,Y_1^m}. \tag{6}$$

For cross-camera tracking tasks, all cameras should work together, i.e., the tasks generated at the same time should be 'as simultaneous as possible' within a certain threshold. Therefore, we expect to find optimal horizontal partition points $\mathbf{p}^* = \{p^{0*}, ..., p^{M-1*}\}$ and the execution nodes $\mathbf{Y}_0^* = \{Y_0^{0*}, ..., Y_0^{M-1*}\}$ and $\mathbf{Y}_1^* = \{Y_1^{0*}, ..., Y_1^{M-1*}\}$ to minimize the following function:

$$\begin{aligned} &\min_{\mathbf{p},\mathbf{Y}_0,\mathbf{Y}_1} T^{\text{Total}} = \min_{\mathbf{p},\mathbf{Y}_0,\mathbf{Y}_1} \{\max_m T^{\text{Total}}_{p^m,Y^m} - \min_{m'} T^{\text{Total}}_{p^{m'},Y^{m'}}\}, \\ &s.t. \quad T^{\text{Total}} < \rho, \end{aligned} \tag{7}$$

where ρ is the time threshold for tasks and $m, m' \in \{0, 1, \ldots, |\mathcal{D}|\}$.

4 Algorithm Design

Framework Overview: Considering the optimization problem in (7), we introduce a Multi-Agent Deep Reinforcement Learning (MADRL) algorithm to provide the selections of partition node and execution computing nodes. Specially. Each agent generates action $\mathbf{A}$ interacting with the environment and observes the environment state $\mathbf{S}$ and rewards $\mathbf{R}$ by duration, and learns policy π to maximize the accumulated reward.

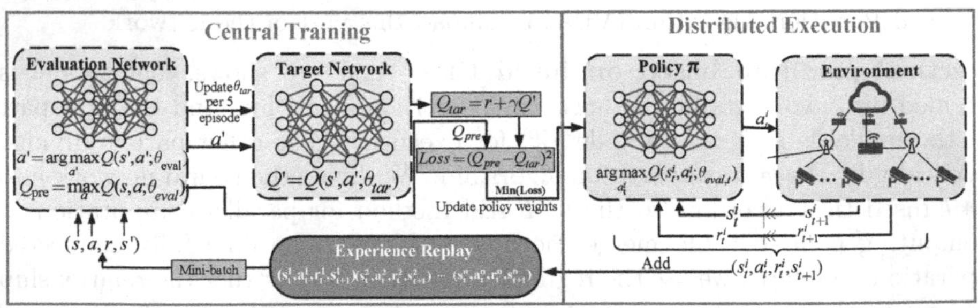

Fig. 6. Workflow of MADRL.

Figure 6 illustrates the structure of our MADRL algorithm, which consists of two main parts: central training and distributed execution. **In the centralized training**, the cloud trains the homogeneous Double DQN (DDQN) network structure of multiple agents. DDQN has an evaluation network and a target network to eliminate the overestimation problem of DQN by decoupling the policy evaluation with action selection and the Q values of the next action. Target network updates weight from evaluation network periodically. **In the distributed execution**, DDQN weights after centralized training is shared with all agents. Each agent has a homogeneous state, action and reward space, which reduces the large number of weights that must be trained [6].

The goal of the agent is to maximize the accumulated discount reward $G_t = R_{t+1} + \gamma Q(S_{t+1}, \arg\max_a Q(S_{t+1}, a; \theta_t), \theta_t^-)$,where γ is the discount factor determines the influence of future reward based on the action. DDQN chooses the best action based on the evaluation network with parameter θ_t, and obtains Q value based on the target network with parameter θ_t^-.

State and Action Space: We model each camera as agent. Each agent $d_m \in D$ selects an action $\mathbf{a_t^m} \in \mathbf{A}$ based on policy π and current state $\mathbf{s_t^m} \in \mathbf{S}$. Each agent observes state s_t^m at time t. Edge device and the cloud follow the **first-come-first-served** principle and need to cooperate to complete multi-agent resource allocation and computation offloading decisions.

Construct the state $\mathbf{s_t^m}$ as follows:

$$s_t^m = (T_{p^m, Y^m}^{Total}, T_{p^m, Y^m}^{tr}, T_{p^m, Y_0^m}, T_{Y_0^m}^{wait}, s(\mathcal{DES})_t^m), \quad (8)$$

where T_{p^m, Y^m}^{Total} denotes the execution time of the detection task of executing the video stream of camera d_m, T_{p^m, Y^m}^{tr} denotes the transmission delay of the detection task between the execution node Y_0^m and the execution node Y_1^m. T_{p^m, Y_0^m} denotes the delay of execution time on the execution node Y_0^m. $T_{Y_0^m}^{wait}$ denotes the waiting time before execution on the execution node Y_0^m. $\mathbf{s(DES)_t^m}$ denotes the state of the base station and the cloud under the influence of other intelligences besides camera, includes the number of tasks waiting to be performed in the device, the amount of tasks, and the expected completion time of the currently performed task.

The action $\mathbf{a_t^m} \in \mathbf{A}$ consists of the selection of the partition point $p^m \in \mathcal{P}$ and offloading computing nodes $Y_0^m, Y_1^m \in \mathcal{E} \cup \mathcal{S}$ in time t, i.e., $a_t^m = \{p^m, Y_0^m, Y_1^m\}$. After interacting with environment, agent d_m obtains reward r_t^m and transitions to a new state s_{t+1}^m.

Reward Design: According to equation (7), The goal of EI-ISS is to minimum total processing time T^{Total} of videos analysis task generated by the camera cluster Periodically. In a distributed scenario, finding an optimal policy π^* for real-time load balancing for multiple continuous video analysis tasks is challenging. EI-ISS not only considers minimizing the processing time of each camera video task but also needs to maintain the time difference between cameras and cameras minimal. The reward of each agent is defined as follows:

$$r_t^m(a_t^0, ..., a_t^{M-1} | s_t^0, ..., s_t^{M-1}) = \begin{cases} Z - T_{p^m, Y^m}^{Total}, & T^{Total} \leq \rho \\ Z - T_{p^m, Y^m}^{Total} - \mu \times T^{Total}, & T^{Total} > \rho \end{cases} \quad (9)$$

where μ denotes the penalty weight, ρ denotes the tolerance of the maximum time difference for all cameras to complete the detection task at moment t, and

Z denotes the maximum time tolerance for completing the detection task of a single camera. T^{Total} is the maximum task completion time difference generated by all camera machines at moment t.

5 Numerical Evaluation

5.1 Experiment Setup

We will evaluate the performance of the proposed algorithms on the real-world dataset. One cloud, six edge devices, and 20 cameras are built into the virtual machine by controlling CPU and GPU usage. The training process is deployed in the Tesla V100 (16G) GPU as a cloud. Unlike other work based on general public datasets, **we innovatively constructed a real dataset based on 20 surveillance cameras deployed on campus and used it in our experiments**. These cameras capture data from different areas of the campus, such as cafeterias, academic buildings, and dormitories and their surrounding areas. We sampled 150 representative 5-minute continuous video clips from 20 cameras at a frame rate of 30 FPS and a resolution of 1920 × 1080P. The codes are implemented using Python3.

We set five model partition selections, i.e., {0, 1, 2, 3, 4} through the regression function in Sect. 3.2, where the selection 0 represents the model that does not perform the vertical partition. In the experimental details, we randomly change the network state with a probability of 0.9 to simulate network fluctuations in a real-world environment. Video tasks generated by 20 cameras per second are continuously transmitted into the system.

5.2 Experimental Result

To show the superiority of our online system in a cooperative multitasking flow scenario, we designed the Epsilon-Greedy algorithm and the stochastic algorithm for comparison. The Epsilon-Greedy algorithm makes optimal decisions based on the current tasks, resource availability, and latency constraints. The discount factor and the minimum greedy rate of Epsilon-Greedy are 0.0001 and 0.3, respectively. And the Random algorithm makes decisions randomly without considering the cooperative relationship among edge devices.

Figure 7 shows the collaboration success rate of MTMCT system. The MTMCT system requires collaborative cross-camera tracking, and excessive time differences in task processing among cameras will result in lost and duplicated tracking targets. Therefore, based on the user-defined time threshold ρ, we designed a metric to determine the success rate of collaboration among tasks. As shown in the figure, the collaboration rate of DDQN gradually increases and converges, and finally reaches the collaboration time requirement proposed by the user. And the collaboration rate of the greedy algorithm ends up fluctuating only around 0.4.

Figure 8 shows the average rewards of DDQN, Epsilon-Greedy, and Random algorithms training on different edge devices. We set the rewards of 700–1300 as

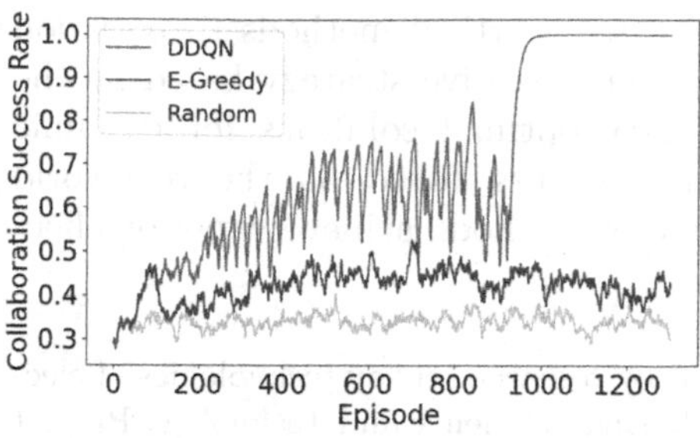

Fig. 7. Collaboration success rate under three algorithms

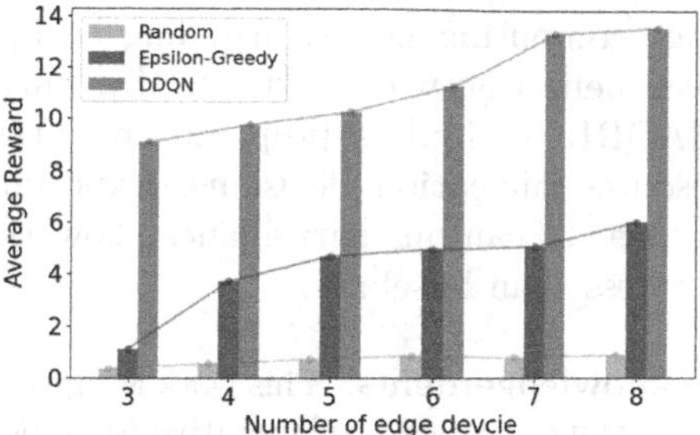

Fig. 8. Number of edge device versus the average reward

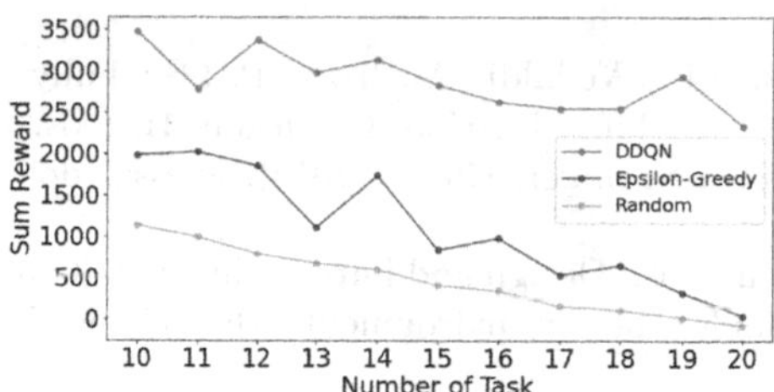

Fig. 9. Number of task versus the sum of reward

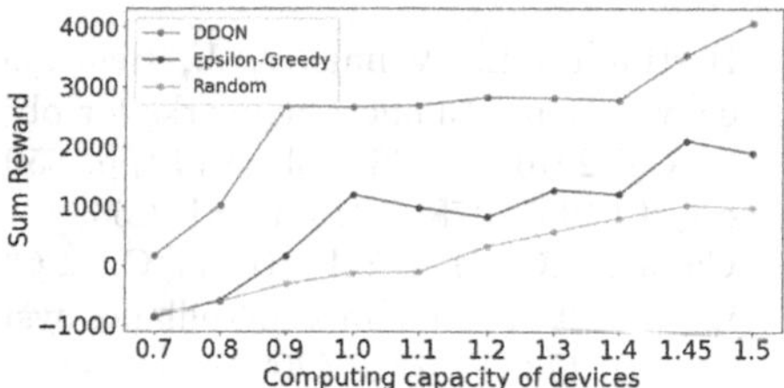

Fig. 10. Computing capacity versus the sum of reward

the average reward, where Epsilon-Greedy and DDQN algorithms end up with a greedy rate of 0.1. As shown in the figure, the performance of DDQN improves significantly as the number of edge devices increases, while the average reward of the Epsilon-Greedy algorithm is flatter.

Figure 9 shows the sum of rewards under the different number of cameras. The sum of the rewards of the random and epsilon-greedy methods decreases rapidly due to the more intensive video tasks in the system as the number of cameras increases. In contrast, the DDQN has the least fluctuation in decline. Further, Fig. 10 shows the sum of rewards for different edge devices' computing capabilities, where DDQN has a faster upward trend compared to the baselines and the Epsilon-Greedy algorithm performs poorly. DDQN outperforms all baselines in terms of computational utilization and collaborative inference and has more stable performance in dynamic environments.

6 Conclusion

With the booming field of ISSs, edge intelligent collaborative inference has great potential for various cross-video stream tracking application scenarios. This paper proposes a lightweight model inference approach for resource-constrained edge computing nodes, including horizontal and vertical methods to partition deep neural networks dynamically. Further, an adaptive strategy based on the MADRL method is proposed to obtain near-optimal solutions for dynamic resource allocation decisions. Experimental results based on the real-world dataset of campus surveillance show that our methods achieve superior effectiveness than baselines.

Acknowledgements. This work is supported by research on key technologies of electrical cloud-edge-end collaborative AI model sharing in Science and Technology Project of State Grid Headquarters (2021, Power base support technology-30, Shandong Electric Power Company, No. 520627210010).

References

1. Bertinetto, L., Valmadre, J., Henriques, J.F., Vedaldi, A., Torr, P.H.S.: Fully-convolutional Siamese networks for object tracking. In: Hua, G., Jégou, H. (eds.) ECCV 2016. LNCS, vol. 9914, pp. 850–865. Springer, Cham (2016). https://doi.org/10.1007/978-3-319-48881-3_56
2. Chang, H.C., Hsu, Y.L., Hsiao, C.Y., Chen, Y.F.: Design and implementation of an intelligent autonomous surveillance system for indoor environments. IEEE Sens. J. **21**(15), 17335–17349 (2021)
3. Chou, Y.S., Wang, C.Y., Chen, M.C., Lin, S.D., Liao, H.Y.M.: Dynamic gallery for real-time multi-target multi-camera tracking, pp. 1–8 (2019)
4. Das, S., Dereniowski, D., Uznanski, P.: Brief announcement: energy constrained depth first search, vol. 107, pp. 1–5 (2018)
5. Girshick, R., Donahue, J., Darrell, T., Malik, J.: Rich feature hierarchies for accurate object detection and semantic segmentation. In: Proceedings of the IEEE Conference on Computer Vision and Pattern Recognition (CVPR) (2014)
6. Gündogan, A., Gürsu, H.M., Pauli, V., Kellerer, W.: Distributed resource allocation with multi-agent deep reinforcement learning for 5g–v2v communication. ACM (2020)
7. He, L., Liu, G., Tian, G., Zhang, J., Ji, Z.: Efficient multi-view multi-target tracking using a distributed camera network. IEEE Sens. J. **20**(4), 2056–2063 (2020)
8. He, W., Guo, S., Guo, S., Qiu, X., Qi, F.: Joint DNN partition deployment and resource allocation for delay-sensitive deep learning inference in IoT. IEEE Internet Things J. **7**(10), 9241–9254 (2020)
9. He, Y., Wei, X., Hong, X., Shi, W., Gong, Y.: Multi-target multi-camera tracking by tracklet-to-target assignment. IEEE Trans. Image Process. **29**, 5191–5205 (2020)
10. Hu, C., Bao, W., Wang, D., Liu, F.: Dynamic adaptive DNN surgery for inference acceleration on the edge. In: IEEE INFOCOM 2019-IEEE Conference on Computer Communications, pp. 1423–1431 (2019)
11. Islam, K.A., Hill, V., Schaeffer, B., Zimmerman, R., Li, J.: Semi-supervised adversarial domain adaptation for seagrass detection using multispectral images in coastal areas. Data Sci. Eng. **5**(2), 111–125 (2020)
12. Kang, Y., et al.: Neurosurgeon: Collaborative intelligence between the cloud and mobile edge. ACM SIG. Comput. Archit. News **45**(1), 615–629 (2017)
13. Ke, R., Zhuang, Y., Pu, Z., Wang, Y.: A smart, efficient, and reliable parking surveillance system with edge artificial intelligence on IoT devices. IEEE Trans. Intell. Transp. Syst. **22**(8), 4962–4974 (2021)

14. Kim, H., Cha, Y., Kim, T., Kim, P.: A study on the security threats and privacy policy of intelligent video surveillance system considering 5G network architecture. In: 2020 International Conference on Electronics, Information, and Communication (ICEIC), pp. 1–4 (2020)
15. Li, D., Zhang, Z., Yu, K., Huang, K., Tan, T.: ISEE: An intelligent scene exploration and evaluation platform for large-scale visual surveillance. IEEE Trans. Parallel Distrib. Syst. **30**(12), 2743–2758 (2019)
16. Li, E., Zhou, Z., Chen, X.: Edge intelligence: on-demand deep learning model co-inference with device-edge synergy. In: Workshop on Mobile Edge Communications, pp. 31–36. ACM (2018)
17. Lu, S., Yao, Y., Shi, W.: Clone: Collaborative learning on the edges. IEEE Internet Things J. **7**, 10222–10236 (2020)
18. Mao, J., Chen, X., Nixon, K.W., Krieger, C., Chen, Y.: MoDNN: local distributed mobile computing system for deep neural network. In: Design, Automation & Test in Europe Conference & Exhibition (DATE), pp. 1396–1401 (2017)
19. Mohammed, T., Joe-Wong, C., Babbar, R., Francesco, M.D.: Distributed Inference Acceleration with Adaptive DNN Partitioning and Offloading. IEEE INFOCOM, pp. 854–863 (2020)
20. Muchtar, K., Afdhal, A., Nasaruddin, N.: Convolutional network and moving object analysis for vehicle detection in highway surveillance videos. In: 2020 3rd International Seminar on Research of Information Technology and Intelligent Systems (ISRITI), pp. 509–513 (2020)
21. Redmon, J., Farhadi, A.: Yolo9000: Better, faster, stronger. In: Proceedings of the IEEE Conference on Computer Vision and Pattern Recognition, pp. 7263–7271 (2017)
22. Tian, B., et al.: Hierarchical and networked vehicle surveillance in its: a survey. IEEE Trans. Intell. Transp. Syst. **16**(2), 557–580 (2015)
23. Winter, K., Wien, J., Molin, E., Cats, O., Morsink, P., van Arem, B.: Taking the self-driving bus: a passenger choice experiment. In: 2019 6th International Conference on Models and Technologies for Intelligent Transportation Systems (MT-ITS) (2019)
24. Xie, J., et al.: Deep learning-based computer vision for surveillance in its: evaluation of state-of-the-art methods. IEEE Trans. Veh. Technol. **70**(4), 3027–3042 (2021)
25. Yoo, H., Kim, K., Byeon, M., Jeon, Y., Choi, J.Y.: Online scheme for multiple camera multiple target tracking based on multiple hypothesis tracking. IEEE Trans. Circuits Syst. Video Technol. **27**(3), 454–469 (2017)
26. You, S., Yao, H., Xu, C.: Multi-target multi-camera tracking with optical-based pose association. IEEE Trans. Circuits Syst. Video Technol. **31**(8), 3105–3117 (2021)
27. Zeng, X., Fang, B., Shen, H., Zhang, M.: Distream: scaling live video analytics with workload-adaptive distributed edge intelligence. In: Proceedings of the 18th Conference on Embedded Networked Sensor Systems, pp. 409–421 (2020)
28. Zhao, Z., Barijough, K.M., Gerstlauer, A.: Deepthings: Distributed adaptive deep learning inference on resource-constrained IoT edge clusters. IEEE Trans. Comput. Aided Des. Integr. Circuits Syst. **37**(11), 2348–2359 (2018)
29. Zhou, X., Xu, X., Liang, W., Zeng, Z., Yan, Z.: Deep-learning-enhanced multitarget detection for end-edge-cloud surveillance in smart IoT. IEEE Internet Things J. **8**(16), 12588–12596 (2021)

Trajectory Optimization for Propulsion Energy Minimization of UAV Data Collection

Juan Xu[1,2(✉)], Di Wu[1], Jiabin Yuan[1], Hu Liu[1], Xiangping Zhai[1], and Kun Liu[1]

[1] College of Computer Science and Technology, Nanjing University of Aeronautics and Astronautics, Nanjing, China
{juanxu,karel1137,jbyuan,liuhu,blueicezhaixp,lk1009}@nuaa.edu.cn

[2] China and Collaborative Innovation Center of Novel Software Technology and Industrialization, Nanjing, China

Abstract. As a flexible communication manner, unmanned aerial vehicle (UAV) communication is a promising technology for wireless communication systems. Considering UAV data collection in wireless sensor network, this paper proposes a novel trajectory optimization scheme to minimize UAV's propulsion energy consumption. The scenario of a fixed-wing UAV flying uniformly at a fixed altitude is considered. Thus the theoretical minimization model of UAV's propulsion energy is derived based on the line-of-sight channel model and reliable communication distance. Then the minimum-degree-prior (MDP) placement algorithm for the minimum clique partitioning problem we just presented in another paper is utilized to deploy the virtual base stations (VBSs) and determine the UAV's waypoints. The trajectory is finally optimized by leveraging the travelling salesman problem with convex optimization technique. Our scheme requires fewer virtual base stations owing to the effectiveness of MDP algorithm as compared with the scheme that first proposed the concept of VBS. The numerical results consequently show that our scheme is superior over the benchmark schemes in the minimization of UAV flight distance and propulsion energy.

Keywords: UAV data collection · Propulsion energy minimization · Trajectory optimization · Minimum-degree-prior placement algorithm · Minimum clique partitioning

1 Introduction

The progress of new technologies promotes the application and development of Internet of Things (IoT) and wireless sensor network (WSN), such as Internet of Vehicles (IoV), blockchain, artificial intelligence (AI) and 5G networks [1–4]. In recent years, with the flexibility and increasing robustness of unmanned aerial vehicles (UAVs), the issue that how to efficiently collect wireless sensor data by the use of UAVs has attracted increasing

This work is supported by the Fundamental Research Funds for the Central Universities (Grant No. NZ2020021), the Aeronautical Science Funds (Grant No. 2020Z073052001) and the National Natural Science Foundation of China (Grant No. 62132008).

B. Li et al. (Eds.): APWeb-WAIM 2022, LNCS 13421, pp. 224–236, 2023.
https://doi.org/10.1007/978-3-031-25158-0_18

interest [5]. Considering the scenario, for example, a large region in harsh environment without available terrestrial communication facilities has deployed a certain number of battery-powered sensor nodes to gather required information (e.g. temperature, gas, and humidity). In this WSN, the nodes close to the sink should forward the collected samples from other nodes, so they will exhaust their energy quickly, leading to an energy hole around the sink nodes [6]. However, if the data collection is assisted by UAV, the energy hole problem could be solved since UAV can fly over all the wireless sensor nodes, collect their data packets and send them back to data center while moving. As a result, the UAV-aided information gathering can balance the energy consumption, reduce the operation power of sensor nodes, so that the entire network lifetime is prolonged [7]. Thus this paper concentrates on the UAV-assisted data collection WSN, and since the UAV can fly close to the nodes so as to shorten the link distance for more energy-efficient data gathering, the system performance critically determined by the UAV trajectories, which is considered to be optimally designed for energy minimization in this paper.

In UAV-assisted WSN, UAV consumes great energy to keep flying and moving, the propulsion energy consumption of UAV accounts for a large proportion of the total energy consumption, far more than the energy required for information transfer [7]. So it becomes an important subject to research the efficient UAV trajectory design, aiming at minimizing the propulsion energy and extending the data collection time for UAV. In 2017, Zeng and Zhang derived a theoretical model on propulsion energy consumption of fixed-wing UAVs as a function of flying speed, direction, and acceleration, and then optimized the trajectory jointly considering both the communication throughput and energy consumption [8]. In 2018, Zhan *et al.* proposed a trajectory design for distributed estimation by applying the classic travelling salesman problem (TSP) and convex optimization methods [9]. In the same year, Zeng *et al.* put forward efficient schemes for the waypoint design based on virtual base station (VBS) placement combined with convex optimization, and achieved significant reduction in mission completion time [10]. In 2019, Zeng *et al.* studied energy minimization of a rotary-wing UAV communicating with multiple ground nodes, and presented efficient algorithms to optimize UAV trajectory and communication time allocation in simple and general cases [11]. In 2021, Jiang and Chen provided an optimal discontinuous UAV trajectory planning based on Q-learning algorithm in reinforcement learning, and improved the energy and data collection efficiency for UAV-assisted data collection [12].

The fixed-wing UAV is considered in this paper. From the analysis of [10], we know that the optimal UAV trajectory only needs to constitute connected line segments, that is, the problem of UAV trajectory optimization can be reduced to finding a set of optimal waypoints. Thus, the novel concept of virtual base station (VBS) was proposed which resembled the standard base station (BS) placement problem for ensuring user coverage with a given coverage distance D. The spiral BS placement algorithm [13] was adopted in [10] and achieved significant performance gains over other schemes. Inspired by this idea, the VBS approach is employed in this paper to optimize the UAV trajectory for information acquisition in WSN aiming at minimizing propulsion energy consumption. Without loss of optimality, we adopt the same assumptions that UAV flies horizontally with a fix altitude H and the UAV-sensor channels are dominated by line-of-sight (LoS) links since the UAV flies close enough to the sensor nodes. And the TSP method with convex optimization technique is also used for further optimization. So the optimal scheme proposed in [10] is taken as the vital benchmark scheme in this paper.

Different from the optimal scheme in [10], this paper utilizes the minimum-degree-prior (MDP) placement algorithm we just proposed in another paper to deploy VBSs, which minimizes the number of base stations by transforming the issue to the minimum clique partitioning problem with the minimum enclosing circle coverage constraint. Furthermore, the scenario of data collection, not information dissemination, is treated here. Thus, for given set of waypoints, we suggest using polling mechanism to save energy of the UAV and the nodes. The numerical results demonstrates that the proposed scheme in this paper is superior over the benchmark schemes in the minimization of UAV flight distance and propulsion energy.

2 Problem Formulation and System Model

In this paper, we consider a wireless communication system consisting of K sensors (generally considered as ground terminal, GT) denoted by the set $\mathcal{K} = \{1, 2, .., K\}$. The locations of all sensors are assumed to be known for the UAV trajectory design, denoted as $w_k, k \in \mathcal{K}$. A fixed-wing UAV flying at a constant altitude H over the sensors is dispatched to collect the data from all the sensors. In practice, H could be chosen as the minimum altitude to ensure safe UAV operations without frequent ascending or descending and shorten the communication distance between the UAV and sensors to save energy. As a result, the channel of UAV-sensor is supposed to be dominated by LoS links whose quality is determined by distance.

Under the LoS channel model, the coverage range of UAV can be formulated to disks with fixed radius, called virtual base station (VBS). To design the UAV trajectory, the number and positions of VBSs are firstly determined according to the distribution of sensors on ground. After that, reliable data transfer with all sensors can be guaranteed as long as the UAV flies through all VBSs. The number of VBSs should be reduced as much as possible to minimize UAV flight distance. The notations used in our scheme are listed in Table 1.

Below we will derive a propulsion-energy-consumption minimization model based on the reliable communication distance.

Table 1. Notations used in our scheme

Symbol	Denotation
$\mathcal{K} = \{1, 2, .., K\}$	Set of sensors
$\mathcal{M} = \{1, 2, .., M\}$	Set of VBSs/MBSs
$\mathcal{D} = \{D_1, D_2, .., D_m\}$	Set of coverage disks
u_m	Horizontal location of mth VBS/MBS

(continued)

Table 1. *(continued)*

Symbol	Denotation
w_k	Horizontal location of kth sensor
d_0	Reliable communication distance
$D(t)$	Reliable communication region
$q(t)$	The UAV position at time t
$G(V, E)$	GT adjacent graph
$V(D_m)$	Set of the vertices in D_m
V_{can}	Set of the candidate vertices

2.1 Reliable Communication Distance

For a given signal modulation scheme (e.g. Binary Phase Shift Keying), the bit error rate of wireless channel ρ_ϵ should be below a exact value (e.g. 10^{-6}) to guarantee the reliable data collection between the UAV and the sensors. So the minimum signal-to-noise ratio at the UAV should be larger than a certain threshold. It can be satisfied when the distance between the UAV and the sensor is less than d_0 based on the LoS channel assumption. Thus d_0 is termed as reliable communication distance, and can be expressed as [10]

$$d_0 \triangleq \sqrt{\frac{\gamma_0}{\rho_\epsilon} - H^2} \geq \|q(t) - w_k\|, 0 < t < T, \tag{1}$$

where γ_0 is signal-to-noise ratio at 1m, ρ_ϵ is bit error rate of wireless channel, H is the flight altitude, $q(t)$ is the UAV position at time t (T is the total flight duration), and w_k is the sensor's horizontal position on the ground. In this section, suppose the transmitting power and minimum signal-to-noise ratio required for all sensors are the same. As a result, the reliable communication distance d_0 for all sensors is the same. The communication range projecting onto the ground can be modeled as a disk when the UAV flies at a fixed height. If the radius of the disk is d_0, the reliable communication region $D(t)$ of UAV can be formulated as [10]

$$D(t) = \left\{q(t) \in \mathbb{R}^{2\times 1} \|\mathrm{q(t)} - \mathrm{w_k}\| \leq \mathrm{d_0}\right\}, 0 < t < T, \tag{2}$$

with all parameters stated above. Assume that data collection time is negligible compared to flight time due to the small amount of data to be transferred from each sensor. Therefore, when the sensor is in the reliable communication range of the UAV, the UAV is considered to have reliable data collection with the sensor. The data collection mission is supposed to be accomplished only when the reliable communication region of the UAV has covered all sensors. Define the coverage indicator function $\hat{I}_n(t)$ and coverage indicator variable I_n as follows [9]:

$$\hat{I}_n(t) = \begin{cases} 1, \text{ if } k_n \in D(t), \\ 0, \text{ otherwise}, \end{cases} \tag{3}$$

$$I_n = \begin{cases} 1, \text{ if} \int_0^T \hat{I}_n(t)dt > 0, \\ 0, \text{ otherwise}, \end{cases} \tag{4}$$

where $\hat{I}_n(t)$ indicates whether the sensor k_n is within $D(t)$ or not at each time instant t, and I_n indicates whether the sensor k_n is covered by $D(t)$ (at least once) during the time horizon T.

2.2 Propulsion Energy Minimization Model

The object of this section is to minimize the propulsion energy while the UAV accesses all sensor nodes to collect data. Define the energy-minimum speed of UAV as V_{me} [8]. Suppose the UAV flies at uniform velocity V_{me} to simplify the energy minimization problem to trajectory minimization problem. Thus we formulate the propulsion energy minimization problem as

$$(\text{P1}) : \begin{cases} \underset{q(t)}{min} \quad E(q(t)) \\ \text{s.t. } \sum_{n=1}^{K} I_n = K, \end{cases} \tag{5}$$

$$q(0) = q_0, q(T) = q_F. \tag{6}$$

In (P1), Constraint (5) indicates that each sensor should be covered by the UAV at least once, and Constraints (6) specifies the initial/final locations for the UAV.

Without loss of optimality, the UAV trajectory consists only of connected line segments [9, 10]. When the reliable communication distance $d_0 = 0$, the UAV needs to fly to the position of each sensor for data collection. In this way, problem (P1) is transformed to the problem of finding the optimal set of ordered path points, which is equivalent to TSP. Otherwise, when $d_0 > 0$, problem (P1) is equivalent to TSPN (Travelling salesman problem with neighborhoods) [14]. However, existing algorithms for solving TSPN require disjoint circles between neighbors, so it cannot be utilized directly to solve (P1). In this paper, we adopt the concept called virtual base station [10] combined with TSPN to determine the initial flight path, and optimize the final path by successive convex approximation technique. The specific scheme of trajectory optimization for propulsion energy minimization is described in the next section.

3 Propulsion Energy Minimization Scheme

3.1 Virtual Base Station Deployment

As deploying virtual base stations, if the radius of coverage disk is d_0, the UAV must fly through the center of the base station to meet the restriction that the distance between the UAV and the sensor should be no more than d_0. On the other hand, if he radius of coverage disk is $d_0/2$, the UAV can travel through any location in each VBS to fulfill the restriction. This offers new freedom degrees in trajectory design. Thus we minimize the number of VBSs to reduce the total length of UAV trajectory on the premise that the VBSs cover all sensors. The specific method of VBS deployment is depicted as follows.

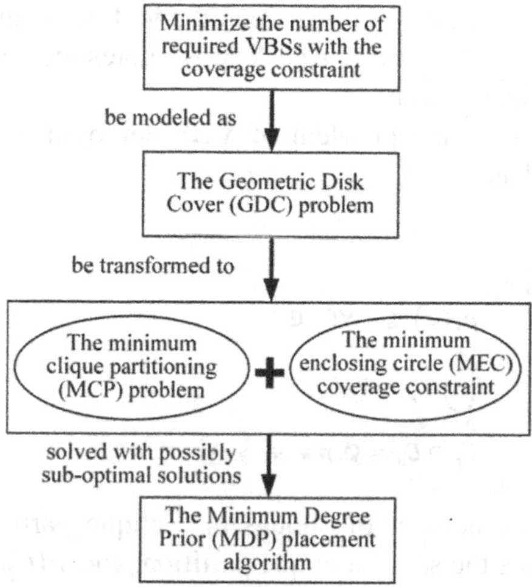

Fig. 1. The solution idea for VBS deployment of MDP algorithm.

Actually, we adopt the minimum-degree-prior (MDP) placement algorithm we just presented in [15] to deploy VBSs. The idea of MDP algorithm is showed in Fig. 1.

Figure 1 Shows that the minimizing the number of required VBSs with coverage constraint has been modeled as the Geometrc Disk Cover (GDC) problem in existing schemes, which can be formulated as [13]

$$\begin{cases} \min\limits_{\{u_m\}_{m\in M}} & |\mathcal{M}|, \\ \text{s.t.} & \min\limits_{m\in M} ||w_k - u_m|| \leq r, \forall k \in K, \end{cases}$$

where |M| is the number of required VBSs, $||w_k - u_m||$ is the horizontal Euclidean distance between GT k and VBSs m.

Since GDC problem is NP-hard, convertion and simplification efforts have been made. Different from these methods, we transform VBS placement problem into the minimum clique partitioning (MCP) problem with the minimum enclosing circle (MEC) coverage constraint and propose a heuristic algorithm with sub-optimal solutions to solve the problem. Here the definition of GT Adjacent Graph is given [15] (GT means ground terminal, and in this paper it refers to wireless sensor).

Definition 1 (GT Adjacent Graph): A GT adjacent graph is an undirected graph $G(V, E)$, where the vertex v_i represents GT i, and e_{ij} represents the edge between the vertices v_i and v_j. The distance between v_i and v_j is the Euclidean distance between GT i and GT j. The construction process of the GT adjacent graph is as follows. Only if the distance between v_i and v_j is not more than $2r$, then there is an edge between v_i and v_j, which can be formulated as follows:

$$\begin{cases} \forall v_i, v_j \in V(G), \text{if distance}\left(v_i, v_j\right) \leq 2r, \exists e_{ij}, & (7) \\ \forall v_i, v_j \in V(G), \text{if distance}\left(v_i, v_j\right) > 2r, \nexists e_{ij}, & (8) \end{cases}$$

where distance (i, j) is the distance between the vertices v_i and v_j. The degree of v_i denotes the number of neighbors of v_i. And $N(v_i)$ denotes the neighbors of v_i, which is the set of the vertices adjacent to v_i.

Based on the definition, the problem of VBS deployment with minimized base stations can be modeled as

$$
(\text{P2}): \begin{cases} \min z \\ s.t. \quad \rho(C_p) \leq r, \forall C_p \in C, & (9) \\ \quad \bigcup_{p=1}^{z} C_p = V(G), & (10) \\ \quad C_p \cap C_q = \emptyset, p \neq q, & (11) \end{cases}
$$

where z indicates the number of cliques in a clique partition of $G(V, E)$, $C = \{C_1, C_2, \cdots, C_z\}$ denotes the set of a clique partition, and $\rho(C_p)$ represents the radius of the minimum enclosing circle for C_p, which is less or equal to r given in Constraint (9). Constraint (10) indicates that the union of all cliques equals $V(G)$, and Constraint (11) means that different cliques have no same vertex.

The proposed algorithm to tackle the VBS deployment problem is called minimum-degree-prior (MDP) algorithm. In this algorithm, vertex v_i with the minimum degree in $G(V, E)$ will be first selected, and a new coverage disk D_m (corresponds to VBS m) is deployed on v_i (corresponds to GT i). If more than one vertex has the minimum degree, one of them will be randomly selected. Then, the location of D_m will be refined to include as many as possible vertices. When D_m is deployed, D_m will be deleted from the GT adjacent graph. Finally, we deploy the next coverage disk D_{m+1} until $G(V, E)$ is empty.

Here is the detail of refining the location of D_m. The vertices not adjacent to v_i will not be considered, since they cannot be covered by the same coverage disk. Thus, the vertices in $N(v_i)$ are the candidate vertices that can be added to D_m. We use V_{can} to denote the set of the candidate vertices. The following steps will continue until V_{can} is empty, at which the deployment of D_m is finished, and D_m will be deleted from $G(V, E)$.

1) Select the vertices within distance r to the center of D_m, and add them to $V(D_m)$. Since they are within the coverage of D_m, remove them from V_{can}.
2) Select the vertex v_j with the minimum degree in V_{can}. If there is more than one vertex with the minimum degree, the vertex with shortest distance to the center of D_m is selected.
3) If v_j is not adjacent to all vertices in D_m, v_j will be not considered. Otherwise, we will execute the MEC subroutine to determine whether v_j can be added to $V(D_m)$.
4) Remove v_j from V_{can}, regardless of whether v_j can be added to $V(D_m)$.

The MDP algorithm described above is summarized in Algorithm 1 [15].

Algorithm 1: Minimum Degree Prior VBS Deployment Algorithm Without The Capacity Constraint

```
Input: GT set 𝒦, coverage radius r.
Output: Coverage disk set 𝒟.
Initialization: G(V,E) ← 𝒦, r; 𝒟 = ∅.
while G(V,E) is not empty do
    Select v_i with the minimum degree in G(V,E).
    Deploy a new coverage disk D_m on v_i.
    Update candidate vertex set V_can ← N(v_i).
    while V_can ≠ ∅ do
        Find all the vertices within distance r to the
         center of D_m. Add them to V(D_m) and
         remove them from V_can.
        Select v_j with the minimum degree in V_can.
        if v_j is adjacent to all vertices in V(D_m) then
            if v_j can be added to V(D_m) via the MEC
             subroutine then
                Refine the location of D_m.
            end
        end
        Remove v_j from V_can.
    end
    𝒟 ← 𝒟 ⋃ {D_m}.
    Remove the vertices in V(D_m) from G(V,E).
end
```

After MDP algorithm is executed, wireless sensors in Fig. 2(a) will be covered by seven VBS as shown in Fig. 2(b).

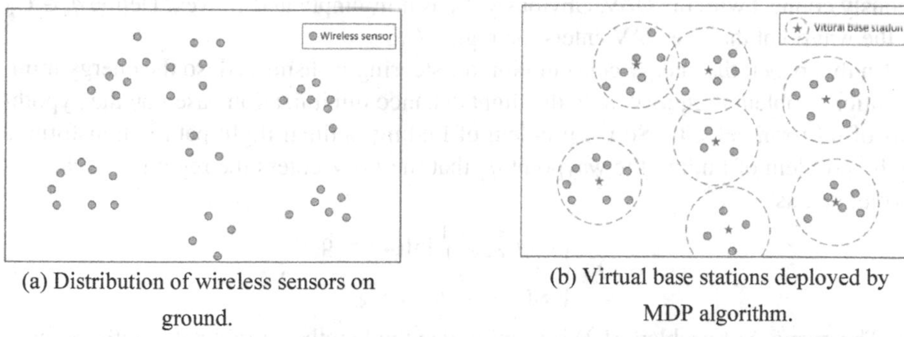

(a) Distribution of wireless sensors on ground.

(b) Virtual base stations deployed by MDP algorithm.

Fig. 2. An illustration of VBS placement after MDP algorithm. Trajectory Initialization and Optimization

3.2 Trajectory Initialization and Optimization

After VBS deployment, the optimal access sequence for VBSs should be determined before initial trajectory design. Suppose the UAV need to fly through the center of all

VBSs, the optimal access sequence will be equivalent to No-Return TSP problem [7]. The initial trajectory obtained by solving the No-Return TSP problem using Matlab toolbox is illustrated in Fig. 3(a). In practice, there is no need for UAV to go through the center of each VBS, so the trajectory can be further optimized by applying convex optimization technique.

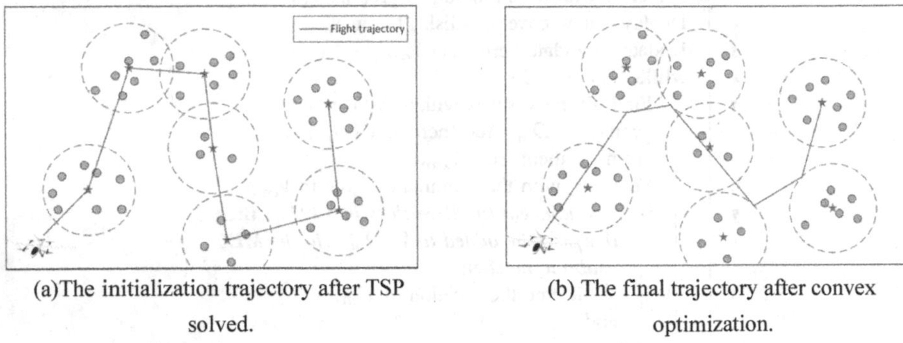

(a)The initialization trajectory after TSP solved.

(b) The final trajectory after convex optimization.

Fig. 3. An illustration of the initialization and optimization of UAV trajectory.

Suppose the GTs are partitioned into G ordered clusters $\mathcal{S}_g\,(g = 1, 2, \cdots, G)$, where $\mathcal{S}_g \subset K$ denotes the subset of GTs covered by the gth VBS. For the gth ordered cluster in $\mathcal{S}_g$, define the following set [10]

$$C_g \triangleq \left\{q \in \mathbb{R}^{2\times 1} \| \mathrm{q} - \mathrm{w_i} \| \leq \mathrm{d_0}, \forall \mathrm{i} \in \mathcal{S}_\mathrm{g}\right\}, \tag{12}$$

where C_g is the set of all possible UAV locations ensuring that all GTs in $\mathcal{S}_g$ simultaneously connect with the UAV. Obviously, C_g is non-empty and convex. Define $q_g \in C_g$ as the waypoint that the UAV enters the region C_g.

In this paper, the energy consumption for steering is dismissed, so the energy minimization problem is equivalent to the flight distance minimization based on the hypothesis of uniform velocity. So the problem of finding optimal flight path is transformed to the problem of finding the waypoint q_g that the UAV enters the region C_g, which is formulated as

$$(\mathrm{P3}): \begin{cases} \underset{q_g}{min} \sum_{g=1}^{G-1} \| \mathrm{q_{g+1}} - \mathrm{q_g} \| \\ \text{s.t.} \qquad q_g \in C_g \end{cases}, \tag{13}$$

The purpose of problem (P3) is to minimize the length of flight path, with a convex optimization objective function and a convex limiting condition. So it can be solved by convex optimization technique. As a result, we can obtain the optimized trajectory shown in Fig. 3(b).

3.3 Polling Mechanism

Based on the feasible waypoints given above, the polling mechanism is suggested to be used in the data collection phase to improve energy efficiency. The polling mechanism can be executed like this [16]:

(1) Since the waypoints are predetermined, the UAV knows the order of VBSs to pass, as well as the sensors belonged to each VBS. As soon as the UAV arrives at the site within the coverage of the first VBS, it sends a wake-up signal to those sensors within this range.
(2) The sensor nodes only periodically turns on their radio in polling period and check for a wake-up signal. Once the wake-up signal is confirmed, the sensor sends its data to the UAV.

The procedure is repeated until the UAV reaches the final location. It is supposed that the polling and data collection process can be completed during the flight time from the UAV reaches the first site within the VBS's coverage to the last one. So the UAV can fly at a fixed speed. If it cannot be completed, the UAV can fly quasi-stationarily around the last site which will not be discussed here.

4 Numerical Results

In this section, numerical results are provided to verify the effectiveness of the proposed scheme. For comparison, three benchmark schemes are chosen as follows:

Scheme 1 (Fly individually): The UAV flies to all sensor positions individually for information acquisition.

Scheme 2 (K-means + COT) [17]: Deploy the virtual base stations using K-means algorithm and design trajectory by taking advantage of the travelling salesman problem (TSR) with convex optimization technique (COT).

Scheme 3 (Spiral+COT) [13]: Deploy the virtual base stations using Spiral algorithm and design trajectory by taking advantage of TSR with COT.

Scheme 4 (MDP+COT): Deploy the virtual base stations using MDP algorithm and design trajectory by solving TSR with COT (the proposed scheme in this paper).

Here we suppose that there are K sensors distributed in a 2000 m*2000 m square area (illustrated in Fig. 4). The main simulation parameters used in this section are showed in Table 2, where $V_{me} = (c_2/3c_1)^{1/4}$, and $P(V) = c_1V^3 + c_2/V$(V is the UAV's fixed flight speed). Besides, the reliable communication distance d_0 between UAV and sensor is determined by sensor transmit power, noise power and error rate requirement in practice, and it is assumed to be $d_0 = 500$ m in this section.

Table 2. The main simulation parameters

Parameter	Value
UAV Flight Altitude H	100 m
UAV Flight Parameter c_1	9.26×10^{-4}
UAV Flight Parameter c_2	2250
Reliable Communication Distance d_0	500 m
Minimum Energy Consumption Flight Speed V_{me}	30 m/s
UAV Flight Power $P(V)$	100 W

The trajectory design results with 50 sensors of the three comparison schemes as well as the proposed scheme are demonstrated in Figs. 4, 5, 6, 7. All the trajectory design algorithms are programmed in Python 3.8, and call Matlab TSP and convex optimization toolbox functions [18, 19] to solve TSP and convex optimization problems. The validation programs are run on the desktop computer with i7–4790 CPU and 16G RAM.

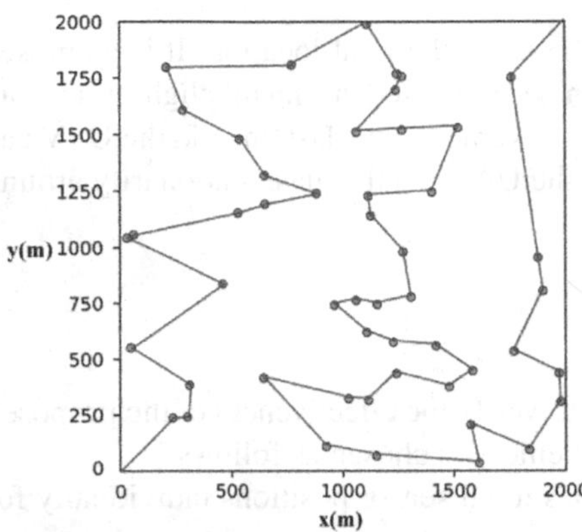

Fig. 4. The trajectory of Scheme 1 (Fly individually). The flight distance is 13448 m, flight time is 448.27 s, and the total energy consumption is 44827 J.

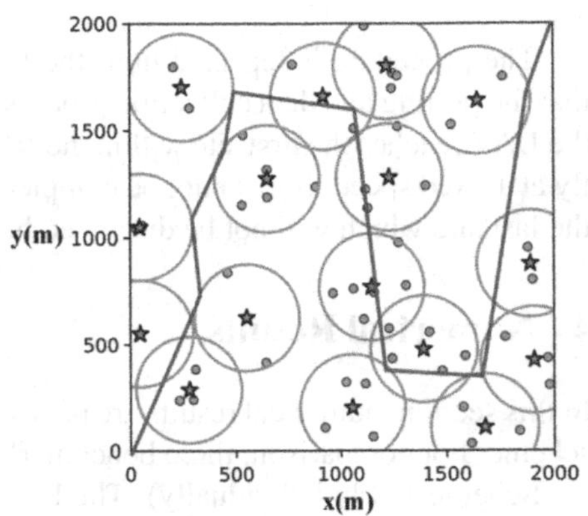

Fig. 5. The trajectory of Scheme 2 (K-means + COT). The flight distance is 5738 m, flight time is 191.27 s, and the total energy consumption is 19127 J.

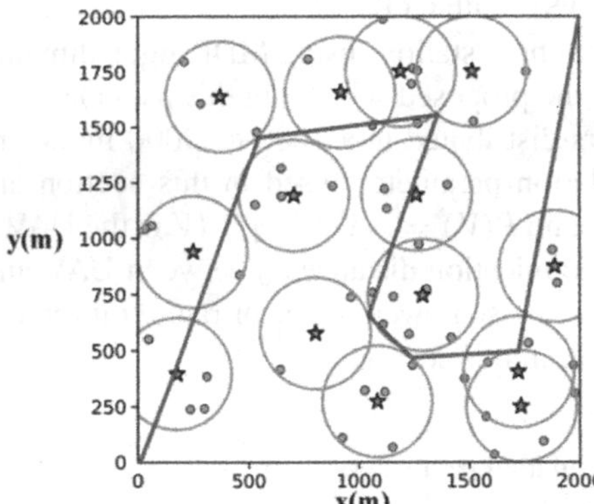

Fig. 6. The trajectory of Scheme 3 (Spiral + COT). The flight distance is 5637m, flight time is 187.9s, and the total energy consumption is 18790J.

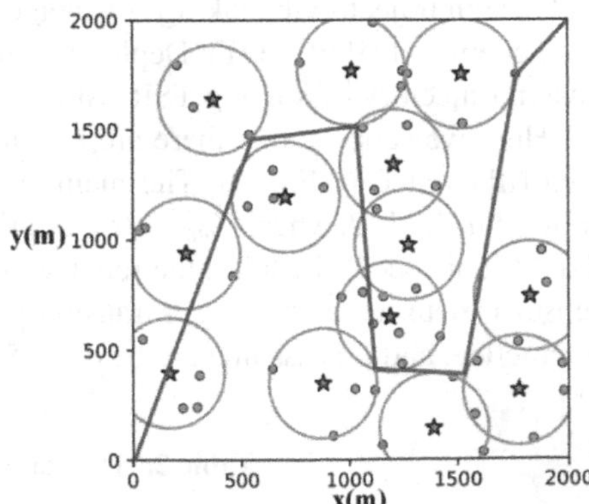

Fig. 7. The trajectory of Scheme 4 (MDP + COT). The flight distance is 5234 m, flight time is 174.47 s, and the total energy consumption is 17447 J.

In Fig. 4, the UAV flies to all sensor positions individually to collect information, underutilizing the facts that the UAV can collect information dependably from the sensors within a certain range. So the flight path of Scheme 1 is quite long and the UAV will consume much energy for data collection. The flight paths of Schemes 2, 3 and 4 are much shorter due to the coverage method. Among these three schemes, MDP algorithm deploys 16 virtual base stations, while K-means algorithm deploys 16 and Spiral deploys 14. Because of the fewest virtual base stations [15], the trajectory obtained by the proposed scheme (MDP + COT) has the shortest flight distance, and the least propulsion energy consumption as compared with the three benchmark schemes.

In order to further verify the effectiveness of our algorithm, the energy consumption values of the four trajectory designs are calculated under several different numbers of sensors. The results are shown in Table 3. It can be seen that when the number of sensors is small, there is little gap between the energy consumption of different schemes; because the number of sensors covered by the UAV at the same time is small. However, when the number of sensors increases, Scheme 1 causes a sharp increase in energy consumption, so the trajectory design of Scheme 1 is only suitable for the case of a small number of sensors. By comparison, the increase of energy consumption of Scheme 2, 3 and 4 is not obvious as the number of sensors increases. Thus the trajectory design methods based on virtual base station can significantly reduce the flight energy consumption of UAV. Finally, we can come to this conclusion again that the trajectory obtained by our scheme consumes the least propulsion energy owing to the least virtual base stations among the four typical schemes.

Table 3. Energy consumption of the four schemes under different number of sensors (Unit: Joule)

Scheme	K (the number of sensors)					
	20	50	80	150	200	300
Scheme 1 (Fly individually)	30567	44827	53463	63383	105930	171933
Scheme 2 (K-means + COT)	18350	19127	20860	21937	22637	23690
Scheme 3 (Spiral + COT)	18037	18790	20143	20667	21077	22507
Scheme 4 (MDP + COT)	17250	17447	19130	19870	20103	21550

5 Conclusion

To minimize the UAV's propulsion energy consumption for data collection in wireless sensor network, this paper optimizes the trajectory by deploying virtual base stations in a new way of coverage. Specifically, firstly the reliable communication range is determined based on the LoS channel model and the reliable communication distance, so that it avoids the UAV flying to the position of sensors one by one for data collection. Then the virtual base stations are deployed by the minimum-degree-prior (MDP) algorithm we just put forward to obtain fewer UAV waypoints than other benchmark algorithms. Finally, the optimized trajectory is designed by leveraging the travelling salesman problem with convex optimization technique. The numerical results show that the proposed scheme can accomplish UAV data collection of all sensors in the wireless network with less propulsion energy consumption as compared with the benchmark schemes.

References

1. Hbaieb, A., Ayed, S., Chaari, L.: A survey of trust management in the Internet of Vehicles. Comput. Netw. **203**(11), 108558 (2022)
2. Li, B., Liang, R., Zhou, W., Yin, H., Gao, H., Cai, K.: LBS meets blockchain: an efficient method with security preserving trust in SAGIN. IEEE Internet Things J. **9**(8), 5932–5942 (2022)
3. Nie, L., et al.: Network traffic prediction in Industrial Internet of Things backbone networks: a multitask learning mechanism. IEEE Trans. Industr. Inf. **17**(10), 7123–7132 (2021)
4. Dwivedi, A.D., Singh, R., Kaushik, K., Mukkamala, R.R., Alnumay, W.S.: Blockchain and artificial intelligence for 5G-enabled Internet of Things: Challenges, opportunities, and solutions. https://doi.org/10.1002/ett.4329. Accessed 14 July 2021
5. Al-Mashhadani, M.A., Hamdi, M.M., Mustafa, A.S.: Role and challenges of the use of UAV-aided WSN monitoring system in large-scale sectors. In: 2021 3rd International Congress on Human-Computer Interaction, Optimization and Robotic Applications (HORA), pp. 1–5 (2021)
6. Ren, J., Zhang, Y., Zhang, K., Liu, A., Chen, J., Shen, X.S.: Lifetime and energy hole evolution analysis in data-gathering wireless sensor networks. IEEE Trans. Industr. Inf. **12**(2), 788–800 (2016)
7. Zeng, Y., Wu, Q., Zhang, R.: Accessing from the sky: A tutorial on UAV communications for 5G and beyond. Proc. IEEE **107**(12), 2327–2375 (2019)
8. Zeng, Y., Zhang, R.: Energy-efficient UAV communication with trajectory optimization. IEEE Trans. Wireless Commun. **16**(6), 3747–3760 (2017)
9. Zhan, C., Zeng, Y., Zhang, R.: Trajectory design for distributed estimation in UAV-Enabled wireless sensor network. IEEE Trans. Veh. Technol. **67**(10), 10155–10159 (2018)
10. Zeng, Y., Xu, X., Zhang, R.: Trajectory Design for completion time minimization in UAV-enabled multicasting. IEEE Trans. Wireless Commun. **17**(4), 2233–2246 (2018)
11. Zeng, Y., Xu, J., Zhang, R.: Energy minimization for wireless communication with rotary-wing UAV. IEEE Trans. Wireless Commun. **18**(10), 2329–2345 (2019)
12. Jiang, B., Chen, H.: Trajectory Planning for unmanned aerial vehicle assisted WSN data collection based on Q-Learning. Computer Eng. **47**(4), 127–134,165 (2021)
13. Lyu, J.B., Zeng, Y., Zhang, R., Lim, T.J.: Placement optimization of UAV-mounted mobile base stations. IEEE Commun. Lett. **21**(3), 604–607 (2017)
14. Dumitrescu, A., Mitchell, J.S.: Approximation algorithms for TSP with neighborhoods in the plane. J. Algorithms **48**(1), 135–159 (2003)
15. Wu, D., Xu, J., Yuan, J., Zhai, X.: A novel deployment method for UAV-mounted mobile base stations. In: The 17th International Conference on Mobility, Sensing and Networking. IEEE, New York (2021). *to be published*
16. Zohar, N.: Efficient data gathering from passive wireless sensor networks. In: 2021 Wireless Telecommunications Symposium (WTS), 1570705088. IEEE, New York (2021)
17. Galkin, B., Kibilda, J., DaSilva, L.A.: Deployment of UAV-mounted access points according to spatial user locations in two-tier cellular networks. In: 8th Wireless Days (WD) Conference. IEEE, NEW YORK, pp. 1–6 (2016)
18. Ghadle, K.P., Muley, Y.M.: Travelling salesman problem with MATLAB programming. Int. J. Advances in Applied Mathematics and Mechanics **2**(3), 258–266 (2015)
19. Grant, M., Boyd, S.: CVX: Matlab software for disciplined convex programming, version 2.1 (2014)

Robust Clustered Federated Learning with Bootstrap Median-of-Means

Ming Xie, Jie MA, Guodong Long(✉), and Chengqi Zhang

Australian Artificial Intelligence Institute, Faculty of Engineering and IT, University of Technology Sydney, Sydney, Australia
{ming.xie-1,jie.ma-5}@student.uts.edu.au,
{guodong.long,Chengqi.Zhang}@uts.edu.au

Abstract. Federated learning (FL) is a new machine learning paradigm to collaboratively learn an intelligent model across many clients without uploading local data to the server. Non-IID data across clients is a major challenge for the FL system because its inherited distributed machine learning framework is designed for the scenario of IID data across clients. Clustered FL is a type of FL method to solve non-IID challenges using a client clustering method in the FL context. However, existing clustered FL methods suffer the challenge of processing client-wise outliers which could be produced by minority clients with abnormal behaviour patterns or be derived from malicious clients. This paper is to propose a novel Federated learning framework with Robust Clustering (FedRoC) to tackle client-wise outliers in the FL system. Specifically, we will develop a robust federated aggregation operator using a bootstrap median-of-means mechanism that can produce a higher breakdown point to tolerate a larger proportion of outliers. We formulate the proposed FL framework into a bi-level optimization problem, and then a stochastic expectation-maximization method is adopted to solve the optimization problem in an alternative updating manner by considering EM steps and distributed computing simultaneously. The experiments on three benchmark datasets have demonstrated the effectiveness of the proposed method that outperforms other baseline methods in terms of evaluation criteria.

Keywords: Federated learning · Robust clustering · Bootstrap median-of-means

1 Introduction

Federated Learning (FL) is a new machine learning paradigm to enable many clients collaboratively learn intelligent models. The vanilla FL, namely FedAvg [28], was proposed to learn a server-side intelligent model using many distributed clients without direct access to their local dataset. This distributed machine learning framework with data locality can greatly mitigate the risk of privacy [19] in contrast to a traditional learning system with centralized data storage.

B. Li et al. (Eds.): APWeb-WAIM 2022, LNCS 13421, pp. 237–250, 2023.
https://doi.org/10.1007/978-3-031-25158-0_19

Due to the heterogeneous nature of such a distributed system, a major challenge for FL is to tackle non-IID data across clients. For example, a smartphone typing tool, GBoard in an Android smartphone, needs to auto-fill the incomplete words by considering the typing context and user's language preference. The user's historical data are usually non-IID across clients, thus the learning system needs to tackle this non-IID challenge in the FL's distributed settings.

To solve the non-IID challenge in FL, one solution is to enhance the robustness of a single model at the server to tackle various distributions across clients. However, this kind of solution can only tackle the scenario with slight differences of data distributions across clients. A recent solution for tackling non-IID issues is personalized FL that aims to optimize each client-wise local model while using the global model as a regularize to exchange shared information and constraint the divergence across client-wise local models. The personalized FL suffers the increased complexity of optimization problems that usually treat client-wise model learning as a joint optimization problem across clients. Moreover, it is impractical to find a proper trade-off between shared knowledge and personalization.

Clustered FL is a trade-off solution between single model FL and personalized FL. It aims to learn multiple global models on the server while each global model is a cluster-wise personalized model for the clients with similar data distribution. In particular, the clustering method is a tool to assign clients to different clusters. Therefore, clustered FL can gain better personalization capability than single model FL, and also can learn a model with better generalization than client-wise personalized FL methods. However, clustering among clients is very sensitive to outliers or adversarial attacks. In general, an Outlier is a data point that primarily differs from other observations. Some examples of outliers in the case of Federated Learning are those systematic mislabelling of data or Byzantine failures [4]. In practice, even a small proportion of outliers can render clustering unreliable, while cluster centres and model parameter estimators can be severely biased. Therefore, tackling client-wise outliers will be a new challenge for clustered FL systems.

To tackle the aforementioned challenge, this paper proposes a novel robust clustered Federated Learning framework to tackle the client-wise outliers which could be a minority of users with abnormal behaviour patterns or could be from malicious clients equipped with Byzantine attack tools, i.e., arbitrarily corrupt the information using some adversarial attache mechanism. In particular, we enhance the federated aggregation mechanism by adopting a bootstrap sampling method and a robust approach based on a median-of-means estimator. We formulate the problem into a bi-level optimization framework for a general form and then use a stochastic EM method to solve the optimization problem in an alternative updating strategy.

The motivation for using bootstrap median-of-mean to implement robust clustered FL is quite straightforward. The clients clustering in the FL system is usually based on the client's local models that usually to be a high dimensional vector derived from deep neural networks, such as CNN, RNN and Transformers.

The high dimensional data exaggerates the outlier problem in clustering, and also most distance-based regularization-based robust clustering is impractical in this scenario. Specifically, most clustering methods use a mean-based estimator to compute the centre of a cluster. However, computing barycenter or mean can be very sensitive to the presence of outliers. In contrast to the mean-based estimator with penalty term, a median point-based estimator will be a better option to implement the robustness of the clustering algorithm.

The paper's contributions are summarized as below.

- We propose a simple yet effective approach, namely FedRoc, to implement robust clustered FL.
- We adopt bootstrap sampling during initialization together with a median-of-means estimator to solve the outlier problem in clustered FL contexts.
- We formulate the problem into a bi-level optimization problem.
- Compared with other FL methods on a few datasets, it shows that FedRoC is computationally competitive and more robust than any other baseline algorithms.

The remainders of this paper is organized as below. We will introduce the related work at Sect. 2, and then introduce the method at Sect. 3. The experiment results has been analyzed in Sect. 4. We make conclusion and discuss future work at Sect. 5.

2 Related Work

2.1 Clustered Federated Learning

FL is designed for specific scenarios that can be further expanded to a standard framework to preserve data privacy in large-scale machine learning systems or mobile edge networks [20,23,24]. For example, [38] expanded FL by introducing a comprehensive, secure FL framework that includes horizontal FL, vertical FL, and federated transfer learning. [19] discussed the advances and open problems in FL. [6] proposed LEAF, a benchmark for federated settings with multiple datasets. [25] proposed an object detection-based dataset for FL.

Traditionally, the data distribution over different workers in decentralised setting is non-IID, which is a natural assumption of real-world applications. However, early FL approaches [28] use only one global model as a single-center to aggregate the information of all users. The stochastic gradient descent (SGD) for single-center aggregation is designed for IID data, and therefore, conflicts with the non-IID setting in FL. Some research work is done which are popular approaches to this problem, clustering, multi-task learning, local adaption, ensemble learning. Federated clustering approaches can be divided into two types: model clustering and data clustering.

CFL [30] and Robust FL in heterogeneous network [16], which claims a novel framework involves three-steps module process, both are identified as FL model clustering methods. CFL is a method that extends existing FedAvg with iterative clustering. Analogous to CFL, Robust FL in heterogeneous network, also identified as FL clusters models, which performs clustering on local empirical risk minimizers, but the difference is that these clustering models are not based on FedAvg and only need to complete local independent training. Moreover, this method incorporate three different modular steps, each of those has not been fully theoretically work well as a system and thus not suitable to sensitive applications. Research [15] proposes a similar algorithm named IFCA, for which a convergence bound is established under the assumption of good initialization and all clients having the same amount of data. The work [26] proposes a unified bi-level optimization framework for CFL and prove the convergence. Typical algorithms include FedAMP [18], which adds an attention-inducing function to the local objective, and pFedMe [11], which formulates the regularization as Moreau envelopes. A highly influential work of multi-task learning in FL is [31]. Multi-task learning has recently emerged as an alternative approach to learn personalized models in the federated setting and allows for more nuanced relations among clients' models. A number of other robust study in the field distributed or Federated Learning [14,39], but they do not have a clustering structure of the nodes. Some other techniques including prototype [32] and graph [7] also be applied into FL to improve its privacy or performance.

In addition, the idea of Clustered FL is to train multiple global statistical models w^k instead of one. The principle is to estimate a worker node cluster identity via finding a center with minimal distance. Here the distance-based clustering method is used, and the metric is often squared Euclidean distance. Throughout the learning process, the worker nodes will have a cluster identity and run local updates within the same cluster. Then each cluster obtains a new global model using a weighted average that is identical to Federated Averaging. FeSEM [35] can be viewed as one of the important techniques to learn an unknown mixture from samples. Though, one drawback of existing EM is poorly performed with dimensions. Given that the local model size can have millions of parameters in practice, the need to have the robustness of high dimensional SEM is naturally obvious. This work focus on the property of robustness to a small number of outliers and improved convergence. In the future, we will robustify the SEM against resistance to malicious data or model attacks such as adversarial federated nodes.

2.2 Model Poisoning Attacks in Federated Learning

In this section, we will introduce our adversarial settings. There are two common threats models in Federated Learning; the first case is data is mislabeled or maliciously injected some wrong data, which is called noise labelling and has been addressed by some work including [34]; the other threat is the attackers aim to manipulate the learning process such that the learned model has a higher testing error rate. Normally, attackers can only inject data into training datasets with the aim to make data poisoning attacks when they have full control of those worker nodes, while the learning process is often deemed as protected. This work will focus on model poisoning attacks and is based on the assumption that attackers have knowledge about the aggregation rule. The most basic aggregation rule uses mean estimator and weighted mean in Federated Learning.

Several model poisoning attack and their variants emerged in recent literatures [3]. In this work, we do not claim Bootstrap of Median of Means is a comprehensive defense measure against heavy model poisoning attacks. Hence, we select the idea of Krum [4] which is simply boosting each iteration of the learned model in some worker nodes to manipulate the learning process such that the learned model achieved label misclassifications. The way of explicit boosting works is to mimic the benign worker nodes during the learning process; the node tries to perform the same number of epochs on the local dataset via the same training objectives to obtain an initial gradient update. Since the malicious node wants to ensure the outcome deviates from the true label, it will have to overcome the scaling effect of gradient updates collected from other nodes. In other words, the final gradient updates the malicious nodes send back are then scaled a factor λ by which the malicious nodes boost the initial update. This attack has proved to negate the combined effect of normal worker nodes.

2.3 Robust Clustering

The robust clustering algorithm is a principal approach to enhance the robustness against the presence of outliers [13]. A significant number of work has been reported in this context, such as [10,29]. Typical robust clustering methods include mixture modeling [36], trimming approach [12]. A number of works in robust clustering have been studied by [1,9,12,17,36,37] from recent literature. Recent works of using bootstrap of classical MOM with K-means are emerging [5]. The median-of-means (MOM) estimator of the mean in dimension one consists in taking the median of some arithmetic means derived from a collection of samples, as in our case, derived from a collection of local model parameters. The bootstrap of median-of-means is thus a collection block b_1^B, which are generated through a random process, that proceeds without replacements (disjoint blocks) and according to the uniform distribution on the remaining data at each step. It is worth noting that bootstrap MOM (bMOM) is a randomized estimator. Also, for any fixed sample size n, we can choose any block size n_B and the number of blocks B to define a bMOM estimator, on the contrary to the classical MOM, where the product of the block size with the number of blocks should be equal to the sample size.

3 Methodology

3.1 Problem Definition

The basic of FL is that many clients in a network collaboratively train a global model. The dataset kept on a local device is only a shard of a much larger dataset. To formulate the FL system, it is composed of m smart devices that has a private dataset $\mathcal{D}_i = \{< \mathcal{X}_i, \mathcal{Y}_i >\}$ for each, where $\mathcal{X}_i$ and $\mathcal{Y}_i$ denote input samples and corresponding labels respectively, and $i \in \{1, ..., m\}$ is the index of a client. Each dataset $\mathcal{D}_i$ on the device will be used to train a local supervised learning model $\mathcal{M}_i : \mathcal{X}_i \rightarrow \mathcal{Y}_i$. $\mathcal{M}$ usually denotes a deep neural model parameterized by weights ω. It is built to solve a specific task, and all devices share the same model architecture. Generally, the problem of FL can be usually denote as minimizing this formula below,

$$\min_{w \in \mathbb{R}^d} \sum_{i=1}^{m} \frac{n_i}{n} \mathcal{L}(\mathcal{M}_i, \mathcal{D}_i, \omega_i) \tag{1}$$

where n_i is the number of samples kept in i-th node and n is the total number of samples on all nodes, $\mathcal{L}(\mathcal{M}_i, \mathcal{D}_i, \omega_i)$ is the local objective function which describes how good are the trained classifier, and the local objective function is different for different classifiers (e.g., logistic regression, neural network). For the i-th device, given a private training set $\mathcal{D}_i$, the training procedure of $\mathcal{M}_i$ is briefly notes as

$$\omega_i^* = \underset{\omega_i}{\operatorname{argmin}} \, \mathcal{L}(\mathcal{M}_i, \mathcal{D}_i, \omega_i), \tag{2}$$

where $\mathcal{L}(\cdot)$ is a general definition of loss function for any supervised learning task, and its arguments include model structure $\mathcal{M}_i$, training data D_i and learnable parameters ω_i whose vector space is d dimensional.

In particular, a master node needs to keep a global model whose parameters w normally is the weighted average of model parameters ω_i of all worker nodes. During the FL learning process, three steps will be performed at each iteration:

- step I: a master node sends the global model parameter to all.
- step II: worker nodes compute update with respect to the global model parameter using local objective function and training data kept on devices then sends back the update.
- step III: a master node aggregates all updates to obtain a new global model using a certain aggregation rule. In the case of Federated Averaging, the rule is the weighted average. Formally noted as: $w = \sum_{i=1}^{m} \frac{n_i}{n} \omega_i$.

To tackle the non-IID challenge in FL, clustered FL is an important variant which can achieve good performance on this. And its formulation can be written as a bi-level objective as below,

$$\underset{C}{\text{minimize}} \ \frac{1}{m} \sum_{k=1}^{K} \sum_{i=1}^{m} r_{i,k} \mathcal{L}(\mathcal{M}, D_i, c_k) \tag{3a}$$

$$\textit{subject to } r_{i,k}, C = \underset{r_{i,k}, C}{\operatorname{argmin}} \sum_{k=1}^{K} \sum_{i=1}^{m} r_{i,k} d(g_i, G_k) \tag{3b}$$

where all clients share the same model structure $\mathcal{M}$, K is the number of clusters, $r_{i,k}$ is the indication assignment matrix of the clustering problem to determine that i-th device belongs to cluster k, $C = \{c_1, \ldots, c_K\}$ represents centroids of K clusters, g_i and G_k are the measure of client i and cluster k, respectively, which can be model parameters or loss, d is the distance function. To simply the formulation, weight of each client is set to be $1/m$. The upper Eq. 3a is the objective of FL, and the lower Eq. 3b is the clustering objective, while a bi-level optimization structure is adopted to connect the FL with clustering.

3.2 Robust Clustered FL with bMOM

While K-means is widely used for clustering, its robustness is limited. It shows a poor convergence rate or is not able to correctly group data when there are outliers and adversarial contamination. Several other robust versions of EM or K-means already existing, such as K-PDTM, trimmed-K-means, K-medians have also been proposed. Compared to the above robust variants of K-means, K-bMOM may be increased the computation complexity at each step of the learning process due to bootstrapping. Yet, K-bMOM holds a number of beneficial effects, such as better break down points. Also, K-bMOM in theory has a higher convergence rate when there is a certain level of outliers in the sample compared to the K-means method.

The breakdown point is a classical measure in the robust statistics literature to measure, which represents the maximum proportion of outliers that leaves the estimator bounded. In bMOM, it implies if the block size n_B is rightly chosen, then the probability that the bMOM remains stable under adversarial contamination tends to be one when the number of blocks tends to infinity [5]. This shows that when the number of blocks is big enough, then the bMOM can have a better breakdown point than empirical means. Overall, to address the outliers and non-IID in FL and make the performance of FL more robust, bMOM estimator is imported to combined with clustered FL. And the loss function of the clustering task can be defined as:

$$\mathcal{R} = \text{med}\{\sum_{k=1}^{K}\sum_{i=1}^{m} r_{i,k}\|\omega_i - c_k^{(b)}\|_2^2 : b \in \{1, ..., B\}, n_B > K\}, \tag{4}$$

where B blocks are bootstrapped from m devices' data with block size $n_B > K$, med is to find the median of B minimum losses, and $c_k^{(b)}$ is the center of cluster k based on b-th block of data.

Combined bMOM with the clustered FL problem, using model parameters to measure the client or cluster, and Euclidean distance to measure the distance of clients and clusters, and k-means as the clustering method, we can formulate the loss as a bi-level optimization problem.

$$\underset{C^{(b_{med})}}{\text{minimize}} \ \frac{1}{m}\sum_{k=1}^{K}\sum_{i=1}^{m} r_{i,k}\lambda_i \mathcal{L}(\mathcal{M}_k, \mathcal{D}_i, c_k^{(b_{med})}) \tag{5}$$

$$\textit{subject to } r_{i,k}, C^{(b_{med})} = \underset{r_{i,k}, c_1^{(b_{med})}, ..., c_K^{(b_{med})}}{\text{argmin}} \mathcal{R} \tag{6}$$

where $C^{(b_{med})} = \{c_1^{(b_{med})}, \ldots, c_K^{(b_{med})}\}$ represents centroids of the block b_{med}, which has the median loss in B blocks, and λ_i is the weight of i-th device.

Given the above formulation and analysis, we believe bMOM is robust against this setting due to the statistical nature of the classical median of means. As with every iteration in the learning process, we sample a list of blocks from the worker nodes. When the number of malicious worker nodes is sufficiently small than the number of blocks, this randomness will nullify the effect caused by malicious worker nodes. As we take the median of a list of empirical risk computed on each cluster, the block that malicious worker node is naturally either larger or small than the median empirical risk of normal block and will be discarded. In the experiment section, this feature of K-bMOM is proved empirically.

3.3 Algorithm

To address the bi-level optimization problem above, we proposed an Robust Clustered FL algorithm called FedRoC. FedRoC starts with K initial model parameters. To initialize FedRoC, firstly we do *Bootstrap*(B, n_B) to sample $n_B > K$ devices with replacement randomly and uniformly for B times to get B blocks $1, \ldots, B$. Then k-means++ initialization [2] is proceeded for each block, and the empirical risks of B blocks are calculated. At last block with median risk and its centroids are got.

For the iterative round, four steps will be performed. At first we still need to do *Bootstrap*(B, n_B), and then we perform the EM algorithm and calculate the empirical risk for each block. The next step is to select the block which has the median clustering loss, and its centroids. And the last step is to perform local update of FL for each cluster in the selected block, and get the updated centroids. Then we iterate these four steps until convergence.

The pseudo code of FedRoC is shown in Algorithm 1.

4 Experiment

4.1 Training Setups

As a proof-of-concept scenario to demonstrate the effectiveness of the proposed method, we experimentally evaluate and analyze the proposed FeRobust on federated benchmarks dataset(Caldas et al. 2018).

Dataset. We employed three publicly-available federated benchmarks datasets introduced in LEAF [6], which is a benchmarking framework for learning in federated settings. The datasets used are Federated Extended MNIST (FEMNIST)[1] [8] and Federated CelebA (FedCelebA)[2] [22], and finally Synthetic dataset which inspired by [21]. We follow the data processing instructions from its official repository. In FEMNIST, the handwritten images are splited according to the writers. For FedCelebA, the face images are extracted for each person and developed an on-device classifier to recognize whether the person smiles or

[1] http://www.nist.gov/itl/products-and-services/emnist-dataset.

[2] http://mmlab.ie.cuhk.edu.hk/projects/CelebA.html.

Algorithm 1: FedRoC

```
Input: {D_1, D_2, ..., D_m}, K
Output: r_{i,k}, C^{(b_med)}
Initialize:
Bootstrap(B, n_B)
for blocks b from 1 to B: do
    K-means++ initialization
    Calculate the empirical risk
end
Select the block b_med get initialized centroids {c_1^{(b_med),0}, ..., c_K^{(b_med),0}}.
Iterate:
while stop condition is not satisfied do
    Bootstrap(B, n_B)
    for blocks b from 1 to B: do
        E-Step:
        Assign each device in b to its closest centroid using updated centroids
        M-Step:
        Recompute the centroids
        Calculate the empirical risk
    end
    Select the block b_med and get centroids {c_1^{(b_med),t}, ..., c_K^{(b_med),t}}
    Federated Learning-Step:
    for each cluster k = 1,... K in b_med do
        Assign c_k^{(b_med)} to every device in Cluster k. for i ∈ C_k do
            for E local epochs do
                c_k^{(b_med),t+1} ← c_k^{(b_med),t} − η∇L(c_k^{(b_med),t}, M_k, D_i)
            end
        end
    end
end
```

not. For Synthetic, the dataset is generated with 1000 nodes and five classes. A statistical description of the datasets is described in Table 1.

Local Model. We use a CNN with the same architecture from [22] for two image classification datasets and multi-class logistic regression for a Synthetic dataset. Two data partition strategies are used: (a) an ideal IID data distribution using randomly shuffled data, (b) a non-IID partition by use a $\mathbf{p}_k \sim Dir_J(0.5)$. Part of the code is adopted from [33]. For FEMINST data, the local model's learning rate is 0.003, and the local epoch is 5. For FedCelebA, the learning rate is 0.1, and the local epochs are 10.

Baselines.

1. **NonFed**: We will conduct the supervised learning task at each device without the FL framework.

Table 1. Statistics of datasets. "# of inst. per dev." represents the average number of instances per device.

DATASET	SYNTHETIC	FEMNIST	FEDCELEBA
# Data points	107553	805,263	200,288
Model	LOG-REG	CNN	CNN
Classes	5	62	2
# of device	1000	3,550	9,343
LR	0.01	0.003	0.1
Epochs	10	5	10

2. **FeSEM**: A clustered FL method that clusters clients by considering the distance between their model parameters. It uses stochasti EM as the optimization algorithm. [35]
3. **FedAvg**: The vanilla FL method [28] proposed by Google in 2017. It is an SGD-based FL with weighted averaging.
4. **FedCluster**: A clustered FL method that is to enclose FedAvg into a hierarchical clustering framework [30].
5. **HypCluster(K)**: A clustered FL method that measure distance using the performance of each client's model, namely hypothesis-based clustered FL [27].
6. **FedRoC** Our proposed algorithm that is robust clustered FL algorithm using bootstrap median-of-mean to tackle outliers in clustering.

Training Settings. We used 80% of each device's data for training and 20% for testing. For the initialization of the cluster centers in FeSEM, we conducted pure clustering 20 times with randomized initialization, and then the "best" initialization, which has the minimal intra-cluster distance, was selected as the initial centers for FeSEM. For the local update procedure of FeSEM, we set N to 1, meaning we only updated ω_i once in each local update.

Evaluation Metrics. Given each global model of a cluster perform differently across numerous devices of a cluster, we evaluated the overall performance of the FL methods. We used classification accuracy and F1 score as the metrics for the two benchmarks. In addition, due to the multiple devices involved, we explored two ways to calculate the metrics, i.e., micro and macro. The only difference is that when computing an overall metric, "micro" calculates a weighted average of the metrics from devices where the weight is proportional to the data amount, while "macro" directly calculates an average over the metrics from devices.

4.2 Experiment Analysis

Convergence. To verify the convergence of the proposed approach, we conducted a convergence analysis by running FedRoC with different cluster numbers K

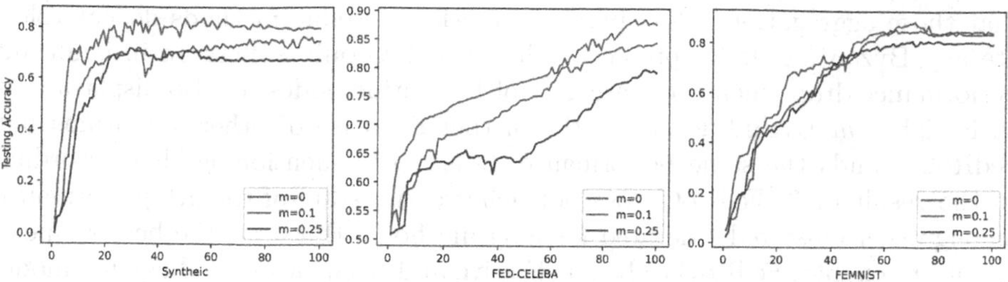

Fig. 1. Convergence Analysis from Benchmarks with Model Poisoning Attack

(from 2 to 5) in 100 iterations by the same set of other hyperparameters. As shown in Fig. 1, robust clustered Federated Learning can efficiently converge on all datasets, and results show that the best performance can be achieved with the cluster number $K = 5$. In this figure, we show the testing accuracy against the number of iterations on three datasets. The Red line shows the FeRobust with no model poisoning attack. The green line shows the FedRoC with 10% of workers are Byzantine nodes, and the Blue line shows the FeRobust with 25% of workers are Byzantine node. The figure display that the testing accuracy of FedRoC dropped by varied of 3.0-11.2 in Synthetic, Feminist and Celeba, while the testing accuracy on Celeba decreases the most.

Table 2. FeSEM v.s. FedRoC

Dataset	Approach	No Attak	m = 0.1	m = 0.25
Synthetic	FeSEM	2.8 ± 1.6	3.6 ± 2.2	3.9 ± 2.0
	FedRoC	1.5 ± 0.6	1.8 ± 0.8	1.0 ± 0.8
FEMNIST	FeSEM	3.0 ± 0.2	4.1 ± 2.4	4.9 ± 2.0
	FedRoC	1.2 ± 0.2	1.1 ± 0.6	1.1 ± 0.8
Celeba	FeSEM	2.4 ± 0.2	3.2 ± 2.0	3.5 ± 2.1
	FedRoC	0.7 ± 0.5	0.7 ± 0.7	0.9 ± 0.7

Comparison Study. We report the experiment of classification on three datasets and start training a global model with/without Byzantine nodes. There are m=0.05 of total clients as Byzantine nodes. Unsurprisingly, the average convergence rate without Byzantine nodes is faster than FedRoC, even without Byzantine nodes. However, figure on the right, we report the case training a global model with Byzantine nodes. Each Byzantine node estimates an update on their auxiliary datasets and before sending it to the server, scaled by a large factor (set to the number of total workers that have sampled to train at the current round), note that in FeSEM and Fedavg, each worker, including Byzantine worker is selected uniformly at the beginning of each round. The figure displays that those aggregation rules operated by other baseline methods do not tolerate any Byzantine nodes presence, while FedRoC only suffers an insignificant performance drop when there are 25% of Byzantine nodes. It also displays that with higher m Byzantine nodes, the further decrease of other algorithms but FedRoC stands the same performance. It is worth mentioning that according to the result of 3, FedRoC does not achieve the state-of-the-art performance among other clustered federated learning methods. However, the better "mean operator" enables FedRoC to be an effective and resilient method against model poisoning attacks. Our empirical result is aligned with the property of FedRoC, and the average similarity measure of each cluster with its most similar cluster in 2 supports this hypothesis.

Table 3. Comparison of our proposed FedRoC(K) algorithm with the baselines on FEMNIST and FedCelebA datasets. Note the number in parenthesis denotes the number of clusters, K.

Datasets	FEMNIST				FedCelebA			
Metrics(%)	Micro-Acc	Micro-F1	Macro-Acc	Macro-F1	Micro-Acc	Micro-F1	Macro-Acc	Macro-F1
NoFed	79.0	67.6	81.3	51.0	83.8	66.0	83.9	67.2
FeSEM	90.3	70.6	91.0	53.4	93	74.8	94.1	69.4
FedAvg	84.9	67.9	84.9	45.4	86.1	**78.0**	86.1	54.2
FedCluster	84.1	64.3	84.2	64.4	86.7	67.8	87.0	67.8
HypCluster	77.7	60.9	74.2	62.4	77.6	55.4	80.4	55.2
FedRoC(5)	88.6	69.3	86.3	62.2	87.2	72.7	90.1	68.3

Clustering Case Study. As a case study, a figure displays that plots all nodes and their assignment in training. The highlight of this table is that, as expected, every iteration of the block, which is the median block among others, shows different risks to those blocks has Byzantine nodes. Those blocks show very different statistical properties, and their gradient updates will not take into the global model. It also shows a successful dense against model poisoning attacks.

5 Conclusion and Remarks

In this paper, we propose a novel clustered FL approach to offset the adversarial worker nodes using the K-bMOM estimator. This algorithm has better breakdown point to address outliers and converges fast. In experiments based on three datasets, the proposed algorithm has similar performance to other baselines while showing much superior clustering performance than baseline methods in all three datasets. We also discussed possible extensions of FedRoC for better distance computation if the local model is a neural network.

References

1. Ana, L.F., Jain, A.K.: Robust data clustering. In: 2003 IEEE Computer Society Conference on Computer Vision and Pattern Recognition, 2003. Proceedings, vol. 2, p. II. IEEE (2003)
2. Arthur, D., Vassilvitskii, S.: k-means++: The advantages of careful seeding. Tech. rep, Stanford (2006)
3. Bhagoji, A.N., Chakraborty, S., Mittal, P., Calo, S.: Analyzing federated learning through an adversarial lens. In: International Conference on Machine Learning, pp. 634–643. PMLR (2019)
4. Blanchard, P., El Mhamdi, E.M., Guerraoui, R., Stainer, J.: Machine learning with adversaries: Byzantine tolerant gradient descent. In: Advances in Neural Information Processing Systems 30 (2017)
5. Brunet-Saumard, C., Genetay, E., Saumard, A.: K-bMOM: A robust Lloyd-type clustering algorithm based on bootstrap median-of-means. Comput. Stat. Data Anal. **167**, 107370 (2022)
6. Caldas, S., et al.: Leaf: a benchmark for federated settings. arXiv preprint arXiv:1812.01097 (2018)
7. Chen, F., Long, G., Wu, Z., Zhou, T., Jiang, J.: Personalized federated learning with graph. arXiv preprint arXiv:2203.00829 (2022)
8. Cohen, G., Afshar, S., Tapson, J., Van Schaik, A.: EMNIST: extending MNIST to handwritten letters. In: 2017 International Joint Conference on Neural Networks (IJCNN), pp. 2921–2926. IEEE (2017)
9. Davé, R.N., Krishnapuram, R.: Robust clustering methods: a unified view. IEEE Trans. Fuzzy Syst. **5**(2), 270–293 (1997)
10. Deshpande, A., Kacham, P., Pratap, R.: Robust k-means++. In: Conference on Uncertainty in Artificial Intelligence, pp. 799–808. PMLR (2020)
11. Dinh, C.T., Tran, N.H., Nguyen, T.D.: Personalized federated learning with moreau envelopes. arXiv preprint arXiv:2006.08848 (2020)
12. García-Escudero, L.A., Gordaliza, A., Matrán, C., Mayo-Iscar, A.: A general trimming approach to robust cluster analysis. Ann. Stat. **36**(3), 1324–1345 (2008)
13. García-Escudero, L.A., Gordaliza, A., Matrán, C., Mayo-Iscar, A.: A review of robust clustering methods. Adv. Data Anal. Classif. **4**(2), 89–109 (2010)
14. Ge, Y.F., Cao, J., Wang, H., Chen, Z., Zhang, Y.: Set-based adaptive distributed differential evolution for anonymity-driven database fragmentation. Data Sci. Eng. **6**(4), 380–391 (2021)
15. Ghosh, A., Chung, J., Yin, D., Ramchandran, K.: An efficient framework for clustered federated learning. arXiv preprint arXiv:2006.04088 (2020)
16. Ghosh, A., Hong, J., Yin, D., Ramchandran, K.: Robust federated learning in a heterogeneous environment. arXiv preprint arXiv:1906.06629 (2019)
17. Guha, S., Rastogi, R., Shim, K.: Rock: a robust clustering algorithm for categorical attributes. Inf. Syst. **25**(5), 345–366 (2000)
18. Huang, Y., et al.: Personalized cross-silo federated learning on Non-IID data. In: AAAI, pp. 7865–7873 (2021)
19. Kairouz, P., McMahan, H.B., et al.: Advances and open problems in federated learning. arXiv preprint arXiv:1912.04977 (2019)
20. Kumagai, A., Iwata, T., Fujiwara, Y.: Transfer metric learning for unseen domains. Data Sci. Eng. **5**(2), 140–151 (2020)
21. Li, T., Sanjabi, M., Smith, V.: Fair resource allocation in federated learning. CoRR abs/1905.10497 (2019). http://arxiv.org/abs/1905.10497

22. Liu, Z., Luo, P., Wang, X., Tang, X.: Deep learning face attributes in the wild. In: Proceedings of the IEEE ICCV, pp. 3730–3738 (2015)
23. Long, G., Shen, T., Tan, Y., Gerrard, L., Clarke, A., Jiang, J.: Federated learning for privacy-preserving open innovation future on digital health. In: Chen, F., Zhou, J. (eds.) Humanity Driven AI, pp. 113–133. Springer, Cham (2022). https://doi.org/10.1007/978-3-030-72188-6_6
24. Long, G., Tan, Y., Jiang, J., Zhang, C.: Federated learning for open banking. In: Yang, Q., Fan, L., Yu, H. (eds.) Federated Learning. LNCS (LNAI), vol. 12500, pp. 240–254. Springer, Cham (2020). https://doi.org/10.1007/978-3-030-63076-8_17
25. Luo, J., et al.: Real-world image datasets for federated learning. arXiv preprint arXiv:1910.11089 (2019)
26. Ma, J., Long, G., Zhou, T., Jiang, J., Zhang, C.: On the convergence of clustered federated learning. arXiv preprint arXiv:2202.06187 (2022)
27. Mansour, Y., Mohri, M., Ro, J., Suresh, A.T.: Three approaches for personalization with applications to federated learning. arXiv preprint arXiv:2002.10619 (2020)
28. McMahan, B., Moore, E., Ramage, D., Hampson, S., y Arcas, B.A.: Communication-efficient learning of deep networks from decentralized data. In: Artificial Intelligence and Statistics, pp. 1273–1282. PMLR (2017)
29. Paul, D., Chakraborty, S., Das, S.: Robust principal component analysis: a median of means approach. arXiv preprint arXiv:2102.03403 (2021)
30. Sattler, F., Müller, K.R., Samek, W.: Clustered federated learning: Model-agnostic distributed multi-task optimization under privacy constraints. arXiv preprint arXiv:1910.01991 (2019)
31. Smith, V., Chiang, C.K., Sanjabi, M., Talwalkar, A.S.: Federated multi-task learning. In: Advances in Neural Information Processing Systems 30 (2017)
32. Tan, Y., et al.: Fedproto: federated prototype learning across heterogeneous clients. In: AAAI Conference on Artificial Intelligence, vol. 1, p. 3 (2022)
33. Wang, H., Yurochkin, M., Sun, Y., Papailiopoulos, D., Khazaeni, Y.: Federated learning with matched averaging. In: International Conference on Learning Representations (2020). https://openreview.net/forum?id=BkluqlSFDS
34. Wang, Z., Zhou, T., Long, G., Han, B., Jiang, J.: FedNoiL: a simple two-level sampling method for federated learning with noisy labels. arXiv preprint arXiv:2205.10110 (2022)
35. Xie, M., et al.: Multi-center federated learning. arXiv preprint arXiv:2108.08647 (2021)
36. Yang, M.S., Lai, C.Y., Lin, C.Y.: A robust EM clustering algorithm for gaussian mixture models. Pattern Recogn. **45**(11), 3950–3961 (2012)
37. Yang, M.S., Wu, K.L.: A similarity-based robust clustering method. IEEE Trans. Pattern Anal. Mach. Intell. **26**(4), 434–448 (2004)
38. Yang, Q., Liu, Y., Chen, T., Tong, Y.: Federated machine learning: concept and applications. ACM Trans. Intell. Syst. Technol. (TIST) **10**(2), 12 (2019)
39. Yin, D., Chen, Y., Kannan, R., Bartlett, P.: Defending against saddle point attack in byzantine-robust distributed learning. In: International Conference on Machine Learning, pp. 7074–7084. PMLR (2019)

SAPMS: A Semantic-Aware Privacy-Preserving Multi-keyword Search Scheme in Cloud

Qian Zhou[1], Hua Dai[1,2(✉)], Zheng Hu[1], Yuanlong Liu[1], and Geng Yang[1,2]

[1] Nanjing University of Posts and Telecommunications, Nanjing 210023, China
daihua@njupt.edu.cn

[2] Jiangsu Security and Intelligent Processing Lab of Big Data, Nanjing 210023, China

Abstract. Most of the traditional privacy-preserving search schemes adopt TF-IDF model which is on the basis of keyword frequency statistics. The embedding semantic association between keywords and documents are not considered. To solve this problem, we propose an efficient semantic-aware privacy-preserving multi-keyword search scheme over encrypted cloud data. The LDA topic model is adopted to generate the topic information-embedded vectors for documents and queried keywords. The homomorphic encryption on vectors is used to perform privacy-preserving semantic relevance score computation between queried keywords and documents. To achieve efficient search processing, a novel tree-based index is designed, which is constructed following the divisive hierarchical clustering algorithm. By using the tree-based index, a depth-first privacy-preserving multi-keyword search algorithm is proposed. The experimental results show that the proposed scheme outperforms the existing schemes in terms of the semantic precision of search results and the search time cost.

Keywords: Cloud computing · Searchable encryption · Multi-keywords search · Semantic-aware · Privacy-preserving

1 Introduction

Nowadays, cloud computing is becoming more and more popular and useful for enterprises, which provides dynamic, easy-scalable and virtualized resources through the Internet. Data and software of enterprises could be outsourced and deployed in the cloud for cost saving. However, the outsourcing of data makes the data owner losing the independent and exclusive control on his/her data, which could lead privacy leakage issues. The cloud service provider could obtain the valuable information of the outsourced data by statistical interfering, etc. To protect data privacy, an easy idea is to encrypt the data before it is outsourced to the cloud. But the encrypted data is difficult to search and use, which severely decreases the availability of the data. It is a challenge to provide searchability

B. Li et al. (Eds.): APWeb-WAIM 2022, LNCS 13421, pp. 251–263, 2023.
https://doi.org/10.1007/978-3-031-25158-0_20

and privacy preservation on the outsourced data, simultaneously. Therefore, the research on searchable encryption schemes is a necessary.

Recently, many searchable encryption schemes have been proposed, such as single keyword search schemes [3,19,22,24,27,28,32,33] and multi-keyword search schemes [2,4–9,11–13,15–18,20,21,25,26,30,31]. Most of these schemes adopt the term frequency-inverse document frequency model (TF-IDF) based vector space model to generate vectors to represent documents and searched keywords, and these vectors are encrypted by the random invertible matrix based vector encryption method [2] for privacy preservation. The encrypted vectors are organized into the sophisticate encrypted indexes which support the privacy-preserving multi-keyword search processing over the encrypted cloud data. However, the TF-IDF model is a statistical model which utilizes the quantized frequency statistics to evaluate the importance of keywords to a set of documents. It ignores the semantic association between keywords and documents, which could lead to unsatisfied results for users. At the same time, due to the large number of keywords in the dictionary extracted from the whole documents, the generated vectors following the TF-IDF model could have an extremely high dimension and sparsity. It costs a lot of storage space and search time. The previous work in [5] firstly proposed a LDA-based semantic-aware search scheme LDA-EMRSE. Documents are trained by using the LDA model and represented by lower dimensional topic vectors. But the semantic relation between documents are not considered when constructing the search index, there is still room of improvement in search precision and efficiency.

In this paper, we proposed a semantic-aware privacy-preserving multi-keyword search scheme (SAPMS). The data owner first uses the LDA model to train documents and generate the topic information-embedded vectors for documents. The vectors are taken as the input of the divisive hierarchical clustering algorithm, and a novel binary clustering tree (BCI-tree) is formed which is used as the index for search processing. The index is encrypted by using the homomorphic matrix encryption, and the encrypted index is outsourced to the cloud server together with the encrypted documents. When starting a multi-keyword search to require k highest relevant documents in semantic, the data user converts the queried keywords into the trapdoor which is submitted to the cloud server as the search command. The cloud server executes the search by using the encrypted index and returns the result to the data user. At last, the data user performs the decryption to obtain the plaintext result. The experimental results show that the proposed scheme has better performance in terms of the search result precision and search time cost.

2 Notations and Preliminaries

2.1 Notations

- F: a set of n documents, $F = \{f_1, f_2, \cdots, f_n\}$, whose encrypted form is $\widehat{F} = \{\widehat{f_1}, \widehat{f_2}, \cdots, \widehat{f_n}\}$.
- D: a dictionary having h keywords, $D = \{w_1, w_2, \cdots, w_h\}$, which is extracted from documents of F.

- T: a set of m topics, $T = \{t_1, t_2, \cdots, t_m\}$.
- Γ: An $n \times m$-dimensional document-topic correlation matrix, whose encrypted form is $\widehat{\Gamma}$.
- Ψ: an $h \times m$-dimensional keyword-topic correlation matrix.
- P_T: an m-dimensional topic probability vector.
- P_W: an h-dimensional keyword probability vector.
- $\mathcal{I}$: a binary tree-based index, whose encrypted form is $\widehat{\mathcal{I}}$.
- Q: a search with multiple queried keywords, $Q \subseteq D$.
- V_Q: an m-dimensional query topic vector, whose encrypted form is $\widehat{V}_Q$.

2.2 Preliminaries

LDA Model. The Latent Dirichlet Allocation model (LDA) is an unsupervised learning algorithm proposed by David Blei [1]. By using the LDA model to train the documents of F, a document-topic correlation matrix Γ and a keyword-topic correlation matrix Ψ will be generated, which describe the numerical topic relevance between topics and documents and keywords, respectively. Each document and keyword can be represented by a document topic vector and a keyword topic vector, respectively. Given a document $f_i \in F$, a topic $t_j \in T$ and a keyword $w_e \in W$, $\Gamma[i]$ is the document topic vector of f_i, where $\Gamma[i][j]$ records the topic relevance score between f_i and t_j; $\Psi[e]$ is the keyword topic vector of w_e, where $\Psi[e][j]$ records the topic relevance score between w_e and t_j.

Sematic Relevance Measurement: In this paper, documents and searches are represented by the vectors embedding the topic semantic information. The sematic relevance score between a document and a search is the inner product of the corresponding vectors. Assuming that $\Gamma[i]$ and V_Q are the document topic vector of f_i and the query topic vector of Q, respectively, the semantic relevance score between f_i and Q is the inner product between $\Gamma[i]$ and V_Q, i.e. $Score(f_i, Q) = \Gamma[i] \cdot V_Q$.

Divisive Hierarchical Clustering. The divisive hierarchical clustering algorithm [23] is one of the hierarchical clustering algorithms, which can divide a large cluster into multiple small sub-clusters recursively and form a binary clustering tree. Each node in the tree represents a cluster and its left and right child nodes represent the generated two sub-clusters.

Secure Inner Product Operation. The secure inner product [2] is implemented by a homomorphic matrix encryption. We assume that x and y are two given n-dimensional vectors and M is a random $n \times n$ invertible matrix. Here, M is used as a private key. The vectors x and y are encrypted by M and the encrypted vectors are denoted as $\widehat{x}$ and $\widehat{y}$, where $\widehat{x} = M^T \cdot x$ and $\widehat{y} = M^{-1} \cdot y$. The inner product of x and y is equal to the inner product of $\widehat{x}$ and $\widehat{y}$ because $\widehat{x} \cdot \widehat{y} = (M^T x) \cdot (M^{-1} y) = (M^T x)^T (M^{-1} y) = x^T M M^{-1} y = x \cdot y$. It indicates that the inner product of two vectors can be obtained without knowing the plaintext values of any dimensions of vectors.

3 Problem Statement

System Model. The system model is the shown in Fig. 1, which involves three different entities: the data owner, the data user and the cloud server. The data owner owns a collection of documents and he/she outsources the encrypted documents and index to the cloud server and shares the secure key and other information with the data user. The data user generates a trapdoor when he/she starts a multi-keyword search. The trapdoor is submitted to the cloud server as the search command. The cloud server performs the search processing by using the encrypted index and returns the search result in ciphertext to the data user. When the data user receives the encrypted result, he/she decrypts it to get the final plaintext search result.

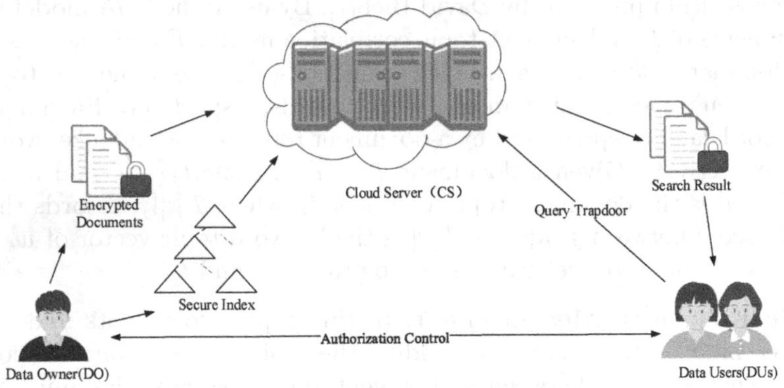

Fig. 1. An example of BCI-tree search.

Threat Model. In this paper, we adopt the same threat model as most of the existing searchable encryption schemes [2–4,8,9,11–13,17–21,24–28,30–32], which is the "honest-but-curious" threat model. In this model, the cloud server is assumed to execute the search processing honestly and returns the correct search results to the data user. However, the cloud server is curious about the information of the outsourced data by statistical analyzing or other methods.

Design Goals. Given a multi-keyword search Q, the proposed search scheme can achieve the following design goals.

1) **Semantic-aware multi-keyword search.** In the scheme, the cloud server can return the k documents which have the highest k semantic relevance scores to the queried keywords.
2) **Search efficiency and accuracy.** The scheme realizes efficient and accurate search by using semantic model, special index and efficient search algorithm.
3) **Privacy preservation.** The scheme can protect the privacy of documents, index and queried keywords.

4 The Proposed Search Scheme

There are three modules in the proposed search scheme, which are the data preprocessing module, index building module and multi-keyword search module.

4.1 Data Preprocessing Module

The data preprocessing module mainly includes the private key generation and the LDA topic model training, which are described as the algorithms *GenKey* and *LDATrain*.

$\boldsymbol{K \leftarrow GenKey(1^{l(n)})}$. The *GenKey* algorithm is to generate a set of private keys $K = \{S, M_1, M_2, k_f\}$, where S is an m-bit random vector, M_1 and M_2 are two random $m \times m$ invertible matrices and k_f is the key used to encrypt documents. K is only shared between the data owner and the authorized data user while the cloud server has no idea of it.

$\boldsymbol{(\Gamma, \Psi, P_T, P_W) \leftarrow LDATrain(F)}$. The *LDATrain* algorithm is to generate the document-topic correlation matrix Γ, the keyword-topic correlation matrix Ψ, the topic probability vector P_T and the keyword probability vector P_W. The algorithm is shown as the following steps.

1) Taking the documents of F as the input, the data owner uses the LDA model to perform the train processing, and the document-topic correlation matrix Γ and the keyword-topic correlation matrix Ψ are generated.
2) The data owner generates the topic probability vector P_T. The ith dimension of P_T stores the prior belief of the topic t_i existing in F. Assuming that documents in F have the same importance, the prior probability of t_i is the average probability that each document of F belongs to t_i.

$$P_T = \frac{1}{n} \cdot \sum_{j=1}^{n} \Gamma[j]. \tag{1}$$

3) The data owner generates the keyword probability vector P_W. The ith dimension of the vector stores the prior belief of the keyword w_i existing in F.

$$P_W = P_T \cdot \Psi^T. \tag{2}$$

In the end of data preprocessing module, the data owner shares K, Ψ, P_T and P_W generated in the above steps to the data user.

4.2 Index Building Module

This section proposes the structure definition and the construction algorithm of a novel binary clustering tree based index which is denoted as the BCI-tree.

Definition 1. Maximized Filter Vector. Given a set of m-dimensional vectors $G = \{V_1, V_2, \cdots, V_x\}$, the maximized filter vector of G is denoted as $MFV(G)$ which is an m-dimensional vector. The ith dimension of $MFV(G)$ stores the maximum of the value set consisting of the ith dimension of each vector in G. Assuming that $V = MFV(G)$, we have $V[i] = max\{V_j[i] | V_j \in G\}$.

Definition 2. BCI-tree. The BCI-tree is a binary tree which is constructed by the divisive hierarchical clustering algorithm. The tree is used as the search index and denoted as $\mathcal{I}$. Each node u in $\mathcal{I}$ is a four-element tuple, $u =< id, lp, rp, vec >$, where id is an identity, vec is an m-dimensional vector, and lp and rp point to the left and right child nodes of u, respectively. The setting of u depends on the position of u in $\mathcal{I}$ as follows:

- if u is a leaf node, then u corresponds to a document which is assumed to be f_i, and the setting of u is $u.lp = u.rp = \varnothing$, $u.id = i$ and $u.vec = \Gamma[i]$.
- if u is not a leaf node, then $u.lp$ and $u.rp$ point to the left and right child nodes of u, $u.id = \varnothing$, and $u.vec$ is the maximized filter vector of $u.lp$ and $u.rp$ where $u.vec = MFV(\{u.lp.vec, u.rp.vec\})$.

On the basis of Definition 2, we give the construction algorithm of the encrypted index, $BuildIndex$, which is performed by the data owner.

$\widehat{\mathcal{I}} \leftarrow \boldsymbol{BuildIndex(\Gamma, K)}$. The data owner first takes the document topic vectors of $\Gamma = \{\Gamma[1], \Gamma[2], \cdots, \Gamma[n]\}$ as the input and uses the divisive hierarchical clustering algorithm to construct the plaintext BCI-tree index. Then, the data owner encrypts the index with the private keys of K. The details of the $BuildIndex$ algorithm is shown as follows:

1) Taking the document topic vectors of Γ as input, the data owner performs the $GenBCItree$ algorithm to construct the plaintext BCI-tree index $\mathcal{I}$. The detail of $GenBCItree$ is shown in Algorithm 1.
2) For each node in $\mathcal{I}$, the data owner uses the m-bit random vector S in K to split the vector V stored in the node into two vectors, V' and V''. If $S[i] = 0$, then $V'[i] = V''[i] = V[i]$; and if $S[i] = 1$, then $V'[i] + V''[i] = V[i]$. Afterwards, the data owner uses the random invertible matrices M_1 and M_2 in K to encrypt V' and V'', respectively, where $\widehat{V} = \{\widehat{V}', \widehat{V}''\} = \{M_1^T V', M_2^T V''\}$. When the vectors of all nodes have been processed, the encrypted index $\widehat{\mathcal{I}}$ is generated.

In the end of the index building module, the data owner uses the key k_f in K to encrypt the documents of F. And then, the encrypted documents are outsourced to the cloud server together with the encrypted index.

4.3 Multi-keyword Search Module

There are two key algorithms in the multi-keyword search module, $GenTrapdoor$ and $MKSearch$. The former is to generate the trapdoor corresponding to a search with multiple queried keywords and the latter is to perform the search processing in privacy-preserving manner.

$\widehat{V}_Q \leftarrow \boldsymbol{GenTrapdoor(Q, \Psi, P_T, P_W, K)}$. We assume that the started search is $Q = \{w_{e_1}, w_{e_2}, \cdots, w_{e_p}\}$, where w_{e_p} is the e_pth keyword in D, and the search requests the k documents with the k highest semantic relevance scores to Q. The data user first generates the corresponding query topic vector V_Q and then uses the private keys of K to encrypt V_Q to get the trapdoor $\widehat{V}_Q$. We describe in detail the generation process of trapdoor as follows:

Algorithm 1: *GenBCItree(C)*

```
Input: C is a cluster of document topic vectors representing a set of
   corresponding documents
Output: the root of a BCI-tree
Create a new BCI-tree node u;
if |C| > 1 then
   u.id = ∅, u.vec = MFV(C);
   Take the cluster C as the input to run the divisive hierarchical clustering
   algorithm, and the output sub-clusters are assumed to be C_1 and C_2;
   u.lp = GenBCItree(C_1);
   u.rp = GenBCItree(C_2);
else if |C| = 1 then
   Assume that the only one document topic vector in C is Γ[i] which
   corresponds to the document f_i;
   u.vec = Γ[i], u.id = i, u.lp = u.rp = ∅;
end if
```

1) The data user generates the query topic vector V_Q according to Eq.(3) where $\circ$ is the Hadamard product operator [29]. V_Q is an m-dimensional vector that represents the probabilities that the topic intent of Q belongs to each topic.

$$V_Q = \sum_{w_{e_i} \in Q} \frac{P_T \circ \Psi[e_i]}{|Q| \cdot P_W[e_i]} \tag{3}$$

2) The data user splits the vector V_Q into two random vectors $\{V'_Q, V''_Q\}$ with the n-bit random vector S in K. The split rule is the same as that in the second step of *BuildIndex*.
3) The data user encrypts $V_Q = \{V'_Q, V''_Q\}$ into $\widehat{V}_Q = \{\widehat{V}'_Q, \widehat{V}''_Q\} = \{M_1^{-1}V'_Q, M_2^{-1}V''_Q\}$ by using the random inverted matrices M_1 and M_2 in K. The encrypted query topic vector $\widehat{V}_Q$ is the trapdoor which is transmitted to the cloud server as the search command.

***R* ← *BCISearch*($\widehat{\mathcal{I}}, \widehat{V}_Q, k, \tau$).** Once receiving the trapdoor $\widehat{V}_Q$, the cloud server performs the *BCISearch* algorithm by using the encrypted index $\widehat{\mathcal{I}}$ to obtain the search result R. The detail of *BCISearch* is shown in Algorithm 2.

According to Algorithm 2, *BCISearch* is a depth-first search algorithm. In the beginning of the algorithm, u is the root of the encrypted BCI-tree $\widehat{\mathcal{I}}$ and $\tau = 0$. The pruning threshold τ can improve the search efficiency. For a sub-tree in $\widehat{\mathcal{I}}$, if the inner product between the vector stored in the root of the subtree and the trapdoor vector is lower than τ, the subtree will be pruned because the inner product between the vector stored in any node of the subtree and the trapdoor vector must be lower than τ. And the documents stored in the leaf nodes of the subtree are definitely not in the search result. There is no need to check any nodes in the subtree.

In addition, the inner product between an encrypted document topic vector $\widehat{\Gamma}[i] = \{M_1^T \Gamma'[i], M_2^T \Gamma''[i]\}$ and the encrypted query topic vector $\widehat{V}_Q = \{M_1^{-1} V_Q', M_2^{-1} V_Q''\}$ equals to the inner product between the corresponding plaintext vectors, i.e. $\widehat{\Gamma}[i] \cdot \widehat{V}_Q = \Gamma[i] \cdot V_Q$. The proof is shown in Eq.(4).

$$\begin{aligned} \widehat{\Gamma}[i] \cdot \widehat{V}_Q &= \{V_Q' M_1^{-1}, V_Q'' M_2^{-1}\} \cdot \{\Gamma'[i] M_1^T, \Gamma''[i] M_2^T\} \\ &= (V_Q' M_1^{-1}) \cdot (\Gamma'[i] M_1^T)^T + (V_Q'' M_2^{-1}) \cdot (\Gamma''[i] M_2^T\})^T \\ &= V_Q' \cdot \Gamma'[i] + V_Q'' \cdot \Gamma''[i] \\ &= V_Q \cdot \Gamma[i]. \end{aligned} \tag{4}$$

It indicates that the cloud server can perform the multi-keyword search correctly even just knowing the encrypted data about documents and index, and the scheme proposed in this paper can achieve semantic-aware privacy-preserving multi-keyword search over encrypted cloud data.

Algorithm 2: *BCISearch*(u, $\widehat{V}_Q$, k, τ)

Input: u is the root of a subtree, $\widehat{V}_Q$ is the trapdoor, k is the number of requested documents, and τ is the pruning threshold

Output: the search result R

if $u \neq \varnothing$ **then**
  if $u.vec \cdot \widehat{V}_Q > \tau$ **then**
    if u is not a leaf node **then**
      $BCISearch(u.lp, \widehat{V}_Q, k, \tau)$;
      $BCISearch(u.rp, \widehat{V}_Q, k, \tau)$;
    else if u is a leaf node **then**
      Add u to R;
      if $|R| > k$ **then**
        Delete the document with the lowest score to $\widehat{V}_Q$ in R;
      end if
      if $|R| = k$ **then**
        Set τ equal to the lowest value in the semantic relevance scores between each document in R and $\widehat{V}_Q$;
      end if
    end if
  end if
end if

5 Performance Evaluation

In this section, we evaluate SAPMS and compare it with the schemes presented in [5,30] which are denoted as TFIDF-MRSE and LDA-EMRSE, respectively.

We implement these schemes by using Python 3.6 in Windows 10 with Intel Core(TM) i5-9300H. The evaluation is based on two metrics, the semantic precision of search result and the search time cost. The dataset used in the evaluation is 20 newsgroups [14]. The default setting of the number of documents n, the number of topics m and the number of required documents k are $n = 8000$, $m = 100$ and $k = 100$, respectively.

5.1 Semantic Precision Evaluation

Assuming that the queried keywords come from the same category and the documents in different categories are semantically irrelevant, we evaluate the semantic precision of three schemes by using the calculation method adopted by [10], which is shown in Eq.(5),

$$Precision = \frac{TP}{TP + FP} \times 100\% \tag{5}$$

where *TP* and *FP* are the numbers of documents in the search result that belong and do not belong to the category of the search intent, respectively.

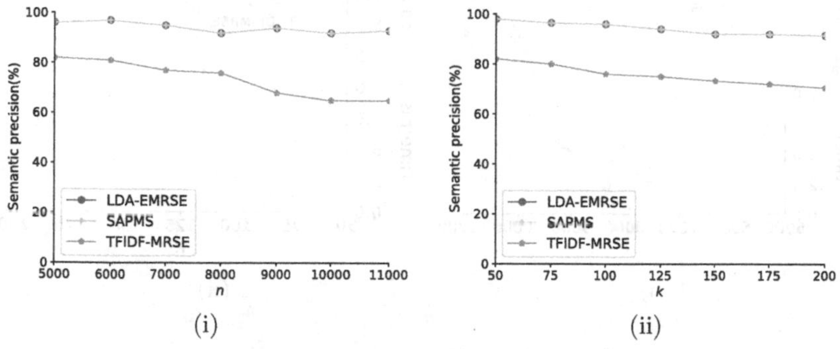

Fig. 2. Semantic precision versus n and k.

1) Fig. 2(i) shows that as the number of documents n increases, the semantic precision of SAPMS and LDA-EMRSE remains stable, while the semantic accuracy of TFIDF-MRSE gradually decreases. The reason is that in SAPMS and LDA-EMRSE, as the number of documents increases, the topics extracted by the LDA model can better reflect the semantic features of the documents but slightly affect the semantic precision of search results.
2) Fig. 2(ii) shows that as the number of documents k to be returned increases, the semantic precision of the three schemes decreases gradually. The reason is that as k increases, documents with lower relevance scores to the search intent are added to the search result, which lead to a slight decrease in semantic precision.

In addition, Fig. 2(i) and 2(ii) both show that the semantic precision of SAPMS and LDA-EMRSE are the same and higher than that of TFIDF-MRSE. The reason is that the former two schemes are based on the LDA model and have the ability to find the same result, thus they have the same semantic precision.

5.2 Search Time Cost Evaluation

Figure 3(i) shows that as the number of documents n increases, the search time cost of all three schemes increases. The reason is that three schemes all need to traverse the nodes of the index trees and perform the inner product to obtain the search result. The larger n is, the more nodes need to be traversed, so the inner product increases, which leads to the increase of search time cost.

Figure 3(ii) shows that as the number of requested documents k increases, the search time cost of all three schemes increases. The reason is that with the increase of k, the search processing of three schemes all need to traverse more nodes. It means that more inner products need to be calculated, which increases the search time cost.

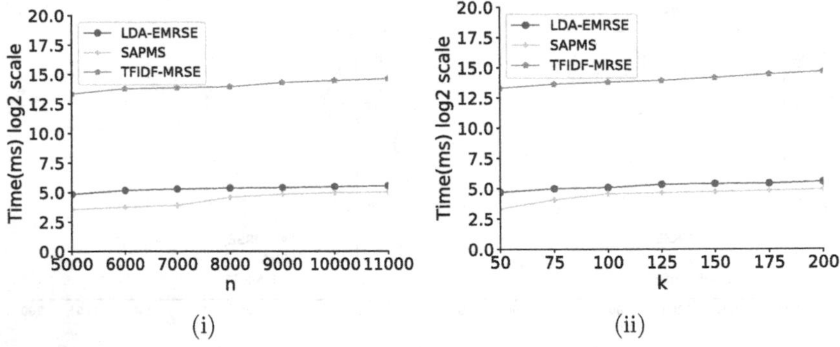

Fig. 3. Time Cost versus n and k.

The experimental results in Fig. 3(i) and 3(ii) both show that the search time cost of TFIDF-MRSE is much higher than the other two schemes. The reason is that the dimension of the vector in TFIDF-MRSE is the same as the scale of the dictionary, which is much higher than the dimension of vectors in SAPMS and LDA-EMRSE. The search time cost of an inner product between vectors with small dimension is obviously faster than that with high dimension. Meanwhile, Fig. 3(i) and Fig. 3(ii) both show that the search time cost of SAPMS is less than that of LDA-EMRSE. The reason is that documents are randomly distributed in leaf nodes in LDA-EMRSE, while documents in SAPMS are well organized in the leaf nodes and documents with high relevance are usually near with each other. Thus, the search efficiency is improved.

6 Conclusion

The searchable encryption in the cloud environment is an effective method to provide multi-keyword searches while protect the privacy of the outsourced data. A semantic-aware privacy-preserving multi-keyword search scheme is proposed in this paper. By using the LDA model and the divisive hierarchical clustering algorithm, an novel tree-based index is presented, which supports semantic-aware multi-keyword searches over encrypted cloud data. The experimental results show that the proposed scheme has better performance in terms of semantic precision of search results and search time cost.

Acknowledgements. This work was supported by the National Natural Science Foundation of China under the grant Nos.61872197,61902199 and 61972209; the Jiangsu Province Postgraduate Scientific Research Innovation Program under Grand No.KYCX22_0984; the "14th Five-Year Plan" Civil Aerospace Pre-research Project of China (D020101).

References

1. Blei, D.M., Ng, A.Y., Jordan, M.I.: Latent dirichlet allocation. J. Mach. Lear. Res. **3**, pp. 993–1022 (2003)
2. Cao, N., Wang, C., Li, M., Ren, K., Lou, W.: Privacy-preserving multi-keyword ranked search over encrypted cloud data. IEEE Trans. Parallel Distrib. Syst. **1**(25), 222–233 (2014)
3. Chang, Y.-C., Mitzenmacher, M.: Privacy Preserving Keyword Searches on Remote Encrypted Data. In: Ioannidis, J., Keromytis, A., Yung, M. (eds.) ACNS 2005. LNCS, vol. 3531, pp. 442–455. Springer, Heidelberg (2005). https://doi.org/10.1007/11496137_30
4. Chen, C., et al.: An efficient privacy-preserving ranked keyword search method. IEEE Trans. Parallel Distrib. Syst. **27**(4), 951–963 (2016)
5. Dai, H., Dai, X., Yi, X., Yang, G., Huang, H.: Semantic-aware multi-keyword ranked search scheme over encrypted cloud data. J. Network Comput. Appl. **147**, 102442 (2019)
6. Dai, H., Yang, M., Yang, G., Xiang, Y., Hu, Z., Wang, H.: A keyword-grouping inverted index based multi-keyword ranked search scheme over encrypted cloud data. IEEE Trans. Sustain. Comput. **7**(3), pp. 561-578 (2022)
7. Dai, X., Dai, H., Rong, C., Yang, G., Xiao, F.: Enhanced semantic-aware multi-keyword ranked search scheme over encrypted cloud data. IEEE Trans. Cloud Comput. **10**(4), 2595–2612 (2022)
8. Dai, Y., Shao, J., Hu, G., Guo, L.: A resource-aware approach for authenticating privacy preserving GNN queries. World Wide Web **22**(2), 437–454 (2019)
9. Fu, Z., Sun, X., Ji, S., Xie, G.: Towards efficient content-aware search over encrypted outsourced data in cloud. In: IEEE INFOCOM 2016-The 35th Annual IEEE International Conference on Computer Communications, pp. 1–9 IEEE (2016)

10. Gabryel, M., Damaševičius, R., Przybyszewski, K.: Application of the Bag-of-Words Algorithm in Classification the Quality of Sales Leads. In: Rutkowski, L., Scherer, R., Korytkowski, M., Pedrycz, W., Tadeusiewicz, R., Zurada, J.M. (eds.) ICAISC 2018. LNCS (LNAI), vol. 10841, pp. 615–622. Springer, Cham (2018). https://doi.org/10.1007/978-3-319-91253-0_57
11. Hozhabr, M., Asghari, P., Javadi, H.: Dynamic secure multi-keyword ranked search over encrypted cloud data. J. Inf. Secur. Appl. **61**(1), 102902 (2021)
12. Ibrahim, A., Jin, H., Yassin, A.A., Zou, D.: Secure rank-ordered search of multi-keyword trapdoor over encrypted cloud data. In: 2012 IEEE Asia-Pacific Services Computing Conference, pp. 263–270. IEEE (2012)
13. Kiayias, A., Oksuz, O., Russell, A., Tang, Q., Wang, B.: Efficient Encrypted Keyword Search for Multi-user Data Sharing. In: Askoxylakis, I., Ioannidis, S., Katsikas, S., Meadows, C. (eds.) ESORICS 2016. LNCS, vol. 9878, pp. 173–195. Springer, Cham (2016). https://doi.org/10.1007/978-3-319-45744-4_9
14. Lang, K.: NewsWeeder: Learning to filter Netnews. In: Machine Learning Proceedings 1995. Elsevier, pp. 331–339. (1995)
15. Li, B., Liang, R., Zhou, W., Yin, H., Gao, H., Cai, K.: LBS meets blockchain: An efficient method with security preserving trust in SAGIN. IEEE Internet Things Journal **9**(8), 5932–5942 (2022)
16. Li, B., Liang, R., Zhu, D., Chen, W., Lin, Q.: Blockchain-based trust management model for location privacy preserving in VANET. IEEE Trans. Intell. Transp. Syst. **22**(6), 3765–3775 (2021)
17. Li, J., Wang, Q., Wang, C., Cao, N., Ren, K., Lou, W.: Fuzzy keyword search over encrypted data in cloud computing. In: 2010 Proceedings IEEE INFOCOM, pp. 1–5. IEEE (2010)
18. Liang, Y., Li, Y., Zhang, K., Ma, L.: DMSE: dynamic multi-keyword search encryption based on inverted index. J. Syst. Archit. **119**, 102255 (2021)
19. Liu, C., Zhu, L., Chen, J.: J. Network Comput. Appl. **86**, 3–14 (2017)
20. Liu, Q., Peng, Y., Wu, J., Wang, T., Wang, G.: Secure multi-keyword fuzzy searches with enhanced service quality in cloud computing. IEEE Trans. Netw. Serv. Manage. **18**(2), 2046–2062 (2020)
21. Orencik, C., Kantarcioglu, M., Savas, E.: A practical and secure multi-keyword search method over encrypted cloud data. In: 2013 IEEE Sixth International Conference on Cloud Computing, pp. 390–397. IEEE (2013)
22. Padhya, M., Jinwala, D.C.: R-OO-KASE: revocable online offline key aggregate searchable encryption. Data Sci. Eng. **5**(4), 391–418 (2020)
23. Roux, M.: A comparative study of divisive and agglomerative hierarchical clustering algorithms. J. Classif. **35**(2), 345–366 (2018)
24. Swaminathan, A., et al.: Confidentiality-preserving rank-ordered search. In: Proceedings of the 2007 ACM workshop on Storage security and survivability, pp. 7–12 (2007)
25. Tseng, C.Y., Lu, C., Chou, C.F.: Efficient privacy-preserving multi-keyword ranked search utilizing document replication and partition. In: 2015 12th Annual IEEE Consumer Communications and Networking Conference (CCNC), pp. 671–676. IEEE (2015)
26. Wang, B., Yu, S., Lou, W., Hou, Y.T.: Privacy-preserving multi-keyword fuzzy search over encrypted data in the cloud. In: IEEE INFOCOM 2014-IEEE Conference on Computer Communications, pp. 2112–2120. IEEE (2014)
27. Wang, C., Cao, N., Li, J., Ren, K., Lou, W.: Secure ranked keyword search over encrypted cloud data. In: 2010 IEEE 30th International Conference on Distributed Computing Systems, pp. 253–262. IEEE (2010)

28. Wang, C., Cao, N., Ren, K., Lou, W.: Enabling secure and efficient ranked keyword search over outsourced cloud data. IEEE Trans. Parallel Distrib. Syst. **23**(8), 1467–1479 (2011)
29. Wang, P., Ravishankar, C.V.: On masking topical intent in keyword search. In: 2014 IEEE 30th International Conference on Data Engineering, pp. 256–267. IEEE (2014)
30. Xia, Z., Wang, X., Sun, X., Wang, Q.: A secure and dynamic multi-keyword ranked search scheme over encrypted cloud data. IEEE Trans. Parallel Distrib. Syst. **27**(2), 340–352 (2015)
31. Xiangyang, Z., Hua, D., Xun, Y., Geng, Y., Xiao, L.: Muse: an efficient and accurate verifiable privacy-preserving multikeyword text search over encrypted cloud data. Secur. Commun. Netw. **2017**, 1–17 (2017)
32. Zerr, S., Olmedilla, D., Nejdl, W., Siberski, W.: Zerber+r: top-k retrieval from a confidential index. In: Proceedings of the 12th International Conference on Extending Database Technology: Advances in Database Technology, pp. 439–449 (2009)
33. Zhou, Q., Dai, H., Shen, W., Liu, Y., Yang, G.: Evss: An efficient verifiable search scheme over encrypted cloud data. World Wide Web pp. 1–21 (2022)

Task Assignment with Spatio-temporal Recommendation in Spatial Crowdsourcing

Chen Zhu[1], Yue Cui[2], Yan Zhao[3](✉), and Kai Zheng[1]

[1] University of Electronic Science and Technology of China, Chengdu, China
chenzhu@std.uestc.edu.cn, zhengkai@uestc.edu.cn
[2] The Hong Kong University of Science and Technology, Hong Kong, China
ycuias@cse.ust.hk
[3] Aalborg University, Denmark, Denmark
yanz@cs.aau.dk

Abstract. With the development of GPS-enabled smart devices and wireless networks, spatial crowdsourcing has received wide attention in assigning location-sensitive tasks to moving workers. In real-world scenarios, workers may show different preferences in different spatio-temporal contexts for the assigned tasks. It is a challenge to meet the spatio-temporal preferences of workers when assigning tasks. To this end, we propose a novel spatio-temporal preference-aware task assignment framework which consists of a translation-based recommendation phase and a task assignment phase. Specifically, in the first phase, we use a translation-based recommendation model to learn spatio-temporal effects from the workers' historical task-performing activities and then calculate the spatio-temporal preference scores of workers. In the task assignment phase, we design a basic greedy algorithm and a Kuhn-Munkras (KM)-based algorithm which could achieve a better balance to maximize the total rewards and meet the spatio-temporal preferences of workers. Finally, extensive experiments are conducted, verifying the effectiveness and practicality of the proposed solutions.

Keywords: Spatial crowdsourcing · Task assignment · Spatio-temporal preference

1 Introduction

With the development of mobile Internet and GPS-equipped smart devices, spatial crowdsourcing (SC) has been booming in both academia and industry. Specifically, task requesters can issue spatial tasks to the SC server, and then the server employs smart device carriers as workers to physically travel to the specified locations and accomplish these tasks. SC platforms recruit a group of available workers to go to a specific location to complete tasks, e.g., taking photos, monitoring traffic conditions, picking up a delivery, etc. The way of SC platform to allocate these spatial tasks to workers is called task assignment.

B. Li et al. (Eds.): APWeb-WAIM 2022, LNCS 13421, pp. 264–279, 2023.
https://doi.org/10.1007/978-3-031-25158-0_21

Research on SC [3,5,16,24] has gained momentum in recent years. Most existing researches focus on task assignment algorithm [1,2,6,9,12,14,15,18,19, 22,23], which aim to maximize the total number of completed tasks [1,9,12,18], the diversity score of assignments [2], the fairness of task assignment [23], or the total rewards of the platform [14,15,19]. An implicit assumption of these works is that workers are willing to perform the tasks assigned to them. However, not all the workers are willing to accept assigned tasks, which may influence the quality of task performing and harm the interests of the platform and workers' initiative. Therefore, it is of essence to consider the preferences of workers in task assignment.

Several preference-based task assignment approaches have been developed in SC. For example, Zhao et al. [21] propose a preference-aware spatial task assignment system based on workers' temporal preferences, which consists of two components: history-based context-aware tensor decomposition for workers' temporal preferences modeling and preference-aware task assignment. Considering the preference of group task assignment, Li et al. [8] propose a framework for group task assignment based on worker groups' preferences, which includes two components: social impact-based preference modeling and preference-aware group task assignment. Nevertheless, they fail to capture workers' preference on the spatio-temoral context of tasks. For example, a worker is happy to go to a restaurant to perform tasks during her lunch break near shopping center (spatio-temoral context) but will definitely refuse to do it in her working hours. Modeling such context can be beneficial since workers could show different preferences in different spatio-temporal contexts, which may influence the initiative of workers to different tasks.

We next illustrate the spatio-temporal preference-based task assignment problem through a toy example. In Fig. 1, there are three workers $\{w_1, w_2, w_3\}$ and four tasks $\{s_1, s_2, s_3, s_4\}$. Each worker has a current location and a reachable distance, e.g., worker w_1 is at $(1,5)$, near the walk street region. Each task is published at a specific POI (Point of Interest) where it will be performed, e.g., task s_2 is published at a milk tea shop $(3, 6.6)$, implying it may be a delivery order. Considering the spatio-temporal constraints between workers and tasks (an assigned task should be located in the reachable range of the corresponding worker, and the worker should arrive at the location of the assigned task before its deadline), w_1 and w_2 have two available tasks, $\{s_1, s_2\}$ and $\{s_3, s_4\}$ respectively, while there is no available task for w_3. Moreover, Fig. 1 also depicts the spatio-temporal preferences of each worker to different tasks, e.g., the preference of w_1 to s_2 is 0.78, which is higher than his preference to s_1 (0.31), since s_2 is associated with the POI of milk tea shop that is a hot spot at the walk street at 9:00 a.m., while s_1 is associated with a cinema that is not popular in this spatio-temporal context. In SC, it is an intuitive move to assign the nearby tasks to workers without violating the spatio-temporal constraints. Therefore, without considering the spatio-temporal preference, we can get a task assignment $\{(w_1, s_1), (w_2, s_4)\}$. However, the total spatio-temporal preference score of this task assignment is only 0.54, and the two workers both have low preferences for their assigned tasks s_1 and s_4. Under such an assignment, workers may refuse to accept tasks or complete tasks with low quality. Moreover, workers

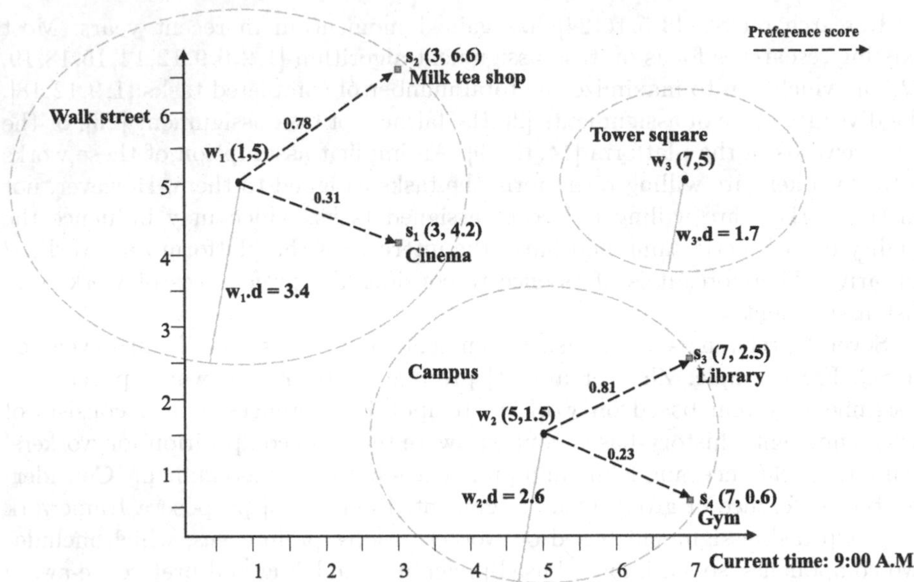

Fig. 1. Running example

may leave the platform if the server always assigns them tasks they dislike. If we assign tasks by giving higher priorities to the workers who are more interested in the tasks, we can get the task assignment result $\{(w_1, s_2), (w_2, s_3)\}$, the total spatio-temporal preference score of which is 1.59.

In this paper, we investigate a Spatio-temporal Preference-aware Task Assignment (SPTA) problem in spatial crowdsourcing. Specially, given a set of workers and a set of tasks, it aims to maximize the total rewards of the task assignment considering the spatio-temporal preferences of workers. In order to tackle this problem, we propose a two-phase framework, which includes a translation-based recommendation phase to capture workers' spatio-temporal preferences, and a task assignment phase. More specifically, in the first phase, we use a Translation-based recommendation (Trans) model to learn spatio-temporal effects from the workers' historical task-performing activities and calculate the spatio-temporal preference scores of workers to each task. In the assignment phase, we design a basic greedy algorithm and a KM-based algorithm to assign a suitable worker to each task by giving higher priorities to workers who are more interested in the tasks.

The contributions made by this paper can be summarized as follows:

- To the best of our knowledge, this is the first work in SC that considers the spatio-temporal preferences among workers.
- We employ a translation-based recommendation model to learn the spatio-temporal effects of workers from the historical task-performing data.
- We propose a greedy algorithm and a KM-based algorithm to perform the task assignment based on spatio-temporal preferences of workers.

- Extensive experiments are conducted to verify the proposed methods that could achieve a better balance to maximize the total rewards and meet the spatio-temporal preferences of workers.

2 Problem Definition

In this section, we will briefly introduce a set of preliminary concepts and then give our problem statement. Table 1 summarizes the major notations used throughout the paper.

Table 1. Summary of notations

Notation	Definition
v	Point of Interest
V	A set of POIs
s	Spatial Task
S	A set of tasks
$s.v$	Associated POI of spatial task s
$s.p$	Publication time of spatial task s
$s.e$	Expiration time of spatial task s
$s.r$	Reward of spatial task s
w	Worker
W	A set of workers
$w.l_0$	Current location of worker w
$w.d$	Reachable distance of worker w
t	A time slot
l	A location/region
tl	A spatio-temporal context $< t, l >$
TL	A set of spatio-temporal contexts
D	A collection of worker task-performing records $D = \{d \mid d = (w, t, l, s)\}$
T	A collection of worker task-performing tripts $T = \{d \mid d = (w, tl, s)\}$
A	A spatial task assignment
$A.r$	Total reward in task assignment A

Definition 1 (POI). *A POI (point of interest) v is defined as a uniquely identified geographical site with some functions (e.g., a restaurant or a hotel), and we use V to denote a set of POIs, i.e., $V = \{v\}$.*

Definition 2 (Spatial Task). *A spatial task, denoted by $s = (s.v, s.p, s.e, s.r)$, is a task to be performed at POI $s.v$, published at time $s.p$, and will expire at $s.e$, with the reward of $s.r$.*

Different from previous work [8], the task may appear at any point in the 2D space. In our work, a task can only be published at POI. Since similar tasks usually appear at similar POIs, e.g., a delivery man (worker) goes to a milk tea shop (POI) to pick up orders (tasks).

Definition 3 (Worker). *A worker, denoted as* $w = (w.l_O, w.d)$*, has a current location* $w.l_O$*, and a reachable distance* $w.d$*. The reachable area of worker* w *is a circular area with* $w.l_O$ *as the center and* $w.d$ *as the radius, in which the worker* w *can accept task assignment.*

A worker can be in an online or offline mode. In our work, we assume all workers are online when assigning tasks and only can process only one task at a certain time instance.

Definition 4 (Spatio-temporal context). *A spatio-temporal context, denoted as* tl *, is a combination of a time slot* t *and a location/region* l *, e.g., (20:00, the High Street).*

In this work, we discretize timestamps associated with task-performing records into time slots, e.g., 24 h in a day. Similarly, we divide the whole spatial space into some geographical regions based on some predefined criteria. In the rest of the paper, we will use the terms regions and locations interchangeably. After the discretization, the number of spatio-temporal context (tl) becomes limited.

Definition 5 (Task-performing activity). *A task-performing activity is a tuple* $d = (w, t, l, s)$*, which means worker* w *performing task* s *at geographical region/location* l *and time* t*.*

Definition 6 (Spatial Task Assignment). *Given a set of workers* W *and a set of tasks* S*, a spatial task assignment is denoted as* A*, consisting of a set of tuples of form* (w, s)*, which means the spatial task* s *is assigned to worker* w*.*

Spatio-temporal Preference-aware Task Assignment (SPTA) Problem Statement Given a set of workers W, a set of tasks S and a dataset $D = \{d|d = (w, t, l, s)\}$ recording a set of workers' historical task-performing activities at the current time on an SC platform, SPTA problem aims to find an optimal task assignment A_o that maximizes the total rewards of all assigned tasks (i.e., $\forall A_i \in \mathbb{A}, A_i.r \leq A_o.r$) considering the spatio-temporal preferences of workers, where $A_i.r$ denotes the total rewards of the task assignment A_i, and $\mathbb{A}$ denotes all the possible ways of assignments.

3 Methodology

3.1 Framework Overview

As shown in Fig. 2, our framework consists of two components: 1) Translation-based recommendation model (Trans) to learn workers' spatio-temporal preferences; and 2) Preference-aware Task Assignment (PTA) based on workers' spatio-temporal preferences.

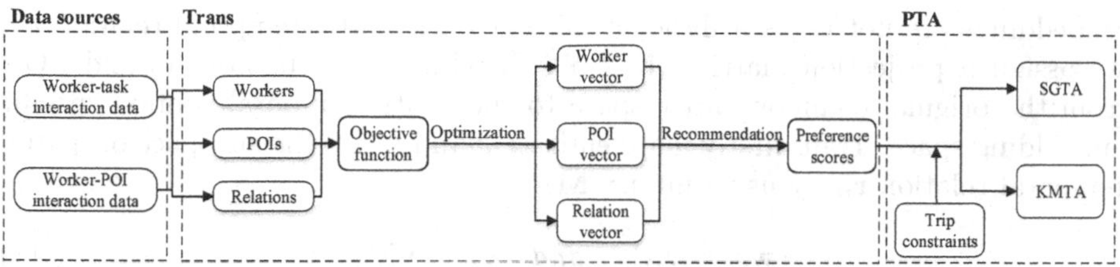

Fig. 2. Framework overview

In the Trans part, inspired by the success of [11] in modeling the spatio-temporal effects in POI recommendation, we use a translation-based recommendation model to learn workers' spatio-temporal preferences in different spatio-temporal contexts, e.g., (7:00, Campus) vs. (20:00, the High Street). Specially, we extract task-performing activities (w, tl, s) from worker-task interaction data and worker-POI interaction data. In order to learn workers' spatio-temporal preferences, we employ spatio-temporal context (tl) as the relation to connect worker (w) and task's (s) associated POI $(s.v)$. Finally, we can get the representation vectors of workers, POIs and relations through joint optimization. Then we could calculate workers' spatio-temporal preference scores by the recommendation model.

In the PTA phase, by considering trip constraints including workers' reachable distance and tasks' expiration time, we optimize the task assignment based on workers' spatio-temporal preference, and propose two algorithms, i.e., Spatio-temporal preference-aware Greedy Task Assignment (SGTA) algorithm and KM-based Task Assignment (KMTA) algorithm.

3.2 Spatio-temporal Preference Modeling

In this subsection, we first introduce our Trans model that aims to capture spatio-temporal effects of context-aware task-performing activities. Then we show how to use this model to calculate workers' spatio-temporal preference scores, which will be used in the task assignment phase.

Model Description. The translation-based model is good at capturing the transition relation [11]. Inspired by the success of [11,13] in learning the spatio-temporal effects of user check-in activities, we employ the translation-based model Trans to model spatio-temporal preferences of workers. Given the worker-task interaction data and worker-POI interaction data, we could extract task-performing activity (w, tl, s), which means that worker w performs task s in the spatio-temporal context tl. Then we could construct a triplet (w, tl, v) from task-performing activity (w, tl, s), where v represents the associated POI of task s. For each triplet (w, tl, v), the worker and task entities are embedded into vectors $\mathbf{w} \in \mathbb{R}^k$ and $\mathbf{v} \in \mathbb{R}^k$ respectively, and relation tl is embedded into $\mathbf{r}_{tl} \in \mathbb{R}^d$. The dimensions of worker and POI embeddings and spatio-temporal relation embeddings are not necessarily identical. For each spatio-temporal relation $\mathbf{r}_{tl}$, we assign a projection matrix $\mathbf{M}_r \in \mathbb{R}^{k\times d}$, which projects workers and POIs from the original common entity space to the spatio-temporal context-specific embedding space. Trans firstly maps entities w and v into the subspace of spatio-temporal relation $\mathbf{r}_{tl}$ by using matrix $\mathbf{M}_r$:

$$\mathbf{w}_r = \mathbf{w}\mathbf{M}_r \quad and \quad \mathbf{v}_r = \mathbf{v}\mathbf{M}_r \tag{1}$$

With the mapping matrix, the representation vector of worker w and POI v could be mapped into the same space as the representation vector of spatio-temporal relation tl. In this space, the POI embedding $\mathbf{v}_r$ should be the nearest neighbor of $\mathbf{w}_r + \mathbf{r}_{tl}$, indicating that worker w prefers performing tasks at POI v in the spatio-temporal context tl. The score function is then defined as:

$$f(w, tl, v) = \parallel \mathbf{w}_r + \mathbf{r}_{tl} - \mathbf{v}_r \parallel_2^2 \tag{2}$$

Given the score function defined in Eq. (2) for a triple (w, tl, v), the following margin-based ranking loss defined in [10] is used for training:

$$L = \sum_{(w,tl,v)\in S} \sum_{(w',tl,v')\in S'} max(0, f(w, tl, v) + \gamma - f(w', tl, v')) \tag{3}$$

where $max(a, b)$ is used to get the maximum between a and b, γ is used to control the margin between positive and negative samples, S and S' are the set of positive and negative triplets, respectively. The positive triplets are extracted from task-performing activities, and the negative triples are generated by replacing the head or tail entities in positive triples with the dissimilar worker or POI. Thus, we corrupt the correct triplet $(w, tl, v) \in S$ to construct incorrect triplets (w', tl, v') by replacing either head or tail entities with other entities from the same group so that:

$$S' = (w', tl, v) \cup (w, tl, v') \tag{4}$$

Preference Calculation. Once we have learned the embeddings for all workers, POIs, and spatio-temporal contexts, given a worker w and its current location l and current time t, we could calculate the preference of worker w to task s in the spatio-temporal context tl:

$$p = \parallel \mathbf{w}\mathbf{M}_r + \mathbf{r}_{tl} - \mathbf{v}\mathbf{M}_r \parallel_2^2 \tag{5}$$

Eq. (5) is the combination of Eq. (1) and Eq. (2), it is also could be seen as the Euclidean distance between $\mathbf{w}\mathbf{M}_r + \mathbf{r}_{tl}$ and $\mathbf{v}\mathbf{M}_r$. Given m workers, n tasks, and their spatio-temporal contexts, we could use a preference matrix $\mathbf{P} \in \mathbb{R}^{m\times n}$ to store workers' spatio-temporal preferences to tasks, where $\mathbf{P}_{ij}$ represents the spatio-temporal preference of worker w_i to task s_j.

3.3 Task Assignment

In this subsection, given a set of workers $W = \{w_1, w_2, ..., w_{|W|}\}$ and a set of tasks $S = \{s_1, s_2, ..., s_{|S|}\}$, we should find an appropriate task assignment that maximizes the total rewards of all assigned tasks while considering workers' spatio-temporal preferences. We firstly generate the available worker sets for each task s based on the trip constraints (i.e., workers' reachable distance and tasks' expiration time), and we then propose the Spatio-temporal preference-aware Greedy Task Assignment (SGTA) algorithm and KM-based Task Assignment (KMTA) algorithm based on spatio-temporal preferences to assign tasks to appropriate workers.

Due to the constraint of workers' reachable distance and tasks' expiration time, each task $s(s \in S)$ can only be completed by a small subset of workers in a time instance. Therefore, we firstly find the set of available workers that can complete task s without violating the constraints, which is denoted as $AW(s)$. The available worker set should satisfy the following conditions: $\forall s \in S, \forall w \in AW(s)$:

1) $t_{now} + t(w.l_0, s.v) \leq s.e$, and
2) $d(w.l_0, s.v) \leq w.d$,

where t_{now} is the current time, $t(w.l_0, s.v)$ is the travel time from $w.l_0$ to $s.v$ and $d(w.l_0, s.v)$ is the travel distance(e.g., Euclidean distance) between $w.l_0$ and $s.v$. The above conditions guarantee that $\forall w \in AW(s)$, worker w could accomplish task s before its expiration time. Here, we assume all the workers share the same speed.

Greedy Task Assignment. We propose Spatio-temporal preference-aware Greedy Task Assignment algorithm (SGTA), which takes workers' preference as the priority of task assignment. Algorithm 1 outlines the major procedure of the SGTA algorithm, which takes the worker set W, task set S, and spatio-temporal contexts TL as input. First we sort all the tasks by their rewards in descending order (line 2). Then the algorithm generates the available worker set $AW(s)$ from W for each task s (line 4). Next, the algorithm will assign worker $w \in AW(s)$ with the maximum preference p to task s (line 5-8). Finally, Algorithm 1 will return a suitable task assignment result (line 9).

KM Task Assignment. It is obvious that the greedy task assignment algorithm cannot get the optimal result for the task assignment problem. Therefore, we transform this problem to a Bipartite Maximum Weight Matching problem [14] and employ KM-based algorithm to solve it. Specially, we construct a bipartite graph $G_{WS} = (W \cup S, E_{WS})$, where W is the worker set and S is the task set to be assigned, $W \cup S$ is the node set of G_{WS}, and E_{WS} represents the edges between workers and tasks. There is an edge $e_{ij} \in E_{WS}$ between worker $w_i \in W$ and task $s_j \in S$ if worker $w_i \in AW(s_j)$, where $AW(s_j)$ is the available worker set of task s_j. Moreover, the weight (w_{ij}) of edge e_{ij} can be set as the

Algorithm 1: SGTA Algorithm

Input: Worker set W, task set S, spatio-temporal contexts TL
Output: Task assignment: A

$A \leftarrow \emptyset$;
$S \leftarrow$ Sort S in descending order of task's reward ;
for *each task* $s \in S$ **do**
  Obtain the available worker set $AW(s)$ from W;
  for *each worker* $w \in AW(s)$ **do**
    calculate spatio-temporal preference p by Equation (5);
  $w_s \leftarrow$ worker $w \in AW(s)$ with the maximum preference p;
  $A \leftarrow A \cup \{(w_s, s)\}$;
return A;

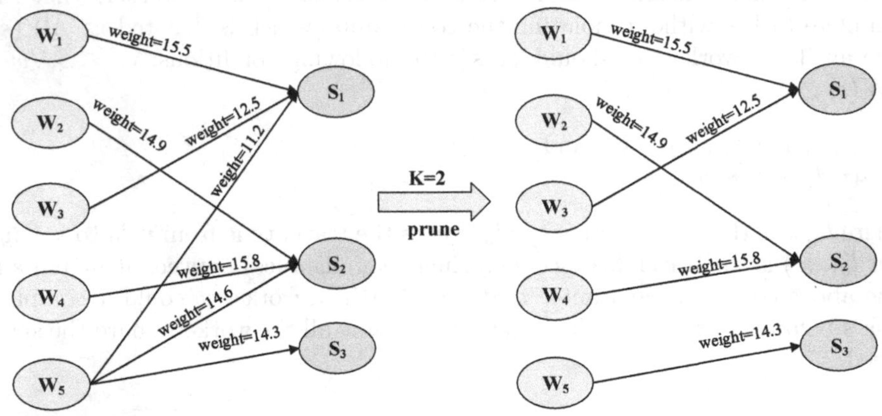

Fig. 3. Worker-task bipartite graph

sum of worker w_i's spatio-temporal preference to task s_j and the reward of task s_j, i.e., $w_{ij} = \mathbf{P}_{ij} + s_j.r$, where $\mathbf{P}_{ij}$ is the workers' preference matrix mentioned in Sect. 3.2. The left part of Fig. 3 depicts an example of such bipartite graph for five workers and three tasks. To improve efficiency, we limit the number of edges in the graph. For each task $s_j \in S$, we prune associated edges with task s_j. Therefore, for each task node, we only retain those associated edges with top K weight. In the example of Fig. 3, we set K to 2. The task assignment problem is now converted into a bipartite maximum weight matching problem. Finally we employ the Kuhn-Munkres (KM) [7] algorithm to solve this problem.

4 Experimental Evaluation

In this section, we conduct extensive experiments on a real-world dataset to evaluate the performance of our proposed algorithms. All the algorithms are implemented on an Intel(R) Xeon(R) CPU E5-2650 v4 @ 2.20 GHz with 256GB RAM, and NVidia GeForce RTX 1080Ti GPU.

Table 2. Experiment parameters

Parameters	Values
Valid time of tasks $e - p$	0.1, 0.15, 0.2, 0.25, <u>0.3</u>
Number of tasks $\|S\|$	1000, 2000, <u>3000</u>, 4000, 5000
Number of workers $\|W\|$	400, 800, 1200, 1600, <u>2000</u>
Reachable distance of workers d	1, 2, 3, 4, <u>5</u>
Coefficient K	10, 20, 30, 40, <u>50</u>

4.1 Experimental Setup

Dataset. We conduct our experiments on a check-in dataset from Foursquare, which provides long-term (about 10 months) check-in data in Tokyo city from 12 April 2012 to 16 February 2013 including 2293 users and 61858 POIs [17]. The dataset is used widely in the evaluation of the SC platform [1]. When using this dataset in our experimental study, we assume the users are the workers in the SC system since users who check in different spots may be good candidates to perform spatial tasks in the vicinity of those spots, and their locations are those of the most recent check-in points. We assume the spots are the tasks of the SC platform. To calculate workers' spatio-temporal preference to each task, we collect check-in data around time 2012–10–13 12:00, employing spot's location and check-in time as the location and publication time of the task. In order to get the spatio-temporal contexts, we use the same method as that in [11], dividing time into 24 time slots that corresponds to 24 h, and the whole geographical space into 100 regions clustered by a standard k-means method. We finally get 2387 spatio-temporal relations.

Since the original dataset does not contain the rewards of tasks, we uniformly generate the reward from the ranges [1,5]. The speed of workers is set to $20km/h$. Table 2 shows our experimental settings where the default values of all parameters are underlined. The margin γ is set to 1 in Eq. (3).

Evaluation Methods. We compare and evaluate the performance of the following methods:

1) SGTA: The Spatio-temporal preference-aware Greedy Task Assignment (SGTA) algorithm.
2) GTA: The Greedy Task Assignment algorithm without considering spatio-temporal preference. The spatio-temporal preference is replaced by the distance between worker and task.
3) SR-KMTA: The KM Task Assignment (KMTA) algorithm considers workers' spatio-temporal preferences and tasks' rewards (SR).
4) S-KMTA: The KM Task Assignment algorithm only considers workers' spatio-temporal preferences (S). The difference with SR-KMTA is algorithm KM's weight.
5) R-KMTA: The KM Task Assignment algorithm only considers tasks' rewards (R). The difference with SR-KMTA is as S-KMTA.

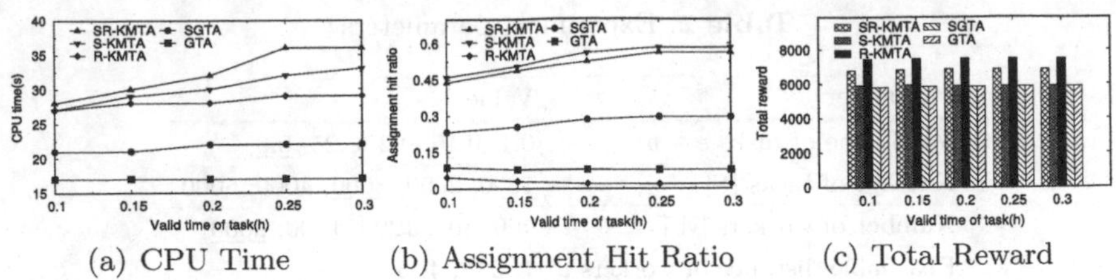

(a) CPU Time (b) Assignment Hit Ratio (c) Total Reward

Fig. 4. Performance of task assignment: Effect of $e - p$

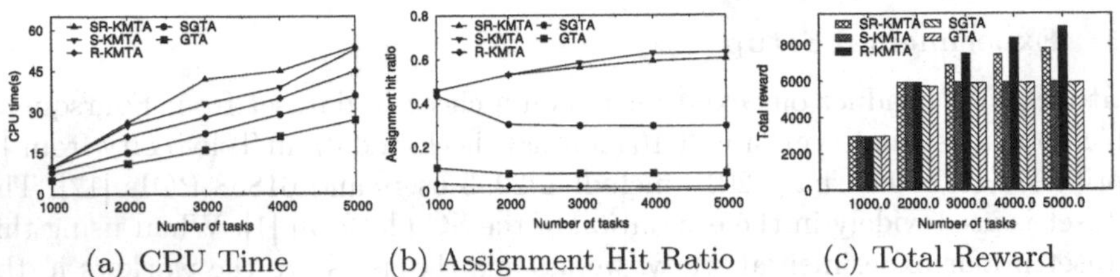

(a) CPU Time (b) Assignment Hit Ratio (c) Total Reward

Fig. 5. Performance of task assignment: Effect of $|S|$

Metrics. Three metrics are compared among the above methods:

- CPU-cost: the CPU time cost for finding the task assignment.
- AHR: Assignment Hit Ratio is the ratio of the number of hit workers to the amount of successful assigned workers. An assigned worker is a hit worker when assigned task s satisfies the spatio-temporal preference of worker w, i.e., $s.v \in D_{train\_w}$ or $s.v \in D_{test\_w}$, where D_{train_w} is the associated training POI set and D_{test_w} is the associated test POI set extracted from task-performing records of worker w. The AHR can be calculated by:

$$AHR = \frac{N(W_{hit})}{N(W_{assign})} \tag{6}$$

  where $N(W_{hit})$ is the number of hit workers, and $N(W_{assign})$ is the number of assigned workers.
- Total reward: the sum of all successful assigned tasks' rewards.

4.2 Experimental Results

In this subsection, in order to demonstrate the effectiveness of our method, we compare our evaluation metrics with other baselines.

Effect of $e - p$. We first study the effect of the valid time $e - p$ of tasks. As depicted in Fig. 4(a), the running time of all methods increases for longer valid times of tasks, since there are more worker-task assignments to process. In terms of the success hit ratio of task assignment, as shown in Fig. 4(b), the SR-KMTA, S-KMTA and SGTA algorithms increase while R-KMTA and GTA decrease since

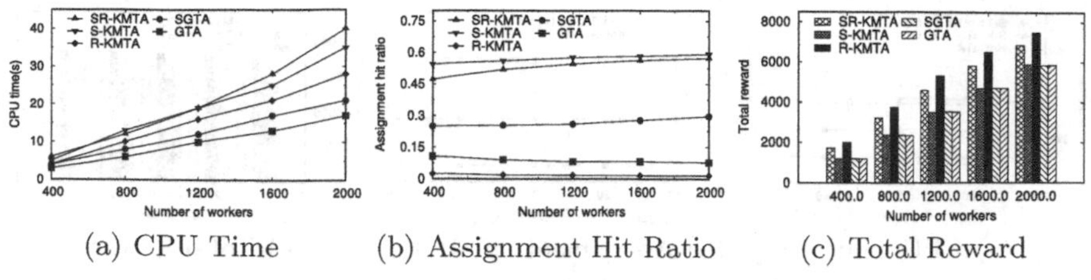

(a) CPU Time (b) Assignment Hit Ratio (c) Total Reward

Fig. 6. Performance of task assignment: Effect of $|W|$

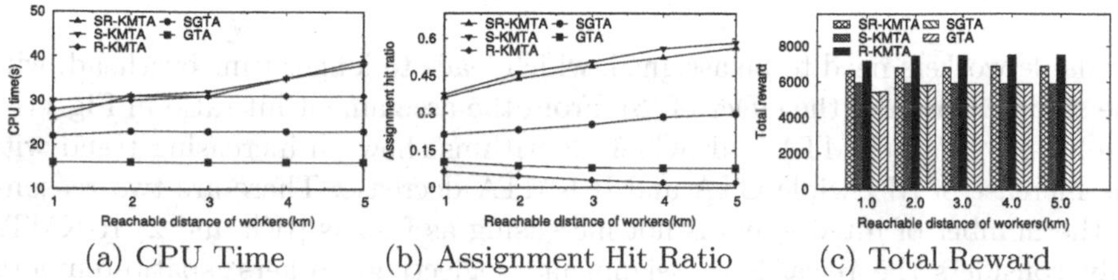

(a) CPU Time (b) Assignment Hit Ratio (c) Total Reward

Fig. 7. Performance of task assignment: Effect of d

they only consider tasks' reward. Moreover, the total rewards of all the methods except for S-KMTA show a not obvious growth trend (see Fig. 4(c)). This is because the weight of S-KMTA is the reward of the assignment when matching workers and tasks. As shown in Fig. 4, all the five methods remain unchanged when $e - p$ exceeds $0.25h$, which may be due to the fact that a majority of the tasks can be completed before 0.25h.

Effect of $|S|$. Next, we study the scalability of the algorithms by changing the number of tasks $|S|$ from 1000 to 5000. It can be seen from Fig. 5(a) that all the methods' CPU time is increasing. At the same time, we can conclude that the CPU time of the GTA-related algorithms is lower than that of the KMTA-related algorithms. This is because with the increase of the number of tasks, there will be more edges between workers and tasks, which may increase the recursion depth of the KMTA-related algorithms, costing more CPU time. As the amount of tasks increases, the total reward of all the methods increases since each worker could have more chances to access higher-reward tasks. We can also find that compared with other algorithms, the total reward of the R-KMTA algorithm is highest while its AHR is lowest (see Fig. 5(b) and Fig. 5(c)) since it only considers the reward of tasks when assigning. In addition, with the increasing $|S|$, the AHR of the SR-KMTA and S-KMTA algorithms increases while that of the other three methods decrease. The reason behind it is the number of hit workers of these three methods (R-KMTA, GTA, SGTA) increases slower than the number of assigned workers.

Effect of $|W|$. To study the effect of $|W|$, we generate 5 datasets containing 400 to 2000 workers by random selection from the dataset. As depicted in Fig. 6(a), the CPU time gets longer when $|W|$ gets larger. This is because more and more

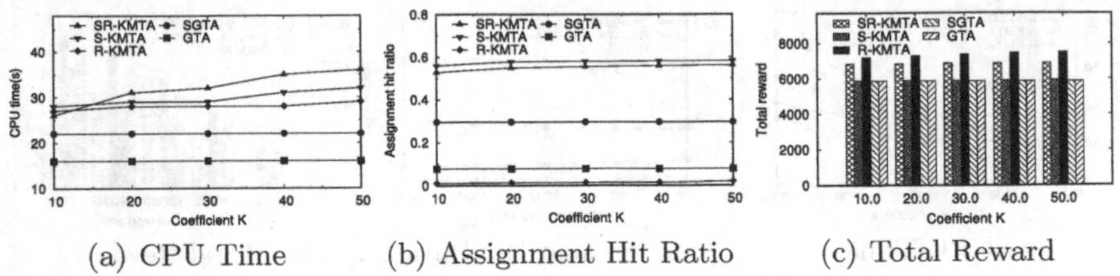

(a) CPU Time (b) Assignment Hit Ratio (c) Total Reward

Fig. 8. Performance of task assignment: Effect of K

available workers need to be assigned, which leads to longer time overhead, with the similar reason of the effect of $|S|$. From the assignment hit ratio in Fig. 6(b), the SR-KMTA, S-KMTA and SGTA algorithms show an increasing trend with the increase of $|W|$ while GTA and R-KMTA decrease. There are two reasons: 1) the number of hit workers is not increasing as fast as $|W|$; and 2) R-KMTA only considers the reward of assignment, neglecting workers' spatio-temporal preferences. In addition, the total rewards of all the methods increase with the increase of $|W|$ (see Fig. 6(c)). At the same time, we could find that the total reward of SR-KMTA and R-KMTA is much better than the other three methods since those methods (GTA, SGTA, and S-KMTA) don not consider the reward of tasks during task assignment. In summary, it can be seen that compared with S-KMTA and R-KMTA, SR-KMTA could have a better balance between workers' spatio-temporal preferences and tasks' rewards.

Effect of d. In this part of the experiment, we further evaluate the effect of the reachable distance d of workers. It can be seen from Fig. 7(a), the CPU time of all the methods gradually increases with d being enlarged. This is because when the reachable distance of workers increases, the number of available workers for each task increases, resulting in a larger search space. In terms of the hit ratio of assignment, as shown in Fig. 7(b), with d increasing, the AHR of SR-KMTA, S-KMTA and SGTA increase. This is similar to $|W|$ since the bigger d could cause more available workers for each task. In addition, the total reward of all the methods except for S-KMTA increases with the increase of d (see Fig. 7(c)), with the similar reason of $e - p$.

Effect of K. Finally, we study the effect of K, which limits the number of edges associated with task vertices in KM-related algorithms. With the increase of K, each worker could associate with more tasks which means each worker could have more opportunities to be assigned to a task. Therefore, the total reward and AHR of R-KMTA, S-KMTA and SR-KMTA increase with the increase of K (see Fig. 8(b) and Fig. 8(c)). At the same time, the bigger K also could cause the competition among workers more intense, leading a higher CPU time (see Fig. 8(a)).

5 Related Work

Task assignment is the core problem in spatial crowdsourcing. Based on the task publishing models, Kazemi et al. [6] divide SC into worker selection task (WST) mode and server assignment task (SAT) mode. For the SAT mode which is the main focus of existing stuies, an SC server is responsible for directly assigning proper tasks to nearby workers to achieve a certain goal of task assignment [12, 20,25]. For the WST mode, the server publishes various spatial tasks online where workers can select any tasks based on their own preferences without the need to coordinate with the server [4]. In this work, we fous on the SAT mode.

In the task assignment problem of SC, most stuies fail to consider workers' preferences when assigning tasks. In order to solve this problem, Zhao et al. [21] propose a preference-aware spatial task assignment system based on workers' temporal preferences. Li et al. [8] use a bipartite graph embedding model and the attention mechanism to learn the social impact-based preferences of the worker groups for different categories of tasks. However, those works ignore the connection between spatial and temporal relations while workers may appear temporal preference as well as spatial preference which is called spatio-temporal preference in our work. Unlike the previous studies which only consider workers' temporal or spatial preference independently, our proposed approach could model workers' spatio-temporal preference simultaneously based on historical task-performing data and better meet the spatio-temporal preferences of workers when assigning tasks.

6 Conclusion

In this paper, we propose and offer solutions to an SC problem called Spatio-temporal Preference-aware Task Assignment (SPTA), which aims to find the optimal task assignment with the maximal total rewards of all assigned tasks while considering spatio-temporal preferences of workers. In order to model spatio-temporal preferences of workers, we employ translation-based recommendation (Trans) model to learn spatio-temporal effects from the workers' historical task-performing data. Based on these spatio-temporal preferences, we propose greedy and optimal algorithms for achieving the task assignment to trade off the efficiency and effectiveness. To the best of our knowledge, this is the first work in SC that considers spatial and temporal preferences simultaneously and performs task assignment based on these preferences. Extensive empirical study based on a real dataset confirms the effectiveness of our proposed solutions.

Acknowledgements. This work is partially supported by NSFC (No. 61972069, 61836007 and 61832017), and Shenzhen Municipal Science and Technology R&D Funding Basic Research Program (JCYJ20210324133607021).

References

1. Cheng, P., Lian, X., Chen, L., Shahabi, C.: Prediction-based task assignment in spatial crowdsourcing. In: ICDE, pp. 997–1008 (2017)
2. Cheng, P., Lian, X., Chen, Z., Fu, R., Chen, L., Han, J., Zhao, J.: Reliable diversity-based spatial crowdsourcing by moving workers. PVLDB **8**(10), 1022–1033 (2015)
3. Cheng, P., Lian, X., Jian, X., Chen, L.: Frog: a fast and reliable crowdsourcing framework. TKDE **31**(5), 894–908 (2018)
4. Deng, D., Shahabi, C., Demiryurek, U.: Maximizing the number of worker's self-selected tasks in spatial crowdsourcing. In: SIGSPATIAL, pp. 324–333 (2013)
5. Dickerson, J.P., Sankararaman, K.A., Srinivasan, A., Xu, P.: Assigning tasks to workers based on historical data: Online task assignment with two-sided arrivals. In: AAMAS, pp. 318–326 (2018)
6. Kazemi, L., Shahabi, C.: Geocrowd: enabling query answering with spatial crowdsourcing. In: SIGSPATIAL, pp. 189–198 (2012)
7. Kuhn, H.W.: The hungarian method for the assignment problem. Naval Res. logistics Q. **2**(1–2), 83–97 (1955)
8. Li, X., Zhao, Y., Guo, J., Zheng, K.: Group Task Assignment with Social Impact-Based Preference in Spatial Crowdsourcing. In: Nah, Y., Cui, B., Lee, S.-W., Yu, J.X., Moon, Y.-S., Whang, S.E. (eds.) DASFAA 2020. LNCS, vol. 12113, pp. 677–693. Springer, Cham (2020). https://doi.org/10.1007/978-3-030-59416-9_44
9. Li, Y., Zhao, Y., Zheng, K.: Preference-aware group task assignment in spatial crowdsourcing: a mutual information-based approach. In: ICDM, pp. 350–359 (2021)
10. Lin, Y., Liu, Z., Sun, M., Liu, Y., Zhu, X.: Learning entity and relation embeddings for knowledge graph completion. In: AAAI, pp. 2181–2187 (2015)
11. Qian, T., Liu, B., Nguyen, Q.V.H., Yin, H.: Spatiotemporal representation learning for translation-based poi recommendation. TOIS **37**(2), 1–24 (2019)
12. Tong, Y., Chen, L., Zhou, Z., Jagadish, H.V., Shou, L., Lv, W.: Slade: a smart large-scale task decomposer in crowdsourcing. TKDE **30**(8), 1588–1601 (2018)
13. Wang, X., Salim, F.D., Ren, Y., Koniusz, P.: Relation Embedding for Personalised Translation-Based POI Recommendation. In: Lauw, H.W., Wong, R.C.-W., Ntoulas, A., Lim, E.-P., Ng, S.-K., Pan, S.J. (eds.) PAKDD 2020. LNCS (LNAI), vol. 12084, pp. 53–64. Springer, Cham (2020). https://doi.org/10.1007/978-3-030-47426-3_5
14. Wang, Z., Zhao, Y., Chen, X., Zheng, K.: Task assignment with worker churn prediction in spatial crowdsourcing. In: CIKM, pp. 2070–2079 (2021)
15. Xia, J., Zhao, Y., Liu, G., Xu, J., Zhang, M., Zheng, K.: Profit-driven task assignment in spatial crowdsourcing. In: IJCAI, pp. 1914–1920 (2019)
16. Xu, L., Zhou, X.: A crowd-powered task generation method for study of struggling search. Data Sci. Eng. **6**(4), 472–484 (2021)
17. Yang, D., Zhang, D., Zheng, V.W., Yu, Z.: Modeling user activity preference by leveraging user spatial temporal characteristics in LBSNS. IEEE Trans. Syst. Man Cybern. Syst. **45**(1), 129–142 (2015)
18. Ye, G., Zhao, Y., Chen, X., Zheng, K.: Task allocation with geographic partition in spatial crowdsourcing. In: CIKM, pp. 2404–2413 (2021)
19. Zhao, Y., Guo, J., Chen, X., Hao, J., Zhou, X., Zheng, K.: Coalition-based task assignment in spatial crowdsourcing. In: ICDE, pp. 241–252 (2021)
20. Zhao, Y., Li, Y., Wang, Y., Su, H., Zheng, K.: Destination-aware task assignment in spatial crowdsourcing. In: CIKM, pp. 297–306 (2017)

21. Zhao, Y., et al.: Preference-aware task assignment in spatial crowdsourcing. In: AAAI, pp. 2629–2636 (2019)
22. Zhao, Y., Zheng, K., Cui, Y., Su, H., Zhu, F., Zhou, X.: Predictive task assignment in spatial crowdsourcing: a data-driven approach. In: ICDE, pp. 13–24 (2020)
23. Zhao, Y., Zheng, K., Guo, J., Yang, B., Pedersen, T.B., Jensen, C.S.: Fairness-aware task assignment in spatial crowdsourcing: Game-theoretic approaches. In: ICDE, pp. 265–276 (2021)
24. Zhao, Y., Zheng, K., Li, Y., Su, H., Liu, J., Zhou, X.: Destination-aware task assignment in spatial crowdsourcing: a worker decomposition approach. TKDE, pp. 2336–2350 (2019)
25. Zhao, Y., Zheng, K., Yin, H., Liu, G., Fang, J., Zhou, X.: Preference-aware task assignment in spatial crowdsourcing: from individuals to groups. TKDE **34**(7), 3461–3477 (2020)

DE-DQN: A Dual-Embedding Based Deep Q-Network for Task Assignment Problem in Spatial Crowdsourcing

Yucen Gao[1], Dejun Kong[1], Haipeng Dai[2], Xiaofeng Gao[1(✉)], Jiaqi Zheng[2], Fan Wu[1], and Guihai Chen[2]

[1] Shanghai Jiao Tong University, Shanghai 200240, China
{guo_ke,kdjkdjkdj99}@sjtu.edu.cn, {gao-xf,fwu}@cs.sjtu.edu.cn
[2] Nanjing University, Nanjing 210008, China
{haipengdai,jzheng,gchen}@nju.edu.cn

Abstract. Along with the rapid development of sharing economy, spatial crowdsourcing has become a hot topic of general interests in recent years. One of its core issues is the task assignment problem, in which tasks are released on the platforms and then assigned to available workers. Due to the various characteristics of tasks and workers, finding the optimal assignment scheme and routing plan are difficult.

In this paper, we define the utility-driven destination-aware spatial task assignment problem (UDSTA), which is proved to be NP-complete. Our goal is to maximize the total utility of workers. We propose a dual-embedding based deep Q-Network (DE-DQN) to sequentially assign tasks to suitable workers. Specifically, we design a utility embedding to reflect the top-k utility tasks for workers and worker-task pairs, and propose a coverage embedding to represent the potential future utility of an assignment action. For the first time, we combine the dual embedding with DQN to realize the multi-task and multi-worker matching, and obtain route plans of workers. Experiments based on both synthetic and real-world datasets indicate that DE-DQN performs well and shows significant advantages over the baseline methods.

Keywords: Dual embedding · Deep Q-Network · Task assignment · Spatial crowdsourcing

1 Introduction

Spatial crowdsourcing is an emergent working mode in recent years, which decomposes sophisticated tasks from task publishers into multiple small and easy tasks, and then assigns these tasks to numerous mobile workers with various backgrounds to finish the tasks efficiently and fast. Along with the rapid

This work was supported by National Key R&D Program of China [2019 YFB2102200]; National Natural Science Foundation of China [61872238, 6187 2178, 61972254], Shanghai Municipal Science and Technology Major Project [2021 SHZDZX0102] and Huawei Cloud [TC20201127009]. Thanks for the help of Wei Liu.

B. Li et al. (Eds.): APWeb-WAIM 2022, LNCS 13421, pp. 280–295, 2023.
https://doi.org/10.1007/978-3-031-25158-0_22

development of mobile intelligent devices, spatial crowdsourcing is extending to many aspects of the society. Many related applications based on spatial crowdsourcing systems are proposed and employed in practice such as Ear-phone [14] for urban noise sensing and Uber [17] for taxis hailing.

Traditional spatial crowdsourcing problems concentrate on task assignment, quality control, incentive system, and privacy protection. Nevertheless, they typically overlook mobility, online situation and tempo-spatial characteristics, which are crucial to many applications. Besides, these features also make the problems more complicated. Therefore, researchers have paid much attention to explore easier, quicker and more robust methods to deal with spatial crowdsourcing.

One of the core issues in spatial crowdsourcing is the task assignment. Task assignment is to allocate tasks with unique attributes such as revenue and location to workers in service who also have specific characteristics such as service scope and ability. Such a process of task-worker matching is realized via a spatial crowdsourcing platform as shown in Fig. 1. The task publishers submit tasks to be completed with various requirements to the platform. The workers join in the platform through recruitment. Then, the platform matches compatible tasks with workers according to the tasks pool and workers pool. Nevertheless, if the scales of tasks and workers increase, the computational complexity of the problem may increase in factorial order. Thus, proposing efficient and fast methods for the situations of large-scale data deserves in-depth research.

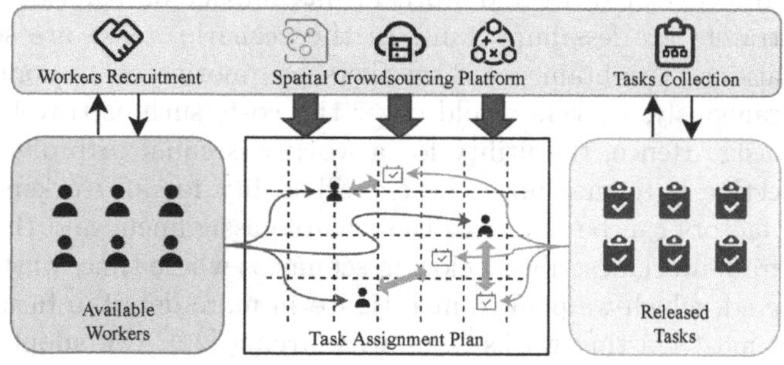

Fig. 1. Illustration of task assignment in spatial crowdsourcing.

In our scenario, we consider the problem of which multiple tasks and workers are distributed in a spatial area. A worker can receive and finish a limited number of tasks, while a task can only be assigned to one worker. A cost emerges when a worker deals with a task, which is proportional to the distance between the worker and the task. When a task is completed, the worker will receive the corresponding reward. The goal of the problem is to maximize the total utility of all the workers given the cost and reward. The problem is named as utility-driven destination-aware spatial task assignment problem (UDSTA). Apparently, there may be conflicts between global optimum and local optimum (*e.g.*, high-reward tasks are preferred by many workers). From the perspective of workers, maximizing their own revenue can be a naive strategy for task selection, which is mainly associated to less working interval and larger task reward. Less working interval leads to preferring neighboring tasks, while larger task revenue leads to preferring tasks with high reward. There may exist a decision dilemma when the two strategies conflict with each other, making the optimization more difficult.

As for UDSTA, traditional algorithms like greedy algorithms and heuristic algorithms may fall into local optimum and cannot obtain optimal solutions. Even worse, when the scale of the problem is large, solving the problem is very time-consuming. Hence, to leverage the advantages of machine learning techniques in finding global optimal solutions and addressing large-scale problems, we propose a new value based deep learning framework called dual-embedding based deep Q-Network (DE-DQN), which innovatively utilizes dual embedding vectors as the input of DQN to learn the optimal strategy of task assignment. DE-DQN can significantly improve the efficiency of problem solving and performs best in large-scale scenarios. What's more, since UDSTA problem is a generic setting for the task assignment problem, the proposed DE-DQN can be easily adapted to address many variant problems such as sweep coverage and vehicle dispatching by adding time-related factors and modifying the optimization goal.

2 Problem Statement and Preliminaries

In the paper, we pay attention to the scenarios such as region monitoring, where location distribution of tasks and workers has impact on solutions and time-related constraints are less important. In the scenario, tasks are scattered in different locations. To obtain profit, workers are motivated to complete some tasks. Simultaneously, workers would cover the costs such as travel costs when completing tasks. Hence, the utility for a worker is equal to profit minus cost and the objective is to maximize the overall utility for all workers. However, time-related factors can be linked to the feasible assignment and thus included in the DQN framework described below in scenarios where time constraints need to be considered, which we plan to investigate in more detail in future work.

Following most existing works in crowdsourcing [22] , we adopt the round-based model. In each round, the platform operates on a set of pended tasks and workers. Basic definitions related to worker, task and utility are introduced here.

Definition 1 (Worker). *A worker w_i is denoted by a tuple $\langle l_i^w, r_i, C_i \rangle$, where l_i^w is the current location of worker w_i; r_i is w_i's route, which consists of tasks assigned to w_i; C_i is w_i's capacity limit of workload.*

Definition 2 (Spatial task). *A spatial task s_j is denoted by $\langle l_j^s, p_j \rangle$, where l_j^s is the geographical location of task s_j; p_j is the profit when task s_j is completed.*

Definition 3 (Worker's utility for accepting task). *The utility of worker w_i to accept task s_j is defined by a function $u(\cdot)$ where*

$$u(w_i, s_j) = p_j - cost(w_i, s_j) \tag{1}$$

Here $cost(w_i, s_j)$ represents the cost for w_i to complete s_j.

Definition 4 (Overall utility). *The overall utility is defined as the sum of workers' utilities, i.e.,*

$$U_{all} = \sum_{s_j^*} p_j - \sum_{w_i} cost(w_i) \tag{2}$$

where s_j^* represents the assigned task and $cost(w_i) = \sum_{s_j \in r_i} cost(w_i, s_j)$ represents the overall cost of worker w_i.

We consider the task assignment mode with several constraints as follows:

- **Uniqueness constraint:** A spatial task can only be assigned to one worker.
- **Spatial constraint:** A spatial task can be completed only if the worker arrives at the location of the task.
- **Capacity constraint:** The number of tasks assigned to w_i cannot exceed the workload capacity limit C_i.

Without loss of generality, we also have the following assumptions.

Assumption 1. *Since a worker should arrive at the location of a task to complete it, the $cost(w_i, s_j)$ may be proportional to $dist(l_i^w, l_j^s)$, which means*

$$cost(w_i, s_j) = \beta_i \cdot dist(l_i^w, l_j^s) \tag{3}$$

where β_i is the cost per unit travel distance for w_i and $dist(\cdot)$ is distance function.

Assumption 2. *According to individual rationality assumption, a worker accepts the task only if the utility is positive, i.e., $u(w_i, s_j) > 0$.*

Based on the aforementioned discussions, we formulate the *Utility-driven Destination-aware Spatial Task Assignment* problem (UDSTA) as follows.

Definition 5 (Utility-driven Destination-aware Spatial Task Assignment problem). *Given a worker set $\mathcal{W}$ and a task set $\mathcal{S}$, the Utility-driven Destination-aware Spatial Task Assignment problem (UDSTA) aims to find a global assignment solution $\mathbb{A}$ to maximize the overall utility for workers, while satisfying the aforementioned assumptions and constraints.*

Figure 2 shows a toy example of the UDSTA problem with the positions of 2 workers and 9 tasks, where the radius of nodes represents the profit of tasks, and the length of edges represents the distance between two locations. Assume the two workers' cost per unit travel distance β_1 and β_2 are both equal to 1, the proper task sequence for worker w_1 with capacity $c_1 = 3$ will be $r_1 = [s_2, s_3, s_5]$. In fact, worker w_1 chooses the task with the highest utility at each step. However, such greedy idea does not always make sense. If worker w_2 with capacity $c_2 = 2$ adopts a greedy policy, he/she will pick up s_6 instead of s_8 at the first step. Hence, the greedy route r_2' will be $[s_6, s_7]$, obtaining a total utility of 6. However, a better route $r_2 = [s_8, s_9]$ obtains a utility of 7, since there exist better neighboring tasks around s_8 compared with s_6. Our proposed DE-DQN method will focus on the important potential future utility.

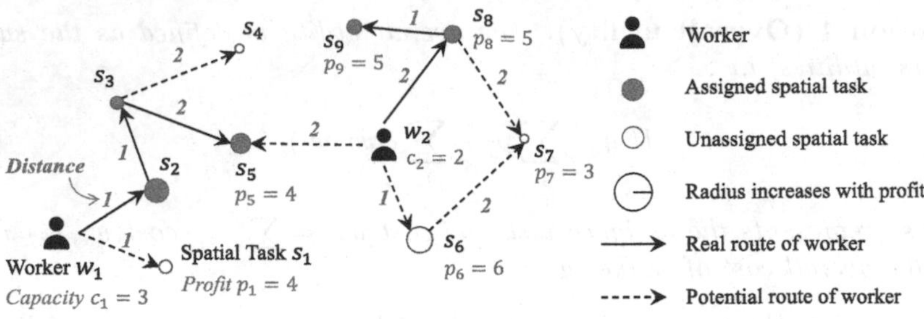

Fig. 2. A toy example with 2 workers and 9 spatial tasks.

NP-completeness of the UDSTA Problem. We prove the NP-completeness of the UDSTA problem with a reduction from a variation of the TSP problem, say, Traveling Salesman Path problem (abbreviated as TSP-Path) without returning to the original source, which has been proved to be NP-complete [6].

Theorem 1. *The UDSTA problem is NP-Complete.*

Proof. It is trivial to check a certificate of the UDSTA problem in polynomial time. Thus the UDSTA is an NP problem. Next, let us consider a reduction from TSP-Path. Given $G(V, E)$ with $|V| = n$, we define $cost(e) = 1, \forall e \in E$. Assume that $n - 1$ nodes represent tasks whose profits are 2, while the remaining node represents the worker with capacity $c = n - 1$. To maximize the overall utility, the worker will accept all the $n - 1$ tasks. We add edges E' to G to make G a complete graph and define $cost(e) = 2, \forall e \in E'$. In this situation, maximizing overall utility means minimizing cost. We consider the decision version of the UDSTA problem, which is "Is there a route of cost at most $n - 1$?". Now we can use Cook's reduction and solve the decision version of the TSP-Path problem by an oracle of UDSTA. Therefore, UDSTA is NP-Complete. □

The important notations in the rest of the paper are summarized in Table 1.

3 DE-DQN: Dual-Embedding Based Deep Q-Network for Task Assignment

Since cost is assumed to be proportional to distance between worker and task, we utilize the graph to model the destination-aware spatial crowdsourcing and propose a neural network-based learning method on the graph to solve the UDSTA problem. We first describe the dual-embedding approach to obtain the vector representations of workers and tasks for flexible computation in the neural network-based learning. Then, we use a Deep Q-Network (DQN) to sequentially assign tasks to proper workers and obtain the global task assignment solution.

Table 1. Definitions and notations

Symbol	Definition
$\mathcal{W}$	worker set, each worker $w_i \in \mathcal{W}$
$\mathcal{S}$	spatial task set, each task $s_j \in \mathcal{S}$
$\mathcal{S}^*$	assigned spatial task set
w_i	a worker $w_i = \langle l_i^w, r_i, C_i \rangle$
s_j	a spatial task $s_j = \langle l_j^s, p_j \rangle$
s_j^*	tasks received by workers
l_i^w	location of worker w_i
r_i	task sequence received by worker w_i
C_i	limit of task received by worker w_i
l_j^s	location of task s_j
p_j	profit when task s_j is completed
$dist(\cdot)$	distance between two locations
β_i	cost per unit travel distance for w_i
$cost(w_i, s_j)$	cost for worker w_i to complete task s_j
$u(w_i, s_j)$	utility for worker w_i to complete task s_j
U_{all}	overall utility for worker set, $U_{all} = \sum_{s_j^*} p_j - \sum_{s_j^*} cost(w_i, r_i, s_j)$

3.1 Dual-Embedding

The crowdsourcing graph is composed of a large number of nodes with worker and task features, and edges with distance information, which makes the graph complex and difficult to directly perform neural network-based learning on it. As is shown in Fig. 2, reasonable consideration of potential future utility makes sense. Since the potential future utility of an assignment can be measured from the breadth and depth perspective, we propose a dual-embedding method for nodes, using the utility embedding to reflect the top-k neighboring tasks' utilities from the breadth perspective and the coverage embedding to represent a worker's potential future utility linked with the remaining capacity from the depth perspective.

Utility Embedding: According to a worker's historical record, current location, and neighboring uncompleted tasks, we construct the utility embedding vector to reflect the top-k neighboring tasks' utilities for the worker, which is helpful to provide a larger search space. To consider the potential utility of a longer route, we similarly construct the utility embedding vector for pair (w_j, s_j) to reflect the top-k neighboring utilities when s_j is assigned to w_i.

Representation Vector for Worker: For worker w_i, we use w_i^k to represent the index of the k_{th} highest utility task for w_i, and then obtain uncompleted task sequence with the top-k highest utilities for w_i: $[s_{w_i^1}, s_{w_i^2}, \cdots, s_{w_i^k}]$. Therefore, the utility embedding for w_i is

$$\mathbf{V}_{w_i}^u = (u(w_i, s_{w_i^1}), u(w_i, s_{w_i^2}), \cdots, u(w_i, s_{w_i^k})) \tag{4}$$

Representation Vector for Pair: An assignment can be seen as a worker-task pair (w_j, s_j), which means that s_j is assigned to w_i and w_i arrives at l_j^s. In the scenario, we define $u_{w_i}(s_j, s_k) = p_k - \beta_i \cdot dist(l_j^s, l_w^s)$ to represent the utility for w_i to complete s_k after completing s_j, and use $(w_i s_j)^k$ to represent the index of the k_{th} highest utility task when w_i arrives at l_j^s. Hence, the utility embedding for pair (w_j, s_j) is

$$\mathbf{V}_{w_i s_j}^u = (u_{w_i}(s_j, s_{(w_i s_j)^1}), u_{w_i}(s_j, s_{(w_i s_j)^2}), \cdots, u_{w_i}(s_j, s_{(w_i s_j)^k})) \tag{5}$$

Coverage Embedding: The coverage embedding reflects the potential future utility for an assignment considering the remaining workload capacity of a worker. Specifically, for pair (w_i, s_j), the potential future utility can be denoted by the sum of the highest neighboring task utilities in the propagation chain with the remaining workload capacity rc_i as length. We define $s_{(w_i s_j)^1}^m$ to represent the m_{th} highest utility task in the propagation chain. For instance, $s_{(w_j s_j)^1}^2 = s_{(w_i s_{(w_i s_j)^1})^1}$. Especially, $s_{(w_i s_j)^1}^0 = s_j$. Hence, for pair (w_i, s_j), we can express the potential future utility of assigning $s_{w_i s_j}^k$ to w_i as

$$u_{w_i}^c(s_j, s_{(w_i s_j)^k}) = u_{w_i}(s_j, s_{(w_i s_j)^k}) + \sum_{m=1}^{rc_i - 1} u_{w_i}(s_{(w_i s_j)^1}^{m-1}, s_{(w_i s_j)^1}^m) \tag{6}$$

Therefore, for pair (w_i, s_j), we concatenate the top-k potential future utilities to form the coverage embedding $\mathbf{V}^c(w_i, s_j)$ as

$$\mathbf{V}^c(w_i, s_j) = (u_{w_i}^c(s_j, s_{(w_i s_j)^1}), \cdots, u_{w_i}^c(s_j, s_{(w_i s_j)^k})) \tag{7}$$

3.2 Framework and Training for DE-DQN

Reinforcement learning is a learning process of studying proper interaction with the environment to maximize a special reward. To solve the UDSTA problem, we adopt the DQN, which is a structure combining Q-learning with the deep neural network, to assign tasks to proper workers. Specifically, the framework of our DE-DQN is shown in Fig. 3.

In the DE-DQN framework, the agents are the decision makers. The environment consists of the worker and task graph and determined task sequences. The agent takes an action to assign a task to the proper worker according to the environment. The environment then gives a reward as feedback to the agent. The goal of the agent is to maximize the long-term gain. These components are described as follows.

State: The state reflects the current task assignment situation in the spatial crowdsourcing network. The state vector concatenates three parts. The first part is the mean of the top-k utilities for all workers $\mathbf{V}_W^u = \frac{1}{|W|}\sum_{w_i \in W} \mathbf{V}_{w_i}^u$. The second part indicates the total remaining workload capacity $rc = \sum_{w_i \in W} rc_i$. The third part records the sum of cumulative utility and cost of completed tasks $uc = (\sum_{s_j \in S^*} p_j, \sum_{w_i} cost(w_i))$. The state vector $x_s = (\mathbf{V}_W^u || rc || uc)$.

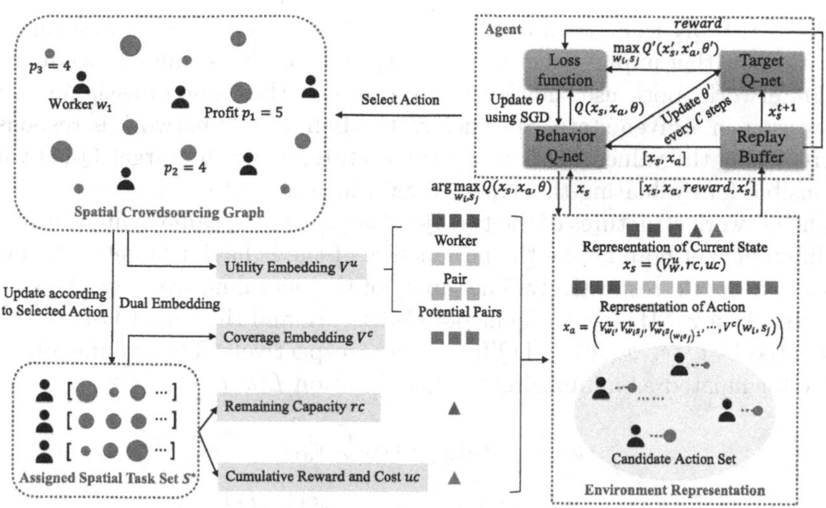

Fig. 3. The illustration of DE-DQN. DE-DQN first calculates the utility embedding $\mathbf{V}^u$ and coverage embedding $\mathbf{V}^c$ for workers and pairs. The state representation x_s is the concatenation of three parts: mean of the utility embedding for workers $\mathbf{V}^u_W$, remaining capacity rc, and cumulative reward and cost uc. The assignment action representation x_a is the concatenation of utility embedding and coverage embedding. DE-DQN selects the proper action based on the current state and candidate action set. After determining a worker-task pair, the agent receives the reward, and then updates the corresponding variables and representations.

Action: The action is to assign a task to a worker. To take the future reward into account, we consider the current task to be assigned and the potential future tasks. Specifically, we concatenate the utility embedding of worker w_i, task s_j, potential future tasks $\{s_{(w_i s_j)^1}, \cdots, s_{(w_i s_j)^k}\}$, and coverage embedding to indicate the action of agent $x_a = (\mathbf{V}^u_{w_i} || \mathbf{V}^u_{w_i s_j} || \mathbf{V}^u_{w_i s_{(w_i s_j)^1}} || \cdots || \mathbf{V}^c(w_i, s_j))$.

Reward: The reward represents the utility obtained. When the agent assigns task s_j to worker w_i, the reward is equal to $p_j - cost(w_i, s_j)$.

Policy: The deep Q-learning is an off-policy algorithm where we employ a neural network to approximate the Q value. Specifically, we use the ϵ-greedy algorithm in the training stage to do more exploration, and the pure greedy algorithm in the testing stage to maximize the overall reward.

The ϵ-greedy policy in the training stage is

$$\gamma_{\epsilon-greedy}(x_s) = \begin{cases} \underset{((w_i,s_j)|s_j \in \mathcal{S}-\mathcal{S}^*)}{\arg\max} Q & \text{w.p. } \epsilon \\ \text{a random feasible pair } ((w_i,s_j)|s_j \in \mathcal{S}-\mathcal{S}^*) & \text{o.w.} \end{cases} \tag{8}$$

The pure greedy policy in the testing stage is

$$\gamma_{greedy}(x_s) = \underset{((w_i,s_j)|s_j \in \mathcal{S}-\mathcal{S}^*)}{\arg\max} Q \tag{9}$$

The DE-DQN framework consists of a behavior Q-network and a target Q-network to further improve the stability of Q-value in the training procedure. The two Q-networks both use the Q-value to evaluate the value of assigning a task to a worker in a given state. Specifically, the behavior Q-network is responsible for predicting the value of action in current state, while the target Q-network is responsible for evaluating the expected gain in next state.

The network structures of the two Q-networks are the same, while parameters are different. Let θ and θ' be the parameters of the behavior Q-network and the target Q-network, respectively. The inputs of the neural network are the state x_s and the action x_a. Hence, the behavior Q-network and the target Q-network can be denoted by $Q(x_s, x_a, \theta)$ and $Q'(x'_s, x'_a, \theta')$, respectively. The training objective can be formulated as minimizing the loss function $L(\theta)$:

$$L(\theta) = \mathbf{E}[(\hat{y^t} - Q(x_s^t, x_a^t, \theta))] \tag{10}$$

$$\hat{y^t} = reward^t + \eta \max_{(w_i, s_j)} Q'(x_s^{t+1}, x_a^{t+1}, \theta') \tag{11}$$

where x_s^t is the state vector at step t, x_a^t is the action vector at step t, and $\eta \in [0, 1]$ is the discount factor.

To reduce the correlation between the predicted Q-value and the target Q-value, the parameters of the target Q-network remains unchanged in C training steps. After C steps, the parameters of the behavior Q-network are copied to update θ' in the target Q-network.

Algorithm 1: DE-DQN Agent Training

Input: Spatial crowdsourcing graph, worker set $\mathcal{W}$, task set $\mathcal{S}$.
Output: Parameters of Q-network.

Initialize replay buffer D;
Initialize parameters of the behavior and the target Q-networks, $\theta = \theta'$;
for *episode* $= 1$ *to* E **do**
    Initialize the assigned spatial task set $\mathcal{S}^* = \emptyset$;
    for $t = 1$ *to* $\sum_{w_i} C_i$ **do**
        Select action by $\gamma_{\epsilon-greedy}$ policy;
        Calculate $reward^t$;
        $x_s^{t+1} \leftarrow f_t(x_s^t, x_a^t)$;
        Store $[x_s^t, x_a^t, reward^t, x_s^{t+1}]$ into D;
        Randomly sample a minibatch data from D;
        if $t = k$ **then**
            $y^t = reward^t$;
        else
            $y^t = reward^t + \eta \max_{(w_i, s_j)} Q'(x_s^{t+1}, x_a^{t+1}, \theta')$;
        Use SGD to take a gradient descent step on $(y^t - Q(x_s^t, x_a^t, \theta))^2$;
        After C steps, update $\theta' = \theta$.

The training process for agent is shown in Algorithm 1. We employ the replay buffer proposed in [12] to reduce the dependency among data. For each training episode, we initialize the assigned spatial task set in Line 4. The episode ends when the number of assigned tasks reaches $\sum_{w_i} C_i$, the sum of workload capacity limit of all workers. In each episode, we adopt $\gamma_{\epsilon-greedy}$ policy to choose an action from candidate sets. After selecting an action, we update the representation of state vector through a state transition function in Line 8. After obtaining the reward and the next state, we save it in the replay buffer. We update the parameters in Q-network by stochastic gradient descent (SGD) method. After C steps, we update the parameters in the target Q-network.

4 Experiments

In this section, we conduct experiments on both synthetic and real-world datasets and evaluate performances of baseline algorithms and our DE-DQN framework.

4.1 Experiment Setups

Dataset Overview: The experiments are carried out on two datasets: Gowalla[1] and synthetic (SYN). Gowalla is an open source real-world dataset [2], which includes a total of 6, 442, 890 check-ins of users over the period of Feb 2009-Oct 2010. We choose the locations where the number of user check-ins is greater than 100 as the initial locations of workers, and the locations where the number of item check-ins is greater than 10 as the locations of tasks by random sampling in the longitude range from −120 to −110 and the latitude range from 30 to 40. The task profit is determined by sampling the order value from Didi Chuxing Chengdu dataset. Table 2 shows the percentage of tasks in different value ranges, which indicates that the number of tasks with high profits is often small in reality. Besides, a distance conversion function is used to calculate the distance between points with longitude and latitude. The worker cost per unit travel distance is obtained from a Poisson distribution with the mean of 0.5.

Table 2. Percentage of Tasks in Different Value Ranges

Range	(0, 0.1]	(0.1, 0.2]	(0.2, 0.3]	(0.3, 0.4]	(0.4, 0.5]	(0.5, 0.6]	(0.6, 0.7]	(0.7, 0.8]	(0.8, 0.9]	(0.9, 1.0]
Ratio	0.75	0.21	$3.5e^{-2}$	$6.3e^{-3}$	$1.3e^{-3}$	$4.5e^{-4}$	$2.0e^{-4}$	$3.8e^{-5}$	$2.9e^{-5}$	$1.9e^{-5}$

To obtain the synthetic dataset, we generate workers and tasks following a uniform distribution within the $2D$ space $[0, 400]^2$. The task profit is obtained from a probability density function of step descent, which is in line with the above observation in reality. Besides, the worker cost per unit travel distance is obtained from a Poisson distribution with the mean of 0.1.

[1] http://snap.stanford.edu/data/loc-Gowalla.html.

Parameter Settings: The parameter settings are shown in Table 3, where the default ones are marked in bold. As for training the DE-DQN, we do the training with 500 episodes on randomly sampled data from Gowalla and SYN.

Table 3. Parameter setting

Parameter	Description	Value
$\|\mathcal{W}\|$	Worker Set	$50, \mathbf{60}, 70, 80$
$\|\mathcal{S}\|$	Spatial Task Set	$800, 900, \mathbf{1000}, 1100$
Capacity Mean	Mean of $\{C_i\}$	$5, 7, \mathbf{10}, 15$

Baseline methods: We take the following baselines for performance evaluation.

- **DisGreedy:** The DisGreedy algorithm is proposed in [20], which takes distance as the highest priority criterion for selecting the next task. It assigns the nearest task to workers.
- **PftGreedy:** The PftGreedy algorithm is also proposed in [20], which takes profit as the highest priority criterion for selecting the next task. It assigns the task with the highest profit to the nearest worker.
- **Utility Priority:** [4] proposes the Utility Priority algorithm, matching the largest utility worker-task pair at each step.

4.2 Performance Analysis

All the experiments are implemented on an Apple M1 chip with 8-core CPU and 16GB unified memory. The DE-DQN and baseline algorithms are implemented with Python 3.6. Tensorflow 2.0 is used to build the machine learning framework. As shown in Fig. 4 and 5, DisGreedy works better than PftGreedy when travel cost is high on Gowalla, while PftGreedy works better than DisGreedy when travel cost is low on SYN. However, DE-DQN and Utility Priority perform well regardless of the change of travel cost.

Effect of $|\mathcal{W}|$: As shown in Fig. 4(a) and 5(a), the performance of DE-DQN is better than that of other baseline algorithms with various $|\mathcal{W}|$. Simultaneously, the performance gap of DE-DQN over Utility Priority increases as $|\mathcal{W}|$ increases, indicating that DE-DQN is suitable for the complex spatial crowdsourcing scenario where there exists a great quantity of workers. We note that the performance of algorithms degrades when $|\mathcal{W}| = 60$ compared with $|\mathcal{W}| = 50$ on Gowalla, indicating that the performance is influenced by data instances.

Effect of $|\mathcal{S}|$: As shown in Fig. 4(b) and 5(b), DE-DQN achieves the highest overall utility with various $\mathcal{S}$. We note that DE-DQN and Utility Priority can discover better tasks as $|\mathcal{S}|$ increases compared with DisGreedy and PftGreedy.

Simultaneously, the performance gap between DE-DQN and Utility Priority decreases as $|\mathcal{S}|$ increases on SYN. This may be because Utility Priority can benefit more from the increase of highly profitable tasks when the total worker workload capacity is constant within a uniform distribution.

Effect of Capacity Mean: As shown in Fig. 4(c) and 5(c), the higher the capacity mean of workers, the better the DE-DQN outperforms the baseline methods when the number of tasks is constant, which indicates that considering future utility is more important as the total worker capacity is closer to $|\mathcal{S}|$.

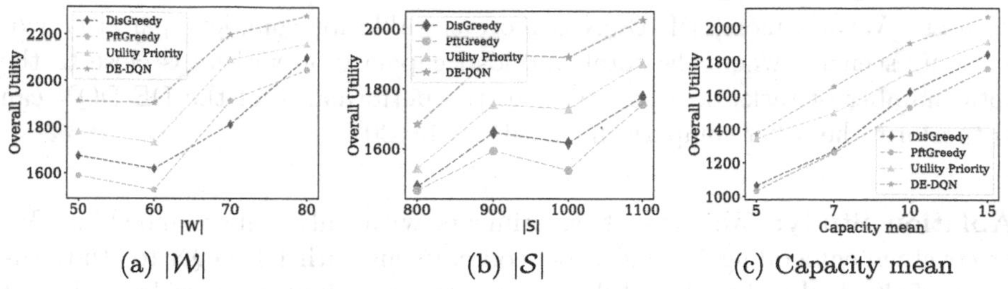

(a) $|\mathcal{W}|$ (b) $|\mathcal{S}|$ (c) Capacity mean

Fig. 4. Comparison with various $|\mathcal{W}|$, $|\mathcal{S}|$, and capacity mean on Gowalla.

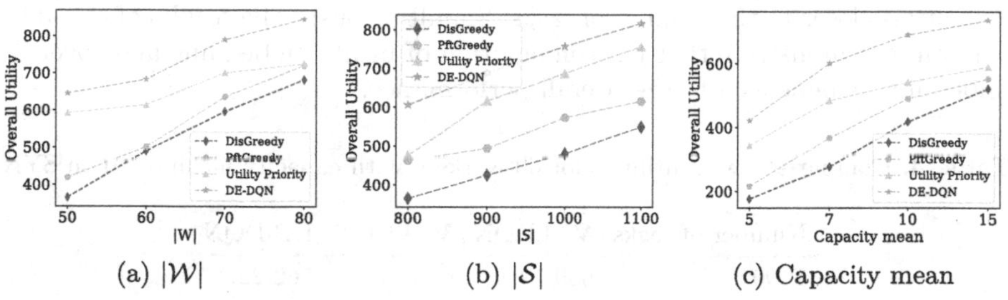

(a) $|\mathcal{W}|$ (b) $|\mathcal{S}|$ (c) Capacity mean

Fig. 5. Comparison with various $|\mathcal{W}|$, $|\mathcal{S}|$, and capacity mean on SYN.

Effect of $|\mathcal{W}| \times |\mathcal{S}|$: As shown in Fig. 6, the overall utility grows with the number of workers and the number of tasks for all the four algorithms. DE-DQN outperforms the baseline algorithms under most cases. When there are a relatively small number of tasks and the total workload capacity of workers is small, the performance gap between DE-DQN and other baseline algorithms is not large. However, when there are many tasks and the total workload capacity of workers is close to the task number, the performance gap is large.

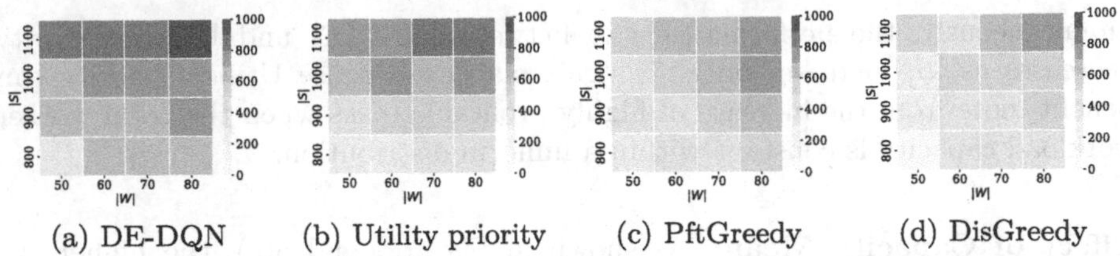

(a) DE-DQN (b) Utility priority (c) PftGreedy (d) DisGreedy

Fig. 6. Comparison with the overall utility w.r.t. $|\mathcal{W}| \times |\mathcal{S}|$ on SYN.

In summary, DE-DQN performs well on both synthetic and real-world datasets. What's more, DE-DQN is more suitable for complex spatial crowdsourcing scenario where the total workload capacity of workers is close to the total number of tasks. In this situation, the performance of the DE-DQN can outperform the other comparison algorithms by 20%.

Ablation Study: We conduct experiments with only utility embedding $\mathbf{V}^u$ or coverage embedding $\mathbf{V}^c$ and make comparisons with DE-DQN to study the effect of the dual-embedding. Table 4 shows the overall utility based on different techniques with 1000 and 1200 tasks. The results are consistent with the results shown in Fig. 4(c), which indicates that the potential future utility reflected by coverage embedding $\mathbf{V}^c$ is more important when the total worker workload capacity is closer to the number of tasks. Simultaneously, the results of the ablation study demonstrate that the combination of utility embedding and coverage embedding is beneficial to the overall performance.

Table 4. Comparison with utilities for 60 workers with capacity mean of 10 on SYN

Number of tasks	$\mathbf{V}^u$+DQN	$\mathbf{V}^c$+DQN	DE-DQN
1000	639.70	696.97	762.22
1200	836.53	804.32	928.67

5 Related Work

Task Assignment Problem: Task assignment problem in crowdsourcing has been extensively paid attention to and researched in the past ten years as most task-worker matching requirements in real world application can be formulated as this problem. Task assignment problem can be classified into offline problem or online problem according to whether all the information of tasks and workers is known before the assignment decision process starts [3]. As for the online task assignment, it can be classified into one side online problem [5] and both sides online problem [21]. A common model employed for solving task assignment problem is bipartite graph [11], in which tasks and workers are abstracted into nodes of both sides. [9] concentrates on ride-sharing routing problem associated to customer selection and proposes a two-stage structure to obtain the optimal solution with pseudo-polynomial time complexity. [4] defines a team-oriented task planning problem, which is actually a multi-task multi-worker matching problem, considering the worker's skill constraint of tasks.

Multiple methods based on different technologies have been studied and mentioned in recent works. [1] explores distributed and centralized algorithms for task selection to improve the quality of task accomplishment, where the distributed one is a game theory based approximation algorithm, the centralized one is a greedy based approximation algorithm. Similarly, [13] respectively define dependency-aware spatial crowdsourcing and cooperation-aware spatial crowdsourcing, and also propose game theory and greedy based approximation algorithms aiming at maximizing the assigned tasks and the cooperation quality separately. Besides traditional methods, [10] first presents a deep reinforcement learning framework on vehicular crowdsourcing problem, both maximizing task accomplishment and coverage fairness, and minimizing the energy cost.

Deep Reinforcement Learning: Reinforcement learning (RL) has become a hotspot in research field recently. Some variants of RL also emerged combined with other methods. Deep reinforcement learning (DRL) combines deep learning methods with traditional RL methods, which is found to be suitable for solving decision optimization problems [16], and therefore has been widely researched for improvement and application. Similar to RL, DRL can also be distinguished by whether it is model-based. As for model-free DRL, Deep Q-Network (DQN) is a popular method based on value function [12], which utilizes convolutional neural network and Q-learning [7], and performs well in many applications including task arrangement in crowdsourcing [8,18]. What's more, [15] proposes auxiliary-task based DRL to achieve multi-objective optimization in participant selection problem of mobile crowdsourcing. [19] introduces geographic partition into spatial crowdsourcing and then conducts the RL method to solve the problem.

6 Conclusion

In this paper, we focus on solving the global task assignment problem in the spatial crowdsourcing. Our goal is to maximize the overall utility for all workers, which is formulated as the utility-driven destination aware spatial task assignment problem and is proved to be NP-complete. To deal with the problem, we propose a DE-DQN framework which balances the current and future utility through the dual-embedding. We construct the proper representations of state, action, and reward of DQN according to our scenario, and use the experience replay buffer to train the Deep Q-Network. The experiments based on both synthetic and real-world dataset verify the effectiveness of the DE-DQN framework. In the future, we will consider the fairness problem where a homogeneity metric is presented to balance the utility among workers.

References

1. Cheung, M.H., Hou, F., Huang, J., Southwell, R.: Distributed time-sensitive task selection in mobile crowdsensing. IEEE Trans. Mob. Comput. **20**(6), 2172–2185 (2021)
2. Cho, E., Myers, S.A., Leskovec, J.: Friendship and mobility: user movement in location-based social networks. In: ACM SIGKDD International Conference on Knowledge Discovery and Data Mining, pp. 1082–1090 (2011)
3. Du, Y., Sun, Y.E., Huang, H., Huang, L., Xu, H., Wu, X.: Quality-aware online task assignment mechanisms using latent topic model. Theoret. Comput. Sci. **803**, 130–143 (2020)
4. Gao, D., Tong, Y., Ji, Y., Xu, K.: Team-oriented task planning in spatial crowdsourcing. In: Asia-Pacific Web and Web-Age Information Management Joint Conference on Web and Big Data, pp. 41–56 (2017)
5. Gao, G., Wu, J., Xiao, M., Chen, G.: Combinatorial multi-armed bandit based unknown worker recruitment in heterogeneous crowdsensing. In: IEEE Conference on Computer Communications, pp. 179–188 (2020)
6. Gutin, G., Punnen, A.P. (eds.): The traveling salesman problem and its variations. Springer Science & Business Media, LLC, Springer (2007). https://doi.org/10.1007/b101971
7. Krizhevsky, A., Sutskever, I., Hinton, G.E.: Imagenet classification with deep convolutional neural networks. In: Advances in Neural Information Processing Systems 25 (2012)
8. Li, B., Zhang, A., Chen, W., Yin, H., Cai, K.: Active cross-query learning: a reliable labeling mechanism via crowdsourcing for smart surveillance. Comput. Commun. **152**, 149–154 (2020)
9. Lin, Q., Deng, L., Sun, J., Chen, M.: Optimal demand-aware ride-sharing routing. In: IEEE Conference on Computer Communications, pp. 2699–2707 (2018)
10. Liu, C.H., et al.: Curiosity-driven energy-efficient worker scheduling in vehicular crowdsourcing: a deep reinforcement learning approach. In: IEEE International Conference on Data Engineering, pp. 25–36 (2020)
11. Liu, Q., Peng, J., Ihler, A.T.: Variational inference for crowdsourcing. In: Advances in Neural Information Processing Systems 25 (2012)
12. Mnih, V., et al.: Human-level control through deep reinforcement learning. Nature **518**(7540), 529–533 (2015)
13. Ni, W., Cheng, P., Chen, L., Lin, X.: Task allocation in dependency-aware spatial crowdsourcing. In: IEEE International Conference on Data Engineering, pp. 985–996 (2020)
14. Rana, R.K., Chou, C.T., Kanhere, S.S., Bulusu, N., Hu, W.: Ear-phone: an end-to-end participatory urban noise mapping system. In: ACM/IEEE International Conference on Information Processing in Sensor Networks, pp. 105–116 (2010)
15. Shen, W., He, X., Zhang, C., Ni, Q., Dou, W., Wang, Y.: Auxiliary-task based deep reinforcement learning for participant selection problem in mobile crowdsourcing. In: ACM International Conference on Information & Knowledge Management, pp. 1355–1364 (2020)
16. Wang, H.N., et al.: Deep reinforcement learning: a survey. Front. Inf. Technol. Electr. Eng. **21**(12), 1726–1744 (2020)
17. Xiong, F., Xu, S., Zheng, D.: An investigation of the uber driver reward system in china-an application of a dynamic pricing model. Technol. Anal. Strat. Manage. **33**(1), 44–57 (2021)

18. Xu, L., Zhou, X.: A crowd-powered task generation method for study of struggling search. Data Sci. Eng. **6**(4), 472–484 (2021)
19. Ye, G., Zhao, Y., Chen, X., Zheng, K.: Task allocation with geographic partition in spatial crowdsourcing. In: ACM International Conference on Information & Knowledge Management, pp. 2404–2413 (2021)
20. Yin, B., Li, J., Wei, X.: Rational task assignment and path planning based on location and task characteristics in mobile crowdsensing. IEEE Trans. Comput. Soc. Syst. **9**(3), 781–793 (2022)
21. Zhao, Y., Zheng, K., Cui, Y., Su, H., Zhu, F., Zhou, X.: Predictive task assignment in spatial crowdsourcing: a data-driven approach. In: IEEE International Conference on Data Engineering, pp. 13–24 (2020)
22. Zheng, L., Cheng, P., Chen, L.: Auction-based order dispatch and pricing in ridesharing. In: IEEE International Conference on Data Engineering, pp. 1034–1045 (2019)

Dynamic Vehicle-Cargo Matching Based on Adaptive Time Windows

Chong Feng[1], Jiajun Liao[1], Jiali Mao[1(✉)], Jiaye Liu[1], Ye Guo[2], and Weining Qian[1]

[1] School of Data Science and Engineering, East China Normal University, Shanghai, China
{51205903030,52215903003,51184501030}@stu.ecnu.edu.cn, {jlmao,wnqian}dase.ecnu.edu.cn

[2] Jing Chuang Zhi Hui (Shanghai) Logistics Technology, Shanghai, China
guoye@jczh56.com

Abstract. The core task of vehicle-cargo matching is to dispatch the cargoes to the trucks. The existing matching policies mainly focus on maximizing the shipping weight for each truck. Due to each cargo is bulky and heavy in bulk logistics area, such strategies cannot ensure maximization of total weight of cargoes to be transported, and lead to a few cargoes be stranded. To tackle this issue, we present an intelligent decision framework for vehicle-cargo matching, called *ILPD*. Based on the limiting rules and features related to loading plan decisions that extracted from historical logistics data, we design a *time window*-based matching policy to achieve the goal of maximizing the total shipping weight and minimizing the quantity of stranded cargoes. Specifically, in each time window, dynamic programming and *Branch-and-Bound* method are leveraged to generate the loading plans of cargoes with the aim of minimization of stranded cargoes' quantities. Then, *Kuhn-Munkres* algorithm is used to make the matching decisions to obtain maximum weight matching. Further, to fit for dynamic changing number of trucks and cargoes, a *time zone*-based *Q-learning* algorithm is proposed to adjust the time window size adaptively. Extensive experimental results on real data sets validate the effectiveness and practicality of our proposal.

Keywords: Steel logistics · Vehicle-cargo matching · Task assignment · Adaptive time window

1 Introduction

Vehicle-cargo matching is a core function for bulk logistics, which is one of key factors for reducing logistics costs. Its main task is to decide the cargo-loading plan for each truck. Take steel logistics as an example, the cargoes in the waybills are first split into several loading plans according to the stocks and the loading plans are matched with chronologically arrived trucks. Due to each cargo is bulky and heavy in bulk logistics area, heavy trucks are widely used in transporting bulk cargoes by road. But the number of heavy trucks is limited and

B. Li et al. (Eds.): APWeb-WAIM 2022, LNCS 13421, pp. 296–312, 2023.
https://doi.org/10.1007/978-3-031-25158-0_23

has uneven distribution in different regions. How to maximize total shipment weights using limited capacity resources for bulk logistics has become a serious challenge. Although vehicle-cargo matching issue can be viewed as a task assignment one, it is distinct from that in the scenario like taxi dispatching where a taxi is matched with a nearest passenger [14–16,18]. It needs to take a series of factors into account, e.g., each type of cargo has its transportation environment requirement for packaging (i.e. loading conditions of truck), each truck driver has his preferred transportation destinations and cargoes, etc. So the solutions for task assignment cannot be directly used to tackle our proposed matching issue.

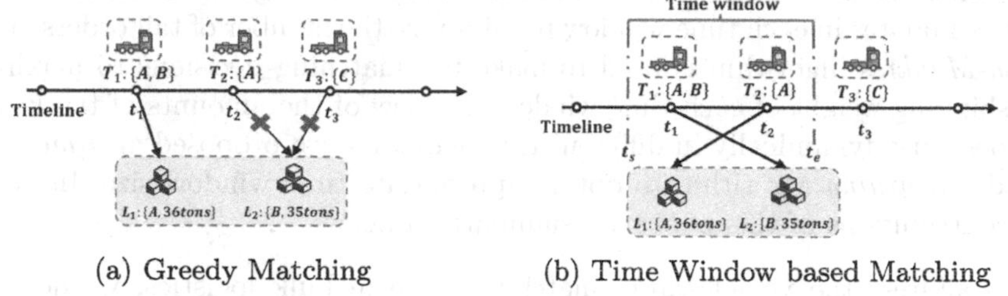

(a) Greedy Matching (b) Time Window based Matching

Fig. 1. Illustration of different matching strategies, here T_i, L_i, t_i denote the truck, loading plan and arrived time of truck respectively, and A and B represent different transporting destinations.

A common solution is to maximize truckloads under the premise of not exceeding each truck's load limit. Recently, some policy considers the priority level of cargoes to be delivered in matching process to tackle the issue of limited transportation capacity, and also incorporates the preferences of drivers on the direction of transportation flow to enhance their satisfactions [11]. But it aims to maximize total shipping weight of higher priority cargoes, such a greedy policy neglects current matched result's impact on future matching decision. As shown in Fig. 1(a), the truck T_1, T_2 and T_3 arrive at different time (i.e. t_1, t_2 and t_3) respectively, the cargo loading plan L_1 and L_2 are ready for transporting before the timestamp t_2 and t_3 separately. According to the greedy policy, L_1 is dispatched to T_1, but L_2 cannot be dispatched to T_2 or T_3 due to no matched transporting destination between loading plan and truck. A workaround to this problem is to have T_1 waited a little of time until L_2 is ready, in such a way that T_1 and T_2 are matched with their preferred cargo loading plans, as illustrated in Fig. 1(b). The policy of instantly matching the loading plan with the arrived truck is changed into that of matching several loading plans with the trucks in a shorter period of time (denoted as *time window*). It is obviously that the latter has higher matching rate.

Nevertheless, the above mentioned policy unavoidably leaves parts of cargoes not exceeding the truck minimal load limit stranded, and the waybills to which these cargoes belong are thus unfinished, called *tail orders*. The cargoes in tail orders have to be packed up with those of other waybills in the next time period and wait for transporting by other trucks. This imperceptibly increases the transporting costs of logistics platform. Therefore, it necessitates to design appropriate matching policy not only to maximize *total shipping weight* to be transported, but also to minimize the number of tail orders in a global view. To the best of our knowledge, none of the researches so far accounts for the optimization goal of minimizing the number of tail orders.

To achieve the aforementioned optimization goals, we develop a vehicle-cargo matching decision framework, called as *Intelligent Loading Plan Decision Framework* (or *ILPD* for short). *ILPD* consists of an offline training part for time window splitting, and an online matching part. At each phase, *Dynamic Programming* and *Bound and Branch* methods are first leveraged to generate loading plan candidates in each time window to minimize the number of tail orders. Then *Kuhn-Munkres* algorithm is used to make the matching decisions to maximize the shipping weight of each time window. In view of the amounts of trucks and cargoes vary dynamically in different time windows, we proposed an *time zone*-based *Q-learning* algorithm to obtain appropriate time window size. In detail, the contributions of this paper are summarized below.

- To address the vehicle-cargo matching issue in bulk logistics, we design a two-phase framework consisting of off-line training for time window cutting and online matching, called *ILPD*.
- We put forward an online matching mechanism involving loading plan generation and matching decision making, in which *Dynamic Programming* and *Bound and Branch* methods are applied to minimize the number of *tail orders*, and *time zone*-based *Q-learning* algorithm is presented to obtain optimal time window size.
- We conduct a comprehensive series of experiments on real and synthetic data sets from logistics platform. Experimental results validate the effectiveness of our proposal in terms of whole revenue of the platform.

Outline. The remainder of this paper is organized as follows. Section 2 reports the state of the art on vehicle-cargo matching issue. In Sect. 3, the problem is defined formally. Section 4 elaborates the details of *ILPD*. Section 5 reports the evaluations of our proposal on both real and synthetic data sets. Finally, the last section presents the conclusion of the work.

2 Related Work

The vehicle-cargo matching issue can be treated as a task assignment one, which is usually solved by online bipartite graph matching. Besides, as stated previously, *Q-learning* method is introduced into our matching strategy to fit for dynamic changing number of trucks and cargoes. Next, we perform a systematic evaluation of works about *Online Bipartite Graph Matching* and *Reinforcement Learning*-based matching.

Online Bipartite Graph Matching. The strategy of online bipartite graph matching aims to match two sets of objects in terms of a preset matching policy. In past few years, it has received considerable attentions [2–7,13,15–17]. These approaches can be grouped into three categories according to whether both sides of bipartite graph change or not with time, including *two-sided static bipartite* graph, *one-sided dynamic* one and *two-sided dynamic* one. Sun et al. studied the vehicle-cargo matching issue using two-sided static bipartite graph, where the information of all the vehicles and cargoes was known before task allocation was conducted [13]. Hu et al. presented a matching method using one-sided dynamic bipartite graph, where only the information of all the cargoes was known in advance [6]. Although the aforementioned approaches focused on the optimization goals of maximizing total number of assignments and minimizing total cost, they are unsuitable for our proposed matching issue which needs to consider the minimization of the number of tail orders.

Reinforcement Learning Based Matching. Recently, a few researches proposed *reinforcement learning*-based matching strategies [12,18]. They attempted to achieve global optimal matching by learning an adaptive strategy. Spivey et al. modeled the online matching issue as a *Markov Decision Process*, which takes each matching behavior as an action [12]. But the aforementioned methods cannot be directly applied to tackle our proposed matching issue due to different optimal goals. Inspired by them, we need to constantly cumulate dynamically generated tasks (i.e. loading plans) and workers (i.e. trucks) into a series of micro-batches, and begin to conduct matching only when the size of each micro-batch (i.e. a time window) is suitable. Consider that the volumes of cargoes and the number of trucks are constantly changing, we design an adaptive time window matching strategy by incorporating *reinforcement learning* technique to obtain a solution much closer to global optimal one.

3 Problem Statement

In this section, we first give explicit definitions of terminologies in the steel logistics scenario, then discuss main challenges we face and the goals we want to achieve.

Our goal is to maximize total shipping weight(denoted as TSW) and minimize the number of tail orders (denoted as O_t). Firstly, we need to pack up the cargoes (denoted as c) of each waybill (denoted as w) into some loading plans (denoted as l), then match them with newly arrived trucks (denoted as T). The matrix $M_l^c = \{d_{ij}\}_{N\times K}$ represents N pieces of cargoes to be distributed to K loading plans, where $1 \leq i \leq N$ and $1 \leq j \leq K$. Since each cargo at most belongs to one loading plan, there is a restriction: $d_{ij} = \begin{cases} 1, & \text{if } c_i \text{ is distributed to } l_j. \\ 0, & \text{else.} \end{cases}$, which subjects to: $\sum_{i=0}^{K} d_{ij} \leq 1$.

Definition 1 (Bipartite Graph). *A bipartite graph can be defined as* $G = (T, LPS, \mathcal{E})$, *where* T *and* LPS *are the nodes of* G, *they represent the set of* *trucks and the loading plan set generated by grouping the cargoes in the waybills respectively, and $\mathcal{E}$ denotes the edge set between T and LPS.*

Each edge $e \in \mathcal{E}$ has a weight which equals to total weight of a loading plan l_i, denoted by $\omega(i)$. G corresponds to the matches between loading plans and trucks in a time window (denoted as $\mathcal{W}$). $\mathcal{W}$ is denoted as $\mathcal{W} = (n_T, n_L, r_W, G)$, here n_T, n_L and r_W represent the number of trucks, loading plans and the time range of each time window respectively. $M_G = \{f_{ij}\}_{K \times M}$ is a matrix that indicates the matching results between T and LPS, where $1 \leq i \leq K$ and $1 \leq j \leq M$. Likewise, one loading plan only corresponds to one truck, so we have: $f_{ij} = \begin{cases} 1, & \text{if loading plan } l_i \text{ is matched with truck } T_j. \\ 0, & \text{else.} \end{cases}$, which subjects to: $\sum_{i=0}^{M} f_{ij} \leq 1$ (Table 1).

Table 1. Summary of notations

Notation	Description	Notation	Description
T_i	The i-th truck in truck set T	C_i	The i-th cargo in cargo set C
w_i	The i-th waybill in waybill set WBS	l_i	The i-th loading plan in loading plan set LPS
W_u^T	The standard upper weight of truck T	W_l^T	The standard lower weight of truck T
M_l^c	The cargo loading plan decision matrix	G	The bipartite graph between T and LPS
$\mathcal{E}$	The set of edges in G	$\omega(i)$	The weight of l_i
O_t^i	The tail order after splitting waybill w_i	$\psi(i)$	The number of tail order O_t^i
Se	The sequence of trucks and waybills	$\mathcal{W}$	The time window corresponds to G
M_G	The matching result matrix of G	β	The max size of time window $\mathcal{W}$

Finally, we summarize the problem as below. Given a sequence of chronologically arrived trucks and waybills, denoted as Se, our goal is to divide Se into multiple sub-sequences $(\mathcal{W}_1, \mathcal{W}_2, ..., \mathcal{W}_n)$ by a time window splitting strategy (denoted as O), and then achieve optimal vehicle-cargo matching with the aim of maximizing total shipping weight while minimizing the number of tail orders. The optimization goals of our framework are formulated as:

$$Se \xrightarrow{O} (\mathcal{W}_1, \mathcal{W}_2, ..., \mathcal{W}_n) \tag{1}$$

$$\max \sum_{k=0}^{n} \sum_{i=0}^{K} \sum_{j=0}^{M} \omega(i) \cdot f_{ij} \tag{2}$$

$$\min \sum_{k=0}^{n} \sum_{i=0}^{N} \psi(i) \tag{3}$$

4 Methodology

As illustrated in Fig. 2, *ILPD* framework consists of offline training part and online matching part. Specifically, at the offline training phase, a *time zone*-based *Q-learning* algorithm is proposed to learn a strategy of dynamical adjusting the size of time window. It helps to achieve a near-optimal matching decision from a global perspective. At online matching phase, the cargoes in waybills are packaged into several loading plans with the aim of minimizing the number of tail orders, and then the matching decisions are made in each time window through using the offline-learned strategy. With the guiding of adjusting strategy for time window size, we split the sequence *Se* into multiple continuous time windows, then assign each loading plan to appropriate truck to acquire a match of maximum weight in each time window.

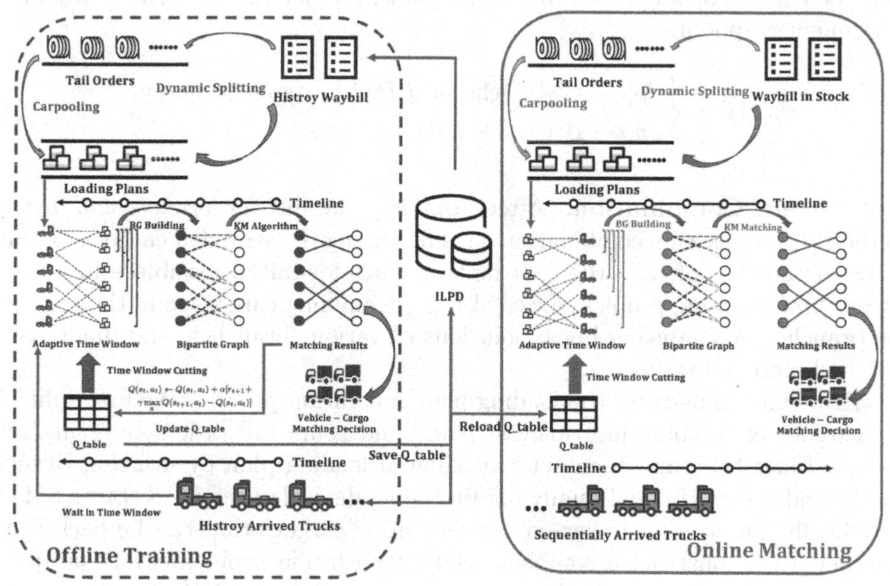

Fig. 2. *ILPD* framework

4.1 Loading Plan Generating

Initially, we need to pack up the cargoes of the waybills in current time window into loading plans. In order to achieve the optimization objective of minimizing the number of tail orders, we apply dynamic programming technology to split the cargoes of the waybills into loading plans, and then employ *branch and bound* algorithm [10] to turn unpackaged cargoes (i.e. stranded cargoes) into newly combined loading plans.

Waybill Splitting. Commonly, the cargoes of any waybill cannot be transported by one truck due to each cargo is bulky and heavy. Thus they need to be split into several transportation tasks. According to the load limit of trucks, the trucks can be grouped into different load types. The load type (denoted as LPT) needs to be considered when the cargoes of waybills are split. For instance, a waybill has 95 tons of cargoes whose unit weight is 5 tons, W_u^T is set as 36 tons and W_l^T is set as 29 tons. Due to LPT is between W_u^T and W_l^T, LPT has two choices, i.e. 30 tons and 35 tons. This may leave some tail order of 5 tons or 20 tons respectively. Thus, improper splitting way will leave more tail orders that cannot be shipped in time. To ensure the minimum of number of tail orders, we use dynamic programming technique to conduct the splitting process. As shown in the following DP matrix, dp[i][j] represents the minimum weight of stranded cargoes when the waybill's weight is j. Then we choose the i-th load type $LPT[i]$ to split the cargoes of waybills. After building DP matrix, we split the cargoes of waybills into loading plans, and let the tail orders to join in subsequent carpooling.

$$dp[i][j] = \begin{cases} dp[i-1][j], & \text{choose } LPT[i] \text{ as split strategy} \\ dp[i-1][j - n \times LPT[i]], & \text{else} \end{cases} \tag{4}$$

Tail Order Combination. After splitting, we utilize *branch and bound* method [10] to search a combinational result among the stranded cargoes to minimize the number of tail orders. It first searches for initial feasible solutions of integer programming problem for combining remaining cargoes, and then prunes the branches by analyzing linear solutions of various branches, and finally gets the combinational result.

Algorithm 1 illustrates the loading plan generating process. We first split all the cargoes of waybills into loading plans and some tail orders by using DP matrix (Lines 1–7), and then get consolidated loading plan by utilizing Branch and Bound algorithm, and finally obtain the loading plan set LPS (Lines 8–14). Specifically, we attempt to screen out remaining cargoes that can be packed up together, then construct a combinational optimization problem which subjects to: the sum of chosen cargoes' weight is between W_u^T and W_l^T. Finally we use *branch and bound* method to solve this optimization problem and get remaining cargoes combination results.

Algorithm 1. Loading Plan Candidate Generation for Each Truck.

Input:
The arrived truck set, T;
The waybill set, w_i;
Output:
The loading plan candidates for T, LPS;
Extract waybills from w_i that can be dispatched to T;
for all w_i such that $w_i \in WBS$ **do**
  Acquire LPT based on unit weight and total weight of w_i;
  Construct DP matrix based on LPT computed before.
  Package cargoes into loading plan set LPS based on DP matrix;
  Join the left cargoes in O_t^i into remaining cargo list RCL
end for
for all r such that $r \in RCL$ **do**
  Acquire the cargo list CL that can be dispatched with r;
  Build combinational optimization problem;
  Search a cargo combination result by using *Branch and Bound* algorithm;
  Join the loading plan that acquired before into LPS
end for
return LPS;

4.2 Adjusting Strategy Learning of Time Window Size

After generating loading plans for the trucks in current time window, the matches between the trucks and loading plans need to be made. Distinct from the greedy matching policy which serves the truck arrived earlier, we make matches between the arrived trucks and loading plans in a time window. This ensures more proper matches between trucks and loading plans. Since the amount of trucks and the number of loading plans vary with time, we need to dynamically adjust the time window size to ensure the global optimal matching result. How to determine an optimal timing to split time window to achieve the maximization of the total shipping weight? To solve this problem, we adopt reinforcement learning technique to learn a strategy of dynamically adjusting the size of time window. It has been proven that can solve the sequential decision making problem effectively. Specifically, we propose a *time zone*-based *Q-Learning* to cut time window, which is a typical model-free reinforcement learning method with relatively low computational complexity. We model the environment and actions with the *time zone*-based Q-Learning method in the following steps.

- **State:** We model the matching system as an agent. The state is expressed as a four-element tuple, namely, $s = (n_T, n_L, d, r_W)$. Elements in s represent the number of the trucks, the number of loading plan candidates, the hour index of current time window, the length of timespan of the current time window $\mathcal{W}$ respectively. It roughly indicates the statuses of cargoes and trucks in time windows. The hour index d is *time zone* in a day, and it is used to meet dramatically changing arrival distribution of the trucks and waybills.

- **Action:** The actions a have two types. One type is to cut the current time window $\mathcal{W}$ and acquire the matching results. The other type does not cut $\mathcal{W}$. The action to determine whether to cut or not depends on the number of a, which represents the expected time window size of agent. When a is not less than r_W, it means that the time window has not reached expected size and needs to be grown. Otherwise, the time window is cut and the matches are obtained.
- **State Transition:** The cutting decision will change the current state into $(n_T - n, \max(0, n_L - n), d, r'_W)$, in which n, r'_W represent the number of trucks that has been matched in $\mathcal{W}$ and the length of time span of $\mathcal{W}$ respectively. Another action will turn the current state into (n'_T, n'_L, d', r'_W), which indicates the next state.
- **Reward:** The instant reward is the sum of discount weight of the matches between trucks and loading plans to maximize TSW from a global perspective. For the reason of the matching process's timeliness and practicality, we bring in a discount factor γ and time-influencing factors ι into the formula of instant reward function. ι denotes the waiting time of the truck nodes, and γ is introduced into controlling the impact of $\mathcal{W}$'s length on instant reward, which is denoted as $R(s,a) = \sum_{\iota=0}^{r_W} \gamma^{\iota} \cdot \frac{R}{r_W}$.

Particularly, *time zone*-based *Q-Learning* algorithm used to approximate Q-value in Q-table is based on Temporal-Difference(TD) rule. As illustrated in Eq. 5, s_t denotes the current state, a_t is the action recommended by Q-function, s_{t+1} is the next state and r_{t+1} is the earned instant reward after taking a_t in the current state of s_t. *Q-Learning*'s bellman equation is denoted by:

$$Q(s_t, a_t) \leftarrow Q(s_t, a_t) + \alpha[r_{t+1} + \gamma \max_{a} Q(s_{t+1}, a) - Q(s_t, a_t)] \tag{5}$$

We add a restriction on the state space in view of infinite matching action required (i.e. the steel factory never ceases production), hence the action space is also infinite. Besides, if the size of a time window is too large, the drivers earlier arrived have to wait until get the matching decision, which is impractical. Hence, we add a length limitation of time window in the learning process, denoted as β. Algorithm 2 details the procedure of *time zone*-based *Q-Learning* method. It first initializes Q table randomly (Line 1), and in each training episode, matching is made according to Q table (Lines 5–19). Then the feedback of rewards is given and Q table is updated using bellman equation. After the training process finishes, the adjusting strategy for the time window size is acquired, which can be recommended by Q table (Line 20).

Algorithm 2. Time zone-based Q-Learning

Input:
learning rate $\alpha \in (0,1]$, discount factor $\gamma \in (0,1]$, *Se*;
Output:
state-action value funciton $Q(s, a)$;
1: initialize $Q(s, a)$ with *Random*(), $\forall$ a $\in A$, s $\in S$;
2: **for all** *episode* such that *episode* $\in$ *episodes* **do**
3: $TSE \leftarrow$ the timespan of episode;
4: $t \leftarrow 0$;
5: **while** $t < TSE$ **do**
6: get new trucks and loading plans nodes from *Se* and update $\mathcal{W}$
7: construct the bipartite graph G corresponds to $\mathcal{W}$
8: update r_W based on the status of $\mathcal{W}$
9: $a = argmax_{a'}\ Q((n_T, n_L, d, r), a')$
10: **if** $a <= r_W$ **then**
11: implement KM algorithm in G and observe s_{t+1}, r_t;
12: $Q(s_t, a) \leftarrow Q((n_T, n_L, d, r_W), a) + \alpha[r_t + \max_{a'} Q(s_{t+1}, a') - Q((n_T, n_L, d, r_W), a)]$
13: Δt $\leftarrow$ the time period between s_t and s_{t+1}
14: $t \leftarrow t + \Delta$t, $l_t \leftarrow \Delta$t
15: **else**
16: increase the time span of $\mathcal{W}$ and observe s_{t+1};
17: $Q(s_t, a) \leftarrow Q((n_T, n_L, d, r_W), a) + \alpha[\max_{a'} Q(s_{t+1}, a') - Q((n_T, n_L, d, r_W), a)]$;
18: Δt $\leftarrow$ the time period between s_t and s_{t+1};
19: $t \leftarrow t + \Delta$t, $l_t \leftarrow l_t + \Delta$t;
20: **end if**
21: **end while**
22: **end for**
23: **return** Q;

4.3 Matching Decision Making

At online decision making phase, we split *Se* into multiple continuous time windows according to the time window size adjusting strategy, and then assign each loading plan to appropriate truck by using *Kuhn-Munkres* algorithm [9] to acquire maximum weight matching of G in the current time window $\mathcal{W}$. Specifically, when the size of current time window $\mathcal{W}$ is big enough but $\mathcal{W}$ has not got matching results, we set the max value of time window size (denoted as β) for the consideration of the performance of algorithm and practicality of matching. When the size of current time window equals to β (Line 7), we assign each loading plan to the truck by implementing *Kuhn-Munkres* algorithm immediately (Lines 8–10). Algorithm 3 shows the details of vehicle-cargo matching algorithm.

Complexity Analysis. Let $N_T, N_L, |D|, |\beta|$ denote the maximal number of truck and loading plan nodes in time window with maximal size β, the number of hour index and maximal time window size respectively. The space complexity of online vehicle-cargo matching is $O(N_T N_L|D||\beta|)$, which equals to the memory size of Q table. The time complexity of each time window is at most $O(KM(N_T, N_L))$, which is the time of executing *Kuhn-Munkres* algorithm on a bipartite graph with at most N_T truck nodes and N_L loading plan nodes. Considering real-time requirements of matching for steel logistics and high time complexity of *Branch and Bound*, we pack up the cargoes in waybills every 20 min when the waybills are updated by logistics platform.

Algorithm 3. Online Vehicle-Cargo Matching

Input:
Q table learned by Algorithm 2, the trucks and waybills sequence Se
Output:
matching results M_G;
while Se is not finished **do**
  initialize the arrived trucks list RTT based on Se;
  initialize the loading plan set LPS based on Se;
  update $\mathcal{W}$ based on RTT and LPS
  update state s_t and time span r_W based on the status of $\mathcal{W}$;
  $a \leftarrow Q(s_t)$;
  if $a <= r_W$ or $a = \beta$ **then**
    $md \leftarrow KuhnMunkres(RTT, LPS)$;
    update RTT and LPS based on md;
    $M_G \leftarrow M_G \cup md$;
  end if
end while
return M_G;

5 Experiments

5.1 Data Description

The real data sets are from a steel logistics company named *JCZH*, including a transporting data set and a waybill data set. The time spans of both data sets range from *Sept. 24th* to *Oct. 23th* in 2020.

Transporting Data Set. It is derived from the transporting records of 5,629 trucks in *RiZhao Steel Plant.* It contains the attributes of *arriving timestamp, transporting destination* and *commodity of preferred cargoes.*

Waybill Data Set. It records all the waybill information of *RiZhao Steel Plant.* It contains the attributes of *transporting destination, total weight, unit weight, timestamp* and *the number of cargoes* for each waybill.

Due to data privacy and company security requirements, our real data sets from *JCZH* cannot be made public. In addition, to evaluate the robustness of our proposed framework, we manually synthesize the transporting data set and waybill data set using different distributions and cardinalities to simulate different arrival distributions and densities of both sides.

Synthetic Data Set. We simulate different arrival distributions by utilizing a normal distribution with different sparsities σ, and obtain different densities by using a uniform distribution with variety of cardinalities δ of truck and loading plan nodes.

5.2 Evaluation Metric

Competitive ratio is an evaluation metric for analyzing the performance of online algorithm in the condition of unknown information. We utilize it as evaluation metric to make a comparison between our proposal and other approaches.

Competitive Ratio: Given an arbitrary online algorithm A, if there exists a constant K for any experiment sequence I_s, $C_A(I_s) \leq \alpha\ C_{opt}(I_s) + K$, we call A is α- competitive. The competitive ratio α_A is the infimum corresponding to worst performance in the s-round experiments, denoted as $\alpha_A = \frac{\min(A(I_s))}{OPT(I_s)}, s = 1, 2, ..., n$.

5.3 Comparative Approaches

To evaluate the performance of our proposal, we compare it with different approaches.

- **Optimal matching(OPT):** It takes all the trucks and waybills information as input to make the maximum weight matching.
- **Greedy algorithm(GR):** It is a simple policy originally used in steel logistics platform. As soon as the truck arrives in steel factory, the max-weight loading plan that the truck can transport will be assigned.
- **Perturbed Greedy algorithm(PG)** [1]**:** It combines greedy algorithm and ranking algorithm [8] to ensure $\frac{e-1}{e}$ competitive ratio.
- **Fixed-Batch algorithm(FB):** It is conducted by using a fixed size of time window. We try to choose all the suitable time window size and regard the best one as the parameter. In each fixed-size time window, *KM* algorithm is implemented to acquire a match of maximum weight.
- **TGOA** [15]**:** *TGOA* combines the greedy policy with fixed-size batch matching strategy to achieve the goal of maximizing global shipping weight, and learns the greedy threshold to decide the usages of both strategies.

5.4 Quantitative Comparison Results

Next, we evaluate the effectiveness of our framework by analyzing the results of experiments on both real and synthetic data sets. As illustrated in Fig. 3(a)(b), after running all the approaches on real data sets for a few days, the results show that *ILPD* performs significantly better than the other methods in terms of the objectives of maximizing *TSW*. In addition, its performance is closest with that of *OPT* method and it can at least achieve a competitive ratio of 0.85. More specifically, *time window* based matching methods like *FB* and *TGOA* have

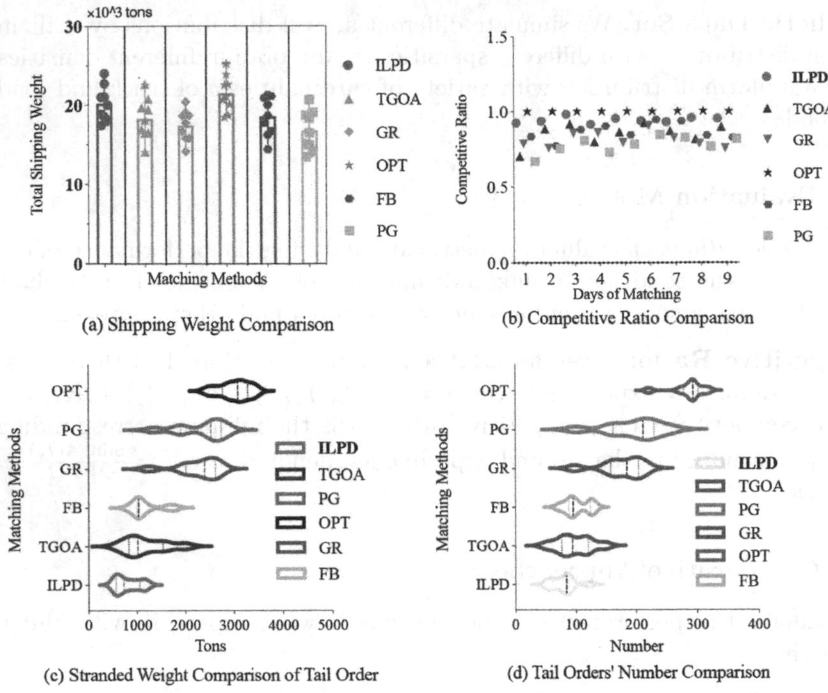

(a) Shipping Weight Comparison (b) Competitive Ratio Comparison

(c) Stranded Weight Comparison of Tail Order (d) Tail Orders' Number Comparison

Fig. 3. Quantitative comparative results

better performance than greedy matching methods like *GR* and *PG*. Because the former can obtain more in-stock information that used to generate loading plan than the latter. Additionally, *TGOA* and *FB* lose the competitiveness for their lackness of taking the current matches' impact into consideration. As for the optimization goal of minimizing the number of tail orders, *ILPD* can leave minimal tail orders in comparison to other methods, as shown in Fig. 3(c) and (d). Also, *time window* based matching methods perform better than greedy ones in terms of minimizing weight and number of tail orders.

Moreover, we utilize synthetic data sets to evaluate the robustness of *ILPD*. Different distributions and cardinalities of trucks and waybills are introduced to simulate the distribution of trucks' and waybills' arrival. As shown in Fig. 4(a)(b), we use normal distribution with a constant cardinality and different sparsities σ to simulate the trucks' and waybills' arrivals in Fig. 4(a), the higher the value of σ means that trucks' and waybills' arrival distributions are more intensive. It can be clearly observed that *ILPD* can acquire the best performance comparing to other online matching methods. Moreover, we leverage uniform distribution with different trucks and loading plans cardinalities δ to simulate the trucks' and waybills' arrivals of someday. As illustrated in Fig. 4(b), *ILPD* also obtains a best performance except *OPT*.

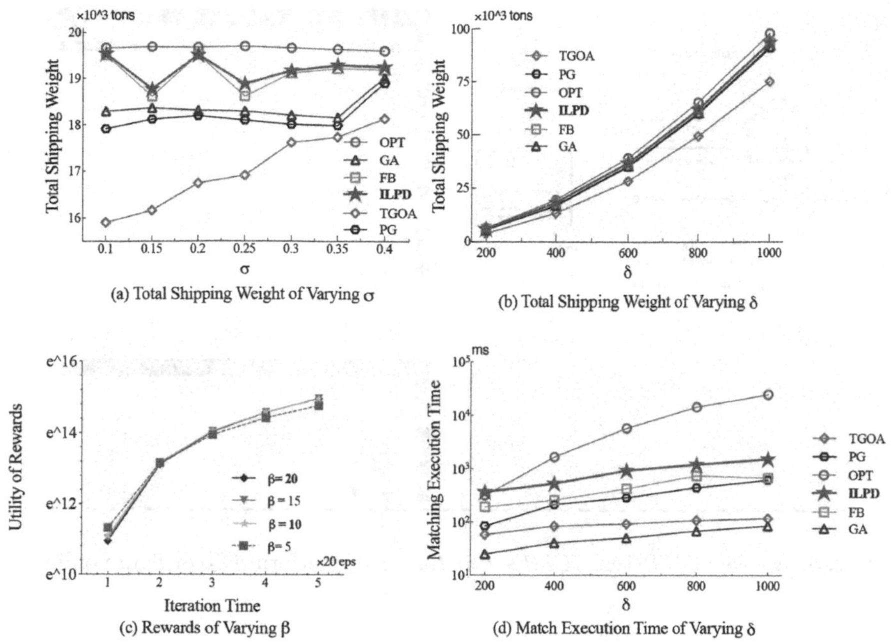

Fig. 4. Results on varying parameters: β, σ, δ

Furthermore, we discuss the choice of the max size of time window β. We use four types of β to train our model and gather the utility of reward. As shown in Fig. 4(c), β which equals to 5 obtains the most utility of rewards in first forth epochs while gets the minimal utility of rewards later comparing to other types of β, which almost stay in a same level while the training process runs deep. In the practical scenario, the bigger value of β means that more trucks wait in the steel factory, which will affect the operational efficiency of the steel plant. Hence, we set β as 10 to gain more rewards in order not to impact the operational efficiency of the steel plant.

Finally, we compare the match execution time of all the matching methods. As shown in Fig. 4(d), the execution time of *ILPD* is higher than any other methods as the truck and waybills cardinality δ increases except OPT. This is due to that *ILPD* splits more time windows than *FB* whose r_W is constant value of 10, which incurs more execution time. Due to high time complexity of *KM* algorithm, the execution time of *ILPD* and *FB* is higher than *GR* and *PG*. The max execution time of *ILPD* is 1.6 s when the truck and waybills cardinality δ is 1000. Though *ILPD* consumes more execution time than other methods, the gap between *ILPD* and other methods is not big enough to affect the operational efficiency. It is acceptable under practical vehicle-cargo matching scenario.

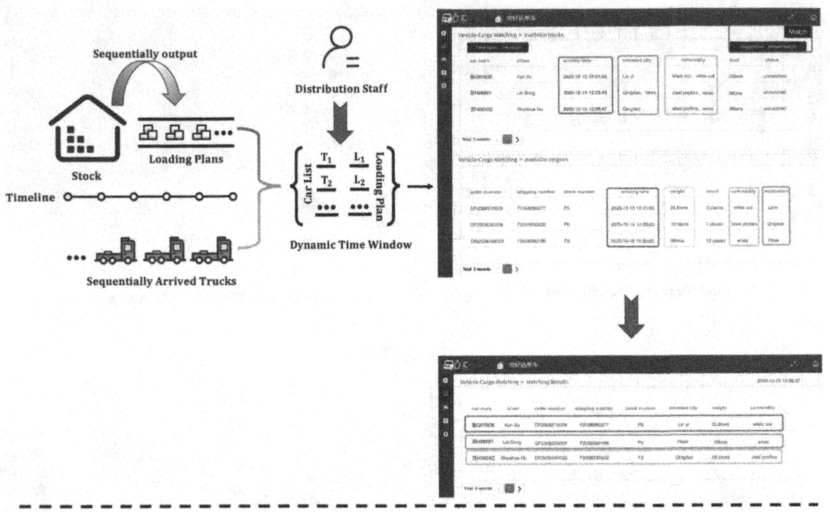

Fig. 5. A case of *ILPD* at *JCZH*'s *Hui Hao Yun* platform (Color figure online)

5.5 Case Study

Furthermore, we verify the practicality of *ILPD* by implementing it on steel logistics platform. The visualization of vehicle-cargo matching process is illustrated in Fig. 5. When the loading plans are generated continuously and trucks sequentially arrives in steel factory, the logistics platform gathers and presents the truck information and stock information in the website. Then the website shows the status of current time window, including all information of each truck and loading plans (highlighted in different colors box). As illustrated in Fig. 5, the website gives a recommendation for vehicle-cargo matching in the current time window. It provides the choices for the dispatcher to determine the matching result of trucks and loading plans.

6 Conclusions

In this paper, we propose an intelligent decision framework for vehicle-cargo matching in bulk logistics field, called *ILPD*. To achieve the goal of maximizing total weight of cargoes and minimizing the number of tail orders, we first conduct the analysis and training on the real logistics data, and then design a two-step online matching mechanism. Further, to achieve a near optimal matching result, we introduce a *time window*-based *Q-learning* algorithm to determine the size of each time window adaptively. The experiment evaluations on the real logistics and synthetic data sets show our proposal performs better than the existing matching policies. In the future, we will apply *ILPD* to more logistics scenarios to improve its efficiency and generalization ability.

Acknowledgments. This research was supported by NSFC (Nos. 62072180, U1911203 and U1811264).

References

1. Aggarwal, G., Goel, G., Karande, C., Mehta, A.: Online vertex-weighted bipartite matching and single-bid budgeted allocations. In: Proceedings of the Twenty-Second Annual ACM-SIAM Symposium on Discrete Algorithms, SODA 2011, San Francisco, California, USA, 23–25 January 2011, pp. 1253–1264 (2011)
2. Ashlagi, I., et al.: Min-cost bipartite perfect matching with delays. APPROX/RANDOM **81**, 1–1 (2017)
3. Azar, Y., Chiplunkar, A., Kaplan, H.: Polylogarithmic bounds on the competitiveness of min-cost perfect matching with delays. In: ACM-SIAM Symposium on Discrete Algorithms, pp. 1051–1061 (2017)
4. Chen, Z., Cheng, P., Zeng, Y., Chen, L.: Minimizing maximum delay of task assignment in spatial crowdsourcing. In: ICDE, pp. 1454–1465 (2019)
5. Dickerson, J.P., Sankararaman, K.A., Srinivasan, A., Xu, P.: Assigning tasks to workers based on historical data: online task assignment with two-sided arrivals. In: AAMAS (2018)
6. Hu, Z., Sheng, Z.: A decision support system for public logistics information service management and optimization. Decis. Support Syst. **59**, 219–229 (2014)
7. Huang, Z., Kang, N., Tang, Z.G., Wu, X., Zhang, Y., Zhu, X.: How to match when all vertices arrive online. In: ACM SIGACT Symposium on Theory of Computing, pp. 17–29 (2018)
8. Karp, R.M., Vazirani, U.V., Vazirani, V.V.: An optimal algorithm for on-line bipartite matching. In: Proceedings of the Twenty-Second Annual ACM Symposium on Theory of Computing, pp. 352–358 (1990)
9. Kuhn, H.W.: The Hungarian method for the assignment problem. Naval Res. Logist. Q. **2**(1–2), 83–97 (1955)
10. Land, A.H., Doig, A.G.: An automatic method for solving discrete programming problems. In: Jünger, M., et al. (eds.) 50 Years of Integer Programming 1958–2008 - From the Early Years to the State-of-the-Art, pp. 105–132. Springer, Heidelberg (2010). https://doi.org/10.1007/978-3-540-68279-0_5
11. Liu, J., et al.: Adaptive loading plan decision based upon limited transport capacity. In: Nah, Y., Cui, B., Lee, S.-W., Yu, J.X., Moon, Y.-S., Whang, S.E. (eds.) DASFAA 2020. LNCS, vol. 12114, pp. 685–697. Springer, Cham (2020). https://doi.org/10.1007/978-3-030-59419-0_42
12. Spivey, M.Z., Powell, W.B.: The dynamic assignment problem. Transp. Sci. **38**(4), 399–419 (2004)
13. Sun, Y., Zhang, J., Tang, Q., Yan, Y.: Research on benefit-risk of vehicle-cargo matching platform based on matching degree. In: ICITE, pp. 580–584. IEEE (2020)
14. Tong, Y., She, J., Ding, B., Chen, L., Wo, T., Xu, K.: Online minimum matching in real-time spatial data: experiments and analysis. VLDB **9**(12), 1053–1064 (2016)
15. Tong, Y., She, J., Ding, B., Wang, L., Chen, L.: Online mobile micro-task allocation in spatial crowdsourcing. In: ICDE, pp. 49–60 (2016)
16. Tong, Y., et al.: Flexible online task assignment in real-time spatial data. VLDB **10**(11), 1334–1345 (2017)

17. Wang, Y., Wong, S.C.W.: Two-sided online bipartite matching and vertex cover: beating the greedy algorithm. In: ICALP, pp. 1070–1081 (2015)
18. Wang, Y., Tong, Y., Long, C., Xu, P., Xu, K., Lv, W.: Adaptive dynamic bipartite graph matching: a reinforcement learning approach. In: ICDE, pp. 1478–1489 (2019)

Microservice Workflow Modeling for Affinity Scheduling to Improve the QoS

Yingying Wen(✉), Guanjie Cheng, ShuiGuang Deng, and Jianwei Yin

Computer Science and Technology, Zhejiang University, Hangzhou, China
{11821102,11821019,dengsg}@zju.edu.cn, zjuyjw@cs.zju.edu.cn

Abstract. Attracted by the flexibility of microservice architecture, many Cloud services are composed of components that communicate with the Remote Procedure Calls (RPC). Considering the high cost of RPC between components running on different machines, this raises a question about how to effectively arrange and place these components to physical machines in Cloud to offer good quality of services. Current co-location strategies mainly consider the resource constraints and performance interference among individual components but ignore the workflow dependencies of components. Though workflow of background jobs has been used as an important modeling element when seeking optimal scheduling schema, the workflow-aware scheduling for microservice applications lacks research efforts. In this paper, we propose a workflow-aware component placement schema for microservice applications to reduce their running time. Specifically, we design a new workflow model based on Directed Acyclic Graph (DAG) and probabilistic theory describing components' calling and time dependency to predict the running time of applications. Based on the proposed model, we quantify the affinity degree of two components that further supports the pod affinity scheduling. The affinity degree considers multiple dimensions, including the critical path, the possible improvement of running time, and the throughput of user requests. The experimental evaluation results prove the accuracy of the workflow model and its application effect on pod affinity scheduling.

Keywords: Microservice architecture · Workflow scheduling · Performance optimization · Affinity scheduling

1 Introduction

The mature microservice architecture and container techniques have caused a revolution in our way of developing and deploying software services. In Cloud, constituent components of microservice applications can be distributed over a set of machines to overcome the resource limit of a single machine, while the container technique supports the sharing of a physical machine among multiple components. These microservice components are composed via standardized

B. Li et al. (Eds.): APWeb-WAIM 2022, LNCS 13421, pp. 313–328, 2023.
https://doi.org/10.1007/978-3-031-25158-0_24

Remote Procedure Calls (RPC) interfaces, such as Google's gRPC [15] or Facebook/Apache's Thrift [1], to cross the container boundaries. The RPC communication cost between two containers would be reduced when they are located at the same domain, which can directly impact the quality of services (QoS). Thus, Cloud administrators face a challenge to seek an optimal service placement schema to offer better quality of services.

Current container orchestration, like the Kubernates[1], has already supported the advanced affinity scheduling since 2017. Pod[2] affinity is one of the supported objects based on pods' key/value labels, which can specify pod affinity rules about how pods should be placed relative to one another. Pod affinity/anti-affinity of Kubernates is very flexible to support various rules. Therefore, in this paper, we focus on the affinity rule design.

The state-of-the-art designs for pod affinity mainly intend to reduce the resource competition/performance interference [2,11] including traffic consideration [5,6]. To the best of our knowledge, no research effort has been devoted to optimizing the quality of web services with joint considerations of microservice workflows. Though workflow scheduling for jobs has lots of solutions, unfortunately, these solutions are unsuitable to web services, caused by the difference of their resource sharing methods. Job scheduling is to place a job's tasks[3] to use limited resources by time slots sharing [4]. After one task is finished, it would release the allocated resources. But long-running web services would not return the allocated resources after finishing a user request and would keep waiting for the subsequent user requests. Thus the web service's workflow scheduling is to change their pods' placement, which requires new design of pod affinity rules different from jobs' workflow scheduling.

In this paper, for improving the quality of web services, we propose a new workflow modeling method to help with the design of pod affinity rules. Considering the inefficiency of estimating the service response time by summing the running time of all participating service components [12,16,19] since the lack of considerations on the parallel execution structures, we profiles the RPC dependencies by DAG graph. But different from the SDG (Service Dependency Graph) [6] that treats a node's direct downstream nodes as being executed parallelly, we model the microservice workflow as time dependency graph (TDG) that considers the synchronous/asynchronous relationships of calls. As the diverse parameters of user requests would incur dynamic microservice workflows even for the same service, we further introduce the probability modeling method into our workflow model. Based on the workflow model, we can predict the distribution of service response time when a pair of microservices are re-located to the same domain. We thus define the pod affinity of two microservices as their impact degree on the service response time when their pod placements are changed.

[1] https://kubernetes.io/.

[2] Pods are the smallest deployable units of computing that you can create and manage in Kubernetes. https://kubernetes.io/docs/concepts/workloads/pods/.

[3] Jobs would be decomposed into smaller tasks to run up.

We further evaluate the proposed workflow model by validating its prediction accuracy and its application effect on pod affinity scheduling. As verified by the experimental evaluation using five services with diverse microservice workflows, our performance model can precisely predict the running time from the mean, median, P99, P95, and P90 dimensions within 10% error bound in average. And the optimization method performs the best when compared to four other affinity rule designs.

2 Background and Related Work

2.1 Affinity Scheduling

Container-based virtualization offers operating system level abstraction and isolation to allow a physical machine to be shared among multiple containers [17]. The containers are the dynamic running instances that manage the resources and run the source code up. Pod is the concept introduced by Kubernates, which is a group of containers as the smallest deployable unit. Scheduling is to manage these containers/pods to offer good quality of services and constrain the resource usage.

About the affinity scheduling, there are two types of affinity that need to be clarified: They are the node affinity and pod affinity. In the literature, node affinity is referred to as the service placement problem (SPP) [3,4,13] and consists in identifying the optimal computation and storage servers for hosting a cloud service's application and data components, while taking into consideration some placement constraints. Pod affinity is referred to as the workload co-location problem about how pods should be placed relative to one another. Performance interference [7,9,10,20] is mostly considered in this research area. Pod affinity is what this paper intend to promote.

Except the performance interference considerations, the former affinity designs [5,6] have intended to complement the pod affinity rules by reducing the overall communication cost to improve the QoS. As the parallel execution structure of microservice workflow, the component pair with most communication cost is not the most influential one for the QoS. We show an illustration example in Fig. 1. For optimizing the overall communication cost, the most costly pair [m_1,m_2] needs to be co-located with highest priority. Contrarily, for reducing the running time from start to end, the pair [m_3, m_4] in the critical path is the one that needs to be co-located with highest priority. Therefore, in this paper, we intend to complement the pod affinity strategies from the impact of running time dimension with workflow-awareness.

2.2 Microservice Workflow

The execution relationship among the constituent services of a composite application is usually expressed as a Directed Acyclic Graph(DAG), sometimes called a workflow [18]. The microservice workflow as the research object in this paper

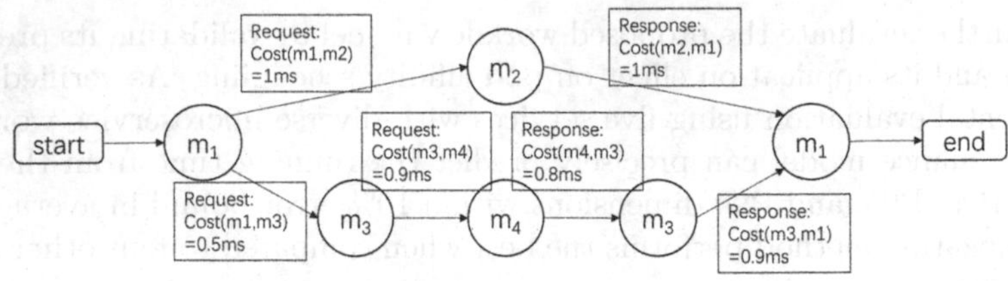

Fig. 1. An example to show the component pair with highest communication cost is not the most influential one for the QoS.

means the dynamic running order that is called as Execution History Graph (EHG) in Qiu's research [11]. In Fig. 2, we show the example of Service Dependency Graph (SDG) [6] and two EHGs, where SDG shows the dependencies (the edges of the graph) between microservice and EHG shows space-time diagram of the distributed execution of a user request. EHG's edges represent the RPC invocations corresponding to RPC send and RPC receive. EHG is better than SDG at showing the running states from time dimension.

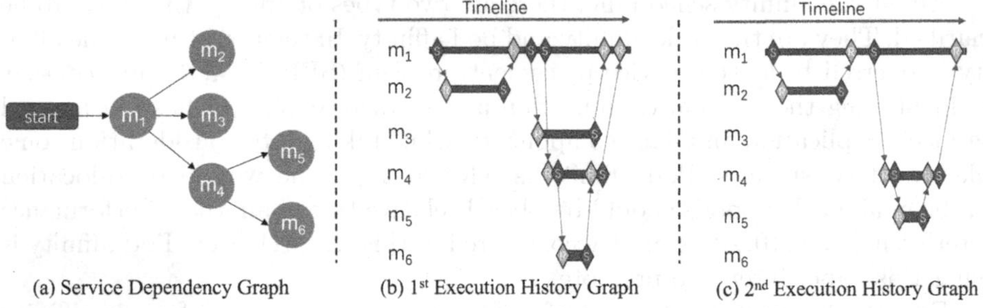

Fig. 2. Microservice workflow overview: (a) is the Service Dependency Graph of an application; (b) and (c) are two possible execution history graphs of this application.

Each user request traverses a subset of vertices in the SDG and forms its own EHG. The EHGs of two user requests for the same application would diverse. The EHG can be constructed using the global view of execution provided by distributed tracing tools, like Jaeger[4], Zipkin[5], and Dapper [14]. However, an EHG only shows the running states of a single user request. For affinity scheduling, we need to consider the overall service quality by unifying all user requests. How to model and unify diverse EHGs of a service is one of the major contributions of this paper.

---

[4] https://www.jaegertracing.io/.
[5] https://zipkin.io/.

3 Microservice Workflow Modeling

3.1 Definition of the Microservice Workflow

The microservice workflow of a microservice application G_m is defined as a weighted Directed Acyclic Graph (DAG):

$$G_m = \{V, E, DT, PN, P, MS, DN, DC\} \tag{1}$$

where

- V is a finite set of $|V|$ vertices $m_1, m_2, \ldots, m_{|V|}$, representing microservice modules;
- $E \subseteq V \times V$ is a finite set of directed edges. The directed edge from $m_u \in V$ to $m_v \in V$, denoted as $e_i = (m_u, m_v)$, represents the interaction between vertex m_u and m_v defined by the business logic;

the V related elements include:

- $DT : V \rightarrow [0, +\infty)$ is the duration of a microservice module;
- $PN : V \rightarrow [0, +\infty)$ is a penalty value denoting the limitation aspect or negative effect of this module to integrate with other modules;

and the E related elements include:

- $P : V \times V \rightarrow [0, 1]$ is the transition probability. The probability of the edge (m_u, m_v), denoted as $P(m_u, m_v)$, represents the probability to call m_v when the workflow reaches to m_u defined by the relative number of invocations of each node in V.
- $MS : V \times V \rightarrow [0, +\infty)$ is the message size delivered from m_u to m_v, denoted as $MS(m_u, m_v)$.
- $DN : V \times V \rightarrow [0, +\infty)$ is a network delay function. $DN(m_u, m_v)$ identifies the delay from m_u to m_v incurred by the network communication.
- $DC : V \times V \rightarrow [0, +\infty)$ is a delay function of serialization/deserialization cost of remote calls. $DC(m_u, m_v)$ identifies this kind of delay of m_u calling for m_v.

Typically, the execution of the microservice workflow starts with a user request from the interaction layer as a trigger. Then the following modules are linked by remote process calls with sending the request by a client and receiving the response from the corresponding server as a complete calling process until the workflow completes the business logic. Therefore, we include a start node m_{str} and end node m_{end} in V, defining the workflow's entry point and ending point, respectively. And we model the request and response of a remote call from m_u to m_v as two edges (m_u, m_v) and (m_v, m_u), set the $P(m_u, m_v)$ related to the number of invocations, and set the $P(m_v, m_u)$ equals 1 to denote the server would certainly response when it has received a request. The edges started from a node is independent, which means the probability sum of these edges needn't equal to 1. Figure 3 shows an example of microservice workflows for an application.

Based on the definition of the microservice workflow, we define the following notations:

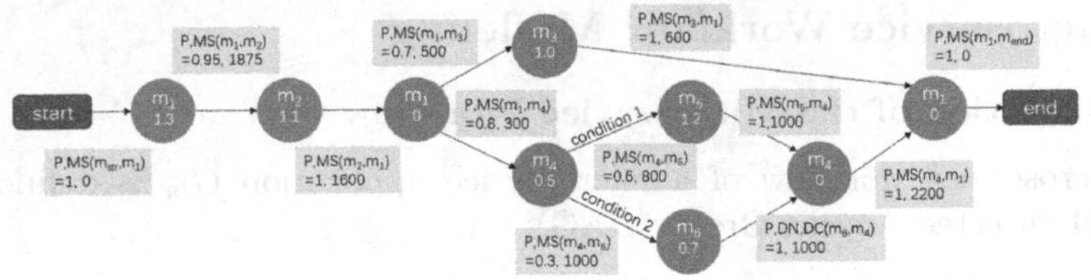

Fig. 3. Workflow of a microservice application composed of 6 modules.

1. A simple path in the workflow is a finite sequence of vertices and edges $s = m_1 e_1 m_2 e_2 \dots e_{n-1} m_n$ satisfying (1) $f_i \in V$ for all $1 \leq i \leq n$, (2) $e_i \in E$ for all integers $1 \leq i \leq n-1$, and (3) $e_i = (m_i, m_{i+1})$ for all $1 \leq i \leq n-1$.
2. The transition probability of the simple path $s = m_1 e_1 m_2 e_2 \dots e_{n-1} m_n$ is defined as

$$TP(s) = \prod_{i=1}^{n-1} P(m_i, m_{i+1}) \tag{2}$$

3. The running time of a simple path $s = m_1 e_1 m_2 e_2 \dots e_{n-1} m_n$ denoted as $RT(s)$, is defined as Eq. (3), namely the sum of the duration of all vertices and the delay, including the network delay and serialization/deserialization delay, incurred by edges in this path.

$$RT(s) = \sum_{i=1}^{n} DT(m_i) + \sum_{i=1}^{n-1} (DN(m_i, m_{i+1}) + DC(m_i, m_{i+1})) \tag{3}$$

4. $AS(m_u, m_v)$ denotes all simple paths between vertex m_u and m_v, which is the set of all possible simple paths in the graph starting from m_u and ending at m_v.
5. $CP(G_m)$ denotes the critical path of a workflow. It is an element coming from $AS(m_{start}, m_{end})$, and satisfies Eq. (4), namely the path with longest running time.

$$\max_{s \in AS(m_{start}, m_{end})} RT(s) = RT(CP(G_m)) \tag{4}$$

6. $out(m_i)$ denotes the set of all edges starting from vertex m_i and $in(m_j)$ denotes the set of all edges ending to vertex m_j.

3.2 Running Time Estimation

We propose a performance model to get the estimated running time of the microservice workflow denoted as $RT(G_m)$. As defined, G_m is a probabilistic DAG, such that

1. G_m has no cycles and loops.
2. Simple path s in G_m has its corresponding transition probability $TP(s)$, which means this path may sometimes not exist.

3. The running time of a workflow depends on the critical path with the longest running time.
4. The critical path $CP(G_m)$ in G_m would not always be the same. A simple path s has its probability of being a critical path CP whose value is different from $TP(s)$.

We define the probability of a path s_i to be the critical path as $PCP(s_i)$. Thus the $RT(G_m)$ can be calculated as

$$RT(G_m) = \sum_{s_i \in AS(m_{start}, m_{end})} PCP(s_i) \cdot RT(s_i) \tag{5}$$

The estimated running time of the workflow is the weighted average running time of all possible critical paths. The following sub-sections describes how the estimation method processes the probability structures in the microservice workflow to get the $RT(G_m)$ value.

Process the Parallels. The parallel edges are formed by two types of structures: One is branch structure, in which the edges are mutually exclusive, and the other is asynchronous structure, in which edges are independent. The main idea of processing parallels is to convert these two types of edges into a single type. Our design converts the independent edges into mutually exclusive edges and then calculates the weighted average running time.

Considering a sub-graph $G_p = \{\{m_u, m_v\}, E_p\}$ where E_p contains n parallel edges that link start node m_u and end node m_v. Two types of edges mix up in E_p. We process the parallels in G_p following Algorithm 1.

Algorithm 1. Processing the parallels

Require: a sub-graph $G_p = \{\{m_u, m_v\}, E_p\}$
Ensure: return the processed sub-graph $\hat{G_p}$

for e_i in E_p **do**
  $C(e_i) = DN(e_i) + DC(e_i)$
end for
E_p = rank(E_p) {according to their $C(\cdot)$ values in ascending order}
for $i = 0 \rightarrow len(E_p)$ **do**
  if e_i is asynchronous type **then**
    reset $P(e_i)$ value as $\hat{P}(e_i)$ according to Eq. (6)
  end if
  if e_i is branch type **then**
    reset $P(e_i)$ value as $\hat{P}(e_i)$ according to Eq. (7)
  end if
end for
return $\hat{G_p}$

There are two processing equations in Algorithm 1. The performance model converts asynchronous e_i to be a branch edge with the new probability value calculated as

$$\hat{P}(e_i) = P(e_i) \times \prod_{e_{aa} \in E^i_{async}} (1 - P(e_{aa})) \times (1 - \sum_{e_{bb} \in E^i_{branch}} P(e_{bb})) \quad (6)$$

where E^i_{async} and E^i_{branch} are the edge lists coming from ordered list $[e_{i+1}, e_{i+2}, \ldots, e_n] \subset E_p$ classified according their types, whose costs are higher than e_i. For branch edge e_i, as the adding of new edge converted from asynchronous edges, its probability also need to be updated as:

$$\hat{P}(e_i) = P(e_i) \times \prod_{e_{aa} \in E^i_{async}} (1 - P(e_{aa})) \quad (7)$$

where E^i_{async} is the asynchronous edge list whose cost are higher than e_i coming from ordered list $[e_{i+1}, e_{i+2}, \ldots, e_n] \subset E_p$. These two equations mean that a simple path can be the critical path only when all the paths with longer running time do not exist in the workflow. After getting the new probability values, the conversion is completed by changing the probability values and adding 'condition' annotations on edges.

Next, we can calculate the weighted running time of G_p as

$$RT(G_p) = \sum_{i=1}^{n} C(e_i) \cdot \hat{P}(e_i) + DT(m_u) + DT(m_v) \quad (8)$$

If there is no other simple path linking the m_{start} and m_{end} after subtracting G_p from workflow graph, the performance model can simplify G_p into a single vertex. The performance model

1. trims G_p by removing the edges in E_p and left two nodes, m_u and m_v;
2. combines these two nodes into a single one $\ddot{m}_u$ and keeps all the edges of the workflow not in E_p unchanged;
3. and replaces the DT value of this new $\ddot{m}_u$ node as $RT(G_p)$.

But if other simple paths link the m_{start} and m_{end} after subtracting G_p from the workflow graph, for the accuracy of the performance model, we need to leave these parallel edges uncombined and process these edges with the serial structure until there is no other simple path except the G_p.

Process the Series. The main idea of processing the series is to reduce the middle vertex. Considering a serial structure $G_s = \{V_s, E_s\}$ with three vertices $V_s = \{m_u, m_v, m_w\}$ and multiple edges linking these vertices, we can simplify this structure by processing the serial edges and trimming the m_v. We denote these edges as $E(m_u, m_v) = \{(m_u, m_v)_1, \ldots, (m_u, m_v)_p\}$ and $E(m_v, m_w) = \{(m_v, m_w)_1, \ldots, (m_v, m_w)_q\}$, where $p \geq 1$ and $q \geq 1$.

After processing these $p + q$ edges, we get $p \cdot q$ edges $E_{new} = \{e_{[1,1]}, \ldots, e_{[1,q]}, e_{[2,1]}, \ldots, e_{[p,q]}\}$ linking m_u and m_w. For an edge $e_{[pp,qq]}, 1 \leq$

Table 1. List of critical paths and their corresponding probabilities

ID	Simple Path	Probability	ID	Simple Path	Probability
cp_1	$m_1m_2m_1$	0.95*0.06=0.057	cp_2	$m_1m_2m_1m_3m_1$	0.95*0.364=0.3458
cp_3	$m_1m_2m_1m_4m_5m_4m_1$	0.95*0.47=0.456	cp_4	$m_1m_2m_1m_4m_1$	0.95*0.024=0.0228
cp_5	$m_1m_2m_1m_4m_6m_4m_1$	0.95*0.062=0.0684	cp_6	m_1	0.05*0.06=0.003
cp_7	$m_1m_3m_1$	0.05*0.364=0.0182	cp_8	$m_1m_4m_5m_4m_1$	0.05*0.48=0.024
cp_9	$m_1m_4m_1$	0.05*0.024=0.0012	cp_{10}	$m_1m_4m_6m_4m_1$	0.05*0.072=0.0036

$pp \leq p, 1 \leq qq \leq q$ in these $p \cdot q$ edges, the probability of $e_{[pp,qq]}$ and its cost value are calculated as

$$\begin{aligned} P(e_{[pp,qq]}) &= P(m_u, m_v)_{pp} * P(m_v, m_w)_{qq} \\ C(e_{[pp,qq]}) &= C(m_u, m_v)_{pp} + C(m_v, m_w)_{qq} + DT(m_v) \end{aligned} \tag{9}$$

The above procedures simplify the workflow graph by trimming a vertex.

Summary, Example, and Analysis. Algorithm 2 implements the estimation algorithm. By detecting the parallels and series iteratively, the algorithm leverages different procedures to process the workflow graph. We can get the critical paths with corresponding probabilities as shown in Table 1 and get the average running time of the application is 10.5516.

Algorithm 2. Running time prediction

Require: a microservice workflow G_m
Ensure: return the expected running time $RT(G_m)$

repeat
  if serial edges exist **then**
    process serials {Eq. (9)}
  else if parallel edges exist **then**
    if independent edges exist **then**
      change all edges to mutually exclusive edges {Eq. (6) and Eq. (7)}
    end if
    if no path between start to end except these parallel edges **then**
      combine parallel edges {Eq. (8)}
    end if
  end if
until left one vertex m_u
return $RT(m_u)$

About the time complexity of the performance model, let us consider the case with a balanced workflow shape. Each vertex m_i in the workflow model calls k vertex, namely $|out(m_i)| = k$, and there are total v vertices (except the vertices for running constraints). There are total $k \cdot v$ probability edges, which means,

Table 2. The list of module pairs appeared in critical paths and their weights

Module Pair	Appeared in CP	Weight
(m_1, m_2)	$cp_1, cp_2, cp_3, cp_4, cp_5$	3.23
(m_1, m_3)	cp_2, cp_7	1.638
(m_1, m_4)	$cp_3, cp_4, cp_5, cp_8, cp_9, cp_{10}$	0.9216
(m_4, m_5)	cp_3, cp_8	1.104
(m_4, m_6)	cp_5, cp_{10}	0.1294

if we iterate all possible conditions, there are $2^{k \cdot v}$ conditions since each edge can exist or not. By contrast, the worst case of the performance model depends on the largest parallel degree between start vertex to end vertex, which equals $O(k^v)$. We can find that the performance model reduces the time complexity from $O((2^k)^v)$ to $O(k^v)$.

3.3 Affinity Degree Calculation

The affinity degree of a pair of microservice components are defined as its weight multiplying with the RPC cost between them. About the weight, we set it as the sum of the probabilities of critical paths where this pair of components appeared. We show the algorithm of affinity degree calculation by Algorithm 3. The weights of module pairs are shown in Table 2.

Algorithm 3. To get weights for module pairs

Require: $CP(G_m)$ and $PCP(G_M)$ for a microservice workflow G_m
Ensure: a list of module pairs *pp* and their corresponding weights $w(m_u, m_v)$.
pp = [] {Initialize a list to record the module pairs}
for s_i in $CP(G_m)$ **do**
  len = $|s_i|$
  for $j = 1$ to $len - 1$ **do**
    m_{from} = the jth vertex in s_i
    m_{to} = the $(j+1)^{th}$ vertex in s_i
    if pair (m_{from}, m_{to}) is in pp **then**
      $w((m_{from}, m_{to})) = w((m_{from}, m_{to})) + PCP(s_i) * C(m_{from}, m_{to})$
    else if pair (m_{to}, m_{from}) is in pp **then**
      $w((m_{to}, m_{from})) = w((m_{from}, m_{to})) + PCP(s_i) * C(m_{from}, m_{to})$
    else
      $w((m_{from}, m_{to})) = PCP(s_i) * C(m_{from}, m_{to})$
    end if
  end for
end for

---

4 Experimental Evaluation

In this section, we utilize the microservice applications to evaluate the accuracy of our performance model and show the performance of the proposed optimization algorithm utilizing the workflow-aware weights by comparing it with three other weight designs.

4.1 Experiment Setting

We deploy our experiment workload to the cloud instances of ByteDance Cloud. About the software stack, the cloud instance is implemented by a framework similar to Kubernetes. We implement the RPC communication layer with the support of Apache's Thrift[6], and a tool based on kitex[7] to help with the development of the RPC layer.

We run up 5 applications for this experiment, denoted as "App1" to "App5" with varying numbers of modules. App1 has 3 modules, App2 has 5 modules, App3 has 11 modules, App4 has 42 modules, and App5 has 73 modules. For each application, we warm up the context by 100 repeated requests and record the running time of the following 1000 requests to be the observed data.

4.2 Accuracy Validation on the Model

Experimental Design. The workflow model is expected to estimate the running time. For validating our model, in this section, we compare the estimated results of the model to the actual running result to show its accuracy.

The current information about the RPC cost of the experiment applications is the message size. While the impact variables on the network cost have been revealed [8], including the impact of the message size and network environment, we profile the relationship between message size and RPC cost for our experiment environment to predict the RPC cost and would not discuss further on the impact of other variables (like the data structure variation and network connection setting) on RPC cost in this paper.

Experimental Result. We set the message that needs to communicate between two microservice modules as a string with a specified length and set the network communication based on a long connection that can be reused without the cost of the TCP three-way handshake. We change the request message size from 0 to 400 KB with the step of 40 KB and record the RPC cost as shown in Fig. 4. The RPC cost does not start from zero because it needs to set connection with the remote machine, and zero string length of the message still needs wrap headers to get through network layers (such as TCP, IP, and MAC). The RPC cost increasing with message size can be caused by increasing network and encoding/decoding cost. We apply the function between message size and RPC cost to predict the running time in the following step.

---
[6] https://github.com/cloudwego/thriftgo.
[7] https://github.com/cloudwego/kitex.

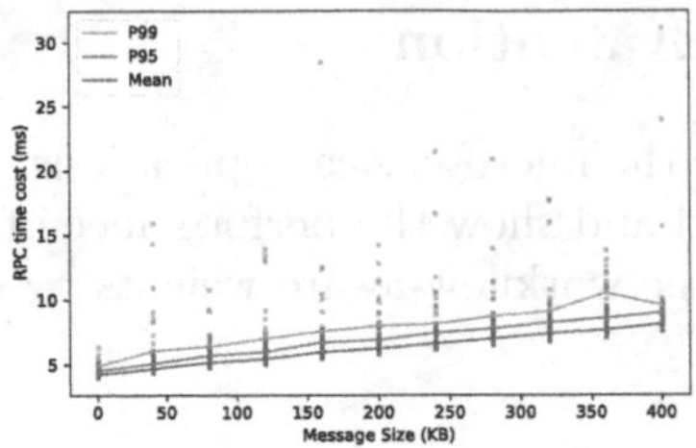

Fig. 4. The impact of message size on the RPC cost

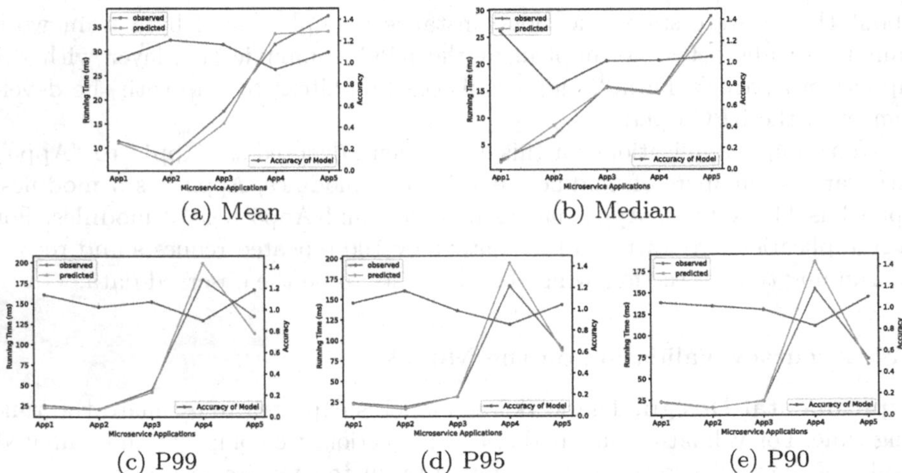

Fig. 5. Comparison between observed running time and predicted values of our performance model from five dimensions

We apply our performance prediction model on these experiment applications to compare with their running result. The comparisons of these dimensions between observed running results and predicted values of our performance model are shown in Fig. 5. The average accuracy values of the estimation method are 103.0% of Median, 109.9% of Mean, 106.4% of P99, 102.3% of P95, and 99.5% of P90. Such results indicate the high accuracy of the proposed performance model.

4.3 Effectiveness Validation on Performance Optimization Method

Experiment Design. The performance optimization method is used to find the most effective combination groups to co-locate them. The formerly used weights for constructing weighted graphs mainly include the number of invocations and the communication cost [5,6]. Thus we compare the effect of our workflow-aware method (W) with four other weight designs: (1) C: the RPC cost between components, (2) Q: the throughput between components quantified by the Queries Per Second (QPS) dimension, (3) QC: the product of these two previous dimensions $QPS * C$, and (4) R: randomly generated value as the baseline.

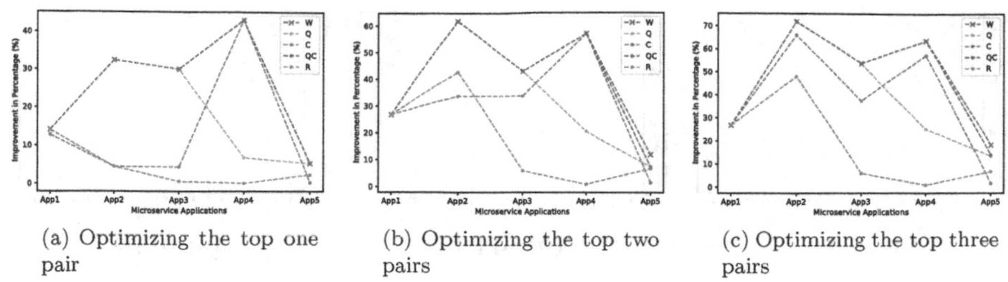

(a) Optimizing the top one pair

(b) Optimizing the top two pairs

(c) Optimizing the top three pairs

Fig. 6. The effect of improvement when optimizing the most effective pairs

We optimize the RPC cost of a component group by compiling them together as a single module, which converts the remote calls to local calls and saves all the RPC costs among this group. The effects of other optimization strategies are not included in this paper. The following running time data is all based on this optimization strategy.

Experiment Result. We denote the optimization effect by the running time improvement in percentage to the original running time for these experiment applications. The improvements of the average running time of these five applications after combining the top one to three most effective pairs are shown in Fig. 6. We can find our design W has the best performance, QC approaches W's performance, W has better performance than C and Q designs, while the R has the worst performance. The reason for QC approaching to W's performance is that the module pair with large throughput and high RPC cost tends to appear in the critical path. The weight design of W contains the considerations of QC design.

But there is still a performance gap between QC and our W design, as shown by the results of "App5" in the Fig. 6b and Fig. 6c. To show the effect of our optimization method on copying with complex structures, we thus explore the optimization further on applications with large number of modules: the "App3", "App4", and "App5". For avoiding too large to fit into Cloud running instance, we set the penalty limit as no more than 4 components in a group. We optimize the most effective module pair for each step chosen according to QC and W. The step-by-step optimization effects comparing the QC and W on these three applications are shown in Fig. 7. We can find that our optimization method has a larger improvement percentage than the QC design.

Furthermore, we find the first several steps contribute most of the improvement. In App3, the first 3 steps bring 58.4% improvement while the left steps bring 2.9% improvement. In App4, the first 5 steps bring 70% improvement while the left 15 steps bring 4.5% improvement. In App5, the first 6 steps bring 30.4% improvement while the left 15 steps bring 3.4% improvement. This result recommends that the optimization focusing on top effective module pairs would gain a high cost-performance ratio.

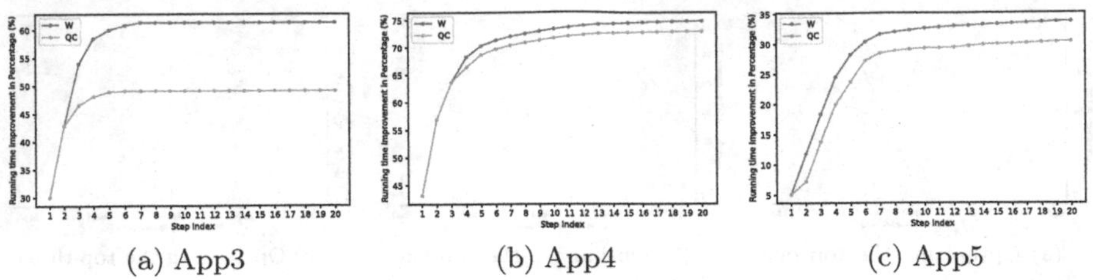

(a) App3 (b) App4 (c) App5

Fig. 7. Comparison between our optimization method W to another good performance design QC on step by step optimization process

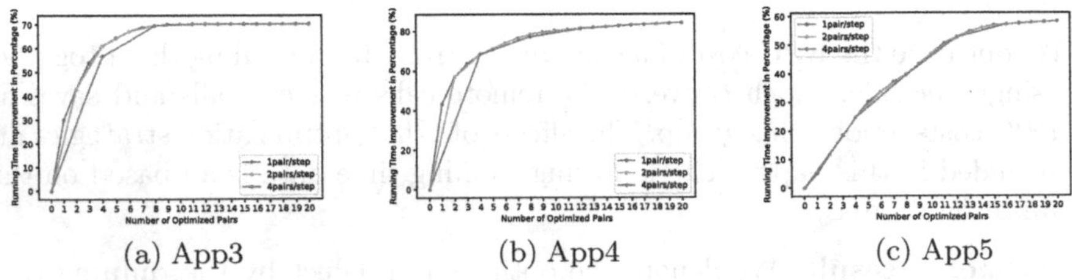

(a) App3 (b) App4 (c) App5

Fig. 8. The effect of changing the number of combination pairs per step

The optimization may change the probability of a module pair appearing in the critical path. The most accurate result is to re-calculate the weights to rank the module pairs after each optimization step (one pair per step). But to reduce the calculation cost, we compare the results of multiple pairs per step. The experiment is conducted on applications with enough modules to support the exploration. The results of "App3", "App4", and "App5" are shown in Fig. 8. We can find that, though there are performance differences like the 4 pairs combination result of "App3" and the 8 pairs combination result of "App4", the increasing number of pairs per step brings little performance degradation. We thus recommend it is no need to re-rank pairs after each optimization step.

5 Conclusion

We model the microservice workflow using the probability DAG graph from the time dependency view, which can effectively estimate the running time distribution of the workflow and proposes the running time optimization schema by co-locating microservice components with the considerations on workflow structure, which outperforms traditional weight designs.

As the first work introducing the time dependency workflow to online services, it emphasizes that workflow is a significant impact not only on computation-intensive jobs but also on online services. This work is an initial version design that calls for further contributions, like a more advanced clustering algorithm, more sophisticated penalty dimension design, and richer application practices, to fill the gap of container placement schema for running time optimization in a large-scale environment.

References

1. Bardhi, F., Arnould, E.J.: Thrift shopping: combining utilitarian thrift and hedonic treat benefits. J. Consum. Behav. Int. Res. Rev. **4**(4), 223–233 (2005)
2. Gollapudi, R.T., Yuksek, G., Ghose, K.: Cache-aware dynamic classification and scheduling for Linux. In: 2019 IEEE Symposium in Low-Power and High-Speed Chips (COOL CHIPS), pp. 1–3 (2019). https://doi.org/10.1109/CoolChips.2019.8721355
3. Hajji, M.A., Mezni, H.: A composite particle swarm optimization approach for the composite SAAS placement in cloud environment. Soft. Comput. **22**(12), 4025–4045 (2018)
4. Hedhli, A., Mezni, H.: A survey of service placement in cloud environments. J. Grid Comput. **19**(3), 1–32 (2021)
5. Hu, Y., de Laat, C., Zhao, Z.: Optimizing service placement for microservice architecture in clouds. Appl. Sci. **9**(21) (2019). https://doi.org/10.3390/app9214663, https://www.mdpi.com/2076-3417/9/21/4663
6. Huang, K.C., Shen, B.J.: Service deployment strategies for efficient execution of composite SAAS applications on cloud platform. J. Syst. Softw. **107**, 127–141 (2015). https://doi.org/10.1016/j.jss.2015.05.050, https://www.sciencedirect.com/science/article/pii/S0164121215001156
7. Iorgulescu, C., et al.: Perflso: performance isolation for commercial latency-sensitive services. In: 2018 USENIX Annual Technical Conference (USENIX ATC 2018), pp. 519–532. USENIX Association, Boston, July 2018. https://www.usenix.org/conference/atc18/presentation/iorgulescu
8. Jayasinghe, M., Chathurangani, J., Kuruppu, G., Tennage, P., Perera, S.: An analysis of throughput and latency behaviours under microservice decomposition. In: Bielikova, M., Mikkonen, T., Pautasso, C. (eds.) ICWE 2020. LNCS, vol. 12128, pp. 53–69. Springer, Cham (2020). https://doi.org/10.1007/978-3-030-50578-3_5
9. Nathuji, R., Kansal, A., Ghaffarkhah, A.: Q-clouds: managing performance interference effects for QoS-aware clouds. In: Proceedings of the 5th European Conference on Computer Systems. EuroSys 2010, pp. 237–250. Association for Computing Machinery, New York (2010). https://doi.org/10.1145/1755913.1755938
10. Patel, T., Tiwari, D.: Clite: efficient and QoS-aware co-location of multiple latency-critical jobs for warehouse scale computers. In: 2020 IEEE International Symposium on High Performance Computer Architecture (HPCA), pp. 193–206 (2020). https://doi.org/10.1109/HPCA47549.2020.00025
11. Qiu, H., Banerjee, S.S., Jha, S., Kalbarczyk, Z.T., Iyer, R.K.: FIRM: an intelligent fine-grained resource management framework for SLO-oriented microservices. In: 14th USENIX Symposium on Operating Systems Design and Implementation (OSDI 2020), pp. 805–825. USENIX Association, November 2020. https://www.usenix.org/conference/osdi20/presentation/qiu
12. Samanta, A., Li, Y., Esposito, F.: Battle of microservices: towards latency-optimal heuristic scheduling for edge computing. In: 2019 IEEE Conference on Network Softwarization (NetSoft), pp. 223–227 (2019). https://doi.org/10.1109/NETSOFT.2019.8806674
13. Shi, T., Ma, H., Chen, G., Hartmann, S.: Location-aware and budget-constrained service deployment for composite applications in multi-cloud environment. IEEE Trans. Parallel Distrib. Syst. **31**(8), 1954–1969 (2020). https://doi.org/10.1109/TPDS.2020.2981306

14. Sigelman, B.H., et al.: Dapper, a large-scale distributed systems tracing infrastructure. Technical report, Google, Inc. (2010). https://research.google.com/archive/papers/dapper-2010-1.pdf
15. Talwar, V.: GRPC: a true internet-scale RPC framework is now 1.0 and ready for production deployments (2016). https://cloud.google.com/blog/products/gcp/grpc-a-true-internet-scale-rpc-framework-is-now-1-and-ready-for-production-deployments
16. Tanković, N., Galinac Grbac, T., Žagar, M.: Elaclo: a framework for optimizing software application topology in the cloud environment. Expert Syst. Appl. **90**, 62–86 (2017) https://doi.org/10.1016/j.eswa.2017.07.001, https://www.sciencedirect.com/science/article/pii/S0957417417304700
17. Wan, X., Guan, X., Wang, T., Bai, G., Choi, B.Y.: Application deployment using microservice and docker containers: framework and optimization. J. Netw. Comput. Appl. **119**, 97–109 (2018). https://doi.org/10.1016/j.jnca.2018.07.003, https://www.sciencedirect.com/science/article/pii/S1084804518302273
18. Wieczorek, M., Hoheisel, A., Prodan, R.: Taxonomies of the Multi-criteria Grid Workflow Scheduling Problem, pp. 237–264. Springer, Boston (2008). https://doi.org/10.1007/978-0-387-78446-5_16
19. Yu, Y., Yang, J., Guo, C., Zheng, H., He, J.: Joint optimization of service request routing and instance placement in the microservice system. J. Netw. Comput. Appl. **147**, 102441 (2019). https://doi.org/10.1016/j.jnca.2019.102441, https://www.sciencedirect.com/science/article/pii/S1084804519303017
20. Zhu, Q., Tung, T.: A performance interference model for managing consolidated workloads in qos-aware clouds. In: 2012 IEEE Fifth International Conference on Cloud Computing, pp. 170–179 (2012). https://doi.org/10.1109/CLOUD.2012.25

Data Mining

SynBERT: Chinese Synonym Discovery on Privacy-Constrain Medical Terms with Pre-trained BERT

Lingze Zeng[1], Chang Yao[1](✉), Meihui Zhang[2], and Zhongle Xie[1]

[1] Zhejiang University, Hangzhou, China
{chang.yao,0620515}@zju.edu.cn
[2] Beijing Institute of Technology, Beijing, China
meihui_zhang@bit.edu.cn

Abstract. Discovering medical synonym sets (i.e.,set of terms referring to a similar medical concept) is an important task in real-world, which can benefit many downstream applications such as medical information retrieval system and clinical decision support system. Recent synonym discovery methods take words as the input unit and leverage raw text as contextual information. However, they are ill-suited in Chinese participle as taking word as the input unit leads to serious Out-of-Vocabulary (OOV) problems. Additionally, it is hard to get large-scaled raw clinical texts in medical domain because of the privacy and security. Therefore, we define a new task discovering Chinese synonym from Privacy-Constrain terms (i.e., only terms without raw corpus) and propose a framework *SynBERT* to solve it. *SynBERT* consists of a binary classifier, inferring whether two term sets can form a synonym set, and two-phase clustering algorithm, applying classifier to cluster given terms into different synonym sets. In particular, *SynBERT* composes term's embedding with character's embedding to address the OOV problems. *SynBERT* introduces a BERT model pre-trained on public large-scaled corpus before to eliminate the need of raw context information. According to our experiment, *SynBERT* outperforms better than baseline methods such as Kmeans, L2C, SynSetMine, etc.

Keywords: Data mining · Information extraction · Information retrieval

1 Introduction

Synonym discovery, which is to identify terms referring to the same or similar real-world concept, is a widely-used technique in text analysis [1], question answering [2] and web searching [3]. In the medical domain, discovering synonyms from clinical texts in Electronic Medical Records (EMRs) is an important task which not only helps researchers to automatically build medical knowledge base, but also benefits many downstream applications such as medical information retrieval system and clinical decision system. For instance, when physicians issue a query (e.g., "flu") to find relevant information, the discovered synonym set (e.g. "influenza", "catarrh") can expand the query and thereby improve the retrieval performance.

B. Li et al. (Eds.): APWeb-WAIM 2022, LNCS 13421, pp. 331–344, 2023.
https://doi.org/10.1007/978-3-031-25158-0_25

In this paper, we study a new problem named *Chinese Synonym Discovery on Privacy-Constrain Data*, which is more restrictive than normal synonym discovery in the following two points.

(1) Out-of-Vocabulary (OOV) problem. Due to the variety of Chinese phrase combinations and the difficulty in Chinese term segmentation, extracting many strange and unseen words from corpus is inevitable. Additionally, transliteration of professional term in specific field will create different Chinese words. Take the *amoxycillin* as an example, Chinese term corresponding to it is translated according to the pronunciation. Thus, "阿莫西林", "安莫西林" are all right. How to precisely represent the meaning of these OOV words makes synonym discovery harder.
(2) Lack of context corpus. Unstructured clinical texts in Electronic Medical Records (EMRs) contain rich medical information, meanwhile, it also includes patient's private information, such as patient-centered narratives and patient's disease treatment. Considering the privacy and security, many institutes only release medical terms that are extracted from the clinical corpus, without the large-scaled raw clinical texts. In this paper, we refer these given terms without context corpus as Privacy-Constrain data.

Regarding to the above two challenges, the problem can be defined specifically: *Given a set of terms extracted from Chinese medical raw texts, cluster these terms into different synonym sets.* Investigating effective methods under this setting is particularly meaningful since data privacy issues are getting more attention with the development of artificial intelligence. How to utilize less sensitive information but accomplish the task with high performance becomes essential in real-world applications.

For synonym discovery, there are two types of common methods. The first type [4,5,21] is directly mapping given terms to standard entities from existing knowledge bases (KBs) and gathering terms linking to the same entity as a synonym set. However, only synonyms related to existing knowledge bases can be discovered while the terms that the knowledge base do not cover can not be grouped to a synonym set. The problem is further severe in the Chinese medical domain, considering the incomplete and outdated knowledge base. Another type of approaches [6,7] focuses on automatically discovering synonyms from an enormous volume of domain corpus. They usually leverage the local context (sentences or paragraphs mentioning terms) to predict whether two terms can form a synonym set. For privacy-constrain data, it can not work well since the large-scaled corpus is unavailable.

In order to break through the aforementioned limitations, we propose a novel framework named *SynBERT* which consists of two modules, a supervised binary-Classifier called SynSet classifier and a synonym generation algorithm. (1) The SynSet classifier is trained to predict whether two synonym set can be combined. In this module, we take Chinese characters as the input unit and combine them to generate term's embedding, which greatly relieves OOV problems. It also works well for detecting synonyms that look similar in character-level. Additionally, we introduce BERT [8] as main body of classifier, which is pre-trained on large-scaled corpus before. It plays a complementary role to represent the context information of terms. Different from previous methods which predict term pairs separately, *SynBERT* emphasizes the holistic meaning of a synonym set, directly compare two synonym sets. (2) The synonym generation algorithm is to detect synonym sets from given terms applying the SynSet classifier. It consists of two phases. The first step is cluster initialization, which enumerates all given terms once and generates initialized synonym sets. The second step is cluster combination, which enumerates all initialized synonym set and combines related synonym sets. To summary, our contribution:

- We come up with a new concept in synonym discovery, Privacy-Constrain data which refer to terms without contextual text. It's meaningful to research methods to mine synonyms from them, as large-scaled data related to people's privacy is in accessible in many cases.
- We propose an effective framework named *SynBERT* to discover Chinese synonym set from Privacy-Constrain Data. *SynBERT* trains the SynSet classifier and groups given terms into different synonym sets. In particular, it works well in the new setting regardless of the OOV problem and lack of large corpus.
- We conduct multiple experiments and case studies on Privacy-Constrain Data in medical domain. The results show that *SynBERT* outperforms all baseline methods and therefore prove the effectiveness of our framework.

2 Concept Introduction

In this part, We introduce basic concepts and their notations, then present the definition of the ultimate task.

Term. A word or a phrase which is a kind of expression of an exact concept in the real world.

Privacy-Constrain Data. A set of terms without contextual text. Usually, these terms are extracted from field-specific raw text involving private information, such as Electronic Medical Records (EMRs). Due to the security and privacy requirement, the contextual text is unavailable.

Synonym Set. A set of terms that refer to the same or similar real-world concept. Take synonym set ("flu","influenza", "H1N1-flu") as an example, terms in it refer to the same real-world disease influenza. "flu" is the alternative name and "H1N1-flu" is a specific influenza.

Synonym Candidate. Synonym Candidate is a pair of two synonym sets. It is the input of the SynSet classifier, which will predict whether they can form a new synonym set. In particular, synonym set is allowed to contain only one term.

Task Definition. Given terms (i.e. Privacy-Constrain Data) V and a group of synonym sets C, the task is to leverage the distant supervision of C to train the SynSet classifier $F(x, y)$, then cluster terms from V into different synonym sets applying this SynSet classifier $F(x, y)$

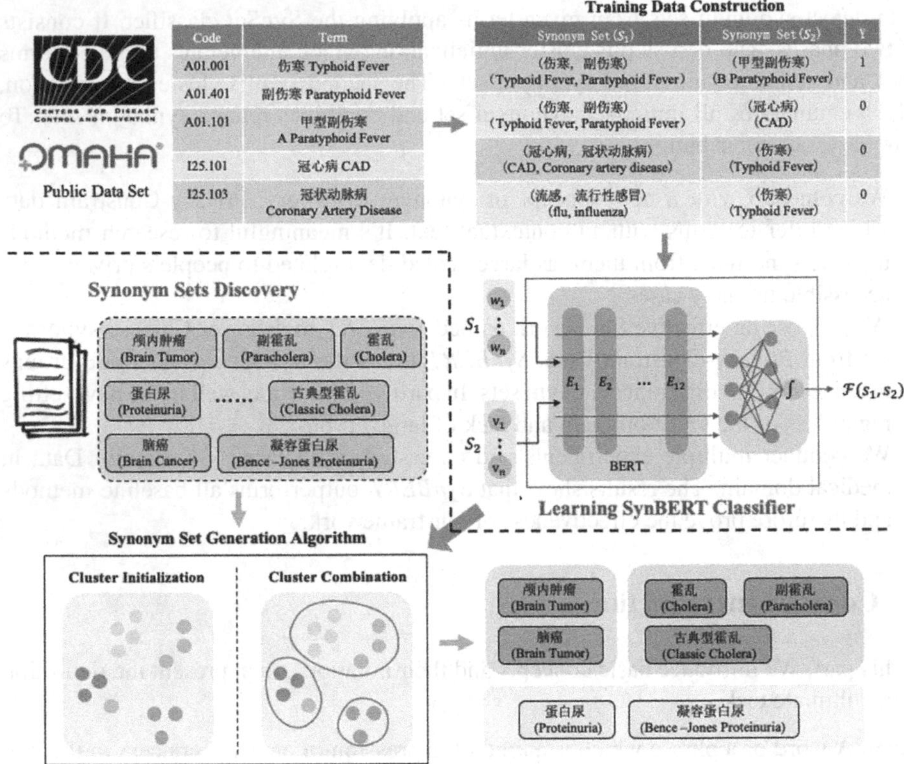

Fig. 1. Framework overview

3 Framework Introduction

We introduce the framework *SynBERT* in this section. Figure 1 shows the workflow of our framework, which contains two stages. In the first stage, a binary classifier called SynSet is trained with supervised data, which can predict whether two synonym sets can form a new synonym set. The second stage is applying the SynSet classifier trained before to detect different synonym sets according to the synonym set generation algorithm. Each stage is elaborated in the following sections.

3.1 Training Binary Classifier

Generation of Training Data. The first step of training SynSet classifier is sampling synonym candidate to construct training data from existing supervised datasets. Existing public datasets, like Chinese Symptom Knowledge Base (CSKB) [9], Open Medical and Healthcare Alliance (OMaha) [10], contain many standard medical terms extracted from the large volume of Electronic Medical Records (EMRs). These terms are manually clustered into synonym sets by research institution. Furthermore, certain datasets, like ICD-10 [11], do not have synonyms, but every term has a standard code [12]. We can simply treat the terms with the same codes as synonym sets. However, these synonym sets can not be used to train SynSet classifier directly. We have to sample synonym candidate from synonym sets C. For each synonym set S, randomly choose a subset S_{sub}, where $1 \leq Size_{S_{sub}} \leq Size_S - 1$. In the left set $S_{left} = S - S_{sub}$, randomly choose a term i as an i_{pos}, consider it as a synonym set $S\{i_{pos}\}$ and construct a positive synonym candidate, denoting it as $(S_{sub}, S\{i_{pos}\}, 1)$. For each positive synonym candidate, generate K negative synonym candidates. Randomly select K negative terms $i_0, i_1 ... i_{k-1}$ in term pool P, where $P = S_0 \cup S_1 ... \cup S_n, \forall j, 0 \leq j \leq n, S_j \in C - S$. Then pair each of these terms with S_{sub} to construct negative synonym candidate, denoting it as $(S_{sub}, S\{i_j\}, 0), 1 \leq j \leq k - 1$.

K controls the ratio of positive and negative synonym candidates. We study how the negative sample size K influences the model performance in later experiments.

Architecture of Classifier. After constructing the training data, we introduce the architecture and the training process of the model. In order to predict whether two synonym sets can form a new synonym set, one key requirement is that the each term from one synonym set should be related to terms from another synonym set. Intuitively, one way [1] is to decompose the set-set prediction into multiple term-term relations, then judge the relations between two terms and finally aggregate all result to get the final decision. However, SynSetMine [7] points out that the holistic information of a synonym set is ignored if following decomposing method. It focuses on the holistic change on synonym set when two synonym set are combined. If the holistic change exceeds a threshold, it regards the two synonym sets can be combined. The disadvantage of SynSetMine [7] is that as the set increases, a newly added term change the holistic meaning of synonym set less, leading to collect unrelated term into synonym set.

In our task, *SynBERT* considers two types of synonym relationships. One is two synonyms are similar in character-level. Some terms are the misspelled form of another

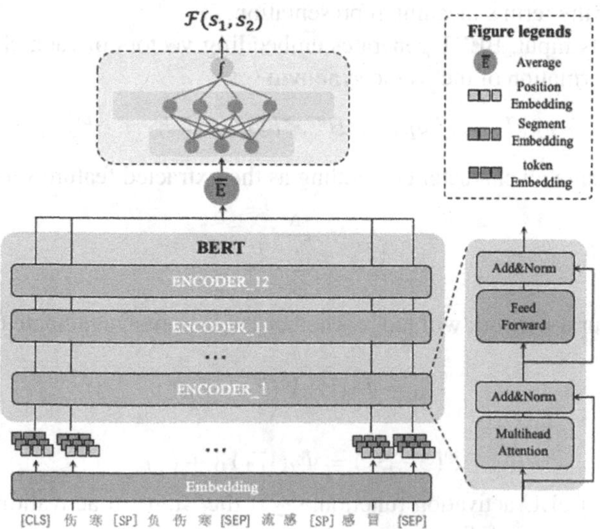

Fig. 2. SynSet classifier

term. *SynBERT* makes term's embedding out of characters, which is beneficial to detect the first type of relationship. When the difference in character-level between two synonyms is small, the similarity of term's embedding is high. Another relationship is that two synonyms are similar in semantics-level while different in character-level, which is more difficult, especially without raw text corpus. Thus, we choose the BERT [8] model as the main component of SynSet classifier. BERT pre-trained on a large generic corpus before, plays a complementary role to initialize the embedding of every character with contextual information.

Understanding the previous practices, we also emphasize the holistic meaning of a synonym set. The transformer module in BERT [8] can be considered as a good feature extraction for a synonym set. Since transformer is insensitive to the order of input data. Occasionally, the order of terms in a synonym set is meaningless. Thereby, not just a character encoder, BERT [8] is also used as a feature extraction for the whole synonym sets. Following the pattern in *Next Sentence Prediction (NSP) Task* proposed in BERT paper [8] to fine-tune BERT model, we regard a synonym set as a sentence of NSP task in training process.

In general, the classifier is composed of BERT model and a neural network, shown in Fig. 2. We join two synonym set with $[SEP]$ and $[CLS]$ to generate input data. In a synonym set, we use a special symbol $[SP]$ to split terms. Each character of a term consists of three embedding vectors, including token embedding, segment embedding, position embedding.

$$E^j = E^j_{token\_emb} + E^j_{segment\_emb} + E^j_{position\_emb} \tag{1}$$

The token embedding represents the initialized meaning of this character. The segment embedding can help BERT to distinguish the character belong to the first synonym set or the second synonym set. It is a tag in NSP task to mark which sentence the word belongs to. The position vector is to represent the different meaning brought by different positions between characters in a term since in one term, the order of characters makes sense for the term's meaning representation.

Based on this input, BERT generates embedding vectors of each character, which contains the information of the whole synonym set.

$$T_{CLS}, T_1, ..., T_{SEP} = BERT(E_{CLS}, E_1, ..., E_{SEP}) \tag{2}$$

We directly average all character embedding as the extracted feature vector.

$$V(T) = \sum_{0}^{N} T_j / N \tag{3}$$

Then the neural network will judge whether the synonym candidate can form a new synonym set

$$Y_0 = \Phi_1(W_0 V(T) + b_0) \tag{4}$$

$$F(S^1, S^2) = \Phi_2(W_1 Y_0 + b_1) \tag{5}$$

where Φ_1 is the ReLU activation function, Φ_2 is the Sigmoid activation function. The calculation details are as follows:

$$\Phi_1(x) = Max(0, x) \tag{6}$$

Algorithm 1. Cluster Initialization algorithm

Input: A SynSet classifier F; Given terms $V = (s_1, s_2, ..., s_{|V|})$; A threshold $\theta \in [0, 1]$
Output: m synonym sets $C = [C_1, C_2, ..., C_m]$ where $C_i \subseteq V, C_i \cap C_j = \emptyset, \forall i \neq j$.

Initialize $C = [s_1]$
for i from 2 to $|V|$ **do**
    $best\_score = 0$; $best\_j = 1$;
    for j from 1 to $|C|$ **do**
        if $F(C_j, s_i) > best\_score$ **then** $best\_score = F(C_j, s_i)$; $best\_j = j$;
    if $best\_score > \theta$ **then** $C_{best\_j}.add(s_i)$;
    else$C.append(s_i)$;
return C;

$$\Phi_2(x) = \frac{1}{1 + e^{-x}} \tag{7}$$

Given a collection of m synonym candidate $\{(S_j^1, S_j^2, y_j)|_{j=0}^N\}$, we learn the SynSet classifier using the Binary Cross Entropy as follows:

$$L_j = \sum_{j=0}^{N} -y_i log(P) - (1 - y_j) log(1 - P) \tag{8}$$

$$P = F(S_j^1, S_j^2) \tag{9}$$

where y_j equals to 1 if this synonym candidate is positive and equals to 0 otherwise.

3.2 Synonym Sets Discovery

In this section, we present how to apply the SynSet classifier to cluster terms into different synonym sets. The whole process is divided into two phases, i.e., cluster initialization and cluster combination.

Cluster Initialization. We present cluster initialization in Algorithm 2. This algorithm takes the above learned SynSet classifier, given terms (e.i., Privacy-Constrain Data) V, and a threshold θ as input and outputs synonym sets from given terms. This algorithm enumerates the vocabulary V once and maintains a pool of all detected sets C. For each word $s_i \in V$, it applies the SynSet classifier F to calculate the probability of adding this term into each detected set in C and finds the best set C_j with the largest probability. If this probability value is larger than the threshold θ, we will add s_i into set C_j. Otherwise, we create a new set s_i with this single term and add it into the set pool C. The entire algorithm stops after one pass of all terms and returns all detected sets in C.

Algorithm 2. Cluster Combination algorithm

Input: A SynSet classifier F; Initialized synonym sets $C = [C_1, C_2, ..., C_m]$; A threshold $\theta \in [0, 1]$

Output: m synonym sets $C^{'} = [C_1{'}, C_2{'}, ..., C_k{'}]$ where $C_i^{'} \cap C_j^{'} = \emptyset, \forall i \neq j$.

Initialize $C^{'} = \emptyset$
while $C \neq \emptyset$ **do**
  Random Choose C_p where $C_p \in C$
  $C.pop(C_p)$; $L = [C_p]$; $C_k^{'} = \emptyset$;
  for all $c \in C$ **do**
    if $F(C_p, c) > \theta$ **then** $L.append(c)$; $C.pop(c)$
  for all $l \in L$ **do**
    $C_k^{'} = C_k^{'} \cup l$; $C^{'}.append(C_k^{'})$
return $C^{'}$;

Cluster Combination. Cluster Initialization only enumerate all terms once. If a new term cluster is not gathered into correct synonym set, it will lead to error accumulation in clustering later terms. Therefore, Cluster Combination plays a double-check for the final result, in which all related synonym sets are gathered into a new synonym set. We present the algorithm in Algorithm 2. This algorithm enumerates the synonym sets C and maintains a pool of new synonym sets. For one synonym set C_p, we apply the SynSet Classifier F on it with other all synonym sets in C. Collect all synonym sets whose result is larger than threshold θ, and group them with C_p into a new synonym set $C_k^{'}$. Add $C_k^{'}$ into pool of new synonym sets C. This algorithm stops after one pass of all synonym sets and returns combined synonym sets.

4 Experiment

4.1 Baseline Methods

We compare **SynBERT** with the following baselines.

KMeans [13] is an unsupervised clustering algorithm. This algorithm requires a predefined cluster number K and we set its value to the oracle number of clusters for each test dataset.

DBSCAN [14] is an unsupervised clustering algorithm based on density, which takes the term embedding as features and discovers sets in spatial space.

GMM [15] is Gaussian Mixture Model, a generalized Kmeans model. We set n_c to the oracle number of clusters for each dataset.

AC [16] is Hierarchical agglomerative clustering algorithm. We set n_c value to the oracle number of clusters for each dataset.

L2C [17] is a supervised clustering method that learns a pairwise similarity prediction neural network and a constrained clustering network on training sets. Then, it applies the mode on test datasets to detect new synonym sets.

SynSetMine [7] is a supervised clustering method which trains a set-instance classifier and integrates it seamlessly into an efficient set generation algorithm.

Table 1. Dataset

Dataset	MSKB	MDKB	HITSyn
Synonym sets in train	167	4615	1365
Terms in train	442	17253	6254
Synonym sets in test	42	136	68
Terms in test	232	1124	531

BERT-SynSetMine [7] is a supervised clustering method based on **SynSetMine**. In implementation, we apply pre-trained model BERT as the scorer part in SynSetMine model.

4.2 Dataset Construction

We evaluate *SynBERT* on three Chinese Privacy-Constrain datasets, shown in Table 1.

- **MSKB** Terms are Chinese medical symptom like "牙龈肿痛 "(Swollen gums) extracted from CSKB [9]
- **MDKB** Terms are Chinese medical disease name like "流行性感冒 "(influenza) extracted from OMaha [10] and ICD-10 [11].
- **HITSyn** Terms related to medical domain are from Chinese Synonym Dictionary.

4.3 Terms Embedding and Parameter Setting

Since the lack of corpus, it is difficult to get word embedding following traditional method like Word2Vec. To get a fair comparison, *SynBERT* choose to use pre-trained BERT model to generate word embedding. Characters of a term are entered into BERT. The $[CLS]$ output is taken as the embedding of the whole term. In our work, we take *bert-wwm-ext* [18] as pre-trained model and the hidden size of the two-layer neural network is 256. We optimize our model using *Adam* with initial learning rate $1e^{-5}$ and apply the dropout technique with a dropout rate of 0.3. For the clustering algorithm, we set the probability threshold θ to be 0.5. We will discuss the influence of these hyper-parameters later.

4.4 Evaluation Metrics

For all the methods, we evaluate them by three standard clustering evaluation metrics.

ARI [19] measures the similarity of two cluster assignments. Given the ground truth cluster assignment C^* and model predicted cluster assignment C, we use $TP(TN)$ to denote the number of element pairs that are in the same (different) cluster in both C^* and C. We denote the total number of element pairs in C^* as N. E_{RI} is the expected RI of random assignments.

$$ARI = \frac{RI - E_{RI}}{max(RI) - E_{RI}}, RI = \frac{(TP + TN)}{N} \tag{10}$$

Table 2. Clustering result

Method	MSKB			MDKB			HITSyn		
Metrics	ARI	NMI	FMI	ARI	NMI	FMI	ARI	NMI	FMI
Kmeans	47.22	81.16	48.64	65.34	90.60	65.69	30.73	74.45	32.45
DBSCAN	44.70	83.94	47.28	52.00	89.91	52.47	15.61	77.95	20.40
GMM	43.49	79.35	45.35	63.92	90.20	64.30	30.73	74.45	32.45
AC	49.84	82.11	51.07	69.13	92.00	69.43	39.26	77.19	40.35
L2C	17.44	71.79	25.14	14.68	62.37	44.45	10.95	50.25	24.80
SynSetMine	18.08	80.59	24.91	59.15	89.26	59.68	25.19	78.98	29.35
SynSetMine*	52.50	88.65	57.81	79.18	95.28	80.22	50.98	87.74	55.54
SynBERT	**71.76**	**91.25**	**72.93**	**85.59**	**96.70**	**85.87**	**61.15**	**88.82**	**62.33**

FMI [20] Fowlkes-Mallows index uses $FP(FN)$ to denote the number of elements pairs that belong to the same cluster in $C^*(C)$ but in different cluster in $C(C^*)$. Calculate FMI as follows.

$$FMI = \frac{TP}{\sqrt{(TP+FP)(TP+FN)}} \tag{11}$$

NMI calculates the normalized mutual information, which is widely used in literature. The calculation details can be found in [22].

4.5 Experiment Results

The clustering performance of all methods are shown in Table 2. *SynBERT*'s performance is far better than other baseline methods across all three datasets. Without supervision from the training set, the unsupervised baseline methods all have stable performance. Obviously, the pre-setting cluster number K brings great improvement. On the contrary, supervised methods perform bad, especially in small dataset. The reason is that the limited expressive power of neural network without enough training data. SynSetMine's result is better than L2C's, since L2C is based on pairwise similarity while SynSetMine has a holistic view of set. BERT-SynSetMine performs far better than SynSetMine, proving that transformer structure is a good feature extractor of synonyms set.

For further analysis, we conduct three different case study experiments.

The result of ablation study is shown in Table 3, which compares the necessity of the different parts of model. Without pre-trained BERT module, *SynBERT*'s performance is reduced by 20%, 5%, 21% in ARI, NMI, FMI, but is still better than all supervised baseline methods. It demonstrates that transformer in BERT [8] can be a good feature encoder for synonym set. With frozen pre-trained BERT module, although the performance is reduced slightly, *SynBERT* still outperforms other baseline method. It proves that the pre-trained process is necessary for our task. The full connected layer improves the performance, but only a little bit.

Table 3. Ablation study

Method	MDKB		
	ARI	NMI	FMI
SynBERT	84.13	96.11	85.32
BERT-No-Pretrained	64.70	91.21	64.97
BERT-Freeze	72.21	93.74	72.46
BERT-No-FCLayers	83.11	96.08	83.50

Table 4. Dataset size study

Dataset size	MDKB			HITSyn		
	ARI	NMI	FMI	ARI	NMI	FMI
1	84.13	96.11	85.32	61.72	88.01	62.3
0.75	85.25	96.59	85.47	61.67	88.50	62.52
0.5	83.22	96.22	84.69	60.95	88.03	61.62
0.25	83.22	96.25	83.5	60.13	87.46	60.27
0.1	81.24	95.74	81.42	55.70	86.85	56.56

In hyper-parameter study, we analyze the influence of the negative sampling size K in generation of synonym candidate and the probability threshold θ. In Fig. 3, generating more negative samples can slightly improve the performance of *SynBERT*. However, when K is larger than 50, it will lead to decrease in experiment result. Figure 3 shows that when the θ is between 0.45 and 0.55, *SynBERT* can get a better result in these two datasets.

To demonstrate that *SynBERT* doesn't rely too much on large distant supervision, We truncate the training datasets *MDKB* and *HITSyn* by percentage and test the performance of *SynBERT*. Results are shown in Table 4. With the decrease of the size of the datasets, SynBERT's performance is reduced slightly. However, compared to other baseline methods, it still has the best results in this task.

5 Related Work

5.1 Medical Entity Linking

Medical entity linking is an essential task in text mining that maps the entity mentions in the medical text to standard entities in a given knowledge base. This task is of great importance in the medical domain. It can be used for merging different medical and clinical ontologies. [4] extracts medical entities from MIMIC-III and automatically link them with the existing biomedical knowledge graphs, including ICD-9 ontology. [23] uses the category of entities in ICD-9 as a constrain to solve Multi-implication's entity normalization. [24] generates entity candidates using a robust and portable candidate

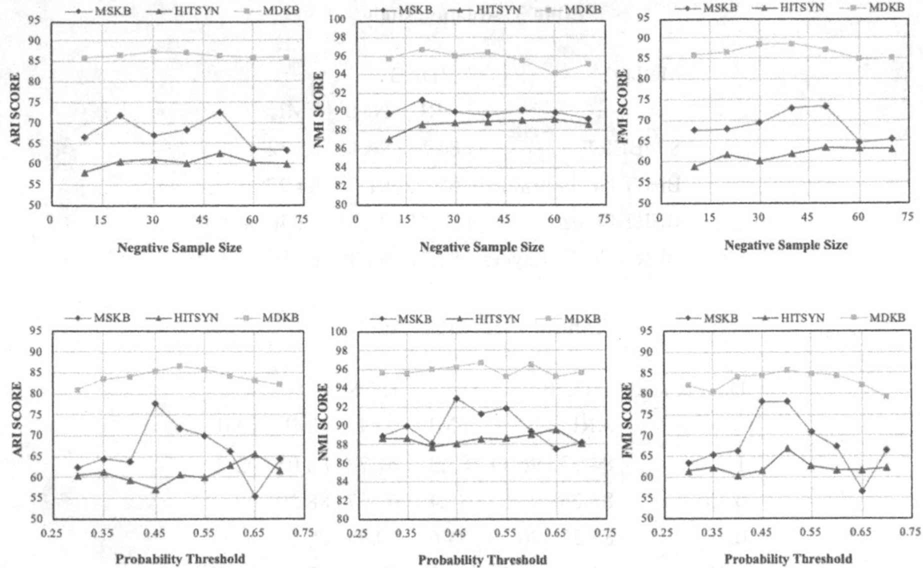

Fig. 3. Hyper-parameter study

generation scheme. then it makes use of the Triplet network to ranking extracted candidates according to a standard entity. These methods attempt to group their entities based on a standard knowledge base. However, the limitation of using these methods to discover synonyms is the difficulty to find a new synonym out of standard entities.

5.2 Entity Synonym Discovery

Entity synonym discovery is a crucial task in NLP, and many efforts have been invested. One straightforward approach is to obtain synonyms from public knowledge database, such as WordNet [5], ConceptNet [25]. Additionally, some works try to discover synonyms from a raw text corpus. For example, SynonymNet [6] proposed a multi-context bilateral matching framework, which extract multi-dimensional context information. SynSetMine [7] generate supervision in knowledge database and train a set-instance classifier to discover synonyms. The challenge to apply these methods in the medical domain is the lack of context information, considering the privacy of medical data.

6 Conclusion

In this paper, we study the problem discovering synonym sets from Privacy-Constrain Data (i.e., terms without contextual text)in Chinese medical domain. We propose a practical framework named *SynBERT*, which leverages supervision to train a SynSet classifier and applies it in an two-phase synonym set generation algorithm to cluster given terms. We conduct extensive experiments and *SynBERT* substantially outperforms various baselines on three real-world datasets. For future work, we will extend *SynBERT* to generic domain and test it on datasets from different fields.

References

1. Wang, Z., Yue, X., Moosavinasab, S., Huang, Y., Lin, S., Sun, H.: SurfCon: synonym discovery on privacy-aware clinical data. In: Proceedings of the 25th ACM SIGKDD International Conference on Knowledge Discovery and Data Mining, pp. 1578–1586 (2019)
2. Zhou, G., Liu, Y., Liu, F., Zeng, D., Zhao, J.: Improving question retrieval in community question answering using world knowledge. In: Twenty-Third International Joint Conference on Artificial Intelligence (2013)
3. Cheng, T., Lauw, H.W., Paparizos, S.: Entity synonyms for structured web search. IEEE Trans. Knowl. Data Eng. **24**(10), 1862–1875 (2011)
4. Wang, M., et al.: PDD graph: bridging electronic medical records and biomedical knowledge graphs via entity linking. In: d'Amato, C., et al. (eds.) ISWC 2017. LNCS, vol. 10588, pp. 219–227. Springer, Cham (2017). https://doi.org/10.1007/978-3-319-68204-4_23
5. Fellbaum, C.: Wordnet. In: The Encyclopedia of Applied Linguistics (2012)
6. Zhang, C., Li, Y., Du, N., Fan, W., Yu, P.S.: Synonymnet: multi-context bilateral matching for entity synonyms. arXiv preprint arXiv:1901.00056 (2018)
7. Shen, J., Lyu, R., Ren, X., Vanni, M., Sadler, B., Han, J.: Mining entity synonyms with efficient neural set generation. In: Proceedings of the AAAI Conference on Artificial Intelligence, vol. 33, pp. 249–256 (2019)
8. Devlin, J., Chang, M.-W., Lee, K., Toutanova, K.: BERT: pre-training of deep bidirectional transformers for language understanding. arXiv preprint arXiv:1810.04805 (2018)
9. Ruan, T., et al.: An automatic approach for constructing a knowledge base of symptoms in Chinese. J. Biomed. Semant. **8**(1), 33 (2017)
10. Martin, K.S., Scheet, N.: The Omaha system. Appl. Commun. Health Nurs. 1992 (2005)
11. Hirsch, J., et al.: ICD-10: history and context. Am. J. Neuroradiol. **37**(4), 596–599 (2016)
12. Quan, H., et al.: Coding algorithms for defining comorbidities in ICD-9-CM and ICD-10 administrative data. Med. Care, 1130–1139 (2005)
13. Hamerly, G., Elkan, C.: Learning the k in k-means. Adv. Neural. Inf. Process. Syst. **16**, 281–288 (2004)
14. Schubert, E., Sander, J., Ester, M., Kriegel, H.P., Xu, X.: DBScan revisited, revisited: why and how you should (still) use DBScan. ACM Trans. Database Syst. (TODS) **42**(3), 1–21 (2017)
15. Liu, J., Cai, D., He, X.: Gaussian mixture model with local consistency. In: Proceedings of the AAAI Conference on Artificial Intelligence, vol. 24, no. 1 (2010)
16. Beeferman, D., Berger, A.: Agglomerative clustering of a search engine query log. In: Proceedings of the Sixth ACM SIGKDD International Conference on Knowledge Discovery and Data Mining, pp. 407–416 (2000)
17. Hsu, Y.-C., Kira, Z.: Neural network-based clustering using pairwise constraints. In: ICLR workshop (2016). https://arxiv.org/abs/1511.06321
18. Cui, Y., et al.: Pre-training with whole word masking for Chinese BERT. arXiv preprint arXiv:1906.08101 (2019)
19. Steinley, D.: Properties of the Hubert-arable adjusted rand index. Psychol. Methods **9**(3), 386 (2004)
20. Fowlkes, E.B., Mallows, C.L.: A method for comparing two hierarchical clusterings. J. Am. Stat. Assoc. **78**(383), 553–569 (1983)
21. Wawrzinek, J., Wiehr, O., Pinto, J M.G., Balke, W.-T.: Exploiting latent semantic subspaces to derive associations for specific pharmaceutical semantics. Data Sci. Eng. **5**(4), 333–345 (2020)
22. McDaid, A.F., Greene, D., Hurley, N.: Normalized mutual information to evaluate overlapping community finding algorithms. arXiv preprint arXiv:1110.2515 (2011)

23. Yan, J., Wang, Y., Xiang, L., Zhou, Y., Zong, C.: A knowledge-driven generative model for multi-implication Chinese medical procedure entity normalization. In: Proceedings of the 2020 Conference on Empirical Methods in Natural Language Processing (EMNLP), pp. 1490–1499 (2020)
24. Mondal, I., et al.: Medical entity linking using triplet network. arXiv preprint arXiv:2012.11164 (2020)
25. Liu, H., Singh, P.: ConceptNet - a practical commonsense reasoning tool-kit. BT Technol. J. **22**(4), 211–226 (2004)

Improving Motor Imagery Intention Recognition via Local Relation Networks

Lin Yue[1], Yuxuan Zhang[2], Xiaowei Zhao[2], Zhe Zhang[2(✉)], and Weitong Chen[3(✉)]

[1] University of Newcaslte, Newcastle, NSW, Australia
[2] Northeast Normal University, Changchun, Jilin, China
zhangz059@nenu.edu.cn
[3] University of Adelaide, Adelaide, SA, Australia
t.chen@adelaide.edu.au

Abstract. Brain-computer interface (BCI) is a new communication and control technology established between human or animal brains and computer or other electronic equipment that does not rely on conventional brain information output pathways. The non-invasive BCI technology collects EEG signals from the cerebral cortex through signal acquisition equipment and processes them into signals recognized by the computer. The signals are preprocessed to extract signal features used for pattern recognition and finally are transformed into specific instructions for controlling external types of equipment. Therefore, the robustness of EEG signal representation is essential for intention recognition. Herein, we convert EEG signals into the image sequence and utilize the Local Relation Networks model to extract high-resolution feature information and demonstrate its advantages in the motor imagery (MI) classification task. The proposed method, MIIRvLR-Net, can effectively eliminate signal noise and improve the signal-to-noise ratio to reduce information loss. Extensive experiments using publicly available EEG datasets have proved that the proposed method achieved better performance than the state-of-the-art methods.

Keywords: Brain-computer Interface (BCI) · Intention recognition · Electroencephalogram (EEG)

1 Introduction

The Brain-Computer Interface (BCI) is a direct link between the human or animal brain and external equipment that enables the information exchange between the brain and equipment. When humans carry out different thinking activities, they activate different neurons. Then neurons in the cerebral cortex produce tiny electric currents. The brain-computer interface technology can

The second author has an equal first-author-level contribution to this work. This work has partially been supported by the Project of Philosophy and Social Sciences of Jilin Province under Project No. 2019C70 and the Thirteenth Five-Year Program for Social Science of Education Department of Jilin Province under Project No. JJKH20201187SK.

B. Li et al. (Eds.): APWeb-WAIM 2022, LNCS 13421, pp. 345–356, 2023.
https://doi.org/10.1007/978-3-031-25158-0_26

directly extract these neural signals in the brain to control external devices. It will build a bridge between people and machines and ultimately promote communication between people [5,6,9,14,17]. The output of the brain-computer interface system may replace the natural output lost due to injury or disease. For example, a person who has lost the ability to speak can output text through a brain-computer interface [20]. For the control field, in addition to manual control methods, brain control methods can also be added to achieve multi-modal control [19].

The non-invasive BCI mainly uses the electrodes on the head-mounted EEG cap to collect EEG signals. This method can monitor the discharge activity of a group of neurons on the scalp and has no risk of damage and infection to the human body. The main disadvantage is that it is not accurate enough. Therefore, the technical problem mainly focuses on EEG signal extraction, representation learning, intention classification and recognition. Compared with other signals, EEG-based motion imagery (MI) has the advantages of spontaneity and essential naturalness. It is more suitable as a decoding object and control signal for brain-computer interfaces. However, the accuracy of motion image classification is relatively low. Although the accuracy of existing CNN based methods meets or exceeds that of traditional machine learning methods [16], it is still insufficient to guarantee a wide range of applications, especially in complex tasks such as multi-dimensional motor intention recognition.

To improve the accuracy of the motor intention recognition task, we used a model that simultaneously learns robust feature representation and classified the raw EEG signal to detect different intentions. In this paper, we sliced, decomposed, and converted the EEG signals into a series of two-dimensional multi-channel images. Inspired by state-of-the-art image classification technology to learn robust representation, these images were enhanced with colour and spatial features using the proposed method MIIRvLR-Net. We evaluated the model on two datasets and obtained satisfactory accuracy and robustness in identifying five instruction intentions with a limited number of EEG electrodes.

2 Related Work

The intention recognition task ends up with getting the classification of different intentions. Alomari et al. [2] utilized a support vector machine-based method for binary classification along with features extracted from multi-resolution EEG signals. Shenoy et al. [12] deployed regularisation to improve the robustness and accuracy of CSP estimation in features extracting processing. Fisher's linear discriminant is used to perform binary tasks. In [10], after decomposition of the raw EEG data using neural networks to extract salient features, a binary classification task of EEG signals is performed. Kim et al. [8] used a random forest classifier for prediction, in which the Mu and Beta rhythms are obtained from the nonlinear EEG signals. Sita et al. [13] extracted features from open source EEG data and solved multiple classification by LDA. Zhang et al. [19] applied deep recurrent neural networks on an open EEG database for multiple classification. Chen et al. [5] proposed multi-task RNNs model (MTLEEG) for motion intention recognition based EEG signals. Chen et al. [6] developed DAMTRNN, a delta attention-based multi-task recurrent neural network for human intention recognition.

Among representation learning methods, CNN-based models have achieved great success on multiple tasks dealing with various data types. These learning methods can also be transferred to EEG for classification tasks. Researchers have made different attempts on EEG preprocessing and deep learning frameworks [3]. Hu et al. [7] adaptively determined aggregation weights based on the compositional relationship of local pixel pairs. This relational approach can composite visual elements into higher-level entities in a more effective way to encourage semantic inference. Compared to other deep neural networks that use bottom-up computation of weights, most of them do not apply to large-scale recognition tasks or only act as a complement to regular convolution. Moreover, these methods do not consider the geometric relationships between pixels but simply aggregate the entire feature space. In our model, we use locality and geometric prior for feature extraction of the EEG transformed image signal. The Local Relation Networks (LR-Net) is able to provide greater modelling capacity than its counterpart built with regular convolution.

3 MI Intention Recognition via LR-Net

This section describes the proposed method MIIRvLR-Net including Data Acquisition and Preprocessing, Generating Image Sequence from EEG Signals, and Architecture. LR-Net is one of the main components in this part used to learn the colour and spatial variations of image objects transformed from EEG signals (Fig. 1).

3.1 Data Acquisition and Preprocessing

The subject generates motor imagery through a dynamic state during which an individual mentally rehearses or simulates a physical action. This type of phenomenal experience implies that the subject feels themselves performing the action. The EEG signals are collected by the multiple electrodes wearable devices, i.e., 64 and 14 electrodes in this work. The n (64/14) electrodes placed on the subject's scalp continuously capture the real-time voltage values. The raw EEG readings obtained are an n-dimensional vector $R_t = [r_t^1, r_t^2, ..., r_t^n]$, where r_t^i is the reading of the ith electrode sensor at time step t.

The EEG consists of multiple time series corresponding to measurements at different spatial locations over the cortex. Like speech signals, the most salient features reside in the frequency domain, usually studied using a spectrogram of the signal. Since different frequency bands of EEG signals correspond to different biological meanings, the collected EEG signals can be subdivided into several variable time series with a frequency range between 0.5 Hz 28 Hz. In this work, the raw EEG signals can be decomposed into four frequency bands of waves, i.e., Delta, Theta, Alpha, and Beta. The raw EEG signals in r_t^i can be divided

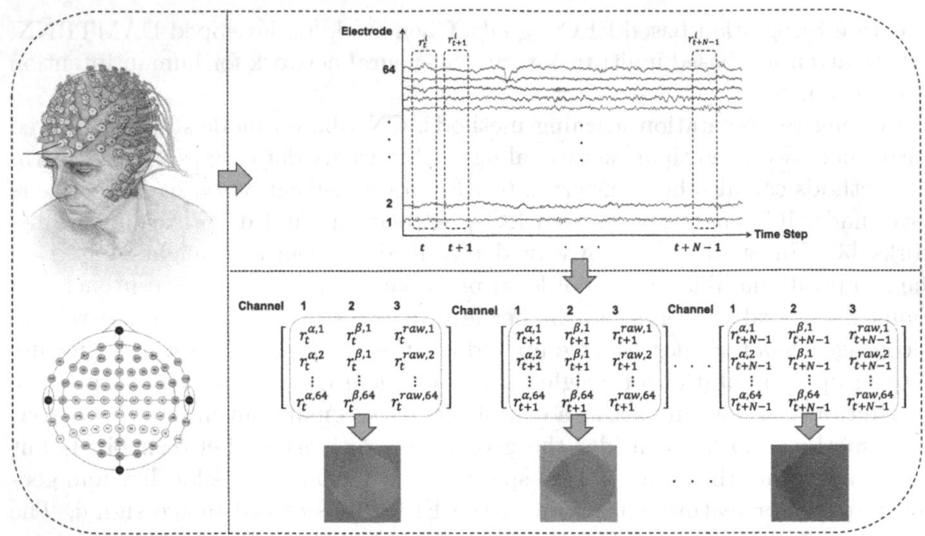

Fig. 1. EEG signal processing.

into different categories of bandwidth c, where $c = (\delta, \theta, \alpha, \beta)$. The decomposed EEG reading of the ith electrode sensor at step t can be expressed as $r_t^{c,i} = [r_t^{\delta,i}, r_t^{\theta,i}, r_t^{\alpha,i}, r_t^{\beta,i}]$. Specifically, the frequency range of the Alpha wave is 8 to 12.5 Hz, which mainly corresponds to the brain activities of relaxation and eye closure. Beta waves have a frequency range of 12.5 28 Hz and correspond to motor behaviour. According to [14,17], combining multi-wavelet or split-wavelet can make the correlation rows between features clearer and improve intention recognition performance. In this work, we select the filtered Alpha, Beta and raw signal on the ith electrode sensor at step t ($r_t^i = [r_t^{\alpha,i}, r_t^{\beta,i}, r_t^{raw,i}]$) as input feature. We set the sliding window size to $[n, 1]$, where n equals 64 or 14; we slice the input feature along the time axis. The resulting data matrix is known as the sliding matrix.

3.2 Generating Images from EEG Signals

All EEG electrodes in wearable devices are distributed on the scalp in three-dimensional space. To convert the spatially distributed activity map into a 32×32 two-dimensional image, we first need to project the electrodes' positions from the three-dimensional space onto the two-dimensional surface. However, this transformation should also maintain the relative distance between neighbouring electrodes. In this case, each pixel point in the 32 × 32 mesh is composed of three channels. The time series of Alpha, Beta, and raw signal that we use are superimposed on each pixel point of the mesh in the form of three channels. Inspired by [1,18], we apply the Clough-Tocher scheme for interpolating the scattered power measurements over the scalp. This procedure is repeated for each interested frequency band, resulting in three maps of cortical activity corresponding to each frequency band. These three spatial maps are then incorporated to form an RGB image with three channels. We refer to the generated RGB images as brain activity maps. The brain activity maps will be provided as input to the classification model, as described in the following section. The length and width of the brain activity map correspond to the spatial distribution of thinking activity in the cerebral cortex. The different electrodes of the brain activity map correspond to the corresponding cortical areas. Distinct cortical aspects, in turn, correspond to unique spatial regions in thinking activity. The activity diagram of the brain represents the temporal distribution of thinking activity.

3.3 Architecture

The classification model consists of two parts, i.e., the image enhancement and the classification network parts. The input of LR-Net is the brain activity map (3 channels), which expresses the time-space EEG information. To obtain better classification results, we processed these images to make their features more distinct. First, brain activity maps were enhanced for colour, size, and location learning. Then the picture features are captured by the local relational network for classification (see Fig. 2).

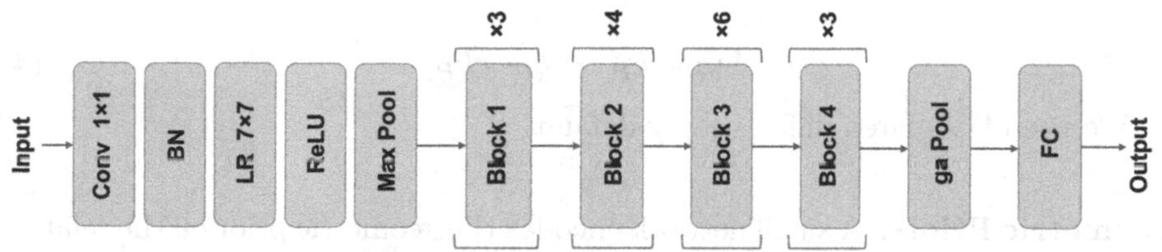

Fig. 2. MIIRvLR-Net architecture.

We employ Local Relation Networks to deal with the inherent structure of EEG data. In this paper, we construct a classification network model combining the Local Relation Layer and CNN network inspired by [7], in which meaningful composition structures can be inferred adaptively between visual elements in local regions. This network adjusts the aggregation weights based on the combinability of local pixel pairs. It has good performance in image classification and exploiting large kernel neighbourhoods.

The local relation network is mainly composed of a local relational layer with a convolutional kernel size of 7×7 and a channel transform layer with a convolutional kernel size of 1×1. Following each local relational layer is a ReLU activation function. These layers with different functions are stacked to form bottleneck blocks. Multiple bottleneck blocks are stacked together, followed by the global average pool layer and the FC layer. The core of this classification network is the local relational layer, in which its aggregation weights are defined as:

$$\omega(P',P) = softmax(\Phi(f_{\theta_q}(X_{P'}), f_{\theta_k}(X_P)) + f_{\theta_g}(P-P')) \quad (1)$$

where the term $\Phi(f_{\theta_q}(X_{P'}), f_{\theta_k}(X_P))$ is a measure of composability between the target pixel P' and a pixel P within its position scope, based on their appearance after transformations f_{θ_q} and f_{θ_k}, following recent works on relation modeling [4]. The term $f_{\theta_g}(P-P')$ defines the composability of a pixel pair (P, P') based on a geometric prior. The geometric term adopts the relative position as input and is translationally invariant.

Appearance Composability. We follow a general approach for relation modelling [4] to compute appearance composability $\Phi(f_{\theta_q(X_{P'})}, f_{\theta_k(X_P)})$, where $X_{P'}$ and X_P are projected to a *query* (by a channel transformation layer f_{θ_q}) and *key* (by a channel transformation layer f_{θ_k}) embedding space, respectively. We use scalars to represent the *query* and *key*. Scalar representation has a better speed-accuracy trade-off. We consider the following instantiations of function Φ: squared difference, absolute difference and multiplication.

$$\Phi(q_{P'}, k_P) = -(q_{P'} - k_P)^2 \quad (2)$$

$$\Phi(q_{P'}, k_P) = -|q_{P'} - k_P| \quad (3)$$

$$\Phi(q_{P'}, k_P) = q_{P'} \cdot k_P \quad (4)$$

We use (4) squared difference by default.

Geometric Priors. A small network encodes the geometric prior on the relative position of P to P'. The small network consists of two-channel transformation layers, with a ReLU activation in between. The small network on relative position treats relative positions as vectors in metric space, while the direct method treats different positions as independent identities.

Note that the inference process with using a small network is the same as directly learning the geometric priors. In fact, during inference, a fixed learning weight θ_g yields a fixed geometric prior $f_{\theta_g}(\Delta P)$ for the relative position ΔP. We use these fixed geometric prior values instead of the original model weights θ_g for more convenient inference.

Weight Normalization. The layer uses softmax for weight normalization.

Channel Sharing. Following [11], the local relation layer uses channel sharing in aggregation computation, where multiple channels share the same aggregation weights. Channel sharing can generally reduce the model size and facilitate GPU memory scheduling for efficient implementation.

In general, EEG has a low signal-to-noise ratio. However, learning models using algorithms with too much expressive power can lead to over-fitting to poor-quality EEG data. In this paper, using a residual strategy can prevent gradient disappearance and handle the problem of overfitting and reduce the risk of gradient loss. Therefore, more complete extraction of more efficient features and better classification results can be achieved compared to traditional convolutional classification networks.

4 Experiments

4.1 Datasets and Settings

We tested the proposed method and all the benchmark methods using 5-fold cross-validation on two datasets. Two public EEG datasets involved in this paper are eegmmidb and emotiv, respectively.

- EEGMMIDB[1], is used as the baseline dataset, and EEG signals are collected using the BCI 2000 system[2], including 64 EEG electrodes, each of which is sampled at 160 times per second, as well as comment channels. The dataset consists of over 1500 one- and two-minute EEG recordings obtained from 109 volunteers, in which each subject performs five kinds of experimental runs. For each one-minute baseline run, five different imagery tasks are shown in Table 1.
- EMOTIV[3], is collected using Emotiv Epoc+ 14 channels headset, which contains 7 participants aged 23 to 27. Each subject has 80640 samples. For each subject, 46080–76800 samples belong to 5 class samples (every 48 s), a total of 240 s of collection data, with a sampling 128 Hz. For each 48-second baseline run, five different imagery tasks are shown in Table 1.

Table 1. Class annotation in two different datasets

Task	EEGMMIDB	EMOTIV
Task 0	Keep eyes closed	Confirm
Task 1	Open and close both feet	Up
Task 2	Open and close both fists	Down
Task 3	Open and close left fist	Left
Task 4	Open and close right fist	Right

[1] https://physionet.org/pn4/eegmmidb/.
[2] http://www.schalklab.org/research/bci2000.
[3] shorturl.at/gqB78.

4.2 Benchmark Methods

For the baseline models, we kept the same structures and settings. We fed baselines with different kinds of features extracted from the same dataset to evaluate the influence of the multi-resolution signals. And these methods include:

- **Alomari et al.** [2]: A support vector machine-based method is used for binary classification task, along with features extracted from multi-resolution EEG signals.
- **Shenoy et al.** [12]: Regularisation is deployed to improve the robustness and accuracy of CSP estimation for feature extraction. Fisher's linear discriminant is used to perform the binary classification task.
- **Rashid et al.** [10]: A neural network is utilized to perform EEG signal binary-class classification, decomposing the raw EEG data to extract significant features.
- **Kim et al.** [8]: Random forest classifier is used for motor imagery classification, in which the Mu and Beta rhythms are obtained from the nonlinear EEG signals.
- **Sita et al.** [13]: Features are extracted from open source EEG data, and multiple classification problem is solved by LDA.
- **Zhang et al.** [19]: A deep recurrent neural network (DRNN) is proposed for multiple classification problems, which is tested on an open EEG database.
- **Chen et al.** [5]: A multi-task RNNs model (MTLEEG) is proposed for motion intention recognition based on EEG signals.
- **Chen et al.** [6]: A delta attention-based recurrent neural network, DAMTRNN, is developed for human intention recognition.

For the binary classification methods, we used One-vs-Rest (OvR)[4] strategy for multiclass classification tasks.

4.3 Results and Discussion

In this section, we present the overall performance and comparative results of our proposed model. To evaluate the performance of our model, we conducted the following experiments: First, we compared our method with the state-of-the-art methods on intention recognition tasks, i.e., simple binary classification and more complex multi-class classification. The overall performance of MIIRvLR-Net and benchmark methods are summarized in Tables 2 and 3. We can be observed that MIIRvLR-Net outperforms other state-of-the-art methods in terms of Accuracy.

Secondly, we illustrated the classification results on particular intention tasks in terms of Accuracy, Precision, Recall, F1-score and AUC (see Table 4 and 5). We presented that MIIRvLR-Net will not be primarily influenced by fewer sensors and a lower sampling rate from this part of the experiments. Even with low-resolution signal data using fewer signal acquisition channels (14 electrodes), we can still obtain high prediction accuracy. That is, fewer features, lower frequency

[4] shorturl.at/nrZ02.

resolution, and higher phase noise existing in data, naturally, will all be solved with the proposed method and get robust performance in intention recognition.

Finally, We used the area under the receiver operating characteristic curve (AUC) between the true positive (TP) and false positive (FP) rates to evaluate the performance of our model. The ROC curves indicate the prediction accuracy situation by true positive rate versus false-positive rate according to the threshold value. Figure 3 shows that the ROC curve in all categories and macro-precision, macro-recall, and macro-F1-score approximate 1, which means the high performance of MIIRvLR-Net.

Table 2. Comparisons of MIIRvLR-Net and benchmark methods on EEGMMIDB dataset

Index	Method	Class	Accuracy
1	Alomari et al. [2]	Binary	0.7490
2	Shenoy et al. [12]	Binary	0.8206
3	Rashid et al. [10]	Binary	0.9200
4	Kim et al. [8]	Multi(3)	0.8050
5	Sita et al. [13]	Multi(3)	0.8724
6	Zhang et al. [19]	Multi(5)	0.9553
7	Chen et al. [5]	Multi(5)	0.9786
8	Chen et al. [6]	Multi(5)	0.9887
9	**MIIRvLR-Net**	**Multi(5)**	**0.9980**

Table 3. Comparisons of MIIRvLR-Net and benchmark methods on EMOTIV dataset

Index	Method	Class	Accuracy
1	Alomari et al. [2]	Binary	0.5627
2	Shenoy et al. [12]	Binary	0.5553
3	Rashid et al. [10]	Binary	0.7538
4	Kim et al. [8]	Multi(3)	0.7695
5	Sita et al. [13]	Multi(3)	0.6985
6	Zhang et al. [19]	Multi(5)	0.7361
7	Chen et al. [5]	Multi(5)	0.8396
8	Chen et al. [6]	Multi(5)	0.8997
9	**MIIRvLR-Net**	**Multi(5)**	**0.9768**

Table 4. Comparisons of each task on EEGMMIDB dataset

Task	Accuracy	Precision	Recall	F1-score	AUC
0	0.9949	0.9984	1.0	0.9992	0.9957
1	0.9979	0.9945	0.9945	0.9945	0.9930
2	0.9993	0.9977	0.9977	0.9977	0.9950
3	0.9991	0.9988	0.9955	0.9972	0.9805
4	0.9989	0.9971	0.9971	0.9971	0.9971
Macro	**0.9980**	**0.9973**	**0.9970**	**0.9971**	**0.9923**

Table 5. Comparisons of each task on EMOTIV dataset

Task	Accuracy	Precision	Recall	F1-score	AUC
0	0.9896	0.9837	0.9577	0.9705	0.9984
1	0.9620	0.9293	0.8762	0.9020	0.9920
2	0.9668	0.9075	0.9364	0.9217	0.9917
3	0.9858	0.9665	0.9619	0.9642	0.9976
4	0.9801	0.9322	0.9778	0.9544	0.9939
Macro	**0.9768**	**0.9438**	**0.9420**	**0.9425**	**0.9949**

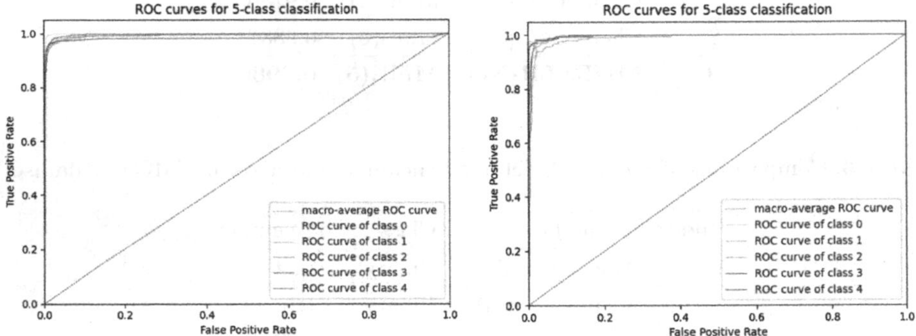

Fig. 3. ROC Curve on two datasets. Left: On eegmmidb; Right: On emotiv.

5 Conclusion

In this paper, we focus on how to improve the motor imagery intention recognition through MIIRvLR-Net, which utilizes the local relation layer based on the combinability of pixel pairs and determines the fusion weight in a bottom-up way providing a more effective spatial combination with analyzing intention related EEG signals. Compared to others, extensive experiments using EEGMMIDB and EMOTIV datasets proved that MIIRvLR-Net achieved better performance than the state-of-the-art methods. In the future, we will put more effort into reducing the computational complexity of the model [15]. And we plan to conduct experimental studies using more abundant data sets to verify the generalizability of the algorithm utilized in this paper.

References

1. Alfeld, P.: A trivariate clough-tocher scheme for tetrahedral data. Comput. Aided Geometric Des. **1**(2), 169–181 (1984)
2. Alomari, M.H., AbuBaker, A., Turani, A., Baniyounes, A.M., Manasreh, A.: EEG mouse: a machine learning-based brain computer interface. Int. J. Adv. Comput. Sci. Appl. **5**(4), 193–198 (2014)
3. Bashivan, P., Rish, I., Yeasin, M., Codella, N.: Learning representations from EEG with deep recurrent-convolutional neural networks. arXiv preprint arXiv:1511.06448 (2015)
4. Battaglia, P.W., et al.: Relational inductive biases, deep learning, and graph networks. arXiv preprint arXiv:1806.01261 (2018)
5. Chen, W., Wang, S., Zhang, X., Yao, L., Yue, L., Qian, B., Li, X.: EEG-based motion intention recognition via multi-task RNNs. In: Proceedings of the 2018 SIAM International Conference on Data Mining, pp. 279–287. SIAM (2018)
6. Chen, W., Yue, L., Li, B., Wang, C., Sheng, Q.Z.: DAMTRNN: a delta attention-based multi-task RNN for intention recognition. In: Li, J., Wang, S., Qin, S., Li, X., Wang, S. (eds.) ADMA 2019. LNCS (LNAI), vol. 11888, pp. 373–388. Springer, Cham (2019). https://doi.org/10.1007/978-3-030-35231-8_27
7. Hu, H., Zhang, Z., Xie, Z., Lin, S.: Local relation networks for image recognition. In: Proceedings of the IEEE/CVF International Conference on Computer Vision, pp. 3464–3473 (2019)
8. Kim, Y., Ryu, J., Kim, K.K., Took, C.C., Mandic, D.P., Park, C.: Motor imagery classification using mu and beta rhythms of EEG with strong uncorrelating transform based complex common spatial patterns. Computational intelligence and neuroscience 2016 (2016)
9. Qiu, Y., Chen, W., Yue, L., Xu, M., Zhu, B.: Stct: spatial-temporal conv-transformer network for cardiac arrhythmias recognition. In: International Conference on Advanced Data Mining and Applications, pp. 86–100. Springer (2022)
10. or Rashid, M.M., Ahmad, M.: Classification of motor imagery hands movement using levenberg-marquardt algorithm based on statistical features of EEG signal. In: 2016 3rd International Conference on Electrical Engineering and Information Communication Technology (ICEEICT), pp. 1–6. IEEE (2016)
11. Sabour, S., Frosst, N., Hinton, G.E.: Dynamic routing between capsules. arXiv preprint arXiv:1710.09829 (2017)
12. Shenoy, H.V., Vinod, A.P., Guan, C.: Shrinkage estimator based regularization for EEG motor imagery classification. In: 2015 10th International Conference on Information, Communications and Signal Processing (ICICS), pp. 1–5. IEEE (2015)
13. Sita, J., Nair, G.: Feature extraction and classification of EEG signals for mapping motor area of the brain. In: 2013 International Conference on Control Communication and Computing (ICCC), pp. 463–468. IEEE (2013)
14. Yue, L., Shen, H., Wang, S., Boots, R., Long, G., Chen, W., Zhao, X.: Exploring BCI control in smart environments: intention recognition via EEG representation enhancement learning. ACM Trans. Knowl. Discovery Data (TKDD) **15**(5), 1–20 (2021)

15. Yue, L., Shi, Z., Han, J., Wang, S., Chen, W., Zuo, W.: Multi-factors based sentence ordering for cross-document fusion from multimodal content. Neurocomputing **253**, 6–14 (2017)
16. Yue, L., Tian, D., Chen, W., Han, X., Yin, M.: Deep learning for heterogeneous medical data analysis. World Wide Web **23**(5), 2715–2737 (2020)
17. Yue, L., Tian, D., Jiang, J., Yao, L., Chen, W., Zhao, X.: Intention recognition from spatio-temporal representation of EEG signals. In: Qiao, M., Vossen, G., Wang, S., Li, L. (eds.) ADC 2021. LNCS, vol. 12610, pp. 1–12. Springer, Cham (2021). https://doi.org/10.1007/978-3-030-69377-0_1
18. Zang, Y., Liu, Y., Chen, W., Li, B., Li, A., Yue, L., Ma, W.: Gisdcn: A graph-based interpolation sequential recommender with deformable convolutional network. In: International Conference on Database Systems for Advanced Applications, pp. 289–297. Springer, Cham (2022). https://doi.org/10.1007/978-3-031-00126-0_21
19. Zhang, X., Yao, L., Huang, C., Sheng, Q.Z., Wang, X.: Intent recognition in smart living through deep recurrent neural networks. In: International Conference on Neural Information Processing, vol. 10635, pp. 748–758. Springer, Cham (2017). https://doi.org/10.1007/978-3-319-70096-0_76
20. Zhang, X., Yao, L., Sheng, Q.Z., Kanhere, S.S., Gu, T., Zhang, D.: Converting your thoughts to texts: Enabling brain typing via deep feature learning of eeg signals. In: 2018 IEEE international conference on pervasive computing and communications (PerCom), pp. 1–10. IEEE (2018)

EAS-GCN: Enhanced Attribute-Aware and Structure-Constrained Graph Convolutional Network

Jijie Zhang[1], Yan Yang[1,2](✉), Shaowei Yin[1], and Zhengqi Wang[3]

[1] School of Computer Science and Technology, Heilongjiang University, Harbin 150080, China
yangyan@hlju.edu.cn

[2] Key Laboratory of Database and Parallel Computing of Heilongjiang Province, Harbin 150080, China

[3] School of Computer Science and Technology, Harbin Engineering University, Harbin 150001, China

Abstract. Recently, graph neural networks (GNNs) have achieved significant success in many graph-based tasks. However, most GNNs are inherently restricted by over-smoothing, which limits performance improvement. In this paper, we propose an Enhanced Attribute-aware and Structure-constrained Graph Convolutional Network (EAS-GCN). Specifically, EAS-GCN first uses degree prediction to incorporate graph local structure information into autoencoder-specific representation. A delivery mechanism is then designed to pass the autoencoder-specific representation to the corresponding GCN layer. Autoencoder mainly assists GCN in learning enhanced attribute information, and node degree prediction assists GCN in learning local structure information. Furthermore, we theoretically analyze autoencoder could help alleviate the over-smoothing in GCN. Experimental results show that EAS-GCN enjoys high accuracy on the node classification task and can better alleviate over-smoothing.

Keywords: Autoencoder · GCN · Over-smoothing · Delivery mechanism

1 Introduction

In recent years, graph neural networks (GNNs) [27] have received a lot of attention. They have achieved great success in many tasks, such as node classification [6], graph classification [25], and link prediction [20]. Traditional GNNs generally take node features and graph structures as input to generate node representations or graph representations required for downstream tasks. Although GNNs have enjoyed great success in node classification, most of them require a certain number of labeled nodes to be trained. GNNs are limited in real-world applications since this labeled information is usually scarce and valuable. Moreover, models like Graph Convolutional Network (GCN) [10] and Graph Attention

B. Li et al. (Eds.): APWeb-WAIM 2022, LNCS 13421, pp. 357–371, 2023.
https://doi.org/10.1007/978-3-031-25158-0_27

Network (GAT) [17] both work best at two layers, and this shallow architecture limits their ability to extract information from high-order neighbors. Stacking more layers and increasing nonlinearity tends to degrade the performance of these models. The main reason is that these models suffer from over-smoothing. It was first pointed out in [13] that each layer of the graph convolutional network is a special kind of Laplacian smoothing. The graph convolution operation updates the node representations by aggregating the representations of the neighbors so that the neighboring nodes will become similar. Thus, stacking multiple layers of graph convolution operations results in all nodes having similar representations. For sparse graphs, it is essential to exploit the high-order information of the nodes. Therefore, this work mainly addresses the semi-supervised learning on graphs and alleviates over-smoothing.

Collecting large amounts of labeled data is often difficult and expensive for a given task. However, unlabeled data is typically rich and easily accessible. For node classification, unlabeled data is utilized by GNNs through a simple information aggregation process. This process may not be sufficient to exploit unlabeled data fully. Therefore, to enable GCN to utilize the information of unlabeled data fully, we use two modules, autoencoder and degree prediction, to assist GCN in learning node representations. After GCN stacks more layers, the obtained structure information tends to be global. Therefore, it is necessary for GCN to learn better local structure information. The key is how to integrate local structure information into GCN better? In this work, we use node degree as the local structure information. Then we assist GCN to learn better local structure information by predicting the degree of nodes. For using autoencoder to assist GCN to learn enhanced attribute information, we naturally think of a question: how to combine GCN and autoencoder elegantly? Autoencoder consists of multiple neural network layers, each learning different potential attribute information. GCN can capture various types of structure information between data. Hence, what is the relation between different structure information in GCN and different potential attribute information in autoencoder? Considering the above problems, we design a delivery mechanism to pass the autoencoder-specific representations to the corresponding GCN layer. In this way, different structure information in GCN are naturally combined with different potential attribute information in autoencoder. The main contributions of this paper are as summarized as follow:

- We design a delivery mechanism to pass the autoencoder-specific representations to the corresponding GCN layer, aiming to assist the GCN to learn enhanced attribute information. Furthermore, we design a degree prediction module to assist GCN in learning local structure information.
- We have conducted a theoretical analysis of EAS-GCN, and based on our theoretical analysis, the designed delivery mechanism will effectively alleviate the over-smoothing in GCN. The theoretical analysis proves that the node representation learned by EAS-GCN is equivalent to the sum of representations with different-order structure information.
- Extensive experiments have demonstrated that the proposed framework can outperform representative baselines, and it can better alleviate over-smoothing.

2 Related Work

2.1 Graph Neural Networks

GNNs [18,23] have boosted research on graph data mining. The key to the success of most GNNs lies in the message passing mechanism, which propagates the neighbors information to a target node in an iterative manner. In the growing number of GNN architectures, the most representative method is GCN [10] and GAT [17]. GCN and GAT follow coupling feature transformation and neighborhood aggregation together for representation learning. Nevertheless, some recent work finds that coupling feature transformation and neighborhood aggregation is unnecessary and causes over-smoothing. APPNP [11], SGC [21], and S^2GC [28] achieve good node classification by separating the two operations. DropEdge [14] is introduced to alleviate the over-smoothing by randomly dropping some edges in the graph during each training epoch. GCNII [3] obtains better results by applying two simple techniques, initial residuals and identity mapping, to graph convolutional networks. Similar to APPNP, we also alleviate over-smoothing by integrating different order information into GCN. The difference from APPNP is that APPNP uses the same representations and different order adjacency matrices to generate node representations; we use different representations and different order adjacency matrices to generate node representations. This makes our model contain more information.

2.2 Auto-Encoder

Currently, more deep models are beginning to be designed for graph-structured data. For example, autoencoders have been extended for node representation learning on graph-structured data [2,19]. These deep graph models facilitate the development of graph data-related tasks in different scenarios beyond GNN capabilities and significantly extend the graph deep learning techniques. The autoencoder architecture, which extracts complex features using only unlabeled data, allows deep learning techniques to be applied to a broader range of domains. The autoencoder can be viewed as an unsupervised learning model for obtaining compressed low-dimensional representations of data samples. SDNE [19] uses the deep autoencoder with multiple non-linear layers to capture the first and second order proximites. SDCN [2] converts raw data into the low-dimensional representations by the encoder and then decodes the low-dimensional representations to reconstruct the raw data. In this work, we use the basic autoencoder [8] to learn representations for raw data.

2.3 Self-supervised Learning

In recent years, deep learning has been applied in many fields with great success. Deep learning requires a large amount of labeled data for superior performance. However, collecting high-quality labeled data often requires a lot of time and resources. Hence, Self-Supervised Learning (SSL) [1,24] was introduced to alleviate the demand for massive labeled data and provide adequate supervision. SSL seeks to design a domain-specific pretext task on unlabeled data and then learns better representation with the pretext task. Although initially applied in computer vision [4] and natural language processing [5], recent interest has been in leveraging SSL to enhance deep learning on graphs. For node classification, unlabeled data has been incorporated by GNNs via the simple message passing process. This process could be insufficient to make use of unlabeled data fully. Hence, the works [9,22] have been systematically designed different self-supervised pretext tasks in GNNs. To fully utilize the structure information of unlabeled nodes, this paper proposes a self-supervised task using node degree prediction to assist the GCN to learn local structure information.

3 The Proposed Model

EAS-GCN's overall framework is illustrated in Fig. 1. The model is divided into three main parts: the autoencoder module, the node degree prediction module and the delivery mechanism. We will describe our proposed model in detail in the following.

3.1 Preliminaries

Let $G = (V, E)$ denote a graph, where $V = \{v_1, v_2, \ldots, v_N\}$ is a node set, $E \subseteq V \times V$ is a set of $|E| = M$ edges between nodes. The raw data is denoted as $\boldsymbol{X} = \{\boldsymbol{x}_1, \boldsymbol{x}_2, \ldots, \boldsymbol{x}_N\}$. $\boldsymbol{A} \in \{0,1\}^{N\times N}$ denotes the adjacency matrix of G, with each $\boldsymbol{A}_{ij} = 1$ indicating there exists an edge between v_i and v_j, otherwise $\boldsymbol{A}_{ij} = 0$.

This work focuses on semi-supervised graph representation learning, in which each node v_i is associated with 1) a feature vector $\boldsymbol{x}_i \in \boldsymbol{X} \in \mathbb{R}^{N\times d}$ and 2) a label vector $\boldsymbol{Y}_i \in \boldsymbol{Y} \in \{0,1\}^{N\times C}$ with C representing the number of classes. For semi-supervised classification, T nodes ($0 < T \ll N$) have observed their labels $\boldsymbol{Y}^L$ and the labels $\boldsymbol{Y}^U$ of the remaining $N - T$ nodes are missing. The objective is to learn a predictive function $f : G, \boldsymbol{X}, \boldsymbol{Y}^L \to \boldsymbol{Y}^U$ to infer the missing labels $\boldsymbol{Y}^U$ for unlabeled nodes.

3.2 Auto-Encoder Module

In our model, the basic autoencoder is used as a feature encoder in order to reduce the complexity of the model. Assume that the encoder and decoder parts have L layers each and ℓ represents the ℓ-th layer. The processing of the encoder is defined as:

$$\boldsymbol{H}_e^{(\ell)} = \sigma\left(\boldsymbol{W}_e^{(\ell)} \boldsymbol{H}_e^{(\ell-1)} + \boldsymbol{b}_e^{(\ell)}\right), \tag{1}$$

where $\sigma(\cdot)$ is the activation function and we apply *Relu*. Here $\boldsymbol{W}_e^{(\ell)}$ and $\boldsymbol{b}_e^{(\ell)}$ are the trainable weight matrix and bias of the ℓ-th layer in the encoder, respectively.

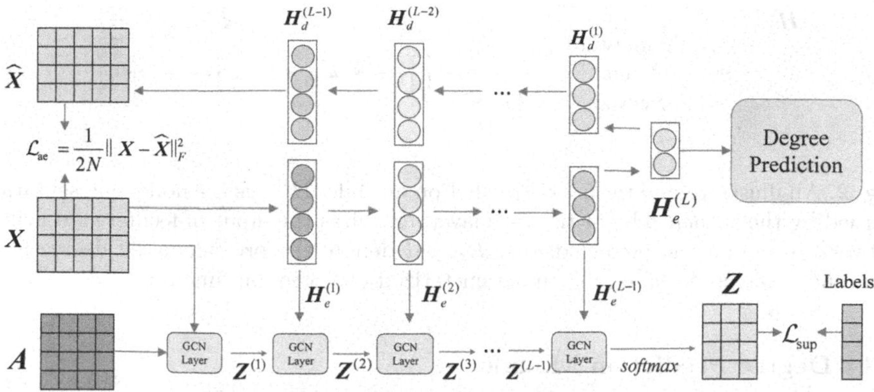

Fig. 1. An illustration of our proposed EAS-GCN. For clarity, we show the process to generate the node prediction. $\boldsymbol{X}$, $\widehat{\boldsymbol{X}}$ are the input data and the reconstructed data. $\boldsymbol{Z}$ are the node prediction probabilities. $\boldsymbol{Z}^{(1)}$, $\boldsymbol{Z}^{(2)}$, $\boldsymbol{Z}^{(3)}$ and $\boldsymbol{Z}^{(L-1)}$ are node representations in different GCN layers. $\boldsymbol{H}_e^{(1)}$, $\boldsymbol{H}_e^{(2)}$, $\boldsymbol{H}_e^{(L-2)}$ and $\boldsymbol{H}_e^{(L-1)}$ are node representations in different encoder layers of autoencoder. $\boldsymbol{H}_d^{(1)}$, $\boldsymbol{H}_d^{(L-2)}$ and $\boldsymbol{H}_d^{(L-1)}$ are node representations in different decoder layers of autoencoder. The input data $\boldsymbol{X}$ first obtains the node embeddings through the autoencoder, and $\boldsymbol{H}_e^{(L)}$ is the input of the degree prediction module. Then the local structure information is integrated into the autoencoder-specific representations. Through the designed delivery mechanism, the autoencoder-specific representations with local structure information is transferred to the corresponding GCN layer, and after multiple GCN layers, the final output node prediction is obtained.

$\boldsymbol{H}_e^{(\ell-1)}$ and $\boldsymbol{H}_e^{(\ell)}$ are node representations of layer $\ell-1$ and layer ℓ in the encoder respectively while $\boldsymbol{H}_e^{(0)}$ is set to raw data $\boldsymbol{X}$. Similarly, the decoder part is defined as:

$$\boldsymbol{H}_d^{(\ell)} = \sigma\left(\boldsymbol{W}_d^{(\ell)} \boldsymbol{H}_d^{(\ell-1)} + \boldsymbol{b}_d^{(\ell)}\right), \tag{2}$$

where the input of decoder part is $\boldsymbol{H}_d^{(0)} = \boldsymbol{H}_e^{(L)}$, $\boldsymbol{W}_d^{(\ell)}$ and $\boldsymbol{b}_d^{(\ell)}$ are the trainable weight matrix and bias of the ℓ-th layer in the decoder, respectively. $\boldsymbol{H}_d^{(\ell-1)}$ and $\boldsymbol{H}_d^{(\ell)}$ are node representations of layer $\ell-1$ and layer ℓ in the decoder respectively. The output of the decoder part is the reconstruction of the raw data $\widehat{\boldsymbol{X}} = \boldsymbol{H}_d^{(L)}$. The corresponding reconstruction loss is defined as:

$$\mathcal{L}_{\text{ae}} = \frac{1}{2N} \sum_{i=1}^{N} \|\boldsymbol{x}_i - \widehat{\boldsymbol{x}}_i\|_2^2 = \frac{1}{2N} \|\boldsymbol{X} - \widehat{\boldsymbol{X}}\|_F^2. \tag{3}$$

By minimizing the reconstruction loss in Eq. (3), the autoencoder can learn good low-dimensional representations from the data itself but ignores the structure information of the node. Therefore, we introduce the node degree prediction module to learn the structure information of the node (Fig. 2).

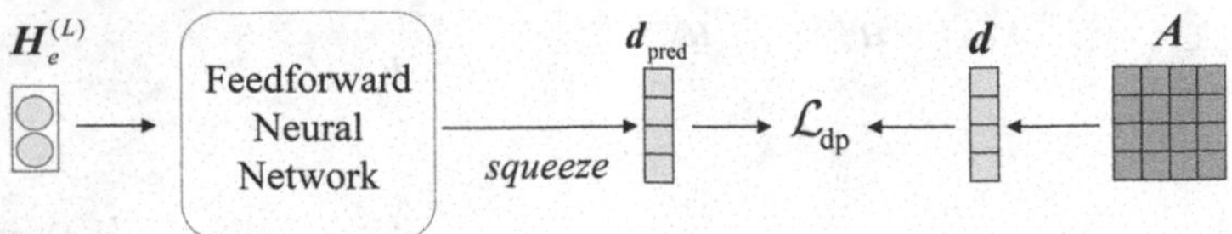

Fig. 2. An illustration of the degree prediction module. $\boldsymbol{H}_e^{(L)}$ is the node representation learned by the autoencoder at the L-th layer, and it as the input of feedforward neural network to obtain the predict degree $\boldsymbol{d}_{\text{pred}}$. $\boldsymbol{d}$ denote the pre-calculated degree from adjacency matrix $\boldsymbol{A}$. *squeeze* is unparameterized compression function.

3.3 Degree Prediction Module

The node degree prediction module aims to encourage the GCN to learn better local structure information. We use node degree as the local structure information. Firstly, we pre-calculate the degree from the adjacency matrix. This pre-calculated degree can be used as the label of the predicted degree. This calculation process is defined as:

$$\boldsymbol{d}_i = \sum_{j=1}^{N} \boldsymbol{A}_{ij} \tag{4}$$

where $\boldsymbol{d}_i$ represents degree of node v_i. Then we will convert autoencoder-specific representations through a feedforward neural network to get the node degree $\boldsymbol{d}_{\text{trans}}$, as:

$$\boldsymbol{d}_{\text{trans}} = \sigma\left(\boldsymbol{H}_e^{(L)}\boldsymbol{W} + \boldsymbol{b}\right) \quad \in \mathbb{R}^{N\times 1}, \tag{5}$$

where $\boldsymbol{W}$ and $\boldsymbol{b}$ are the trainable weight matrix and bias of the feedforward neural network.

Because the dimension of the converted node degree is different from the pre-calculated label dimension, we use the compression function to change the dimension of the converted node degree:

$$\boldsymbol{d}_{\text{pred}} = squeeze\left(\boldsymbol{d}_{\text{trans}}\right) \quad \in \mathbb{R}^{N}, \tag{6}$$

where *squeeze* is unparameterized compression function, i.e., $f : \boldsymbol{X} \in \mathbb{R}^{N\times 1} \rightarrow \boldsymbol{X}_{comp} \in \mathbb{R}^{N}$. After the compression function, matrix can be converted into a vector.

The corresponding node degree prediction loss is given by

$$\mathcal{L}_{\text{dp}} = \frac{1}{N}\sum_{i=1}^{N}\left(\boldsymbol{d}_{\text{pred}}[i] - \boldsymbol{d}_i\right)^2. \tag{7}$$

By optimizing the above loss function, the graph structure information can be incorporated into the autoencoder-specific representations. In order for GCN to obtain better local structure information, we need to pass the autoencoder-specific representations with local structure information to GCN, so we design a delivery mechanism.

3.4 Delivery Mechanism

The last module is a delivery mechanism for passing the autoencoder-specific representations with local structure information to the corresponding GCN layer. Standard GCN's operation is given by

$$\boldsymbol{Z}^{(\ell)}=\sigma\left(\widehat{\boldsymbol{D}}^{-\frac{1}{2}}\widehat{\boldsymbol{A}}\widehat{\boldsymbol{D}}^{-\frac{1}{2}}\boldsymbol{Z}^{(\ell-1)}\boldsymbol{W}^{(\ell-1)}\right), \tag{8}$$

where $\widehat{\boldsymbol{A}}=\boldsymbol{A}+\boldsymbol{I}$ is the adjacency matrix obtained by adding the self-loops, $\widehat{\boldsymbol{D}}$ is the degree matrix, i.e., $\widehat{\boldsymbol{D}}_{ii}=\sum_{j=1}^{N}\widehat{\boldsymbol{A}}_{i,j}$, $\boldsymbol{W}^{(\ell-1)}$ is a learnable matrix of GCN.

Considering that the node representations $\boldsymbol{H}_e^{(\ell-1)}$ learned by the autoencoder can reconstruct the data itself while obtaining extra valuable attribute information, we design a delivery mechanism to pass $\boldsymbol{H}_e^{(\ell-1)}$ to the representation $\boldsymbol{Z}^{(\ell-1)}$ learned by GCN and combine them to obtain a more complete and efficient representation, as:

$$\widehat{\boldsymbol{Z}}^{(\ell-1)}=(1-p)\boldsymbol{Z}^{(\ell-1)}+p\boldsymbol{H}_e^{(\ell-1)}, \tag{9}$$

where $\boldsymbol{Z}^{(0)}=\boldsymbol{H}_e^{(0)}=\boldsymbol{X}$, p is the transfer parameter to balance the ratio between the autoencoder-specific representations and the representations learned by GCN, set to 0.5 by default. We will perform sensitivity analysis on this parameter later in the experimental section.

We then use $\widehat{\boldsymbol{Z}}^{(\ell-1)}$ as the input to the ℓ-th layer GCN for generating $\boldsymbol{Z}^{(\ell)}$ which is defined as:

$$\boldsymbol{Z}^{(\ell)}=\sigma\left(\widehat{\boldsymbol{D}}^{-\frac{1}{2}}\widehat{\boldsymbol{A}}\widehat{\boldsymbol{D}}^{-\frac{1}{2}}\widehat{\boldsymbol{Z}}^{(\ell-1)}\boldsymbol{W}^{(\ell-1)}\right). \tag{10}$$

The raw data $\boldsymbol{X}$ is the input of GCN at first layer, as Eq (11).

$$\boldsymbol{Z}^{(1)}=\sigma\left(\widehat{\boldsymbol{D}}^{-\frac{1}{2}}\widehat{\boldsymbol{A}}\widehat{\boldsymbol{D}}^{-\frac{1}{2}}\boldsymbol{X}\boldsymbol{W}^{(1)}\right) \tag{11}$$

After obtaining $\boldsymbol{Z}^{(L)}$, the node prediction probabilities is obtained through the softmax multi-classification layer, as Eq (12).

$$\boldsymbol{Z}=\operatorname{softmax}\left(\boldsymbol{Z}^{(L)}\right) \tag{12}$$

The cross-entropy loss for T labeled nodes can be calculated as

$$\mathcal{L}_{\text{sup}}=-\sum_{i=0}^{T-1}\sum_{k=1}^{C}\boldsymbol{Y}_i[k]\ln\boldsymbol{Z}_i[k], \tag{13}$$

where $\boldsymbol{Z}_i[k]$ denotes the probability of node i belongs to class k. In total, we define the final loss of EAS-GCN as the combination of three losses above:

$$\mathcal{L}=\mathcal{L}_{\text{sup}}+\alpha\mathcal{L}_{\text{ae}}+\beta\mathcal{L}_{\text{dp}}, \tag{14}$$

where α and β is the weighting coefficient to balance these losses.

We provide a detailed description of EAS-GCN in Algorithm 1.

Algorithm 1. EAS-GCN

Input: Adjacency matrix $\boldsymbol{A}$, raw data $\boldsymbol{X}$, Number of layers L, Transfer parameter p, learning rate η, balance coefficient α and β. A model combining autoencoder and GCN $f(\boldsymbol{X}, \boldsymbol{A}, \Theta)$.

Output: Prediction $\boldsymbol{Z}$.

for $\ell = 1$ to L **do**
  Initialize $\boldsymbol{W}_e^{(\ell)}$, $\boldsymbol{b}_e^{(\ell)}$, $\boldsymbol{W}_d^{(\ell)}$, $\boldsymbol{b}_d^{(\ell)}$, $\boldsymbol{W}^{(\ell)}$.
end for
while not convergence **do**
  for $\ell = 1$ to $L-1$ **do**
    Generate autoencoder-specific representations $\boldsymbol{H}_e^{(\ell)}$ via Eq (1) and $\boldsymbol{H}_d^{(\ell)}$ via Eq (2).
    Generate representations $\widehat{\boldsymbol{Z}}^{(\ell)}$ via Eq (9).
    Generate the next GCN layer representations $\boldsymbol{Z}^{(\ell+1)}$ via Eq (10).
  end for
  Calculate $\mathcal{L}_{\text{ae}}$, $\mathcal{L}_{\text{dp}}$, $\mathcal{L}_{\text{sup}}$ via Eq (3), Eq (7) and Eq (13) respectively.
  Update the parameters Θ by gradients descending: $\Theta = \Theta - \eta \nabla_{\Theta} (\mathcal{L}_{\text{sup}} + \alpha \mathcal{L}_{\text{ae}} + \beta \mathcal{L}_{\text{dp}})$
end while

3.5 Theory Analysis

Theorem 1. *The representation $\boldsymbol{Z}^{(\ell)}$ learned by EAS-GCN is equivalent to the sum of the representations with different-order structure information.*

Proof. Let us assume that $\sigma(x) = x$, $\boldsymbol{b}_e^{(\ell)} = 0$, $\boldsymbol{W}_e^{(\ell)} = \boldsymbol{I}$ and $\boldsymbol{W}^{(\ell)} = \boldsymbol{I}$, $\forall \ell \in [1, 2, \cdots, L]$. We can rewrite Eq (10) as

$$
\begin{aligned}
\boldsymbol{Z}^{(\ell+1)} &= \widetilde{\boldsymbol{A}} \widehat{\boldsymbol{Z}}^{(\ell)} \\
\boldsymbol{Z}^{(\ell+1)} &= (1-p) \widetilde{\boldsymbol{A}} \boldsymbol{Z}^{(\ell)} + p \widetilde{\boldsymbol{A}} \boldsymbol{H}_e^{(\ell)},
\end{aligned} \tag{15}
$$

where $\widetilde{\boldsymbol{A}} = \widehat{\boldsymbol{D}}^{-\frac{1}{2}} \widehat{\boldsymbol{A}} \widehat{\boldsymbol{D}}^{-\frac{1}{2}}$. After L-th propagation step, the result is

$$
\boldsymbol{Z}^{(L)} = (1-p)^L \widetilde{\boldsymbol{A}}^L \boldsymbol{X} + p \sum_{\ell=1}^{L} (1-p)^{\ell-1} \widetilde{\boldsymbol{A}}^\ell \boldsymbol{H}_e^{(\ell)}. \tag{16}
$$

The output of the standard GCN is $\widetilde{\boldsymbol{A}}^L \boldsymbol{X}$, which pays attention to high-order structure information and is prone to over-smoothing. Moreover, if $L \to \infty$, the left term of Eq. (16) tends to 0 and the data representation is dominated by the right term of Eq. (16). EAS-GCN can alleviate the over-smoothing in GCN, because we can clearly see that the right term of Eq. (16) is the sum of representations with different-order structure information.

4 Experiments

To evaluate the effectiveness of our proposed EAS-GCN, we conduct extensive experiments on node classification tasks. First, we introduced the datasets, experimental environment and parameter settings. Then, we compare EAS-GCN with the previous state-of-the-art baseline to prove the superiority of EAS-GCN. Finally, we conducted some performance studies to validate the proposed model further.

4.1 Datasets

We use the same train/validation/test splits as [10] for citation datasets (Cora, CiteSeer, and PubMed). For the other datasets [15], we randomly select 20 labeled nodes per class as the training set, 30 labeled nodes per class as the validation set, and the rest as the test set. The statistics of datasets are summarized in Table 1.

Table 1. Statistics of datasets.

Dataset	Nodes	Edges	Features	Classes	Training nodes	Validation nodes	Test nodes
Cora	2708	5278	1433	7	20 per class	500	1000
CiteSeer	3327	4552	3703	6	20 per class	500	1000
PubMed	19717	44324	500	3	20 per class	500	1000
Amazon Photo	7487	119043	745	8	20 per class	30 per class	Rest Nodes
Amazon Computers	13381	245778	767	10	20 per class	30 per class	Rest Nodes
Coauthor CS	18333	81894	6805	15	20 per class	30 per class	Rest Nodes
Coauthor Physics	34493	247962	8415	5	20 per class	30 per class	Rest Nodes

Table 2. Hyper-parameters specifications.

DataSet	α	β	p	L	Learning rate	Weight decay	Dropout rate	Hidden dimension
Cora	0.5	0.3	0.6	5	1e−3	1e−2	0.8	256
CiteSeer	1	1	0.6	5	1e−2	1e−2	0.8	256
PubMed	1	0.1	0.1	5	1e−2	1e−3	0.8	128
Amazon Photo	0.4	1	0.1	5	5e−3	5e−3	0.8	128
Amazon Computers	1	0.6	0.2	5	5e−3	5e−3	0.8	128
Coauthor CS	0.3	0.1	0.6	5	5e−3	5e−3	0.8	256
Coauthor Physics	0.3	1	0.2	5	1e−2	5e−3	0.8	128

4.2 Implementation and HyperParamter

The experiments are conducted on a machine with AMD Ryzen 7 5800H CPU @ 3.20 GHz, and a single NVIDIA GeForce RTX 3070 with 8GB GPU memory. As for software versions, we use Python 3.8, Pytorch 1.9.1, Pytorch Geometric 2.0.1 [7], and CUDA 11.1. The hyper-parameters in each baseline are set according to the original paper if available. We perform a grid search to tune the hyper-parameters for EAS-GCN based on the accuracy on the validation set. Adam optimizer are used on all datasets. α, β and p is obtained from a search of range 0.1 to 1 with step 0.1, L is set to 5, the dropout rate is set to 0.8, the learning rate and weight decay is chosen from $\{1e-3, 5e-3, 1e-2\}$, the hidden dimension is chosen from $\{128, 256\}$. The detailed hyper-parameter settings for EAS-GCN is in Table 2.

4.3 Classification Results

We choose the following baseline methods: GCN [10], GAT [17], ResGCN [12], JK-Net [26], SGC [21], APPNP [11], S^{2}GC [28], DropEdge [14], and GCNII [3]. To alleviate the influence of randomness, we repeat each method 100 times and report the mean performance and the standard deviations. The experimental results are summarized in Table 3. On Cora, EAS-GCN performs worse than GCNII. The reason may be due to the richness of information contained in the original features of the nodes in the Cora dataset, while GCNII can incorporate certain original features of the nodes through residuals. On the other datasets, EAS-GCN performs better than the representative baselines by significant margins and outperforms GCN by a margin of 1.3% to 4.0%. While most baselines use a multilayer perceptron (MLP) for feature transformation, our proposed framework uses a autoencoder for feature transformation, which allows us to better extract information from the data itself. EAS-GCN passes the autoencoder-specific representations incorporating local structural information to GCN, which is significantly better than other baselines in extracting attribute and structural information.

Table 3. Results on all datasets in terms of classification accuracy (in percent).

Model	Cora	CiteSeer	PubMed	Amazon computers	Amazon Photo	Coauthor CS	Coauthor Physics
GCN	81.3 ± 0.6	71.1 ± 0.8	78.9 ± 0.5	82.4 ± 0.4	91.2 ± 0.6	90.7 ± 0.2	92.4 ± 0.9
GAT	83.1 ± 0.4	70.8 ± 0.5	79.0 ± 0.3	80.1 ± 0.6	90.8 ± 0.9	90.5 ± 0.6	89.8 ± 1.1
ResGCN	82.2 ± 0.6	70.8 ± 0.7	78.3 ± 0.6	81.1 ± 0.7	91.3 ± 0.9	87.9 ± 0.6	92.2 ± 1.5
JK-Net	81.8 ± 0.5	70.7 ± 0.7	78.8 ± 0.5	81.9 ± 0.8	91.6 ± 0.7	89.8 ± 0.7	92.1 ± 0.5
SGC	81.0 ± 0.1	71.3 ± 0.3	78.9 ± 0.2	82.1 ± 0.7	91.5 ± 0.8	90.3 ± 0.5	91.6 ± 0.9
APPNP	83.3 ± 0.5	71.8 ± 0.6	80.1 ± 0.2	81.7 ± 0.3	91.3 ± 0.4	91.8 ± 0.4	92.7 ± 0.9
S^{2}GC	82.7 ± 0.4	73.0 ± 0.1	79.9 ± 0.4	83.0 ± 0.4	91.5 ± 0.7	91.6 ± 0.6	92.9 ± 0.8
DropEdge	82.8 ± 0.2	72.3 ± 0.4	79.6 ± 0.3	82.4 ± 0.7	91.3 ± 0.5	91.5 ± 0.8	92.5 ± 0.5
GCNII	**85.5 ± 0.5**	73.4 ± 0.6	80.3 ± 0.4	81.9 ± 0.3	92.1 ± 0.5	92.0 ± 0.5	92.8 ± 0.7
EAS-GCN	84.3 ± 0.2	**73.9 ± 0.5**	**80.6 ± 0.1**	**85.7 ± 0.5**	**93.1 ± 0.5**	**92.9 ± 0.4**	**93.6 ± 0.5**

4.4 Training Time Comparison

We compared the training time of some baselines on citation datasets. We marked the top two models with the fastest training time and highest classification accuracy. The results are shown in Table 4. We can see that EAS-GCN achieves the fastest training time among all mehods except SGC. The excellent training time is attributed to the model's ability to converge quickly. On citation datasets, the model achieves good classification accuracy with 50 epochs. SGC is a simplified graph convolution model, which removes the feature transformation in the propagation phase so that the training time will be faster. EAS-GCN outperforms standard GCN in both classification accuracy and training time. Remarkably, EAS-GCN has a competitive effect compared to the current state-of-the-art method GCNII, while achieving up to 103×, 54×, and 57× training speedups on Cora, CiteSeer, and PubMed. This is because EAS-GCN has a faster convergence rate than GCNII.

Table 4. Training time Comparison on Cora, CiteSeer, and PubMed.

Model	Cora		CiteSeer		PubMed	
	time(s)	ACC(%)	time(s)	ACC(%)	time(s)	ACC(%)
GCN	1.03	81.3	1.12	71.1	0.85	78.9
GAT	4.07	83.1	4.27	70.8	5.84	79.0
SGC	**0.65**	81.0	**0.35**	71.3	**0.74**	78.9
APPNP	0.89	83.3	1.05	71.8	0.86	80.1
GCNII	87.38	**85.5**	52.18	**73.4**	39.33	**80.3**
EAS-GCN	**0.85**	**84.3**	**0.96**	**73.9**	**0.69**	**80.6**

4.5 Ablation Study

We conduct an ablation study to examine the contributions of different components in EAS-GCN. Specifically, we build the following variants:

- **Without Degree Prediction (DP):** We only use the autoencoder to assist GCN in learning enhanced attribute information. Since the degree prediction module is removed, GCN will not be enough to learn good local structure information with the stacking of layers.
- **Without Auto-Encoder (AE):** Since the degree prediction module needs the hidden representation of the autoencoder, when the autoencoder module is removed, the model degenerates to the standard GCN.

Table 5. Ablation study results in terms of accuracy of node classification.

Ablation	Cora	CiteSeer	PubMed	Amazon Computers	Amazon Photo	Coauthor CS	Coauthor Physics
EAS-GCN	**84.3 ± 0.2**	**73.9 ± 0.5**	**80.6 ± 0.1**	**85.7 ± 0.5**	**93.1 ± 0.5**	**92.9 ± 0.4**	**93.6 ± 0.5**
w/o DP	83.8 ± 0.3	73.1 ± 0.8	79.7 ± 0.6	84.7 ± 1.3	92.6 ± 0.4	92.2 ± 0.5	93.1 ± 0.7
w/o AE	81.3 ± 0.6	71.1 ± 0.8	78.9 ± 0.5	82.4 ± 0.4	91.2 ± 0.6	90.7 ± 0.2	92.4 ± 0.9

In Table 5, we have two observations. First, all EAS-GCN variants show a significant performance degradation compared to the full model, indicating that each component contributes to the performance of EAS-GCN. Second, compared with learning sufficient local structure information, GCN is more deficient in learning effective attribute information.

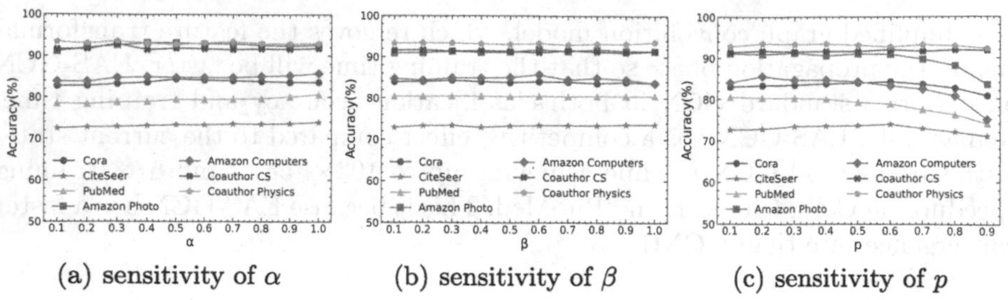

(a) sensitivity of α (b) sensitivity of β (c) sensitivity of p

Fig. 3. The performance of EAS-GCN with varying different hyper-parameters on all datasets.

4.6 Parameter Sensitivity Analysis

We perform parameters analysis on all datasets for three parameters: α, β and p. The results on the validation set are shown in Fig. 3. We find that EAS-GCN is robust to the α and β. α and β takes smaller values to achieve good results, indicating that the GCN does not need to excessively extract information from the data itself and does not need to incorporate too much local structure information. As p increases, the performance of the model decreases on most datasets, indicating that the attribute information passed from the autoencoder to GCN should be appropriate. If too much attribute information is transferred to GCN, it will affect the effectiveness of the GCN instead.

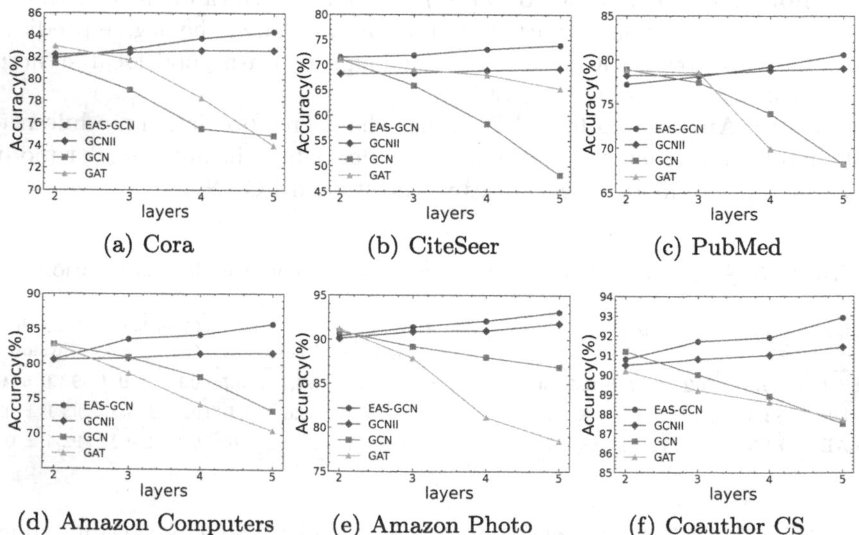

(a) Cora (b) CiteSeer (c) PubMed

(d) Amazon Computers (e) Amazon Photo (f) Coauthor CS

Fig. 4. Over-smoothing comparison

4.7 Over-Smoothing Analysis

For the target node with fewer neighbors, it is necessary to obtain more information by expanding the receptive field of the node. However, many GNNs face the problem of over-smoothing when stacking many layers, making nodes with different labels indistinguishable. We study the EAS-GCN's ability to alleviate over-smoothing by using the classification results in the case of stacking different layers. Figure 4 shows the classification accuracy of different models. This suggests that EAS-GCN can better alleviate over-smoothing while existing representative GNNs show significant performance degradation as the layer deepen. Because after stacking multiple layers, GCN focus on high-order structure information. And the GCN module in EAS-GCN is concerned with different order structure information. On most datasets, both EAS-GCN and the current state-of-the-art method GCNII show an increase in performance with deeper layers. This demonstrates the effectiveness of deep models. It is worth mentioning that the optimal effect reported in the original paper of GCNII is achieved at dozens of layers, and our framework can exceed its optimal effect at shallow layers.

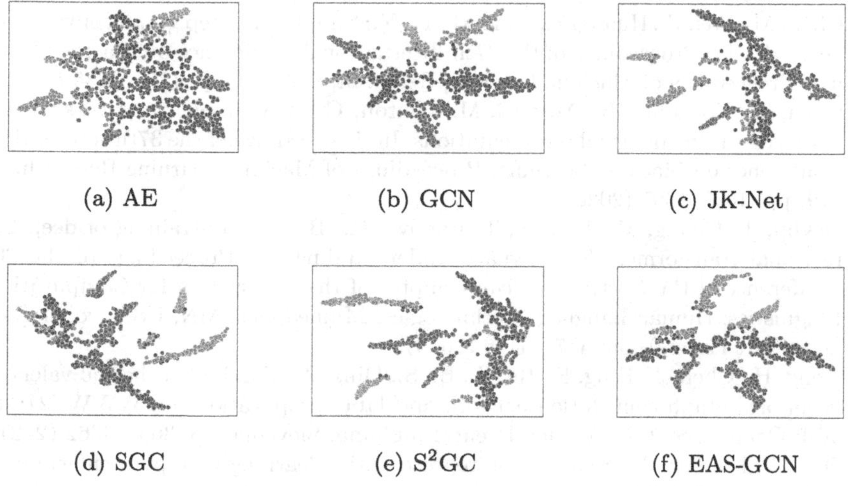

Fig. 5. Visualization on Cora

4.8 Visualization

To provide a more intuitive understanding of the learned node representations, we visualize node representations of autoencoder (AE), GCN and EAS-GCN by using t-SNE [16]. Each point represents a test node on Cora, and the color represents the node label. The results are shown in Fig. 5. We clearly observe that the nodes are better classified in EAS-GCN than AE, which means that EAS-GCN captures more detailed class information. Compared with GCN, the classification boundary of EAS-GCN is clearer.

5 Conclusions

In this paper, we propose an enhanced attribute-aware and structure-constrained graph convolutional network. Because GCN does not learn enough structure and attribute information and is prone to over-smoothing, we design a delivery mechanism to assist GCN to learn the better structure and attribute information through degree prediction and autoencoder. Extensive experiments on seven benchmark graph datasets demonstrate the high accuracy and efficiency of EAS-GCN against the state-of-the-art GNNs.

References

1. Abburi, H., Parikh, P., Chhaya, N., Varma, V.: Fine-grained multi-label sexism classification using a semi-supervised multi-level neural approach. Data Sci. Eng. **6**(4), 359–379 (2021)
2. Bo, D., Wang, X., Shi, C., Zhu, M., Lu, E., Cui, P.: Structural deep clustering network. In: WWW '20: The Web Conference 2020, Taipei, Taiwan, pp. 1400–1410 (2020)
3. Chen, M., Wei, Z., Huang, Z., Ding, B., Li, Y.: Simple and deep graph convolutional networks. In: Proceedings of the 37th International Conference on Machine Learning. Proceedings of Machine Learning Research, vol. 119, pp. 1725–1735 (2020)
4. Chen, T., Kornblith, S., Norouzi, M., Hinton, G.E.: A simple framework for contrastive learning of visual representations. In: Proceedings of the 37th International Conference on Machine Learning. Proceedings of Machine Learning Research, vol. 119, pp. 1597–1607 (2020)
5. Devlin, J., Chang, M., Lee, K., Toutanova, K.: BERT: pre-training of deep bidirectional transformers for language understanding. In: Proceedings of the 2019 Conference of the North American Chapter of the Association for Computational Linguistics: Human Language Technologies, Minneapolis, MN, USA, vol. 1 (Long and Short Papers), pp. 4171–4186 (2019)
6. Dong, H., Chen, J., Feng, F., He, X., Bi, S., Ding, Z., Cui, P.: On the equivalence of decoupled graph convolution network and label propagation. In: WWW '21: The Web Conference 2021, Virtual Event/Ljubljana, Slovenia, pp. 3651–3662 (2021)
7. Fey, M., Lenssen, J.E.: Fast graph representation learning with pytorch geometric. CoRR abs/1903.02428 (2019)
8. Hinton, G.E., Salakhutdinov, R.: Reducing the dimensionality of data with neural networks. Science **313**, 504–507 (2006)
9. Jin, W., et al.: Self-supervised learning on graphs: Deep insights and new direction. CoRR abs/2006.10141 (2020)
10. Kipf, T.N., Welling, M.: Semi-supervised classification with graph convolutional networks. In: 5th International Conference on Learning Representations, ICLR (2017)
11. Klicpera, J., Bojchevski, A., Günnemann, S.: Predict then propagate: Graph neural networks meet personalized pagerank. In: 7th International Conference on Learning Representations, ICLR (2019)
12. Li, G., Müller, M., Thabet, A.K., Ghanem, B.: Deepgcns: Can gcns go as deep as cnns? In: 2019 IEEE/CVF International Conference on Computer Vision, ICCV 2019, Seoul, Korea (South), pp. 9266–9275 (2019)

13. Li, Q., Han, Z., Wu, X.: Deeper insights into graph convolutional networks for semi-supervised learning. In: Proceedings of the Thirty-Second AAAI Conference on Artificial Intelligence, New Orleans, Louisiana, USA, pp. 3538–3545 (2018)
14. Rong, Y., Huang, W., Xu, T., Huang, J.: Dropedge: towards deep graph convolutional networks on node classification. In: 8th International Conference on Learning Representations, ICLR (2020)
15. Shchur, O., Mumme, M., Bojchevski, A., Günnemann, S.: Pitfalls of graph neural network evaluation. CoRR abs/1811.05868 (2018)
16. Van Der Maaten, L., Hinton, G.: Visualizing data using t-sne. J. Mach. Learn. Res. **9**(2605), 2579–2605 (2008)
17. Velickovic, P., Cucurull, G., Casanova, A., Romero, A., Liò, P., Bengio, Y.: Graph attention networks. In: 6th International Conference on Learning Representations, ICLR (2018)
18. Waikhom, L., Patgiri, R.: Graph neural networks: Methods, applications, and opportunities. CoRR abs/2108.10733 (2021)
19. Wang, D., Cui, P., Zhu, W.: Structural deep network embedding. In: Proceedings of the 22nd ACM SIGKDD International Conference on Knowledge Discovery and Data Mining, San Francisco, CA, USA, pp. 1225–1234 (2016)
20. Wang, X., He, X., Wang, M., Feng, F., Chua, T.: Neural graph collaborative filtering. In: Proceedings of the 42nd International ACM SIGIR Conference on Research and Development in Information Retrieval, Paris, France pp. 165–174 (2019)
21. Wu, F., Jr., A.H.S., Zhang, T., Fifty, C., Yu, T., Weinberger, K.Q.: Simplifying graph convolutional networks. In: Proceedings of the 36th International Conference on Machine Learning, Long Beach, California, USA. Proceedings of Machine Learning Research, vol. 97, pp. 6861–6871 (2019)
22. Wu, S., Zhang, Y., Gao, C., Bian, K., Cui, B.: GARG: anonymous recommendation of point-of-interest in mobile networks by graph convolution network. Data Sci. Eng. **5**(4), 433–447 (2020)
23. Wu, Z., Pan, S., Chen, F., Long, G., Zhang, C., Yu, P.S.: A comprehensive survey on graph neural networks. IEEE Trans. Neural Networks Learn. Syst. **32**(1), 4–24 (2021)
24. Xie, Y., Xu, Z., Wang, Z., Ji, S.: Self-supervised learning of graph neural networks: a unified review. CoRR abs/2102.10757 (2021)
25. Xu, K., Hu, W., Leskovec, J., Jegelka, S.: How powerful are graph neural networks? In: 7th International Conference on Learning Representations, ICLR (2019)
26. Xu, K., Li, C., Tian, Y., Sonobe, T., Kawarabayashi, K., Jegelka, S.: Representation learning on graphs with jumping knowledge networks. In: Proceedings of the 35th International Conference on Machine Learning, Stockholmsmässan, Stockholm, Sweden. Proceedings of Machine Learning Research, vol. 80, pp. 5449–5458 (2018)
27. Zhang, Z., Cui, P., Zhu, W.: Deep learning on graphs: a survey. IEEE Trans. Knowl. Data Eng. **34**(1), 249–270 (2022)
28. Zhu, H., Koniusz, P.: Simple spectral graph convolution. In: 9th International Conference on Learning Representations, ICLR (2021)

Mining Frequent Patterns with Counting Quantifiers

Yanxiao He, Xin Wang(✉), Yuji Sha, Xueyan Zhong, and Yu Fang

Southwest Petroleum University, Chengdu, China
{202022000283,202025000059}@stu.swpu.edu.cn,
{xinwang,zhongxueyan,fangyu}@swpu.edu.cn

Abstract. Frequent Pattern Mining (FPM) is a classical graph mining task. In recent years, we have witnessed wide applications of FPM. However, emerging applications keep calling for more expressive patterns to capture more complex structures in a large graph. In light of this, we investigate the problem of *mining frequent patterns with counting quantifiers*. We first introduce quantified graph patterns (QGPs), which are the patterns incorporated with counting quantifiers. We then develop an algorithm along with an optimization technique to mine QGPs in a single large graph. On real-life graphs, we verified the performances of our (optimization) techniques.

Keywords: Quantified graph patterns · Frequent pattern mining

1 Introduction

The problem of FPM has been a core task in the field of graph mining, due to its wide applications. Facing more complicated scenarios, more expressive patterns are required, especially those with counting quantifiers (CQs). The introduction of CQs can not only extend the expressive ability of semantics but also capture complex structures in the underlying graphs [13].

Example 1. Graph patterns with CQs reflect the regularity of connections between entities in social networks. Patterns Q_1 and Q_2 of Fig. 1 show that:

∘ *if (i) person X_0 is in a music club, and (ii) among the people whom X_0 follows, at least m of them like Shakira's album, then X_0 may also like the album.*

∘ *if (i) m people Z whom X_0 follows, (ii) Z recommends Lenovo computer, then X_0 may buy a Lenovo computer.*

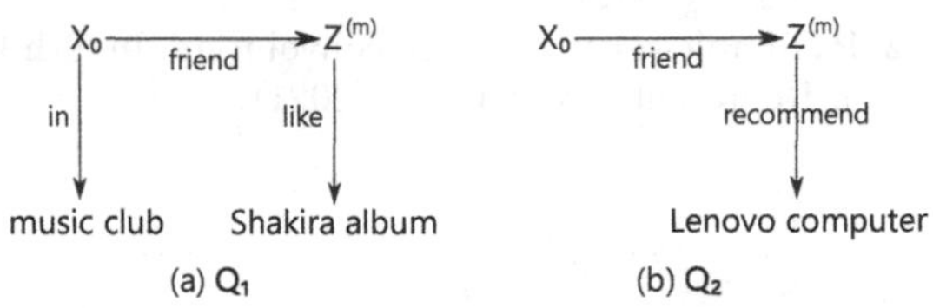

Fig. 1. Quantified graph patterns

B. Li et al. (Eds.): APWeb-WAIM 2022, LNCS 13421, pp. 372–381, 2023.
https://doi.org/10.1007/978-3-031-25158-0_28

Intuitively, the above graph patterns take quantifiers m, and achieve quantitative aggregation through m. Compared with the traditional graph pattern, the graph pattern with CQs has even richer semantics. Indeed, formulating a marketing strategy based on the above patterns can achieve better performance than traditional marketing methods. Since empirical studies suggest that "90% of customers believe in peer recommendations, while only 14% of customers trust advertisements [2]", and "the peer influence from one's friends leads to more than 50% increase in the probability of buying a product [7]". Despite the wide application, we are not aware any prior work for discovering such patterns.

Contributions. Our contributions in this paper can be summarized as follows.

(1) We extend the traditional graph pattern and propose QGPs (Sect. 2) for capturing more complex structures. (2) We propose QGPM, a framework for mining frequent QGPs in large single graphs. Based on QGPM, corresponding optimization strategies are proposed to improve the performance of the algorithm (Sect. 3). (3) Using real-life and synthetic graphs, we experimentally verify the performance of our algorithms (Sect. 4).

Related Work. We categorize related work as follows.

Graph Pattern Mining. [11] presented GraMi, a novel framework for frequent subgraph mining in a single large graph. [15] proposed FSSG, an algorithm that uses graph invariant properties and symmetries present in a graph to generate candidate subgraphs. FSSG reduces the generation of a large number of candidate subgraphs, thereby reducing the complexity of candidate generation and frequency counting. [17] introduced the SOCMI algorithm, its core idea is to store the appearance of patterns with pathgraph, which makes it easier for SOCMI to extend patterns and calculate frequency during mining. [6,16] proposed to mine weighted frequent subgraphs in a weighted single large graph. StreamFSM [21] and IncGM+ [4] are mining algorithms based on dynamic graphs.

Pattern Semantic Extension. SPARQLog extends SPARQL with first-order logic (FO) rules, including existential and universal quantification of nodes in the graph [10]. For social networks, SocialScope [5] and SNQL [20] are algebraic languages with numerical aggregations on node and edge sets. Furthermore, social recommendation rules are studied in [18], and support counts are introduced as constraints. Prior work show that existing extensions still suffer from limitations and in emerging applications, patterns with complex features, notably counting quantifiers (CQs), predicates and negation are required [19]. Among these features, CQs have received more attention [8,12,14,19].

2 Preliminaries

2.1 Conventional Graph Pattern Mining

Graph. A *graph* is defined as $G = (V, E, L)$, where (1) V is a set of nodes; (2) $E \subseteq V \times V$ is a set of edges; and (3) each node v in V carries a tuple

$L(v) = (A_1 = a_1, \cdots, A_n = a_n)$, where $A_i = a_i (i \in [1, n])$ represents that the node v has a value a_i for the attribute A_i, and is denoted as $v.A_i = a_i$, *e.g.*, v.job_title = "*DBA*".

A graph $G' = (V', E', L')$ is a *subgraph* of $G = (V, E, L)$, denoted by $G' \subseteq G$, if $V' \subseteq V$, $E' \subseteq E$, and moreover, for each $v \in V'$, $L'(v) = L(v)$.

Pattern. A *pattern* Q is defined as a graph (V_p, E_p, L_v), where V_p and E_p are the set of nodes and edges, respectively; for each u in V_p, it is associated with a predicate $L_v(u)$ defined as a conjunction of atomic formulas of the form '$A = a$' such that A denotes an attribute of the node u and a is a value of A.

Graph Pattern Matching. Consider graph $G = (V, E, L)$ and pattern $Q = (V_p, E_p, L_v)$, a node v in G satisfies the search conditions of a pattern node u in Q, denoted as $v \sim u$, if for each atomic formula '$A = a$' in $L_v(u)$, there exists an attribute A in $L(v)$ such that $v.A = a$. A match of pattern Q in graph G is a *bijective function* f from the nodes of Q to the nodes of a subgraph G, such that (1) for each node $u \in V_p$, $f(u) \sim u$, and (2) (u, u') is an edge in Q if and only if $(f(u), f(u'))$ is an edge in G. We denote by $Q(G)$ the set of matches of Q in G. Consider Q_1 and G in Fig. 2, a match of Q_1 in G is $\{v_3, v_1, v_4\}$.

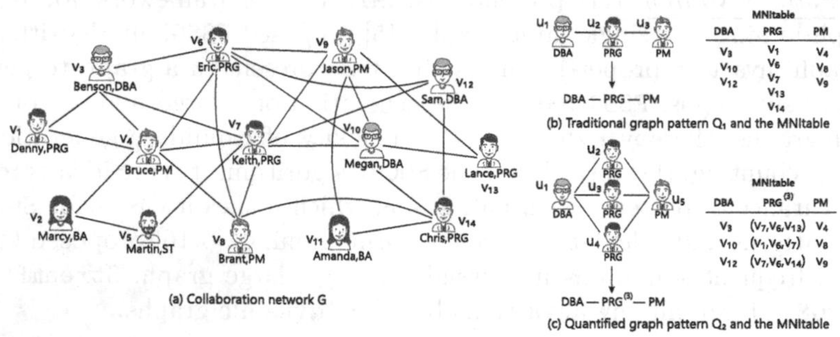

Fig. 2. Graph & Traditional graph pattern & Quantified graph pattern(QGP)

Support. The support of a pattern Q in a graph G, denoted by $\mathbf{sup}(G, Q)$, indicates the occurrence frequency of Q in G. It is crucial to develop an anti-monotone support metric for search space pruning. Several anti-monotone support metrics have been introduced, we adopt the MNI [9] in this paper, as it can be calculated more efficiently.

Let f be the set of isomorphic mappings of pattern $Q = (V_p, E_p, L_v)$ in G. The length of the set f is m, while $F(u) = \{f_1(u), f_2(u), \cdots, f_m(u)\}$ is the mapping function for each u on the set f after de-duplication, where $u \in V_p$. The minimum image based support (MNI) of Q in G, denoted by $\mathbf{sup}(G, Q)$, is defined as $\mathbf{sup}(G, Q) = \mathbf{min}\{t | t = |F(u)|, u \in V_p\}$.

We use *MNItable* to represent the MNI field of pattern Q. An *MNItable* consists of a set of MNIcol. An MNIcol(u_i) contains a list of distinct nodes, *i.e.*, matches in G of pattern node u_i in Q. Figure 2(b) shows a *MNItable*, where each `MNIcol` is given.

2.2 Quantified Graph Patterns

Quantified Graph Patterns (QGPs). A QGP $Q_{(u_0)}^{(m)}$ is defined as (V_p, E_p, L_u, f_u), where (1) V_p and E_p are the set of pattern nodes and edges, respectively; (2) L_u is a function that assigns a label to a node u ($u \subseteq V_p$); and (3) f_u assigns a CQ m (a natural number) to the specified quantifier-constrained node u_0 such that $f_u(u_0) = m$.

Intuitively, the quantified graph pattern extends the traditional graph pattern by incorporating a CQ. It supports counting on a single node, *i.e.*, for two or more nodes with the same label, if they all connect to the same set of nodes, they can be shrunk into a single node with a CQ. *e.g.*, in Fig. 2(c), Q_2 is a quantified graph pattern and can be represented as $DBA - PRG^{(3)} - PM$, *i.e.*, PRG is associated with a superscript "(3)" as a quantifier.

In particular, we only consider the case where a single node in the pattern has a CQ. For a QGP, when the quantifier constraints are not considered, the corresponding pattern is called a quantifier-free pattern, and is denoted as Q^w. *e.g.*, Q_1 in Fig. 2(b) is the quantifier-free pattern of Q_2 in Fig. 2(c).

Support of QGP. The support of QGP is defined along the same line its traditional counterpart, *e.g.*, for Q_2 in Fig. 2(c), `MNIcol`($PRG^{(3)}$)=$\{(v_7, v_6, v_{13}), (v_1, v_6, v_7), (v_7, v_6, v_{14})\}$, and it can be easily verified that $\mathtt{sup}(G, Q_2) = 3$.

Theorem 1. *Given a QGP* ${Q_1}_{(v)}^{(m)}$*, if there is another QGP* ${Q_2}_{(v)}^{(m')}$*, they have the same* Q^w*, but* $m > m'$*, then* ${Q_1}_{(v)}^{(m)}$ *implies* ${Q_2}_{(v)}^{(m')}$*, denoted as* ${Q_1}_{(v)}^{(m)} \Rightarrow {Q_2}_{(v)}^{(m')}$.

Example 2. Consider two QGPs, $Q_{(B)}^{(2)}$: $\mathtt{A} - \mathtt{B}^{(2)} - \mathtt{C}$ and $Q_{(B)}^{(3)}$: $\mathtt{A} - \mathtt{B}^{(3)} - \mathtt{C}$, we will only focus on the latter, because (1) $Q_{(B)}^{(3)} \Rightarrow Q_{(B)}^{(2)}$. (2) According to the "anti-monotone" property, if $Q_{(B)}^{(3)}$ is frequent, $Q_{(B)}^{(2)}$ must also be frequent.

For frequent patterns with the same Q^w, the one with the largest CQ is considered.

1-itemset. 1-itemset refers to the structure of $\mathtt{X} - \mathtt{Y}^{(m)}$ $(m \geq 1)$ in the graph, where X and Y are quantifier-free node and quantifier-constrained node, respectively. It is frequent when the support of a *1-itemset* meets the threshold.

Star- QGP. A star-QGP preserves the "star" topology, and a quantifier-constrained node is deemed as the star center, other nodes are connected to the star center.

Forward & Backward extension. The forward extension on a pattern Q essentially introduces a new edge from one node in Q; while the backward extension includes a new edge from two existing nodes. Interested readers can refer to [22].

3 Frequent QGPs Mining

In this section, we develop techniques to address the frequent QGPs mining problem.

Problem Statement. Given a graph G, support threshold τ, the *frequent* QGP*mining* (QPM) problem is to discover a set $\mathbb{S}$ of frequent QGPs Q of G such that $\text{sup}(G, Q) \geq \tau$ for any Q in $\mathbb{S}$. However, the problem is NP-hard and computationally expensive.

3.1 Identifying Frequent QGPs

Algorithm. We now provide an algorithm QGPM for mining QGPs. Unless otherwise stated, the patterns mentioned below refer to the QGP.

Initialization. QGPM first initializes a set of parameters: set *result* for keeping track of frequent QGPs (line 1), sets *fEdge* and *fItemSet* are used to maintain frequent single-edge patterns and frequent *1-itemsets*, respectively (lines 2–3). Note that the *1-itemset* here is the case with the largest CQ in the same Q^w.

Frequent QGPs Mining. Based on *fEdge* and *fItemSet*, QGPM is divided into different stages to obtain results. In the first stage, for each *1-itemset* (from *fItemSet*), QGPM applies the procedure *QuanExt* to generate a set of frequent Star-QGPs (lines 4–6); the second is forward expansion, QGPM calls the procedure *FWExt* to extend the Star-QGP produced in the first stage by adding new edges (from *fEdges*) (lines 7–8); finally, QGPM employs the procedure *BWExt* for backward extension (lines 9–10). In addition, to exclude the already generated extensions, we adopt the VF2-based graph isomorphism detection algorithm. Next, we describe the details of each stage.

(I) Star-QGP generation. QGPM employs procedure *QuanExt* to generate Star-QGPs. More specifically, *QuanExt* takes a QGP S as input and tries to extend it with the *1-itemset* of *fItemSet* (lines 2–5). That is, when S and the current *1-itemset* have the same quantifier-constrained node label, it means that they can be merged to generate a new candidate pattern. All applicable extensions that have not been previously considered are stored in *candidateSet* (line 6). Then, *QuanExt* (lines 7–11) evaluates all candidate patterns generated. If the candidate is frequent, the program continues to extend and evaluates it recursively. Otherwise, decrement the CQ for the current pattern, and check whether the pattern after the decrement is frequent.

Example 3. Recall Fig. 3 with τ=2 and G in Fig. 2(a). QGPM first initializes the sets *fEdge* and *fItemSet* (shown in Fig. 3(a)), as their support is greater than 2, and then QGPM works step-by-step. For example, considering Q_1 and Q_2 (Fig. 3(a)), QGPM applies procedure *QuanExt* to extend them, and generates a candidate Star-QGP Q_3 ($\text{sup}(G, Q_3) = 3$).

(II) Forward expansion. QGPM applies *FWExt* (not shown) for forward expansion, which is aimed at the quantifier-free nodes in QGPs. *FWExt* takes QGP S

as input, and then extends S by adding new edges (from *fEdges*) to generate candidate patterns. After that, *FWExt* evaluates the support of candidate patterns and eliminates the members of the candidate set that do not satisfy the support threshold τ. Finally, *FWExt* is recursively executed to further expand the frequent subgraphs.

Algorithm 1: QGPM

```
Input: graph G, support threshold τ.
Output: All QGPs Q of G such that sup(G, Q) ≥ τ.
result ← ∅;
Let fEdges be the set of frequent single edge patterns in G;
Let fItemSet be the set of frequent 1-itemset in G;
for each item ∈ fItemSet do
    result ← result ∪ QuanExt(item,G,τ,fItemSet);
    Remove item from fItemSet;
for each fwitem ∈ result do
    result ← result ∪ FWExt(fwitem,G,τ,fEdges);
for each bwitem ∈ result do
    result ← result ∪ BWExt(bwitem,G,τ,fEdges);
return result;
```

Algorithm 2: QuanExt

```
Input: QGP S, graph G, support threshold τ, frequent itemset fItemSet.
Output: frequent Star-QGPs.
result ← S, candidateSet ← ∅;
for each item ∈ fItemSet do
    if item.QnodeLabel equals S.QnodeLabel then
        Let G'_s be the extension of S with item;
        if G'_s was not generated then
            candidateSet ← candidateSet ∪ {G'_s};
for each c ∈ candidateSet do
    if ISFrequent(c, G, τ) then
        result ← result ∪ QuanExt(c, G, τ, fItemSet);
    else
        candidateSet ← candidateSet ∪ CQdec(c, G, τ);
return result;
```

(III) Backward expansion. `QGPM` applies *BWExt* (not shown) for backward expansion. Similar to *FWExt*, *BWExt* takes `QGP` S as input and then extends S by adding new edges (from *fEdges*), but the difference is that *BWExt* does not introduce new nodes (i.e. two nodes of the extended edge already exist in pattern S).

Example 4. Continuing with Example 3, *FWExt* generates candidate pattern Q_4 by enlarging Q_3 with frequent single-edge pattern $PM - BA$. Then Q_5 is generated by extending Q_4 through *BWExt*.

During the pattern extension phase, QGPM uses the procedure *CQdec* (not shown) to reduce the CQ value of a pattern on hand. Specifically, if a QGP is infrequent but its CQ $m > 2$, we reduce m and then continue to determine if the pattern is frequent.

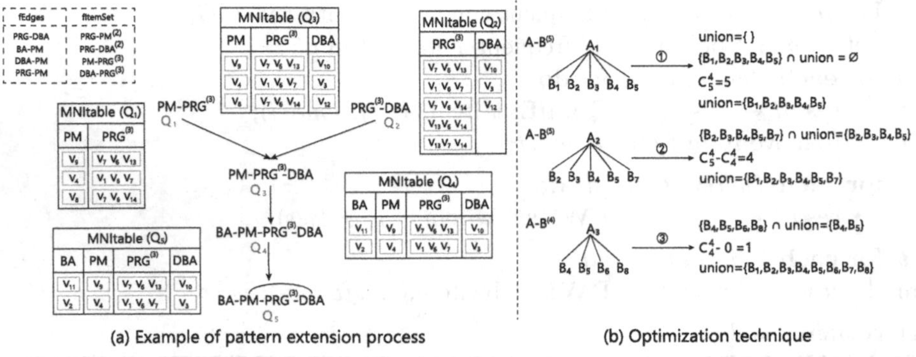

Fig. 3. Pattern extension & Optimization

Support Evaluation. Recall Sect. 2, a QGP S is frequent in G (*i.e.*, $\mathtt{sup}(G, Q) \geq \tau$) if there exist at least τ nodes in each domain ($\mathtt{MNIcol}(v_1), \cdots, \mathtt{MNIcol}(v_n)$) that are valid variable assignments (*i.e.*, are part of a match) for the corresponding variables $(v_1, \cdots, v_n)$. To evaluate support, a procedure *ISFrequent*(not shown) is employed by QGPM. It returns *true* iff S is a frequent QGP in G or *false* otherwise. In a nutshell, *ISFrequent* considers each match and places nodes assigned to variables in the corresponding domains. If all domains have at least τ nodes, then S is frequent in G. Otherwise, *ISFrequent* continues to consider other matches.

We refer interested readers to [1] for algorithm details.

Optimization. To further improve the performance of QGPM, we propose the following optimization. The technique is based on the observation that the support evaluation does not need to obtain the exact result, instead, one only needs to know whether the number of valid matches of each pattern node meets the threshold τ. We hence introduce a technique for support estimation.

Figure 3(b) shows the specific process. An empty set union (recording the matches that have already appeared) is initialized first, and then the existing matches are traversed in turn. For the first match, there are 5 matches of $Q^{(4)}_{(B)}$, and then update the *union*. For the second match, the intersection with the union is $\{B_2, B_3, B_4, B_5\}$, indicating that a match formed by the intersection has been generated before, and if such a match occurs again, it must be subtracted, so the number of matches $Q^{(4)}_{(B)}$ formed by the second match is 4, and the *union* is then updated as well. For the third match, its intersection with *union* is $\{B_4, B_5\}$, and the cardinality of the intersection is less than the number of quantifiers, indicating that the previous match will not be generated, so there is 1 match. The final match count is the sum of the matches obtained each time, *i.e.*, the matches of $Q^{(4)}_{(B)}$ is 10.

4 Experimental Study

Using real-life data, we conducted a set of experiments to evaluate the performance of QGPM and its optimization technique.

Experimental Setting. We used six graphs, (a) CiteSeer [11] represents a graph consisting of publications (nodes) and citations between them (edges). (b) lasftm_asia [3], a social network consisting of users (nodes) and following

Table 1. Description of graphs ($|L|$ indicates no. of node labels).

DataSets	$\|V\|$	$\|L\|$	$\|E\|$
Citeseer	3312	6	4519
Lasftm_asia	7624	12	27806
Mico	100000	29	1080298
DBLP	425957	25	1049866
Amazon	334863	80	925872
AstroPh	18772	50	396160

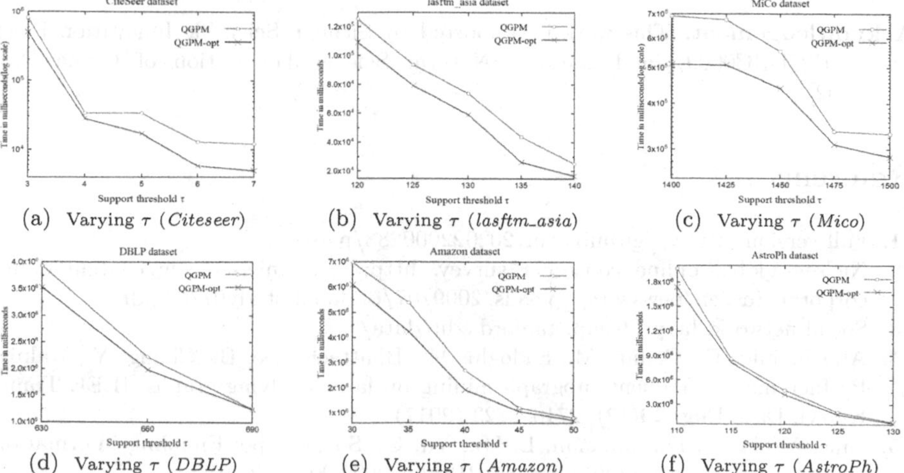

(a) Varying τ (*Citeseer*) (b) Varying τ (*lasftm_asia*) (c) Varying τ (*Mico*)

(d) Varying τ (*DBLP*) (e) Varying τ (*Amazon*) (f) Varying τ (*AstroPh*)

Fig. 4. Performance of QGPM & QGPM-opt on real-life graphs

relationships (edges) between them. (c) Mico [11] models the Microsoft co-authorship information. (d) DBLP [3] represents a collaborative network consisting of authors (nodes) and partnerships (edges) between them. (e) Amazon [3], is a product co-purchasing network. (f) AstroPh [3] covers scientific collaboration between author papers submitted to the Astrophysics category. Table 1 shows the statistics of each graph.

Experimental Results. We next report our findings.

Exp-1: Performance of QGPM & QGPM-opt. The results presented in Fig. 4(a)–4(f) show the results of the algorithm on CiteSeer, lasftm_asia, MiCo, DBLP, Amazon, and AstroPh, respectively, and demonstrate the following. (1) All algorithms take longer with a small τ because more candidate patterns and their matches need to be verified. (2) QGPM-opt performs better than QGPM, as unnecessary computations are avoided owing to the strategy QGPM-opt applied.

Frequency Threshold. As is shown, the support threshold τ is a crucial factor. When the threshold decreases, there might be an exponential number of candidate pattern that leads to exponential run time. For a given time budget, an efficient algorithm should be able to solve mining problems with low τ values.

5 Conclusion

Incorporating counting quantifiers into traditional graph patterns can capture more complex structural relationships in graph data. Accordingly, we have studied the problem of mining Quantified Graph Pattern (QGP) in a single large graph and propose QGPM to identify QGPs in the graph that satisfy threshold conditions. Our experimental study validates the efficiency and effectiveness of our techniques.

Acknowledgement. This work is supported by Sichuan Scientific Innovation Fund (No. 2022JDRC0009), and National Natural Science Foundation of China [No. 62172102].

References

1. Full version. https://github.com/202022000283/paper
2. Nielsen global online consumer survey. http://www.nielsen.com/content/dam/corporate/us/en/newswire/uploads/2009/07/prglobal-study07709.pdf
3. Social network. http://snap.stanford.edu/data/
4. Abdelhamid, E., Canim, M., Sadoghi, M., Bhattacharjee, B., Chang, Y., Kalnis, P.: Incremental frequent subgraph mining on large evolving graphs. IEEE Trans. Knowl. Data Eng. **29**(12), 2710–2723 (2017)
5. Amer-Yahia, S., Lakshmanan, L.V.S., Yu, C.: Socialscope: Enabling information discovery on social content sites. CoRR abs/0909.2058 (2009)
6. Ashraf, N., et al.: Wefres: weighted frequent subgraph mining in a single large graph. In: ICDM 2019, pp. 201–215. ibai Publishing (2019)

7. Bapna, R., Umyarov, A.: Do your online friends make you pay? a randomized field experiment on peer influence in online social networks. Manag. Sci. **61**(8), 1902–1920 (2015)
8. Blau, H., Immerman, N., Jensen, D.: A visual language for querying and updating graphs. University of Massachusetts Amherst Computer Science Technical Report **37**, 2002 (2002)
9. Bringmann, B., Nijssen, S.: What is frequent in a single graph? In: Washio, T., Suzuki, E., Ting, K.M., Inokuchi, A. (eds.) PAKDD 2008. LNCS (LNAI), vol. 5012, pp. 858–863. Springer, Heidelberg (2008). https://doi.org/10.1007/978-3-540-68125-0_84
10. Bry, F., Furche, T., Marnette, B., Ley, C., Linse, B., Poppe, O.: SPARQLog: SPARQL with rules and quantification. In: de Virgilio, R., Giunchiglia, F., Tanca, L. (eds.) Semantic Web Information Management, pp. 341–370. Springer, Heidelberg (2009). https://doi.org/10.1007/978-3-642-04329-1_15
11. Elseidy, M., Abdelhamid, E., Skiadopoulos, S., Kalnis, P.: GRAMI: frequent subgraph and pattern mining in a single large graph. Proc. VLDB Endow. **7**(7), 517–528 (2014)
12. Fan, W.: Graph pattern matching revised for social network analysis. In: 15th International Conference on Database Theory, ICDT 2012, Berlin, Germany, 26–29 March, 2012, pp. 8–21. ACM (2012)
13. Fan, W., Wang, X., Wu, Y., Xu, J.: Association rules with graph patterns. Proc. VLDB Endow. **8**(12), 1502–1513 (2015)
14. Fan, W., Wu, Y., Xu, J.: Adding counting quantifiers to graph patterns. In: Proceedings of the International Conference on Management of Data, pp. 1215–1230. ACM (2016)
15. Kavitha, D., Haritha, D., Padma, Y.: Optimized candidate generation for frequent subgraph mining in a single graph. In: Proceedings of International Conference on Computational Intelligence and Data Engineering, pp. 259–272. Springer, Singapore (2021). https://doi.org/10.1007/978-981-15-8767-2_23
16. Le, N., Vo, B., Nguyen, L.B.Q., Fujita, H., Le, B.: Mining weighted subgraphs in a single large graph. Inf. Sci. **514**, 149–165 (2020)
17. Li, L., Ding, P., Chen, H., Wu, X.: Frequent pattern mining in big social graphs. IEEE Trans. Emerging Topics Comput. Intell., 1–11 (2021)
18. Lin, W., Alvarez, S.A., Ruiz, C.: Collaborative recommendation via adaptive association rule mining. Data Min. Knowl. Disc. **6**(1), 83–105 (2000)
19. Mahfoud, H.: Expressive top-k matching for conditional graph patterns. Neural Computing and Applications, pp. 1–17 (2021)
20. Martín, M.S., Gutiérrez, C., Wood, P.T.: SNQL: a social networks query and transformation language, vol. 749 (2011)
21. Ray, A., Holder, L., Choudhury, S.: Frequent subgraph discovery in large attributed streaming graphs. In: Proceedings of the 3rd International Workshop on Big Data, pp. 166–181. PMLR (2014)
22. Wang, X., Xu, Y., Zhan, H.: Extending association rules with graph patterns. Expert Syst. Appl. **141** (2020). https://doi.org/10.1016/j.eswa.2019.112897

MST-GNN: A Multi-scale Temporal-Enhanced Graph Neural Network for Anomaly Detection in Multivariate Time Series

Zefei Ning, Zhuolun Jiang, Hao Miao, and Li Wang(✉)

Data Science College, Taiyuan University of Technology, Jinzhong 030600, Shanxi, China
wangli@tyut.edu.cn

Abstract. Anomaly detection in time is an important task in many applications. Sensors are deployed in the industrial site to monitor the condition of different attributes or different places in real time, which generate multivariate time series. Recently, many methods were proposed to detect anomalies with multivariate time series, but they focused on the sequence attributes or spatial and temporal correlation, ignoring the characteristic of single sensor time series. In this paper, we propose a novel model MST-GNN that builds each sensor representation from Multi-Scale Temporal (MST) view, and use Graph Neural Network to mine their latent correlation to improve the performance of anomaly detection. In the MST representation, shapelets learning is introduced to extract its distinguishing features, a recurrent-skip neural network is used to extract the local temporal dependence relationship, and the raw data retains the original features of time series. These three features are fused to form the multi-scale temporal-enhanced features. Subsequently, the graph neural network is adopted to capture the potential interdependencies between multivariate time series and obtain the optimal representation of time series. Finally, bias assessment and anomaly detection are carried out. Extensive experiments on real-world datasets show that MST-GNN outperforms other state-of-the-art methods consistently, which provides an effective solution for anomaly detection in multivariate time series.

Keywords: Anomaly detection · Mutlivariate time series · Multi-scale temporal-enhanced · Shapelets

1 Introduction

Anomaly detection is an important task in time series analysis, which aims to identify abnormal states to help eliminate abnormal events. At present, anomaly detection has been widely used in finance, aerospace, medical and many other fields [1].

Multi-sensor multi-point monitoring provides characteristics from different attributes and locations of time series, effectively improving the accuracy of scene state judgment in multivariate time series analysis. Generally, the algorithm of anomaly detection obtains the normal law of time series through a learning mechanism, and then evaluates the deviation between observed and expected normal values to determine

B. Li et al. (Eds.): APWeb-WAIM 2022, LNCS 13421, pp. 382–390, 2023.
https://doi.org/10.1007/978-3-031-25158-0_29

anomalies. Shapelets are subsequences that express the discriminative features of time series, which can be used to represent time series [2]. At present, the introduction of deep learning makes remarkable progress on anomaly detection in multivariate time series. Deng et al. [3] introduced graph structure to express the relationship between multivariate time series, captured correlations between sensors and detected anomalies based on Graph Neural Networks (GNN). These studies provide solutions for anomaly detection in terms of multivariate time series fusion and reconstruction. However, most methods either focus on the time series features of single time series, or focus on the correlations of multivariate time series, ignoring the importance of the expression of single time series for discovering the correlation between multivariate time series.

In this paper, we propose a novel Multi-Scale Temporal-enhanced Graph Neural Network (MST-GNN) to detect anomalies in multivariate time series. Specifically, MST-GNN utilizes shapelets learning to extract spatial features that are discriminative from other sensors based on full historical data. Recurrent-skip neural network is introduced to extract temporal features with temporal structure discrimination, which are used to enhance the temporal features of single sensor. Subsequently, the two spatiotemporal features and the raw data are fused to form the multi-scale temporal-enhanced representation. Then the enhanced representation is added with a learnable embedding, which is used to calculate the dependency degree between sensors and based on the relationship, a sensor graph is created. Further, the ultimate graph representation is obtained based on graph attention network. Finally, anomaly detection is performed by evaluating deviations. The main contributions are summarized as follows:

- A multi-scale temporal enhanced representation for multivariate time series is proposed, which extracts the spatiotemporal representation of time series and preserves the raw characteristics of time series.
- An improved graph deviation network with multi-scale temporal-enhanced representation is proposed, which takes the multi-scale temporal-enhanced representation as prior knowledge, to guide the construction of graph and the calculation of graph attention, so as to obtain the final sensor representation.
- Extensive experiments are conducted that show our model outperforms the state-of-the-art methods in anomaly detection, especially in the anomaly recall ability.

2 Related Work

2.1 Traditional Anomaly Detection

The traditional anomaly detection methods of multivariate time series include linear model-based method, distance-based method, probabilistic and density estimation-based method, etc. Linear model-based method trains a linear classifier to detect whether sample points are abnormal [4]. Distance-based methods use nearest neighbor distance to define anomaly [5]. Probabilistic model-based and density estimation-based methods pay more attention to the distribution of data [6]. However, none of these methods can consider the temporal correlation of time steps.

In the task of anomaly detection in time series, feature expression of time series is crucial. Ye et al. [2] proposed shapelets, which are characterized by representative and differentiated subsequences of raw data. Cheng et al. [7] constructed a shapelets evolution graph to learn the representation of time series for anomaly detection. The use of shapelets in the analysis of time series has proved promising in a variety of studies.

2.2 Deep Learning-Based Anomaly Detection

Recently, deep learning has shown outstanding strengths in the analysis of multivariate time series. As a popular deep learning model, Autoencoder (AE) reconstructs multivariate time series and detects its reconstruction deviation to capture anomalies. Long Short-Term Memory network (LSTM) models long-term dependencies more effectively by adding hidden layer units that store long-term states. However, these methods encapsulate the interaction between variables as a globally hidden state.

When analyzing multivariate time series, the correlation between time series should be considered. Recently, GNN shows high ability in dealing with relation dependence. Deng et al. [3] constructed a graph with sensors as nodes according to the distance between multivariate time series, predicted time series and detected anomalies through graph attention network. Ray et al. [8] adopted GNN to learn the complex dependencies between multivariate time series to detect anomalies. Although existing methods have achieved significant improvements, most of them only focus on the relationship between multivariate time series, but ignore the temporal dependence within time series.

3 Methodology

3.1 Task Definition and Notations

We are interested in the task of multivariate time series anomaly detecting. Given time series data $\mathbf{X} = (\mathbf{x}_1, \mathbf{x}_2, ..., \mathbf{x}_n)^T \epsilon \mathbf{R}^{n \times T}$, where n is the dimension, T is the length of the time series. The task is to detect exceptions in the test dataset, and the output of the model is represented as a vector $\mathbf{Y} = [\mathrm{y}_{T+1}, \mathrm{y}_{T+2}, ..., \mathrm{y}_L]$, $\mathrm{y}_i \in \{0, 1\}$, where $\mathrm{y}_i = 1$ indicates that the ith timestamp is anomalous. The data from the sensors of the system under attack is considered anomalous.

3.2 Model Architecture

To solve this problem, we propose a novel framework for anomaly detection in multivariate time series, Multi-Scale Temporal-enhanced Graph Neural Network, MST-GNN for brevity (Fig. 1). In this framework, we adopt shapelets learning to extract features with spatial structure differentiation based on full historical data. A recurrent-skip neural network is introduced to extract the long-term temporal dependencies. Then, the spatiotemporal features are fused with the raw data. The enhanced representation is added to calculate the dependencies between nodes and construct the graph structure. After that, the ultimate graph representation is obtained based on graph attention network. Finally, the deviation is evaluated and abnormal detection is performed.

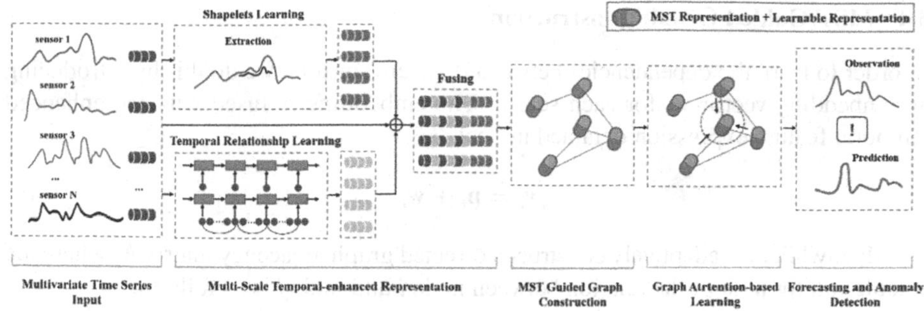

Fig. 1. The framework of MST-GNN.

3.3 Multi-scale Temporal-Enhanced Representation

In order to enhance the feature representation of sensors, we propose a multivariate time series representation method based on multi-scale temporal feature enhancement.

Shapelets Learning. In this model, shapelets [2] are used to find a sequence with the shortest distance from the raw time series as shapelets feature, so as to output time series data features with spatial differentiation.

S_i is defined as shapelets of time series with length T. In this paper, a regression classification model is constructed to obtain shapelets sequence in the process of optimizing the model according to [9].

Temporal Relationship Learning. This module uses Gate Recurrent Unit (GRU) and Recurrent-Skip Block to learn the temporal relationship.

Firstly, we use GRU as a recursive layer to learn the "importance pattern" of different combinations along all timestamps. Then, we introduce a recurrent-skip block against long-term extraction and expression of the sequential dependencies to extend the time span of the information flow [10]. The process is as follows:

$$\mathbf{z}^t = \sigma(\mathbf{W}_{xz}x^t + \mathbf{W}_{hz}\mathbf{h}^{t-p} + \mathbf{b}_z) \tag{1}$$

$$\mathbf{r}^t = \sigma(\mathbf{W}_{xr}x^t + \mathbf{W}_{hr}\mathbf{h}^{t-p} + \mathbf{b}_r) \tag{2}$$

$$\mathbf{c}^t = \text{RELU}(\mathbf{W}_{xc}x^t + \mathbf{z}_t(\mathbf{W}_{hc} \odot \mathbf{h}^{t-p}) + \mathbf{b}_c) \tag{3}$$

$$h_S^t = (1 - \mathbf{r}_t) \odot \mathbf{h}_S^{t-p} + \mathbf{r}_t \odot c^t \tag{4}$$

where p is the number of hidden cells skipped through.

Finally, the three features of space, time and raw are concatenated to form enhanced temporal features, where $\mathbf{u}_i^t$ is raw data, and $\oplus$ represents the concatenation operation.

$$\mathbf{p}_i^t = \mathbf{s}_i^t \oplus \mathbf{h}_i^t \oplus \mathbf{u}_i^t \tag{5}$$

3.4 MST Guided Graph Construction

In order to learn the dependencies between different sensors, we do this by introducing an embedding vector $\mathbf{w}_i$ for each sensor. The embedding is fused with the enhanced temporal feature expression obtained in Sect. 3.3.

$$\mathbf{v}_i = \mathbf{p}_i + \mathbf{w}_i \tag{6}$$

Meanwhile, we adaptively construct a directed graph adjacency matrix $\mathbf{A}$, where the calculation formula of the relation between node i and node j is as follows:

$$\mathrm{e}_{ij} = \frac{\mathrm{v}_i^T \mathrm{v}_j}{\|\mathrm{v}_i\| \cdot \|\mathrm{v}_j\|} \tag{7}$$

3.5 Graph Attention-Based Learning

To capture the relationships between sensors, we introduce a graph attention-based feature extractor to deal with the dependency of nodes in the graph.

$$\mathbf{z}_i^t = \mathrm{ReLU}(\mathbf{v}_i^t + \sum_{j \in \mathrm{N}(i)} \alpha_{i,j} \mathbf{v}_j^t) \tag{8}$$

where $\mathrm{N}(i) = \{j|\mathbf{A}_{ij} > 0\}$ is the set of neighbors of node i obtained from the learned adjacency matrix A, and the attention coefficients $\alpha_{i,j}$ are computed as follows:

$$\pi(i,j) = \mathrm{LeakyReLU}\left(\boldsymbol{a}^T\left(v_i^t \oplus v_j^t\right)\right) \tag{9}$$

$$\alpha_{i,j} = \frac{\exp(\pi(i,j))}{\sum_{k \in \mathrm{N}(i)} \exp(\pi(i,k))} \tag{10}$$

where $\mathbf{a}$ is a vector of learned coefficients for the attention mechanism.

3.6 Forecasting and Anomaly Detection

Through the above feature extractor, we get the representation $\{\mathbf{z}_1^{(t)}, \cdots, \mathbf{z}_N^{(t)}\}$ of N nodes. This feature vector is passed to MLP to make the final prediction $\hat{y}_i$:

$$\hat{y}_i = f_\theta([\mathbf{w}_1 \odot \mathbf{z}_1^t, \mathbf{w}_2 \odot \mathbf{z}_2^t \cdots, \mathbf{w}_N \odot \mathbf{z}_\mathrm{N}^t]) \tag{11}$$

We use the MeanSquaredError between the predicted $\hat{y}_i^{(t)}$ and the observed $y_i^{(t)}$ as the minimum loss function. Meanwhile, our model calculates individual anomaly scores for each sensor and combines them to obtain the overall anomaly scores at the moment. We follow the idea mentioned in [3] to generate the smoothed scores $Score_s(t)$. A time tick t is labelled as anomaly if $\mathrm{Score}_s(t)$ exceeds a fixed threshold.

4 Experiments

4.1 Datasets and Evaluation Metrics

Datasets. In this experiment, we used three datasets to validate our model, which are MSL (Mars Science Laboratory rover), SWaT (Secure Water Treatment) and WADI (Water Distribution). In order to observe the performance of the model in scenes with scarce data and abnormal frequency, the MSL data set used is a part of all the data. Table 1 shows the overall statistics.

Table 1. Overall statistics of 3 datasets in the experiments.

Statistics	MSL	SWaT	WADI
Number of Dimensions	27	51	127
Training Set Size	1565	47520	118800
Training Set Size	2049	44991	17280
Anomaly Rate (%)	78.087	12.209	6.038

Evaluation Metrics. We evaluate the performance of our model and the baseline model using the Precision ($Prec = \frac{TP}{TP+FP}$), Recall ($Rec = \frac{TP}{TP+FN}$) and F1-Score ($F1 = 2 \times \frac{Prec \times Rec}{Prec+Rec}$), where TP is the correctly detected anomaly, FP is the falsely detected anomaly and FN is the falsely assigned normal.

4.2 Performance Comparison

We compare our model with five anomaly detection methods for multivariate time series, including two traditional linear model-based and distance-based anomaly detection methods, and three deep learning-based anomaly detection methods.

We performed anomaly detection on three datasets. Ten sets of experiments were conducted for each method, and the average of the ten sets of results was used as the final experimental results, whose specific results are shown in Table 2.

Reference data with * comes from [13]. Reference data with # comes from [3]. For more objective comparison, other comparative data are from our reproduction experiment.

The results show that our model performs better on three datasets, with the F1-score achieving the best results on the MSL and WADI, which are higher than the second best results by 2.27% and 16.33%, and 6.67% higher on SWaT than the results in the GDN replication experiments. Given that the application of our model in a real-world task is anomaly capture, it is important that the model is able to detect all attacks, even though there may be some false anomaly alerts. Therefore, we consider recall to be the main metric used to measure anomaly detection performance. In such a scenario, our model achieves great results with 99.99%, 66.32% and 42.47% on the three datasets.

Table 2. The performance of MST-GNN with other baseline methods.

Method	MSL			SWaT			WADI		
	Prec	Rec	F1	Prec	Rec	F1	Prec	Rec	F1
PCA [4]	25.93	22.63	0.24	33.99	29.36	0.32	29.95	12.96	0.18
	—	—	—	24.92#	21.63#	0.23#	39.53#	5.63#	0.10#
KNN [5]	8.31	7.99	0.08	10.32	11.79	0.11	6.98	9.13	0.08
	—	—	—	7.83#	7.83#	0.08#	7.76#	7.75#	0.08#
LSTM-VAE [11]	56.13	87.84	0.68	97.15	51.23	0.67	78.53	13.10	0.22
	52.57*	95.46*	0.68*	96.24#	59.91#	0.74#	87.79#	14.45#	0.25#
MAD-GAN [12]	83.24	93.31	0.88	98.55	58.14	0.73	34.67	35.12	0.35
	85.17*	89.91*	0.87*	98.97#	63.74#	0.77#	41.44#	33.92#	0.37#
GDN [3]	78.75	99.81	0.88	99.14	60.88	0.75	80.51	35.26	0.49
	—	—	—	99.35#	**68.12#**	**0.81#**	**97.50#**	40.19#	**0.57#**
MST-GNN	81.51	**99.99**	**0.90**	**99.44**	66.32	0.80	85.27	**42.47**	**0.57**

4.3 Discussion

To show our model's performance in real-world scenarios, we choose two cases on the WADI and SWaT and compare our model with GDN [3].

Figure 2 and Fig. 3 show the changes of each sensor value over a period of time, in the red area of time the system is abnormal, in the non-red area manually supplemented with three status labels used to display abnormal judgment, respectively (1) attack, the real anomalies, 0 means normal, 1 means abnormal; (2) GDN_result, the judgment of the GDN; (3) MST-GNN_result, the judgment of our model.

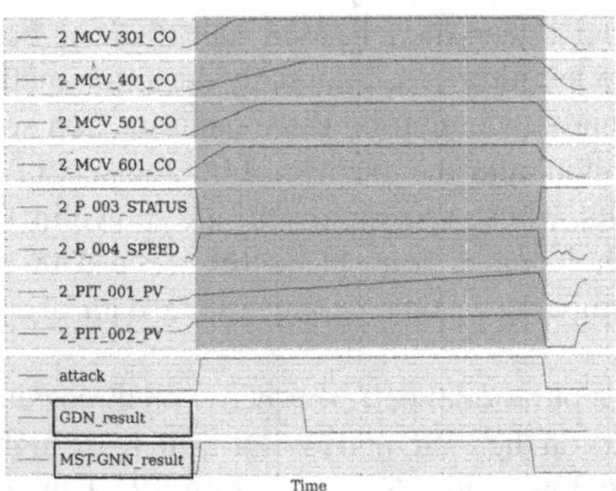

Fig. 2. The case on WADI

In the case of Fig. 2, the GDN misjudges the state as abnormal before the attack occurs. The anomaly judgement curve of our model is similar to the real situation,

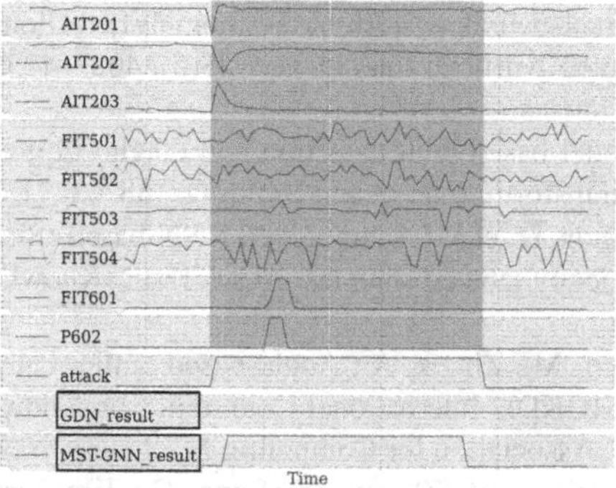

Fig. 3. The case on SWaT

indicating that our model makes timely judgements when anomalies occur and stop. It is worth noting that the fluctuations of some of the curves have flattened out in the middle and post time of the attack, but are numerically well above normal, and we speculate that the GDN incorrectly judges this smooth trend as a return to normal, but our model captures this anomaly at the numerical level.

Figure 3 shows the case on SWaT, we find that the GDN does not respond to the anomalies, while the result curves of our model match the real anomalies, which indicates that our model successfully captures the subtle anomalies by detecting the performance and slight transient fluctuations in some sensors. This phenomenon also corresponds to the high recall rate of our model in the experiment, which indicates that the shapelets learning in our model structure can retain the differential characteristics of sensors in the multivariate time series.

5 Conclusion

In this paper, we propose a novel Multi-Scale Temporal-enhanced Graph Neural Network for anomaly detection in time series, which models the correlation and temporal dependence among multivariate time series. Our method outperforms other state-of-the-art methods on three real-world datasets. Future work considers capturing the evolution of time series patterns to improve the effectiveness of anomaly detection.

Acknowledgments. This work was supported by the National Natural Science Foundation of China (No. 61872260) and National key research and development program of China (No. 2021YFB3300503).

References

1. Pang, G., Shen, C., Cao, L., Hengel, A.V.D.: Deep learning for anomaly detection: a review. ACM Comput. Surv. **54**(2), 1–38 (2021)
2. Ye, L., Keogh, E.: Time series shapelets: a new primitive for data mining. In: 15th ACM SIGKDD International Conference on Knowledge Discovery and Data Mining, pp. 947–956. Association for Computing Machinery, Paris (2009)

3. Deng, A., Hooi, B.: Graph neural network-based anomaly detection in multivariate time series. In: AAAI Conference on Artificial Intelligence. **35**(5), 4027–4035. AAAI Press, New York (2021)
4. Li, S., Wen, J.: A model-based fault detection and diagnostic methodology based on PCA method and wavelet transform. Energy and Buildings. **68**, 63–71 (2014)
5. Ying, S., Wang, B., Wang, L., et al.: An improved KNN-based efficient log anomaly detection method with automatically labeled samples. ACM Trans. Knowl. Discov. Data **15**(3), 1–22 (2021)
6. Kriegel, H.P., Schubert, M., Zimek, A.: Angle-based outlier detection in high-dimensional data. In: 14th ACM SIGKDD International Conference on Knowledge Discovery and Data Mining, pp. 444–452. Association for Computing Machinery, New York (2008)
7. Cheng, Z., Yang, Y., Wang, W., Zhuang, Y., Song, G.: Time2graph: revisiting time series modeling with dynamic shapelets. In: AAAI Conference on Artificial Intelligence. **34**(04), 3617–3624. AAAI Press, New York (2020)
8. Ito, T., Tsubouchi, K., Sakaji, H., Yamashita, T., Izumi, K.: Contextual sentiment neural network for document sentiment analysis. Data Science and Engineering **5**(2), 180–192 (2020)
9. Grabocka, J., Schilling, N., Wistuba, M., Schmidt-Thieme, L.: Learning time-series shapelets. In: 20th ACM SIGKDD International Conference on Knowledge Discovery and Data Mining, pp. 392–401. ACM Press, New York (2014)
10. Lai, G., Chang, W. C., Yang, Y., Liu, H.: Modeling long-and short-term temporal patterns with deep neural networks. In: 41st International ACM SIGIR Conference on Research & Development in Information Retrieval, pp. 95–104. ACM Press, Michigan (2018)
11. Park, D., Hoshi, Y., Kemp, C.C.: A multimodal anomaly detector for robot-assisted feeding using an lstm-based variational autoencoder. IEEE Robotics and Automation Letters. **3**(3), 1544–1551 (2018)
12. Li, D., Chen, D., Jin, B., Shi, L., Goh, J., Ng, S.-K.: MAD-GAN: multivariate anomaly detection for time series data with generative adversarial networks. In: Tetko, I.V., Kůrková, V., Karpov, P., Theis, F. (eds.) ICANN 2019. LNCS, vol. 11730, pp. 703–716. Springer, Cham (2019). https://doi.org/10.1007/978-3-030-30490-4_56
13. Zhao, H., Wang, Y., Duan, J., et al.: Multivariate time-series anomaly detection via graph attention network. In: IEEE ICDM, pp. 841–850. IEEE Press, Sorrento (2020)

Category Constraint Spatial Keyword Preference Query Based Spatial Pattern Matching

Yi Li[1] and Shaopeng Wang[1,2,3,4](✉)

[1] College of computer science-college of software, Inner Mongolia university, Hohhot, China
wangsp@imu.edu.cn

[2] Engineering Research Center of Ecological Big Data Ministry of Education, Hohhot, China

[3] Inner Mongolia Engineering Laboratory for Cloud Computing and Service Software, Hohhot, China

[4] Inner Mongolia discipline inspection and supervision big data laboratory, Hohhot, China

Abstract. Spatial Pattern Matching (SPM) is the pattern-based spatial keyword query. It can fit user's intention on text keywords, distances and exclusion relationships of spatial objects at the same time. However, SPM still suffer from the following issues. Firstly, in some application scenario, SPM cannot meet the user's preference query requirements for different attributes (such as price, service, etc.) of spatial objects. In addition, SPM only follows strict text keyword matching and ignores category during the query process, which may make the object categories in the query results incorrect, compromising query effectiveness. In view of these drawbacks, this paper formulates a novel query, i.e., category constraint spatial keyword preference query, or CSKPQ in short. The query integrates attributes of spatial objects and categories into SPM, finding one most suitable collection of spatial objects for user. Further, we propose an efficient query processing algorithm which use a new hybrid index structure. Extensive empirical experiments on real datasets demonstrate the effectiveness of our proposed algorithm.

Keywords: Spatial keyword query · Category constraint · Spatial pattern matching · Skyline algorithm

1 Introduction

Spatial pattern matching is a new way of querying spatial objects, which allows users to restrict the keywords, distances and mutual relationships of spatial objects [4]. However, the SPM has some problems which may lead to query results that not satisfy user's query preferences well.

Table 1 lists the detailed spatial object information in the dataset D and the corresponding location is shown in Fig. 1 (a). The query pattern P shown in Fig. 1

B. Li et al. (Eds.): APWeb-WAIM 2022, LNCS 13421, pp. 391–398, 2023.
https://doi.org/10.1007/978-3-031-25158-0_30

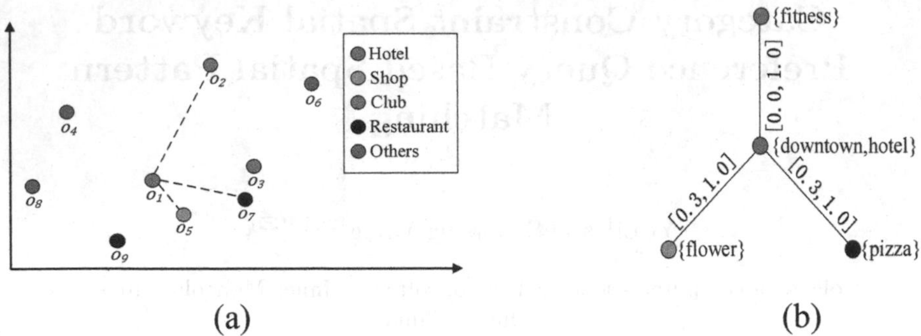

Fig. 1. An example of CSKPQ

Table 1. Detail information of dataset D.

Object	Category	keywords	Numerical attributes		
			a1	a2	a3
o_1	Hotel	Days, In, Downtown	0.53	0.22	0.45
o_2	Club	LA, Fitness	0.32	0.46	0.70
o_3	Hotel	Best, Downtown, Hotel	0.66	0.31	0.52
o_4	School	Future, Stars	0.70	0.12	0.54
o_5	Shop	Flower, Plant	0.42	0.26	0.73
o_6	Hospital	Health, Central, Care	0.66	0.51	0.22
o_7	Restaurant	Neighborhood, Pizza	0.83	0.32	0.65
o_8	Station	Broadway, Downtown, Station	0.52	0.26	0.77
o_9	Restaurant	McDonald's	0.36	0.41	0.72

(b) indicates that user wants to find a downtown hotel and specify that the hotel should not far (e.g., [0.0 km, 1.0 km]) from a fitness club. At the same time, user also requires that the flower shop and a pizza restaurant are least 0.3 km from the hotel to avoid the noise of crowds, but the distance should be limited to 1 km. After SPM query, three sets of objects S_1, S_2, S_3 satisfying pattern P are obtained. The $S_1 = \{o_2, o_5, o_7, o_8\}$ is incorrect because the category of o_8 is "Station" instead of "Hotel". In addition, o_1 and o_3 differ in attributes. The user may has preference for a dimension-specific attribute, such as a1. We assume that the smaller the value is, the more in line with the user's query preference. Thus, compared to $S_2 = \{o_2, o_3, o_5, o_7\}$, the $S_3 = \{o_1, o_2, o_5, o_7\}$ connected in dashed line is the optimal set. In summary, these shortcomings of SPM cause the quality of the result set to decrease.

2 Related Works

There exist few studies related to our work. Most existing SKQ studies focus on retrieving single object, which is close to the query locations and satisfy the query keywords [9]. With the complexity of users' query requirements, the result of SKQ is no longer a single spatial object, but a group of objects, such as spatial group keyword query [1], m-Closest Keyword query [2], minSK [3]. These objects collectively match the query conditions.

SPM are similar to them, but SPM query captures users' requirements better. There are some studies [4,6] on SPM recently. The SPM problem is NP-hard [5,6], and the paper [4] proposed an IR-Tree-based MSJ algorithm to solve SPM queries. SPM does not take categories and users' preference into account and this paper proposes a new query method to solve these weaknesses.

Furthermore, users may also have corresponding preferences for the attributes of different dimensions of object, such as lower prices or better services. The literature [7] was first defined attributes aware spatial keyword query (ASKQ) and proposed query processing methods. ASKQ uses a two-layer hybrid index structure, QDR-tree, to efficiently process query through skyline-based filtering method. An AIR-Tree index structure is used in literature [8] to achieve personalization of query. However, neither QDR-tree nor AIR-Tree can be directly applied to effectively deal with the distance interval and category constraint of spatial objects in SPM.

3 Problem Definition

The dataset $D = \{o_1, o_2, o_3, \ldots, o_n\}$, one spatial object o_i is represented by a tuple $< Loc, Word, Attr, Cate >$, where $o_i.Loc = (lng, lat)$ is the location of o_i; $o_i.Word$ represents the text keyword of the object; numerical attributes $o_i.Attr = [a_1, a_2, ..., a_n]$, each a_i is normalized into [0,1]. The smaller value, the more suitable attributes with query preferences; $o_i.Cate$ is the category of object; $|o_i, o_j|$ denotes spatial euclidean distance of two objects.

The **Category constraint Spatial Pattern (CSP)** is a simple graph composed of n vertices and m edges. CSP $= (V, E, C, W, Q)$, in which: (1) V and E correspond to the vertex set and edge set of CSP, respectively; (2) For each edge $(v_i, v_j) \in E$, it corresponds to a distance interval $[L_{i,j}, U_{i,j}]$. $L_{i,j}$ and $U_{i,j}$ denote the lower and upper distance limits between two spatial objects matching with v_i and v_j respectively; (3) C is the set of categories C_i corresponding to vertex v_i; $Q = \{qW_1, qW_2, , , qW_n\}$, where qW_i is the set of query keywords represented in vertex v_i; (4) $W = [w_1, w_2, , , w_d]$, $\forall w_i \in W, w_i \in [0,1] (i = 1, 2, ..., d)$, $\sum_{i=1}^{d} w_i = 1$. d is the number of dimensions of attributes. The larger the value of w_i, the more preferences the user shows to the attribute a_i of the spatial object;

Objects o_i and o_j in dataset D form the **spatial match** of edge (v_i, v_j) if and only if: (1) $C_i(C_j) = o_i.Cate(o_j.Cate)$; (2) $\exists w \in qW_i(qW_j)$, $w \in o_i.Word(o_j.Word)$; (3) $|o_i, o_j| \in [L_{i,j}, U_{i,j}]$;

A group of spatial objects form a **pattern match collection** S of CSP: $\forall o_i, o_j \in S$, there is one and only one spatial match constituted by $o_i$and o_j for each edge (v_i, v_j) in pattern;

Category constraint Spatial Keyword Preference Query (CSKPQ) refers to determining pattern match collection that best meet the user's query intention. The evaluation function of object o is defined as follows:

$$Score_o = \alpha \times S_W + (1-\alpha) \times S_A \quad (1)$$

S_W and S_A respectively represent the keyword relevance and the matching degree in attribute of the object, shown as Eqn. (2) and (3). α trades off the importance between S_W and S_A, and the default value is 0.7, which means text keywords are more important than multi-dimensional attributes relatively.

$$S_W = \frac{|qW_i \cap o_i.Word|}{|o_i.Word|} \quad (2)$$

$$S_A = 1 - \sum_{i=1}^{|W|} w_i * a_i \, (a_i \in o_i.Attr, w_i \in W) \quad (3)$$

Assuming $W = [0.3, 0.5, 0.2]$ and $qW_1 = \{downtown, hotel\}$ in example. For object o_1, $S_W = \frac{1}{3}$ and $S_A = 1 - (0.3 \times 0.53 + 0.5 \times 0.22 + 0.2 \times 0.45) = 0.641$, thus $Score_{o_1} = 0.7 \times S_W + 0.3 \times S_A \approx 0.43$. Similarly, $Score_{o_3} \approx 0.63$. In summary, the S_2 is concluded the query result of CSKPQ.

4 CSF Algorithm

CSIR-Tree is a kind of tree that: (1) Each tree node holds a category vector (denoted by CV) in the form of a bitmap. CV is used to characterize the category information of the spatial objects contained in all child nodes that take it as the root. (2) Each leaf node has two additional contents: invert list(InvList) [10] and skyline(SK) [11]. InvList is used to record the corresponding relationship between each text keyword and spatial object. While the SK is a set of spatial objects contained in the leaf node.

On the basis of CSIR-Tree, we propose effective **category skyline filter (CSF)** algorithm. The CSF consists of three sub-algorithms.

a. Find Matching Nodes (FMN) algorithm. First, the FSM algorithm finds all matching nodes in an up-bottom way. Two same level nodes x and y constitute **matching nodes** if they meet the categories C_i and C_j respectively and $[lw, up] \cap [L_{i,j}, U_{i,j}] \neq \emptyset$. lw and up represent the minimum and maximum distance between any two spatial objects contained in the minimum bounding rectangle (MBR) of the node.

The input of the algorithm is the root node $Root$ and a certain edge (v_i, v_j). θ is a mapping that stores matching nodes, the key is a certain node, and the value is the set of all nodes that form a matched node with the key. Firstly, $Root$ and itself constitute matching nodes (line 2). Next, find the matching nodes of each level through the loop (line 3–17). θ is updated continuously until the leaf level. The time complexity of FMN is $O(mc^2)$, where m is the number of edge in the pattern and c stands for the number of different categories in dataset. Each edge has at most c^2 matching nodes.

b. The order of join Second, we use the number of matched leaf nodes to estimate the number of spatial matches and perform the greedy algorithm shown in [4] to obtain the join order γ.

c. Find Spatial Match (FSM) algorithm. KC is a set of objects that meet the text keywords and categories at the same time. The candidate set *Cand* of the leaf node is the intersection of KC and SK (if the intersection is not empty).

Theorem 1. *The object p in the intersection must be better than object q not in the intersection.*

Proof. Firstly, objects in SK dominate other objects that are not in the SK. Then $\forall p \in SK, q \notin SK, S_W(p) > S_W(q)$. Secondly, only the objects in InvList contain the query keyword w, $w \in qW_i$. Obviously, $\forall p \in KC, q \notin KC, S_A(p) > 0$ and $S_A(q) = 0$. Thus, $Score_p > Score_q$ and object p is better than object q.

Algorithm 1: FMN

```
   input : Root,(v_i, v_j)
   output: θ ←all the matched leaf node of(v_i, v_j)
1  θ ← ∅, C_i, C_j, L_{i,j}, U_{i,j} ← (v_i, v_j);
2  μ.add(Root), θ.add(Root, μ);
3  for each level do
4      θ̂ ← ∅;
5      for x ∈ θ.key() do
6          μ̂ ← ∅;
7          for x̂ ∈ x.CV(C_i) do
8              for y ∈ θ.key() do
9                  for ŷ ∈ y.CV(C_j) do
10                     lw ← minDis(x̂.MBR, ŷ.MBR);
11                     up ← minDis(x̂.MBR, ŷ.MBR);
12                     if up < L_{i,j} then
13                         continue;
14                     if lw < U_{i,j} then
15                         μ̂.add(ŷ);
16         θ̂.add(x̂, μ̂);
17     θ ← θ̂;
18 return θ;
```

We use the FSM algorithm to find all the spatial matches in turn, and join them into a pattern match collection. The input of FSM is a matching leaf nodes (L_i, L_j). At first, we get the set KC (Line 3). If intersection of KC and SK is not empty, $Cand_i$ equal to intersection (Line 4). Otherwise, $Cand_i$ is KC (Line 5). And $Cand_j$ is similar (Line 8,9). We decide if the distance condition is satisfied and store spatial matches information (Line 11–13). The time complexity of FSM is $|D|^n$, where n is the number of vertex in the pattern and $|D|$ is the size of

Algorithm 2: FSM

input : (L_i, L_j)
output: $\beta \leftarrow$all the spatial match of edge (v_i, v_j)

$Cand_i \leftarrow \emptyset, Cand_j \leftarrow \emptyset, (v_i, v_j) \leftarrow (L_i, L_j)$;
$C_i, C_j, qW_i, qW_j, L_{i,j}, U_{i,j} \leftarrow (v_i, v_j)$;
$KC_i(KC_j) \leftarrow$ objects contained keyword and matched with category;
if $KC_i \cap SK_i \neq \emptyset$ **then** $Cand_i = KC_i \cap SK_i$;
else $Cand_i = KC_i$;
for $\forall o_i \in Cand_i$ **do**
    $\alpha \leftarrow \emptyset$;
    if $KC_j \cap SK_j \neq \emptyset$ **then** $Cand_j = KC_j \cap SK_j$;
    else $Cand_j = KC_j$;
    for $\forall o_j \in Cand_j$ **do**
        if $|o_i, o_j| \in [L_{i,j}, U_{i,j}]$ **then**
            $\alpha.add(o_j)$
    $\beta.add(o_i, \alpha)$;
return β;

dataset. From the above analysis, the total number of spatial matches could be $|D|^n$.

d. Best Matched Collection (BMC) Algorithm. Finally, the BMC algorithm is used to determine the optimal pattern match collection. For each collection in δ, the score S of it is the sum of $Score_o$ of each object in collection (Line 1–4). Sorting collection and output the collection with the highest S (Line 5–6). The time complexity of BMC is $O(kd)$, where d is the dimension of the attribute of the spatial object and k is the number of collections in δ.

Algorithm 3: BMC

input : $\delta \leftarrow$ List of all spatial collection
output: $C \leftarrow$the best matched collection

1 **foreach** *collection* C *in* δ **do**
2     $S \leftarrow 0$;
      `// each object in` C
3   **for** $\forall o \in C$ **do**
4         $S \leftarrow S + Score_o$
5 $\delta \leftarrow$ sorted by S of each collection in list in descending order;
6 **return** the first collection in δ;

e. CSF Algorithm The overall complexity of CSF is $O(mc^2 + |D|^n + kd)$

5 Experiments

Experiment Setup. We use two real open-source datasets: Yelp[1] and UK[2] dataset. The UK dataset's numerical attribute randomly generat based on a uniform distribution and range from 0 to 1. We generate the query patterns based on the structure of pattern in [4]. The query keywords, categories and distance intervals in the pattern are all randomly generated based on relevant statistical data. The fan-out of CSIR-Tree is 100.

In the experiment, CSF algorithm will mainly carry out comparative experiments with MSJ-A algorithm. We make adjustment to the MSJ algorithm in lecture [4], so that MSJ-A can also find the results that meet the user's query requirements for category and preference. A series of experiments are conducted on the PC of AMD R7 CPU, 3.20 GHz, and 16GB memory in java JDK 1.8.

Effectiveness. We verify that the result of the CSF is consistent with the result of MSJ-A. Table 2 shows the results of the 50 comparisons. The "Y" demonstrates that the result of the CSF is the same as MSJ-A and the effectiveness of the CSF algorithm be proved.

Efficiency. We compared our method with baseline method MSJ-A. YP(UK) and YP-A(UK-A) record the query time using CSF and MSJ-A respectively in dataset Yelp(UK).

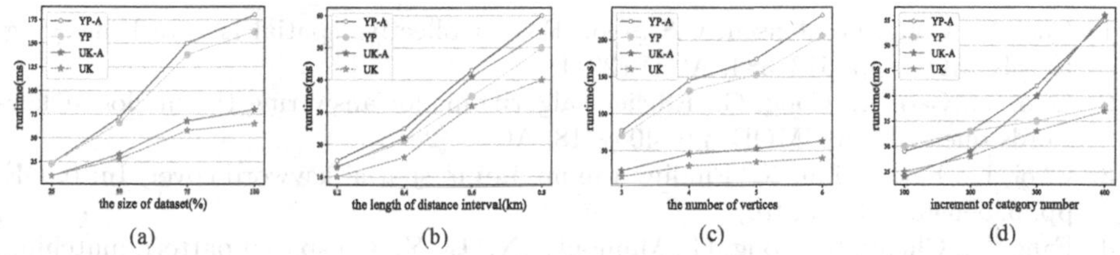

Fig. 2. Efficiency results of different parameter

a. The size of dataset Four different sizes of datasets are used and the experimental results are shown in the Fig. 2 (a). We can find that the larger dataset, the longer query time of the two algorithms. And we also find that the query time of CSF is shorter than MSJ-A.

b. The length of distance interval The Fig. 2 (b) shows the query time of CSF and MSJ-A with the change of distance interval. First, we can find that the larger of interval, the longer of query time will be. And we can still get the same conclusion as above: CSF has a shorter runtime than MSJ-A.

c. The number of vertices in pattern The Fig. 2 (c) shows the query time of CSF and MSJ-A with the change of the number of vertices. We can find that

[1] https://www.yelp.com/dataset.
[2] www.pocketgpsworld.com.

the larger of number, the longer query time will be. And the MSJ-A is more time-consuming than CSF.

d. The number of categories in dataset The experimental result of Fig. 2 (d) shows that CSF is more stable. And we can get the same conclusion as above: CSF algorithm has higher query efficiency than MSJ-A.

6 Conclusion

This paper focus on deficiencies in spatial pattern matching, and proposes a category constraint spatial keyword preference query. Next, we propose category skyline filter (CSF) algorithm to solve CSKPQ. The CSF algorithm based on index structure CSIR-Tree, in which category vector and skyline set are used to improve search efficiency. Finally, empirical experiment demonstrated the effectiveness and efficiency of our algorithm.

Acknowledgment. This research is supported by: the National Natural Science Foundation of China (Grant nos. 61862047, 62066034), Inner Mongolia discipline inspection and supervision big data laboratory open project fund (No. IMDBD2020011)

References

1. Cao, X., Cong, G., Jensen, C.S., Ooi, B.C.: Collective spatial keyword querying. In: SIGMOD, pp. 373–384. ACM (2011)
2. Guo, T., Cao, X., Cong, G.: Efficient algorithms for answering the m-closest keywords query. In: SIGMOD, pp. 405–418. ACM (2015)
3. Choi, D., Pei, J., Lin, X.: Finding the minimum spatial keyword cover. In: ICDE, pp. 685–696. IEEE (2016)
4. Fang, Y., Cheng, R., Cong, G., Mamoulis, N., Li, Y.: On spatial pattern matching, pp. 293–304 (2018). https://doi.org/10.1109/icde.2018.00035
5. Carletti, V., et al.: Challenging the time complexity of exact subgraph isomorphism for huge and dense graphs with VF3. TPAMI (2017)
6. Chen, H., Fang, Y., Zhang, Y., Zhang, W., Wang, L.: ESPM: efficient spatial pattern matching. TKDE **32**(6), 1227–1233 (2020)
7. Zang, X., Hao, P., Gao, X., Yao, B., Chen, G.: QDR-Tree: an efficient index scheme for complex spatial keyword query. In: Hartmann, S., Ma, H., Hameurlain, A., Pernul, G., Wagner, R. (eds.) DEXA 2018. Lecture Notes in Computer Science, vol. 11029, pp. 390–404. Springer, Cham (2018). https://doi.org/10.1007/978-3-319-98809-2-24
8. Meng, X., Li, P., Zhang, X.: A personalized and approximated spatial keyword query approach. In: Access, pp. 44889–44902. IEEE (2020)
9. Chen, L., Shang, S., Yang, C., Li, J.: Spatial keyword search: a survey. GeoInformatica **24**(1), 85–106 (2019). https://doi.org/10.1007/s10707-019-00373-y
10. Zhang, C., Zhang, Y., Zhang, W., Lin, X.: Inverted linear quadtree: efficient top-k spatial keyword search. TKDE **28**(7), 1706–1721 (2016)
11. Lee, J., Hwang, S.: Toward efficient multidimensional subspace skyline computation. In: VLDB 2014, pp. 129–145 (2014)

Many-to-Many Pair Trading

Yingying Wang[1], Xiaodong Li[1(✉)], Pangjing Wu[1], and Haoran Xie[2]

[1] College of Computer and Information, Hohai University, Nanjing, China
{yingying.wang,xiaodong.li}@hhu.edu.cn

[2] Department of Computing and Decision Sciences, Lingnan University, Hong Kong, China
hrxie@ln.edu.hk

Abstract. Pair trading is a market neutral strategy commonly used in hedge funds. There are two main phases of the pair trading: pair formation phase and trading phase. Finding profitable stock pairs during the pair formation phase is an important issue. Previous one-to-one pair trading mostly limits the pair formation phase to a small number of stocks. In this paper, we make an innovative improvement by extending one-to-one pair trading to many-to-many pair trading. We propose a framework that involves association rule algorithms and the OPTICS clustering algorithm, and uses the bipartite graph partition algorithm to form many-to-many pairs. Experimental results show that the framework proposed in this paper is effective in selecting many-to-many pairs for pair trading and more trading opportunities are obtained compared with traditional pair trading.

Keywords: Many-to-many pair trading · Association rule mining · Unsupervised learning · Bipartite graph · Cointegration · Distance

1 Introduction

The traditional pair trading model involves trading in a pair of a single stock. These two stocks have a large similarity in historical prices, and pair trading is a way of leverage the short separation of these two stocks to make a risk-free profit. However, due to the growing popularity of pair trading, trading opportunities are becoming less and less available.

Generally, pair trading can be divided into two main phases. In the pair formation phase, the main task is to select a number of pairs of stocks by using similarity measurement methods such as the distance method [4] and the cointegration method [2]. The pair with the highest similarity is then traded in the second phase. In the trading phase, the many-to-many pairs formed in the former phase will be backtested. The most classic trading model is based on fixed threshold which is first proposed in [4] and we use it in this paper.

This work was supported in part by the National Natural Science Foundation of China under Grant No. 61602149, and in part by the Fundamental Research Funds for the Central Universities, China under Grant No. B210202078.

B. Li et al. (Eds.): APWeb-WAIM 2022, LNCS 13421, pp. 399–407, 2023.
https://doi.org/10.1007/978-3-031-25158-0_31

Growing popularity of pair trading leads to the lose of trading opportunities. In order to increase trading opportunities and improve performance of the traditional pair trading strategy, we propose a many-to-many pair trading strategy (MMPT) that focuses on the pair formation phase. In the traditional one-to-one pair trading strategy, only one pair may have a certain high level of similarity, thus limiting trading opportunities. However, in the MMPT proposed in this paper, several pairs at the same high level of similarity can be obtained through a weighted combination of multiple stocks.

2 Related Work

Pair trading is the operation of buying and selling between two stocks in which a long-term equilibrium relationship exists to make a risk-free profit. Krauss et al. [10] reviewed the literature on pair trading frameworks and classified pair trading techniques into five categories: distance-based, cointegration-based, time series-based, stochastic control-based, and other methods. The main techniques used to select promising stock portfolios from the candidate portfolios are the distance method, Dynamic Time Warping (DTW), the cointegration method, etc. Since Gatev et al. [4] has proposed the distance method [6,7], it is widely used in the financial field. As for the cointegration technique, there are two main methods for identifying cointegration relationship between stocks: Engle-Granger's method [1,2,9] and Johansen's method [8]. Gatev et al. [4] proposed the most classic fixed threshold-based trading model. The stochastic control approach likewise focuses on the trading phase, which originates from [13]. Li et al. [12] applied the stochastic control approach to pair trading. More and more machine learning techniques were also applied to pair trading strategies. Baek et al. [1] used the Error Correction Model framework (ECM) and the Support Vector Machine (SVM) algorithm based on Engle-Granger's method to determine hedging ratios and identify cointegrated pairs. Fallahpour et al. [3] focused on reinforcement learning-based pair trading strategy optimization.

In the existing literature, researchers had focused on one-to-one pair trading, i.e., a single stock to a single stock pair trading. However, research on the MMPT is little. [11] proposed the copula method to complete the formation of many-to-many pairs. [5] used multivariate pair trading for multi-objective optimization. But neither of them gave the specific construction process of the many-to-many pairs. In order to improve practicality and find more trading opportunities, this paper proposes a new framework of many-to-many pair trading which focuses on stock portfolio selection in the pair formation phase, extending one-to-one pair trading to many-to-many pair trading.

3 Many-to-Many Pair Formation for Pair Trading

In this section, the algorithms and models proposed in MMPT are presented. A formal definition of the many-to-many pair formation problem is given first, followed by a description of the algorithms used in the MMPT.

3.1 Problem Description

Definition 1. *Pair: In pair trading, a pair is a combination of two components, and each component is made of one stock or multiple stocks.*

For the pair formation problem, this paper proposes a novel framework based on the OPTICS clustering algorithm and the association rule algorithm, selecting frequent itemsets in each cluster by the Apriori algorithm and introducing the bipartite graph partition algorithm to form many-to-many pairs. The many-to-many pair formation problem is formally defined as a tuple of five elements $\{S, L, P^M, P, E\}$:

- $S = \{s_i | s_i \text{ is a stock}\}$. The MMPT model divides stocks in S into individual stock portfolios for pair trading.
- $L = \{L_i | L_i \text{ are frequent itemsets}\}$. L are the itemsets selected from candidate itemsets (denoted by C) by using the Apriori algorithm.
- $P^M = \{p_i^M | p_i^M$ is a pair of stocks made from L_i, $p_i^M = \{s_m^M, s_n^M\}$, $s_m^M \cap s_n^M = \varnothing\}$. P^M denotes the many-to-many pairs.
- $P = \{p_i | p_i = \{s_m, s_n\}, s_m \in S, s_n \in S, s_m \neq s_n\}$. p_i is a pair of stocks selected according to similarity measurement methods.
- $E = \{e_i | e_i$ is the value of similarity of two components in p_i and $p_i^M\}$. The value of similarity is calculated by similarity measurement methods such as the sum of the squared differences (SSD) and the cointegration method.

Similar many-to-many pairs can be formed by using the SSD and the cointegration method. Stocks in each many-to-many pair can belong to either long or short component in pair trading which are represented by the Boolean vector $s_i^g \in \mathbb{B}$ and $g \in (L, S)$ means the long or short component. For n stocks in each many-to-many pair, a column vector represents weights $\boldsymbol{\omega}^g = (\omega_1^g, \omega_2^g, \ldots, \omega_n^g)^\top$, where $0 \le \omega_i^g \le 1$. We regard $\mathbf{1}$ as a vector of ones. The sum of all weights in each component must equal to one like $(\omega_i^L)^T \mathbf{1} = 1$ and $(\omega_i^S)^T \mathbf{1} = 1$ where the length of $\mathbf{1}$ is equal to the length of the ω_i^L and ω_i^S.

The set $E = [e_1, e_2, \ldots, e_k]$ represents the value of similarity of two components in each pair. The following formulation shows the objective function of the MMPT.

$$\begin{aligned} \max \quad & E = [e_1, e_2, \ldots, e_k] \\ \text{s.t.} \quad & s_i^g \in \mathbb{B}, \\ & g \in (L, S), \\ & s_i^L + s_i^S \le 1, \\ & 0 \le \omega_i^g \le 1, \\ & (\omega_i^L)^T \mathbf{1} = 1, \\ & (\omega_i^S)^T \mathbf{1} = 1. \end{aligned} \tag{1}$$

3.2 Pair Formation Framework

First, some clusters can be generated based on the application of PCA followed by the OPTICS clustering algorithm. One-to-one pairs will be obtained from each cluster by combining any two stocks. The method of generating many-to-many pairs is to firstly obtain a group of multiple stocks by using the one-to-one pairs with the application of the Apriori algorithm, then divide the group into two combinations, and weight the price data of these two combinations into two price series. Figure 1 shows the overall framework proposed in the paper.

We use Algorithm 1 to show the process of forming groups of stocks. When $n = 0, 1, 2, \ldots$, the Apriori algorithm forms many groups of multiple stocks by following procedures:

1) Select L_{n+2} from C_{n+2}: a C_{n+2} is considered to be a L_{n+2} if any combination of $n+1$ stocks in a C_{n+2} is a L_{n+1}.
2) Generate C_{n+3} by using the existing combinations: combine a L_{n+1} and a L_{n+2} to generate a C_{n+3}. To improve the efficiency of generating candidate itemsets: if there are $n+1$ stocks that are the same in two L_{n+2}, these $n+1$ stocks and the one remaining stock of each L_{n+2} can form a C_{n+3}.
3) Add 1 to n and go to step 1) until the L_{n+2} or C_{n+3} has no elements.

Algorithm 1. Association rule mining-frequent itemset

Inputs: P, min-support, N.
Ouput: Frequent itemset L.
count the number of occurrences of each s_i in P;
if the number of occurrences of s_i >= min-support **then**
  s_i is a frequent itemset in L_1;
end if
for $i = 2 \rightarrow N$ **do**
  if any $i-1$ stocks of the i stocks in C_i belong to L_{i-1} **then**
    C_i is a frequent itemset in L_i;
  end if
  obtain C_{i+1} from L_i;
  if there are $i-1$ stocks that are same in two L_i **then**
    form a C_{i+1} by the two sets;
  end if
end for

Then, the bipartite graph partition algorithm divides the multiple stocks into two groups. Two pre-definitions can be made: 1) Any L_n can be traced back to L_2. So the combination of any two stocks in a L_n is a L_2. 2) Treat the n stocks in a L_n as n vertices in a graph. There is an edge connecting two vertices represented by two stocks.

A complete graph of the L_n can be formed by the steps above. The bipartite graph partition process will be carried out on the complete graph to divide all vertices in the complete graph into two disjoint sets (U, V). Algorithm 2 shows the bipartite graph partition algorithm. The variable-new_vertex is defined to indicate the number of new vertices added by each one-to-one pair (the sum of the new_vertex of all pairs in selected_pairs is the number of vertices n). Theorem 1 and the following proof show that a complete graph of L_n can always be transformed into a bipartite graph.

Algorithm 2. Split the stocks of frequent itemset into two assets-bipartite graph partition algorithm

Require: All the frequent itemset L_n, E.
Ensure: The two assets of each frequent itemset.

sort the L_2 in descending order of similarity in E;
select the pairs in L_2 by the order until the selected pairs cover all stocks in the frequent itemset and denote the pairs as selected_pairs;
define new_vertex to represent the number of new vertices added by each pair in selected_pairs;
for each pair in selected_pairs **do**
  if new_vertex = 0 **then**
    delete the pair which means deleting the edge in the graph;
  end if
  if new_vertex = 2 **then**
    place the two stocks in pair separately into asset1,asset2;
  else
    put the new vertex into the asset that is opposite to the asset of the other vertex is in;
  end if
end for

Theorem 1. *A complete graph of a L_n can be transformed into a bipartite graph when $n = 3, 4, 5, \ldots, k$.*

Proof. When $n = 3$, suppose the L_3 is (s_1, s_2, s_3). Let the order of similarity values is (s_1, s_2) > (s_1, s_3) > (s_2, s_3). Select the one-to-one pairs in order and (s_1, s_2) and (s_1, s_3) are selected. After that, these two pairs cover all three stocks in the L_3, so (s_1, s_2) and (s_1, s_3) are recorded as selected_pairs.

Two stocks in (s_1, s_2) are placed into asset1, asset2 respectively. In (s_1, s_3), as s_1 is in asset1, s_3 is put into asset2. Then the L_3 is divided into two groups (s_1, (s_2, s_3)). Therefore the complete graph of the L_3 can be transformed into a bipartite graph.

Let $n = k-1$, suppose that the complete graph of a L_{k-1} can be transformed into a bipartite graph.

Let $n = k$, prove the complete graph of a L_k can be transformed into a bipartite graph. Since the complete graph of the L_{k-1} can be transformed into a bipartite graph, $k-1$ vertices can be divided into two disjoint sets of vertices (U, V). Suppose that L_k can be gained by adding a new vertex to the L_{k-1}, and we denote the new vertex as s_k. Hence that s_k with $s_i \in U$, $s_j \in V$ has a relative order of the similarity values, we can suppose that s_k with $s_m \in U$ has the greatest value of similarity. Then two operations can be performed: add s_k

to the set of vertices V and add a new edge (s_k, s_m) to the bipartite graph of the L_{k-1}. Thus the set of vertices U and V are mutually disjoint. So the complete graph of L_k can be transformed into a bipartite graph. □

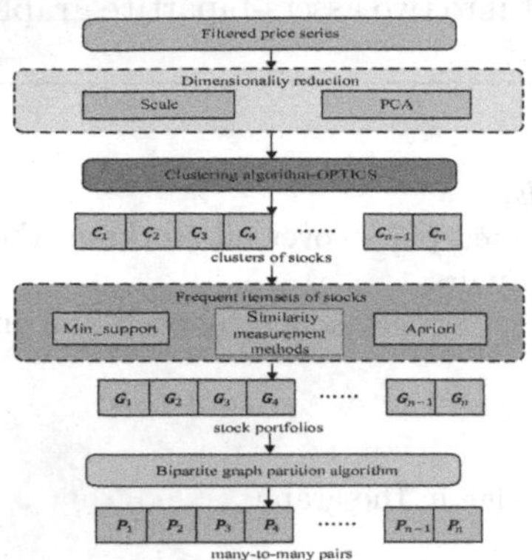

Fig. 1. Overall framework of our proposed model.

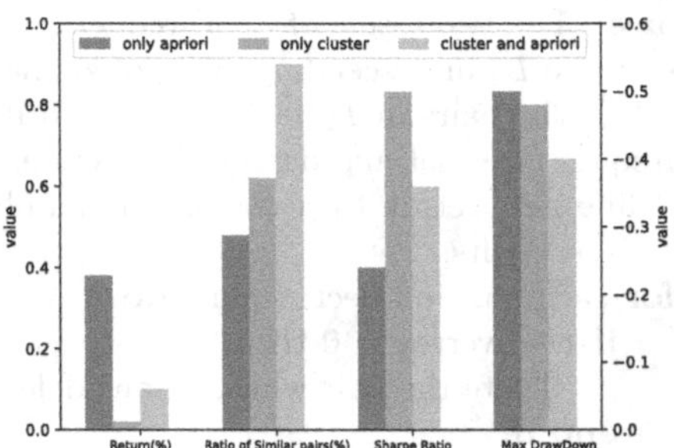

Fig. 2. The average values of the evaluation indicators in three search techniques.

4 Experiments and Discussions

4.1 Datasets and Experimental Settings

In the experiment, we use three datasets to compare their performance with each other: China Securities Index (CSI300), Shanghai Stock Exchange (SSE180) and Shanghai-Hongkong (SH-HK) Stock Connect. The time span of the datasets used in the experiment is during 2013/01/01-2017/12/30. The numbers of three datasets are 300, 180, 581. The industries of stocks in the dataset are various and cover most industries in the market. Stock data obtained from tushare[1] is used in our experiment. Based on the length of the pair formation phase and trading phase in [14], one year of data is used for pair formation, and a six-month trading phase is set after the many-to-many pairs are obtained. The sliding step is set to one month. In our experiment, a comparison between three stock combinations' search techniques is performed (using only the Apriori algorithm, using only the OPTICS clustering algorithm, and using the OPTICS clustering algorithm and the Apriori algorithm).

4.2 Results and Analysis

The similarity of two components in many-to-many pairs is calculated by the similarity measurement methods to determine whether the MMPT is effective in

[1] https://tushare.pro/.

obtaining many-to-many pairs with great similarity. The average values of the evaluation indicators in three search techniques are shown in Fig. 2, in which the values of return and ratio are indicated by the left Y axis. Tables 1, 2 and 3 show the trading performance, the number of total many-to-many pairs, and the ratio of the pairs that satisfy the similarity requirement in three stock combinations' search techniques of each dataset. The first and second best performances are bold and underlined respectively. From the ratio of similar pairs in Fig. 2, we conclude that the three techniques is effective in selecting many-to-many pairs with great similarity, in which the technique which uses the OPTICS clustering algorithm and the Apriori algorithm is the most. As shown in Tables 1, 2 and 3, the SSD-based method has larger ratio of the similar many-to-many pairs than the cointegration-based method. Also, cointegration-based method achieves higher return and SR than SSD-based method mostly. On the contrary, SSD-based method achieves lower MDD than cointegration-based method.

Table 1. Trading performance of stock combinations' search techniques in CSI300.

Search technique	only apriori[a]		only cluster[b]		cluster and apriori[c]	
Similarity measurement method	Cointegration	SSD	Cointegration	SSD	Cointegration	SSD
Return(%)	−0.51	0.01	**0.22**	−0.05	<u>0.17</u>	−0.05
Sharpe Ratio	<u>−0.14</u>	−0.23	−0.17	−0.43	**0.06**	−0.6
Max DrawDown	−0.93	**−0.19**	−0.67	−0.52	<u>−0.42</u>	−0.5
Total pairs	**360**	111	122	122	<u>204</u>	143
Ratio of Similar pairs (%)	0.27	0.5	0.33	<u>0.88</u>	0.77	**0.9**

[a] Search technique which uses only the Apriori algorithm.
[b] Search technique which uses only the OPTICS clustering algorithm.
[c] Search technique which uses the OPTICS clustering algorithm and the Apriori algorithm.

Table 2. Trading performance of stock combinations' search techniques in SSE180.

Search technique	only apriori		only cluster		cluster and apriori	
Similarity measurement method	Cointegration	SSD	Cointegration	SSD	Cointegration	SSD
Return (%)	**0.62**	−0.16	−0.2	0.2	<u>0.27</u>	0.2
Sharpe Ratio	−0.35	−0.69	<u>0</u>	−0.92	**0.47**	−0.86
Max DrawDown	−0.79	**−0.26**	−0.73	−0.28	−0.48	<u>−0.27</u>
Total pairs	**194**	79	99	99	<u>190</u>	108
Ratio of Similar pairs (%)	0.26	0.59	0.36	<u>0.9</u>	**0.92**	**0.92**

Table 3. Trading performance of stock combinations' search techniques in SH/SZ-HK Stock Connect.

Search technique	only apriori		only cluster		cluster and apriori	
Similarity measurement method	Cointegration	SSD	Cointegration	SSD	Cointegration	SSD
Return (%)	**2.6**	−0.26	−0.17	0.12	0	0.02
Sharpe Ratio	**0.61**	−0.63	−0.69	−0.77	−0.21	−0.99
Max DrawDown	−0.45	−0.36	−0.35	−0.35	−0.4	**−0.34**
Total pairs	**310**	118	153	153	131	162
Ratio of Similar pairs (%)	0.66	0.59	0.35	0.89	**0.99**	0.91

5 Conclusion and Future Work

There is little research in the existing literature on MMPT. To fill this gap, this paper proposes a novel many-to-many pair trading framework which uses the OPTICS clustering algorithm and the association rule algorithm to select candidate stocks. Then, the bipartite graph partition algorithm is used to form many-to-many pairs. Our experimental results show that the framework proposed in the paper is effective in forming many-to-many pairs with great similarity. Future work could optimize the pair formation phase by considering new factors like risk, and explore the application of deep reinforcement learning techniques to MMPT.

References

1. Baek, S., Glambosky, M., Oh, S.H., Lee, J.: Machine learning and algorithmic pairs trading in futures markets. Sustainability **12**(17), 6791 (2020)
2. Engle, R.F., Granger, C.W.: Co-integration and error correction: representation, estimation, and testing. Econ. J. Econ. Soc. **55**, 251–276 (1987)
3. Fallahpour, S., Hakimian, H., Taheri, K., Ramezanifar, E.: Pairs trading strategy optimization using the reinforcement learning method: a cointegration approach. Soft. Comput. **20**(12), 5051–5066 (2016). https://doi.org/10.1007/s00500-016-2298-4
4. Gatev, E., Goetzmann, W.N., Rouwenhorst, K.G.: Pairs trading: performance of a relative-value arbitrage rule. Rev. Financ. Stud. **19**(3), 797–827 (2006)
5. Goldkamp, J., Dehghanimohammadabadi, M.: Evolutionary multi-objective optimization for multivariate pairs trading. Expert Syst. Appl. **135**, 113–128 (2019)
6. Gupta, K., Chatterjee, N.: Selecting stock pairs for pairs trading while incorporating lead–lag relationship. Phys. A: Stat. Mech. Appl. **551**, 124103 (2020)
7. Huck, N.: The high sensitivity of pairs trading returns. Appl. Econ. Lett. **20**(14), 1301–1304 (2013)
8. Johansen, S.: Statistical analysis of cointegration vectors. J. Econ. Dyn. Control **12**(2–3), 231–254 (1988)
9. Kim, S., Heo, J.: Time series regression-based pairs trading in the Korean equities market. J. Exp. Theoret. Artif. Intell. **29**(4), 755–768 (2017)
10. Krauss, C.: Statistical arbitrage pairs trading strategies: review and outlook. J. Econ. Surv. **31**(2), 513–545 (2017)

11. Lau, C., Xie, W., Wu, Y.: Multi-dimensional pairs trading using copulas. In: European Financial Management Association 2016 Annual Meetings June (2016)
12. Li, Z., Tourin, A.: A finite difference scheme for pairs trading with transaction costs. Comput. Econ. **60**, 1–32 (2021). https://doi.org/10.1007/s10614-021-10159-w
13. Mudchanatongsuk, S., Primbs, J.A., Wong, W.: Optimal pairs trading: a stochastic control approach. In: 2008 American Control Conference, pp. 1035–1039. IEEE (2008)
14. Rad, H., Low, R.K.Y., Faff, R.: The profitability of pairs trading strategies: distance, cointegration and copula methods. Quant. Finan. **16**(10), 1541–1558 (2016)

Proximity Preserving Graph Convolutional Networks

Zhenglin Yu, Hui Yan(✉), and Ling Guo

School of Computer Science and engineering, Nanjing University of Science and Technology, Nanjing, China
{yuzhenglin,yanhui,guoling}@njust.edu.cn

Abstract. Graph Neural Networks (GNNs) have achieved promising performance in a wide range of graph-based tasks. Their key is to capture the smoothness of labels or features over nodes manifested in graph structure when stacking multiple layers. However, as the number of layers increases, most GNNs suffer from the over-smoothing issue. To solve this issue, we propose proximity preserving graph convolutional networks (PPGCNs) in this paper. Different from previous methods finetuning graph convolutional operator, we resort to modified graph Laplacian regularization in loss function. Specially, a novel optimization objective function with multi-order proximity preserving regularization (PP-Reg) is designed to guide an end-to-end framework. The theoretical evidence w.r.t. such optimization objective function verifies it can further smooth node features but inhibit over-smoothness. Moreover, empirical evidence verifies 1st-order PP-Reg is more suitable for assortative graphs, and 2nd-order PP-Reg is more suitable for disassortative graphs. Empirical evidence verifies PPGCNs achieve state-of-the-art results in the task of semi-supervised node classification on 8 assortative or disassortative benchmark datasets. The code is available on Github (https://github.com/GNN-zl/PPGCNs).

Keywords: Graph convolutional networks · Graph laplacian · Over-smoothing

1 Introduction

Recently, graph neural networks (GNNs) have made great progress in many graph learning tasks and strengthened the representation power of graph embedding [6,11,18]. Their success is partially owing to that GNNs actually perform Laplacian smoothing on node features to perserve proximity based on the homophily hypothesis that connected nodes tend to have the same label or similar features, where DeepWalk [12] and Node2vec [4] has made attempts to preserve proximity in early graph machine learning. GNNs, e.g., graph convolutional network (GCN) [7], update node representations by propagating and transforming information within topological neighborhoods in each convolutional layer. After multi-layer

B. Li et al. (Eds.): APWeb-WAIM 2022, LNCS 13421, pp. 408–416, 2023.
https://doi.org/10.1007/978-3-031-25158-0_32

graph convolutional operations, the effect of Laplacian smoothing makes node representations become more and more similar, eventually becoming indistinguishable. This phenomenon is called over-smoothing [9].

Several works try to tackle the over-smoothing issue by modifying the graph convolutional operator. For example, DropEdge [13] randomly removes edges from the input graph to make node connections more sparse. Simplifying graph convolution (SGC) [15] attempts to get rid of non-linear activation function during the process of convolution. Approximately personalized propagation of neural predictions (APPNP) [8] replaces the power of graph convolutional matrix with the personalized pagerank matrix. Simple spetral graph convolution (SSGC) [19] aggregates multi-step diffusion matrices based on SGC. Deep adaptive graph neural network (DAGNN) [10] learns weights for each layer's embedding generating by SGC, and fuses multi-layers' embeddings to update the target node embedding.

Besides the above modification approaches, there is another factor, i.e., modified loss function, that has been ignored. In this paper, we resort graph Laplacian regularization to design a novel optimization objective function to boost the classification performance. Our contributions are summarized as follows:

We define k-order smoothness of graphs as a variant of graph Laplacian regularization, which is called proximity preserving regularization (PP-Reg). Then, we combine PP-Reg with the prevailing empirical risk with a handful of labeled data and propose proximity preserving graph convolutional networks (PPGCNs). Note that our innovation seems inconsistent with GCN [7], which argues that using graph Laplacian regularization term in the loss function might be invalid and even limited to modeling capacity. However, our theoretical and empirical evidence demonstrates: 1) PP-Reg preserves k-order proximity between the node pairs. It can bridge the gap between labeled data and unlabeled data, enabling the supervised signal spread over the whole graph. 2) Combining PP-Reg with empirical risk in a semi-supervised task can further smooth features, but avoid over-smoothing issue aroused from increasing the number of layers. 3) 1st-order PP-Reg is more suitable for assortative datasets and 2nd-order PP-Reg is more suitable for disassortative datasets, where assortative datasets have high probability that edges exist between node pairs with the same label, while disassortative datasets are opposite.

2 Problem Formalization

Given an undirected attributed graph $\mathcal{G} = (\mathcal{V}, \mathcal{E}, \mathbf{X})$, where $\mathcal{V} = \{v_1, v_2, \cdots, v_n\}$ consists of a set of nodes with $|\mathcal{V}| = n$, $\mathcal{E}$ is a set of edges between nodes, and $\mathbf{X} = [x_1; x_2; \cdots; x_n] \in \mathbb{R}^{n \times d}$ denotes the feature matrix of all the nodes, where $x_i \in \mathbb{R}^d$ is the real valued feature vector associated with node v_i, and d is the number of dimension of the feature vector. The topological structure of graph $\mathcal{G}$ is represented by an adjacency matrix $\mathbf{A} \in \{0,1\}^{n \times n}$, where $\mathbf{A}_{ij} = 1$ indicates the existence of an edge between nodes v_i and v_j, otherwise $\mathbf{A}_{ij} = 0$. $\mathbf{D}$ is a diagonal degree matrix with $\mathbf{D}_{ii} = \sum_j \mathbf{A}_{ij}$. $\widetilde{\mathbf{A}}$ stands for the adjacency matrix with self-loops, $\widetilde{\mathbf{A}} = \mathbf{A} + \mathbf{I}_n$. $\mathbf{I}_n$ denotes the $n \times n$ sized identity matrix. $\widetilde{\mathbf{D}}$ is the diagonal matrix with $\widetilde{\mathbf{D}}_{ii} = \sum_j \widetilde{\mathbf{A}}_{ij}$. $\widetilde{\mathbf{L}}$ stands for the normalized Laplacian matrix, i.e., $\widetilde{\mathbf{L}} = \mathbf{I}_n - \widetilde{\mathbf{D}}^{-1/2}\widetilde{\mathbf{A}}\widetilde{\mathbf{D}}^{-1/2}$. $\mathbf{Z} = [z_1; z_2; \cdots; z_n] \in \mathbb{R}^{n \times c}$ stands for the embedding matrix where c denotes the number of the classes.

3 Proximity Preserving Graph Convolutional Networks

3.1 k-order Smoothness of Graphs

In graph-related task, each row of the feature matrix $\mathbf{X}$ can be considered as a graph signal. A widely adopted assumption for graph signals is that values change smoothly across adjacent nodes, and the more smoothing the graph signals, the higher the proximity in the graph, and vice versa [16]. Besides the immediate proximity is widely adopted to describe smoothness [7], the higher-order proximity between nodes plays a key role in capturing the structure information of the graph [5,6]. We define k-order smoothness of graphs as:

$$\mathcal{L}_{PP}^{k} = trace(\mathbf{X}^T \widetilde{\mathbf{L}}^k \mathbf{X}) \tag{1}$$

where k is the order of the matrix[1] $\widetilde{\mathbf{L}}$, $\mathcal{L}_{PP}^{k}$ is called proximity preserving regularization (PP-Reg). The formula of k-order smoothness is essentially a k-order form of graph-based Laplacian regularization.

Supposing that we are given noisy graph features, to reconstruct the original features $\mathbf{Z}$ with graph smoothness, we design a general framework based on minimizing $\mathcal{L}_{PP}^{k}$:

$$\arg\min_{\mathbf{Z}} trace(\mathbf{Z}^T \widetilde{\mathbf{L}} \mathbf{Z}) \quad s.t., \quad \varphi(\mathbf{Z}) \tag{2}$$

where $\varphi(\mathbf{Z})$ is a constraint term.

For Eq. (2), we take the constraint term $\varphi(\mathbf{Z}) = \min ||\mathbf{X} - \mathbf{Z}||_F^2$. Considering the specific case that the balance parameter is $1/2$, an auto-encoder with $\mathcal{L}_{PP}^{k}$ is:

$$\arg\min_{\mathbf{Z}} \mathcal{L}^1(\mathbf{Z}) := ||\mathbf{X} - \mathbf{Z}||_F^2 + \frac{1}{2} trace(\mathbf{Z}^T \widetilde{\mathbf{L}^k} \mathbf{Z}) \tag{3}$$

The 1-step gradient descent at $\mathbf{X}$ with the learning rate 1 and $k = 1$ of the above auto-encoder is equivalent to the 1-layer graph convolution in GCN:

$$\mathbf{X} - \nabla \mathcal{L}^1(\mathbf{X}) = (\widetilde{\mathbf{D}}^{-1/2} \widetilde{\mathbf{A}} \widetilde{\mathbf{D}}^{-1/2}) \mathbf{X} \tag{4}$$

Likewise, k-step gradient descent is equivalent to a k-layer GCN. When $k \to \infty$, we have $\mathbf{z}^j = (\widetilde{\mathbf{D}}^{-1/2} \widetilde{\mathbf{A}} \widetilde{\mathbf{D}}^{-1/2})^k \mathbf{x}^j = \gamma \widetilde{\mathbf{D}}^{-1/2} \mathbf{1}$ ($\forall j \in \{1, 2, ..., d\}$), where $\mathbf{x}^j \in \mathbb{R}^{n \times 1}$ and $\mathbf{z}^j \in \mathbb{R}^{n \times 1}$ denote the j-th column of the feature matrix $\mathbf{X}$ and $\mathbf{Z}$, respectively. $\mathbf{1}$ is a vector with all elements 1 and γ is a constant [2]. It

[1] Laplacian matrix $\widetilde{\mathbf{L}}$ could be replaced by other variants, such as the adjacency matrix and the similarity matrix, where the replacement is sparse and symmetric.

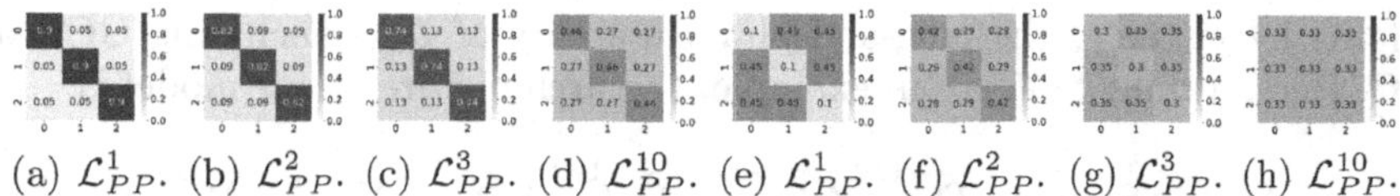

(a) $\mathcal{L}_{PP}^1$. (b) $\mathcal{L}_{PP}^2$. (c) $\mathcal{L}_{PP}^3$. (d) $\mathcal{L}_{PP}^{10}$. (e) $\mathcal{L}_{PP}^1$. (f) $\mathcal{L}_{PP}^2$. (g) $\mathcal{L}_{PP}^3$. (h) $\mathcal{L}_{PP}^{10}$.

Fig. 1. Class-wise similarity matrix. In first four $h = 0.9$ and in the rest $h = 0.1$.

indicates nodes with the same degree have the same embedding, and the final features of the nodes with different classes become indistinguishable. Thereby, infinite-step gradient descent encounters over-smoothing issue as well as infinite-layer convolution in GCN. It means that to alleviate over-smoothing issue, the only alternative is to employ gradient descent with a small amount of steps.

Kipf et al. [7] verify that performing 1-step gradient descent optimization, i.e., 1-layer convolution to obtain the local optimal solution is beneficial for the classification task on real-world graphs. Except for applying proximity, i.e., $\mathcal{L}_{PP}^k$, to message propagation, e.g., convolution in GCN, we can also apply it to the loss function. To alleviate over-smoothing issue, the intuitive approach is to avoid the loss function being dominated entirely by the smoothness term $\mathcal{L}_{PP}^k$.

Properties of $\mathcal{L}_{PP}^k$. Firstly, $\mathcal{L}_{PP}^k$ as a variant of graph Laplacian regularization provides extra supervised information, and it bridges the gap between labeled data and unlabeled data, enabling the supervised signal spread over the whole graph, especially in semi-supervised learning. The effectiveness of graph Laplacian regularization in semi-supervised learning has been shown in Table 1.

Secondly, GCN strengthens smoothness depending on the increase in convolutional layers, which is a discrete integer process. Such process only reflects a crude change. However, as shown in Fig. 2, minimizing $\mathcal{L}_{PP}^k$ strengthens smoothness by finetuning a continuously changing regularization term coefficient, which is precise relatively, and it skillfully avoids the over-smoothing problem.

Lastly, $\mathcal{L}_{PP}^k$ with a flexible coefficient k controls the proximity of different orders. Taking two specific datasets as examples, i.e., an assortative dataset with homophily value as 0.9 and a disassortative dataset with homophily value as 0.1, we suppose the number of classes as 3. As shown in Fig. 1, on assortative datasets, intra-class similarity in 1st-hop neighborhoods is higher than that in 2nd-hop neighborhoods, while the phenomenon is opposite on disassortative datasets. Thus, $\mathcal{L}_{PP}^1$ is more suitable for assortative datasets, while $\mathcal{L}_{PP}^2$ is more suitable for disassortative datasets. However, as the hop of neighborhood increases, e.g., $k \geq 3$, the gap between the similarity of intra-class and inter-class becomes smaller and smaller, that is, k becomes less sensitive to homophily than 1st-hop and 2nd-hop. Besides, the computation cost can be very high with the increase in k. Thus, there is no need to further increase the hop of neighborhood.

3.2 Proximity Preserving Graph Convolutional Networks

In this paper, we adopt a 2-layer graph convolution. We take $\mathcal{L}_{PP}^k$ as a regularization term with the final embedding $\mathbf{Z}$, and add it into GCN and GCN's variants by virtue of the plug-and-play capability of $\mathcal{L}_{PP}^{k}$, including GCN and SGC. Here, Taking SGC as an example, the architecture of our model is:

$$\mathbf{X}^{(2)} = \text{Softmax}(\widetilde{\mathbf{D}}^{-1/2}\widetilde{\mathbf{A}}\widetilde{\mathbf{D}}^{-1/2}(\widetilde{\mathbf{D}}^{-1/2}\widetilde{\mathbf{A}}\widetilde{\mathbf{D}}^{-1/2}\mathbf{X}\mathbf{W}^{(0)})\mathbf{W}^{(1)}) \quad (5)$$

where $\mathbf{W}^{(0)} \in \mathbb{R}^{d\times d'}$ and $\mathbf{W}^{(1)} \in \mathbb{R}^{d'\times c}$ are weight matrixes with d' as the dimension of the hidden layer and c as the number of classes. Softmax function is used to obtain a probability distribution of different classes.

Then the whole model is trained through the composition loss:

$$\mathcal{L} = \mathcal{L}_{cls} + \alpha\mathcal{L}_{PP}^{k} \quad (6)$$

where α is a hyper-parameter to tune the proportion of regularization loss, and $\mathcal{L}_{cls}$ is a widely used cross-entropy loss [7].

3.3 Connections to Squared-Error P-reg

Yang et al. [17] propose a variant of graph Laplacian-based regularization, called Propagation-regularization (P-reg). P-reg and our PP-Reg seem to be similar. We show that P-reg is a special case of our PP-Reg in what follows:

$$\begin{aligned}\mathcal{L}_{P-reg} =& \sum_{i=1}^{n} \|(\widetilde{\mathbf{D}}^{-1}\widetilde{\mathbf{A}}\mathbf{Z})_i^T - \mathbf{Z}_i^T\|_2^2 \\ =& trace(\mathbf{Z}^T\hat{\mathbf{L}}^T\hat{\mathbf{L}}\mathbf{Z})\end{aligned} \quad (7)$$

where $\hat{\mathbf{L}} = \mathbf{I}_n - \widetilde{\mathbf{D}}^{-1}\widetilde{\mathbf{A}}$ is a row-normalized Laplacian matrix. If we replace $\hat{\mathbf{L}}$ with $\widetilde{\mathbf{L}}$, P-reg is proportional to our PP-Reg with $k = 2$. Here, the row-normalization on $\widetilde{\mathbf{A}}$ ignores the effect of the degree of the adjacent node on the target node.

Moreover, P-reg is inflexible in that it fixes the k value, and omits the adaptability of graph regularization to assortative datasets and disassortative datasets. Yang et al. only introduce supervised information to smooth node embeddings over the entire graph topology.

4 Experiments

4.1 Datasets and Experimental Settings

For assortative benchmark datasets, we use Cora, Citeseer, Pubmed, and Photo[2]. For the first three datasets, we employ the same data split in previous works [7]. On Photo 20/30/rest labeled nodes per class are used for training/validation/test. For disassortative benchmark datasets, we use Chameleon, Squirrel, Wisconsin and Cornell[3]. On these datasets 60%/20%/20% nodes are

[2] https://github.com/shchur/gnn-benchmark.

[3] https://github.com/graphdml-uiuc-jlu/geom-gcn.

used for training/validation/test. In our experiments, We combine PP-Reg with two typical models GCN and SGC, and call these PPGCN models GCN-$\mathcal{L}_{PP}^k$ and SGC-$\mathcal{L}_{PP}^k$ ($k = 1, 2$), respectively. Then, we compare our PPGCNs with Node2vec [4], DeepWalk [12], GAT [14], APPNP [8], DropEdge [13], GCN [7], SGC [15], ChebNet [3], SSGC [19], and DAGNN [10].

Table 1. Results (%) with the state-of-art methods on semi-supervised node classification (- means out of memery). The top accuracy is highlighted.

	Assortative datasets				Disassortative datasets			
Method	Cora	Citeseer	Pubmed	Photo	Chameleon	Squirrel	Wisconsin	Cornell
Node2vec	76.7 ± 1.1	51.2 ± 1.7	76.4 ± 1.1	84.1 ± 1.0	53.7 ± 2.2	33.8 ± 0.6	46.5 ± 1.7	54.1 ± 0.1
DeepWalk	69.4 ± 1.4	47.1 ± 1.9	71.4 ± 1.2	84.3 ± 0.8	55.7 ± 1.4	34.3 ± 1.5	45.8 ± 0.5	52.7 ± 1.1
GCN	81.5 ± 0.8	68.2 ± 1.0	76.6 ± 1.5	86.6 ± 1.4	57.9 ± 0.8	40.7 ± 0.6	62.7 ± 1.6	56.8 ± 1.5
SGC	82.5 ± 0.4	72.7 ± 0.2	79.8 ± 0.5	86.4 ± 0.6	55.9 ± 0.6	33.7 ± 0.1	72.5 ± 1.6	59.5 ± 1.2
GAT	83.1 ± 0.2	71.4 ± 0.1	-	-	55.5 ± 0.5	-	72.5 ± 1.0	61.2 ± 0.1
ChebNet	80.6 ± 0.7	69.3 ± 1.2	79.0 ± 0.7	86.1 ± 1.3	60.0 ± 0.1	41.1 ± 0.5	58.8 ± 0.1	54.1 ± 2.7
DropEdge	81.5 ± 0.3	68.8 ± 0.8	78.8 ± 0.5	88.3 ± 0.4	56.4 ± 1.4	41.3 ± 0.8	60.8 ± 1.1	54.1 ± 0.1
APPNP	83.3 ± 0.3	71.8 ± 0.4	80.1 ± 0.3	87.7 ± 0.5	55.3 ± 0.2	35.6 ± 0.5	**74.5** ± 1.1	61.7 ± 1.3
SSGC	83.0 ± 0.7	73.1 ± 0.3	79.8 ± 0.1	87.4 ± 0.2	51.1 ± 0.3	34.8 ± 0.1	70.6 ± 0.6	**62.2** ± 0.8
DAGNN	80.4 ± 0.3	71.0 ± 0.4	78.6 ± 0.7	89.1 ± 1.2	55.5 ± 0.6	38.1 ± 0.4	70.6 ± 1.1	59.5 ± 2.2
GCN-$\mathcal{L}_{PP}^2$	82.9 ± 1.1	69.3 ± 1.8	78.0 ± 0.8	89.4 ± 1.5	**62.7** ± 0.6	**41.7** ± 1.4	68.6 ± 1.1	59.5 ± 1.5
SGC-$\mathcal{L}_{PP}^2$	83.5 ± 0.8	**73.4** ± 0.3	79.8 ± 0.6	88.2 ± 1.2	57.0 ± 2.1	34.0 ± 0.6	**74.5** ± 0.9	**62.2** ± 0.2
GCN-$\mathcal{L}_{PP}^1$	**84.0** ± 0.5	70.2 ± 1.2	78.6 ± 1.2	**89.8** ± 0.7	62.1 ± 1.1	41.5 ± 0.8	66.7 ± 2.4	59.3 ± 1.8
SGC-$\mathcal{L}_{PP}^1$	84.0 ± 0.8	**73.4** ± 0.2	**80.3** ± 1.2	88.0 ± 1.7	56.8 ± 0.9	33.9 ± 0.8	72.5 ± 1.1	61.8 ± 0.2

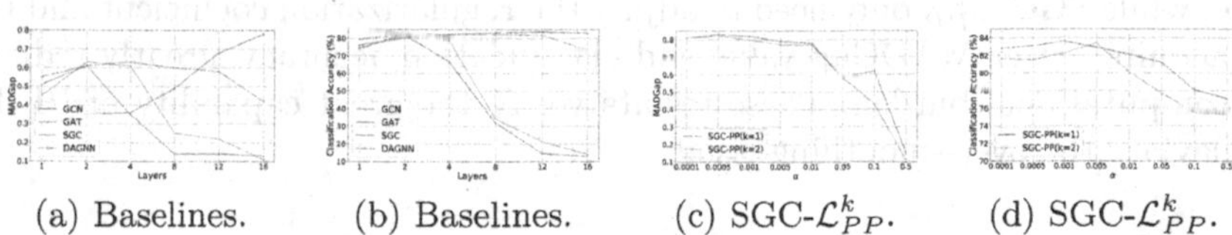

(a) Baselines. (b) Baselines. (c) SGC-$\mathcal{L}_{PP}^k$. (d) SGC-$\mathcal{L}_{PP}^k$.

Fig. 2. The capabilty to relieving over-smoothing issue of PP-Reg.

We set dropout rate as 0.6, hidden dimension as 64, weight decay as 0.0005, the number of convolutional layers as 2 for all methods, and tune the learning rate in $[0.1, 0.2, 0.5, 0.01, ..., 0.09, 0.001, ..., 0.009]$, the regularization coefficient α in $[0.001, ..., 0.009, 0.0001, ..., 0.0009, 0.00001, ..., 0.00009]$. We use the Adam optimizer with 1500 epochs and early stopping with a patience of 100 epochs. We report the mean classification accuracy and standard deviation after 10 runs.

4.2 Node Classification Accuracy of PPGCNs

As shown in Table 1, our PPGCNs performs better than the representative baselines by significant marigins. The results demonstrate the effectiveness of our PPGCNs. On Photo, Chameleon, and Squirrel, the best performance is achieved with GCN-based models, e.g., GCN comparing to SGC and GCN-$\mathcal{L}_{PP}^k$ comparing to SGC-$\mathcal{L}_{PP}^k$. It indicates that these datasets are more complex and need the non-linear activation function ReLU to fit more subtly.

Compared with GCN, GCN-$\mathcal{L}_{PP}^1$ has significant improvement to different degrees on different datasets, and so has SGC-$\mathcal{L}_{PP}^1$ compared with SGC. These results verify the superiority of adding PP-Reg on existing GNNs, which can promote smoothness without encountering over-smoothing issue.

On assortative datasets, the best performance is achieved by PPGCNs with 1st-order PP-Reg, while on disassortative datasets, it is achieved by PPGCNs with 2nd-order PP-Reg. It indicates that 1st-order PP-Reg, i.e., $\mathcal{L}_{PP}^1$, can extract more useful information on assortative datasets, and 2nd-order PP-Reg, i.e., $\mathcal{L}_{PP}^2$, can extract more useful information on disassortative datasets.

Measuring and Evaluating Smoothing Performance. We use MADGap [1] to measure over-smoothness of node embeddings. A smaller MADGap value indicates the more indistinguishable node embeddings. Figure 2 shows the MADGap values and classification results w.r.t. different models. Both the increase in layers and α stands for the increase in smoothness. Figure 2(a) and Fig. 2(b) indicate that some delicate designs in SGC [15] and DAGNN [10] can alleviate the over-smoothing issue. Figure 2(c) and Fig. 2(d) indicate that as α increases, the MADGap value maintains at a relatively high level in a large range until $\alpha > 0.1$, especially when $\alpha = 0.5$, the MADGap value drops to 0.1, while the accuracy still maintains at a high level. Then, We compare SGC-$\mathcal{L}_{PP}^k$ with SGC and DAGNN. The plots show that both metrics in SGC-$\mathcal{L}_{PP}^k$ are higher than them in SGC, and the MADGap of DAGNN needs 16 convolution to reach a relatively high value, while SGC-$\mathcal{L}_{PP}^k$ only need to adjust the regularization coefficient and they can acquire better MADGap value and classification accuracy, greatly reducing the computational burden. These results verify the great capability of PP-Reg to alleviate the over-smoothing issue.

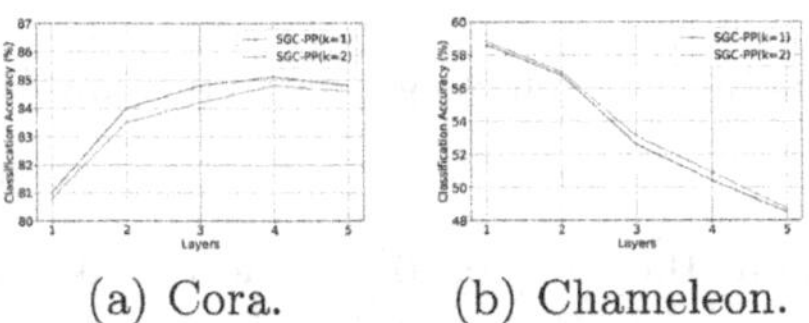

(a) Cora. (b) Chameleon.

Fig. 3. The relation between the number of layers and the order of PP-Reg.

The Relation between The Number of Layers and The Order of PP-Reg. We conduct experiments for SGC-$\mathcal{L}_{PP}^1$ and SGC-$\mathcal{L}_{PP}^2$ w.r.t. different layers on two representative datasets with homophily, i.e., Cora and Chameleon. The results are illustrated in Fig. 3. Intuitively, there is no essential connection between the number of convolutional layers and the order of PP-Reg, e.g., 1-layer convolution matching 1st-order PP-Reg and 2-layer convolution matching 2nd-order PP-Reg. However, the number of convolutional layers should be limited to a certain range to avoid the over-smoothing issue caused by stacking convolutional layers, and the order of PP-Reg should be limited to a certain range to avoid excessive computation complexity as well.

5 Conclusion

In this paper, firstly, we propose proximity preserving regularization (PP-Reg) by introducing k-order graph smoothness into loss function, which avoids over-smoothing issue skillfully. Besides, in our empirical analysis, 1st-order PP-Reg is more suitable for assortative datasets and 2nd-order PP-Reg is more suitable for disassortative datasets. According to comprehensive experiments, PPGCNs (combining PP-Reg with typical GNNs like GCN and SGC) achieve a better performance than current state-of-the-art models on most benchmark datasets.

References

1. Chen, D., Lin, Y., Li, W., Li, P., Zhou, J., Sun, X.: Measuring and relieving the over-smoothing problem for graph neural networks from the topological view. In: Proceedings of the AAAI Conference on Artificial Intelligence. vol. 34, pp. 3438–3445 (2020)
2. Chung, F.R., Graham, F.C.: Spectral graph theory. No. 92, American Mathematical Soc. (1997)
3. Defferrard, M., Bresson, X., Vandergheynst, P.: Convolutional neural networks on graphs with fast localized spectral filtering. In: Advances in Neural Information Processing Systems, vol. 29, pp. 3844–3852 (2016)
4. Grover, A., Leskovec, J.: node2vec: Scalable feature learning for networks. In: Proceedings of the 22nd ACM SIGKDD International Conference on Knowledge Discovery and Data Mining, pp. 855–864 (2016)
5. Jin, W., Derr, T., Wang, Y., Ma, Y., Liu, Z., Tang, J.: Node similarity preserving graph convolutional networks. In: Proceedings of the 14th ACM International Conference on Web Search and Data Mining, pp. 148–156 (2021)
6. Kim, D., Oh, A.: How to find your friendly neighborhood: graph attention design with self-supervision. In: International Conference on Learning Representations (2020)
7. Kipf, T.N., Welling, M.: Semi-supervised classification with graph convolutional networks. arXiv preprint arXiv:1609.02907 (2016)
8. Klicpera, J., Bojchevski, A., Günnemann, S.: Predict then propagate: graph neural networks meet personalized pagerank. arXiv preprint arXiv:1810.05997 (2018)
9. Li, Q., Han, Z., Wu, X.M.: Deeper insights into graph convolutional networks for semi-supervised learning. In: Thirty-Second AAAI Conference on Artificial Intelligence (2018)
10. Liu, M., Gao, H., Ji, S.: Towards deeper graph neural networks. In: Proceedings of the 26th ACM SIGKDD International Conference on Knowledge Discovery & Data Mining, pp. 338–348 (2020)
11. Pei, H., Wei, B., Chang, K.C.C., Lei, Y., Yang, B.: Geom-GCN: geometric graph convolutional networks. arXiv preprint arXiv:2002.05287 (2020)

12. Perozzi, B., Al-Rfou, R., Skiena, S.: Deepwalk: online learning of social representations. In: Proceedings of the 20th ACM SIGKDD International Conference on Knowledge Discovery and Data Mining, pp. 701–710 (2014)
13. Rong, Y., Huang, W., Xu, T., Huang, J.: Dropedge: towards deep graph convolutional networks on node classification. arXiv preprint arXiv:1907.10903 (2019)
14. Veličković, P., Cucurull, G., Casanova, A., Romero, A., Lio, P., Bengio, Y.: Graph attention networks. arXiv preprint arXiv:1710.10903 (2017)
15. Wu, F., Souza, A., Zhang, T., Fifty, C., Yu, T., Weinberger, K.: Simplifying graph convolutional networks. In: International Conference on Machine Learning, pp. 6861–6871. PMLR (2019)
16. Yang, C., Wang, R., Yao, S., Liu, S., Abdelzaher, T.: Revisiting over-smoothing in deep GCNs. arXiv preprint arXiv:2003.13663 (2020)
17. Yang, H., Ma, K., Cheng, J.: Rethinking graph regularization for graph neural networks. arXiv preprint arXiv:2009.02027 (2020)
18. Zang, Y., et al.: GISDCN: a graph-based interpolation sequential recommender with deformable convolutional network. In: Bhattacharya, A. et al. (eds.) DASFAA 2022. LNCS, vol. 13246 pp. 289–297. Springer (2022). https://doi.org/10.1007/978-3-031-00126-0_21
19. Zhu, H., Koniusz, P.: Simple spectral graph convolution. In: International Conference on Learning Representations (2020)

Discovering Prevalent Weighted Co-Location Patterns on Spatial Data Without Candidates

Vanha Tran[1], Lizhen Wang[2](✉), Muquan Zou[2,3], and Hongmei Chen[2]

[1] FPT University, Hanoi 155514, Vietnam
hatvl4@fe.edu.vn
[2] Yunnan University, Kunming 650091, China
{lzhwang,hmchen}@ynu.edu.cn
[3] Kunming University, Kunming 650214, China

Abstract. Prevalent co-location patterns (PCPs) expose the latent relationships between spatial instances and features on spatial data. Traditional PCPs treat spatial instances and features equally. However, in real life, spatial instances and features have different importance depending on their significance or meanings, this leads to traditional PCPs may lose their usefulness and meaningfulness. To address this deficiency, prevalent weighted co-location patterns (PWCPs) are proposed. In PWCP mining, each spatial instance is considered with its varied importance as its weight. The weight of a co-location pattern is determined by the weights of spatial features and instances involved in the pattern. Unfortunately, the weight metric does not satisfy the downward closure property, mining PWCPs becomes very expensive since unnecessary candidates cannot be pruned. To tackle this, an efficient PWCP mining framework is developed based on maximal cliques. This framework improves mining performance by reducing the searching PWCP candidates efficiently. The effectiveness and efficiency of the proposed method are proved by the mining results on both spatial synthetic and real data sets.

Keywords: Prevalent co-location patterns · Prevalent weighted co-location patterns · Maximal cliques

1 Introduction

As one of the important branches of data mining, co-location patterns expose the hidden relationships between spatial instances and spatial features in space. Co-location patterns have been proven to be a powerful tool for discovering potentially useful knowledge from spatial data and it has been applied to many fields such as transportation [11], location-based services [15], public health [5], public security [3], and so on. Figure 1 shows an example of the process of discovering traditional PCPs [13]. The superscript assigned to each instance indicates the weight of it. The line that connects between instances denotes these instances

B. Li et al. (Eds.): APWeb-WAIM 2022, LNCS 13421, pp. 417–425, 2023.
https://doi.org/10.1007/978-3-031-25158-0_33

having a neighbor relationship. The mining process starts from size 2 candidates (that are groups of two Boolean spatial features, e.g., {A, C}) and continues until no higher size candidates are generated. As can be seen, in the traditional PCPs, only the prevalence of instances is considered, while their importance is ignored.

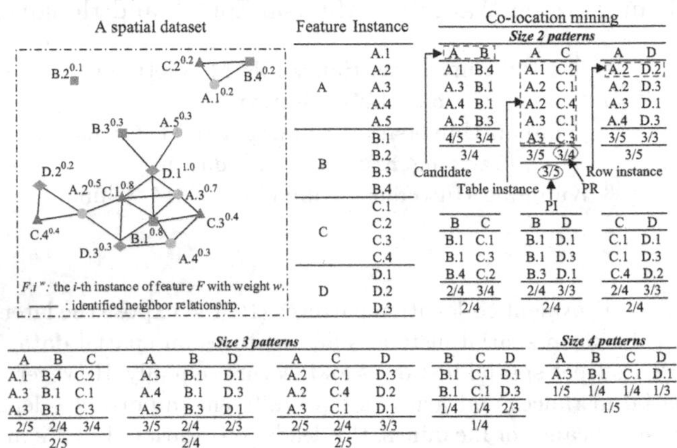

Fig. 1. An illustration of the traditional PCP mining process.

However, in reality, each instance in space has its influence on neighboring areas to indicate its importance. If the factor is not considered, the traditional PCPS may lose its meaningness. To tackle this shortcoming, this paper considers the importance of each instance and introduces the prevalent weighted co-location pattern (PWCP) mining notion. Each instance is assigned a weight value to reflect its importance, and then a weight metric is developed to evaluate the interest of a pattern. Unfortunately, the weight metric does not hold the downward closure property, discovering PWCPs becomes very expensive since a large number of unnecessary candidates is needed to examine their interesting degrees. An efficient PWCP mining framework is proposed without candidates.

2 Related Work

As one of the important branches of data mining, PCPS mining has received more and more attention. Join-based [4] that uses an expensive join operation to collect table instances of candidates, is the first algorithm developed to mine PCPs. Then many algorithms without join operations have been developed such as join-less [14], candidate pattern tree [13], overlapping clique-based [8]. Moreover, some algorithms have been proposed to provide a concise representation of PCPs, e.g., maximal [7], closed [13], non-redundant PCPs [9]. To face big data, parallel mining PCP algorithms have been developed, e.g., Hadoop and Map-reduce [6], the graphic processing unit (GPU) [1], and NoSQL [12].

All of the above mining methods, the importance of spatial instances and features is ignored. A high utility co-location pattern (HUCP) notion is introduced in which each instance has specified a utility [10]. This work considered the difference between features and instances that belong to the same feature and proposed a utility participation index (PUI) metric to evaluate the interest of a pattern. However, PUI does not meet the downward closure property. Although some pruning strategies were proposed, the mining efficiency is still very low, especially on large or/and dense datasets.

Our work overcomes the shortcoming of the traditional PCPs by assigning a weight for each instance to reflect its importance. We focus on mining prevalent weighted co-location patterns on spatial data and show that the proposed patterns are more useful and meaningful than the other inserting measures.

3 Method

3.1 Related Definitions

Definition 1. *(Weight of a spatial instance). Each spatial instance* o_i *in the spatial data set is assigned a value* $w_i \in (0, 1]$ *to reflect its importance depending on its significance or meanings, and denoted as* $o_i^{w_i}$.

Definition 2. *(Weight of a spatial feature). The weight of a feature* f_t *is the sum weight of all instances that belong to the feature type and is denoted as*

$$w(f_t) = \sum o_i^{w_i} \tag{1}$$

Definition 3. *(Co-location pattern). Given a set of spatial instances,* $S = \{o_1^{w_1}, ..., o_n^{w_n}\}$*, these instances belong to a set of boolean spatial features* $F = \{f_1, ..., f_m\}$ *each instance is formed by a four-element vector, <feature type, instance identification, location, weight>, and a neighbor relationship R over on S. A spatial co-location pattern is a subset of the feature set,* $c = \{f_1, ..., f_k\} \subseteq F$ *whose instances frequently form a clique under R. k is the size of c.*

Definition 4. *(Row instance and table instance). A row instance of a pattern* $c = \{f_1, ..., f_k\}$ *is a set of spatial instances,* $I = \{o_1, ..., o_k\}$*, that contains the instances of all feature types in c and they form a clique under R. The table instance of c is the set of all its row instances and is denoted as* $T(c)$.

Definition 5. *(Participation ratio (PR) and participation index (PI)). The participation ratio of a spatial feature* $f_t \in c = \{f_1, ..., f_k\}$ *is denoted by*

$$PR(f_t, c) = |T(c)|_{f_t} \Big/ |S|_{f_t} \tag{2}$$

where $|T(c)|_{f_t}$ *and* $|S|_{f_t}$ *are the number of distinct instances of feature* f_t *in the table instance* $T(c)$ *of pattern c and the given spatial data S, respectively. The minimum of the participation ratios of c is its participation index, i.e.,* $PI(c) = \min\{PR(f_t, c)\}, f_t \in c$.

Definition 6. *(Prevalent co-location pattern, PCP). If the participation index of a co-location pattern c is not smaller than a user-specified prevalence threshold, μ, i.e., $PI(c) \geq \mu$, c is a prevalent co-location pattern.*

As can be seen that in PCP mining, all spatial instances and features are treated similarly and their weights are ignored. The PI metric only considers the presence of distinct instances that contribute to a pattern.

Definition 7. *(Participating instance). The participating instances of a feature $f_t \in c$ is the distinct instances that contribute to c and is denoted by*

$$PaI(f_t, c) = \{o_i^{w_i} | o_i^{w_i} \in T(c) \wedge feature\ type\ of\ o_i^{w_i}\ is\ f_t\} \tag{3}$$

Definition 8. *(Participating weight). The participating weight of a feature $f_t \in c$ is the sum of the weight of all the participating instances of $f_t \in c$ and is denoted by*

$$w(f_t, c) = \sum o_i^{w_i} \Big/ w(f_t), o_i^{w_i} \in PaI(f_t, c) \tag{4}$$

Definition 9. *(Weight of a co-location pattern). The weight of a co-location pattern c is the mean weight of the spatial features that are involved in c and is denoted by*

$$w(c) = \sum w(f_t) \Big/ |c|, f_t \in c \tag{5}$$

Definition 10. *(Prevalent weighted co-location pattern, PWCP). If the weight of a co-location pattern $c = \{f_1, ..., f_k\}$ is not smaller than a user-specified threshold, μ, i.e., $w(c) \geq \mu$, c is a prevalent weighted co-location pattern.*

Lemma 1. *The weight of a co-location pattern is not anti-monotonicity with the increase of the size of patterns.*

Proof. As shown in Fig. 1, we can calculate $w(\{A, B, D\})=0.672 \geq w(\{A, B\})=0.67$ and $w(\{A, B, C\})=0.561 \leq w(\{A, B\})$. The weight of a pattern can be smaller or larger than its supersets.

Since the downward closure property does not meet the weight metric, we cannot find PCPs firstly, and then pick out the PWCPs based-on the PCPs. And besides, if we apply the generating candidate-based framework as in mining PCPs [13,14] to discover PWCPs, a huge number of unnecessary candidates needs to be evaluated, the mining performance is very inefficient. Therefore, a querying mining framework without candidates is proposed in this work.

3.2 Querying Mining Framework Without Candidates

The key to filtering PWCPs is collecting participating instances of patterns and the participating instances are gathered based on the neighbor relationship R. As shown in Fig. 1, after materializing the neighbor relationship, an undirected connected graph G is generated, the neighbor relationship of all spatial instances is included in maximal cliques that are enumerated from G. If we enumerate these maximal cliques first and organize them in a special data structure, using a query machine to collect the participating instances of a pattern instead of from row instances. Therefore, we utilize maximal cliques to improve the efficiency of mining PWCPs. An efficient enumerating maximal clique algorithm [2] is used in our work. However, we can not apply directly in mining PWCPs since if spatial data is large, the intersection in this algorithm becomes very expensive. In addition, it needs a huge storage space to perform the recursive process. We design a graph partition method to address these two problems.

Definition 11. *(Star neighbor instance). The star neighbor instances of $o_i^{w_i} \in S$ whose feature type is f_t is a set of neighboring instances that is denoted by*

$$SNI(o_i^{w_i}) = \{o_j^{w_j} | feature\ type\ of\ o_j^{w_j} \neq f_t \wedge R(o_i^{w_i}, o_j^{w_j})\} \tag{6}$$

Definition 12. *(Induced subgraph and subgraph). An instance $o_i^{w_i}$ and its star neighbor instances $SNI(o_i^{w_i})$ form an induced subgraph, $G_I(o_i^{w_i})$. A subgraph $G_S(o_i^{w_i})$ is an extension of the induced subgraph by adding the star neighbor instances of $o_i^{w_i}$, i.e., $G_S(o_i^{w_i}) = G_I(o_i^{w_i}) \cup SNI(o_i^{w_i})$.*

Algorithm 1 shows the pseudocode of listing maximal cliques. For each instance $o_i^{w_i} \in S$, the algorithm first gets $G_I(o_i^{w_i})$ (Steps 2–4) and $G_S(o_i^{w_i})$ (Step 5). After that, the algorithm iterates over each instance $o_j^{w_j}$ in the star neighbor instances of $o_i^{w_i}$ and put the star neighbor instances of $o_j^{w_j}$ in to $G_S(o_i^{w_i})$ (Steps 6–7). Next, $o_j^{w_j}$) is removed from S (Step 8). Finally, $G_S(o_i^{w_i})$ and the star neighbor instances are taken into the enumerating maximal clique algorithm [2].

Algorithm 1. Enumerating maximal cliques

Input: S, SNI
Output: a set of maximal cliques, MC

1: **while** $S \neq \emptyset$ **do**
2: $\quad o_i^{w_i} \leftarrow S$.pop()
3: $\quad G_I$.add($o_i^{w_i}$)
4: $\quad G_I = G_I \cup SNI(o_i^{w_i})$
5: $\quad G_S \leftarrow G_I$
6: $\quad$ **for** $o_j^{w_j} \in SNI(o_i^{w_i})$ **do**
7: $\qquad G_S = G_S \cup SNI(o_j^{w_j})$
8: $\qquad S$.remove($o_j^{w_j}$)
9: $\quad MC \leftarrow$ Algorithm in [2]
10: **return** MC

Definition 13. *(Clique hash table). A clique hash table is a two-level hash map structure with the key is the set of features that are formed by the instances in the maximal cliques and the value is a hash map with its key is the feature type and its value is the set of instances in the maximal cliques, respectively.*

Figure 2 shows the clique hash table constructed from the enumerated maximal cliques listed from the data set shown in Fig. 1. The pseudocode of the proposed querying framework for mining PWCPs is described in Algorithm 2. It first gets all keys in the clique hash table and stores them in KEY (Steps 1–2). Next, while KEY is not empty, all keys in KEY are sorted in descending order of size (Step 5). Each pattern $c \in KEY$, the algorithm iterates each item in the clique hash table to find the keys that are supersets of c to query and collect the participation instances (Steps 7–10). After that, the weight of the pattern is computed (Step 11). If the weight of c is not smaller than the user-specified threshold, c is a PWCP and is put into RES. Finally, a direct subset of c is generated and put into KEY as new patterns that need to be examined.

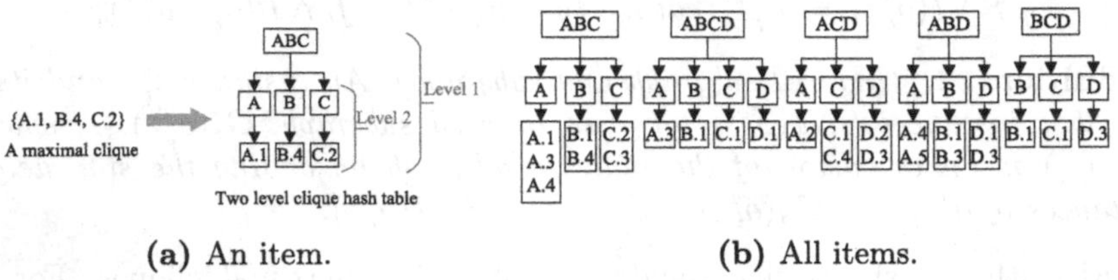

(a) An item. (b) All items.

Fig. 2. The constructed clique hash table.

Algorithm 2. Querying framework for mining PWCPs

Input: a clique hash table, CHT; **Output:** a set of PWCPs, RES

1: **for** $item \in CHT$ **do**
2: $\quad KEY \leftarrow item$.key
3: **while** $KEY \neq \emptyset$ **do**
4: $\quad KEY \leftarrow$ sortBySize(KEY)
5: $\quad c \leftarrow KEY$.pop()
6: $\quad$ **for** $item \in CHT$ **do**
7: $\quad\quad$ **if** $c \subseteq item$.key **then**
8: $\quad\quad\quad PaI(c) \leftarrow item$.value
9: $\quad w(c) \leftarrow$ calculateWeight($PaI(c)$)
10: $\quad$ **if** $w(c) \geq \mu$ **then**
11: $\quad\quad RES$.add(c)
12: $\quad subc \leftarrow$ generateSubset(c)
13: $\quad KEY$.add($subc$)
14: **return** RES

4 Evaluation

This section examines the effectiveness and efficiency of the proposed method. The candidate pattern tree [13] and candidate-based [10] algorithms are chosen as the representations of the traditional PCP and HUCP mining, respectively.

All algorithms are coded in C++ and performed on Intel(R) Core i7-3770 and 16 GB main memory of a computer. A set of synthetic data sets that are generated by a generator [14]. Points of interest of Beijing and Shanghai are two real data sets used in our experiment [8]. The weight of the instances in these data sets is generated according to an exponential distribution under a hypothesis that the number of the low weight instances is much more than the high weight instances.

4.1 The Number of Mined Patterns

We first examine the number of patterns discovered by different interesting metrics. Figures 3 shows the number of patterns mined on different distance and user-specified thresholds. The number of patterns increases with the increase of distance thresholds and decreases with the increase of user-specified thresholds. However, the number of PWCPs is always larger than the others.

4.2 The Efficiency of the Querying Mining Framework

The efficiency of the querying mining framework in improving the performance of discovering PWCPs is evaluated in this experiment. The execution time of the compared algorithms with the increase of distance thresholds is shown in Fig. 4(a). As can be seen, the proposed framework shows good performance, especially at large distance thresholds. Moreover, Fig. 4(b) shows the execution time on different user-specified thresholds. The proposed querying mining framework shows better performance than the generating candidate-based mining framework when the user-specified threshold is set to small.

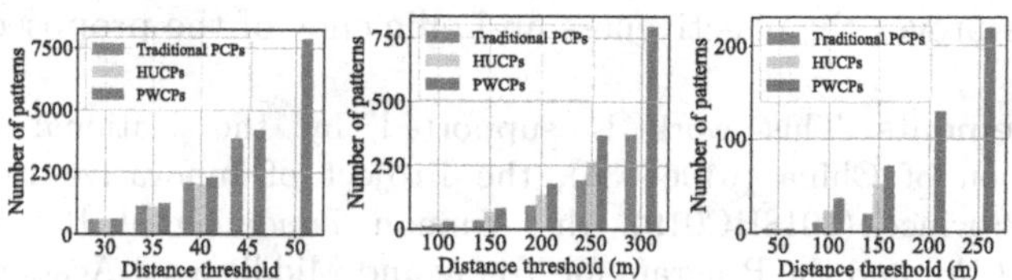

(a) Different distance thresholds on synthetic, Bejing, and Shanghai data sets (μ=0.2).

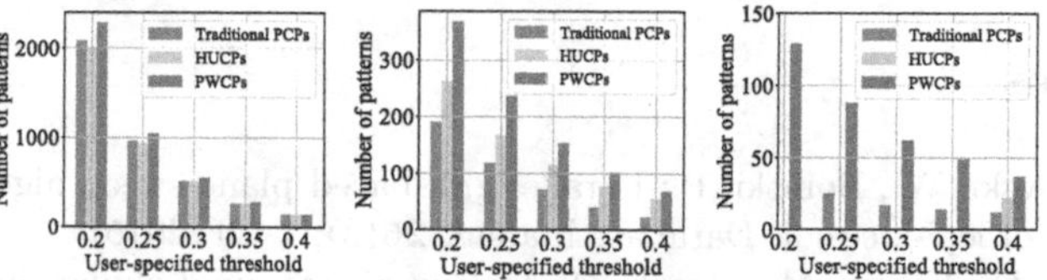

(b) Different user-specified thresholds on synthetic, Bejing, and Shanghai data sets.

Fig. 3. The number of patterns discovered by the different interesting metrics.

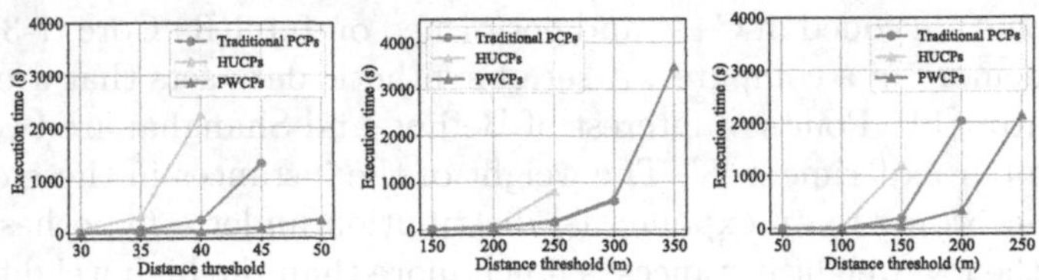

(a) Different distance thresholds on synthetic, Bejing, and Shanghai data sets (μ=0.2).

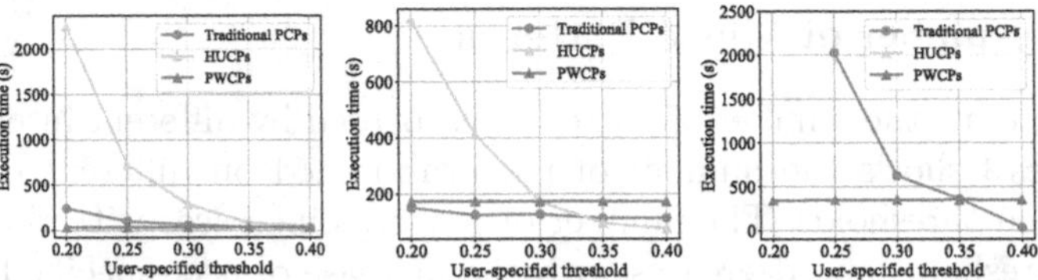

(b) Different user-specified thresholds on synthetic, Bejing, and Shanghai data sets.

Fig. 4. The number of patterns discovered by the different interesting metrics.

5 Conclusion

In this paper, the prevalent weighted co-location pattern notion is introduced to address the shortcomings of the traditional PCPs. In PWCPs, spatial instances and features are considered with their varied importance as their weight. The interest of a pattern is evaluated by its weight which is determined by the weights of spatial features and instances involved in the pattern. To improve the mining performance when the interesting metric does not satisfy the closed property, a querying mining framework is proposed. The experiment on both synthetic and real data sets proves the effectiveness and efficiency of the proposed method.

Acknowledgements. This work is supported by the National Natural Science Foundation of China (61966036), the Project of Innovative Research Team of Yunnan Province (2018HC019), the Yunnan Fundamental Research Projects (202201AS070015), and the Program for Young and Middle-aged Academic and Technical Reserve Leaders of Yunnan Province (202205AC160033).

References

1. Andrzejewski, W., Boinski, P.: Parallel gpu-based plane-sweep algorithm for construction of icpi-trees. J. Database Manag. **26**(3), 1–20 (2015)
2. Eppstein, D., Löffler, M., Strash, D.: Listing all maximal cliques in sparse graphs in near-optimal time. In: Cheong, O., Chwa, K.-Y., Park, K. (eds.) ISAAC 2010. LNCS, vol. 6506, pp. 403–414. Springer, Heidelberg (2010). https://doi.org/10.1007/978-3-642-17517-6_36
3. He, Z., Deng, M., Xie, Z., Wu, L., Chen, Z., Pei, T.: Discovering the joint influence of urban facilities on crime occurrence using spatial co-location pattern mining. Cities **99**, 102612 (2020)

4. Huang, Y., Shekhar, S., Xiong, H.: Discovering colocation patterns from spatial data sets: a general approach. IEEE Trans. Knowl. Data Eng. **16**(12), 1472–1485 (2004)
5. Li, J., Adilmagambetov, A., Mohomed Jabbar, M.S., Zaïane, O.R., Osornio-Vargas, A., Wine, O.: On discovering co-location patterns in datasets: a case study of pollutants and child cancers. GeoInformatica **20**(4), 651–692 (2016)
6. Sheshikala, M., Rao, R.D.: A map-reduce framework for finding clusters of colocation patterns-a summary of results. In: IACC, pp. 129–131. IEEE (2017)
7. Tran, V., Wang, L., Chen, H., Xiao, Q.: Mcht: a maximal clique and hash table-based maximal prevalent co-location pattern mining algorithm. Expert Syst. Appl. **175**, 114830 (2021)
8. Tran, V., Wang, L., Zhou, L.: A spatial co-location pattern mining framework insensitive to prevalence thresholds based on overlapping cliques. In: Distributed and Parallel Databases, pp. 1–38 (2021)
9. Wang, L., Bao, X., Zhou, L.: Redundancy reduction for prevalent co-location patterns. IEEE Trans. Knowl. Data Eng. **30**(1), 142–155 (2017)
10. Wang, L., Jiang, W., Chen, H., Fang, Y.: Efficiently mining high utility co-location patterns from spatial data sets with instance-specific utilities. In: Candan, S., Chen, L., Pedersen, T.B., Chang, L., Hua, W. (eds.) DASFAA 2017. LNCS, vol. 10178, pp. 458–474. Springer, Cham (2017). https://doi.org/10.1007/978-3-319-55699-4_28
11. Yao, X., Jiang, X., Wang, D., Yang, L., Peng, L., Chi, T.: Efficiently mining maximal co-locations in a spatial continuous field under directed road networks. Inf. Sci. **542**, 357–379 (2021)
12. Yoo, J.S., Boulware, D., Kimmey, D.: Parallel co-location mining with mapreduce and nosql systems. Knowl. Inf. Syst. **62**(4), 1433–1463 (2020)
13. Yoo, J.S., Bow, M.: A framework for generating condensed co-location sets from spatial databases. Intell. Data Anal. **23**(2), 333–355 (2019)
14. Yoo, J.S., Shekhar, S.: A joinless approach for mining spatial colocation patterns. IEEE Trans. Knowl. Data Eng. **18**(10), 1323–1337 (2006)
15. Yu, W.: Spatial co-location pattern mining for location-based services in road networks. Expert Syst. Appl. **46**, 324–335 (2016)

A Deep Looking at the Code Changes in OpenHarmony

Yuqing Niu, Lu Zhou(✉), and Zhe Liu

Nanjing University of Aeronautics and Astronautics, Nanjing, China
{nyq1997,lu.zhou,zhe.liu}@nuaa.edu.cn

Abstract. Open-source software has attracted more and more attention from practitioners. At the same time, many Chinese tech companies are embracing open source and opting for open source projects. Most previous research has focused on international companies such as Microsoft or Google, while the actual value of open source projects by Chinese tech companies remains unclear. To address this issue, we conducted an empirical study examining patch information for HarmonyOS. We collected a total of 234,791 commits from 2008 to March 2021. We manually screened out 3,616 submissions about bug fixes. The key findings include: 1) the percentage of bugs decreases rapidly as the number of modified source files increases. 2) About 70% of all bug types of modifications contain only one hunk. 3) About 50% of the bugs can be fixed by modifying one line of code.4) The programmer did not add any files to fix more than 80% of the bugs, Also did not delete any files to fix more than 90% of bugs. Our findings provide direction for software engineering researchers and practical recommendations for software developers and Huawei Technologies.

Keywords: Patch · Code change · HarmonyOS · Empirical study

1 Introduction

Open-source software has become one of the cornerstones of modern software development practices [1]. The open-source movement has dramatically reduced the cost of building and deploying software [2]. Today, more and more organizations and developers rely on open source solutions to sustain and accelerate the development of their software projects. At the same time, the emergence of modern platforms for developing and maintaining open source projects has propelled the open-source movement forward [2,3]. The most famous platform in China is Gitee; developers ([4,5]) use git operations to create a copy of the repository in Gitee, and they then improve the copy by submitting pull requests to the project maintainers.

In recent years, more and more Chinese technology companies have participated in and selected open-source projects, attracting more and more software

This research is supported by the National Key R&D Program of China (No. 2020YFB1005500).

B. Li et al. (Eds.): APWeb-WAIM 2022, LNCS 13421, pp. 426–434, 2023.
https://doi.org/10.1007/978-3-031-25158-0_34

developers to participate and participate. Huawei is a well-known technology company in China and a world-leading information and communication technology (ICT) solution provider, focusing on the ICT field. HUAWEI HarmonyOS is an operating system officially released by Huawei at the Huawei Developer Conference in Dongguan on August 9, 2019. Huawei Hongmeng system is a new distributed operating system for all scenarios, creating a world of interconnected super virtual terminals, organically linking people, devices, and scenarios, and connecting consumers with various types of life.

Despite the above facts, most studies only investigate the activities of software companies such as Microsoft or Google (e.g., Microsoft [6] or Google [7]). This paper takes the first step and conducts an empirical study of patching for HarmonyOS. This research will bring the following benefits: (1) Research results will provide insights into HarmonyOS bug fixes. (2) The findings will provide insights into how existing approaches to HarmonyOS can be improved. And (3) The findings will provide insights into new research directions for fixing HarmonyOS vulnerabilities.

Despite the above benefits, due to the following challenges: (1) It is challenging to conduct such empirical research. First, it requires extensive bug fixes to ensure representative results. Second, It requires high-quality bug fixes to reduce superficial conclusions. (2) Ensuring the accuracy of experimental results is challenging. Existing analysis tools cannot meet our needs due to the limitations of platforms or the inability to analyze multiple languages. To address the first two challenges, we use a manual collection method to screen all HarmonyOS submissions and classify the filtered bugs. We conducted the first empirical study to investigate the above research questions. This paper identifies the following key insights: **(1) Fault location.** Our results show that it is reasonable to assume that it only needs to fix a few files to fix the bug. Furthermore, about 70% of these files contain only one hunk block, and about 50% of the bugs can be fixed by modifying one line of code. **(2) Code fixes bugs.** Our results suggest that it is reasonable to focus on modified source files. These source files are mainly c and c++ files. The programmers did not add any files to fix more than 80% of the bugs and did not delete any files to fix more than 90% of the bugs. **(3) Non-source bugs.** Our results show that about 10% of bugs are in non-source files. Most of these files are configuration files and natural language documents. There are many bug fixes for non-code elements like comments, even in source files.

2 Methodology

This section describes the methodology of conducting our empirical study and the dataset used in this study. To answer these research questions, we should analyze thousands of bug fixes. To ensure the accuracy of our data, we manually identify and fix classification bugs.

2.1 Dataset

1) Data Collection: we manually retrieve all commits in its source code repository. Each commit has a message. We checked the messages and found that when

programmers commit changes, programmers may write a message describing the fix. For example, in third-party-cJSON, there is a commit message "fix encode-string-as-pointer." We identify this commit as a bugfix. Our results show that about 20% of commits are bug fixes. We collected the dataset in March 2021.

2) Categorization: To observe the types of bugs in HarmonyOS open source projects, we manually categorize commits about bugs - by manually looking at the full name, description, and read content to categorize the collected commits. We manually clustered the collected submissions into different groups to facilitate subsequent analysis.So we got 68 small categories, we divided them into five major categories, namely: memory bug(memory leaks, buffer overflows, etc.), compilation bug, function bug(variable type bugs, initialization problems, etc.), I /O bug, document fixes, Bugfixes (fixed specific bugs that are not in the above four categories).

2.2 Research Questions

We consider the following research questions (RQs):

- **RQ-1**. To what extent are the bugs localized in HarmonyOS?
- **RQ-2.** What types of files need to be modified to fix bugs in HarmonyOS?
- **RQ-3.** How many files need to be added or deleted to fix bugs in HarmonyOS?

2.3 Descriptive Statistics on the Data

We start by providing statistics on how developers submit different patches over time. The temporal distribution of patches may shed some light on system maintainers. On the other hand, we also classify patches and study their spatial distribution.

Temporal Distribution of Patches. We count patch distributions from July 2008) until March 2021 and outline them in Fig. 1(a).

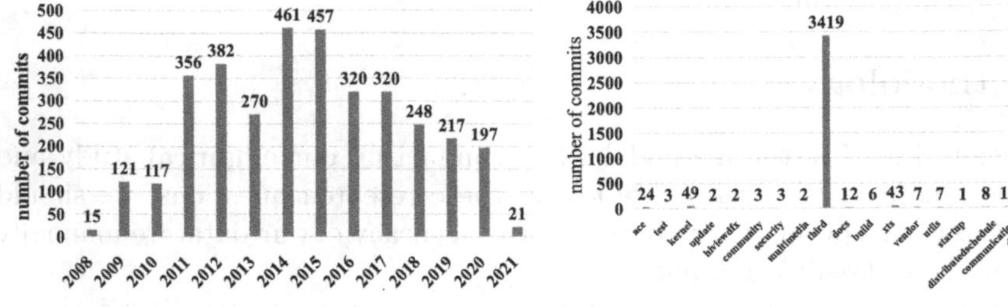

(a) Temporal distributions of patches.

(b) Number of patches committed by each patch process projcets

Fig. 1. Distributions of patches.

Overall, from 2008 to 2014, the overall number of programmers commits for HarmonyOS bug fixes was on an upward trend, with only a drop in 2010 and 2013. After peaking in 2014, the number of bugfix commits began to decline because the system was primarily functional.

Spatial Distribution of Patches. We compute the spatial distribution of patches in the Harmony system. The result is shown in Fig. 1(b). Most patches are for third program code and early development (i.e., in thrid) that are not yet part of the running kernel. It is worth noting that the collection of OpenHarmony ecological certification test suites (XTS/) is another focus for developers.

3 Empirical Results

3.1 [RQ-1:] Fault Distribution

We count the number of modified files fixed per bug type. Figure 3 shows the distribution. The horizontal axis represents the number of modified source files, and the vertical axis represents the percentage of corresponding bugs. The results lead to the following findings:

In general, the programmer modifies one or more source files to fix all kinds of bug types. The number of erroneous bug files is an important parameter for failure prediction models (e.g. [8,9]). Our results show that the percentage of bugs decreases rapidly as the number of modified source files increases.

We also found that programmers did not modify any source files to fix about 10% of the bugs. Yin et al. [10] showed that both open-source and commercial projects contain many bugs in configuration files. During this research, we found bug fixes in configuration files. For example, kernel-liteos-as bug report said that developers need to add QEMU arm virt CFI flash support and enable rootfs and userspace applications, the modified line is shown in Fig. 2:

```
diff --git a/tools/build/config/debug/qemu_arm_virt_ca7_clang.config
          b/tools/build/config/debug/qemu_arm_virt_ca7_clang.config
index b56d817..0ea33d9 100644
--- a/tools/build/config/debug/qemu_arm_virt_ca7_clang.config
+++ b/tools/build/config/debug/qemu_arm_virt_ca7_clang.config
@@ -34,6 +34,7 @@ LOSCFG_ARCH_CPU="cortex-a7"
+LOSCFG_PLATFORM_ROOTFS=y
```

Fig. 2. Patch diff for fixing bug *Chart-2*.

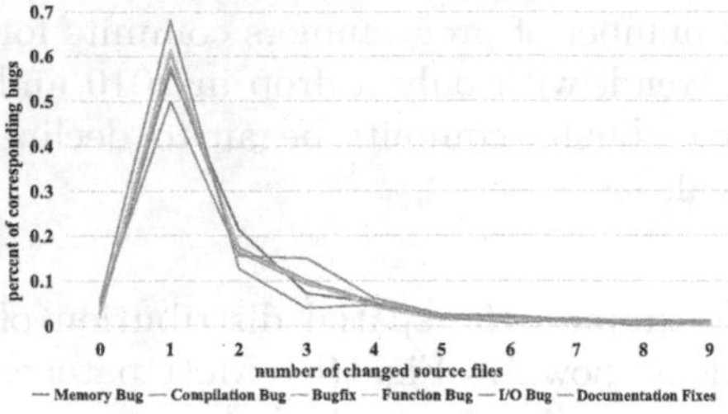

Fig. 3. The fault distribution at the bug level

Fig. 4. The distribution of repair actions

Researchers [8,9,11] proposed models to predict false reports of destructive source files. Since source files are very different from non-source files, it needs extensions to predict problematic non-source files. Furthermore, since non-source files are generally not executable, it requires significant extensions for fault location to locate fault lines of non-source files.

We count the number of hunks of modified files fixed per minute, as shown in Fig. 5(a), where the horizontal axis represents the number of four categories, and the vertical axis represents the bug type of each category. We further investigate patch locality regarding the number of code blocks (i.e., a set of contiguous lines of code) that the patch changes. In practice, code files can be large, and patches may propagate changes to varying degrees within the file, representing some degree of complexity in a fix. We can see from Fig. 5(b) that about 70% of all bug-type modifications contain only one hunk. Bug fixes that contain four or more hunk blocks are less than 10%.

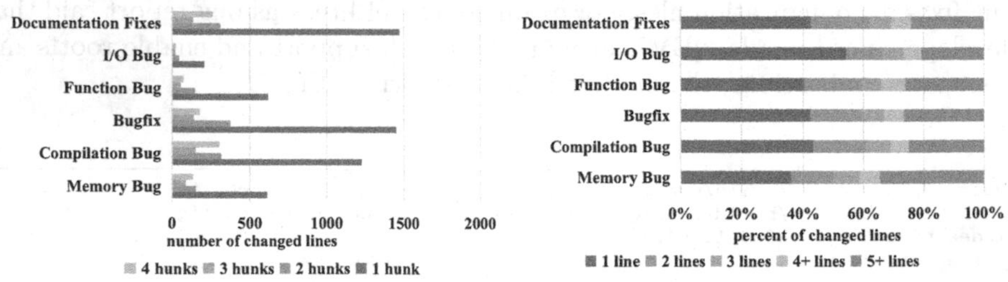

(a) Distribution of patch sizes in terms of hunks.

(b) Distribution of patch sizes in terms of lines.

Fig. 5. Distribution of patch sizes.

At the same time, we have also carried out a more detailed in-depth, we calculated the distribution of the number of lines of the modified file for each bug fix, as shown in Fig. 5(b), the horizontal axis represents the percentage of five categories, and the vertical axis represents each Kind of bug type. We compute the locality of the patch based on the number of rows affected by the change.

Such studies correlate with the proportion of orphaned changes (i.e., single-line changes) that fix bugs. Figure 5 shows that about 50% of bugs can be fixed by modifying one line of code, and about 20% of bugs can be fixed by modifying five or more lines of code.

RQ-1: *our results show that it is reasonable to assume that it only needs to fix a few files to fix the bug. And in these files, about 70% contain only one hunk block, and about 50% of the bugs can be fixed by modifying one line of code.*

3.2 [RQ-2:] File Type for Modifications

We counted file types for all modified files, We use enough file names to represent file types. To save space, we do not offer file types smaller than 1%. From the results, we have the following findings.

The most common modified files are C source files. The result is not surprising because the underlying HarmonyOS is mainly developed in the C language. Programmer did not modify the C source files to fix at least 10% of the bugs. So, on average, programmers modify fewer files than they do when the bug does not involve a C source file. The results show that dependencies between C source files may be higher than other files (such as standard configuration files).

The two most common types of modified non-source files are configuration files and natural language documents. The most found configuration files have names ending in yml or config, and we found three typical types of such files. The first type defines the parameters of the build tool. The second type defines run-time parameters. For example, bug reports in the third-party-jerryscript library ensure that tests are always executed in the appropriate time zone. The modified line is shown in Fig. 6:

```
diff --git a/.travis.yml b/.travis.yml index 27f7c3ea..8ce9861f 100644
--- a/.travis.yml
+++ b/.travis.yml
@@ -72,7 +72,6 @@ matrix:
-         - TZ=America/Los_Angeles
```

Fig. 6. Patch diff for fixing bug *Chart-3.*

The third type defines parameters for third-party tools. For example, the bug report in the third-party-unity library shows: Ensure fresh Python 2.7 on Trusty and compatible Intervaltree.The modified line is shown in Fig. 7:

```
diff --git a/.travis.yml b/.travis.yml
index 8b0b27d8..ad8bda71 100644
--- a/.travis.yml
+++ b/.travis.yml
@@ -135,6 +135,8 @@ matrix:
+       language: python # NOTE: only way to ensure python>=2.7.10 on Trusty image
+       python: 2.7
```

Fig. 7. Patch diff for fixing bug *Chart-4.*

RQ-2: *Our results suggest that it is reasonable to focus on modified source files. These source files are mainly c and c++ files. At the same time, there are also multi-language files such as python, js, and html. About 10% of bugs are in non-source files. Most of these files are configuration files and natural language documents.*

3.3 [RQ-3:] Additions and Deletions

We counted the number of repair operations on each code element, and Fig. 4 shows the results. Its vertical axis shows the name of each type of repair. We do not provide code elements with less than 1% of repair operations to save space. Its horizontal axis shows the number of repair operations, including categories such as additions, deletions, and modifications. The results lead to the following findings:

Overall, changes to files accounted for 90% of all kinds of fixes. The programmers did not add files to fix more than 80% of the bugs, nor did they delete files to fix more than 90% of the bugs. Hattori and Lanza [12] analyzed the nature of commits, and they found that commit size correlates with commit type. In particular, tiny commits are more related to bug fixes, while large commits are more related to new features. Our results reflect another nature of commits that at the file level revisions may be more relevant to bug fixes.

RQ-3:*our results show that programmers did not add any files to fix more than 80% of the bugs, nor did they delete any files to fix more than 90% of the bugs.*

4 Related Work

Bugfix Commit Research: Various research mines software repositories to analyze commits. Purushothaman and Perry [13]studied patch-related commits in terms of size of bug fixes and type of fix operations to study the impact of small source code changes. German [14] analyzed the characteristics of modification records (that is, source code changes in software version control systems) from three aspects: authorship, number of documents, and modification coupling of documents.

Using Real Patch Fixes: Keyunku et al. investigate the practice of patch building to study the impact of different patch generation techniques in Linux kernel development. Long et al. proposed a new system, Genesis, which processes patches to automatically infer code transformations for automatic patch generation.

Empirical research on bug fixes: Yin et al. [15] showed that bug fixes might introduce new bugs. Nguyen et al. [16] showed that reproducibility is common in small-scale bug fixes. Bird et al. show that many projects do not carefully maintain the link between bug reports and bug fixes.

5 Conclusions

This paper reports the results of our exploratory study of patches for the HarmonyOS open source project. We conducted an empirical patch study to gain insight into how many bugs were localized (RQ-1), what types of files needed to be modified in HarmonyOS (RQ-2), and how many files were added or removed to fix bugs (RQ-3). We collected 234,791 HarmonyOS commits from Gitee, mined 3,616 commits for bug fixes, and analyzed them. We study the manual patches of HarmonyOS from the number of file modifications, the number of hunks, and the increase or decrease of the number of lines, etc., to deeply study HarmonyOS at the vulnerability level, and confirm and supplement the field of patch empirical research at the same time.

References

1. Dias, L.F., Steinmacher, I., Pinto, G.: Who drives company-owned oss projects: employees or volunteers. In: V Workshop on Software Visualization, Evolution and Maintenance, VEM, p. 10 (2017)
2. Coelho, J., Valente, M.T.: Why modern open source projects fail. In: Proceedings of the 2017 11th Joint Meeting on Foundations of Software Engineering, pp. 186–196 (2017)
3. Eghbal, N.: Roads and bridges: The unseen labor behind our digital infrastructure. Ford Foundation (2016)
4. Gousios, G., Pinzger, M., van Deursen, A.: An exploratory study of the pull-based software development model. In: Proceedings of the 36th International Conference on Software Engineering, pp. 345–355 (2014)
5. Kalliamvakou, E., Gousios, G., Blincoe, K., Singer, L., German, D.M., Damian, D.: An in-depth study of the promises and perils of mining github. Empir. Softw. Eng. **21**(5), 2035–2071 (2016)
6. Kalliamvakou, E., Damian, D., Blincoe, K., Singer, L., German, D.M.: Open source-style collaborative development practices in commercial projects using github. In: 2015 IEEE/ACM 37th IEEE International Conference on Software Engineering, vol. 1, pp. 574–585. IEEE (2015)
7. Jaspan, C., et al. Advantages and disadvantages of a monolithic repository: a case study at google. In: Proceedings of the 40th International Conference on Software Engineering: Software Engineering in Practice, pp. 225–234 (2018)
8. Zhou, J., Zhang, H., Lo, D.: Where should the bugs be fixed? more accurate information retrieval-based bug localization based on bug reports. In: 2012 34th International Conference on Software Engineering (ICSE), pp. 14–24. IEEE (2012)
9. Kim, D., Tao, Y., Kim, S., Zeller, A.: Where should we fix this bug? a two-phase recommendation model. IEEE Trans. Softw. Eng. **39**(11), 1597–1610 (2013)
10. Yin, Z., Ma, X., Zheng, J., Zhou, Y., Bairavasundaram, L.N., Pasupathy, S.: An empirical study on configuration errors in commercial and open source systems. In: Proceedings of the Twenty-Third ACM Symposium on Operating Systems Principles, pp. 159–172 (2011)
11. Ye, X., Bunescu, R., Liu, C.: Learning to rank relevant files for bug reports using domain knowledge. In: Proceedings of the 22nd ACM SIGSOFT International Symposium on Foundations of Software Engineering, pp. 689–699 (2014)

12. Hattori, L.P., Lanza, M.: On the nature of commits. In: 2008 23rd IEEE/ACM International Conference on Automated Software Engineering-Workshops, pp. 63–71. IEEE (2008)
13. Purushothaman, R., Perry, D.E.: Toward understanding the rhetoric of small source code changes. IEEE Trans. Softw. Eng. **31**(6), 511–526 (2005)
14. German, D.M.: An empirical study of fine-grained software modifications. Empir. Softw. Eng. **11**(3), 369–393 (2006)
15. Yin, Z., Yuan, D., Zhou, Y., Pasupathy, S., Bairavasundaram, L.: How do fixes become bugs? In: Proceedings of the 19th ACM SIGSOFT Symposium and the 13th European Conference on Foundations of Software Engineering, pp. 26–36 (2011)
16. Nguyen, H.A., Nguyen, A.T., Nguyen, T.T., Nguyen, T.N., Rajan, H.: A study of repetitiveness of code changes in software evolution. In: 2013 28th IEEE/ACM International Conference on Automated Software Engineering (ASE), pp. 180–190. IEEE (2013)

GADAL: An Active Learning Framework for Graph Anomaly Detection

Wenjing Chang[1,2], Jianjun Yu[1(✉)], and Xiaojun Zhou[1]

[1] Computer Network Information Center, Chinese Academy of Sciences, Beijing, China
{changwenjing,xjzhou}@cnic.cn, yujj@cnic.ac.cn
[2] University of Chinese Academy of Sciences, Beijing, China

Abstract. Graph Neural Networks (GNNs) have been widely used in graph-based anomaly detection tasks, and these methods require a sufficient amount of labeled data to achieve satisfactory performance. However, the high cost for data annotation leads to some well-designed algorithms in low practicality in real-world tasks. Active learning has been used to find a trade-off between labeling cost and model performance, while few prior works take it into anomaly detection. Therefore, we propose GADAL, a novel Active Learning framework for Graph Anomaly Detection, which employs a multi-aspects query strategy to achieve high performance within a limited budget. First, we design an abnormal-aware query strategy based on the scalable sliding window to enrich abnormal patterns and alleviate the class imbalance problem. Second, we design an inconsistency-aware query strategy based on the effective degree to capture the most specificity nodes in information aggregation. Then we provide a hybrid solution for the above query strategies. Empirical studies demonstrate that our query strategy significantly outperforms other strategies, and GADAL achieves a comparable performance to the state-of-art anomaly detection methods within less than 3% of the budget.

Keywords: Graph anomaly detection · Active learning · Graph neural networks

1 Introduction

Graph-structured data has been widely used in real-world applications, such as social networks, financial networks, and infrastructure networks, to model a wide range of complex systems. Discovering the rare occurrences in networks has a high impact on the applications.

For graph-based anomaly detection tasks, there are two general challenges: class imbalance problem and feature inconsistency problem. Many effective methods [2,6,7] have been proposed to alleviate these problems while they require a sufficient amount of labeled data [11], which is costly in terms of budget and experts. As a result, these well-designed methods are challenging to practice because manually labeling all instances is not feasible in many real-world scenarios. Given a limited annotation budget, it is intriguing to select the

B. Li et al. (Eds.): APWeb-WAIM 2022, LNCS 13421, pp. 435–442, 2023.
https://doi.org/10.1007/978-3-031-25158-0_35

best-performed instances from the unlabeled data to label. To address the labeling problem, Active Learning (AL) [10] provides the idea of selecting examples that can benefit more to model training. Traditional AL strategies are designed for independent and identically distributed data, whereas the graph-structured data is not i.i.d since the information shared along edges. Furthermore, graph-based methods do not face the aforementioned challenges that exist in anomaly detection.

From the above observations, a specific AL query strategy which considers the graph structural features and the challenges in anomaly detection at the same time. The contributions of this paper are summarized as follows:

- We propose an abnormal-aware query strategy that can enrich the abnormal patterns and alleviate the class imbalance problem. We propose an inconsistency-aware query strategy that can discover the disguised anomalies and the most specificity normal nodes.
- We propose GADAL, a novel Active Learning framework for Graph Anomaly Detection, which employs a multi-aspects query strategy to achieve high performance within a limited budget.
- We conduct extensive experiments to verify the effectiveness of our query strategy and show a trade-off solution between budget and performance.

2 Related Works

Anomaly Detection on Graphs. The graph anomaly detection task aims to detect anomalous patterns from various behaviors and relationships on complex networks. Player2Vec [14] adopts an attention mechanism in aggregation process. Semi-GNN [12] applies a hierarchical attention mechanism to better correlate different neighbors and different views. GraphConsis [7] investigates three inconsistency problems in anomaly detection. CARE-GNN [2] employs a similarity-aware neighbor selector to enhance the aggregation process against camouflages. PC-GNN [6] incorporates label distribution information into the sampling process to choose balance train samples. All the above methods require a sufficient number of labeled nodes in model training, which limits the practicality of the model. Different from them, our method focuses on detecting anomalies within a limited labeling budget through AL strategies.

Active Learning on Graphs. Active learning aims to select best-performed instances for model training to label. The core is the query heuristics such as uncertainty, information density and query-by-committee. However, traditional active learning algorithms focus on i.i.d data, which is not feasible for graph-structured data. To measure the informativeness of each node, several graph-based AL methods [1,3,13] have been proposed. They either design a selection criterion based on the graph characteristics [13] or adaptively combine several selection criteria [1,3]. Though the above methods are designed for graph-structured data, they cannot handle the high imbalance and inconsistency problems in graph-based anomaly detection. Therefore, we propose to design an AL strategy dedicated to the problems of graph-based anomaly detection.

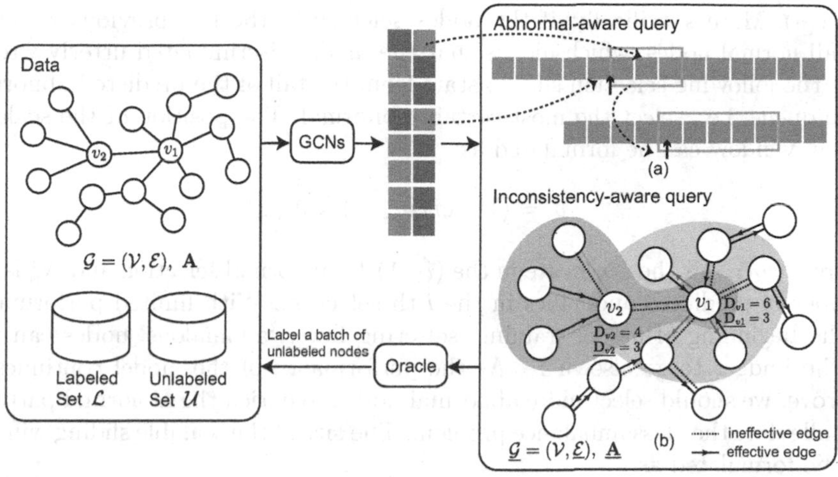

Fig. 1. The workflow of the GADAL framework.

3 Problem Formulation

The problem of active learning for graph-based anomaly detection is defined on the imbalanced graph $\mathcal{G} = (\mathcal{V}, \mathcal{E})$. Denote the set of labeled nodes as $\mathcal{L}$ and the set of unlabeled node as $\mathcal{U}$. Given an annotation budget B, the key of active learning for graph anomaly detection is to design an AL query strategy $\phi(v_i; \Theta)$, which can actively select B nodes from the unlabeled nodes set $\mathcal{U}$ to annotate and add to $\mathcal{L}$. The selected nodes are more conducive to model training than other nodes. Furthermore, the objective of this work is to maximize the performance of the model for anomaly detection tasks within a limited budget.

4 Proposed Method

4.1 Abnormal-aware Query Strategy

In anomaly detection tasks, the proportion of abnormal nodes and normal nodes varies greatly, which causes the minority but more critical class to be neglected. For the generic AL query strategy, the minority class instance has a lower probability of being selected in the query, resulting in the labeled set lying in unbalanced status, which is not suitable for model training. Besides, selecting fewer abnormal nodes also leads to the absence of anomalous patterns.

We propose to use the anomaly discriminate accuracy in the last previous selection to measure the current performance of the model and determine the starting position of the current sampling. If the discriminative error rate of the abnormal node selected last time is high, it indicates that the model has a weak ability to detect an abnormality, and the current sampling boundary should be moved to the position with a higher abnormality score, as shown in

Fig. 1(a). More specifically, if the nodes selected in the last previous selection are all normal nodes, which means that the model discriminated utterly wrong, then the following selection should start from the tail of the predicted abnormal node queue, i.e., select the most certain abnormal. The position of the scalable sliding window can be formulated as:

$$p_i = (1 - error_{i-1}) \times N_p^i \tag{1}$$

where $error_{i-1}$ is the error rate in the $(i-1)$-th abnormal selection and N_p^i is the number of predicted anomalies in the i-th selection. With limited performance at the beginning of model training, selecting fewer 'anomalous' nodes can free up the budget to go elsewhere. As the performance of the model continues to improve, we should select more abnormal nodes to enrich the abnormal patterns and alleviate the class imbalance problem. The size of the scalable sliding window can be formulated as:

$$N_{abs}^i = max(\frac{i}{s}, \alpha) \times b_i \tag{2}$$

where α is the upper limit of the anomaly sampling ratio, s is the current number of selections, and b_i is the budget in the i-th choice. The nodes selected in abnormal-aware query process can be formulated as:

$$L_{abs}^i = \{v, v \in Anomaly[p_i, p_i + N_{abs}^i]\} \tag{3}$$

where *Anomaly* is the ordered list of predicted anomalies.

With the scalable sliding window, we can pick out the relatively high uncertain anomalies in the training of model, thereby enriching the patterns of labeled nodes and alleviating the class imbalance problem in anomaly detection.

4.2 Inconsistency-Aware Query Strategy

The basic idea of GNN is to aggregate the information in the neighborhood along the edges. Measuring the effectiveness of a node in a local graph is important. In previous studies, the degree of a node is used as the centrality evaluation metric of graph. However, each edges have different weights for information propagation whereas the degree-based metrics generally see them equal. Motivated by the down-sampling process, which can filter out dissimilar neighbors based on feature similarity, we propose to calculate the degree matrix after down-sampling as node effectiveness. More specifically, if node v is selected as the down-sampled neighbor of node u, there is information propagating among the connection (u, v), which we define as an effective connection. The effective adjacency matrix can be formulated as:

$$\underline{A_{ij}} = \begin{cases} 1, & v_j \in \mathcal{N}(v_i), v_i \in \underline{\mathcal{N}}(v_j) \\ 0 \end{cases} \tag{4}$$

where $\mathcal{N}(v_i)$ represents the neighbor of v_i and $\underline{\mathcal{N}}(v_j)$ represents the down-sampling neighbor of nodes v_j. If node v_i is not the member of its neighbor v_j

down-sampling neighborhood, the edge between them is ineffective, and $\underline{A_{ij}} = 0$. The effective degree matrix can be formulated as:

$$\underline{D_{ii}} = \sum_j \underline{A_{ij}} \tag{5}$$

If a node has too many redundant connections, it is highly likely to be an anomaly that disguised itself among normal nodes or be a high specificity normal node. These two kinds of nodes are both essential to enrich the feature patterns. Therefore, we design an inconsistency-aware query strategy that calculates effective degree as a selection criterion and picks the most specificity nodes in the neighborhood. The inconsistency scores can be calculated as:

$$\phi_{inconsistency}(v_i) = D_{ii} - \underline{D_{ii}} \tag{6}$$

Through the inconsistency-aware query strategy, both the disguised anomalies and high specific normal nodes can be picked. Discovering these high specificity nodes in the query process both enriches the characteristic patterns and improves the model interpretability.

4.3 Integrating Query Strategies

We propose a hybrid batch-mode query strategy to better select nodes for anomaly detection tasks from specificity and generality. In each selection, we utilize α budget to pick abnormal and specificity nodes by abnormal-aware and inconsistency-aware query strategy, respectively. For the abnormal-aware query strategy based on the scalable sliding window, the budget is linear to its window size N^i_{abs} as Eq. (5). And the remaining budget is used to select the most specificity nodes based on inconsistency scores, and the budget can be formulated as:

$$N^i_{ins} = \alpha b_i - N^i_{abs} \tag{7}$$

Since the above two query strategies tend to select highly specificity nodes in a graph, we need to select some generic nodes as a complement to keep the balance of instances. Therefore, we adopt a random strategy in the hybrid query strategy, which can select nodes according to data distribution, and the budget can be formulated as:

$$N^i_{ran} = b_i \times (1 - \alpha) \tag{8}$$

After annotating the select nodes, compute the error rate in the abnormal-aware query process, which is used to adjust the position of the scalable sliding window, as follows:

$$error_i = \frac{\sum_{v \in L^i_{abs}} 1}{N^i_{abs}} \tag{9}$$

where N^i_{abs} and L^i_{abs} are the sampling size and selected nodes by abnormal-aware query in the i-th selection.

5 Experiments

In this section, we perform empirical evaluations to demonstrate the effectiveness of the proposed model GADAL on two public data sets, which are the YelpChi [9] review dataset and Amazon [8] review dataset. We leverage comprehensive metrics to measure the performance of all the compared methods, namely ROC-AUC, F1-macro, Recall, and GMean [6]. Furthermore, we compare the selected anomalous and normal node ratios (SIR) with other AL strategies.

First, in comparison with other AL strategies, Table 1 shows that GADAL significantly outperforms the other baselines. Compared to Random, the information entropy of selecting the most uncertain nodes performs worse on AUC. It suggests that only studying difficult samples will limit the improvement of model performance. Compared to other single AL criteria, AGE has an unstable performance because the simple linear combination of different criteria cannot adjust the parameters. All baselines perform worse than GADAL due to they are affected significantly by feature inconsistency and class imbalance problems, suggesting that GADAL is more suitable for anomaly detection problems. The experimental results also show that higher SIR is beneficial to model training, which also confirms that the class imbalance problem does affect model training. Compared to all baselines, GADAL achieves at least a 10% improvement on SIR, indicating its ability to alleviate the class imbalance problem.

Table 1. Performance comparison with AL baselines on YelpChi and Amazon.

	Method	Metric				
		AUC	Recall	F1-Macro	GMean	SIR(A/N)
Amazon	Random	0.9207	0.8327	0.8183	0.8213	0.13(23/177)
	Entropy	0.9001	0.8625	0.7250	0.8619	0.06(11/189)
	Degree	0.8878	0.8184	0.6664	0.8175	0.05(10/190)
	PageRank	0.8852	0.8195	0.6590	0.8191	0.05(10/190)
	AGE	0.8474	0.8245	0.7626	0.8161	0.09(17/183)
	Ours	**0.9386**	**0.8820**	**0.8520**	**0.8774**	**0.23(37/163)**
YelpChi	Random	0.6538	0.5460	0.5522	0.3659	0.17(29/171)
	Entropy	0.5566	0.5078	0.5070	0.3875	0.11(20/180)
	Degree	0.6447	0.5743	0.5677	**0.5059**	0.22(36/164)
	PageRank	0.5452	0.5160	0.5161	0.3410	0.10(19/181)
	AGE	0.7340	0.5730	0.5891	0.4181	0.26(41,159)
	Ours	**0.7542**	**0.6032**	**0.6235**	0.4922	**0.48(65/135)**

Then, we compared GADAL with the state-of-art anomaly detection methods, which are well-designed for anomaly detection but without considering the balance between performance and labeling budget. The corresponding results are shown in Fig. 2. GADAL significantly outperforms GCN [5], GraphSAGE [4], and Player2Vec with less than 3% budget. CARE-GNN and PC-GNN can deal with the imbalance problem, and their performance significantly outperforms other baselines. These two methods represent the top performance in graph-based anomaly detection, and GADAL also achieves a near performance in some metrics. For example, GADAL achieves comparable performance with PC-GNN, and improves 4.4% on GMean on the Amazon dataset. It suggests that GADAL strikes a balance between better performance and less budget.

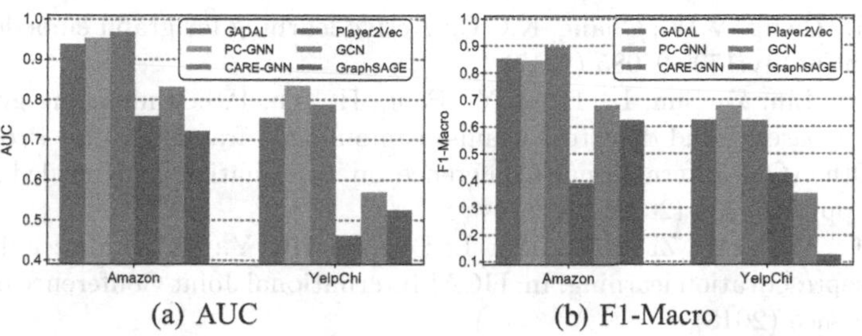

(a) AUC (b) F1-Macro

Fig. 2. Performance comparison with GNN-based anomaly detection methods.

We conduct an ablation study to better examine the contribution of each key component in proposed selection strategy by disabling one component at a time. The performance shows a large decrease if we remove any components in our strategy, which represents each component is effective on its focus problem, and they can be combined in a union strategy effectively (Table 2).

Table 2. The impact of the different components in GADAL.

Dataset	Amazon				YelpChi			
Method	AUC	△	Recall	△	AUC	△	Recall	△
No AbQ	90.24	−3.62	85.06	−3.14	67.15	−8.27	58.38	−1.94
No InQ	89.25	−4.61	83.88	−4.32	71.72	−3.70	54.28	-6.04
GADAL	**93.86**	−	**88.20**	−	**75.42**	-	**60.32**	−

6 Conclusion

In this paper, we proposed a novel active learning framework for anomaly detection that balances model performance and labeling budget. First, to enrich abnormal patterns and alleviate the class imbalance problem, a scalable sliding window is used to select current relatively uncertain anomalies. Second, to capture high specificity patterns, an inconsistency-aware query strategy is devised based on the ineffectiveness of nodes. Comprehensive experiments on real-world anomaly detection datasets show that GADAL outperforms existing AL techniques by significantly improving and achieves comparable performance with the state-of-art anomaly detection methods in less than a 3% budget.

Acknowledgements. This research is supported by a grant from MOE Social Science Laboratory of Digital Economic Forecasts and Policy Simulation at UCAS and CAS 145 Informatization Project CAS-WX2022GC-0301.

References

1. Cai, H., Zheng, V.W., Chang, K.C.C.: Active learning for graph embedding. arXiv preprint arXiv:1705.05085 (2017)
2. Dou, Y., Liu, Z., Sun, L., Deng, Y., Peng, H., Yu, P.S.: Enhancing graph neural network-based fraud detectors against camouflaged fraudsters. In: Proceedings of the 29th ACM International Conference on Information & Knowledge Management, pp. 315–324 (2020)
3. Gao, L., Yang, H., Zhou, C., Wu, J., Pan, S., Hu, Y.: Active discriminative network representation learning. In: IJCAI International Joint Conference on Artificial Intelligence (2018)
4. Hamilton, W., Ying, Z., Leskovec, J.: Inductive representation learning on large graphs. Adv. Neural Inf. Process. Syst. **30**, 1–11 (2017)
5. Kipf, T.N., Welling, M.: Semi-supervised classification with graph convolutional networks. arXiv preprint arXiv:1609.02907 (2016)
6. Liu, Y., et al.: Pick and choose: a gnn-based imbalanced learning approach for fraud detection. In: Proceedings of the Web Conference 2021, pp. 3168–3177 (2021)
7. Liu, Z., Dou, Y., Yu, P.S., Deng, Y., Peng, H.: Alleviating the inconsistency problem of applying graph neural network to fraud detection. In: Proceedings of the 43nd International ACM SIGIR Conference on Research and Development in Information Retrieval (2020)
8. McAuley, J.J., Leskovec, J.: From amateurs to connoisseurs: modeling the evolution of user expertise through online reviews. In: Proceedings of the 22nd International Conference on World Wide Web, pp. 897–908 (2013)
9. Rayana, S., Akoglu, L.: Collective opinion spam detection: bridging review networks and metadata. In: Proceedings of the 21th ACM Sigkdd International Conference on Knowledge Discovery and Data Mining, pp. 985–994 (2015)
10. Settles, B.: Active learning literature survey (2009)
11. Tuteja, S., Kumar, R.: A unification of heterogeneous data sources into a graph model in e-commerce. Data Sci. Eng. **7**(1), 57–70 (2022)
12. Wang, D., et al.: A semi-supervised graph attentive network for financial fraud detection. In: 2019 IEEE International Conference on Data Mining (ICDM), pp. 598–607. IEEE (2019)
13. Zhang, W., Shen, Y., Li, Y., Chen, L., Yang, Z., Cui, B.: Alg: fast and accurate active learning framework for graph convolutional networks. In: Proceedings of the 2021 International Conference on Management of Data, pp. 2366–2374 (2021)
14. Zhang, Y., Fan, Y., Ye, Y., Zhao, L., Shi, C.: Key player identification in underground forums over attributed heterogeneous information network embedding framework. In: Proceedings of the 28th ACM International Conference on Information and Knowledge Management, pp. 549–558 (2019)

Fake News Detection Based on the Correlation Extension of Multimodal Information

Yanqiang Li[1,2], Ke Ji[1,2](✉), Kun Ma[1,2], Zhenxiang Chen[1,2], Jin Zhou[1,2], and Jun Wu[3]

[1] School of Information Science and Engineering, University of Jinan, Jinan 250022, China
ise_jik@ujn.edu.cn

[2] Shandong Provincial Key Laboratory of Network Based Intelligent Computing, University of Jinan, Jinan 250022, China

[3] School of Computer and Information Technology, Beijing Jiaotong University, Beijing 100044, China

Abstract. Online social media is characterized by a large number of users that creates conditions for large-scale news generation. News in multimodal form (images and text) often has a serious negative impact. Existing multimodal fake news detection methods mainly explore the relationship between images and texts by extracting image features and text features. However, these methods typically ignore textual content in images and fail to explore the relationship between news and image texts further. We propose a new fake news detection method based on correlation extension multimodal (CEMM) information to solve this problem. The correlation between multimodal information is extended and the relationship between the extended image information and the news text is explored further by extracting text and statistical features from the image. This CEMM-based detection method consists of five parts, which can discover the relevant parts of news and optical character recognition (OCR) text and the features of fake news images and relevant parts of news text, and combine the information of the news itself to detect fake news. Experimental results proved the effectiveness of our approach.

Keywords: Fake news detection · Multimodal fusion · Deep learning

1 Introduction

News dissemination has gradually changed from the use of traditional electronic equipment, such as TV and radio, to the Internet, such as online social media. Compared with traditional electronic devices, news in the Internet environment spreads faster and has a wider impact, and the authenticity of the news cannot be guaranteed. In this case, fake news is a serious threat that will likely demonstrate negative effects.

Many people have been affected by fake news and suffered emotional or other types of harm or loss [1]. News with pictures is reposted 11 times more than

B. Li et al. (Eds.): APWeb-WAIM 2022, LNCS 13421, pp. 443–450, 2023.
https://doi.org/10.1007/978-3-031-25158-0_36

the average number of reposts of text-only news [6], thereby indicating that multimodal news can cause greater harm. The importance of multimodal forms of fake news detection has been demonstrated [5]. Therefore, exploring and solving the problem of detecting multimodal forms of fake news are necessary. Many approaches [13,14] have been proposed to solve this problem from a multimodal perspective. However, most of them use pre-trained models to extract image features, but ignore the text content on the image and the relationship between the text on the image and the news text.

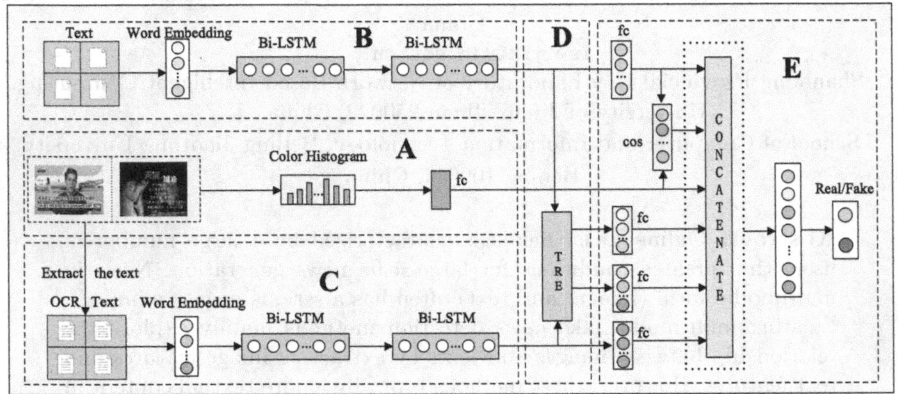

Fig. 1. Overview of our approach.

We propose a multimodal fake news detection model based on the optical character recognition (OCR) text in this study. Our model is shown in Fig. 1 and consists of five parts: (A) visual feature extractor, (B) news text feature extractor, (C) OCR text feature extractor, (D) text correlation extractor (TRE), and (E) feature integration classifier. Our method is significantly better than other approaches and validated using existing datasets.

We introduce the structure of the rest of the paper: Sect. 2 describes related work on fake news detection and text and images, Sect. 3 describes our approach, Sect. 4 presents the dataset, baseline, parameter settings, and comparison results, and Sect. 5 concludes the paper.

2 Related Work

2.1 Image Analysis and Text Classification

Image analysis is the description of the image content and the process of transforming the image described in pixels into data form by means of segmentation or feature extraction. Image analysis has used artificial intelligence or pattern recognition methods in recent years to describe, classify, and interpret images on the basis of image content. The text classification problem has formed the two subproblems of feature engineering and classifier with the development of machine learning. With the development of deep learning, deep learning related methods [4,9] have replaced traditional methods in feature engineering and classifier selection.

2.2 Fake News Detection

The detection was initially based on the text content of news. The content of pictures plays an important role in determining the authenticity of news. Therefore, detecting fake news from the perspective of images is challenging. An image is analyzed from the perspective of an event, and fake news was detected by extracting visual and statistical features of the image in the event [6]. Quality features and semantic information are extracted from the images, and the two features are fused using an attention mechanism to detect fake news images [11]. The attention mechanism is utilized to fuse image and text features to detect the authenticity of news [2,7,12]. A variational autoencoder is utilized to obtain a shared representation of text and images and utilize it to detect fake news [8]. The similarity between text and image is calculated, and features of text and image are added as the basis for fake news detection [17].

The utilization of image information (OCR text and statistical features) is insufficient in previous work, we further explore the relationship between image information and news text.

3 Our Approach

3.1 Problem Definition

Ω represents the dataset, and ω_k represents a sample in the dataset. $y_k \in (0, 1)$, where y_k is the label corresponding to the sample ω_k, 0 represents the true news, and 1 represents the fake news. Particularly, ω_k is composed of text and images.

3.2 Visual Feature Extractor

Images attached to fake news are typically visually impactful [11], and color is an intuitive indication of visual impact. The color histogram is a statistical feature of images for spam classification tasks [15]. This shows that color histogram features are effective in determining fake news images, therefore, we use color histogram and obtain the feature representation V of the image through a 32-dimensional fully connected layer.

3.3 News Text Feature Extractor

The news text T_{news} typically consists of words t^i_{news}. The pretrained language model (BERT) has been successfully used in many tasks. BERT can learn contextual word embedding and demonstrates enhanced generalization capabilities for other tasks after pretraining a large number of corpora. Therefore, we use the pretrained model BERT as our word embedding framework in part B of Fig. 1. We first determine the word embedding w_i that corresponds to the word. And we used two Bi-LSTMs to model the news text to obtain each word and its relationship n_i.

3.4 OCR Text Feature Extractor

We use a feature on the basis of images called OCR text feature (part C in Fig. 1). A given item of fake news includes text and image. The text on the image is crucial in the picture content and an important basis for judging whether the news is fake. A tesseract is an open-source OCR engine used for extracting the text from images and removing spaces and carriage return to obtain the OCR text. However, only some images contain text. We use 10 zero-padding techniques for images without text. And we use the same processing as news text to get OCR text features o_j.

3.5 Text Correlation Extractor

We obtain news and OCR text features extracted by the two layers of Bi-LSTM in the first two parts. We determine the relationship between OCR and news texts in part D of Fig. 1, that is, if news and OCR texts are related, then news and picture texts will likely describe the same thing. Therefore, we use soft alignment [3] on news and OCR texts to calculate the similarity matrix e_{ij} between them and then utilize it as the weight of the attention mechanism as follows [10]:

$$e_{ij} = n_i^{\mathrm{T}} o_j. \tag{1}$$

We use the method of [10] to determine the relationship between the two texts as follows:

$$\overline{n}_i = \sum_{j=1}^{l_o} \frac{exp(e_{ij})}{\sum_{k=1}^{l_o} exp(e_{ik})} o_j, \forall i \in [1_1, \cdots, l_n], \tag{2}$$

$$\overline{o}_j = \sum_{i=1}^{l_n} \frac{exp(e_{ij})}{\sum_{k=1}^{l_n} exp(e_{kj})} n_i, \forall j \in [1_1, \cdots, l_o]. \tag{3}$$

We intuitively find the word that is semantically similar to the news text from the OCR text using (2) and use $\overline{n}_i$ to indicate it. We first calculate the weight of elements in the i-th row and the j-th column in the i-th row in e_{ij} and multiply it with o_j, that is, multiply the weight of the element e_{ij} in e_{i*} with o_j. Finally, the above calculation process is performed for each element in e_{i*} and the sum is obtained to represent the degree of correlation between n_i (i-th word in the news text) and the OCR text. The other formula has the same meaning.

3.6 Feature Integration Classifier

We first summon previous visual, two text, and correlation features into features of the same dimension through the fully connected layer for fusion in part E of Fig. 1. We calculate the cosine similarity between V and n and express it with cos to determine the relationship between news text and images. This method has been effectively utilized in many tasks [16,17].

We then determine whether the news is true or false on the basis of existing information using the following formula:

$$Y = fnc([n_i, o_j, \overline{n}_i, \overline{o}_j, cos, V]), \tag{4}$$

where fnc is the fake news classifier, [ ] is the concatenate operation, and Y is the prediction label.

4 Experiments

4.1 Dataset

Table 1. Details of the dataset.

Category	News	Image	OCR
Fake news	19285	10768	3286
Real news	19186	11064	5276
Total	38471	21832	8562

The dataset[1] in Table 1 we used was obtained from the AI competition platform BIENdata and jointly published by the Institute of Computer Technology of the Chinese Academy of Sciences and Beijing Academy of Artificial Intelligence.

4.2 Baseline

Relevant experiments are performed on the existing dataset to verify the effectiveness of our method.

Bi-LSTM-Att: We use word2vec to convert news text into word embeddings and use Bi-LSTM and attention mechanism to classify news text.

Color Histogram: We employ the color histogram feature to classify using the BPNN model.

OCR Text: The OCR text is processed in the same way as part D in Fig. 1 to obtain the classification results.

att-RNN: We removed the social features in att-RNN [7] to get the classification results.

MVAE: We use MVAE [8] to get the classification results.

[1] https://www.biendata.xyz/competition/falsenews/

Table 2. Experimental results.

Category	Method	Accuracy	Fake news			Real news		
			Precision	Recall	F1	Precision	Recall	F1
News text	Bi-LSTM-Att	0.874	0.900	0.843	0.871	0.850	0.905	0.877
Visual	Color Histogram	0.639	0.615	0.759	0.679	0.679	0.517	0.587
OCR text	Bi-LSTMs	0.545	0.530	0.851	0.653	0.609	0.236	0.340
Multimodal	att-RNN	0.907	0.926	0.886	0.905	0.889	0.928	0.908
	MVAE	0.881	0.880	0.885	0.882	0.882	0.877	0.880
	CEMM	**0.964**	**0.974**	**0.954**	**0.964**	**0.955**	**0.974**	**0.964**

4.3 Parameter Setting

Bi-LSTM consists of two layers with a cell size of 32 and is utilized for processing text. The dimension of the color histogram is 1024. The classifier consists of two fully connected layers. One layer presents a cell size of 64, and the other layer contains a sigmoid activation function. The learning rate is set to 0.0002, and the optimizer uses Adam.

4.4 Comparison Results

Table 2 Results Analysis: The comparison of the experimental results demonstrated that the accuracy and precision of our model are better than those of other models. This finding indicated that our method performs better than other approaches in the task of identifying fake news. The performance of MVAE is similar to Bi-LSTM-Att because the proportion of news containing images in our dataset is only 57%, which may lead to the inability of MVAE to learn a multimodal shared representation.

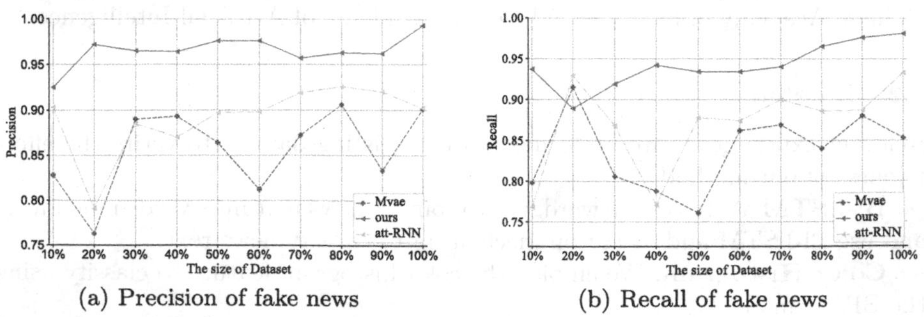

(a) Precision of fake news (b) Recall of fake news

Fig. 2. Variations of different proportions of indicators in the training dataset.

Figure 2 shows the changes in evaluation indicators of different methods using various training data ratios. And Fig. 2 shows that the performance of our method is significantly higher than that of other models.

Table 3. Performance of the model after removing each part.

Method	Accuracy	Fake news			Real news		
		Precision	Recall	F1	Precision	Recall	F1
CEMM	0.964	0.974	0.954	0.964	0.955	0.974	0.964
CEMM^{-bert}	0.925	0.928	0.920	0.924	0.921	0.929	0.925
CEMM$^{-bert\ tre}$	0.923	0.936	0.908	0.922	0.911	0.939	0.925
CEMM$^{-bert\ ocr\ tre}$	0.914	0.960	0.863	0.909	0.876	0.964	0.918
CEMM$^{-bert\ cos}$	0.923	0.943	0.899	0.921	0.904	0.946	0.925
CEMM$^{-bert\ ch}$	0.922	0.923	0.920	0.922	0.920	0.924	0.922

Table 3 Experimental Settings: Table 3 shows the effect of removing certain parts of CEMM.

CEMM^{-bert}: We remove BERT on CEMM and use word2vec for word embedding.

CEMM$^{-bert\ tre}$: We remove the text relevance extractor (tre) on the basis of removing BERT.

CEMM$^{-bert\ ocr\ tre}$: We remove the OCR text and the text relevance extractor on the basis of removing BERT.

CEMM$^{-bert\ cos}$: We remove the cosine similarity on the basis of removing BERT.

CEMM$^{-bert\ ch}$: We remove the color histogram (ch) after removing BERT.

Table 3 Results Analysis: The performance of our method is better than other methods (att-RNN and MVAE) after replacing the pretrained BERT model with the word2vec model. On this basis, the reduced performance of our model indicated that every part of our model exerts a certain effect after the removal of a certain part of the model.

5 Conclusion

The problem of multimodal detection of fake news is investigated in this study. Through the extracted OCR text, the correlation between OCR text and news text is explored, and the similarity between image statistical features and news text is calculated, thereby further extending the correlation between multimodal information. CEMM overcomes the problem of underutilization of image information (OCR text and statistical features) in the fake news detection problem. Experimental results have demonstrated the effectiveness of CEMM.

Acknowledgements. This work is supported by National Science Foundation of China No. 61702216, 61772231, and Higher Educational Science and Technology Program of Jinan City under Grant with No. 2020GXRC057, 2018GXRC002.

References

1. Allcott, H., Gentzkow, M.: Social media and fake news in the 2016 election. J. Econ. Perspectives **31**(2), 211–36 (2017)

2. Chen, J., Wu, Z., Yang, Z., Xie, H., Wang, F.L., Liu, W.: Multimodal fusion network with latent topic memory for rumor detection. In: 2021 IEEE International Conference on Multimedia and Expo (ICME), pp. 1–6. IEEE (2021)
3. Chen, Q., Zhu, X., Ling, Z.H., Wei, S., Jiang, H., Inkpen, D.: Enhanced lstm for natural language inference. In: Proceedings of the 55th Annual Meeting of the Association for Computational Linguistics (Volume 1: Long Papers), vol. 1, pp. 1657–1668 (2017)
4. Devlin, J., Chang, M.W., Lee, K., Toutanova, K.: Bert: Pre-training of deep bidirectional transformers for language understanding. arXiv preprint arXiv:1810.04805 (2018)
5. Hameleers, M., Powell, T.E., Van Der Meer, T.G., Bos, L.: A picture paints a thousand lies? the effects and mechanisms of multimodal disinformation and rebuttals disseminated via social media. Polit. Commun. **37**(2), 281–301 (2020)
6. Jin, Z., Cao, J., Zhang, Y., Zhou, J., Qi, T.: Novel visual and statistical image features for microblogs news verification. IEEE Trans. Multimed. **19**(3), 598–608 (2017)
7. Jin, Z., Cao, J., Guo, H., Zhang, Y., Luo, J.: Multimodal fusion with recurrent neural networks for rumor detection on microblogs. In: Proceedings of the 25th ACM international conference on Multimedia, pp. 795–816 (2017)
8. Khattar, D., Goud, J.S., Gupta, M., Varma, V.: Mvae: Multimodal variational autoencoder for fake news detection. In: The World Wide Web Conference on, pp. 2915–2921 (2019)
9. Mikolov, T., Chen, K., Corrado, G.S., Dean, J.: Efficient estimation of word representations in vector space. In: ICLR (Workshop Poster) (2013)
10. Parikh, A.P., Tackstrom, O., Das, D., Uszkoreit, J.: A decomposable attention model for natural language inference. In: Proceedings of the 2016 Conference on Empirical Methods in Natural Language Processing, pp. 2249–2255 (2016)
11. Qi, P., Cao, J., Yang, T., Guo, J., Li, J.: Exploiting multi-domain visual information for fake news detection. IEEE (2019)
12. Qian, S., Wang, J., Hu, J., Fang, Q., Xu, C.: Hierarchical multi-modal contextual attention network for fake news detection. In: Proceedings of the 44th International ACM SIGIR Conference on Research and Development in Information Retrieval, pp. 153–162 (2021)
13. Singhal, S., Kabra, A., Sharma, M., Shah, R.R., Chakraborty, T., Kumaraguru, P.: Spotfake+: a multimodal framework for fake news detection via transfer learning (student abstract). In: Proceedings of the AAAI Conference on Artificial Intelligence, vol. 34, pp. 13915–13916 (2020)
14. Singhal, S., Shah, R.R., Chakraborty, T., Kumaraguru, P., Satoh, S.: Spotfake: a multi-modal framework for fake news detection. In: 2019 IEEE fifth international conference on multimedia big data (BigMM), pp. 39–47. IEEE (2019)
15. Soranamageswari, M., Meena, C.: Statistical feature extraction for classification of image spam using artificial neural networks. In: 2010 Second International Conference on Machine Learning and Computing, pp. 101–105 (2010)
16. Xue, J., Wang, Y., Tian, Y., Li, Y., Shi, L., Wei, L.: Detecting fake news by exploring the consistency of multimodal data. Inf. Process. Manage. **58**(5), 102610 (2021)
17. Zhou, X., Wu, J., Zafarani, R.: [... formula...]: Similarity-aware multi-modal fake news detection. Advances in Knowledge Discovery and Data Mining 12085, 354 (2020)

SPSTN: Sequential Precoding Spatial-Temporal Networks for Railway Delay Prediction

Junfeng Fu[1,2], Limin Zhong[2,3], Changyu Li[1], Hao Li[1], Chuiyun Kong[2,3(✉)], and Jie Shao[1(✉)]

[1] University of Electronic Science and Technology of China, Chengdu 611731, China
{fujunfeng,changyulve,hao_li}@std.uestc.edu.cn, shaojie@uestc.edu.cn

[2] The Center of National Railway Intelligent Transportation System Engineering and Technology, Beijing 100081, China

[3] Institute of Computing Technologies, China Academy of Railway Sciences Corporation Limited, Beijing 100081, China
chykong@163.com

Abstract. Predicting the delay time of trains is an important task in intelligent transport systems, as an accurate prediction can provide a reliable reference for passengers and dispatchers of the railway system. However, due to the complexity of the railway system, interactions of various spatio-temporal variables make it difficult to find the rules of delay propagation. We introduce a Sequential Precoding Spatial-Temporal Network (SPSTN) model to predict the delay of trains. SPSTN consists of a Transformer encoder that captures long-term dependencies in time series, and spatio-temporal graph convolution blocks that model delay propagation at both temporal and spatial levels. Experiments on a subset of the British railway network show that SPSTN performs favorably against the state-of-the-art, which verifies that the combination of sequential precoding and spatio-temporal convolution can effectively model delay propagation on railway networks.

Keywords: Intelligent transport systems · Railway delay · Graph convolutional networks

1 Introduction

Delay time is an important metric for rail users, but it is often difficult for train operating companies to predict it as the railway delay time is affected by many factors. Various delay prediction methods have been proposed, but they still have certain defects. Traditional methods such as linear regression and multilayer perceptron ignore spatial features of the railway network. Spatio-Temporal Graph Convolutional Network (STGCN) [9] takes the spatial features of the railway

J. Fu and L. Zhong—These authors contributed equally to this work and should be considered co-first authors.

B. Li et al. (Eds.): APWeb-WAIM 2022, LNCS 13421, pp. 451–458, 2023.
https://doi.org/10.1007/978-3-031-25158-0_37

network into consideration, but it performs poorly with long sequence as input due to the existence of gated linear unit. This makes STGCN difficult to achieve optimal performance in railway systems with rich historical data.

In this paper, we propose a novel model to address these problems, by combining the Transformer encoder with spatio-temporal convolution blocks. The spatio-temporal convolution block composed of graph convolution and time-gated convolution enables the model to capture spatio-temporal features simultaneously. Self-attention in the Transformer encoder precodes time series. Due to the existence of precoding, the information loss of spatio-temporal convolution in temporal dimensions is greatly reduced. This enables our method to model long-term dependencies. Therefore, we can predict more accurately with the help of rich historical records. In summary, our contributions are as follows:

- We propose a model named Sequential Precoding Spatial-Temporal Network (SPSTN), which is used to predict train delays on railway networks.
- SPSTN utilizes self-attention to precode time series before combining the spatial and temporal features of the railway network. This enables spatio-temporal modeling under long-term dependencies.
- The proposed method is successfully applied to the Britain railway data. The better performance over the state-of-the-art method in experiments demonstrates its effectiveness.

2 Related Work

Traditional approaches to solve the railway delay prediction problem are usually divided into two types: establishing analytical models and establishing simulation models [7]. However, it is difficult to directly build these models at a complex system. Data-driven approach is an alternative approach for railway delay prediction. However, a test on a section of the railroad in Tennessee, USA shows that support vector regression and deep neural networks perform similarly [1]. This is because these methods only focus on information about delays in temporal dimensions and ignore the structure of the railway network.

Thanks to Bruna et al. [2], convolutional neural networks can be transferred to graph structures. The model named Graph Convolutional Network (GCN) proposed by Kipf and Welling [4] is the most widely used graph neural network. Yu et al. [9] proposed a framework named Spatio-Temporal Graph Convolutional Network (STGCN) to capture correlations of the spatial-temporal dimension. Heglund et al. [3] moved the model on the railway network to predict delays of trains. Li et al. [5] added a long-term encoder, a short-term encoder and an attention-based output layer after GCN.

Transformer [8] was first proposed in translation tasks in natural language processing and has been widely adopted in other fields. Its entire network structure of the Transformer consists only of self-attention. Time series, as natural sequence data, are very suitable to be processed by Transformer. Therefore, there are also may studies in this field [6,10].

3 Methodology

3.1 Problem Formulation

In general, stations and the railways connecting two stations in the railway network are regarded as vertices and edges. However, under this structure, unknown information of stations becomes node-wise features, which is not suitable for delay prediction. Thus, it is better to invert this configuration to define railway links as vertices. In this paper, we use the following definition for graph structure:

Definition 1 (Railway Graph). *$RG = (V, E)$, $V = (v_1, v_2, ..., v_m)$ where v_i represents the i-th pair of two stations for which there exists a trace between them, and $E = (e_1, e_2, ..., e_n)$ where e_j represents the j-th pair (v_p, v_q) for which there exists a station in both v_p and v_q.*

The track information contains the spatial information of the train in the railway system, while the delay record contains the time information. Also, we have the structure from the railway graph defined above. What we want is to calculate the delay time at the k_{future}-th step according to k_{past} previous delay time. This is a conditional time series regression problem formulated as:

$$\hat{y}_{d_{t+k_{future}}} = \underset{d_{t+k_{future}}}{\arg\max} \log P(d_{t+k_{future}} \mid d_{t-k_{past}}, \ldots, d_{t-1}, d_t, RG), \quad (1)$$

where $d_t \in \mathbb{R}^{N \times F}$ is a tensor that represents the delay at time t. $N = |V|$ is the number of vertices and F is the number of features in the tensor.

3.2 Our Proposed Model

The overall structure of Sequential Precoding Spatial-Temporal Network (SPSTN) is shown in Fig. 1. In addition to delays, the input also contains the adjacency matrix of railway graph, denoted as A.

Spatial-Temporal Graph Convolution for Railway Delay. To predict the future delay on the defined railway graph, the model STGCN [9] is suitable because it considers both spatial and temporal patterns of the rail network data.

It is obvious that railway graph is a non-Euclidean structure. The widely accepted graph convolution formula [4] is:

$$\Theta *_{RG} H^l = \sigma\left(\hat{D}^{-\frac{1}{2}}\hat{A}\hat{D}^{-\frac{1}{2}}H^{(l)}W^{(l)}\right). \quad (2)$$

$\hat{A} = A+I$, which means a self-loop, is added to the adjacency matrix by adding to an identity matrix. $W^{(l)}_{F^i \times F^{i+1}}$ is the weight of the l-th layer, and its dimension is $F^i \times F^{i+1}$. $\hat{D}$ is the degree matrix of $\hat{A}$, so $\hat{D}^{-\frac{1}{2}}\hat{A}\hat{D}^{-\frac{1}{2}}$ is the symmetrical normalization of $\hat{A}$. σ is a nonlinear activation function (such as ReLu).

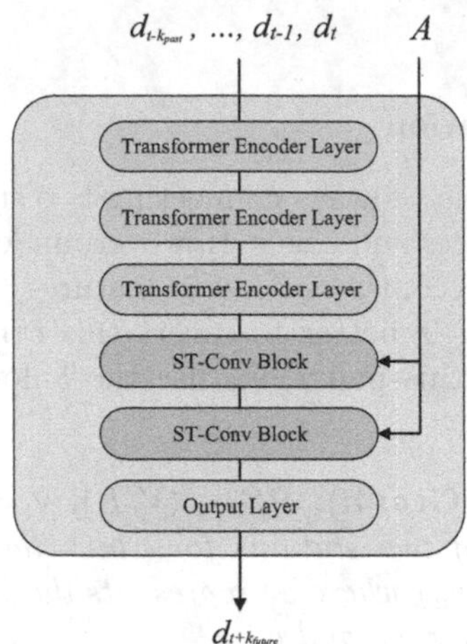

Fig. 1. Structure of the proposed SPSTN model.

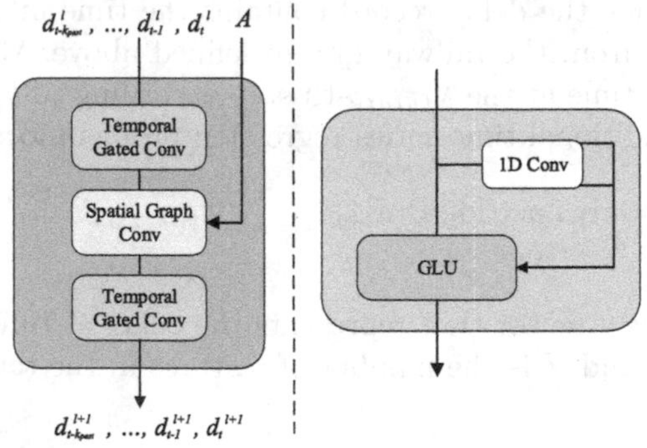

Fig. 2. The left part is the structure of a ST-Conv block. The right part represents a temporal gated conv block.

Figure 2 shows the main structure of the spatial-temporal convolution block and temporal gated convolution block. Each spatial-temporal convolution block consists of three sub-blocks: two temporal gated convolution blocks and a spatial graph convolution block. Each temporal gated convolution block contains a 1D causal convolution and a Gated-Linear Unit (GLU) for activation. By adding nonlinear activation similar to Gated Recurrent Unit (GRU), the block can capture the effect of past delay time series on future delay through its gating. Spatial convolution is defined as:

$$\Gamma *_T H^l = P \odot \sigma(Q), \tag{3}$$

where P and Q are resulted from splitting the result after 1D convolution, i.e., $1DConv(H^l) = concat(P, Q)$.

According to Eq. 2 and Eq. 3, the calculation of spatial-temporal convolution block can be written as:

$$H^{l+1} = \Gamma_2^l *_T \left(\Theta^l *_{RG} \left(\Gamma_1^l *_T H^l\right)\right), \tag{4}$$

where $\Theta^l *_{RG}$ is the spectral kernel of the graph convolution, and Γ_1^l and Γ_2^l are the temporal kernels in block l.

Transformer for Sequential Precoding. Only the encoder part of the Transformer model is needed for the interaction of information from past and future time steps in the delay sequence. The encoder consists of several encoder layers, each containing a self-attention layer and a free forward network [8]. Self-attention can be formulated as:

$$Attention(E^l) = Softmax\left(\frac{Q^l K^{lT}}{\sqrt{d_k^l}}\right) V^l, \tag{5}$$

where $\left(Q\ K\ V\right) = E^l \left(W^q\ W^k\ W^v\right)$ and d_k^l is used for the score normalization of the l-th layer. Here, the first layer's input $\boldsymbol{d_{past}}$ is $\left(d_{t-k_{past}}, \cdots, d_{t-1}, d_t\right) = E^0 \in \mathbb{R}^{N \times F \times T}$. T is the number of time steps in $\boldsymbol{d_{past}}$. The encoder model is shown in Fig. 3. Since Transformer can only process one single feature, we use a parallel structure to contain multiple encoders.

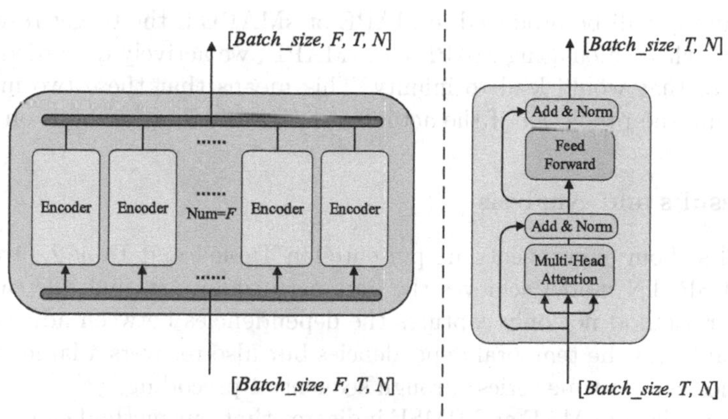

Fig. 3. The left part is one layer of Transformer's encoder without embedding. The right part contains the details of the encoder of the left one.

4 Experiments

4.1 Evaluation Data and Metrics

The dataset used in this work comes from Darwin, the Great Britain rail industry's official train running information engine. Darwin provides 5 data sources on its website and we utilize the Historical Service Performance (HSP) API. In this work, Didcot Parkway and London Paddington are selected as the gateway stations because train delays in this segment occur more frequently. Data for all time types are accurate to the minute. We selected all train journeys from 5:30 AM to 12:00 PM on weekdays from 2016 to 2017. To make the model converge and reduce training difficulty, we remove stations that serve less than one train per day on average. Since historical delay records are needed to predict delays on the future rail network, we obtain the arrival delay by subtracting the scheduled arrival time from the actual arrival time. We sampled at 10-min intervals on the railway routes. All sampled data are normalized by z-score before being used for model training. More details can be referenced from [3].

We compare our model with three methods: Linear Regression (LR), Multilayer Perceptron (MLP), and Spatial-Temporal Graph Convolutional Network (STGCN). The arithmetic mean of the prediction errors at each vertex of the railway graph is taken as the overall error of the model prediction. To measure prediction errors, we use the following four common metrics:

- Mean Absolute Error: $MAE = \frac{1}{n}\sum_{i=1}^{n}(|y_i - \hat{y}_i|)$,
- Root Mean Square Error: $RMSE = \sqrt{\frac{1}{n}\sum_{i=1}^{n}(y_i - \hat{y}_i)^2}$,
- Mean Absolute Percentage Error: $MAPE = \frac{1}{n}\sum_{i=1}^{n}\frac{|y_i - \hat{y}_i|}{|y_i|}$,
- symmetric Mean Absolute Percentage Error: $sMAPE = \frac{1}{n}\sum_{i=1}^{n}\frac{|y_i - \hat{y}_i|}{\frac{1}{2}(|y_i| + |\hat{y}_i|)}$.

An infinity will be produced in MAPE or sMAPE if the target result has 0. Therefore, when calculating MAPE and sMAPE, we actively discarded the part of the data that would lead to infinity. This means that these two metrics do not measure the prediction if the actual situation of train arrival is on time.

4.2 Results and Analysis

The results of our experiments are presented in Table 1 and Table 2. Overall, our proposed SPSTN model achieves the best performance in multiple cases. This proves our method not only captures the dependencies between adjacent nodes in the graph and the temporal dependencies but also recovers a large amount of information loss in time series through sequential precoding.

The large lead in MAE and RMSE indicates that our method can achieve the smallest absolute error. As we mentioned, MAPE and sMAPE only evaluate the part of the data that actually occur delays. Therefore, the results show that LR and MLP, especially LR, are biased in delay prediction. For all data, they are biased to believe that there will be large delays in the future. On the contrary, SPSTN can make accurate predictions for both cases of delay and on time, and our model has smaller errors.

In the case of 30-min and 60-min sampling data, most results are generally inferior to those with 10-min sampling. This is most likely due to a decrease in the amount of data. Since Transformer cannot achieve the best results in small

Table 1. Accuracy under MAE and RMSE on railway delay data.

Models	Metrics					
	MAE			RMSE		
$k_{past} = 6$	10 min	30 min	60 min	10 min	30 min	60 min
LR	0.304	0.365	0.360	0.690	0.834	0.847
MLP	0.341	0.362	0.364	0.966	0.915	1.096
STGCN	0.256	0.311	0.302	0.625	0.803	0.755
SPSTN (ours)	**0.241**	**0.311**	**0.301**	**0.500**	**0.582**	**0.561**
$k_{past} = 12$	10 min	30 min	60 min	10 min	30 min	60 min
LR	0.279	0.337	0.338	0.590	0.753	0.785
MLP	0.331	0.340	0.327	0.982	0.896	0.931
STGCN	0.250	**0.282**	0.327	0.539	0.713	0.669
SPSTN (ours)	**0.221**	0.321	**0.321**	**0.444**	**0.585**	**0.577**

Table 2. Accuracy under MAPE and sMAPE on railway delay data.

Models	Metrics					
	MAPE			sMAPE		
$k_{past} = 6$	10 min	30 min	60 min	10 min	30 min	60 min
LR	**0.828**	1.138	1.289	1.043	1.308	1.288
MLP	0.975	1.076	1.345	1.422	1.545	1.511
STGCN	0.938	0.968	1.085	**0.907**	1.217	1.447
SPSTN (ours)	0.933	**0.929**	**1.003**	0.938	**1.021**	**1.036**
$k_{past} = 12$	10 min	30 min	60 min	10 min	30 min	60 min
LR	**0.767**	1.126	1.154	0.957	1.269	1.106
MLP	0.863	1.083	1.173	1.221	1.510	1.382
STGCN	1.048	1.029	1.166	0.948	1.309	1.234
SPSTN (ours)	0.877	**1.004**	**1.145**	**0.924**	**0.981**	**0.977**

datasets, the performance of SPSTN will also be restricted to a certain extent. Even so, SPSTN performs best in most cases on this dataset.

In addition, k_{past} represents the time steps. In the case of 10-min sampling data, STGCN does not achieve as good performance as SPSTN when increasing the length of time series. This proves that STGCN has insufficient temporal information acquisition in long sequence data. In contrast, our method significantly increases the prediction accuracy when the sequence length is doubled. This illustrates that adding sequential precoding can enable spatio-temporal networks to model long-term dependencies.

5 Conclusion

To predict train delays on railway networks, we propose a new framework named Sequential Precoding Spatial-Temporal Network (SPSTN). SPSTN combines a spatio-temporal graph convolutional network with a Transformer encoder. The encoder is used to precode time-series data, while spatio-temporal convolution blocks are used to predict delays in spatial and temporal dimensions. Experiments show that the prediction accuracy of SPSTN is much higher than that of classical delay prediction methods. SPSTN outperforms the model without sequential precoding. This proves that sequential precoding can effectively deal with the problem of long-term dependencies in spatio-temporal convolution.

Acknowledgements. This work is supported by The Center of National Railway Intelligent Transportation System Engineering and Technology (Contract No. RITS2021KF08), China Academy of Railway Sciences (Contract No. 2021YJ195).

References

1. Barbour, W., Samal, C., Kuppa, S., Dubey, A., Work, D.B.: On the data-driven prediction of arrival times for freight trains on U.S. railroads. In: 21st International Conference on Intelligent Transportation Systems, ITSC 2018, pp. 2289–2296 (2018)
2. Bruna, J., Zaremba, W., Szlam, A., LeCun, Y.: Spectral networks and locally connected networks on graphs. In: 2nd International Conference on Learning Representations, ICLR 2014 (2014)
3. Heglund, J.S.W., Taleongpong, P., Hu, S., Tran, H.T.: Railway delay prediction with spatial-temporal graph convolutional networks. In: 23rd IEEE International Conference on Intelligent Transportation Systems, ITSC 2020, pp. 1–6 (2020)
4. Kipf, T.N., Welling, M.: Semi-supervised classification with graph convolutional networks. In: 5th International Conference on Learning Representations, ICLR 2017 (2017)
5. Li, J., Xu, X., Shi, R., Ding, X.: Train arrival delay prediction based on spatial-temporal graph convolutional network to sequence model. In: 24th IEEE International Intelligent Transportation Systems Conference, ITSC 2021, pp. 2399–2404 (2021)
6. Li, S., et al.: Enhancing the locality and breaking the memory bottleneck of transformer on time series forecasting. In: Advances in Neural Information Processing Systems: Annual Conference on Neural Information Processing Systems 2019, NeurIPS 2019, vol. 32, pp. 5244–5254 (2019)
7. Murali, P., Dessouky, M., Ordóñez, F., Palmer, K.: A delay estimation technique for single and double-track railroads. Transp. Res. Part E: Logist. Transp. Rev. **46**(4), 483–495 (2010)
8. Vaswani, A., et al.: Attention is all you need. In: Advances in Neural Information Processing Systems: Annual Conference on Neural Information Processing Systems 2017, vol. 30, pp. 5998–6008 (2017)
9. Yu, B., Yin, H., Zhu, Z.: Spatio-temporal graph convolutional networks: a deep learning framework for traffic forecasting. In: Proceedings of the Twenty-Seventh International Joint Conference on Artificial Intelligence, IJCAI 2018, pp. 3634–3640 (2018)
10. Zhou, H., et al.: Informer: beyond efficient transformer for long sequence time-series forecasting. In: Thirty-Fifth AAAI Conference on Artificial Intelligence, AAAI 2021, pp. 11106–11115 (2021)

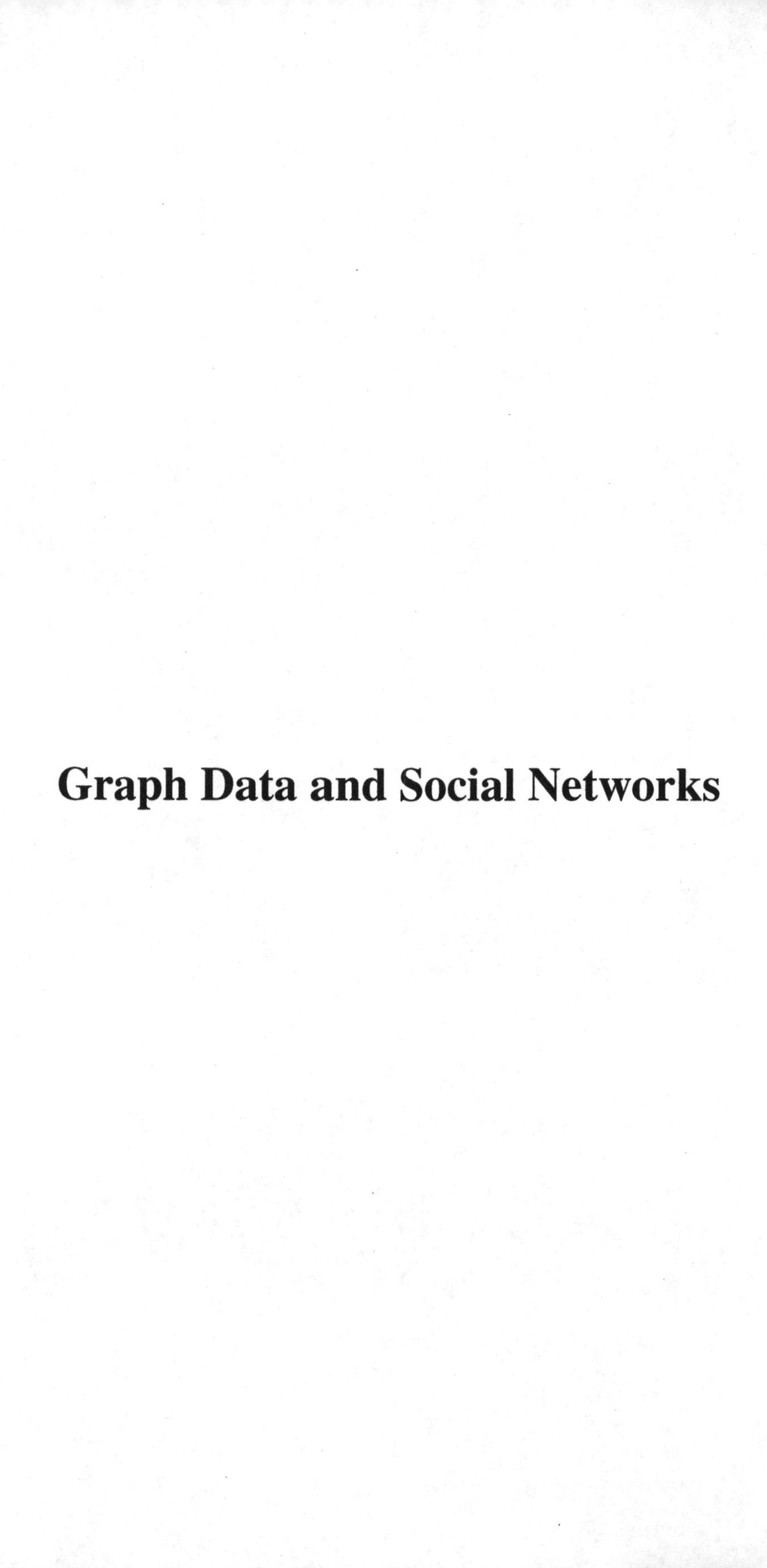

Graph Data and Social Networks

Mining Periodic k-Clique from Real-World Sparse Temporal Networks

Zebin Ren[1], Hongchao Qin[1], Rong-Hua Li[1](✉), Yongheng Dai[2], Guoren Wang[1], and Yanhui Li[3]

[1] Beijing Institute of Technology, Beijing, China
lironghuabit@126.com
[2] Diankeyun Technologies Co., Ltd., Beijing, China
[3] Chongqing Jiaotong University, Chongqing, China

Abstract. In temporal networks, nodes and edges are associated with time series. To seeking the periodic pattern in temporal networks, an intuitive method is to searching periodic communities in them. However, most existing studies do not exploit the periodic pattern of communities. The only few works left do not take the sparse propriety of real-world temporal networks into consideration, such that (i) the answers searched for are few, (ii) the computation suffers from poor performance. In this paper, we propose a novel periodic community model in temporal networks, σ-periodic k-clique, and an efficient algorithm for enumerating all σ-periodic k-cliques in real-world sparse temporal networks. We first design a new data structure to store temporal networks in main memory, which can reduce the maintaining cost and support dynamic deletion of nodes and edges. Then, we propose several efficient pruning rules to eliminate unpromising nodes and edges that do not belong to any σ-period k-clique to reduce graph size. Next, we propose an algorithm that directly enumerates σ-periodic k-cliques on temporal graph to avoid redundant computation. Finally, extensive and comprehensive experiments show that our algorithm runs one to three orders of magnitudes faster and requires significantly less memory than the baseline algorithms.

Keywords: Temporal networks · Periodic community · k-clique

1 Introduction

Real-world networks are usually temporal networks, in which their edges are associated with timestamps, indicating the time periods in which they exist. For example, in an email communication network, edge (u, v, t) means that user u send a message to user v in time t. Many real-world networks are temporal networks such as phone-call networks, social-media interaction networks and research collaboration networks. Many studies have been done to mine significant patterns in temporal networks such as finding fast-changing components [29], detecting information flow [12], mining dense subgraphs [20,27,30] and reachability testing [28]. With

B. Li et al. (Eds.): APWeb-WAIM 2022, LNCS 13421, pp. 461–476, 2023.
https://doi.org/10.1007/978-3-031-25158-0_38

the existence of timestamps, there are many new community models for temporal networks [1,3,18,19,23,31]. Periodicity is a very important phenomenon in time series analysis. Communities that occur periodically often indicate strong patterns in real-world, such as weekly group discussion, monthly friends parties and yearly research collaboration. Therefore, finding the periodically occurring communities can be helpful for finding dense connections between nodes that persist over time or predicting future events.

Mining dense subgraphs in static networks has been widely studied during the past few decades and they are widely used in real world applications such as finance [8], biology [10,22], search engine [11] and queries on graphs [6,15]. As shown below, periodic clique is an important community pattern in temporal networks. However, existing studies on finding periodic communities in temporal networks don't dive in to the problem of how to represent temporal networks in main memory nor take the sparse propriety of real-world temporal networks into account [23,31]. As far as we know, there is no study on one of the most basic periodic community, periodic k-cliques.

In this paper, we propose a new periodic dense subgraph model called σ-period k-clique, which is a k-clique that occurs σ times with some period. The main contributions are summarized as follows:

1. We design a new data structure, *Augmented Adjacency Array*, based on adjacency array, to represent temporal networks in memory. The new data structure has a low maintaining cost and supports dynamic deletion of vertices and edges, which is used in the pruning algorithms.
2. We propose a new method to enumerate periodic k-clique efficiently on real-world sparse temporal networks. We propose several efficient pruning rules to prune the temporal network efficiently, which is much more efficient than the baseline algorithm and performs nearly as well as the complex pruning algorithms [23] in real-world sparse networks. Then, we propose a new algorithm which directly enumerates periodic k-cliques on temporal networks. This algorithm avoids the costly computation of most periods and it prunes vertices and edges as enumeration to reduce unnecessary computation. It also avoids redundant computation of overlapping periods.
3. We conduct extensive experiments on five real-world networks to show the efficiency and effectiveness of our algorithm. The experiments show that our algorithm is one to three orders of magnitudes faster than the baseline algorithm.

2 Preliminaries

An undirected temporal graph is defined as $\mathcal{G} = (\mathcal{V}, \mathcal{E})$, where $\mathcal{V}$ is the set of vertices and $\mathcal{E}$ is the set of temporal edges. Each temporal edge $e \in \mathcal{E}$ is a triplet (u, v, t), where $u, v \in \mathcal{V}$, t is the time when this edges exists. Without loss of generality, we assume t is an integer, because timestamps in real world are usually represented as integer. The de-temporal graph $G = (V, E)$ is a graph that ignores all the timestamps, where $V = \mathcal{V}$ and $E = \{(u, v)|(u, v, t) \in \mathcal{E}\}$. A snapshot of $\mathcal{G}$ at time i is a static graph defined as $G_i = (V_i, E_i)$ where $V_i = \mathcal{V}$

and $E_i = \{(u,v)|(u,v,t) \in \mathcal{E}, t = i\}$. $N_v(G)$ is the neighbor of v. $D_v(G)$ is the degree of v. Let $N_u(\mathcal{G})$ or $N_u(G)$ be the neighbors of u.

A graph $\mathcal{G}' = (\mathcal{V}', \mathcal{E}')$ is called a subgraph of $\mathcal{G}$ if $\mathcal{V}' \in \mathcal{V}$ and $\mathcal{E}' \in \mathcal{E}$. Similarly, graph $G' = (V', E')$ is called a subgraph of G if $V' \in V$ and $E' \in E$. Given a set of vertex $\mathcal{V}_s$, a subgraph $\mathcal{G}_s = (\mathcal{V}_s, \mathcal{E}_s)$ is called a vertex induced subgraph if $\mathcal{S}_s \in \mathcal{V}$ and $\mathcal{E}_s = \{(u,v,t)\ |u,v \in \mathcal{V}_s, (u,v,t) \in \mathcal{E}\}$. We call $G_i = (V_i, E_i)$ a snapshot of temporal graph $\mathcal{G}$ where $V_i = \{u|(u,v,t) \in \mathcal{E}, t = i\}$ and $E_i = \{(u,v)|(u,v,t) \in \mathcal{E}, t = i\}$. Fig-1 shows an example of a temporal graph.

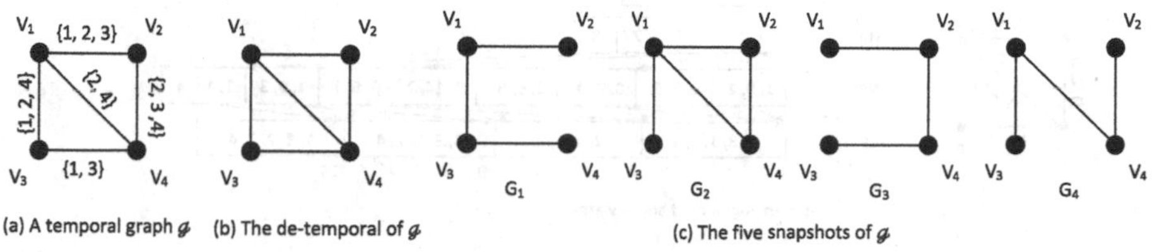

Fig. 1. An example of temporal graph

Definition 1 (time support set). *Given a temporal graph* $\mathcal{G} = (\mathcal{V}, \mathcal{E})$ *and its de-temporal graph* G. *The time support set of vertex* $v \in V$ *is* $TS(v) = \{t|(v,u,t) \in \mathcal{E}, u \in \mathcal{V}\}$; *the time support set of edge* $(u,v) \in E$ *is* $TS(u,v) = \{t_i|(u,v,t) \in \mathcal{E}\}$, *we will refer to the size of* $TS(u,v)$, $|TS(u,v)|$ *as an edge's weight for simplicity. And the time support set of a subgraph of* G, $S = (V_s, E_s)$ *is* $TS(S) = \{t|\forall(u,v) \in E_s, t \in TS(u,v)\}$

Definition 2 (periodic time support set). *Given a sorted time support set* $P = \{ts_1, ts_2, \ldots, ts_k\}$, *if the difference of any two adjacent timestamp* $ts_i - (ts_i - 1)$ *is a constant, then* P *is called a periodic time support. For convenience, we will call periodic time support as period and the difference between any two adjacent timestamps DIFF(P). If* $|P| = \sigma$, P *is called a* σ*-periodic time support set, also called* σ*-period.*

Definition 3 (σ-periodic k-clique). *Given a temporal graph* $\mathcal{G} = (\mathcal{V}, \mathcal{E})$ *and a subgraph of* $\mathcal{G}$, $\mathcal{C} = (\mathcal{V}_c, \mathcal{E}_c)$ *and its de-temporal graph* C *is a* σ*-periodic* k*-clique with period* P *if (1) The de-temporal graph of* $\mathcal{C}$ *is a* k*-clique and (2) for any* $u, v \in \mathcal{V}_c$, $(u,v,t) \in \mathcal{E}_c$ *and (3) for any* $(u,v,t) \in \mathcal{E}_c$, $t \in TS(C)$.

Problem Statement. Given a temporal graph $\mathcal{G}$ and two parameters σ and k, our goal is to find all the σ-periodic k-cliques in $\mathcal{G}$. We refer this problem as the (σ, k)-PC problem.

We simply modify the algorithm of enumerating periodic maximal clique in [23] to find all the σ-periodic k-cliques. The algorithm is referred as PKC-B and it involves three parts: firstly, it prunes the graph using k-core and periodicity of vertices and edges; secondly, it translates the temporal graph into a static graph, the details of step 1 and 2 can be found in [23]; thirdly, it enumerates σ-periodic k-cliques on the translated static graph using the algorithm in [5].

Challenges. The challenges of searching the σ-periodic k-cliques by PKC-B are:

1. Existing studies don't dive into the memory representation of temporal graphs and use trivial methods to represent temporal graphs such as a combination of map and lists [23]. Maintaining such a data structure is costly. It is challenging to design a new data structure which supports fast access to vertices, edges and temporal edges and dynamic deletion of vertices and edges to satisfy the needs of the enumeration algorithm.

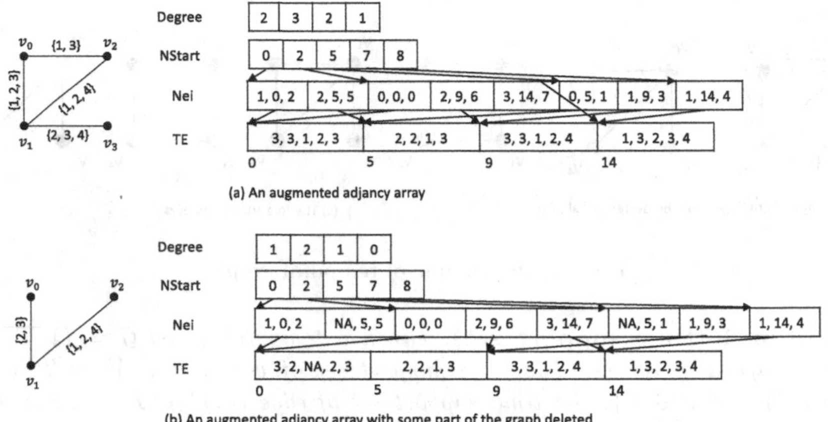

Fig. 2. The data structure of augmented adjacency array

2. In the pruning algorithm of the baseline algorithm, the periods of every vertices and edges in the graph are computed, which is costly. The real-world networks are always sparse. It is challenging to design a more efficient algorithm with the pruning power close to the complex pruning rules in the baseline algorithm [23].
3. The baseline algorithm first converts the pruned temporal graph to a static graph, then it applies clique enumeration algorithm to the static graph. The periods of all the vertices and edges is needed to do this conversion, which is costly to compute. Besides, there are overlapping between periods, which will lead to redundant computation. The problem is can we enumerate periodic k-cliques directly on the temporal graph and also avoid redundant computation?

3 The Proposed Algorithms

In this section, we present our algorithm in three parts. Firstly, we present a new data structure, called augmented adjacency array, an in-memory data structure for temporal networks which supports dynamic deletion of nodes and edges. Then, we propose an efficient pruning algorithm to prune unpromising vertices and edges that cannot be in any σ-periodic k-clique. Finally, we devise a σ-periodic k-clique enumeration algorithm which directly enumerates periodic cliques on temporal graphs.

3.1 Augmented Adjacency Array

To implement the pruning and enumerate algorithm efficiently, a data structure which can provide fast access of vertices and edges and support deletion of vertices and edges is needed. So we design an efficient in-memory data structure, called Augmented Adjacency Array, to store temporal networks in main memory that fits the access pattern of our algorithms. All the following algorithms in this section are based on this data structure since it provides fast access and deletion of the network. The data structure is efficient in four aspects (1) it has a low maintaining cost compared with previous map-based implementation [23], (2) when the temporal graph is scanned sequentially, the data structure preserves cache locality, (3) the deletion of edges and temporal edges can be done in constant time with a few operations and (4) our data structure is space efficient.

The augmented adjacency is composed of four arrays: *Degree*, *NStart*, *Nei*, *TE*. For each vertex v, *Degree*[v] stores the degree of vertex v. *NStart*[v] stores the start position of its neighbors in edges, so *NStart*[$v+1$] stores the position that just after the end of its neighbors in *Nei*. The elements between *NStart*[v] and *NStart*[$v+1$] in *Nei* are the neighbors of vertex v. Each element in *Nei* is composed by three parts: *Id*, *TsIdx* and *REIdx*. *Id* is the vertex id of the neighbor, *TsIdx* points to the start position of the current edge's timestamps in *TE*, and *REIdx* (Reverse Edge Index) points to location of the reverse direction edge in *Nei*, specially edge (v, u) in *Nei*. The *REIdx* is used to delete an edge in O(1) time. The timestamps of each non-temporal edge are stored in *TE* (Temporal Edge), each edge's timestamps are arranged in three parts, *L*, *TSize* and *TS*. *L* is the length of each edge's *TS* part in memory, counted *INVALID* timestamps, *TSize* is the size of time supports for the edge. *TS* is the sorted list of time supports of that edge.

Operations. This data structure supports dynamic deletion for the pruning procedure. We use special value, *INVALID*, to mark if an edge or a temporal edge is deleted. To delete a vertex, we should traverse its neighbor and delete all its edges, then set it's degree to 0. To delete an edge (u, v), mark the *Id* of $v's$ block in $u's$ neighbor and the *Id* of $u's$ block in $v's$ neighbor to be *INVALID*. To delete a temporal edge, we mark the timestamp in *TS* to be *INVALID* and decrement its *TSize*.

Figure 2 is an example of augmented adjacency array. (a) shows the in-memory representation of the left graph. (b) shows when vertex v_3, edge (v_0, v_2) and temporal edge $(v_0, v_1, 1)$ is deleted from the origin graph.

Complexity. The time complexity of constructing the data structure from unordered edges is $O(|\mathcal{E}| + |\mathcal{E}| ln\frac{|\mathcal{E}|}{|E|})$ and the time complexity of deleting edges or temporal edges is $O(1)$, the time complexity of deleting a vertex is $O(D_v(G))$. The space complexity of the data structure is $O(2|V| + 8|E| + |\mathcal{E}|)$. The can be easily computed, thus it is omitted due to space limitation.

3.2 The Proposed Pruning Techniques

Although [23] uses reverse index to avoid re-computation of periods caused by the deletion of a vertex, it is still inefficient because it needs to compute the periods of all vertices and edges and the cost of computing periods can not be omitted. We notice that real-world temporal networks are always sparse, so we proposed several pruning rules that can be efficiently implemented and perform nearly as well as the complex pruning rules.

We first devise several lemma and the pruning rules induced from these lemma. Then, we apply these pruning rules to our pruning algorithm.

Lemma 1. *Given a temporal graph $\mathcal{G} = (\mathcal{V}, \mathcal{E})$, its de-temporal graph $G = (V, E)$ and edge $e_{u,v} = (u, v) \in E$, if $e_{u,v}$ belongs to some σ-periodic k-clique then the time support set of edge $e_{u,v}$, $TS(u, v)$ must contains a σ-period.*

With Lemma 1, given a temporal graph $\mathcal{G} = (\mathcal{V}, \mathcal{E})$ and its de-temporal graph $G = (V, E)$. We can prune any edge $e_{u,v} = (u, v)$, we can have the flowing pruning rule:

Pruning Rule 1. If $|TS(u, v)| < \sigma$, we can delete edge e without reducing the number of σ-period k-cliques in graph $\mathcal{G}$.

Pruning Rule 2. If (1) $|TS(u, v)| = \sigma$, and (2) $TS(u, v)$ is not a σ-period, we can delete edge $e_{u,v}$ without reducing the number of σ-period k-cliques in graph $\mathcal{G}$. Note we only check the edge (u, v) where $|TS(u, v)| = \sigma$ for efficiency.

Lemma 2. *Given a temporal graph $\mathcal{G}$ and its de-temporal graph G. For any σ-periodic k-clique $\mathcal{C}$ in $\mathcal{G}$, it's de-temporal S must be a subgraph of the k-core of G.*

Pruning Rule 3. Given a temporal graph $\mathcal{G} = (\mathcal{V}, \mathcal{E})$ and its de-temporal graph $G = (V, E)$. We can prune any vertex v and it associated edges where v doesn't belong to the k-core of G without reducing the number of σ-period k-cliques in graph $\mathcal{G}$.

Lemma 3. *Given a temporal graph $\mathcal{G}$, for any temporal edge $e = (u, v, t)$, if edge e belongs to any σ-periodic k-clique, then in snapshot G_t, we have $D_u(G_t) \geq k$ and $D_v(G_t) \geq k$.*

Pruning Rule 4. Given a temporal graph $\mathcal{G} = (\mathcal{V}, \mathcal{E})$, we can prune all the temporal edges (u, v, t) where $D_u(G_t) < k$ or $D_v(G_t) < k$ without reducing the number of σ-period k-cliques in $\mathcal{G}$.

The Proof of Lemma 1, 2 and 3 is straightforward, so they are omitted due to space limitation.

Algorithm 1 prunes the input graph with a combination of the four rules with parameter $k = 3$ and $\sigma = 3$. It first computes the degrees of each vertex with pruning rule 1 and 2 (line 5–6). And it then counts the degree with invalid edge deleted (line 7). After that, it computes the k-core on the graph pruned by rule

1 and 2 (lines 11–15). Then it applies pruning rule 4 (lines 16–25). Finally it computes the degree of each vertex at each snapshot (line 19) and the edges do not following pruning rule 4 are deleted (line 23).

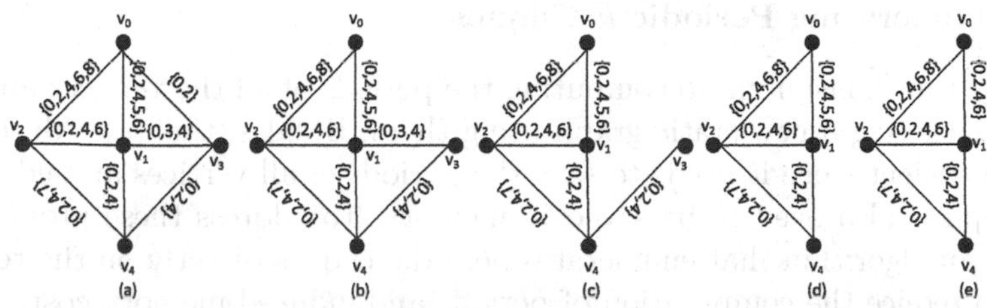

Fig. 3. Illustration of the Pruning algorithms

Figure 3 shows an example of our pruning rules, (a) is the origin graph. (b) is the graph pruned by rule 1, edge $(0, 3)$ is deleted. (c) shows the graph pruned by rule 2 based on (b), edge $(1, 3)$ is deleted. (d) shows the graph pruned by rule 3 based on (c), vertex v_3 and the edges associated with it are deleted. (e) shows the graph pruned by rule 4, temporal edges $(0, 1, 5)$, $(0, 2, 8)$ and $(2, 4, 7)$ are deleted.

Algorithm 1: AggregatedPrune

```
Data: temporal graph 𝒢, its de-temporal graph G, parameter σ, k
Result: pruned temporal graph 𝒢 = (𝒱, ℰ), an array of the degree left Degree
1  Q = ∅;
2  Initialize Degree to an array of 0;
3  foreach u ∈ G do
4      foreach v ∈ u.neighbor do
5          if (u, v).ts.size() < σ then Delete edge (u, v) ;
6          else if |TS(u, v)| = σ AND |TS(u, v)| is not a σ-period then Delete edge (u, v) ;
7          else Degree[u] ← Degree[u]+1 ; // This edge is valid
8      end
9      if Degree[u] < k-1 then Q ← Q ∪ {u} ;
10 end
11 while Q ≠ ∅ do
12     v = Q.pop()
13     if Degree[v] == 0 then continue ;
14     Delete vertex v;
15 end
16 foreach u ∈ 𝒱 do
17     SD ← ∅;
18     foreach v ∈ u.neighbor do
19         foreach t ∈ TS(u, v) do SD[t].append([u, v, t]) ;
20     end
21     foreach t ∈ SD.keys do
22         if |SD[t]| < k-1 then
23             foreach (u, v, t) ∈ SD[t] do Delete temporal edge (u, v, t) ;
24     end
25 end
```

Complexity. The worst case of our pruning algorithm occurs when the time support set of any two edges is a σ-period, then our pruning algorithm will iterate over all the timestamps. So the worst case time complexity is $O(|\mathcal{E}|)$.

3.3 Enumerating Periodic k-Cliques

It is very time-consuming to computing the periods of all the vertices and edges and constructing a new static graph using them. Besides, this procedure is also space inefficient since it needs to save the periods of all vertices and edges. The new graph is also need to be saved in memory. To address these problems, we propose an algorithm that enumerates periodic cliques directly on the temporal graph to reduce the computation of periods and induced memory cost.

Our enumeration algorithm improves the performance in several ways. Firstly, it directly enumerates periodic k-cliques on the temporal graph, reducing memory cost. Secondly, we adopt a new way to process the periods to avoid

Algorithm 2: ComputeMaximalPeriod(TS, σ)

```
Data: An array of timestamps TS, parameter σ
Output: An array of σ-maximal periods in TS
P ← ∅, MP ← ∅, S ← ∅;
foreach t ∈ TS do
    ED ← ∅;
    foreach p ∈ P do
        if t − p.s = p.l ∗ p.d then
            p.l ← p.l + 1, ED ← ED ∪ {p.d};
        else if (t − p.s) > p.l ∗ p.d then
            if p.l ≥ σ then
                MP ← MP ∪ {[s ← p.s, d ← p.d, l ← p.l]};
            P ← P\{p};
    end
    foreach s ∈ S do
        if (t − s)(σ − 2) ≤ max(TS) − t AND (t − s) ∉ ED then
            P ← P ∪ {[s ← s, d ← t − s, l ← 1]};
    end
    S ← S ∪ {t};
end
foreach p ∈ P do
    if p.l ≥ σ then
        MP ← P ∪ {[s ← p.s, d ← p.d, l← p.l]};
end
return MP;

Procedure UnionPeriod(MP, TS)
CMP ← ∅;
foreach p ∈ MP do
    s ← p.s, l ← 0, next ← p.s;
    foreach i ∈[0, p.l-1] do
        if next ∉ TS then
            if l ≥ σ then  CMP ← CMP ∪ {[s ← s, d ← p.d, l ← l]} ;
            s ← next + p.d, l ← 0;
        else  l ← l + 1 ;
        next ← next + p.d;
    end
    if l ≥ σ then  CMP ← CMP ∪ {[s ← s, d ← p.d, l ← l]} ;
end
return CMP;
```

most computation of periods. Thirdly, we introduce a "prune while enumeration" strategy. During enumeration of periodic cliques, the processed vertices and edges are deleted to avoid re-enumeration. We push this deletion on step forward. The deletion of vertices and edges might cause the left graph violates the pruning rules presented in the previous subsection. So, we also delete the vertices that violate the pruning rules due to a deletion of a processed vertex or edge. At last, the overlaps between periods can cause redundant computation. We use "maximal period" in enumeration to prevent the redundant computation caused by overlapping periods.

Definition 4 (σ-maximal period). *Given a time support set TS, a σ-maximal period of TS, $\sigma - MP$, is a period of TS where (1) $|\sigma - MP| \geq \sigma$ and (2) there exist no period P such that $DIFF(P) = DIFF(\sigma - MP)$ and $P \subset \sigma - MP$.*

Definition 5 (σ-maximal support period). *Given a temporal graph $\mathcal{G} = (\mathcal{V}, \mathcal{E})$ and a subgraph $\mathcal{S} = (\mathcal{V}_c, \mathcal{E}_c)$. The ($\sigma$, k)-maximal support period of $\mathcal{S}$, denoted as $MSP_{\sigma,k}(\mathcal{S})$, is a set of σ-maximal period of S where (1) for any $P \in MPS_{\sigma,k}(\mathcal{S})$, P is a σ-maximal period and (2) for any σ-maximal period, P, of subgraph $\mathcal{S}$, $P \in MPS_{\sigma,k}(\mathcal{S})$.*

Algorithm 2 computes the maximal periods of the given sorted timestamps. A period is represented by three parts, s, d, l. s is the smallest timestamp in the

Algorithm 3: KPC-A

```
Data: temporal graph G, parameter k and σ
Result: a vector of σ-periodic k-cliques
Prune the graph using the algorithm in 3.2 ;
cliques ← ∅;
foreach u ∈ V do
    if Degree[u] < k then Delete vertex u, continue ;
    foreach v ∈ u.neighbors do
        PS = ComputeMaximalPeriod(TS(u, v));
        if |PS| ≠ 0 then
            PC ← {u, v};
            CV ← u.neighbor ∩ v.neighbor;
            if |CV| ≥ k − 2 then kcliqueRec(PS, PC, CV) ;
            Delete edge (u, v);
            if Degree[u] < k then Delete vertex u; ;
            if Degree[v] < k then Delete vertex v; ;
    end
    Delete vertex v;
end
Procedure KcliqueRec(PS, PC, CV)
if |PC| = k then cliques ← cliques ∪ PC, return ;
if |PC| + |CV| < k then return ;
for u ∈ CV do
    CV ← CV\{u};
    nCV ← CV ∩ u.neighbor;
    nPS ← PS;
    foreach v ∈ PC do
        nPS ← UnionPeriod(nPS, TS(u, v));
        if |nPS| = 0 then continue ;
    end
    KcliqueRec(nPS, CV ∪ u, nCV);
end
```

period, d is the difference between two adjacent timestamps and l is the size of the period. The algorithm first initializes some data structures (line 1): P is the set of periods that might be extended, MP is maximal periods that can not be extended and S is the set of timestamps that has been traveled, which used to add new periods. It then travels through the sorted timestamps (line 2). ED is used to avoid adding a non-maximal period (line 3). Then, for each candidate period that might be extended, extend it if the new timestamp t can be used to extend the candidate period (line 5–6). If the new timestamp t cannot be used to extend the candidate period and it is larger than the next expected timestamp for it, which means that the current period cannot be extended, so it is added to the result (line 8, 9) and the candidate is erased (line 10). After that, the algorithm will add new maximal period candidates where the current processing timestamp is the second timestamp in the period (line 12 to 14). Note the new candidates' difference cannot be in ED or it can't be the start of a maximal period. After all the timestamps are processed, the left candidates are checked, periods with size not smaller than σ are added to the final result (line 18–21).

Given a σ-maximal support period MP and a time support set TS, UnionPeriod will find the σ-maximal support period of $TS' = \{t|t \in p, p \in MP\} \cup \{t|t \in TS\}$. It travels through each maximal period (line 25). For each maximal period P, it checks the overlapping part of TS and P (line 27–32).

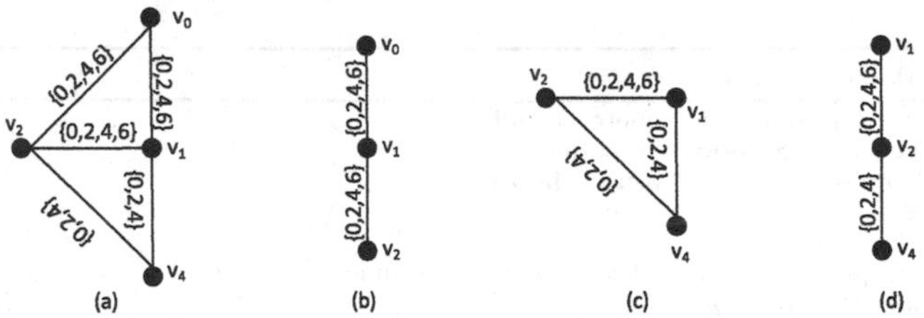

Fig. 4. Example of the Enumeration algorithm

We propose a σ-periodic k-clique enumeration algorithm based on Chiba and Nishizeki's algorithm for listing k-cliques [5]. For each vertex in u, it forms a 1-clique, called a partial clique, then the algorithm finds the vertex that can be added to the partial clique and compute the maximal support recursively. The support period set for the new partial clique can be computed incrementally with the algorithms in algorithm 4 efficiently.

The details of periodic clique enumeration is shown in Algorithm 3. The cliques are enumerated in a recursive manner. For each vertex at the root of the recursive tree, select a vertex in its neighbor (line 5). If the new vertex can be added to the partial clique (line 7), compute the new common neighbor of the partial clique (line 9) and try to add new vertex in the common neighbor to the partial clique recursively. After all the cliques start from a root vertex u and its neighbor v are checked, edge (u, v) is deleted to avoid replicate enumeration (line 11) and vertex u, v is checked if it can be deleted recursively. After all the cliques start with v have been enumerated, vertex v is deleted. The recursive algorithm will first check if it has found a σ-periodic k-clique, if it has found one, return (line 18). And then it selects a vertex u from u's neighbors, compute the new common neighbor (line 22), the new maximal support period is computed when u is added (line 24–27) and then a new vertex is added from the common neighbor of the new partial clique recursively.

Figure 4 shows the enumeration of 3-periodic 3-clique on the pruned graph in Fig. 2. (a) is the pruned graph, (b) is the recursive tree with root v_0, after root v_0 and edge (v_0, v_1) is processed, edge (v_0, v_1) is deleted, causing v_0 deleted. The graph is shown in (c). Then we select v_1 as root, the recursive tree is shown in (d). After that, the enumeration is finished.

Complexity. The worst case complexity of ComputeMaximalPeriod is $O(l_t^3)$ for a time support of length l_t. And the complexity of UnionPeriod is $O(l_p)$ where l_p is the total length of periods in parameter MP. We assume that the degrees and temporal edges are distributed evenly in the graph, the time complexity of the enumeration procedure is $O(2|V|t_a C_{d_a}^{\sigma})$, where d_a is the average degree $|V|/|E|$ and t_a is the average time support $|\mathcal{E}/E|$.

4 Experiments

Algorithms. We implement four algorithms in our experiments. KPC-B is the baseline algorithm shown in Sect. 4.1. KPC-O1 is the implementation of our basic enumeration algorithm without any optimizations. KPC-O2 is the implementation of our enumeration algorithm which uses the pruning algorithm to prune the graph. KPC-A is the implementation of our enumeration algorithm with all the optimizations. All the algorithms are implemented with C++ and compiled

Table 1. Datasets

Dataset	$\|V\|$	$\|E\|$	$\|\mathcal{E}\|$	d_{max}	$d_{tempmax}$	Time scale
PS	242	8,317	32,079	134	503	Hour
LKML	26,885	159,996	328,092	2,989	14,172	Month
Enron	86,978	297,456	499,983	1,726	4,311	Month
DBLP	1,729,816	8,546,306	12,007,380	3,815	5,980	Year
WikiTalk	2,863,439	8,146,544	10,268,684	146,311	277,833	Month

Table 2. Running time(s) on different datasets

	σ, k = 3,3			σ, k = 4,4			σ, k = 5,5		
	KPC-B	KPC-A	Speedup	KPC-B	KPC-A	Speedup	KPC-B	KPC-A	Speedup
PS	0.36	0.04	9.41	0.32	0.02	15.43	0.24	0.01	19.97
LKML	34.32	1.02	31.91	18.27	0.07	27.48	10.31	0.34	30.08
Enron	16.120	0.360	44.21	6.54	0.20	33.43	3.01	0.09	32.20
DBLP	1071.14	2.82	379.53	151.77	0.77	197.90	53.17	0.31	172.00
WikiTalk	9609.68	12.46	771.32	5047.87	4.41	1145.20	2752.41	1.88	1465.06

using GCC version 11.1.0 with optimization level set to O3. All the experiments are carried out on a PC with Linux kernel 5.12.14 running on an Intel i5-9400 with 32 GB memory. The source code can be found in https://www.dropbox.com/s/1xcf8fh7srt0n2b/periodic_k_clique.zip?dl=0.

Datasets. We use 5 real-life temporal networks in our experiments, which are PS, LKML, ENRON, DBLP, WikiTalk. PS is a graph of face-to-face contacts between students and teachers in a French primary school. An temporal edge (u, v, t) in PS means there is face-to-face communication between u and v in time period t. Both LKML and Enron are email communication networks. A temporal edge (u, v, t) in LKML and Enron means u sent an email to v in time period t. Their timescale is rescaled to month. DBLP is a scientific collaboration graph generated from the DBLP dataset. An temporal edge (u, v, t) in DBLP means u and v published a paper together in year t. WikiTalk is a communication network between English wiki users. It's time is also rescaled to month. The time scale represents the time length of each timestamp. Their sizes are shown in Table 1.

Exp-1: Running Time of Different Datasets. Table 2 shows the running time of different datasets with three parameter settings, (σ, k) = (3, 3), (σ, k) = (4, 4) and (σ, k) =(5, 5). We can see that our algorithm outperforms the baseline algorithm in all five datasets and three parameter settings. The speedup grows as the graph size grows. Our algorithm can finish enumeration in DBLP and WikiTalk in a few seconds using a single thread. And it shows a maximum 2822 times speed up in DBLP where σ=3 and k = 3. For each dataset, the running time and speedup grow as parameter σ and k grow, mainly because the effectiveness of pruning algorithm grows as σ and k grows, we can see that in later experiments.

Exp-2: Running Time on Varies Parameters. Figure 5 shows the running time of DBLP with different parameters where both σ and k varies from 3 to

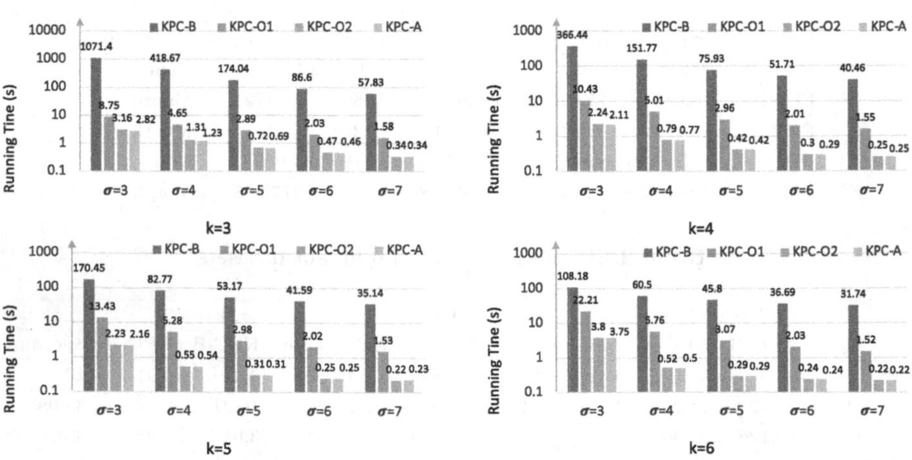

Fig. 5. Running time(s) on different parameters (on DBLP)

6. From the figure we can see that our algorithm outperforms the three baseline algorithms in all the parameter settings. The running time decreases with σ increases consistently, due to the effectiveness of our pruning algorithm. The running time of KPC-O1 and KPC-O2 are close since the enumeration procedure will also prune the graph as it enumerates. With the addition of elimination of overlapping periods, there is a significant improvement in KPC-A.

Exp-3: Pruning Power on Different Datasets. Table 3 shows the pruning power of our pruning algorithm with different datasets when $\sigma = 5$ and $k = 5$. It doesn't performs well in PS since PS is a small dense face-to-face communication graph and there tend to be no obvious large central vertex. The pruning algorithm is very effective in the four real-world large sparse networks and the pruning power grows with the size of graph grows. It can prune out over 97% vertices in four graphs and it even prune 99.79% vertices in DBLP. Despite the simplicity of our pruning algorithm, it performs pretty well in sparse networks.

Table 3. Pruning power with on different datasets ($\sigma = 5$ and k = 5)

	Vertices			Edges			Temporal edges		
	Origin	Left	Percent	Origin	Left	Percent	Origin	Left	Percent
PS	242	232	95.86%	8,317	1,822	21.91%	32,079	15988	49.8%
LKML	26,885	649	2.41%	159,996	7745	4.84%	328,092	77185	23.5%
Enron	86,978	1,003	1.51%	297,456	6601	2.21%	499,983	39639	7.9%
DBLP	1,729,816	3,678	0.21%	8,546,306	10,475	0.12	12,007,380	50438	0.4%
WikiTalk	2,863,439	7,015	0.24%	8,146,544	63,303	0.77%	10,268,684	394728	3.84%

Exp-4: Pruning Power on Different Parameters. Figure 6 shows the pruning power of our pruning algorithm on DBLP with different parameters where both σ and k varies from 3 to 7. We can see that our algorithm is very efficient even $\sigma = 3$ and $k = 3$. And the size of the pruned graph decreases dramatically as σ or k increases. With each increment σ or k, the size of left vertices, edges and temporal edges are nearly halved. And the increment of pruning power are directly reflected of total execution time presented in Experiment 1 and Experiment 2.

Exp-5: Memory Overhead. Figure 7 shows the memory usage of KPC-B and KPC-A. We can see that our algorithm costs significantly less memory than KPC-B. Our algorithm achieves high space-efficiency by using the augmented adjacency array and avoiding store the periods of all edges and vertices in memory.

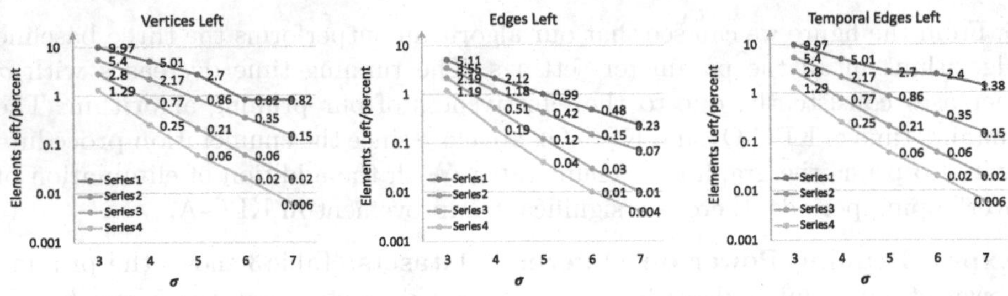

Fig. 6. Pruning power on different parameters on DBLP

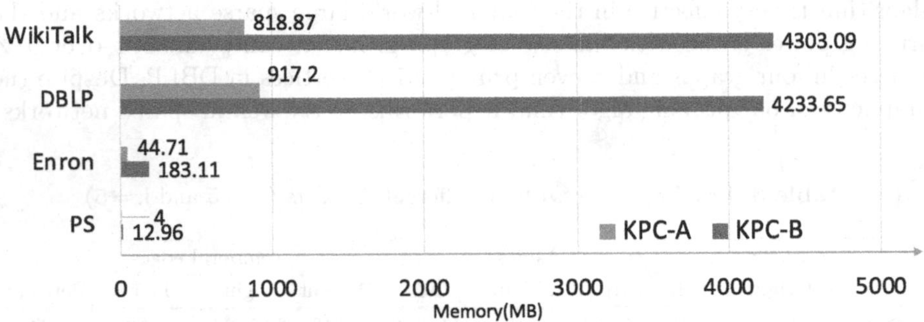

Fig. 7. Memory usage

5 Related Work

k-Clique Listing and Counting. Enumerating k-clique in static graph has been studied for decades [26]. The first practical algorithm of enumerating k-cliques is the algorithm of Chiba and Nishizeki [5]. Danisch et al. [7] give an edge-parallel algorithm to enumerate k-clique and use DAG to avoid re-enumeration. [13] propose an approximately k-clique counting algorithm based on Turan Shadow. Li et al. [17] propose a k-clique listing algorithm based on graph coloring. Jain and Seshadhri [14] propose a k-clique counting algorithm based on the pivot technology and it is much faster than enumeration-based algorithms.

Maximal Clique Enumeration. Many techniques in enumerating maximal cliques can also be used in listing k-cliques with minor change [25]. The best known algorithm in listing maximal cliques is the Bron-Kerbosch algorithm [2]. There are many variants of the Bron-Kerbosch algorithm [4,9,21,25].

Temporal Graph Data Mining. Most previous studies on temporal graph data mining focus on finding dense subgraphs that persist over time or analyzing the evolve of communities. Li et al. [16] propose an algorithm to detect dense subgraphs which persist over a threshold. Ma et al. [20] give an algorithm to find dense subgraphs where edge weights vary with timestamps. Lin et al. [18] propose an algorithm to analyze the evolution of communities through a unified process. Yang et al. [29] propose a method to evaluate the change speed of subgraphs and give an algorithm to detect the fast-changing subgraphs. Rossetti et al. [24] propose a method to extract and track the evolution of overlapping communities using an online iterative procedure. Qin et al. [23] address the problem of finding periodic maximal cliques in temporal cliques. And Zhang et al. [31] propose a new periodic subgraph model called seasonal-periodic subgraph.

6 Conclusion

In this paper, we study the problem of how to find dense subgraphs that periodically occurs in temporal networks, and then propose σ-periodic k-clique to model it. Firstly, we design a new data structure to store temporal networks in main memory as well as support dynamic deletion for the pruning algorithms. Secondly, we propose several effective and efficient rules to prune the graph to reduce search space. Thirdly, we design a search algorithm to enumerate all the σ-periodic k-clique based on the Bron-Kerbosch algorithm for static graphs. Finally, we conduct extensive experiments on five real-world temporal networks to show that our algorithm works efficiently in practice.

Acknowledgments. This work was partially supported by (i) National Key Research and Development Program of China 2020AAA0108503, (ii) NSFC Grants 62072034, 62002036, (iii)Natural Science Foundation of Chongqing CSTC cstc2021jcyj-msxm X0859.

References

1. Agarwal, M.K., Ramamritham, K., Bhide, M.: Real time discovery of dense clusters in highly dynamic graphs: identifying real world events in highly dynamic environments. arXiv preprint arXiv:1207.0138 (2012)
2. Bron, C., Kerbosch, J.: Algorithm 457: finding all cliques of an undirected graph. Commun. ACM **16**(9), 575–577 (1973)
3. Chen, Z., Wilson, K.A., Jin, Y., Hendrix, W., Samatova, N.F.: Detecting and tracking community dynamics in evolutionary networks. In: ICDMW, pp. 318–327 (2010)
4. Cheng, J., Zhu, L., Ke, Y., Chu, S.: Fast algorithms for maximal clique enumeration with limited memory. In: SIGKDD, pp. 1240–1248 (2012)
5. Chiba, N., Nishizeki, T.: Arboricity and subgraph listing algorithms. SIAM J. Comput. **14**(1), 210–223 (1985)
6. Cohen, E., Halperin, E., Kaplan, H., Zwick, U.: Reachability and distance queries via 2-hop labels. SIAM J. Comput. **32**(5), 1338–1355 (2003)
7. Danisch, M., Balalau, O., Sozio, M.: Listing k-cliques in sparse real-world graphs. In: WWW, pp. 589–598 (2018)
8. Du, X., Jin, R., Ding, L., Lee, V.E., Thornton Jr., J.H.: Migration motif: a spatial-temporal pattern mining approach for financial markets. In: SIGKDD, pp. 1135–1144 (2009)

9. Eppstein, D., Strash, D.: Listing all maximal cliques in large sparse real-world graphs. In: International Symposium on Experimental Algorithms, pp. 364–375 (2011)
10. Fratkin, E., Naughton, B.T., Brutlag, D.L., Batzoglou, S.: Motifcut: regulatory motifs finding with maximum density subgraphs. Bioinformatics **22**(14), e150–e157 (2006)
11. Gibson, D., Kumar, R., Tomkins, A.: Discovering large dense subgraphs in massive graphs. Proc. VLDB Endow. 721–732 (2005)
12. Gurukar, S., Ranu, S., Ravindran, B.: Commit: a scalable approach to mining communication motifs from dynamic networks. In: SIGMOD, pp. 475–489 (2015)
13. Jain, S., Seshadhri, C.: A fast and provable method for estimating clique counts using turán's theorem. In: WWW, pp. 441–449 (2017)
14. Jain, S., Seshadhri, C.: The power of pivoting for exact clique counting. In: WSDM, pp. 268–276 (2020)
15. Jin, R., Xiang, Y., Ruan, N., Fuhry, D.: 3-hop: a high-compression indexing scheme for reachability query. In: SIGMOD, pp. 813–826 (2009)
16. Li, R.H., Su, J., Qin, L., Yu, J.X., Dai, Q.: Persistent community search in temporal networks. In: ICDE, pp. 797–808 (2018)
17. Li, R., Gao, S., Qin, L., Wang, G., Yang, W., Yu, J.X.: Ordering heuristics for k-clique listing. Proc. VLDB Endow. (2020)
18. Lin, Y.R., Chi, Y., Zhu, S., Sundaram, H., Tseng, B.L.: Facetnet: a framework for analyzing communities and their evolutions in dynamic networks. In: WWW, pp. 685–694 (2008)
19. Liu, P., Wang, M., Cui, J., Li, H.: Top-k competitive location selection over moving objects. Data Sci. Eng. **6**(4), 392–401 (2021)
20. Ma, S., Hu, R., Wang, L., Lin, X., Huai, J.: Fast computation of dense temporal subgraphs. In: ICDE, pp. 361–372 (2017)
21. Makino, K., Uno, T.: New algorithms for enumerating all maximal cliques. In: Scandinavian Workshop on Algorithm Theory, pp. 260–272 (2004)
22. Presson, A.P., et al.: Integrated weighted gene co-expression network analysis with an application to chronic fatigue syndrome. BMC Syst. Biol. **2**(1), 1–21 (2008)
23. Qin, H., Li, R., Yuan, Y., Wang, G., Yang, W., Qin, L.: Periodic communities mining in temporal networks: concepts and algorithms. IEEE TKDE (2020)
24. Rossetti, G., Pappalardo, L., Pedreschi, D., Giannotti, F.: Tiles: an online algorithm for community discovery in dynamic social networks. Mach. Learn. **106**(8), 1213–1241 (2017)
25. Takeaki, U.: Implementation issues of clique enumeration algorithm. Special issue: Theor. Comput. Sci. Discrete Math. Progress Inform. **9**, 25–30 (2012)
26. Tsourakakis, C.: The k-clique densest subgraph problem. In: WWW, pp. 1122–1132 (2015)
27. Wu, H., et al.: Core decomposition in large temporal graphs. In: IEEE Conference on Big Data, pp. 649–658 (2015)
28. Wu, H., Huang, Y., Cheng, J., Li, J., Ke, Y.: Reachability and time-based path queries in temporal graphs. In: ICDE, pp. 145–156 (2016)
29. Yang, Y., Yu, J.X., Gao, H., Pei, J., Li, J.: Mining most frequently changing component in evolving graphs. WWW, vol. 17, no. 3, pp. 351–376 (2014)
30. Yang, Y., Yan, D., Wu, H., Cheng, J., Zhou, S., Lui, J.C.: Diversified temporal subgraph pattern mining. In: SIGKDD, pp. 1965–1974 (2016)
31. Zhang, Q., Guo, D., Zhao, X., Li, X., Wang, X.: Seasonal-periodic subgraph mining in temporal networks. In: CIKM, pp. 2309–2312 (2020)

LSM-Subgraph: Log-Structured Merge-Subgraph for Temporal Graph Processing

Jingyuan Ma[1], Zhan Shi[1(✉)], Shang Liu[2], Wang Zhang[1], Yutong Wu[1], Fang Wang[1], and Dan Feng[1]

[1] Wuhan National Laboratory for Optoelectronics, Huazhong University of Science and Technology, Wuhan, China
{jyma,zshi,zhangtw,yutongwu,wangfang,dfeng}@hust.edu.cn
[2] Graduate School of Informatics, Kyoto University, Kyoto, Japan
liu.shang.33s@st.kyoto-u.ac.jp

Abstract. Temporal graphs, with a time dimension, are attracting increasing interest from research communities. Existing temporal graph storage formats mainly include copy-based models, log-based models, and hybrid models that have emerged in recent years. Neither the copy-based model nor the log-based model can trade-off storage and query time well. Hybrid models try to find a compromise between the above two models, but existing models do not consider the skewness of vertex degree in temporal graphs is changing over time. Based on these considerations, we propose LSM-Subgraph, a hybrid storage format that only stores snapshots divided by the fluctuation-aware method and in-between logs. First, LSM-Subgraph uses a PMA-based snapshot creation model to store snapshots based on packed memory arrays (PMA), avoiding rebuilding the whole data structure. Second, LSM-Subgraph uses a select-timepoint method based on fluctuation-aware to divide shards during the update, which achieves a good tradeoff between storage overhead and query time cost. Extensive experimental evaluations over various real-world graphs illustrate that LSM-Subgraph outperforms state-of-the-art temporal graph systems in both memory and time consumption.

Keywords: Temporal graph · Packed-memory array · Graph storage · Graph analysis

1 Introduction

Graph, as a fundamental data model, has been widely used in many domains [9,13,21]. Real-world graphs are not only large in size but also often evolve over time as the connections and relationships change. For example, Facebook [23] social network has more than one billion nodes. Among them, 86,000 objects are changing and 2.5 emails are sending out per second on average. These time-evolving graphs are called temporal graphs [5], whose vertex/edge insertion and deletion can happen over time.

B. Li et al. (Eds.): APWeb-WAIM 2022, LNCS 13421, pp. 477–494, 2023.
https://doi.org/10.1007/978-3-031-25158-0_39

Temporal graph processing is challenging in how to trade-off between storage and query time. This is because as a graph evolves constantly, it produces a large number of snapshots as time flies. Each graph snapshot in history should be available. Storing all the snapshots consume memory resource excessively, while disk storage is slow for graph accessing and processing. Besides, how to switch snapshot quickly is not trivial.

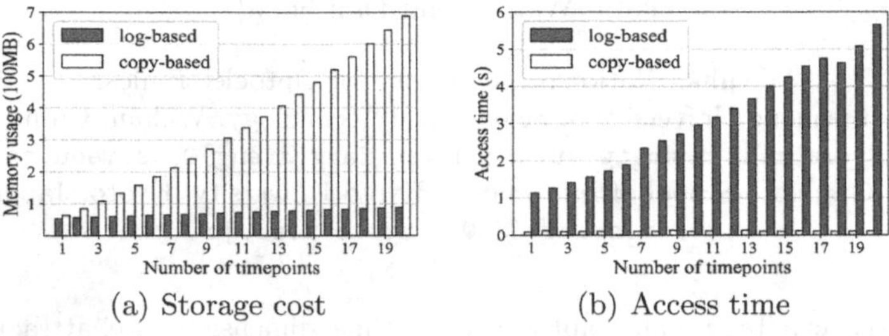

(a) Storage cost (b) Access time

Fig. 1. Overhead comparison

Modern out-of-core graph processing systems store graphs in persistent storage [10]. By loading graphs at desired time points into memory and replacing the previously processed graphs, they finish temporal graph processing. Such graph loading is costly for temporal graph processing. Existing temporal models to handle this problem fall into two types: copy-based models [3,7,12,15] and log-based models [4,8].

Copy-Based Model: It consists of a sequence of snapshots, e.g. *FVF* [15]. Each of which represents a specific state at a single time point. This model generates and stores the copies of the graph at different time points. It stores the graph structure completely. Hence it favors structure locality well and does well in queries. However, there is high redundancy in consecutive snapshots storage, which is inefficient in memory for large-scale dynamic graphs processing.

Log-Based Model: It is composed of a series of events where can be insertions or deletions of nodes and edges, e.g. DeltaGraph [8]. Log-based model stores simple updates at each time point. This model stores only incremental updates of neighbouring snapshots, thus avoiding the storage redundancy of copy-based model. However, it introduces extra overhead at query time for reconstructing snapshots as of specified time points.

We examine the performance of the copy-based model and the log-based model as shown in Fig. 1. We evaluate these two models using the Wiki-talk dataset [11], which is the temporal graph of users editing talk pages on Wikipedia. Each directed edge has a timestamp t. The first 40% is treated as the initial snapshot (e.g. Snapshot 0), and the rest is divided into 20 time-points' updates evenly. The experimental results show that the copy-based model incurs up to 7× memory cost of the log-based model (Fig. 1(a)), while the access time of the log-based model is up to 5× more than that of the copy-based model (Fig. 1(b)). There is an obvious tradeoff between memory and time consumption. We need a new temporal model to support efficient queries with lower storage overhead.

To solve the above problems, hybrid models have been proposed, such as Pensieve [22]. Based on the skewness of vertex degree, Pensieve combines the copy and log model. The log model is used for high-degree vertex data, and the copy model is used for low-degree vertex data. However, the skewness in real-world graphs usually changes over time. In Fig. 2, we examine the Hits@100 of the real-world temporal graph datasets from SNAP [11]. Hits@100 is the proportion of 100 high-degree vertices in the initial snapshot that are still high-degree vertices after being updated. Figure 2 shows that some vertices will no longer be high-degree vertices over time. The degree of a vertex may either grow or shrink in the future, sometimes changing drastically, which is not concerned in Pensive. As a result, Pensieve's design may cause vertex's degree unfit for the model over time and affect the system's performance ultimately.

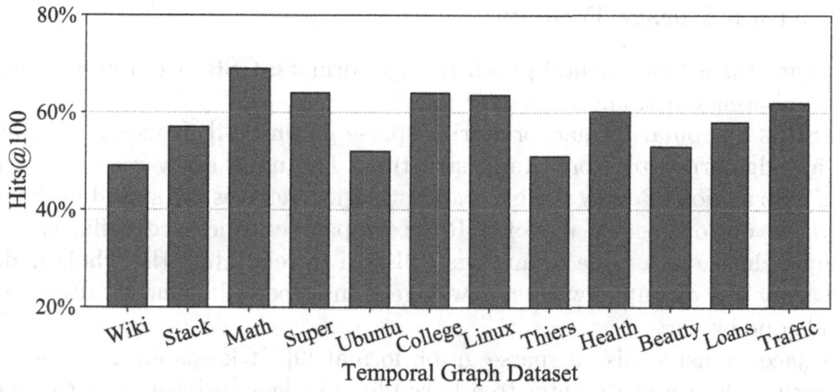

Fig. 2. Hits@100 with real-world temporal graph dataset

The skewness of vertex degree is considered in many systems, but they are based on this static characteristic to design corresponding data structures. However, the skewness of vertex degree in temporal graphs is changing over time, and the data structure designed for the degree skewness is not applicable. Therefore, the main challenge is how to design a data structure according to the changing skewness of vertex degree and how to choose the time point for inserting the key snapshot.

In this work, we propose a temporal graph storage model based on log-structured merge-subgraphs, called LSM-Subgraph. Three factors contribute to the efficiency of LSM-Subgraph. First, it splits all logs into multiple data shards which are composed of a log array and an adjacency array. Specifically, the log array keeps the updates in the edge list format and is designed to accelerate data ingestion. At the same time, the adjacency array based on Packed Memory Arrays [1,2] stores a snapshot of the initial moment which is to solve the problem of snapshot reconstruction caused by the update. Second, by using a fluctuation-aware method, LSM-Subgraph sets a snapshot according to the size change during the evolution of the graph. The next snapshot will be created when the relative change rate of graph size reaches the threshold. Third, LSM-Subgraph reduces the size of logs during the query by utilizing a log-merging method. The edge list format logs will firstly merge in memory and then apply to the adjacency array, which can avoid repeated operations.

We implement our system on top of Ligra [16], a famous graph processing system. For comparison, we evaluate it against two traditional temporal graph processing systems: GraphPool and Chronos, and a hybrid model temporal graph system: Pensieve. The experiment results show that LSM-Subgraph supports faster queries with lower storage overhead than existing systems.

2 Background and Related Work

2.1 Graph Storage Formats

There are three fundamental graph storage formats: CSR (Compressed Sparse Row), adjacency list, and adjacency matrix.

CSR is a popular format for storing sparse graphs [14]. It uses a node array and an edge array to store graph structure. The node array records starting index information of every vertex, and destination vertices are stored in the edge array. The obvious advantages of CSR are compactness and access efficiency. But a simple change of a node or an edge will lead to rebuilding the whole node or edge array. For example, when a new edge is inserted, all elements after it need to slide one by one.

Adjacency list is also a sparse graph format [9]. It keeps an array of nodes and each node stores a pointer to a linked list of edges. Adjacency list supports insertions or deletions well. For example, when a new edge needs to be inserted, we just update the corresponding edge list instead of changing the whole graph. Though it supports fast insertions, the search efficiency is not high because each element is unsorted.

Adjacency matrix stores an $n \times n$ matrix for a graph with n nodes. It excels in storing dense graphs. However, real-world graphs are generally sparse, so it wastes more storage than both CSR and adjacency list. To make it worse, any update of elements will bring the reconstruction of the whole graph. Hence, adjacency matrix is not suitable for dynamic graphs.

2.2 Packed Memory Arrays

Packed Memory Arrays (PMA), a sparse-array structure, has been widely used in many applications [1,6,17,19]. For instance, APMA [1] gives the first adaptive packed-memory array that adjusts to the input pattern automatically. It performs better than traditional PMA, and achieves only $O(logN)$ amortized element moves. RMA [6] proposes an improved version of sparse arrays, Rewired Memory Array, which fixes its major flaws and adds a new technique of memory rewiring into its rebalancing mechanism.

PMA stores N elements in a sorted array, interleaved with empty spaces or gaps. These gaps serve the purpose of providing extra space to fast insert new items in an arbitrary position of the array, instead of moving existing elements. The insertion, deletion, and rebalance operations are as follows.

To insert a new element, we have to find its target position using a binary search. If the cell has been occupied by the other element e, we need to move e towards the nearest empty gaps. To delete an arbitrary element, we just mark this cell as NIL. When element insertion or deletion causes the data structure to be locally dense or sparse, PMA adjusts the position of empty elements in the data structure to meet the local threshold setting.

To examine the performance of PMA, we compare it with traditional data structure CSR. From experimental results, when 70 million data is inserted, the update time of CSR is 40 s while that of PMA is only 2 s. Therefore, PMA is beneficial to design the temporal graph storage format.

However, unlike classical dynamic datasets, the temporal graph is not just skewed, its skewness also changes with time, which makes it hard to determine the number and distribution of empty elements in PMA, and the traditional rebalancing operations are inapplicable. Thus PMA cannot be applied directly. In response to these challenges, we design a PMA-based adjacency array storage model.

2.3 Temporal Graph Storage

There is a growing interest in analyzing and computing temporal graphs. Every temporal system designs and improves temporal graph storage formats for their own applications. For example, GraphOne [9] stores temporal graphs using copy+log approach. It is a unified graph data store abstraction that offers data access at different granularity for various real-time analytics and queries, while supporting high arrival velocity of fined-grained updates simultaneously. But the updated data is stored in adjacency list periodically, losing the temporal property. Pensieve [22] proposes a skewness-aware multi-version graph processing system. It uses a differentiated graph storage strategy that stores high degree vertices using log-based scheme, while stores low degree vertices using copy-based scheme. However, temporal graph is changing over time, and the degree of vertices is not predicted in the future. Huanhuan Wu [20] proposes an equal-weight damped time window model, to tradeoff between the required storage space and the information loss. It is clear that this system achieves excellent storage compactness with the sacrifice of accuracy. DeltaGraph [8] is a distributed hierarchical structure that enables compact storage of the historical trace of a network. It uses the log-based scheme to change the full graph structure to partial storage for modification. This system reduces storage overhead, but suffers

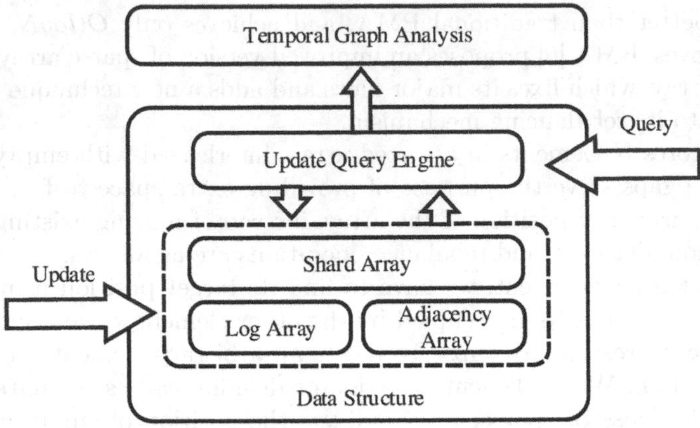

Fig. 3. Architecture of LSM-Subgraph

longer construction time. SAMS [18] presents a novel approach to execute algorithms concurrently for multiple graph snapshots, which is a copy-based model. By leveraging discrepancies between the analyzed graph snapshots, it accelerates temporal analysis. But it just focuses on multiple snapshots' analysis, not on temporal graph storage.

In a word, existing methods fail to achieve a good tradeoff between storage cost and query time overhead during temporal graph processing. None of them consider the characteristic that the skewness of vertex degree is changing over time, which is nonnegligible for improving the performance of temporal graph processing.

3 LSM-Subgraph Design

In this section, we begin with an overview of LSM-Subgraph architecture. Then we present details of the system design.

3.1 Overview

LSM-Subgraph uses a hybrid graph storage format that consists of key snapshots and in-between logs. Specially, we select a few key time points to store snapshots using the copy-based model and store in-between updates in the log-based model. There are two main challenges to deal with: 1) How to create a designated snapshot with the key snapshot and logs instead of reconstructing; 2) How to select the key time points to store graphs in the copy-based model.

To handle them, we elaborate a PMA-based adjacency array model to store snapshots. Then, there is a crucial tradeoff between storage cost and query time overhead when selecting key time points. We design a fluctuation-aware snapshot creation method to determine which format to store graphs at the designated time point (discussed in Sect. 3.3). Finally, We propose a method to reduce the number of log data merged in the query process.

Figure 3 depicts an overview of LSM-Subgraph architecture. LSM-Subgraph mainly includes LSM-Subgraph data structure and update query engine. LSM-Subgraph splits all logs into multiple data shards, and each shard has a log array and an adjacency array based on Packed-Memory Array (PMA). The adjacency array based on PMA handles the reconstruction problem by reserving some space for updating. The function of the update query engine is to accomplish some operations during the update and query. For the coming updated edges, it first computes the discrepancy between the last snapshot and the temporary one currently (defined as TD). If TD goes above the threshold, the temporary snapshot is stored in the copy-based model, and corresponding logs are stored in the log-based model. In the query phase, LSM-Subgraph computes and finds the latest time point first. And then it loads the key snapshot and corresponding logs. After merging logs into the key snapshot, LSM-Subgraph constructs the target snapshot for temporal graph analysis.

3.2 PMA-Based Adjacency Array Model

The PMA-based adjacency array consists of a vertex array and the corresponding adjacency edge array, as shown in Fig. 4. The maximum number of vertices in the current snapshot ID_{max} determines the size of the vertex array because the number of vertices in real-world datasets is much smaller than the number of edges. The vertex array stores all vertex IDs and the degree of each vertex, that is, $0 \sim ID_{max}$. If the vertex does not exist, the corresponding degree is assigned a value of 0. When the vertex degree Deg is greater than 0, the vertex pointer points to the corresponding adjacent edge array. All target vertices in the adjacent edge array are arranged according to the size of ID. In order to ensure the efficient insertion and deletion of updated data, certain empty gaps are reserved for the edge array elements to avoid the elements moving backward as a whole to reduce update overhead. The PMA-based adjacency array is introduced in detail below.

The Size of the Reserved Space: According to the size of memory space and requirements of the graph processing task, we specify the total number of empty elements and then assign empty elements to each vertex according to the degree of each vertex. The calculation process is as Eq. 1:

$$Gap_i = \frac{Gap_i}{Deg_{sum}} * Deg_i \tag{1}$$

Among them, Gap_i is the number of empty gaps allocated to the i vertex adjacency array, Gap_{sum} is the sum of empty gaps, Deg_{sum} is the sum of the degrees of all vertices. If there are no special instructions, in this experiment, Gap_{sum} is set to Deg_{sum}, that is, the number of empty gaps in the adjacent array of each vertex is equal to the size of the vertex's out-degree.

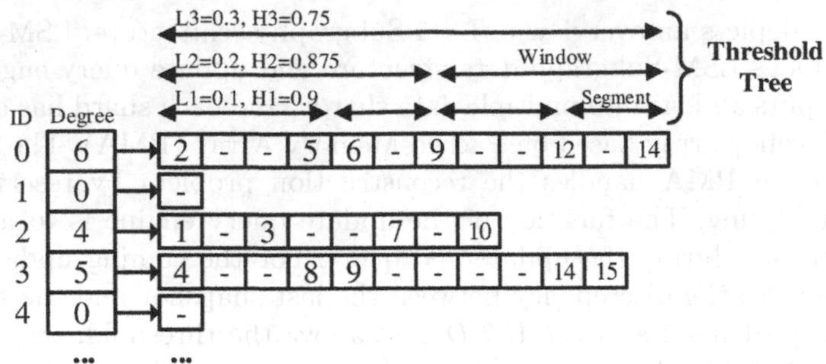

Fig. 4. Sample layout of the key snapshot

The Distribution of Empty Gaps: The traditional methods of determining the distribution of empty gaps TPMA and APMA do not consider the time-evolving feature of temporal graphs, and this paper proposes a new empty gaps distribution method, as shown in Fig. 4. The number of empty gaps between adjacent elements is determined by the difference between the vertex numbers and the total number of empty gaps, as shown in the Eq. 2:

$$Num_{gap} = \frac{Gap_i}{ID_{max} - ID_{min}} * (ID_2 - ID_1) \tag{2}$$

where Gap_i is the total number of empty gaps in the adjacent array of vertices i, ID_{max} and ID_{min} are the maximum and minimum non-empty elements of the existing element numbers in the edge array respectively. ID_1 and ID_2 are the numbers of adjacent non-empty elements before and after the target insertion position. The main reason for using this method is that when the updated log of the temporal graph merges into the initial snapshot, the state of an edge that has been updated several times in the period only saves the latest state. Therefore, taking into account the irregularity of the temporal graph update, LSM-Subgraph reserves appropriate storage space for all possible adjacent edges.

Rebalance: When the update logs apply to the snapshot, the local density in the adjacent array could be too high or too low, and the entire snapshot needs to rebalance. The local density of a specified area is defined in Eq. 3:

$$Density = \frac{Num_{element}}{Num_{element} + Num_{gap}} \tag{3}$$

Among them, $Num_{element}$ and Num_{gap} represent the number of non-empty elements and empty gaps in the local area, respectively. To better apply to the update of temporal graphs, LSM-Subgraph has improved the existing rebalancing operations. As shown in Fig. 4, LSM-Subgraph constructs a 3-level calibration tree for the adjacency array of each vertex. Each layer has a corresponding threshold. When the updated edge is applied to the snapshot and causes the local density to be too high or too low, the rebalancing operation will be triggered.

Two adjacent blocks will merge to meet the threshold requirement. The worst case is that too many insert operations cause the entire array can not meet the threshold requirements. The array needs to copy to new storage space.

When querying other snapshots at arbitrary time points, LSM-Subgraph first finds the latest key time point. After loading the key snapshot and corresponding logs, it displays the updates on the key snapshot using logs. This process of switching can be represented by Formula 4,

Algorithm 1. LSM-Subgraph construction

Input: Set of updated edges U
Output: Set of key snapshots S, Set of in-between logs L

```
for edge e in U do
   if snapshot_i.NotfindNode(e.src) then
      snapshot_i.addNode(e.src);
   end if
   snapshot_i.addEdge(e);
   log_i.addEdge(e);
   if computeDiscrepancy(snapshot_i,snapshot_{i-1})> β then
      S.addSnapshot(snapshot_i);
      L.addLog(log_i);
   else
      continue;
   end if
end for
return S, L
```

Algorithm 2. LSM-Subgraph query

Input: Set of key snapshots S, Set of in-between logs L, Target time point i
Output: $snapshot_i$

```
for timepoint t in L do
   if |t − i|.isMinimum() then
      break;
   end if
end for
if t < i then
   snapshot_t.addEdge(L, t, i);
else
   snapshot_t.deleteEdge(L, t, i);
end if
return snapshot_i
```

$$S_k + L_{ki} \rightarrow S_i, \tag{4}$$

where we use the key snapshot S_k and corresponding L_{ki} to generate the target snapshot S_i.

Algorithm 1 shows the process of snapshot construction. For each edge in the updated set, LSM-Subgraph first constructs a temporary snapshot. When discrepancy exceeds threshold, LSM-Subgraph stores a key snapshot with corresponding logs. Algorithm 2 presents the procedure of query. LSM-Subgraph first loads the latest key snapshot and corresponding logs. Then it merges them into a target snapshot.

3.3 Fluctuation-Aware Snapshot Creation Method

Besides the PMA-based adjacency array model, it is also important to decide when to create snapshots. There is a crucial tradeoff on selecting the interval between snapshots. If the interval is too large, the log will be too large to query effectively. On the contrary, there will be more redundancy between adjacent snapshots, which wastes space.

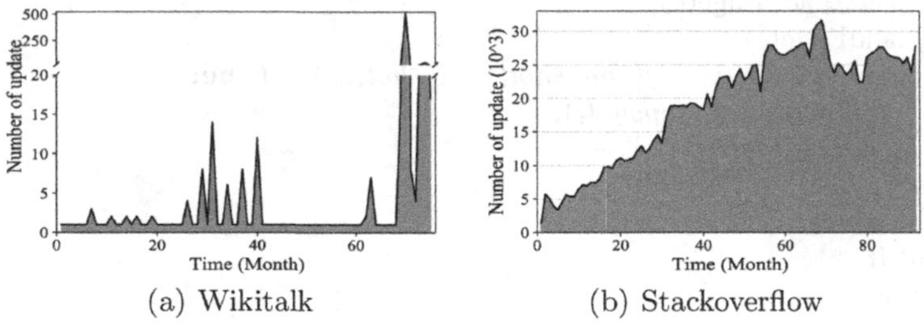

(a) Wikitalk (b) Stackoverflow

Fig. 5. The updated characteristic of temporal graphs

We have two alternative methods to solve this problem. One obvious method is to select key snapshots based on time period, for example, the next key snapshot is created every ten minutes or one hour. However, we have to consider a terrible situation where there are many update operations in the current interval, but few or no update operations in the next one. As shown in Fig. 5, every real-world temporal graph owns itself characteristic over time and this changing characteristic can not be concluded or predicted. The imbalance of distribution will increase the overhead of storage and query. The other is to set up key snapshots based on the size of logs, that's to say, the next snapshot is created when the size of logs reaches the threshold. This scheme can avoid the imbalance of the method based on period. But both insertions and deletions of the same nodes or edges may be in the same period. The next key snapshot would still be created even if the absolute size of logs is less than the threshold, where the absolute size is the number of edges when the system merges all additions and deletions of the same elements. It will result in structural redundancy between adjacent snapshots.

Inspired by FVF [15], we design a fluctuation-aware snapshot creation method to select key time points based on the above discussions. The degree of difference between adjacent snapshots can be defined as:

$$TD(K_1, K_2) = \frac{|E_G|}{|E_{K_1}| + |E_{K_2}|} \tag{5}$$

Among them, $|E_G|$ represents the number of all edges in graph G. $|E_{K_1}|$ and $|E_{K_2}|$ are the number of all edges in the two snapshots. When LSM-Subgraph stores updated data, it first maintains a temporary snapshot K_t in the system. This snapshot adopts the PMA-based adjacency array storage model and then calculates the temporary snapshot and the initial snapshot K_l in the latest data shard in LSM-Subgraph. If $TD(K_l, K_t) > \beta$, where β is a user-defined threshold, create a new data shard and save the temporary snapshot as the initial snapshot, otherwise continue to store the log and update the temporary snapshot.

In order to guarantee the query efficiency and space-saving, we have to find the optimal threshold β, for it affects the tradeoff between storage overhead and access time. We compare the performance over various temporal graphs with different parameters(present in Sect. 4.4). We analyze how the memory usage and access time change with the threshold β in real-world graph dataset. Let M_k represents the average memory overhead of data shards, N represents the number of data shards and M_l represents the memory overhead of the log array. The memory overhead of LSM-Subgraph is $Memory = N * M_k + M_l$. The log array in LSM-Subgraph stores all update edge data, so $M_l = O(E)$, for the same temporal dataset, M_l can be regarded as a constant. As the threshold β increases, N continues to decrease, M_k continues to increase, and finally $Memory$ continues to decrease(see Sect. 4.4 for detail), indicating that N has a much

Algorithm 3. Log Merge

Input: Set of edges list to be updated U
Output: Set of edges list to be updated U'

```
U.sort();
flag1=U.first();flag2=U.second();
while flag1<=U.end()&&flag2<=U.end() do
    if flag1 == flag2 then
        flag1.delection();
        flag1 = flag2;
    else if flag1 == -flag2 then
        flag1.delection;
        flag1 = flag2.next();
        flag2.delection;
    else
        flag1 = flag2;
    end if
    flag2 = flag1.next();
end while
return U'
```

greater impact on the memory overhead than M_k, so we can regard M_k as a constant, and $N \propto \frac{1}{\beta}$. As the threshold β increases, the memory overhead of LSM-Subgraph will continue to decrease. The relationship between $Memory$ and β can be defined as:

$$Memory = M_k * f(\frac{1}{\beta}) + M_l \tag{6}$$

For access time, it consists of the time T_k to find the corresponding data shard and the query time T_l within the data shard. The access time of LSM-Subgraph is $AccessTime = T_k + T_l$. As the PMA-based adjacency array is used to store the snapshot, the time to find the corresponding data shard is much less than average the query time within the data shard, so T_k can be regarded as a constant. As the threshold β increases, the memory overhead by a single snapshot increases. Therefore β and T_l are positively correlated. The relationship between $AccessTime$ and β can be defined as:

$$AccessTime = g(\beta) + T_k \tag{7}$$

According to the above analysis, we know access time and storage usage always intersect in one point. We find that setting the threshold at 0.03 is optimal in most cases(present in Sect. 4.4). Hence, we recommend the threshold β is 0.03. If not specified, we set the $\beta = 0.03$ in the following experiments.

3.4 Log-Merging Method

The query process needs to find the initial snapshot and log data according to the time point, and then merge to generate the snapshot data at the target

Table 1. Details of temporal graph datasets

Datasets	Nodes (M)	Edges (M)	Description
Youtube	1.13	2.99	static, Youtube users
Wikitalk	1.14	7.83	dynamic, Wikipedia users
Wiki-topcat	1.79	28.51	static, Wikipedia hyperlinks
Soc-pokec	1.63	30.62	static, Pokec connections
Stackoverflow	2.60	63.50	dynamic, Stackoverflow interactions
Livejournal	4.84	68.99	static, LiveJournal friendships
Wikipedia	2.1	66.9	static, Wikipedia links (fa)
Orkut	3.07	117.19	static, Orkut connections
Twitter	61.6	1470	static, Twitter connections

time. In the temporal graph update process in the real world, the same element (including vertices and edges) will be updated multiple times at different times. Therefore, we propose a log-merging method to further reduce the amount of log merged data. Assume that the update log within a period time is $Log = e_{1,1}, e_{2,2}, e_{3,3}, e_{3,4}, e_{3,5}, e_{4,6}, -e_{4,7}, ..., e_{n,t}$, where t is the update time, n is the number of the updated edge, e indicates the edge insertion operation, and $-e$ indicates the edge deletion operation. It can be seen that the edge e_3 is continuously and repeatedly inserted at the time t = 3, 4, 5, and the edge e_4 is inserted and deleted at the time t = 6, 7. Algorithm 3 presents the Log-Merging method. For the same update operation continuously, the snapshot only records the latest update status, that is, the update at time t = 5. For reciprocal update operations, the final update operation performed is 0. So the above update log can be expressed as $Log = e_{1,1}, e_{2,2}, e_{3,5}, ..., e_{n,t}$.

4 Evaluation

4.1 Experiment Setup and Datasets

We evaluate LSM-Subgraph on a commodity multi-core machine. It is equipped with a 16-core 1.60 GHz Intel(R) Xeon(R) CPU E5-2603, 128 GB of memory. The program is compiled with g++ version 11.0.

We use nine real-world graph datasets of different scales from SNAP [11], Twitter and Konect. The details of graphs describe in Table 1. Wikitalk and Stackoverflow are temporal graph datasets, and their periods are 2320 days and 2774 days, respectively. Stackoverflow is the largest temporal network that we can find on SNAP. To store all snapshots of copy-based models in the limited memory, we split the dataset evenly into updates of 60's time points based on days. In terms of static graph datasets, we generate random updates by adjusting the parameters including update rate, add rate. Update rate denotes the proportion of updated edges in the graph. Add rate indicates the proportion of additional edges in the update. The default update rate and add rate are 0.001 and 0.9, respectively.

We use the following algorithm as a benchmark to test the performance of the system: **2-hop:** 2-hop neighbour query, **BFS:** Breadth first search, **CC:** Find Connected Components, **BC:** Compute betweenness centrality of vertices, **PR:** Pageranks of vertices.

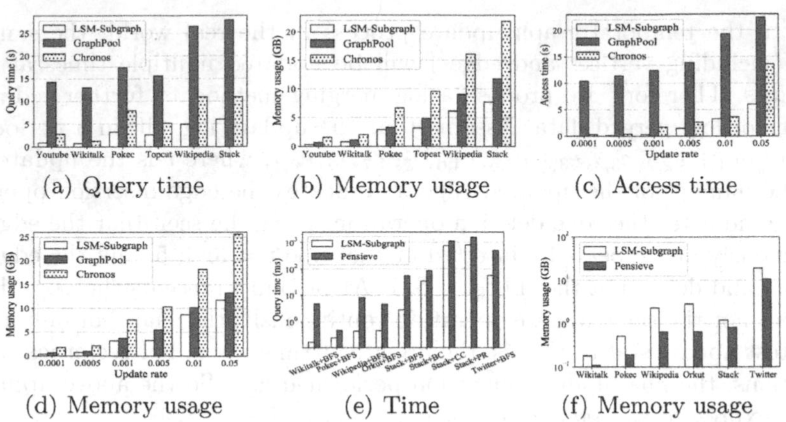

(a) Query time (b) Memory usage (c) Access time

(d) Memory usage (e) Time (f) Memory usage

Fig. 6. Overall performance comparison

4.2 Overall Performance Comparison

Compare with Traditional Models: We select temporal graph systems Chronos [3] and GraphPool [8] to compare against LSM-Subgraph. In terms of basic graph storage formats, Chorons is a copy-based model and GraphPool is a log-based model. We select the 2-hop neighbour query to examine the query performance, where the query time includes both snapshot construction time and 2-hop neighbour compute time.

First, we test datasets of different sizes of temporal graphs in the same update scenario, and the results show in Fig. 6(a) and 6(b). Compared with GraphPool, the query efficiency of LSM-Subgraph is improved by an average of 86%, and the memory overhead can be reduced by 9% to 57%. That is because GraphPool needs to build a dedicated snapshot for each designated moment when querying, and a bitmap is created for each update edge, which increases storage costs. The query efficiency of LSM-Subgraph is 53% higher than that of Chronos on average, and the memory overhead is much smaller than it. That is because Chronos needs to traverse all the bitmaps of the edge when querying and it stores data in the form of snapshots at every moment, resulting in spatial redundancy.

Second, for different update scenarios, we test the memory overhead and query time on the same dataset Wikitalk. The results are shown in Fig. 6(c) and 6(d). Compared with GraphPool, the query time cost of LSM-Subgraph is reduced by 90% on average. The main reason is that the PMA-based adjacency array model of LSM-Subgraph reserves a certain storage location. A small number of updates can be done directly using these empty locations, avoiding Data structure changes and even reconstruction. At the same time, memory overhead has been reduced by an average of 26%. Compared with Chronos, LSM-Subgraph's query time has been reduced by an average of 58%. Memory overhead has been reduced by 64% on average. The main reason is that LSM-Subgraph divides all logs into several data shards, and only one initial snapshot is kept in each data shard, thereby reducing memory overhead.

Compare with Hybrid Model: We compare the performance of LSM-Subgraph with Pensieve using six real-world datasets. Figure 6(e) compares the average querying time of different systems with different datasets and graph algorithms. The result shows that the average querying time of Pensieve is much larger than LSM-Subgraph, up to 12.5×. The reason is that the Pensieve'snapshot needs to be reconstructed for each query. The further away from the moment of the root node, the longer it will take to rebuild the snapshot. The query time of Pensieve at the time point far away from the root node is several times that of the near root node. The average memory overhead of LSM-Subgraph is 3.2× of Pensieve because it uses a fluctuation-aware method to store multiple snapshots. Moreover, the increase in memory is within a tolerable range and is far less than the cost of query reduction.

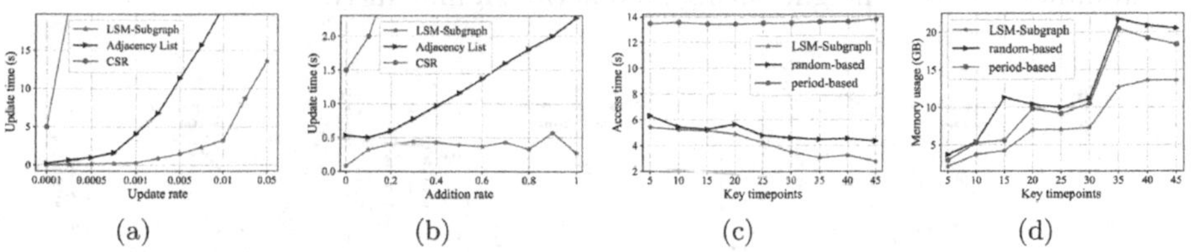

Fig. 7. Performance comparison within LSM-Subgraph

4.3 Comparison Within LSM-Subgraph

We examine the efficiency of each strategy in LSM-Subgraph, including PMA-based adjacency array model and fluctuation-aware snapshot creation method. As for PMA-based snapshot creation model, we examine the update efficiency of LSM-Subgraph against traditional graph storage formats. In terms of the fluctuation-aware snapshot creation method, we evaluate fluctuation-based against period-based and random-based. In this part, all experiments are conducted on temporal dataset Stackoverflow. When update rate varies, all addition rate is set to 0.9, and all update rate is fixed to 0.001 while addition rate changes.

Figure 7(a) shows update performance under different update rates. We can find that the update time of three models increases with the growth of the update rate. When update rate is more than 0.0001, CSR takes unbearable time to rebuild graphs. And when update rate is over 0.0075, the reconstruction time of adjacency list is intolerable. When the number of updated edges is not rather large, LSM-Subgraph uses slots to complete these insertions or deletions. As the update rate grows, slots in segment may be not suitable for updates, incurring some rebalancing operations and increasing the update time.

We further evaluate the influence of addition rate, which represents the skewed distribution of insertions and deletions in updated edges. Figure 7(b) compares the update time on different addition rates. CSR and Adjacency list are sensitive to the number of additional edges. Particularly, we can not compute the update time of CSR when addition rate is more than 0.1. The update time of adjacency list grows linearly as the addition rate increases. In contrast, LSM-Subgraph stays stable whatever the size of additions is.

Then we analyze the performance of key snapshot choice. Figure 7(c) depicts the access time of various key timepoints, where different timepoints mean the number of key snapshots in system. We can find that it costs the least time for LSM-Subgraph to access the graph at arbitrary time point against other strategies. And the period-based method costs up to 3× time of LSM-Subgraph and the number of key timepoints has little influence on it.

In Fig. 7(d), we compare the memory usage of different strategies. The memory usage of them increases as the number of timepoints grows, because system needs more space to store key snapshots and logs. Whatever the number of timepoints is, the storage overhead of LSM-Subgraph is less than both random-based strategy and period-based strategy. This is because LSM-Subgraph reduces the redundancy between neighbouring snapshots significantly.

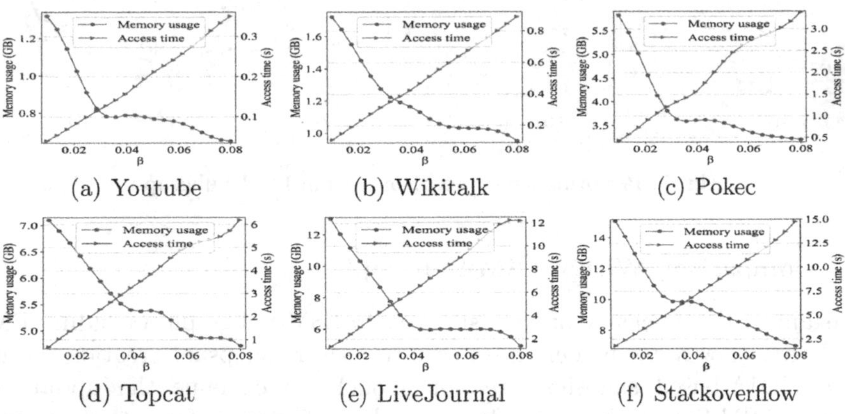

(a) Youtube (b) Wikitalk (c) Pokec

(d) Topcat (e) LiveJournal (f) Stackoverflow

Fig. 8. Performance of different parameters over various graphs

4.4 System Design Parameters

We further verify and analyze the rationality of the selection of the difference threshold during the update process. Figure 8 shows the comparison of memory overhead and query time cost under different threshold parameters when LSM-Subgraph processes different real graph datasets. It can be observed from the figure that when β is approximately 0.03, LSM-Subgraph can achieve a good tradeoff between storage and query. Therefore, we set the fluctuation threshold as 0.03 in our experiments.

5 Conclusion

In this work, we observe that the skewness of vertex degree in temporal graphs is changing over time. Based on our findings, we propose LSM-Subgraph, a temporal graph storage model based on log-structured merge-subgraph. LSM-Subgraph uses a PMA-based adjacency array model to store vertex data and leverages a Fluctuation-Aware snapshot creation method to create snapshots at key time points. Through experiments, we determined the threshold β, under which a comprehensive experiment was carried out to evaluate the performance of the design. Results show that LSM-Subgraph supports faster queries with lower storage overhead than existing systems.

References

1. Bender, M.A., Hu, H.: An adaptive packed-memory array. In: Proceedings of the Twenty-Fifth ACM SIGMOD-SIGACT-SIGART Symposium on Principles of Database Systems, pp. 20–29. PODS (2006)
2. De Leo, D., Boncz, P.: Packed memory arrays - rewired. In: 2019 IEEE 35th International Conference on Data Engineering, pp. 830–841 (2019)
3. Han, W., et al.: Chronos: a graph engine for temporal graph analysis. In: Proceedings of the Ninth European Conference on Computer Systems. EuroSys (2014)
4. Haubenschild, M., Then, M., Hong, S., Chafi, H.: Asgraph: a mutable multi-versioned graph container with high analytical performance. In: Proceedings of the Fourth International Workshop on Graph Data Management Experiences and Systems, pp. 1–6 (2016)
5. Holme, P., Saramäki, J.: Temporal networks. Phys. Rep. **519**(3), 97–125 (2012)
6. Itai, A., Konheim, A.G., Rodeh, M.: A sparse table implementation of priority queues. In: Proceedings of the 8th Colloquium on Automata, Languages and Programming, pp. 417–431 (1981)
7. Ju, X., Williams, D., Jamjoom, H., Shin, K.G.: Version traveler: fast and memory-efficient version switching in graph processing systems. In: 2016 {USENIX} Annual Technical Conference ({USENIX}{ATC} 2016), pp. 523–536 (2016)
8. Khurana, U., Deshpande, A.: Efficient snapshot retrieval over historical graph data (2013)
9. Kumar, P., Huang, H.H.: Graphone: a data store for real-time analytics on evolving graphs. ACM Trans. Storage (2020)
10. Kyrola, A., Blelloch, G., Guestrin, C.: Graphchi: large-scale graph computation on just a PC. In: 10th USENIX Symposium on Operating Systems Design and Implementation (OSDI 2012), pp. 31–46, October 2012
11. Leskovec, J., Krevl, A.: SNAP datasets: Stanford large network dataset collection, June 2014
12. Macko, P., Marathe, V.J., Margo, D.W., Seltzer, M.I.: Llama: efficient graph analytics using large multiversioned arrays. In: 2015 IEEE 31st International Conference on Data Engineering, pp. 363–374 (2015)
13. Mariappan, M., Vora, K.: Graphbolt: dependency-driven synchronous processing of streaming graphs. In: Proceedings of the Fourteenth EuroSys Conference 2019. EuroSys (2019)
14. Nilakant, K., Dalibard, V., Roy, A., Yoneki, E.: Prefedge: SSD prefetcher for large-scale graph traversal. In: Proceedings of International Conference on Systems and Storage, pp. 1–12. SYSTOR (2014)
15. Ren, C., Lo, E., Kao, B., Zhu, X., Cheng, R.: On querying historical evolving graph sequences. VLDB, 726–737 (2011)

16. Shun, J., Blelloch, G.E.: Ligra: a lightweight graph processing framework for shared memory. In: Proceedings of the 18th ACM SIGPLAN Symposium on Principles and Practice of Parallel Programming, pp. 135–146 (2013)
17. Sikos, L.F., Philp, D.: Provenance-aware knowledge representation: a survey of data models and contextualized knowledge graphs. Data Sci. Eng. **5**(3), 293–316 (2020)
18. Then, M., Kersten, T., Günnemann, S., Kemper, A., Neumann, T.: Automatic algorithm transformation for efficient multi-snapshot analytics on temporal graphs. VLDB, 877–888 (2017)
19. Toss, J., Pahins, C.A.L., Raffin, B., Comba, J.L.D.: Packed-memory quadtree: a cache-oblivious data structure for visual exploration of streaming spatiotemporal big data. Comput. Graph. **76**(NOV.), 117–128 (2018)
20. Wu, H., Zhao, Y., Cheng, J., Yan, D.: Efficient processing of growing temporal graphs. In: DASFAA, pp. 387–403 (2017)
21. Yang, J., Yao, W., Zhang, W.: Keyword search on large graphs: a survey. Data Sci. Eng. **6**(2), 142–162 (2021)
22. Ying, T., Chen, H., Jin, H.: Pensieve: skewness-aware version switching for efficient graph processing. In: Proceedings of the 2020 ACM SIGMOD International Conference on Management of Data, pp. 699–713 (2020)
23. Zuckerberg, M.: Facebook (2004). http://www.facebook.com

ForGen: Autoregressive Generation of Sparse Graphs with Preferential Forest

Yao Shi(✉), Yu Liu, and Lei Zou

Peking University, Beijing, China
{shall,dokiliu,zoulei}@pku.edu.cn

Abstract. Graph is a natural way to model interactions between objects, such as in the field of biology and social science. Over the past decades, modeling and generating graphs have been a popular research topic, largely inspired by observed properties of real-world graphs. Since traditional approaches relying on hand-crafted mechanisms are only capable of capturing some specific graph properties, recent focus has been shifted to deep neural methods. However, the task is still challenging in terms of efficiency. To address this issue, we observe that the connectivity (i.e., degree) of nodes follows scale-free distribution for most real-world graphs, which can be utilized to accelerate the generation process. We propose ForGen, a Forest-based Generation model that contains a graph-level and an edge-level autoregressive generator. Specifically, for the edge-level model, motivated by the skewed distribution of node degree and the Huffman tree, we design a forest-like data structure to accelerate edge connection via shallow tree searches and better parallelism. Experiments on both synthetic and real-world graph datasets show that ForGen is two times faster than the current state-of-the-art method for graph generation, and guarantees better generated graph quality.

Keywords: Graph generation · Autoregressive model · Neural network

1 Introduction

Graphs are ubiquitous in the real world, which can represent relational information such as social networks, traffic networks, knowledge bases and molecule structures. To tap the underlying value of complicated structures in these graphs [2], there are mainly two types of approaches [9]: 1) predicting and analyzing patterns on given graphs. 2) modeling and generating graphs, while we focus on the second one.

Since the early work by Erdos and Rényi in 1960 [5], many graph generative models have been proposed due to its wide range of applications which includes mining patterns in social networks [1,8,14], drug design [16,20,29], knowledge graph completion [26], and architecture search [27]. Traditional generative models for graphs (e.g. Barabási-Albert model [1], Kronecker graphs [14], small-world networks [25], and stochastic block models [21]) are based on prior assumptions

B. Li et al. (Eds.): APWeb-WAIM 2022, LNCS 13421, pp. 495–510, 2023.
https://doi.org/10.1007/978-3-031-25158-0_40

and random graph theory, which usually have clear mathematical properties and computational efficiency but lack the ability to model complex dependencies due to their hand-crafted nature. Also, prior assumptions limit the generalization ability and versatility of the model.

To tackle these limitations, recent studies have turned to deep generative models using neural networks. Instead of making prior assumptions, deep graph generators directly learn the distribution from given data to capture latent structural information. Despite the great progress achieved in the molecular area (graph size is small), modeling graphs in the general domain has specific challenges. As a graph with n nodes has an n^2 sized adjacency matrix, most non-autoregressive models [20,29] trying to generate the full matrix fail to scale up to graphs with hundreds of nodes. Moreover, methods like autoencoder-based graph generators [20] treat a graph as a combination of independent components or structures with weak dependency to speed up the generation process. However, this assumption ignores the complex structure dependencies in real-world graphs which can sacrifice the quality of the generated graphs [17].

For the reasons mentioned above, autoregressive models have gained special attention for modeling general real-world graphs (not just molecules) [4,7,17,30], whose basic idea is to sequentially add nodes and edges or small motifs into the graph. Thus new decisions are made based on previous information and complex dependencies can be introduced into the model. Generally, there are two types of methods according to the generation framework they choose. The first way is to use hierarchical architectures, for instance, GraphRNN [30] adopts a graph level RNN combined with an edge level RNN. Another type of approaches leverages massage passing mechanism to bring more neighborhood information into the process [17]. However, the lack of features (semantic information) in the problem setting may hinder GNN's performance and the scalability of GNN and simple sequential models can also be a bottleneck of efficiency.

So far, BiGG [4] is the most scalable autoregressive graph generator and can achieve state-of-the-art sample quality in both synthetic and real-world datasets. Based on the observation that most real graphs are sparse, BiGG attempts to generate the non-zero entries in the adjacency matrix instead of the whole matrix. As GraphRNN does, BiGG adopts a two-level (node and edge) architecture but it reduces the time complexity from $O(n^2)$ to $O((n+m)\log n)$ for a graph with n nodes and m edges. More specifically, BiGG uses a binary tree data structure as the edge level to reduce the time complexity of generating an edge from $O(n)$ to $O(\log n)$, which is the key to efficiency improvement.

It is well studied that the vertex connectivities commonly follow a scale-free distribution in complex real networks [3,19,23]. However, we observed that existing studies have not paid close attention to this characteristic in both experimental setup and model design. As a matter of fact, the degree distributions of graph datasets used in previous work (such as grid graphs and protein graphs [4,17]) are more close to the normal distribution rather than the scale-free distribution, which does not match most real-world application scenarios. Through experiments, we found that it is harder for deep models to learn from scale-free networks than other graph datasets under the same settings, which indicates their complex latent patterns.

When it comes to model design, BiGG treats all the nodes in the same way during the edge generation process. Nevertheless, it is well-known that node degree of numerous real-world graphs (approximately) follows heavy tailed (e.g., scale-free) distribution, and exhibits interesting phenomenons such as "rich-get-richer". This indicates that one should treats nodes *differently* in generating graphs. To make a further improvement on the two-level autoregressive framework, we propose ForGen (Forest-based Generation Model) to model these properties explicitly. Instead of a balanced tree, ForGen conducts the edge level as a hierarchical forest-like data structure inspired by the Huffman tree [10] and the "rich-get-richer" phenomenon, which enables shallow tree search. Also, we improve the traversal-based learning algorithm used in the edge level for better scalability and quality. Thereby, we can generate an edge in $O(1)$ time for the best case which has a higher frequency, and $O(\log n)$ time for the worst case.

To summarize, we have made the following contributions:

- To the best of our knowledge, it is the first time that the skewed distribution of node degree in real-world graphs has been considered in graph generation with deep neural network.
- We propose a two-level deep autoregressive model ForGen with an imbalanced forest-like structure for graph generation, which exploits the "rich-get-richer" phenomenon in scale-free networks.
- Empirical evaluation on both scale-free and non-scale-free networks establishes that compared with the state-of-the-art technique, ForGen can achieve better quality on four datasets and a twice faster sample speed.

2 Related Work

In this section, we will briefly introduce the existing methods in three parts. Table 1 summarizes several notable deep graph generators and their characteristics.

Traditional Graph Models. In the Erdos Rényi random graph model [5], each edge is included with a fixed probability p independently and the degree distribution is proved to be Poisson for large n and $p = c/n$, where c is a constant. However, Barabási et al. [3] find that the degree distribution is commonly a scale-free distribution for realistic networks and propose the Barabási-Albert (BA) model based on the preferential attachment assumption. Though there is various work following the preferential attachment model [1,15,28], they only focus on the degree distribution while neglecting other structural features.

Autoregressive Deep Graph Generators. Li et al. [16] propose a node-by-node method that includes two variants, namely MolMP and MolRNN, to model the probabilities for adding nodes and termination. Similarly, You et al. [30] propose GraphRNN based on two RNNs (one for graph level and one for

Table 1. Existing deep graph generators. n and m denote the number of nodes and edges respectively. M is calculated from the data.

Method	Technology roadmap	Complexity
GraphVAE [20]	Auto-encoder (GNN + MLP)	$O(n^4)$
GCPN [29]	Reinforcement Learning	$O(m^2)$
Graphite [8]	Auto-encoder + GNN	$O(n^2)$
GraphRNN [30]	Autoregressive-based + Two level	$O(nM)$
GRAN [17]	Autoregressive-based + GNN	$O((m+n)n)$
BiGG [4]	Autoregressive-based + Two level	$O((m+n)\log n)$
ForGen (ours)	Autoregressive-based + Two level	$O((m+n)\log n)$

edge level). Subsequently, Liu et al. [18] combine GraphRNN with a vanilla Transformer decoder. Liao et al. [17] propose a GNN-based approach that uses graph neural network with attentive massages to generate one block of nodes and associated edges at each step. Apart from node-based methods, Goyal et al. [7] adopt LSTM to generate edge sequences which are converted from a graph using the minimum DFS code. Recently, Dai et al. [4] propose BiGG, a two-level model with tree-based architecture which achieves the best performance in both quality and scalability.

Non-autoregressive Deep Graph Generators. There exist various work that focuses on generating graphs in a non-autoregressive way. For example, GraphVAE [20] constructs its encoder and decoder based on GCN and MLP respectively, which needs a costly graph matching algorithm to train. Graphite [8] parameterizes variational autoencoders with a GNN, and uses an iterative graph refinement strategy inspired by low-rank approximations for decoding, which can only learn from one input graph. You et al. [29] formulate the molecular graph generation as a Markov Decision Process and adopt PPO to train an RL agent. However, these non-autoregressive graph generators are mainly designed for molecular generation which may suffer from the scalability issue and lack the ability to model complex dependencies.

3 Methodology

3.1 Preliminaries

Graph Generation. A graph is defined by the tuple $G = (V, E)$, where $V = \{v_1, ..., v_n\}$ is the graph's node set and $E = \{(v_i, v_j)|v_i, v_j \in V\}$ denotes its edge set, with $|V| = n$ and $|E| = m$. We represent a graph with an adjacency matrix $A \in \{0,1\}^{n \times n}$, while a graph has up to $n!$ different matrix representations with an ordering set Π over the nodes. More precisely, given an ordering $\pi \in \Pi$, $(\pi(v_1), ..., \pi(v_n))$ is a permutation of $(v_1, ..., v_n)$ and a graph is mapped to a matrix A^{π}. Here, we focus on undirected graphs with no self-loops or attributes.

Our goal is to learn the distribution $p_{model}(\mathbb{G})$ from a given graph set $\mathbb{G} = \{G_1, ..., G_N\}$. Assuming that the observed graphs $\mathbb{G}$ are sampled from an underlying data distribution $p_{data}(\mathbb{G})$, an effective model should be capable of sampling graphs from the learned distribution which are similar to the real ones, i.e., $p_{model}(\mathbb{G}) \approx p_{data}(\mathbb{G})$. Therefore, we maximize the log-likelihood of observed graphs. Following GRAN [17], we use a single canonical ordering $\pi(G)$ to get a valid lower bound of $\log p(G)$ and significantly improve training efficiency:

$$\log p(G) = \sum_{\pi \in \Pi} \log p(A^{\pi}) \simeq \sum_{\pi \in \Theta \subset \Pi} \log p(A^{\pi}) \simeq \log p(A^{\pi(G)}) \tag{1}$$

Table 2. Main notations used in the paper.

Notation	Description
G, V, E	A graph, its node set, its edge set
n, m	The number of nodes and edges respectively
π	A node ordering for a graph G
A^{π}	The adjacency matrix under π
S_t^{π}	The t-th row of A^{π}'s lower triangle part
$L(t)$	The number of trees for node v_t
b	The depth of the first tree
x	A learnable vector $\in \mathbf{R}^d$
g, h	A tuple of embedding vectors $\in \mathbf{R}^d$

Important Characteristics of Real-World Graphs. The heavy-tailed degree distribution in real networks is first found by [19], i.e., the number of nodes of degree k being in inverse proportion to k^{γ} ($\gamma \in (1, \infty)$). The class of networks exhibiting a scale-free degree distribution (at least asymptotically) is called "scale-free network". Moreover, in real-world graphs, there is a higher probability that more and more nodes will link themselves to those with large degrees. In another word, some nodes (called hubs) can have much more connections than others and the existence of this mechanism (denoted as "Preferential Attachment" [3]) has been proved in many real-world graphs [11,13].

3.2 Overview

As shown in Fig 1, ForGen is a two-level autoregressive model based on tree structures. The iterative generation process consists of a top-down decoding stage and a bottom-up encoding stage. The decoding stage starts from the graph level which provides the initial embedding for a new node. To leverage the skewed degree distribution of real graphs, inspired by the preferential attachment, an imbalanced forest-like structure is used as the edge level. After the non-zero entries of A^{π} are decoded from the embedding, new edges are generated. Then the bottom-up stage encodes the edges into a vector within the edge level, which is utilized to update the graph level information. As shown in Table 2, we use $\boldsymbol{h}$ to represent the embeddings in the decoding stage and $\boldsymbol{g}$ in the encoding stage.

Formally, We represent the t-th row of A^π's lower triangle part as S_t^π and the graph level can be factorized in an autoregressive way as

$$p(A^\pi) = \prod_{t=1}^{n} p(S_t^\pi | S_1^\pi, ..., S_{t-1}^\pi) \tag{2}$$

For edge level, instead of generating edges sequentially [4,30] or all at once [17], we divide S_t^π into $L(t)$ groups $S_{t,j}^\pi$ and organize them in a forest-like way:

$$p(S_t^\pi | S_1^\pi, ..., S_{t-1}^\pi) = \prod_{j=1}^{L(t)} p(S_{t,j}^\pi | S_{t,1}^\pi, ..., S_{t,j-1}^\pi, S_1^\pi, ..., S_{t-1}^\pi) \tag{3}$$

Based on this assumption, we generate the non-zero entries inside each group which can introduce more dependencies between groups into the model and take advantage of node diversity in real-world graphs to improve efficiency.

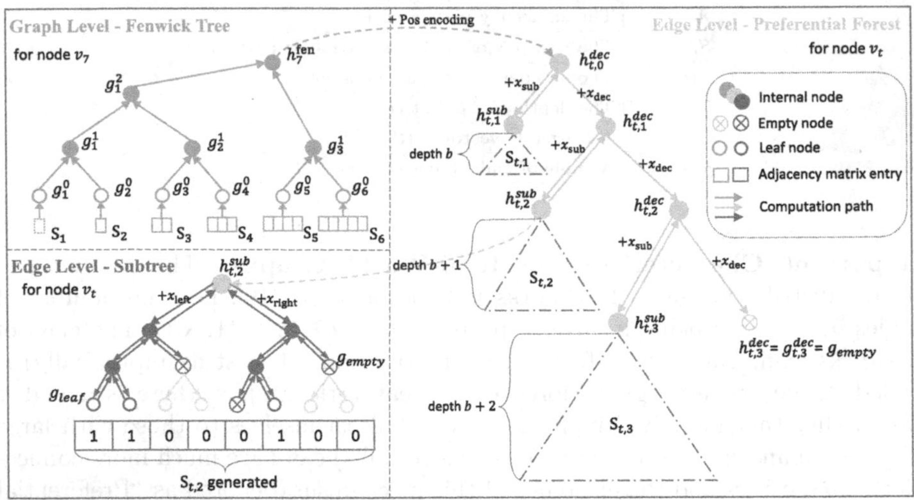

Fig. 1. An example of ForGen's generation process. The generation starts from the graph level computation (block on the top left) and goes to edge level (right part) following the dotted line. Edge level's Preferential forest is modeled in an inorder traversal way, while the subtrees of the forest (the blue part) are built from top to bottom and summarize the generated edges from leaves to the root. (Color figure online)

3.3 A Hierarchical Tree Structure for Edge Generation

The edge level of ForGen is responsible for generating all the edges between node v_t and nodes already generated at timestamp t in an autoregressive way, which is equivalent to finding v_t's neighbors $\mathcal{N}(v_t)$ from a subset of nodes $\{v_1, ..., v_{t-1}\}$. Compared with the graph level, this stage is more critical to improving model efficiency. In this section, we will describe ForGen's edge generation strategy.

Generating with a Standard Binary Tree. BiGG [4] modeled the edge level as an edge binary tree. The binary tree is balanced for the reason that the maximum height difference between the left and right subtrees of each node is one. As illustrated in Fig 2(a), each leaf node represents a candidate for v_t's neighbors and the selection process is decomposed into a series of left/right decisions until reaching the bottom. However, in such a model structure, all the nodes are treated equally which contradicts the nature of real networks.

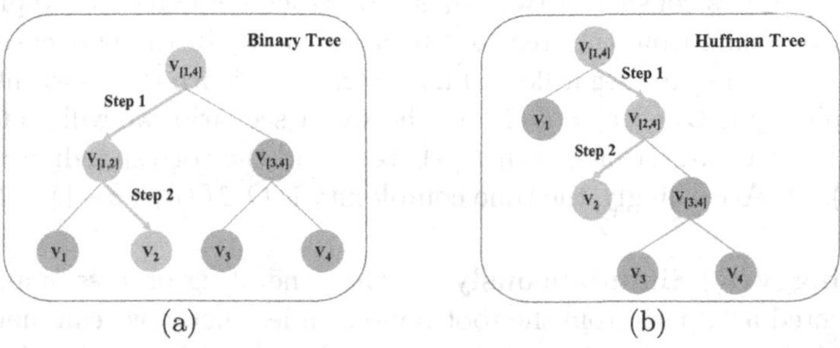

Fig. 2. Simplified edge generation process for node v_5.

Edge Generation Considering Preferential Attachment. Instead of a balanced binary tree, a tree structure that takes nodes' access frequencies into account, will provide a better fit to the scale-free nature of real-world networks. Inspired by the Huffman tree [10], a frequency-sorted binary tree, we organized the tree in almost a sequential way. As Fig. 2(b) shows, all the right children are leaf nodes and represent the candidates. Thereby, the distance between hubs (nodes with large degrees) and the root will be $O(1)$ instead of $O(\log_2 n)$. Since hubs will be chosen frequently, it can significantly improve efficiency. We call it Preferential Tree since nodes closer to the root are preferentially treated.

However, there are many nodes far from the root. Though they may have much lower visiting frequency than the hubs, the structure can be optimized to avoid the $O(n)$ tree depth. Here, we introduce Preferential Forest, which replaces the leaf nodes with binary trees and all the trees make up a forest.

More specifically, the whole Preferential Forest built for node v_t represents the corresponding row S_t^{π}, and each tree in the forest (denoted as $S_{t,j}^{\pi}$) further separates S_t^{π} into different parts. Noted that nodes with higher degrees are fewer and there exists communities of distinct sizes in real graphs, we should keep tree sizes different. We define the j-th tree as the one whose root node is in the j-th level of the whole preferential forest, and its depth is $b + j - 1$, where b is a hyperparameter and $j \in \{1, 2, ..., L(t)\}$. Therefore, the size of the j-th group $|S_{t,j}^{\pi}| = 2^{b+j-1}$, which means the number of hub nodes are small and there exists many nodes with small degrees.

Lemma 1. *Considering the generation of node t's neighbors, the number of trees in the preferential forest* $L(t) = \lceil \log_2 (t + 2^b - 1) - b \rceil = O(\log n)$.

Proof. The number of nodes in $L(t)$ trees is $\sum_{j=1}^{L(t)} 2^{b+j-1} = 2^{b+L(t)} - 2^b$. Since the last tree may be filled with some virtual nodes, we have $\sum_{j=1}^{L(t)-1} 2^{b+j-1} < t - 1 \leq \sum_{j=1}^{L(t)-1} 2^{b+j-1}$. Therefore, $L(t) = \lceil \log_2 (t + 2^b - 1) - b \rceil = O(\log n)$.

The groups are generated sequentially based on Eq. 3 and the tree of each group is generated recursively until reaching the leaf node.

Theorem 1. *In a graph with n nodes, the time complexity of generating one edge with ForGen is $O(1)$ in the best case and $O(\log n)$ in the worst case.*

Proof. The process consists of two parts, i.e., reaching a tree in the preferential forest and traveling from the tree root to a leaf node. In the best case, the first stage costs only one step regardless of n and the tree depth is a constant b. So the time complexity is $O(1+b) = O(1)$. In the worst scenario, we will go to the last tree, which is the largest one, with $L(n)$ steps and the corresponding tree depth is $b+L(n)-1$. Accordingly, the time complexity is $O(2L(n)+b-1) = O(\log_2 n)$.

Generating $\mathcal{N}(v_t)$ Simultaneously. As the generating process of an edge can be represented as a path from the root node to a leaf node, we can merge all the paths belonging to v_t to utilize their overlapping parts for overhead reduction. In other words, we convert the generation of $\mathcal{N}(v_t)$ into pruning v_t's preferential forest so that only the path to its neighbors' leaves are retained.

3.4 Two-Level Autoregressive Model

In this section, we will present the full version of ForGen in detail (see Fig. 1). First, we will discuss how to embed a new node v_t at graph level. Then we will decode its embedding into $\mathcal{N}(v_t)$ via massage passing on the preferential forest.

Generating Initial Embedding h_t for v_t. Following BiGG, we adopt the Fenwick tree [6], a data structure that maintains the prefix sum efficiently, as the graph level. The i-th leaf node of the tree $\boldsymbol{g}_i^0$ is the embedding of S_i generated before. With a Binary TreeLSTM cell [22], the tree is built from bottom to top and the i-th node in the j-th level

$$\boldsymbol{g}_i^j = \text{TreeLSTM}^{\text{fen}}(\boldsymbol{g}_{i*2-1}^{j-1}, \boldsymbol{g}_{i*2}^{j-1}) \tag{4}$$

Thus, we can obtain v_t's initial embedding $\boldsymbol{h}_t$ by computing the prefix sum of $\{\boldsymbol{g}_1^0, ..., \boldsymbol{g}_{t-1}^0\}$ with LSTM (go to the top left part of Fig. 1 for an example):

$$\boldsymbol{h}_t^{fen} = \text{LSTM}^{\text{fen}}([\boldsymbol{g}^k_{\lfloor \frac{t-1}{2^k} \rfloor}, \text{where } (t-1)\,\&\,2^k = 2^k]) \tag{5}$$

Note that & is bitwise AND, the sequence length will be $O(\log n)$ and so is the time complexity of a graph level update.

Generating Preferential Forest. For this part, we need to generate edges between v_t and $\{v_1, ..., v_{t-1}\}$ based on the prefix sum result of the Fenwick tree and then encode the edges into a new embedding $\boldsymbol{g}_t^0$ to update the graph level.

As previously described in 3.3, the preferential forest consists of several binary subtrees which are always the left children (yellow part in Fig. 1), and the right children maintain the information of the generated graph. The root node $\boldsymbol{h}_{t,0}^{dec}$ is the sum of $\boldsymbol{h}_t^{fen}$ and a positional encoding which is the same as [24] to identify different steps and improve generation quality:

$$\boldsymbol{h}_{t,0}^{dec} = \begin{cases} \boldsymbol{h}_t^{fen} + \sin((\mathrm{n}-\mathrm{t})/10000^{\mathrm{i}/\mathrm{d}}), i \bmod 2 = 0 \\ \boldsymbol{h}_t^{fen} + \cos((\mathrm{n}-\mathrm{t})/10000^{(\mathrm{i}-1)/\mathrm{d}}), i \bmod 2 = 1 \end{cases} \tag{6}$$

where the dimension $i \in \{0, 1, ..., d-1\}$.

Then we decode the information by inorder traversal using TreeLSTM and LSTM. First, we need to obtain the root embedding for the subtree in the next level and predict if it has children via a Bernoulli distribution:

$$\boldsymbol{h}_{t,i}^{sub} = \mathrm{LSTMcell}^{\mathrm{toSub}}(\boldsymbol{h}_{t,i-1}^{dec}, \boldsymbol{x}_{sub}) \tag{7}$$

$$p(has\_subtree|\boldsymbol{h}_{t,i}^{sub}) = \sigma(\mathrm{MLP}^{\mathrm{sub}}(\boldsymbol{h}_{t,i}^{sub})) \tag{8}$$

where $\boldsymbol{x}_{sub}$ is a learnable vector to integrate direction information and σ is the sigmoid function.

Based on the $\boldsymbol{h}_{t,i}$, we need to generate the binary subtree before going back to the parent node. However, instead of a traversal way, we build the tree with a bisection method to significantly reduce the sequence length. For the reason that the forest can introduce more complex interactions between node groups, such a partial autoregressive way will not influence the effectiveness but can improve the LSTM-based model's scalability. The process can be represented as follows:

$$p(has\_pos|\boldsymbol{h}_u^{sub}) = \sigma(\mathrm{MLP}^{\mathrm{pos}}(\boldsymbol{h}_u^{sub})) \tag{9}$$

$$\boldsymbol{h}_{\tau(u,pos)}^{sub} = \mathrm{LSTMcell}^{\mathrm{sub}}(\boldsymbol{h}_u^{sub}, \boldsymbol{x}_{pos}) \tag{10}$$

where $pos \in \{left, right\}$ and $\tau(u, pos)$ means the pos child of u in the subtree. We will decide whether a child is an empty node based on the probability This process will continue recursively until no non-empty children or reaching the leaf node where we will add an edge between v_t and the corresponding node.

After all leaves of the subtree are generated, we need to summarize the neighborhood information from bottom to top. Thus, we represent leaf nodes and empty nodes with learnable parameters $\boldsymbol{g}_{leaf}$ and $\boldsymbol{g}_{empty}$ respectively and the bottom-top merging follows the rule:

$$\boldsymbol{g}_u^{sub} = \mathrm{TreeLSTMcell}^{\mathrm{sub}}(\boldsymbol{g}_{\tau(u,left)}^{sub}, \boldsymbol{g}_{\tau(u,right)}^{sub}) \tag{11}$$

Given $\boldsymbol{g}_{t,i}^{sub}$, we can complete the traversal of the preferential forest:

$$\boldsymbol{h}_{t,i}^{tmp} = \mathrm{TreeLSTMcell}^{\mathrm{dec}}(\boldsymbol{h}_{t,i}^{sub}, \boldsymbol{g}_{t,i}^{sub}) \tag{12}$$

$$\boldsymbol{h}_{t,i}^{dec} = \mathrm{LSTMcell}^{\mathrm{dec}}(\boldsymbol{h}_{t,i}^{tmp}, \boldsymbol{x}_{dec}) \tag{13}$$

When it goes to the empty node $\boldsymbol{h}_{t,L(t)}^{dec}$, all subtrees $S_{t,j}$ are generated which means S_t is available. To encode S_t, we adopt a bottom-top method:

$$\boldsymbol{g}_{t,i}^{dec} = \mathrm{TreeLSTMcell}^{\mathrm{forest}}(\boldsymbol{g}_{t,i+1}^{sub}, \boldsymbol{g}_{t,i+1}^{dec}) \tag{14}$$

where $\boldsymbol{g}_{t,L(t)}^{dec} = \boldsymbol{g}_{empty}$ and $i \in \{0, 1, ..., L(t)-1\}$.

The final result $\boldsymbol{g}_t^0 = \boldsymbol{g}_{t,0}^{dec}$ will be added to the Fenwick tree so as to generate the next node v_{t+1} starting from the graph level.

Training Optimization. In previous sections, we describe the inference stage of ForGen. However, if we train the model based on the full $O(n \log n)$ steps as the inference stage does, it will be too time-consuming. Since we have complete adjacency matrix data, we can prepare the training targets concurrently rather than step-by-step. The overall flow path is bottom-top-bottom. First, we gather all subtrees of graphs in the batch and align them from leaf nodes so that subtrees can be built synchronously. Then we present every preferential forest as a sequence and process them together starting from the empty node. After that, all node embeddings based on ground truth are ready with $O(\log n)$ steps and the top-bottom part can be done in a reverse procedure.

4 Experiments

In this section, we compare ForGen to state-of-the-art baselines on both synthetic and real graph datasets. The empirical results demonstrate the effectiveness and efficiency of ForGen.

4.1 Datasets

To make a more comprehensive comparison, we experiment on both scale-free and non-scale-free networks, and part of the benchmark is obtained from [4]. The description of the four datasets is as follows: (1) Grid: standard 2D grid graphs with $100 \leq |V| \leq 400$. (2) BA: random graphs using Barabási-Albert preferential attachment [3] which exhibit a clear scale-free distribution, with $50 \leq |V| \leq 150$. We experiment with different graph densities for further verification. (3) Youtube: 2-hop ego graphs extracted from the social network of YouTube users and their friendship connections [12] with $305 \leq |V| \leq 477$ and $684 \leq |E| \leq 5674$. (4) Amazon: 2-hop ego graphs extracted from the co-purchase network of Amazon [12] with $304 \leq |V| \leq 476$ and $510 \leq |E| \leq 1377$. Each dataset consists of 100 graphs of which 80 for training and 20 for test.

4.2 Settings

Baselines. Following previous work, we adopt the Erdos-Renyi random graph model [5] for traditional approaches and its parameter of graph density is estimated from the training set. For deep graph generators, we compare with GraphRNN[1], GRAN[2] and BiGG[3]. GraphRNN uses a two-level LSTM model while GRAN incorporates GNN with attention mechanism. BiGG is the state-of-the-art model which is designed for sparse graphs. For all the baselines, we use the original code released by their authors.

[1] https://github.com/snap-stanford/GraphRNN.
[2] https://github.com/lrjconan/GRAN.
[3] https://github.com/google-research/google-research/tree/master/bigg.

Evaluation Metrics. We use the same metric as proposed by [30], which is also used in [4,7,17]. Specifically, we use three graph statistics to describe the graph in different levels, i.e., node degree distribution, clustering coefficient distribution of nodes, and orbit count distribution (the number of occurrences of all orbits with 4 nodes). To compare the distribution between the graph statistics of generated graphs and test graphs, we use the Maximum Mean Discrepancy (MMD) with total variation (TV) distance.

4.3 Model Effectiveness

Generation Quality. The first experiment compares the sample quality between ForGen and other baselines across three different metrics on the four datasets. Table 3 shows the experiment results and the best performance under each metric-dataset pair is highlighted in bold font. On the whole, our method, ForGen, outperforms other methods over 3 metrics except for the clustering coefficient of BA dataset which is still the second-best. Also, the result of grid graphs shows that our method can be extended to non-scale-free graphs and achieve good performance.

Table 3. Performance comparison of different methods on the two synthetic datasets and two real-world datasets. For all MMD metrics, the smaller the better.

Datasets		Methods				
		Erdos-Renyi	GraphRNN	GRAN	BiGG	ForGen (ours)
Grid	Degree	0.79	$1.12e^{-2}$	$8.23e^{-4}$	$4.00e^{-4}$	$\mathbf{1.74e^{-4}}$
	Clustering	2.00	$7.73e^{-5}$	$3.79e^{-3}$	$7.00e^{-5}$	**0**
	Orbit	1.08	$1.03e^{-3}$	$1.59e^{-3}$	$5.00e^{-4}$	$\mathbf{2.23e^{-4}}$
BA	Degree	0.17	$9.33e^{-2}$	$1.73e^{-2}$	$1.42e^{-3}$	$\mathbf{1.25e^{-3}}$
	Clustering	0.53	0.31	0.39	**0.14**	0.15
	Orbit	1.16	0.70	0.30	0.36	**0.20**
Youtube	Degree	0.39	$1.60e^{-2}$	$4.40e^{-2}$	$1.30e^{-2}$	$\mathbf{5.84e^{-3}}$
	Clustering	0.75	0.33	0.15	0.12	**0.11**
	Orbit	0.22	0.11	**0.10**	**0.10**	**0.10**
Amazon	Degree	$8.61e^{-2}$	$1.13e^{-2}$	$1.18e^{-2}$	$1.09e^{-2}$	$\mathbf{1.01e^{-3}}$
	Clustering	1.00	0.48	0.18	0.37	**0.11**
	Orbit	0.93	0.13	0.11	$9.83e^{-2}$	$\mathbf{9.57e^{-2}}$

Sample Quality vs. Graph Scale. As shown in Table 3, BiGG (better than other baselines in most cases) performs close to our method in some instances. For further comparison, we test the sample quality from another aspect. Note that it is quite difficult to train on datasets of large graphs due to the limitation of GPU memory, model's ability of expansion is important. To be specific, we will sample graphs of increasing sizes with models trained on the previously mentioned datasets. As such, model performance may diverge and different characteristics of models are amplified. What's more, it offers us an efficient way to get larger graphs with structural information maintained.

However, GraphRNN stops the generation process based on its output which cannot be intervened manually and GRAN can only generate graphs no larger than their training graphs. So, only Erdos-Renyi model, BiGG and ForGen are included in the test based on the Amazon dataset. Figure 3 shows the log-log plot of the MMD results on three graph statistics with graph sizes ranging from 0.5k to 10k. Notably, BiGG's performance on degree distribution declines continuously as the graph size grows while the other two methods are more stable. For clustering coefficient and orbit count, ForGen has consistently lower MMD than BiGG, and ER model is the worst.

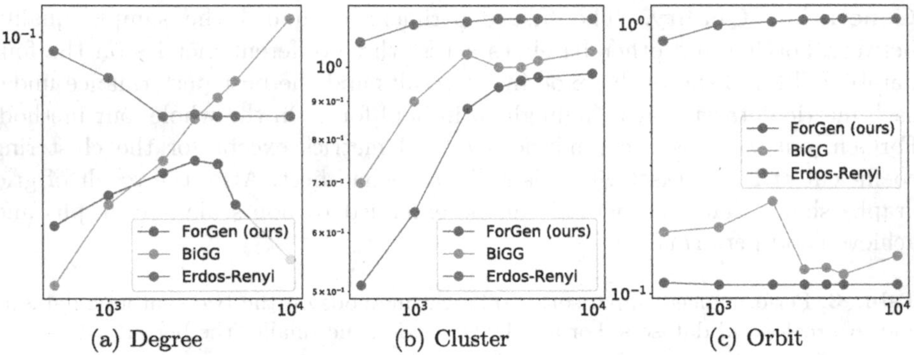

Fig. 3. The log-log plot of sample quality. The vertical axis represents MMD on different graph statistics, while the horizontal axis represents the size of sampled graphs.

Varying Graph Density. The four datasets mentioned in Sect. 4.1 are sparse graphs with an average connection of less than 6 (usually 2). To see what will happen when the graphs become denser, we experiment on BA graphs which can set the number of edges to attach from a new node to existing nodes (denoted as τ). While we have already tested on $\tau = 2$, we adopt $\tau \in \{10, 20\}$ to generate new datasets. We choose the best baseline BiGG and our model ForGen to be trained and tested on the denser graphs. Also, we report the ground truth BA model. Table 4 shows that even though denser graphs may contain more noises which can hinder learning-based methods, ForGen can still achieve the best results across three metrics.

4.4 Model Efficiency

Since BiGG is the most efficient deep autoregressive model and its lower time complexity has been proved, we compare ForGen with it in both training speed and inference speed. Table 5 reports the results of all five scale-free graph datasets we have. The first number of each item is the seconds model takes to finish a training iteration under the same batch size and the latter number is the inference time to generate 1k edges in a graph. Compared with BiGG's, the training

Table 4. Quality test with increasing graph density.

Datasets	Methods	Degree	Clustering	Orbit
BA (τ=2)	BiGG	$1.42e^{-3}$	0.14	0.36
	ForGen (ours)	$1.25e^{-3}$	0.15	0.20
	Ground Truth	$1.13e^{-3}$	0.18	0.29
BA (τ=10)	BiGG	$7.45e^{-3}$	0.14	0.13
	ForGen (ours)	$5.50e^{-3}$	0.11	0.10
	Ground Truth	$4.72e^{-3}$	$8.07e^{-2}$	$6.31e^{-2}$
BA (τ=20)	BiGG	$1.48e^{-2}$	0.17	0.11
	ForGen (ours)	$1.40e^{-2}$	0.16	0.10
	Ground Truth	$1.54e^{-2}$	0.12	$6.25e^{-2}$

Table 5. Training/Inference speed on scale-free graphs.

	BA ($\tau = 2$)	BA ($\tau = 10$)	BA ($\tau = 20$)	Youtube	Amazon
BiGG	**0.79**/90.72	0.99/55.03	1.3/40.29	1.11/83.96	**0.66**/113.6
ForGen (ours)	1.07/**41.44**	**0.97/30.8**	**1.19/23.88**	**1.07/39.21**	0.76/**57.42**

speed of ForGen is faster when the graph is denser since more computation can be synchronized. When it comes to graph generation, ForGen can achieve about twice BiGG's speed on scale-free networks owing to its preferential forest design.

4.5 Ablation Study

Different Node Orderings. We choose four canonical node orderings mainly based on graph properties: the node degree descending ordering, BFS/DFS ordering (nodes with larger degree first), and the default ordering of the data. Also, we test the combination of BFS and DFS. Table 6 shows the results of the Amazon dataset. We can see that both BFS and DFS can achieve good performance since they combine the topological information with preferential information (degree). However, node orderings lacking either information can hinder the learning process and the combination of different orderings with the same training strategy cannot help to improve the performance either. In summary, BFS or DFS ordering is recommended.

Table 6. Performance of models trained with different node orderings.

Node ordering	Degree	Clustering	Orbit
BFS	$1.08e^{-3}$	0.11	0.11
DFS	$1.01e^{-3}$	0.11	$9.57e^{-2}$
Degree descent	$1.51e^{-2}$	0.24	0.11
DFS+BFS	$2.08e^{-3}$	0.12	0.10
Default	0.23	0.35	0.11

Tree Size b**.** We next evaluate how the size of the first tree in the preferential forest (denoted as b) affects the quality. Since too small b will pay too much attention to noise and large b will lead to groups with fewer dependencies, we suggest $b \in \{5, 6, 7\}$. The experiment is conducted on the Youtube dataset. The results in Table 7 suggest that the selection of b within a reasonable range will not influence the effectiveness.

Table 7. Ablation study on tree size.

Tree size	Degree	Clustering	Orbit
$b = 5$	$5.83e^{-3}$	0.11	0.10
$b = 6$	$8.06e^{-3}$	0.12	$9.90e^{-2}$
$b = 7$	$6.34e^{-3}$	0.12	0.11

5 Conclusion

In this paper, we propose a forest-based graph generation model ForGen. First, we verify the wide existence of scale-free networks and skewed degree distributions on graphs of different sizes. To incorporate this important topological feature of real-world graphs, we design a multilevel data structure, which we call Preferential Forest, for the edge generation. We prove that the time complexity of generating an edge is $O(1)$ in the best case (which occurs frequently in scale-free networks) and $O(\log n)$ in the worst case. Together with a graph-level Fenwick tree, an autoregressive generation model that can generate graphs of any size is completed. By synchronization and an optimized tree traversal procedure, model scalability is further improved. We conduct extensive experiments on both synthetic and real graphs and prove ForGen's state-of-the-art sample quality from three different aspects. Furthermore, the generation speed of ForGen is two times faster than the most scalable model. In summary, the two-level tree-based model ForGen outperforms other autoregressive graph generators (like GraphRNN and BiGG) in both quality and efficiency.

References

1. Albert, R., Barabási, A.L.: Statistical mechanics of complex networks. Rev. Mod. Phys. **74**(1), 47 (2002)
2. Barabási, A.: Network Science. Cambridge University , Cambridge (2016)
3. Barabási, A.L., Albert, R.: Emergence of scaling in random networks. Science **286**(5439), 509–512 (1999)
4. Dai, H., Nazi, A., Li, Y., Dai, B., Schuurmans, D.: Scalable deep generative modeling for sparse graphs. In: International Conference on Machine Learning, pp. 2302–2312. PMLR (2020)
5. Erdos, P., Rényi, A., et al.: On the evolution of random graphs. Publ. Math. Inst. Hung. Acad. Sci **5**(1), 17–60 (1960)

6. Fenwick, P.M.: A new data structure for cumulative frequency tables. Softw. Pract. Exp. **24**(3), 327–336 (1994)
7. Goyal, N., Jain, H.V., Ranu, S.: Graphgen: a scalable approach to domain-agnostic labeled graph generation. In: Proceedings of The Web Conference 2020, pp. 1253–1263 (2020)
8. Grover, A., Zweig, A., Ermon, S.: Graphite: iterative generative modeling of graphs. In: International Conference on Machine Learning, pp. 2434–2444. PMLR (2019)
9. Guo, X., Zhao, L.: A systematic survey on deep generative models for graph generation. arXiv preprint arXiv:2007.06686 (2020)
10. Huffman, D.A.: A method for the construction of minimum-redundancy codes. Proc. IRE **40**(9), 1098–1101 (1952)
11. Jeong, H., Néda, Z., Barabási, A.L.: Measuring preferential attachment in evolving networks. EPL (Europhys. Lett.) **61**(4), 567 (2003)
12. Kunegis, J.: KONECT - The Koblenz network collection. In: Proceedings International Conference on World Wide Web Companion, pp. 1343–1350 (2013)
13. Kunegis, J., Blattner, M., Moser, C.: Preferential attachment in online networks: measurement and explanations. In: Proceedings of the 5th Annual ACM Web Science Conference, pp. 205–214 (2013)
14. Leskovec, J., Chakrabarti, D., Kleinberg, J., Faloutsos, C., Ghahramani, Z.: Kronecker graphs: an approach to modeling networks. J. Mach. Learn. Res. **11**(2) (2010)
15. Leskovec, J., Kleinberg, J., Faloutsos, C.: Graphs over time: densification laws, shrinking diameters and possible explanations. In: Proceedings of the Eleventh ACM SIGKDD International Conference on Knowledge Discovery in Data Mining, pp. 177–187 (2005)
16. Li, Y., Zhang, L., Liu, Z.: Multi-objective de novo drug design with conditional graph generative model. J. Cheminform. **10**(1), 1–24 (2018). https://doi.org/10.1186/s13321-018-0287-6
17. Liao, R., et al.: Efficient graph generation with graph recurrent attention networks. In: Proceedings of the 33rd International Conference on Neural Information Processing Systems, pp. 4255–4265 (2019)
18. Liu, C.C., Chan, H., Luk, K., Borealis, A.: Auto-regressive graph generation modeling with improved evaluation methods. In: Proceedings of NeurIPS Workshop Graph Representation Learning (2019)
19. Price, D.J.D.S.: Networks of scientific papers. Science, 510–515 (1965)
20. Simonovsky, M., Komodakis, N.: GraphVAE: towards generation of small graphs using variational autoencoders. In: Kůrková, V., Manolopoulos, Y., Hammer, B., Iliadis, L., Maglogiannis, I. (eds.) ICANN 2018. LNCS, vol. 11139, pp. 412–422. Springer, Cham (2018). https://doi.org/10.1007/978-3-030-01418-6_41
21. Snijders, T.A., Nowicki, K.: Estimation and prediction for stochastic blockmodels for graphs with latent block structure. J. Classif. **14**(1), 75–100 (1997)
22. Tai, K.S., Socher, R., Manning, C.D.: Improved semantic representations from tree-structured long short-term memory networks. In: ACL, no. 1 (2015)
23. Tuteja, S., Kumar, R.: A unification of heterogeneous data sources into a graph model in e-commerce. Data Sci. Eng. **7**(1), 57–70 (2022)
24. Vaswani, A., et al.: Attention is all you need. In: Advances in Neural Information Processing Systems, pp. 5998–6008 (2017)
25. Watts, D.J., Strogatz, S.H.: Collective dynamics of 'small-world' networks. Nature **393**(6684), 440–442 (1998)

26. Xiao, H., Huang, M., Zhu, X.: TransG: a generative model for knowledge graph embedding. In: Proceedings of the 54th Annual Meeting of the Association for Computational Linguistics (Volume 1: Long Papers), pp. 2316–2325 (2016)
27. Xie, S., Kirillov, A., Girshick, R., He, K.: Exploring randomly wired neural networks for image recognition. In: Proceedings of the IEEE/CVF International Conference on Computer Vision, pp. 1284–1293 (2019)
28. Yi, P., Li, J., Choi, B., Bhowmick, S.S., Xu, J.: Flag: Towards graph query autocompletion for large graphs. Data Sci. Eng. **7**(2), 175–191 (2022)
29. You, J., Liu, B., Ying, R., Pande, V., Leskovec, J.: Graph convolutional policy network for goal-directed molecular graph generation. In: Proceedings of the 32nd International Conference on Neural Information Processing Systems, pp. 6412–6422 (2018)
30. You, J., Ying, R., Ren, X., Hamilton, W., Leskovec, J.: GraphRNN: generating realistic graphs with deep auto-regressive models. In: International Conference on Machine Learning, pp. 5708–5717. PMLR (2018)

Iterative Deep Graph Learning with Local Feature Augmentation for Network Alignment

Jiuyang Tang, Zhen Tan(✉), Hao Guo, Xuqian Huang, Weixin Zeng, and Huang Peng

Laboratory for Big Data and Decision, National University of Defense Technology, Changsha, China
{tanzhen08a,guo_hao,huangxuqian15,phmail}@nudt.edu.cn

Abstract. Networks are structures that naturally capture relations between entities in different data sources and information systems. To establish the connections among different networks, the task of network alignment is proposed and intensively studied in network-related research field. Most network alignment methods are based on the representation learning of network structure, which rely only on network topology and are susceptible to structural noise. In addition, such methods focus on the global features, while largely neglect local structure features and fail to take account of the data sparsity issue in real networks. To address these pivotal issues, we propose a novel network alignment method based on iterative deep network learning and local feature augmentation. We first design an iterative deep graph learning model to learn high-quality network structural representation and reduce the structural noise. Furthermore, we embed knowledge representation learning method into the alignment process, which helps to characterize better local structure and alleviate the data sparsity issue. Experiments on real-world network datasets demonstrate that our proposed model achieves state-of-the-art alignment results.

Keywords: Network alignment · Graph learning · Knowledge representation learning

1 Introduction

Networks are natural but powerful structure that capture the relationships between different entities in many domains, such as social networks, referral networks, and bioinformatics networks [12,27]. Network analysis, also known as network science, has received a lot of attention for decades and remains an attractive field [1]. While the analysis of individual network is critical for a variety of applications (e.g., link prediction; community detection), it cannot sufficiently

Supported by Ministry of Science and Technology of China under grants No. 2020AAA0108800, NSFC under grants Nos. 71971212 and 61902417.

B. Li et al. (Eds.): APWeb-WAIM 2022, LNCS 13421, pp. 511–526, 2023.
https://doi.org/10.1007/978-3-031-25158-0_41

address the tasks that require considering the relationships between graphs (e.g., graph clustering; graph alignment). As thus, a subfield of network science is put forward to analyze the relationships between different graphs. In this paper, we aim to solve one of the fundamental problems in comparative graph analysis - network alignment (NA).

The goal of network alignment is to identify the corresponding nodes in different networks. For example, there are a large number of users with accounts in different social networks [23], and network alignment can help identify the same users in different social networks, as shown in Fig. 1. The user correspondence established by network alignment can alleviate the sparsity issue of a single social network, benefiting applications such as link prediction [5] and cross-domain recommendation [14]. Besides, network alignment can also help build a more compact knowledge graph based on existing vertical or cross-language knowledge bases, and enable better knowledge inference. In addition, in bioinformatics, aligning protein-protein interaction networks from different species has also been extensively studied to identify common functional structures [7].

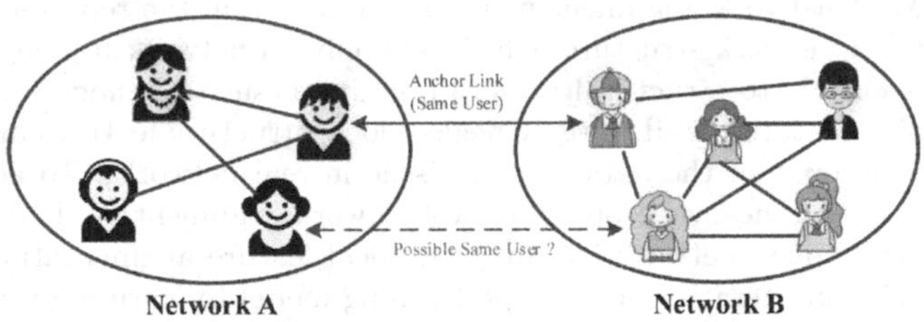

Fig. 1. An example of network alignment. The black lines between two networks are anchor links. And the dash lines are potential aligned nodes.

However, network alignment faces three primary challenges: network noise, data sparsity, and alignment efficiency.

Alignment Efficiency. Some NA work describes network alignment as the maximum matching problem of a binary graph, such as the largest common subgraph problem, but these are all NP-difficult problems [2]. Therefore, many methods employ matrix decomposition formulas such as IsoRank [18], FINAL [28], and REGAL [8]. But the approaches cannot handle very large networks because the computational effort required will grow rapidly as the size of the network increases.

Network Noise. Due to the inevitable error of data measurement or collection, real-world networks are generally noisy or even incomplete. The network noise originates from both the topology of the network and the feature matrix of the nodes.

Data Sparsity. Similar to other types of large-scale data, large-scale networks all obey the long-tail distribution and have severe data sparsity problems [26].

For long-tail nodes, only a few paths are associated with them, so their semantic or inference representation is extremely inaccurate.

To improve alignment efficiency, methods based on the network representing learning are proposed, such as PALE [15], and DeepLink [30]. These alignment techniques can take advantage of the scalability of graph embedding to handle large networks, but these methods rely only on topology information and are susceptible to structure noise, making the models lack generalization capabilities. To overcome the network structure noise, we generate better network structure representation basing on iterative deep graph learning framework.

Moreover, most network alignment methods pay more attention to the global features of the network without valuing the local structure features. Particularly in sparse networks, such methods lead to the poor performance because of the long-tail distribution. While knowledge representation learning methods, such as TransE [3], TransH [24], DistMult [25], ComplEx [21], and RotatE [20] models, can well characterize the local structure of networks. In view of that, to address the data sparsity, we use various graph representation learning methods for network alignment to obtain high-quality local features.

In this paper, we propose **I**terative **D**eep Graph Learning with **L**ocal **F**eature **A**ugmentation for Network Alignment (IDLFA). The model is composed of two parts: encoder module and decoder module. The encoder module learns node structure embedding by iterative deep graph learning model. The decoder module integrates the knowledge representation learning method into the alignment method to augment local feature. And in the training process, the bootstrapping algorithm is applied to add the newly generated alignment nodes to the training set to further alleviate the data sparsity issue. The contributions of this paper can be summarized as follows:

1) We propose a unified network alignment framework, which combines encoder module and decoder model for solving network structure noise and data sparsity.
2) At encoder module, we leverage Iterative Deep Graph Learning for Graph Neural Networks to obtain better node structural embedding and reduce networks noise.
3) At decoder module, for easing data sparsity, we integrate knowledge representation methods to augment local feature and apply bootstrapping algorithm to produce newly alignments for model training.
4) Experiments on real-world datasets demonstrate that the network alignment method based on iterative deep graph learning outperforms state-of-the-art models and is highly robust on alignment tasks.

2 Related Work

In our work, the network alignment techniques are divided into two categories, the **spectral method** and the **network representation learning method**. The goal of the spectral approach is to align the two networks based on the adjacency matrix operation. While, the network representation learning method requires an intermediate step where the nodes in the network are represented as embedding.

2.1 Spectral Method

Many spectral methods [11,17], using matrix factorization, aim to directly calculate the alignment matrices. Assuming that the input graph is in the form of an adjacency matrix, the spectral alignment technique defines the model in the form of a loss function, which considers the adjacency matrix of the source network and the target network. The node features are constants and the alignment matrices are variables. During the alignment process, the alignment matrix is learned by optimizing the loss function based on the structure or property consistency assumptions.

IsoRank [18] is one of the popular and typical techniques in the category which utilizes only topological information. The main idea is that two nodes in two networks are similar if their neighbors are similar, but the technique is highly sensitive to structural noise. BigAlign [10] uses only the attribute information to align the nodes. FINAL [28] is different from previous methods that only used topological or attribute information, and chooses to use both of them to better capture the information of the network nodes. REGAL [8] models the alignment matrix by topology and feature similarity and then employs low-rank matrix approximation to speed up calculation.

2.2 Network Representation Learning Method

Network representation learning approaches [6,16,29] solve the network alignment problem by exploiting graph embedding. It includes two steps: embedding generation and alignment matrix generation. At first, a graph embedding technique is used to represent nodes and the two embedding matrices of graphs are obtained separately. Then, the alignment matrix is designed to map the source network's embeddings to the target network's embedding space.

PALE [15] involves a pre-process step where a priori mapping between two networks is used to populate the missing edges available in one map but not available in the other. DeepLink [30] has the same method as the PALE to construct graph embeddings, but its mapping function varies by considering the mapping direction. DeepLink adopts unbiased random walk to generate embeddings and uses a linked-dual learning process to improve its quality. IONE [13] uses the same mapping function as PALE, its embedding function is more complex because it considers the neighborhood of the node. IONE aims to meet two goals: close nodes in each graph should have similar node embedding and the nodes with close embedding are good candidates.

3 Preliminaries

In this section, we first formally define the task of network alignment, and then we briefly review the knowledge representation model.

3.1 Problem Formulation

Network alignment is the task of identifying corresponding nodes between two different networks. Given two networks: source network $G_s = (V_s, E_s)$)and target network $G_t = (V_t, E_t)$) , where V_s and V_t are sets of network nodes, E_s and E_t are sets of network edges. **Anchor links** represent a node pair(v, v') , where$v \in V_s$, $v' \in V_t$, and v and v' are aligned. The goal of network alignment is to predict all potential anchor links.

3.2 Knowledge Representation Model

In this section, we introduce TransE [3], TransH [24], DistMult [25], ComplEx [21], and RotatE [20] models, which are adapted to our network alignment framework. u and v denote the node embedding; e is the edge embedding.

TransE. The idea of TransE is that the embeddings of the nodes in the source network can be close enough to the corresponding embeddings in the target network through the edge embedding, so that the score function of the TransE model can be represented by the following formula:

$$f_{TransE}(u + e, v) = \|u + e - v\| . \tag{1}$$

TransH. To overcome the defects of TransE in edge modeling, the nodes have a distributed representation when different edges are involved. For an edge, the model positions an edge-specific translation vector d_e in the hyperplane w_e of a specific edge rather than in the space nodes embedded:

$$f_{TransH}(u, v) = \left\|(u - w_e^T u w_e) + d_e - (v - w_e^T v w_e)\right\|_2^2 . \tag{2}$$

DistMult employs bilinear encoding, and the embeddings of nodes and edges can be learned through a neural network. The first layer projects a pair of input nodes onto a low-dimensional vector, and the second layer combines the two vectors onto a scalar. With edge-specific parameters B_e, the score function is:

$$f_{DistMult}(u, v) = u^T B_e v. \tag{3}$$

ComplEx. The model introduces the complex vector space into the embedding, and its score function is:

$$f_{ComplEx}(e, u, v; \Theta) = Re(< e, u, \overline{v} >) = Re(\sum_{k=1}^{K} e_k u_k \overline{v}_k), \tag{4}$$

where $Re(\cdot)$ denotes imaginary; $\overline{v}_k$ and v_k are conjugate.

RotatE. Similar to complEx model, RotatE models the nodes and edges in the complex vector space. The difference is that the RotatE limits the modulus of the edge vector to 1, so that it becomes the rotation vector from the source network node to the corresponding node of the target network. Therefore, its score function is expressed as:

$$f_{RotatE}(u, v) = \|u \odot e - v\|, \tag{5}$$

where $\odot$ denotes Hadamard product; $\|e_i\| = 1$ denotes that the modulus of the edge vector are set to 1.

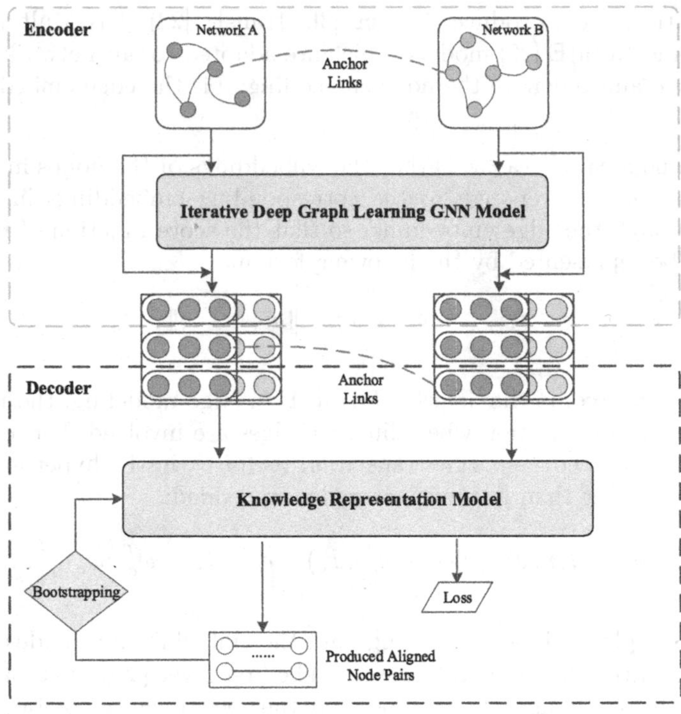

Fig. 2. Overview of IDLFA framework

4 Method

In this section, we present our approach IDLFA, which consists of encoder module and decoder module. The framework is shown in Fig. 2. In encoder module, the network structure is learned according to the iterative deep graph learning framework. In decoder module, we combine the knowledge representation model with the loss function of the IDGL model and adapt it to the network alignment, which helps to learn better local features. To further alleviate the data sparse problem, we use Bootstrapping algorithm [19] to add the predicted aligned node pairs to training data.

4.1 IDGL-Based Node Embedding

Iterative Deep Graph Learning model [4] (IDGL) is an end-to-end graph learning framework, which can jointly and iteratively learning the graph structure and graph embedding. In view of the advantage of obtaining better network representation, we transform it to alignment networks. Figure 3 is the overall architecture of IDGL framework. The brief introduction to the model is following.

Similarity Metric Learning. Without loss of generality, the IDGL model designs a weighted cosine similarity as metric function, $s_{ij}^{p} = cos(w \odot v_i, w \odot v_j)$, where $\odot$ denotes the Hadamard product, and w is a learnable weight vector which has the same dimension as the input vectors v_i and v_j. Note that the two input vectors could be either raw node features or computed node embeddings.

Graph Node Embeddings and Prediction. Both the learned graph structure A and the original graph topology $A^{(0)}$ are helpful to formulate an optimized graph for GNNs. IDGL combines the learned graph with the initial graph,

$$\tilde{A}^{(t)} = \lambda L^{(0)} + (1-\lambda)\{\eta f(A^{(t)}) + (1-\eta) f(A^{(1)})\}, \tag{6}$$

where $L^{(0)} = {D^{(0)}}^{-1/2} A^{(0)} {D^{(0)}}^{-1/2}$ is the normalized adjacency matrix of the initial graph. $A^{(t)}$ and $A^{(1)}$ are the two adjacency matrices computed at the t-th and 1-st iterations, respectively. $A^{(1)}$ is computed from the raw node features X, and $A^{(t)}$ is computed from the previously updated node embeddings $Z^{(t-1)}$ that is optimized toward the downstream prediction task. Hyperparameter η is used to combine the advantages of both; λ is used to balance the learned graph structure and the initial one.

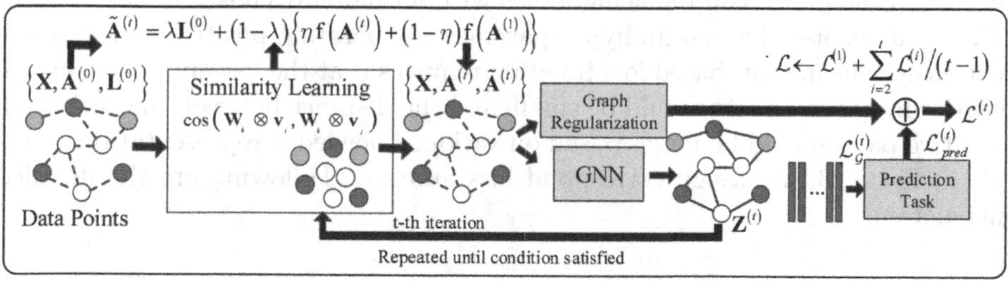

Fig. 3. Overall architecture of the proposed IDGL framework. Dashed lines (in data points on left) indicate the initial noisy graph topology A.

We follow the setting of IDGL model, and adopt a two-layered GCN [9] where the first layer (denoted as GNN_1) maps the raw node features **X** to the intermediate embedding space, and the second layer (denoted as GNN_2) further maps the intermediate node embeddings **Z** to the output space.

$$\mathbf{Z} = \text{ReLU}\left(\mathbf{MP}(\mathbf{X}, \tilde{\mathbf{A}})\mathbf{W}_1\right), \quad \hat{\mathbf{y}} = \sigma\left(\mathbf{MP}(\mathbf{Z}, \tilde{\mathbf{A}})\mathbf{W}_2\right), \quad \mathcal{L}_{pred} = \ell(\hat{\mathbf{y}}, \mathbf{y}), \tag{7}$$

where $\sigma(\cdot)$ and $\mathcal{L}(\cdot)$ are task-dependent output function and loss function, respectively. $\mathbf{MP}(\cdot, \cdot)$ is a massage passing function.

Joint Learning with a Hybrid Loss. IDGL model proposes to jointly and iteratively learning the graph structure and the GNN parameters by minimizing a hybrid loss function combining both the task prediction loss and the graph regularization loss, namely:

$$\mathcal{L} = \mathcal{L}_{pred} + \mathcal{L}_{\mathcal{G}}, \tag{8}$$

where $\mathcal{L}_{\mathcal{G}}$ is the graph regularization loss. At each iteration, a hybrid loss is computed. After all iterations, the overall loss is back-propagated through all previous iterations to update the model parameters.

4.2 Model Training

$\mathcal{L}_{pred}$ is a task-dependent loss function. For network alignment, the object is to embed equivalent nodes as closely as possible in the vector space. So the model training process is performed by minimizing the following margin-based ranking loss function:

$$\mathcal{L}_{pred} = \sum_{(u,v)\in S} \sum_{(u',v')\in S'_{(u,v)}} \left[f\left(u,v\right) + \gamma - f\left(u',v'\right)\right]_+, \tag{9}$$

where $[x]_+ = \max\{0, x\}$; $(u,v) \in S$ represents the set of anchor links used to train the model; $S'_{(u,v)} \in S'_{(u,v)}$ denotes the set of negative instances constructed by corrupting (u,v), i.e., replacing u or v with a randomly chosen nodes in G_s or G_t; $\gamma > 0$ denotes the margin hyper-parameter separating positive and negative instances. The margin-based loss function requires that the distance between the entities in positive pairs should be small, and the distance between the entities in negative pairs should be large. Based on various knowledge representation methods at Sect. 3.3, we design correspond loss function. Following are the detailed instruction.

TransE. We merge it with margin-based ranking loss function:

$$\mathcal{L}_{\text{TransE}} = \sum_{(u,v)\in S} \sum_{(u',v')\in S'_{(u,v)}} \left[f_{TransE}\left(u+e,v\right) + \gamma_1 - f_{TransE}\left(u'+e,v'\right)\right]_+, \tag{10}$$

where $\gamma_1 > 0$ the specific boundary hyperparameter separating the positive and negative node alignment in the TransE model.

TransH. We merge it with margin-based ranking loss function:

$$\mathcal{L}_{\text{TransH}} = \sum_{(u,v)\in S} \sum_{(u',v')\in S'_{(u,v)}} [f_{TransH}(u,v) + \gamma_2 - f_{TransH}(u',v')]_+, \quad (11)$$

where $\gamma_2 > 0$ is the specific boundary hyperparameter.

DistMult. Based on DistMult model, the loss function $\mathcal{L}_{pred}$ can be adjusted to:

$$\mathcal{L}_{\text{DistMult}} = \sum_{(u,v)\in S} \sum_{(u',v')\in S'_{(u,v)}} [f_{DistMult}(u,v) - f_{DistMult}(u',v') + 1]_+. \quad (12)$$

ComplEx. We minimize the negative log-likelihood of the logical model, and train the model using small-batch stochastic gradient descent and AdaGrad. By regularizing the parameters of the considered model, we adjust the learning rates:

$$\mathcal{L}_{\text{ComplEx}} = \sum_{(u,v)\in S} \log\left(1+\exp\left(-\mathbf{Y}_{euv} f_{\text{ComplEx}}(u,e,v;\Theta)\right)\right) + \lambda\|\Theta\|_2^2, \quad (13)$$

where $\mathbf{Y}_{euv} = 1$ when the node pairs are positive; and else $\mathbf{Y}_{euv} = -1$. λ is a weight parameter.

RotatE. Different from above models, RotatE adopts self-adversarial loss function basing on negative sampling for training:

$$\mathcal{L}_{\text{RotatE}} = -\log\sigma\left(\gamma_3 - f_{\text{RotatE}}(\mathbf{u},\mathbf{v})\right) - \sum_{i=1}^{n} p\left(u_i',e,v_i'\right)\log\sigma\left(f_{\text{RotatE}}\left(\mathbf{u}_i',\mathbf{v}_i'\right) - \gamma_3\right), \quad (14)$$

where γ_3 is the specific boundary hyperparameter; σ is the sigmoid function; (u_i', e, v_i') is the i-th negative alignment nodes; $p_{(u_i',e,v_i')}$ can be defined as:

$$p\left(u_j',e,v_j' \mid \{(u_i,e_i,v_i)\}\right) = \frac{\exp \alpha f_{\text{RotatE}}\left(\mathbf{u}_j',\mathbf{v}_j'\right)}{\sum_i \exp \alpha f_{\text{RotatE}}\left(\mathbf{u}_i',\mathbf{v}_i'\right)}, \quad (15)$$

where α denotes the sample weight.

The loss function of the above models will learn a better network structure representation. At the same time, the model further adds the new alignment nodes into the training set through the Bootstrapping algorithm, which helps to alleviate the data sparse problem and further improve the performance.

4.3 Alignment Prediction

We predict the alignment results based on the distance between learned nodes representations from two networks.

The Euclidean distance and Manhattan distance are commonly used distance measures in the Euclidean space. For entities u_i in G_s and v_j in G_t, the distance is defined as:

$$D(u_i, v_j) = \frac{f(\mathbf{u}_i, \mathbf{v}_j)}{d}, \tag{16}$$

where $f(x, y) = \|x - y\|_1$, $\|\cdot\|_1$ is the L_1 norm; d denotes the dimension of embedding. The distance is expected to be small for equivalent entities and large for non-equivalent ones. For a specific entity u_i in G_s, our approach computes the distances between u_i and all the entities in G_t, and returns a list of ranked entities as candidate alignments.

5 Experiment

In the section, we conduct extensive experiments to verify the effectiveness of the model. First, we introduce the experimental settings and then evaluate the performance of our method. For further analysis, ablation study and parameter analysis are performed.

5.1 Experiment Setup

Datasets. This section conducts experiments on 2 real-world datasets (4 real-world networks). The detailed information is shown in Table 1.

Flickr and Myspace datasets: The two subnetworks of Flickr and Myspace are collected in the paper [27] and then processed according to the method in the paper [28]. Flickr's subnet contains 6714 nodes, and Myspace's subnet contains 10,733 nodes. The gender of the user is used to represent node attributes, and only part of the ground truth is available for alignment.

Allmovie and Imdb datasets: The Allmovie network is constructed from Rotten Tomatoes website[1]. Two films have an edge connecting them if they have at least one common actors. Imdb network is constructed in a similar way from Imdb website[2]. The alignment output is constructed by the identity of the film, containing 5176 anchor links.

Evaluation Metrics. We use both *Success@q* [28] and *MAP* (Mean Average Precision) [15] to evaluate the effectiveness of our proposed model. *Success@q* denotes whether a true positive match appears in the previous q candidate. For ranking perspective, *MAP* is also known as Mean Reciprocal Rank under pair-wise setting. Considering that the network alignment is a bidirectional task, we use the average value of $G_s \rightarrow G_t$ and $G_t \rightarrow G_s$ to present the experimental results.

---

1 https://www.kaggle.com/ayushkalla1/rotten-tomatoes-movie-database.

2 https://www.kaggle.com/jyoti1706/imdbmoviesdataset.

Table 1. Statistics of 4 real-world networks.

Network	Nodes	Edge	Attribute
Flickr	6714	7333	3
Myspace	10733	11081	3
Allmovie	6011	124709	14
Imdb	5731	119073	14

Comparison Methods. Our proposed model IDLFA with its variants and the state-of-the-art baseline methods for comparison are listed as following:

PALE [15]: is a network representation technique that learns node embedding by maximizing the co-occurrence likelihood of edge nodes, and then applies a linear or multi-layer perceptron as a mapping function.

REGAL: is a spectral method that models the alignment matrix by topology and feature similarity of nodes, and then accelerates with a low-rank matrix [8].

IsoRank: is a spectral approach and a global alignment method initially with application to protein interaction networks [18].

FINAL: is a spectral method designed for attributed networks, which considers graph structure, node feature, and edge feature [28].

GAlign: is the state-of-the-art alignment mode and proposes a completely unsupervised network alignment framework based on a multi-order GCN embedding model [22].

Hyperparameter Tuning. The margin hyper-parameters γ, γ_1, γ_2 and γ_3 in the relevant loss function are set to 1. And the value of λ is validated in set $\{0.1, 0.03, 0.01, 0.003, 0.001\}$. The embedding dimension is set to 100 and will be further evaluated in the later. We optimize the model with Stochastic Gradient Descent algorithm.

Machines and Repeatability. The results are averaged over 10 runs to mitigate randomness. All experiments are conducted on 8 3.6 GHz Intel Cores with 64 GB RAM and 1 GeForce RTX2080Ti graphic cards. Our proposed algorithm is programmed in Python.

5.2 Experiment Result

To verify the effectiveness of our proposed model, we compares the models with several state-of-the-art models on two real-world datasets, and the experimental results are shown in Table 2. Bold numbers indicate optimal results, and underlined numbers indicate sub-optimal results. The results are obtained with 80% of the anchor nodes as training set and the rest for testing.

Table 2. The performance of network alignment on real-world datasets.

Dataset	Metrics	Ours	GAlign	PALE	REGAL	IsoRank	FINAL
Allmovie-Imdb	*MAP*	**0.9320**	0.8496	0.7601	0.1888	0.5271	0.8459
	Success@1	**0.9068**	0.8214	0.6947	0.0953	0.4653	0.7647
	Success@10	**0.9710**	0.9003	0.7159	0.3869	0.6427	0.9609
Flickr-Myspace	*MAP*	0.1245	**0.1608**	0.0059	0.0090	0.0085	0.0429
	Success@1	0.0556	**0.0774**	0.0000	0.0464	0.0000	0.0206
	Success@10	0.2615	**0.3127**	0.0206	0.1950	0.0275	0.0722

In general, our proposed IDLFA model outperforms all the baselines on datasets Allmovie-Imdb in terms of *MAP*, *Success*@1 and *Success*@10. In terms of *Success*@1, IDLFA achieves more than 90%, exceeds GAlign by 8% and exceeds FINAL by nearly 15%. In terms of *MAP*, IDLFA outperforms the second over 8%. In addition, the *Success*@10 of IDLFA is about 97%.

In Flickr-Myspace, our proposed model achieves the sub-optimal alignment accuracy. Compared with FINAL, *MAP* of IDLFA increases by more than 0.07, and *Success*@10 is about 15% higher. Compared with the SOTA model GAlign, our proposed method is nearly 2% lower in *Success*@1 and about 0.04 lower in terms of *MAP*. Although our model IDLFA does not exceed GAlign, we only use structural information, while GAlign uses additional attribute information.

Weakly Supervised Condition. Table 3 further gives the detailed model comparison when the ratio of training set to test set is 0.2:0.8. Our proposed model is compared with the previous SOTA GAlign in Allmovie-Imdb and Flickr-Myspace datasets. Our method IDLFA still has *Success*@1 $\approx$ 78% better than GAlign in Allmovie-Imdb dataset. In Flickr-Myspace dataset, IDLFA outperforms GAlign by a large margin. The results show that the model is still robust and well-performed in a weakly supervised manner. And our proposed model performs better in Allmovie-Imdb than Flickr-Myspace, which indicates that the model has more prominent performance on datasets with abundant structure information in a weakly supervised manner.

Table 3. The performance of our proposed model IDLFA on 20% anchor links.

Dataset	Metrics	Ours	GAlign
Allmovie-Imdb	*MAP*	**0.8185**	0.7925
	Success@1	**0.7785**	0.7399
	Success@10	**0.9023**	0.8667
Flickr-Myspace	*MAP*	**0.0395**	0.0177
	Success@1	**0.0140**	0.0044
	Success@10	**0.0514**	0.0327

Ablation Study. The local feature argument mechanism is based on knowledge representation model, which is designed for sparse datasets. Although the IDGL model can learn better structure representation, the premise is that the nodes have rich topology information. It can be seen from Table 2 that the alignment accuracy of the same model on different datasets varies greatly. What is the reason for this phenomenon? It can be seen from Table 1 that the network Flickr and Myspace are relative sparse, whose average edges of nodes is about 2. That's to say, these networks have a large number of long-tail nodes. For long-tail nodes, GNN-based models are limited to learn their structure features [26]. So, to verify the effectiveness of our proposed method on sparse datasets, we conduct ablation study on Flickr-Myspace.

IDNA represents our proposed model without local feature augmentation module. IDNA+TransE, IDNA+TransH, IDNA+DistMult, IDNA+ComplEx and IDNA+RotatE denote fusing IDNA with various knowledge representation method to learn better local feature. Table 4 presents the result with 4 metrics. From the table, we can clearly see that local feature augmentation module obtains better performance in terms of *MAP*, *Success*@1, *Success*@3 and *Success*@10, which indicates the local feature augmentation module works well and learns better structure representation. When the percentage of anchor links is 0.8, IDNA+RotatE outperforms other method and *Success*@1 improves by 2% compared to IDNA. When the percentage of anchor links is 0.2, IDNA+ComplEx performs better and has 1.2% improvement in terms of *MAP*.

Table 4. The result of ablation study on Flickr-Myspace.

	Model	*Success*@1	*Success*@3	*Success*@10	*MAP*
Anchor links = 0.8	IDNA	0.0371	0.1111	0.2408	0.1165
	IDNA+TransE	0.0185	0.0556	0.1760	0.0800
	IDNA+TransH	0.0370	0.0370	0.1760	0.0875
	IDNA+DistMult	0.0370	0.0926	0.1945	0.1050
	IDNA+ComplEx	0.0185	0.0278	0.1667	0.0735
	IDNA+RotatE	**0.0556**	**0.1311**	**0.2615**	**0.1245**
Anchor links = 0.2	IDNA	0.0047	0.0140	0.0421	0.0270
	IDNA+TransE	0.0047	0.0094	0.0538	0.0270
	IDNA+TransH	0.0070	0.0280	**0.0631**	0.0355
	IDNA+DistMult	0.0070	0.0210	0.0561	0.0340
	IDNA+ComplEx	**0.0140**	**0.0304**	0.0514	**0.0395**
	IDNA+RotatE	0.0047	0.0140	0.0584	0.0285

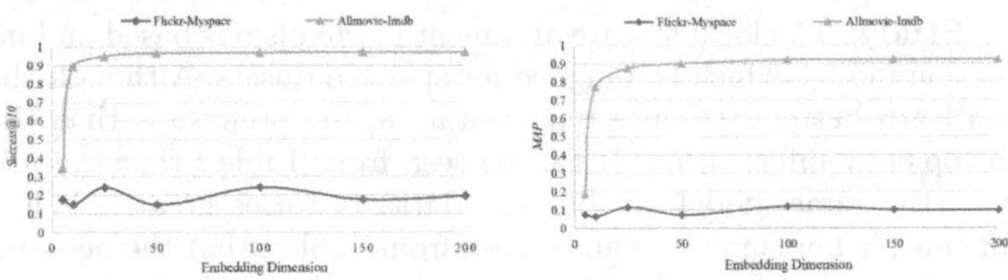

Fig. 4. The relationship between alignment results and embedding dimension.

5.3 Hyperparameter Sensitivity

Figure 4 studies the sensitivity of the embedding dimension. In general, users should not choose a high number of dimensions as it does not increase the performance (*Success*@1) significantly while the time and space complexity definitely become larger. In Flickr-Myspace dataset, as the embedding dimension increases, the model performance will fluctuate to a certain extent, but in general, the effect is better when the embedding dimension is 100. In Allmovie-Imdb, the initial performance increases rapidly over dimension and remains stable when dimension reaches about 100.

6 Conclusion

In this paper, we propose a novel network alignment framework IDLFA, which learns better network structure representation and further solves the networks data sparsity. Comprehensive empirical studies on two pairs of popular real-world datasets show that IDLFA can significantly improve the performance for social network alignment tasks in comparison with existing solutions. On part datasets, such as Allmovie-Imdb, our model shows the superiority, whose *Success*@1 comes to 90%, and can be adopt to practical applications. Our model doesn't take attribute information into consideration. In the future works, we will study relative framework for attributed networks and the proposed framework can be applied to other tasks, e.g., cross-lingual knowledge graph task.

References

1. Albert, R., Barabási, A.L.: Statistical mechanics of complex networks. Rev. Mod. Phys. **74**(1), 47 (2002)
2. Bayati, M., Gerritsen, M., Gleich, D.F., Saberi, A., Wang, Y.: Algorithms for large, sparse network alignment problems. In: 2009 Ninth IEEE International Conference on Data Mining, pp. 705–710. IEEE (2009)
3. Bordes, A., Weston, J., Collobert, R., Bengio, Y.: Learning structured embeddings of knowledge bases. In: Twenty-Fifth AAAI Conference on Artificial Intelligence (2011)

4. Chen, Y., Wu, L., Zaki, M.: Iterative deep graph learning for graph neural networks: better and robust node embeddings. Adv. Neural. Inf. Process. Syst. **33**, 19314–19326 (2020)
5. Gao, H., Zhang, Y., Li, B.: Improving the link prediction by exploiting the collaborative and context-aware social influence. In: Li, J., Wang, S., Qin, S., Li, X., Wang, S. (eds.) ADMA 2019. LNCS (LNAI), vol. 11888, pp. 302–315. Springer, Cham (2019). https://doi.org/10.1007/978-3-030-35231-8_22
6. Goyal, P., Ferrara, E.: Graph embedding techniques, applications, and performance: a survey. Knowl.-Based Syst. **151**, 78–94 (2018)
7. Guzzi, P.H., Milenković, T.: Survey of local and global biological network alignment: the need to reconcile the two sides of the same coin. Brief. Bioinform. **19**(3), 472–481 (2018)
8. Heimann, M., Shen, H., Safavi, T., Koutra, D.: Regal: representation learning-based graph alignment. In: Proceedings of the 27th ACM International Conference on Information and Knowledge Management, pp. 117–126 (2018)
9. Kipf, T.N., Welling, M.: Semi-supervised classification with graph convolutional networks. arXiv preprint arXiv:1609.02907 (2016)
10. Koutra, D., Tong, H., Lubensky, D.: Big-align: fast bipartite graph alignment. In: 2013 IEEE 13th International Conference on Data Mining, pp. 389–398. IEEE (2013)
11. Levy, O., Goldberg, Y.: Neural word embedding as implicit matrix factorization. In: Advances in Neural Information Processing Systems, vol. 27 (2014)
12. Liu, J., Shao, Y., Su, S.: Multiple local community detection via high-quality seed identification over both static and dynamic networks. Data Sci. Eng. **6**(3), 249–264 (2021)
13. Liu, L., Cheung, W.K., Li, X., Liao, L.: Aligning users across social networks using network embedding. In: IJCAI, pp. 1774–1780 (2016)
14. Man, T., Shen, H., Jin, X., Cheng, X.: Cross-domain recommendation: an embedding and mapping approach. In: Proceedings of the 26th International Joint Conference on Artificial Intelligence, pp. 2464–2470 (2017)
15. Man, T., Shen, H., Liu, S., Jin, X., Cheng, X.: Predict anchor links across social networks via an embedding approach. In: IJCAI, vol. 16, pp. 1823–1829 (2016)
16. Ou, M., Cui, P., Pei, J., Zhang, Z., Zhu, W.: Asymmetric transitivity preserving graph embedding. In: Proceedings of the 22nd ACM SIGKDD International Conference on Knowledge Discovery and Data Mining, pp. 1105–1114 (2016)
17. Qiu, J., Dong, Y., Ma, H., Li, J., Wang, K., Tang, J.: Network embedding as matrix factorization: unifying Deepwalk, Line, PTE, and Node2Vec. In: Proceedings of the Eleventh ACM International Conference on Web Search and Data Mining, pp. 459–467 (2018)
18. Singh, R., Xu, J., Berger, B.: Global alignment of multiple protein interaction networks with application to functional orthology detection. Proc. Natl. Acad. Sci. **105**(35), 12763–12768 (2008)
19. Sun, Z., Hu, W., Zhang, Q., Qu, Y.: Bootstrapping entity alignment with knowledge graph embedding. In: IJCAI, vol. 18, pp. 4396–4402 (2018)
20. Sun, Z., Deng, Z.H., Nie, J.Y., Tang, J.: Rotate: knowledge graph embedding by relational rotation in complex space. In: International Conference on Learning Representations (2018)
21. Trouillon, T., Welbl, J., Riedel, S., Gaussier, É., Bouchard, G.: Complex embeddings for simple link prediction. In: International Conference on Machine Learning, pp. 2071–2080. PMLR (2016)

22. Trung, H.T., Van Vinh, T., Tam, N.T., Yin, H., Weidlich, M., Hung, N.Q.V.: Adaptive network alignment with unsupervised and multi-order convolutional networks. In: 2020 IEEE 36th International Conference on Data Engineering (ICDE), pp. 85–96. IEEE (2020)
23. Viswanath, B., Mislove, A., Cha, M., Gummadi, K.P.: On the evolution of user interaction in Facebook. In: Proceedings of the 2nd ACM Workshop on Online Social Networks, pp. 37–42 (2009)
24. Wang, Z., Zhang, J., Feng, J., Chen, Z.: Knowledge graph embedding by translating on hyperplanes. In: Proceedings of the AAAI Conference on Artificial Intelligence, vol. 28 (2014)
25. Yang, B., Yih, W.T., He, X., Gao, J., Deng, L.: Embedding entities and relations for learning and inference in knowledge bases. arXiv preprint arXiv:1412.6575 (2014)
26. Zeng, W., Zhao, X., Wang, W., Tang, J., Tan, Z.: Degree-aware alignment for entities in tail. In: Proceedings of the 43rd International ACM SIGIR Conference on Research and Development in Information Retrieval, pp. 811–820 (2020)
27. Zhang, J., Philip, S.Y.: Multiple anonymized social networks alignment. In: 2015 IEEE International Conference on Data Mining, pp. 599–608. IEEE (2015)
28. Zhang, S., Tong, H.: Final: fast attributed network alignment. In: Proceedings of the 22nd ACM SIGKDD International Conference on Knowledge Discovery and Data Mining, pp. 1345–1354 (2016)
29. Zhou, C., Liu, Y., Liu, X., Liu, Z., Gao, J.: Scalable graph embedding for asymmetric proximity. In: Proceedings of the AAAI Conference on Artificial Intelligence, vol. 31 (2017)
30. Zhou, F., Liu, L., Zhang, K., Trajcevski, G., Wu, J., Zhong, T.: Deeplink: a deep learning approach for user identity linkage. In: IEEE INFOCOM 2018-IEEE Conference on Computer Communications. pp. 1313–1321. IEEE (2018)

OntoCA: Ontology-Aware Caching for Distributed Subgraph Matching

Yuzhou Qin[1], Xin Wang[1(✉)], Wenqi Hao[1], Pengkai Liu[1], Yanyan Song[1], and Qingpeng Zhang[2]

[1] College of Intelligence and Computing, Tianjin University, Tianjin, China
{yuzhou_qin,wangx,haowenqi,liupengkai,songyanyan1895}@tju.edu.cn

[2] School of Data Science, City University of Hong Kong, Hong Kong, China
qingpeng.zhang@cityu.edu.hk

Abstract. With the growing applications of knowledge graphs in diverse domains, the scale of knowledge graphs is dramatically increasing. Based on the fact that a high percentage of queries in practice is similar to previous queries, extensive caching methods have been proposed to accelerate subgraph matching queries by reusing the results of previous queries. However, most existing methods show poor performance when dealing with distributed subgraph matching queries, as numerous intermediate results from the caching should be transmitted to the worker nodes for further validation, leading to extra communication and computation overhead. In this paper, we propose a novel ontology-aware caching method, called OntoCA, which leverages ontology information for efficient distributed queries. Unlike the existing caching methods, our approach fully employs semantic reasoning to filter intermediate results at an early stage, thus improving the query performance. Furthermore, a workload-adaptive prefetching strategy is proposed to increase the hit ratio of OntoCA. The experimental results show that our proposed OntoCA and prefetching strategy outperforms the existing state-of-the-art distributed query method, reducing the query times by 56.16%.

Keywords: Subgraph matching · Ontology · Caching · Partial evaluation

1 Introduction

The Resource Description Framework (RDF) [1] is a W3C recommendation for representing knowledge graphs, and SPARQL Protocol and RDF Query Language (SPARQL) [2] has become a de-facto standard query language to retrieve information from RDF. With the growing applications of knowledge graphs in diverse domains, the volume of RDF data published on the Web is increasing, and complex subgraph matching queries for analyzing RDF data have been emerging, asking for efficient methods to accelerate queries.

As a high percentage of queries in practice has a similarity to some extent with previous queries, numerous caching methods have been proposed to optimize subgraph matching queries by reusing the results of previous queries [3].

B. Li et al. (Eds.): APWeb-WAIM 2022, LNCS 13421, pp. 527–535, 2023.
https://doi.org/10.1007/978-3-031-25158-0_42

However, existing methods focus on queries and do not consider the characteristics of distributed computating. Numerous intermediate results from the caches still need to be transmitted to the worker nodes for further validation, which introduces a significant communication and computation overhead, leading to poor query performance in a distributed environment.

RDFS [4] (RDF Schema) is an extension of the basic RDF, which provides ontology description and supports the inference of semantic information. Based on RDFS, to fully consider the characteristics of distributed computing, we proposed a novel caching method, called OntoCA, which overcomes the shortcomings of the traditional cache mechanism by exploiting the ontology information defined in RDFS to filter intermediate results at an early stage.

Our contributions can be summarized as follows:

(1) We design a novel caching method, called OntoCA, which exploits ontology information to prune the intermediate results, thus improve the communication and computation performance.
(2) A workload-adaptive prefetching strategy aiming at increasing the hit ratio of OntoCA has been proposed, which can further improve the performance of queries.
(3) The extensive experiments have been conducted to verify the efficiency of our methods. The experimental results show that OntoCA with the prefetching strategy outperforms the state-of-the-art methods.

The rest of this paper is organized as follows. Section 2 reviews related works. In Sect. 3, we introduce preliminary definitions. In Sect. 4, we describe the OntoCA schema, workload-adaptive prefetching strategy, and the query processing in detail. Section 5 shows the experimental results, and we conclude in Sect. 6.

2 Related Work

In practice, numerous subgraph matching queries share common structures of query patterns. Based on this fact, extensive caching approaches aiming to accelerate subgraph matching queries have been proposed [3]. Papailiou et al. [5] propose a query canonical labeling algorithm to identify and request queries from the caches, which can be integrated with a dynamic programming planner to generate the optimal execution plan. To limit the overhead space, Zhang et al. [6] propose cache frequently accessed triples, i.e., hot triples, instead of subquery results. Liang et al. [7] propose a workload-aware subgraph querying framework that can exploit query workload for rewriting queries, reusing partial results, and filtering false positive results. Bok et al. [8] propose a two-level caching scheme based on graph connectivity and statistics of subgraph patterns.

However, these methods do not suitable for distributed computing and do not exploit the rich ontology information provided by RDF datasets. In this paper, we take full account of the characteristics of distributed computing and propose OntoCA to accelerate subgraph matching queries by filtering false positive results earlier according to ontology information.

3 Preliminaries

In this section, we introduce the definitions of relevant background knowledge.

Definition 1 (RDF Graph). *Given an RDF dataset as a finite set of triples of the form (s, p, o), its corresponding RDF graph is $G = (V, E, \Sigma)$, where V, E, and Σ denote the set of vertices, edges, and edge labels in G, respectively.*

Definition 2 (BGP Query). *A BGP (Basic Graph Pattern) query is defined as $Q = (V^Q, E^Q, \Sigma^Q)$, where V^Q is a subset of vertices or variables, $E^Q \subseteq V^Q \times V^Q$ is the edge set in query Q, and for each edge $e \in E^Q$, its label is either a variable or in Σ^Q.*

Definition 3 (Subgraph Match). *For a connected query graph Q that has n vertices $\{v_1, ..., v_n\}$, a subgraph match M is a function that maps from $\{v_1, ..., v_n\}$ to $\{u_1, ..., u_m\}$ $(n \geq m)$, where $\{u_1, ..., u_m\} \subseteq V$. M is a result of Q if and only if the following conditions are satisfied:*

1) if v_i is a variable, $M(v_i) \in \{u_1, ..., u_m\}$ should be satisfied; otherwise v_i must be as the same as $M(v_i)$ which is either a URI or a literal value .
2) if there exists an edge $e_1 \in E^Q$ between v_i and v_j in Q, there also exists an edge $e_2 \in E$ between $M(v_i)$ and $M(v_j)$ in G, then e_1 can match e_2. Note that e_1 can match any edge in G if e_1 is a variable edge.

4 Ontology-Aware Caching

In this section, we propose an ontology-aware caching method, called OntoCA. First, we formally define the structure of OntoCA. Then a workload-adaptive prefetching strategy is presented to increase the hit ratio of OntoCA. Finally, we explain how we employ OntoCA in query processing.

4.1 Caching Schema

To filter intermediate results at an early stage, our ontology caching mechanism employs ontology information defined in RDFS. ρdf is a subset of the RDFS vocabulary, that is, {`sp`,`sc`,`type`,`domain`,`range`} (`sp`,`sc` is the abbreviation of `subPropertyOf`, `subClassOf`, respectively). The RDFS inference rules in Table. 1 are reliable and complete [9].

The region encoding is employed to encode the hierarchy ontology information of each class and property in RDFS. Region encoding for classes or properties in the ontology hierarchy is a pair $\langle pre, post \rangle$, where *pre* and *post* indicate the ordinal number of the nodes in the pre-order and post-order traversal sequence of the hierarchy. Formally, for two classes or properties i and j, after region encoding, i `isChildOf` j holds iff $pre_i \geq pre_j \wedge post_i \leq post_j$.

A caching item (CI) is a pair $(Q, \mathcal{M})$, where Q is a subgraph matching query, $\mathcal{M}$ is a result set that for each $M \in \mathcal{M}$, M is a result of Q.

Table 1. Part of RDFS Reasoning Rules

ID	Antecedent	Consequent
R1	p **domain** x, s p o	s **type** x
R2	p **range** x, s p o	o **type** x
R3	p **sp** q, q **sp** r	p **sp** r
R4	s p o, p **sp** q	s q o
R5	s **type** x, x **sc** y	s **type** y
R6	x **sc** y, y **sc** z	s **sc** z

Assume $\mathcal{CIS}$ is a set of CI, and there exist four mappings I, I_r, I_d, and μ, where I maps all entities or properties in the RDF graph to their region encoding, I_r and I_d map from property to its domain and range, respectively. μ maps vertices in the query results of OntoCA to its stored fragment identifier. Formally, OntoCA can be defined as $OntoCA = (\mathcal{CIS}, I, I_r, I_d, \mu)$.

4.2 Workload-Adaptive Prefetching Strategy

Hit ratio is a crucial factor in determining caching performance. Algorithm 1 presents the prefetching strategy designed to preload potentially useful data. Since most of queries have a low number of triples (from 0 to 2) in practice for the majority of the datasets [10], our prefetching strategy mainly focuses on patterns that have no more than two edges and frequently occur in previous queries, with frequent occurrences indicate that the patterns are more likely to be hit by subsequent queries.

Algorithm 1: Prefetching strategy

Input: Last K queries: $\mathcal{Q}^{\mathcal{H}}$, maximum support data size: $maxS$
Output: Caching Item set: $\mathcal{CIS}$

foreach $Q \in \mathcal{Q}^{\mathcal{H}}$ **do**
  for $s \in subgraph(Q, 2)$ **do**
    `// traverse all subqueries of` Q `with no more than two edges`
    $count[s] \leftarrow count[s] + 1$;
while $count \neq \emptyset \wedge size < maxS$ **do**
  $Q \leftarrow$ `top`$(count)$; `// retrieve the most frequent subquery`
  $\mathcal{M} \leftarrow$ `doQuery`(Q) ;
  $size \leftarrow size + |\mathcal{M}| \cdot |V^Q|$; `//` $|V^Q|$ `is the estimated size of a result`
  if $size < maxS$ **then**
    $\mathcal{CIS} \leftarrow \mathcal{CIS} \cup \{(Q, \mathcal{M})\}$
return $\mathcal{CIS}$;

Complexity Analysis. The time complexity of the prefetching algorithm is $O(|V^Q| \cdot |V| \cdot S + |V^Q|^3 \cdot K)$, where $|V^Q|$ is the average vertices number of the last K queries, $|V|$ is the vertices number of the RDF data set. S is employed to represent the average number of generated queries.

4.3 Query Processing with OntoCA

Here we describe query processing with OntoCA. This query processing is implemented based on the *partial evaluation and assembly* [11,12] framework, which is a distributed method that partitions RDF data into different fragments, then processes queries in each fragment in parallel to obtain partial results, and finally assembles these partial results in the master node to form the final results.

As shown in Fig. 1, for example, Q_2 = `(?a,worksFor,?b)` $\wedge$ `(?c,advisor,?a)`, to exploit the results in OntoCA, (1) we first lookup the caching item $CI_i = (Q_i, \mathcal{M}_i)$, where Q_i can match Q_2 or is a subgraph of Q_2. In Fig. 1 Q_i is equal to `(?a,worksFor,?b)`. (2) Then we exclude the invalid intermediate results in $\mathcal{M}_i$ through semantic reasoning. (3) After that, only two pairs, which are (`Professor1`, `University1`) and (`Professor2`, `University2`) need to be transmitted to worker nodes for further computation because we know only `professor` can be someone's `advisor` according to ontology information. (4) Finally, partial results in each worker node will be transmitted to the master node for assembly to obtain the final results, which will be stored in OntoCA to accelerate subsequent queries.

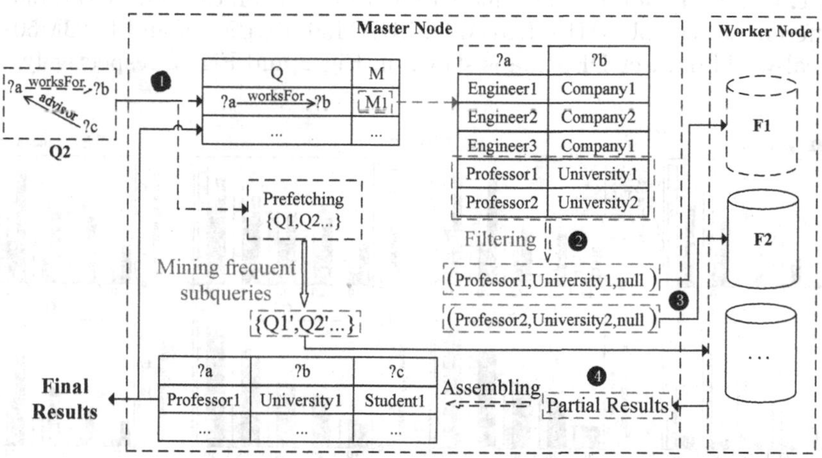

Fig. 1. The query processing with OntoCA

5 Experiments

In this section, we implement our caching method and verify its effectiveness compared with the baselines on several data sets.

5.1 Experimental Settings

The proposed OntoCA was implemented under the *partial evaluation and assembly* framework [11,12], deployed on a 3-node cluster, each node in a docker. The three dockers were deployed on a machine with 16 cores of Intel Xeon Silver 4216 2.10 GHz processors, 512GB of RAM, and 1.92 TB SSD, running 64-bit CentOS 7.7 operating system.

Data Sets. Our experiments were conducted on the real-world dataset DBpedia and the LUBM [13] synthetic datasets of five different scales. 16 benchmark queries[1] on LUBM and DBpedia were created, respectively.

Baselines. To verify the efficiency and effectiveness of OntoCA, we compared the following methods: OntoCA, which is our ontology-aware index; general caching, caching method without ontology; gStoreD [11,12], which is the state-of-the-art distributed RDF store that is based on *partial evaluation and assembly* framework.

Evaluation process. For Exp. 1, we employed the prefetching method to build the ontology caching and then tested the queries based on the caches. For Exp. 2, one set of our experiments applied the prefetching method before testing, and the other did not.

5.2 Experimental Results

Exp 1. Query Efficient. The query execution time on the five LUBM datasets of different scales (LUBM10, LUBM20, LUBM30, LUBM40, and LUBM50) and the real-world dataset DBpeida is shown in Fig. 2 and Fig. 3, respectively.

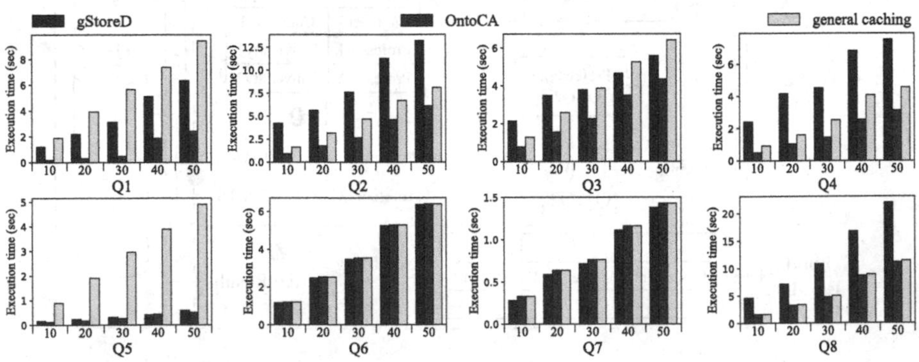

Fig. 2. The results of query execution time on different LUBM datasets

As shown in Fig. 2 and Fig. 3, for most queries, semantic reasoning in OntoCA has a significant impact on speeding up query processing. There are two main

[1] https://github.com/rainboat2/OntoCA.

reasons for this result: (1) for most of our queries that hit caches (Q_1-Q_5, Q_8), a significant number of false positive intermediate results can be filtered earlier by semantic reasoning, which significantly reduces communication overhead. In particular, for Q_1 and Q_5, OntoCA can filter intermediate results directly from the caches by **subClassOf** and **subPropertyOf** reasoning, without the need for communication; (2) as part of the results are known, a high percentage of computations cost to read disk is reduced. For example, in the query process of Q_3 and Q_8, OntoCA forwards more intermediate results than gStoreD, but it still takes less time, as a large number of disk read computations are reduced.

For queries that do not hit caches (Q_6, Q_7), As shown in Fig. 2 and Fig. 3, it can be seen that OntoCA takes slightly more time than gStoreD. There are two reasons: (1) caching lookup takes time, but OntoCA cannot accelerate those queries that miss hit; (2) caching building takes extra time. To add the query results to the caches, we need time to construct the mapping μ defined in OntoCA.

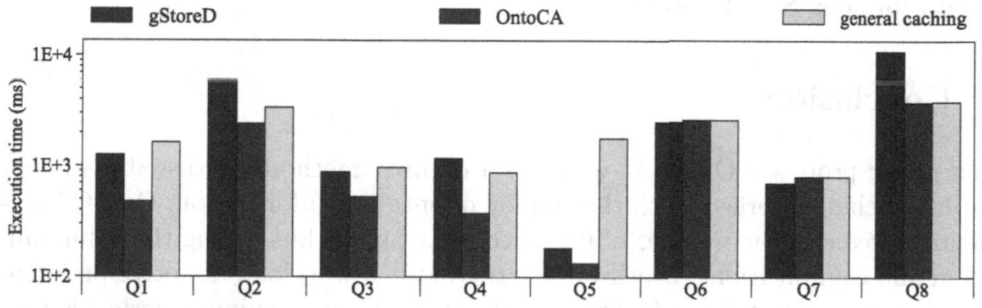

Fig. 3. The results of query execution time on DBpedia (logarithmic scale)

It is clear that OntoCA does not improve the speed of query processing when the caching is missed, which is one of the reasons why we propose the prefetching strategy, so we set up Exp. 2 to further illustrate this issue.

Exp 2. Prefetching Efficient. As shown in Fig. 4(a), we conducted further experiments on different sizes of the LUBM datasets to record the average processing time of eight queries above under different settings. The difference of average query time between OntoCA and gStoreD is insignificant due to the low hit ratio. However, for OntoCA, with the proposed prefetching strategy, we obtain a significant reduction in the average query time because most queries (Q_1-Q_5, Q_8) benefit from the data loaded by our prefetching method.

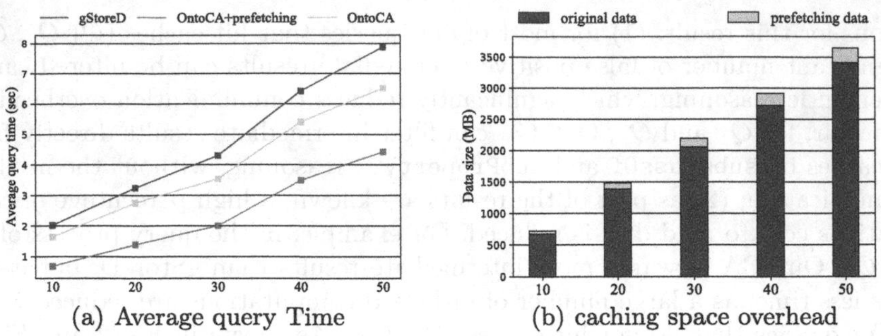

(a) Average query Time (b) caching space overhead

Fig. 4. Average query Time and caching space overhead on different LUBM datasets.

Figure 4(b) depicts the space overhead utilized by the prefetching method for varying data set sizes. As shown in Fig. 4(b), our prefetching strategy can improve the query performance dramatically by preloading a small percentage of data (approx. 8% of database size).

6 Conclusion

This paper proposes OntoCA, which is a caching method for distributed subgraph matching queries. With the employ of ontology information, OntoCA significantly reduces the number of intermediate results, thus saving the communication and computational overhead. Furthermore, a workload-adaptive prefetching strategy, which takes advantage of the statistics of previous queries, is proposed to increase the hit ratio of OntoCA. The extensive experiments were conducted to verify the effectiveness and efficiency of OntoCA and workload-adaptive prefetching strategy. The experimental results show that OntoCA outperforms the existing state-of-the-art distributed query method, reducing the query times by 56.16%.

Acknowledgments. This work is supported by the National Key Research and Development Program of China (2019YFE0198600).

References

1. World Wide Web Consortium et al. RDF 1.1 concepts and abstract syntax (2014)
2. World Wide Web Consortium et al. SPARQL 1.1 query language (2013)
3. Ali, W., Saleem, M., Yao, B., Hogan, A., Ngomo, A.-C.N.: A survey of RDF stores & SPARQL engines for querying knowledge graphs. VLDB J. 1–26 (2021)
4. Brickley, D., Guha, R.V., McBride, B.: RDF schema 1.1. W3C Recommendation 25, 2004–2014 (2014)
5. Papailiou, N., Tsoumakos, D., Karras, P., Koziris, N.: Graph-aware, workload-adaptive SPARQL query caching. In: Proceedings of the 2015 ACM SIGMOD International Conference on Management of Data, pp. 1777–1792(2015)

6. Zhang, Wei Emma, Sheng, Quan Z.., Taylor, Kerry, Qin, Yongrui: Identifying and caching hot triples for efficient RDF query processing. In: Renz, Matthias, Shahabi, Cyrus, Zhou, Xiaofang, Cheema, Muhammad Aamir (eds.) DASFAA 2015. LNCS, vol. 9050, pp. 259–274. Springer, Cham (2015). https://doi.org/10.1007/978-3-319-18123-3_16
7. Yi, J., Li, P., Choi, S.S., Bhowmick, B., Xu, J.: FLAG: towards graph query autocompletion for large graphs. Data Sci. Eng. **7**, 175–191 (2022)
8. Bok, K., Yoo, S., Choi, D., Lim, J., Yoo, J.: In-memory caching for enhancing subgraph accessibility. Appl. Sci. **10**(16), 5507 (2020)
9. Muñoz, Sergio, Pérez, Jorge, Gutierrez, Claudio: Minimal deductive systems for RDF. In: Franconi, Enrico, Kifer, Michael, May, Wolfgang (eds.) ESWC 2007. LNCS, vol. 4519, pp. 53–67. Springer, Heidelberg (2007). https://doi.org/10.1007/978-3-540-72667-8_6
10. Zhao, Y., Yunfei, H., Yuan, P., Jin, H.: Maximizing influence over streaming graphs with query sequence. Data Sci. Eng. **6**(3), 339–357 (2021)
11. Peng, P., Zou, L., Tamer Özsu, M., Chen, L., Zhao, D.: Processing SPARQL queries over distributed RDF graphs. VLDB J. **25**(2), 243–268 (2016)
12. Peng, P., Zou, L.,, Guan, R.: Accelerating partial evaluation in distributed SPARQL query evaluation. In: 2019 IEEE 35th International Conference on Data Engineering (ICDE), pp. 112–123. IEEE (2019)
13. Guo, Y., Pan, Z., Heflin, J.: LUBM: a benchmark for owl knowledge base systems. J. Web Semant. **3**(2–3), 158–182 (2005)

A Social-Aware Deep Learning Approach for Hate-Speech Detection

George C. Apostolopoulos, Panagiotis Liakos(✉), and Alex Delis

University of Athens, 15703 Athens, Greece
g.c.apostolopoulos@gmail.com, {p.liakos,ad}@di.uoa.gr

Abstract. Despite considerable efforts to automatically identify hate-speech in online social networks, users still face an uphill battle with toxic posts that seek to sow hatred. In this paper, we initially observe that there is a great deal of social properties transcending both hateful passages and respective authors. We then exploit this observation by *i)* developing deep learning neural networks that classify online posts as either hate or non-hate based on their content, and *ii)* proposing an architecture that may invigorate any such text-based classifier with the use of additional social features. Our combined approach considerably enhances the classification accuracy of previously proposed state-of-the-art models and our evaluation reveals social attributes that are the most helpful in our classification effort. We also contribute the first publicly-available dataset for hate-speech detection that features social properties.

Keywords: Social features · Deep learning · Twitter · User profile

1 Introduction

With more than half of the world's population actively using social media, much of the communication among individuals is now taking place online [8]. Massive online interactions on *social networks* (*SN*s) have incentivized malicious players to exploit pertinent infrastructures in pursuit of illicit actions. The well-documented correlation between violent acts and *SN*s inflammatory speech have necessitated the need for censoring such content [14]. In the past, the problem of automatic hate speech detection has been predominantly approached as a supervised document classification task. Such efforts mainly entail: *i)* traditional approaches employing feature engineering [4], and *ii)*deep learning approaches that automatically learn features from raw data using neural networks [16].

Feature engineering approaches have relied in the use of *surface* features including *dictionaries* of insults and swear-words [10], *N-grams* [6], as well as *URLs*, mentions, hashtags and capitalization [3]. Besides simple *surface* features, existing methods have employed sentiment analysis to identify negative polarity [9], *part of speech (POS)* to detect the role of each word in the context of a sentence [2], and meta-information related to the author of a passage and her past activity [7,15]. Using the above features, approaches have deployed *SVM*

B. Li et al. (Eds.): APWeb-WAIM 2022, LNCS 13421, pp. 536–544, 2023.
https://doi.org/10.1007/978-3-031-25158-0_43

and *Naive Bayes* classification algorithms to train models for hate-speech detection [3,7].

In [1], the *CNN* and *LSTM* network architectures are examined along the use of *random* and *GloVe* word embeddings [12] to obtain vector representations of terms. [1] shows that deep neural network architectures outperform traditional classifiers such as Logistic Regression, SVM and Gradient Boosted Decision Trees, offering greater classification accuracy. Although linear classifiers and deep learning models have certainly helped attain noteworthy accuracy in identifying hate-speech, these methods have strictly focused on the textual content posted by online users, without considering whether *social properties* can improve the performance of the proposed models.

In this paper, we advocate that using social properties is very beneficial to the task of identifying hate-speech in social media content. We focus on `Twitter` and apply deep learning methods that use *both* the post *content and* the *social properties* related to the user and the post itself. Our main contributions are:

- We propose a deep learning model that exploits the network structure, as well as individual social reactions, e.g., likes, to help identify hate-speech.
- We generate and make publicly available an annotated dataset of `Twitter` posts that do or do not feature hate-speech and is generated by the activity of different individuals. Previous available datasets are *biased* as they involve activity of a very limited number of users, and lead to training models that may penalize a particular writing style. This is the first hate-speech related dataset to include numerous social attributes associated with each post.
- We outperform state-of-the-art methods in terms of classification accuracy through our *combined approach* and demonstrate that network relationships and social activity do enhance the accuracy of hate-speech detection models. Moreover, our approach is orthogonal to text-based hate-speech detection as our architecture can complement classifiers focusing strictly on text.

The rest of the paper is organized as follows: in Sect. 2, we introduce our text-based models and discuss how we can enhance their effectiveness with a novel social-based deep learning approach. Section 3 describes the dataset we collected. Next, we present our experiments in Sect. 4, followed by the conclusion in Sect. 5.

2 Methodology

Our approach commences with a preprocessing step whose objective is to normalize the content of our dataset. This step first removes URLs, emojis, usernames and numbers. Moreover, we split hashtags on upper case letters, as oftentimes they are used to compose sentences and we need to come up with the individual words. Also, we split words appended with slashes, and we mark punctuation repetitions and elongated words. Finally, we transform all upper case letters to lower case for consistency. As an example, given the tweet:

> *@TheLeoTerrell: Good News!!! Leo 2.0 just received message from Team Trump. Going to be used in Campaign. Cannot wait to help*

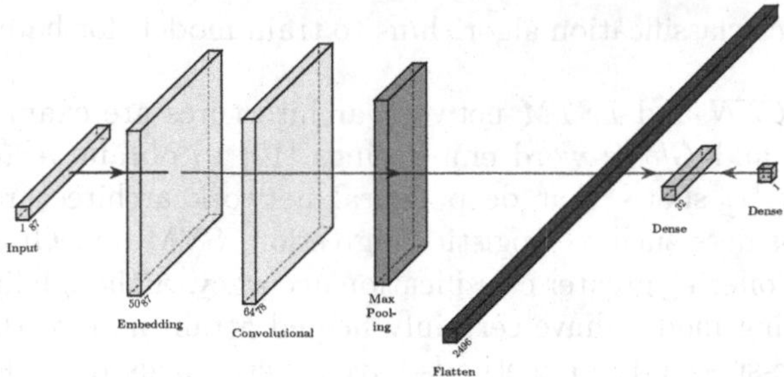

Fig. 1. Our convolutional neural network approach.

we come up with the following text which is ready for our vectorization process:

> *<user>: good news! leo <number> just received message from team trump. going to be used in campaign. cannot wait to help*

2.1 CNN Architecture

The first model we train is a CNN (Convolutional Neural Network), which is known in the literature to extract features effectively [11]. Figure 1 depicts the functional layout of the CNN in question whose components we outline below.

Input Layer: Our CNN model initially features an Input layer, to which we feed the vectors list. The maximum vector produced after the preprocessing of the tweets in our dataset has a length of 87 tokens.

Embedding Layer: The next layer of our architecture is an embedding layer, which allows words with similar meaning to have a similar representation. Word embeddings are one of the key breakthroughs for the impressive performance of deep learning methods on challenging natural language processing problems [5]. Individual words are represented as real-valued vectors in a predefined vector space. We use a pre-trained dimensional embedding called GloVe [12]. In our approach, we exploit a 50 dimensional pre-trained GloVe vector that is created using a total of 2 billion tweets, 27 billion tokens with an overall vocabulary of 1.2 million words. Our embedding layer produces a 87×50 representation.

Convolutional Layer: The output of the embedding layer is then fed into a convolutional layer that learns to extract salient features from the word embeddings. We use a $1D$ convolutional layer with 64 filters and a sliding window size of 10. These are the same initialization values used in the model of [1]. Our convolutional layer uses the rectified linear unit function for activation. This process convolves the input feature space into a 78×64 representation.

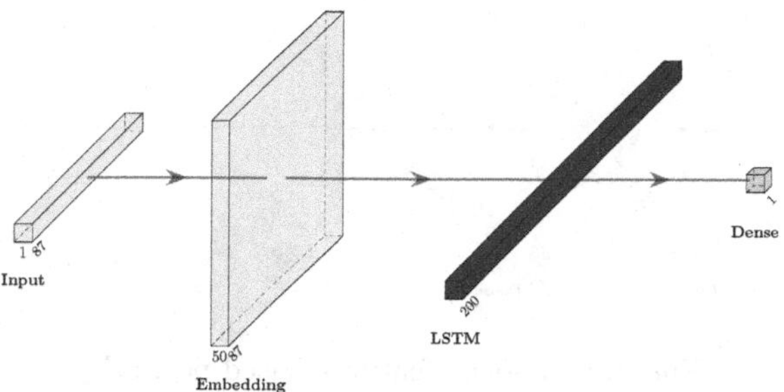

Fig. 2. Our LSTM network.

Pooling Layer: Next, we down-sample the incoming vectors using a pooling layer. Here, we use maximum pooling, which calculates the maximum, or largest, value in each patch of each feature map. This process halves the layer's input, producing a 39 × 64 representation.

Flatten and Dense Layers: The down-sampled vectors produced by the pooling layer are then flattened to a $1D$ array of size 39 × 64 with a Flatten layer. Next, follows a standard dense layer of 32 neurons. Finally, we classify the input as hate or not-hate, with the use of the sigmoid activation function, that is frequently used for binary classification.

2.2 LSTM Architecture

Our second model is based on an LSTM (Long Short Term Memory) network, and is shown in Fig. 2. The input and embedding layers are identical to our CNN approach. The vector length of our input layer is 87, defined by the length of the longest input vector in our dataset. The embedding layer is a 50 dimensional pre-trained GloVe vector, that produces a 87 × 50 output representation.

After the embedding layer, our model features an LSTM layer with a dropout probability. We use a bidirectional LSTM layer of 100 units to preserve information from both past and future, and better capture the context of sentences. Moreover, we set the dropout probability to be equal to 0.5. This leads to ignoring half of the input in each repetition, which is a well-known form of regularization that prevents neural networks from overfitting [13]. The size of the LSTM layer and the dropout probability have been selected after extensive experimentation and in accordance to earlier findings [13].

Finally, our LSTM based model features a dense layer that classifies the input with the use of the sigmoid activation function.

2.3 Social-Aware Deep Learning Network

Our CNN and LSTM networks perform text-based classification using the content of a user's post. Similar networks have been studied in the past and are

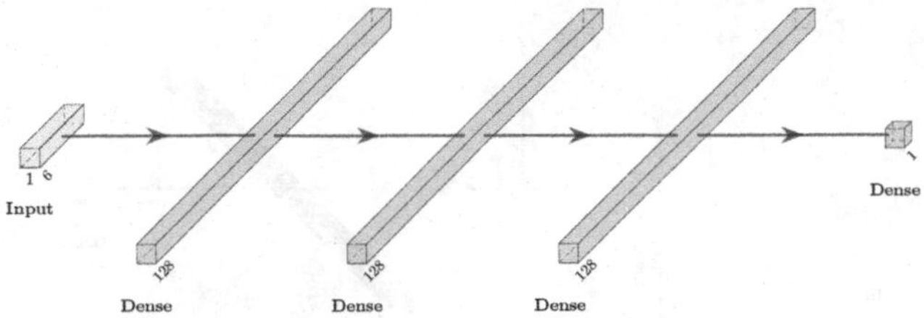

Fig. 3. Our social features based network

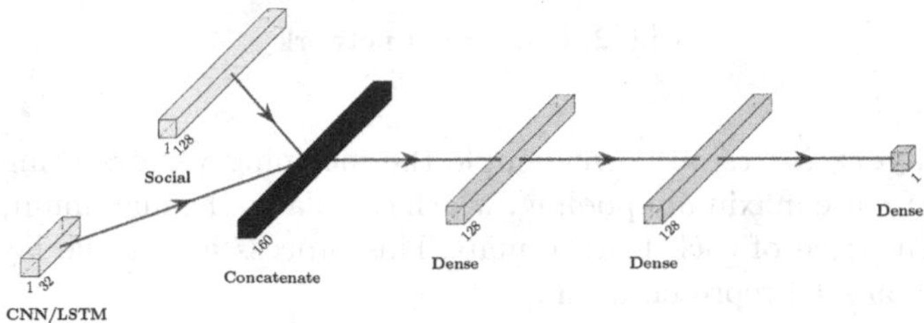

Fig. 4. Combining text-based models with our social features based networks.

shown to outperform traditional classifiers [1]. Here, we discuss how we can additionally exploit information related to the post. In particular, we enhance our deep learning approach with a second network that takes as an input additional social properties about the user and the post itself. Our working hypothesis is that this meta-information can help build more accurate models. Thus, we focus on the following six key features: *i)* user followers, *ii)* user followees, *iii)* user posts, e.g., tweets, *iv)* post shares, e.g., retweets, *v)* post likes, and *vi)* whether the post is a reply. Our goal is to uncover which of the above can enhance the performance of our text-based models. To this end, we build a neural network that comprises a series of consecutive dense non-linear layers and takes as input any combination of the above features; this network is depicted in Fig. 3.

The network of Fig. 3 is ultimately combined with our text-based models as portrayed in Fig. 4. Here, we deploy another network of hidden non-linear layers that takes as input the concatenated output of the network in Fig. 3 and any text-based models, such as those depicted in Figs. 1 and 2. This architecture allows for exploiting the social information through an agnostic –with regard to text classification– approach. Consequently, our contribution is orthogonal to previous approaches [1,16] that exclusively use the textual content of a post to detect hate-speech.

3 Dataset

Our dataset focuses on former US president, Donald J. Trump. The collection process took place just before the 2020 Presidential US elections, a time at

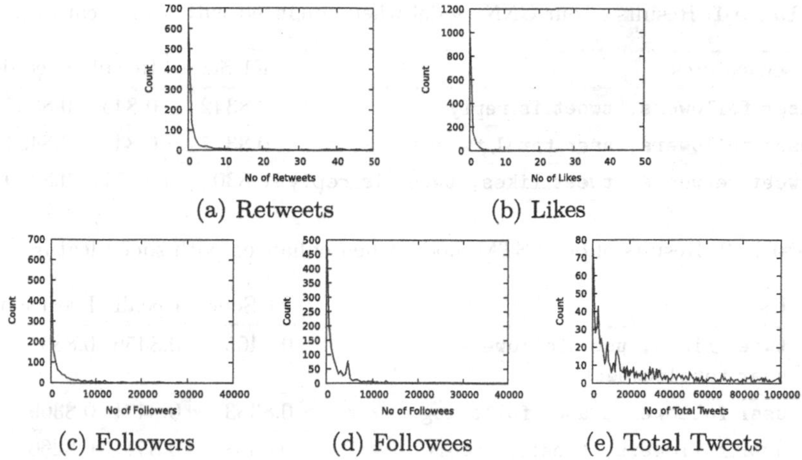

(a) Retweets (b) Likes

(c) Followers (d) Followees (e) Total Tweets

Fig. 5. Analysis of the social features of our dataset.

which media focused heavily on Trump, as a candidate. We followed a collection approach similar to that of [15,16]. In particular, we initially utilized the `Twitter` *API* through the `tweepy`[1] Python library to collect the tweets containing the word `Trump`. Naturally, most tweets did not feature hate speech. After a collection period of one month that ended 3 days before the elections, we had collected a total of 1,311 tweets, out of which 219 are labeled as hateful.

We also present a useful analysis of our social features in Fig. 5. We can see that most posts exhibit a small number of retweets or likes, with the former being more frequent than the latter. Moreover, our users include accounts who have neither followers nor followees as they are posting their very first tweets as well as people with significant activity and thousands of followers.

4 Experiments and Results

We implemented our models using Python `Keras` with the `TensorFlow`-backend and the `scikit-learn` library. Our implementation and dataset are publicly available.[2] For all models discussed in this section, we split the dataset into 80:20 to use 80% with cross-validation to tune learning epochs. We test the optimized model on the 20% held-out data and report average results of multiple executions using weighted precision, recall and *F1*-scores, as in [1,16]. The evaluation of our models answers the following key questions: (i) Does the use of social features improve the effectiveness of our models? (ii) What are the social features that prove to be most helpful?

We begin with an investigation of the potential of *social* features to enhance the performance of our models. There are a total of 63 combinations of features

[1] https://www.tweepy.org/.
[2] https://github.com/giorgos-apo/hate-speech-detection-using-user-attributes/.

Table 1. Results of our CNN model when enhanced with social features.

	User features	F1 Score	Recall	Precision
1	`user_followers, tweet_is_reply`	0.8342	0.8463	0.8387
2	`user_followers, user_total_tweets`	0.8327	0.8494	0.8404
3	`tweet_retweets, tweet_likes, tweet_is_reply`	0.8301	0.8517	0.8339

Table 2. Results of our LSTM model when enhanced with social features.

	User features	F1 Score	Recall	Precision
1	`tweet_likes, user_followers, user_total_tweets`	0.8405	0.8456	0.839
2	`user_followers, user_following`	0.8383	0.8471	0.8369
3	`user_followers, tweet_is_reply`	0.8318	0.841	0.8266

that we can utilize to enhance the performance of our models. We experiment with all and report in Table 1 the ones with the most significant gains, when combined with our CNN model. We obtain the best results when combining the CNN model with a network that takes as input the author's number of followers and whether the tweet is a reply or not. This combination improves the *F1*-score by almost 2 points.

Table 2 lists the combinations of social features that are most effective when combined with our LSTM network. We see that our combined approach manages again to increase the classification accuracy of this network even more than the case was with the CNN model. The most notable improvement is achieved when combining the number of user followers with the user's total number of tweets and the tweet's likes. This synthesis helps us increase the F1-score of the LSTM network by more than 4 points.

Overall, our combined approach shows significant improvement with regards to all precision, recall and accuracy for both our text-based deep-learning models. We note here that our architecture allows for any text-based classifier to be plugged into our model. We have experimented using a CNN and an LSTM network that are trained using the content of user posts. However, our contribution is orthogonal to text-based hate-speech detection and can be exploited by any such approach. More importantly, we show here that the use of meta-information regarding social properties of posts and their respective authors is very beneficial for the task of hate-speech detection. The results of Tables 1 and 2 quantify the importance of various contributing social features as they clearly point out that the number of user followers is very helpful when it comes to identify hate-speech content.

5 Conclusion

We present a novel approach that enhances classification accuracy of hate-speech detection models through the use of *social properties* related to content posted online. Our model exploits meta-information for each post we want to have classified as hate or non-hate, in addition to the actual text of the post. Our experimentation ascertains the importance of social properties in significantly improving the effectiveness of all text-based models we have deployed. Moreover, we investigate and quantify the importance of various social features with regard to detecting hate-speech. Our findings point out that the number of followers a user has is a very helpful property to consider when building hate-speech detection classification models. Combinations of social properties that include this feature improve our text-based classifiers significantly with regards to *F1*-score. Last but not least, we make available a new dataset, complementing existing ones by considering social features instead of exclusively focusing on text content.

References

1. Badjatiya, P., Gupta, S., Gupta, M., Varma, V.: Deep learning for hate speech detection in tweets. In: WWW 2017 (Companion), pp. 759–760. Perth, Australia (2017)
2. Burnap, P., Williams, M.L.: Cyber hate speech on twitter: an application of machine classification and statistical modeling for policy and decision making. Policy Internet **7**(2), 223–242 (2015)
3. Davidson, T., Warmsley, D., Macy, M.W., Weber, I.: Automated hate speech detection and the problem of offensive language. In: ICWSM 2017, pp. 512–515 (2017)
4. Fortuna, P., Nunes, S.: A survey on automatic detection of hate speech in text. ACM Comput. Surv. **51**(4), 1–30 (2018)
5. Deep Learning for Natural Language Processing. Apress, Berkeley (2018). https://doi.org/10.1007/978-1-4842-3685-7_5
6. Greevy, E., Smeaton, A.F.: Classifying racist texts using a support vector machine. In: SIGIR 2004, pp. 468–469. Sheffield, United Kingdom (2004)
7. He, J., Liu, H.: Bi-labeled LDA: inferring interest tags for non-famous users in social network. **5**, 27–47 (2020)
8. Liakos, P., Papakonstantinopoulou, K.: On the impact of social cost in opinion dynamics. In: Proceeding of the Tenth International Conference on Web and Social Media, Cologne, Germany, pp. 631–634 (2016)
9. Liu, S., Forss, T.: Combining N-gram based similarity analysis with sentiment analysis in web content classification. In: IC3K 2014, p. 530–537. Rome, Italy (2014)
10. Liu, S., Forss, T.: New classification models for detecting hate and violence web content. In: IC3K 2015, pp. 487–495 (2015)
11. Ordóñez, F.J., Roggen, D.: Deep convolutional and LSTM recurrent neural networks for multimodal wearable activity recognition. Sensors **16**(1), 115 (2016)
12. Pennington, J., Socher, R., Manning, C.D.: Glove: global vectors for word representation. In: EMNLP 2014. pp. 1532–1543. Doha, Qatar (2014)
13. Srivastava, N., Hinton, G., Krizhevsky, A., Sutskever, I., Salakhutdinov, R.: Dropout: a simple way to prevent neural networks from overfitting. J. Mach. Learn. Res. **15**(1), 1929–1958 (2014)
14. Vosoughi, S., Roy, D., Aral, S.: The spread of true and false news online. Science **359**(6380), 1146–1151 (2018)

15. Waseem, Z., Hovy, D.: Hateful symbols or hateful people? Predictive features for hate speech detection on Twitter. In: Proceedings of the NAACL Student Research Workshop, pp. 88–93. ACL, San Diego, CA, June 2016
16. Zhang, Z., Robinson, D., Tepper, J.: Detecting hate speech on Twitter using a convolution-GRU based deep neural network. In: ESWC 2018, pp. 745–760 (2018)

Community Detection Based on Deep Dual Graph Autoencoder

Zhiyuan Jiang[1], Kai Xu[1], Zhixiang Wu[1], Zhenyu Wang[1(✉)], and Hui Zhu[2]

[1] South China University of Technology, Guangzhou, China
{201921043738,Sekxu,202021045997}@mail.scut.edu.cn, wangzy@scut.edu.cn

[2] College of Economics and Trade Guangdong Mechanical and Electrical Polytechnic, Guangzhou, China

Abstract. Recently, researchers try to use graph neural networks (GNNs) to solve community detection, which is a fundamental task in social network analysis. We can promote targeted products and detect abnormal users by mining the community structure in social network. In this paper, we propose the Community Detection based on Deep Dual Graph Autoencoder (CDDGA). Our model consists of the deep dual graph autoencoder module and clustering layer module. The autoencoder module can simultaneously decode the graph structure and node content. The extension path between encoder and decoder is helpful to learn higher-order structural features. We use clustering layer module to achieve better clustering performance. Finally, both modules are jointly optimized to divide communities. The experimental results show that our algorithm outperforms several state-of-the-art community detection methods.

Keyword: Community detection · Graph autoencoders · Graph neural network

1 Introduction

Community detection, also known as graph clustering, is a fundamental task in social network analysis. A common definition of community detection is to partition the nodes in the graph into some disjoint groups. We can effectively promote targeted products, detect some abnormal users and identify terrorist organizations [2] by mining the community structure in social networks.

To effectively process graph-structured data, researchers try to use GNNs to solve community detection, such as GAE [3], MGAE [3], GALA [6], ARGA [5], etc. These models decode only graph structure or node content, which will weaken the learning of the graph structure or node content. Thus Wang et al. proposed GASN [11]. To alleviate the influence of over-smoothing problem caused by the multi-layer network, Hu et al. proposed GCLN [1]. The above methods are all two-steps framework which firstly achieves the node embeddings through GNN, and then uses the K-means or Spectral clustering method for community detection. Wang et al. proposed DAEGC [9], which designs a graph clustering layer to learn community structural features.

B. Li et al. (Eds.): APWeb-WAIM 2022, LNCS 13421, pp. 545–552, 2023.
https://doi.org/10.1007/978-3-031-25158-0_44

Motivated by the above observations, we propose the deep dual graph autoencoder model CDDGA for community detection in this paper. Our contributions can be summarized as follows:

- We propose a novel deep dual graph autoencoder model CDDGA to decode graph structure and node content simultaneously.
- We stack multiple layers and use the extension path to learn higher-order neighbor features.
- We jointly optimize the dual graph autoencoder and clustering layer to achieve better clustering performance.
- Our algorithm shows state-of-the-art performance compared with other baseline techniques on community detection tasks.

2 Related Work

There are a lot of non-Euclidean data in real life, such as social networks, academic citation networks, etc. It is difficult for traditional deep learning methods to model non-Euclidean data. To effectively represent non-Euclidean data, graph neural networks have been developed.

Kipf et al. proposed GCN in 2017 [4], which can aggregate the features of first-order neighbors. However, the coefficients are the same when the GCN aggregates neighbor features, but the central node's emphasis on neighbor is different in real life. Noticing this, Veličković et al. proposed GAT [8], which uses a single-layer linear neural network to learn the attention coefficients of neighbor nodes. Although the graph attention network has achieved great results in the processing of graph data, there are still some limitations in the understanding of graph structure. Thus Xu et al. proposed GIN [14], which uses multi-layer perceptron on GCN to learn a single injection function, which has a simple structure but is very effective.

However, these methods are two-step frameworks for community detection tasks, which first use GNN to learn node embeddings, and then use K-means or Spectral to divide communities. This framework cannot integrate community structural features into node embeddings, which may lead to suboptimal clustering performance.

3 Preliminary

We consider the task of community detection on attributed graph in this paper. An attribute network $G = V, E, X$ is consists of n nodes $V = \{v_1, v_2, ..., v_n\}$ and m edges $E = \{e_{ij}\} \subseteq V \times V$. Each node v_i has a vector x_i of q dimension that describes the node content information, and the vectors of these nodes constitute the attribute matrix $X \in \mathbb{R}^{n \times q}$. The topology of G can be defined as the adjacency matrix $A = (a_{ij})_{n \times n}$, if $e_{ij} \in E$, then $a_{ij} = 1$, otherwise $a_{ij} = 0$. We consider non-overlapping community, which is defined as $C = \{C_1, C_2, ..., C_k\}$, $C_i \cap C_j = \varnothing$, $\forall i, j$. Here, C_i denotes the i-th community. Given the graph G, community detection is to divide each node v_i in G into one of k communities through a mapping function $\mathcal{F}$.

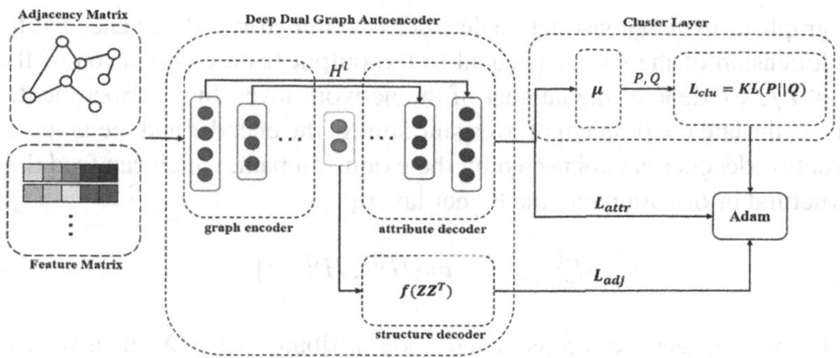

Fig. 1. Overall framework of CDDGA

4 Proposed Method

In this section, we present the Community Detection based on Deep Dual Graph Autoencoder (CDDGA) for community detection tasks. Our model consists of the deep dual graph autoencoder module and clustering layer module. The framework of the whole model is shown in Fig. 1.

4.1 Deep Dual Graph Autoencoder

The deep dual graph autoencoder module includes the graph encoder and the graph decoder. Inspired by the GAE [3], we use GCN as the layer of the graph encoder which can effectively encode both the graph structure and node content into the node embeddings. The l-th layer of the encoder is defined as follows:

$$H^{(l+1)} = \sigma\left(\tilde{D}^{-\frac{1}{2}}\tilde{A}\tilde{D}^{-\frac{1}{2}}H^{(l)}W^{(l)}\right) \tag{1}$$

where $\tilde{A} = A + I_N$, A denotes the adjacency matrix of the graph, and I_N denotes the identity matrix. $\tilde{D}_{ii} = \sum_j \tilde{A}_{ij}$, $W^{(l)}$ is the weight matrix of the l-th layer of the encoder. $\sigma(\cdot)$ means an activation function, here we use $ReLU(\bullet) = max(0, \bullet)$. $H^{(l)} \in \mathbb{R}^{n \times d}$ denotes input feature matrix of the l-th layer of the encoder, $H^{(0)} = X$.

The graph decoder, which simultaneously reconstruct the graph structure A and node content X, consists of the graph structure reconstruction decoder and the graph attribute reconstruction decoder.

The node embeddings learned by the graph encoder part contain structure and content information, if the value of $x_u \bullet x_v$ is larger, there is a high probability that an edge is connected between node u and node v, thus we choose a simple inner product decoder as the graph structure reconstruction decoder to predict the links between nodes.

$$\hat{A} = sigmoid\left(ZZ^T\right) \tag{2}$$

where $Z \in \mathbb{R}^{n \times d}$ denotes the node embeddings of the encoder output. We use the cross-entropy loss function to measure the difference between A and $\hat{A}$.

$$L_{adj} = -\frac{1}{n^2}\sum_i^n \sum_j^n a_{ij} log\left(\hat{a}_{ij}\right) \tag{3}$$

The graph attribute reconstruction decoder is symmetric with the encoder, the input feature dimension of the l layer is equal to the output feature dimension of the $L - l$ layer ($l \leq L/2$), L denotes the number of the network layer. In the experiment, $L = 8$. In order to alleviate the problem of over-smoothing, the encoder and the graph attribute reconstruction decoder are connected by the extension path, which can feed the lower-order structural information into the higher layer [1].

$$H_{expand}^{(l)} = sum\left(H^{(l)}, H^{(L-l)}\right) \tag{4}$$

Finally, we can get the reconstructed node attribute matrix $\widehat{X}$, then we use mean square error loss function to measure the difference between X and $\widehat{X}$.

$$L_{attr} = \frac{1}{n}\sum_i^n \|x_i - \hat{x}_i\|^2 \tag{5}$$

4.2 Clustering Layer

In this clustering layer, it is necessary to consider how to assign the nodes to different communities, when the embeddings of the nodes is obtained. The t-distribution can be used as a kernel function [13] to measure the similarity between the embedding h_i of node i and the embedding μ_j of cluster center j.

$$q_{ij} = \frac{\left(1+\|h_i-\mu_j\|^2/v\right)^{-\frac{v+1}{2}}}{\sum_{j'}\left(1+\|h_i-\mu_{j'}\|^2/v\right)^{-\frac{v+1}{2}}} \tag{6}$$

where v denotes the degrees of freedom in the t-distribution, we set $v = 1$. q_{ij} is a probability of assigning node i to community j, i.e. soft assignment probability. After obtaining the probability distribution $Q = \left[q_{ij}\right]$, we need to find a high-confidence probability distribution P, so that the model can optimize the node representations in the process of Q constantly approaching P. We use the method proposed by Xie [13] to calculate P by Q.

$$P_{ij} = \frac{q_{ij}^2/f_j}{\sum_{j'} q_{ij'}^2/f_{j'}} \tag{7}$$

Here, $f_j = \sum_i q_{ij}$ denotes the frequency of the soft cluster. Finally, we can use the KL divergence between the two probability distributions to obtain the following objective function.

$$L_{clu} = KL(P\|Q) = \sum_i \sum_j p_{ij} log\frac{p_{ij}}{q_{ij}} \tag{8}$$

To avoid instability in the optimizing process, we consider updating Q for several iteration before updating P in our experiment.

4.3 Joint Optimization

We jointly optimize the dual graph autoencoder and the clustering layer, we define total objective function as:

$$L = L_{adj} + \beta L_{attr} + \delta L_{clu} \tag{9}$$

where L_{adj} denotes the structure reconstruction loss, L_{attr} denotes the attribute reconstruction loss and L_{clu} denotes the clustering loss. β and δ are hyperparameter. We could gain our clustering result directly from the last optimized Q:

$$c_i = \underset{j}{\operatorname{argmax}} q_{ij} \tag{10}$$

5 Experiment

5.1 Experimental Settings

Datasets. We used three standard citation networks (Cora, Citeseer, and Pubmed) for experiments. The summary of each dataset is presented in Table 1.

Table 1. Benchmark graph datasets

Dataset	Nodes	Edges	Dims	Clusters
Cora	2708	5429	1433	7
Citeseer	3327	4732	3703	6
Pubmed	19717	44338	500	3

Baselines. We compared ten algorithms with our method in experiments.

K-means&Spectral are the most widely-used clustering algorithm.

DeepWalk [7] is a widely-used structural representation learning methods based on random walk.

M-NMF [12] is a novel modularity non-negative matrix factorization model that uses both microscopic graph structure and macroscopic community structure.

GAE&VGAE [3] combine graph convolutional network with the (variational) autoencoder to learn embeddings.

MGAE [10] introduces noise in the graph structure and node attribute.

ARGA&ARVGA [5] are adversarially regularized autoencoder, which can integrate features of graph structure and node content.

DAEGC [9] is a goal-directed graph clustering approach employing an attention network.

Metrics. In our experiment, we use three metrics to evaluate the community detection results: clustering accuracy (ACC), normalized mutual information (NMI) and adjusted rand index (ARI).

Parameter Settings. For the baseline algorithms, we carefully select the parameters for each algorithm, following the procedures in the original papers. We run the K-means algorithm 50 times to get an average value for all embedding learning baseline methods for fair comparison. For our method, we set the total number of network layers $L = 8$. For Cora and Citeseer we set the encoder neurons to be 1024–512-256–16 respectively, for Pubmed we set to 256–128-64–16. We uniformly set $\beta = 4$ and $\delta = 24$ for all the datasets. The update period of the probability distribution P is set to 3 for the stability of the optimization process. We train our model for 200 iterations using the Adam optimizer with the learning rate of 0.00001.

Table 2. Experimental results on datasets

Methods	Info	Cora			Citeseer			Pubmed		
		ACC	NMI	ARI	ACC	NMI	ARI	ACC	NMI	ARI
K-means	F	0.503	0.317	0.244	0.544	0.312	0.285	0.580	0.278	0.246
Spectral	G	0.398	0.297	0.174	0.308	0.090	0.082	0.496	0.147	0.098
DeepWalk	G	0.484	0.327	0.243	0.337	0.089	0.092	0.543	0.102	0.088
M-NMF	G	0.423	0.256	0.161	0.336	0.099	0.070	0.470	0.084	0.058
GAE	F&G	0.611	0.482	0.302	0.456	0.221	0.191	0.632	0.249	0.246
VGAE	F&G	0.592	0.408	0.347	0.467	0.261	0.206	0.619	0.216	0.201
MGAE	F&G	0.681	0.489	0.436	0.669	0.416	0.425	0.593	0.282	0.248
ARGA	F&G	0.640	0.449	0.352	0.573	0.350	0.341	0.681	0.276	0.291
ARVGA	F&G	0.638	0.450	0.374	0.544	0.261	0.245	0.513	0.117	0.078
DAEGC	F&G	0.704	0.528	**0.496**	**0.693**	0.397	0.410	0.671	0.266	0.278
CDDGA	F&G	**0.714**	**0.540**	0.492	0.689	**0.429**	**0.437**	**0.686**	**0.292**	**0.305**

5.2 Experiment Result

The experimental results of different methods on different datasets are summarized in Table 2, where the values marked in bold are the best among all methods. C, S and C&S indicate if the algorithm uses only content, structure, or both content and structure information, respectively.

We can see that our method is significantly better than most of the comparison methods in evaluation metrics, which shows the effectiveness of the algorithm on the task of community detection. The methods of GAE, MGAE, ARGA and DAEGC only decode graph structure and ignore higher-order structural features. Our method decodes both the graph structure and node content, which can better fuse structure and content information. The multiple network and extension path help learn higher-order structural features in our model. Therefore, the performance of CDDGA is better than most of the baseline methods.

Table 3. The results of ablation experiment

Methods	Cora			Citeseer			Pubmed		
	ACC	NMI	ARI	ACC	NMI	ARI	ACC	NMI	ARI
CDDGA/clu	0.680	0.532	0.442	0.669	0.410	0.404	0.652	0.295	0.270
CDDGA/adj	0.683	0.539	0.464	0.657	0.383	0.401	0.649	0.279	0.262
CDDGA/path	0.608	0.438	0.368	0.582	0.300	0.302	0.603	0.176	0.173
CDDGA	**0.714**	**0.540**	**0.492**	**0.689**	**0.429**	**0.437**	**0.686**	**0.292**	**0.305**

5.3 Ablation Study

In this section, in order to analyze the effectiveness of different modules in our model, we conduct ablation experiments, The experimental results are summarized in Table 3. The comparison experiments are described as follows.

CDDGA/clu: the clustering layer module is removed.

CDDGA/adj: the graph structure reconstruction decoder is removed.

CDDGA/path: the extension path is removed.

CDDGA: complete model.

Among the three ablation experiments, CDDGA/path performed worst, because of the problem of over-smoothing. CDDGA/clu achieves great results, but it is worse than the complete model because it does not learn the features related to community structure. CDDGA/adj also lower than the complete model, because the model will focus more on the content information of nodes. According to the results, every part in the model is important.

6 Conclusion and Further Work

In this paper, we propose a novel deep dual graph autoencoder framework for community detection task. A comparison of the experimental results with several state-of-the-art algorithms validate CDDGA's community detection performance. In the future, we will try to study community detection in heterogeneous graph.

Acknowledgement. This work is funded by NSF of Guangdong (No. 2019A1515011792) ;13th five year plan project of philosophy and social sciences of Guangdong (No.GD20XYJ20); key soft science project of Guangdong electromechanical vocational and Technical College (No.YJZD20210002).

References

1. Hu, R., Pan, S., Long, G., Lu, Q., Zhu, L., Jiang, J.: Going deep: graph convolutional ladder-shape networks. Proceedings of the AAAI Conference on Artificial Intelligence **34**, 2838–2845 (2020)

2. Jin, D., et al.: A survey of community detection approaches: from statistical modeling to deep learning. IEEE Transactions on Knowledge and Data Engineering, Early Access Article (2021)
3. Kipf, T.N.,Welling, M.: Variational Graph Auto-Encoders. arXiv preprint arXiv:1611.07308 (2016)
4. Zang, Y., et al.: GISDCN: a graph-based interpolation sequential recommender with deformable convolutional network. In: International Conference on Database Systems for Advanced Applications. Springer, Cham (2022). https://doi.org/10.1007/978-3-031-00126-0_21
5. Pan, S., Hu, R., Long, G., Jiang, J., Yao, L., Zhang, C.: Adversarially Regularized Graph Autoencoder for Graph Embedding. arXiv preprint arXiv:1802.04407 (2018)
6. Park, J., Lee, M., Chang, H.J., Lee, K., Choi, J.Y.: Symmetric graph convolutional autoencoder for unsupervised graph representation learning. In: Proceedings of the IEEE/CVF International Conference on Computer Vision, pp. 6519–6528 (2019)
7. Perozzi, B., Al-Rfou, R.,Skiena, S.: Deepwalk: Online learning of social representations. In: Proceedings of the 20th ACM SIGKDD International Conference on Knowledge Discovery and Data Mining, pp. 701–710 (2014)
8. Veličković, P., Cucurull, G., Casanova, A., Romero, A., Lio, P., Bengio, Y.: Graph Attention Networks. arXiv preprint arXiv:1710.10903 (2017)
9. Wang, C., Pan, S., Hu, R., Long, G., Jiang, J., Zhang, C.: Attributed Graph Clustering: A Deep Attentional Embedding Approach. arXiv preprint arXiv:1906.06532 (2019)
10. Wang, C., Pan, S., Long, G., Zhu, X., Jiang, J.: Mgae: Marginalized graph autoencoder for graph clustering. In: Proceedings of the 2017 ACM on Conference on Information and Knowledge Management, pp. 889–898 (2017)
11. Wang, J., Liang, J., Yao, K., Liang, J., Wang, D.: Graph convolutional autoencoders with co-learning of graph structure and node attributes. Pattern Recogn. **121**, 108215 (2022)
12. Wang, X., Cui, P., Wang, J., Pei, J., Zhu, W., Yang, S.: Community preserving network embedding. In: Thirty-first AAAI Conference on Artificial Intelligence, **31**(1) (2017)
13. Xie, J., Girshick, R.,Farhadi, A.: Unsupervised deep embedding for clustering analysis. In: International conference on machine learning, pp. 478–487. PMLR (2016)
14. Xu, K., Hu, W., Leskovec, J., Jegelka, S.: How Powerful are Graph Neural Networks? arXiv preprint arXiv:1810.00826 (2018)

SgIndex: An Index Structure Supporting Multiple Graph Queries

Shibiao Zhu, Yuzhou Huang, Zirui Zhang, and Xiaolin Qin(✉)

Nanjing University of Aeronautics and Astronautics, Nanjing, China
qinxcs@nuaa.edu.cn

Abstract. With the rise of social networks, traffic navigation and other fields, graph applications become increasing extensive. To improve query efficiency, indexes are built to manage large-scale graph data. However, these indexes cost large memory space and can only support one single graph operation. We propose a two-layered index structure SgIndex, in which the first layer stores subgraph information, and the second layer stores adjacency information to support multiple path operations and subgraph matching queries. We propose a subgraph matching algorithm based on path join, which completes subgraph matching by searching SgIndex twice. The experimental results show that SgIndex achieves better performance on path queries and subgraph matching than existing index structures, and reduces memory overhead.

Keywords: Index · Graph data · Path query · Subgraph matching

1 Introduction

Large-scale graph data with complex internal structure and diverse query requirements have emerged in different fields and various data operations have appeared in the application of large-scale graph data [2,11,15]. On the one hand, some algorithms have difficulties in adapting to large-scale graphs. On the other hand, most of the existing index structures lack generality. In practical applications, multiple indexes need to be established for the same graph to response to different query requirements. Therefore, this paper proposes a subgraph-based index structure SgIndex: for large-scale graph data, establish an index structure that meets various query requirements.

First, graph operations are expensive on large-scale graph data. Compared with traditional relational data or XML tree, graph data lacks structural constraints and is complicated to operate. Second, flexible and diverse query requirements for large-scale graph data mean that multiple indexes need to be built to meet these requirements. Furthermore, graph data operations usually require loop iterations, which is also the part that our index needs to support.

We have made the following contributions in this paper: 1. We propose an index that supports large-scale datasets, which reduces storage space by optimizing the design of the index; 2. The index structure can implement various

B. Li et al. (Eds.): APWeb-WAIM 2022, LNCS 13421, pp. 553–561, 2023.
https://doi.org/10.1007/978-3-031-25158-0_45

queries on graph data such as path query and subgraph query, including two-point path query, path query with limited path length, path query with limited path hops, and subgraph query.

2 Related Work

Graph is already a widely used data structure, but querying large-scale graph data has always been a difficult problem [5]. A popular solution is to build an index that does not waste too much storage space, but also produces effective positive feedback on the query process under various circumstances. However, in practical applications, there are often multiple queries that need to be satisfied for the same graph data [2,14,15] (Fig. 1). Given a directed graph G and two vertices s and t in it, a reachability query asks G if there is a path from s to t. Reference [13] proposes a set of tools to quantify and query network reachability, and uses decision graphs as data structures to represent reachability matrices. Based on two classical shortest path algorithms, the shortest path problem can be decomposed into a linear complex problem [1,8], or the target solution can be optimized in terms of distance or time [7,9,10], or one or more preprocessing steps to speed up the shortest path query time [4,12]. Subgraph is one of the basic concepts of graph theory, which refers to a graph in which the vertex set and the edge set are subsets of the vertex set and edge set of a certain graph, respectively. Reference [3] adopts a left-side deep join ranking strategy, which

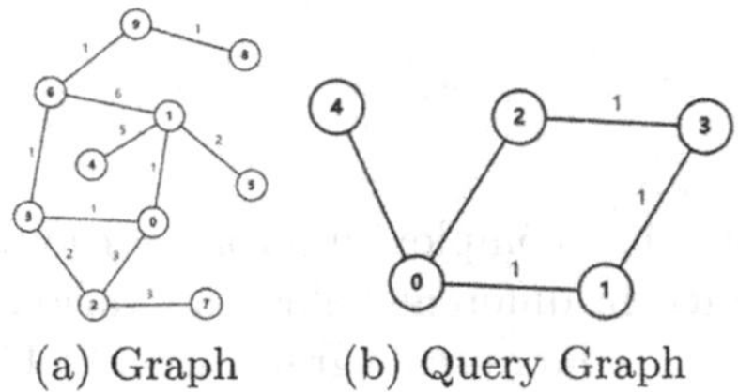

(a) Graph (b) Query Graph

Fig. 1. Graph example and query graph example

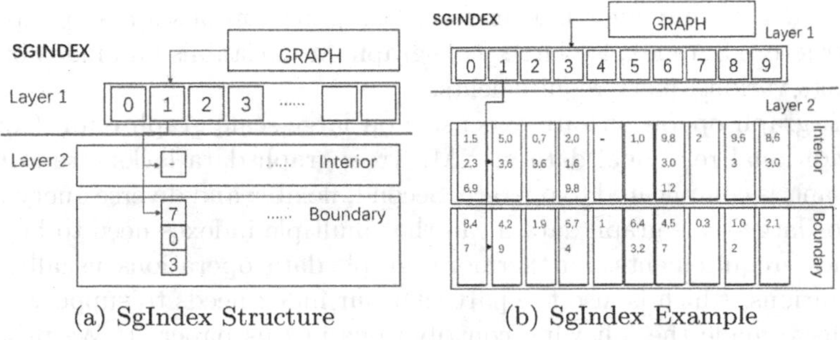

(a) SgIndex Structure (b) SgIndex Example

Fig. 2. SgIndex structure diagrame

models the enumeration process as a join problem. Subtree features are also used for indexing, and they take less time to index than more general subgraph features. BINDEX [6] is a secondary index with excellent query efficiency. It consists of Filter Layer and Refine Layer.

3 Index Based on Subgraph

3.1 Index Structure and Establishment Method

Algorithm 1. Create SgIndex

Input: Graph data $G(V_G, E_G)$; Vertex set V_G; Edge set E_G; Threshold r;
Output: Index I;

```
i ← 0;
while i < |E_G| do
    if e_i.weight < r then
        Put e_i.end into the interior collection of e_i.start;
    else
        Put e_i.end into the boundary collection of e_i.start;
    i++;
i ← 0;
while i < |V_G| do
    Get interior collection of v_i as IN_i;
    j ← 0;
    while j < |IN_i| do
        Get interior node of v_j;
        Calculate the sum of the weights w;
        if w < r then
            Put the interior node of v_j into the interior collection of v_i;
        else
            Put the interior node of v_j into the boundary collection of v_i;
        j++;
    i++;
```

SgIndex is a subgraph-based index structure. This index has two layers. First, we need to locate each vertex of the graph to the head of the adjacency list by hash. Secondly, for each vertex, we store interior vertex information and boundary vertex information for it, as shown in Fig. 2(a). Both types of vertex information include vertex value, path weight, and path hop. For each vertex, its internal vertices are vertices in the local subgraph, and the vertices directly adjacent to subgraph are classified as boundary vertices. The scale of the subgraph is denoted as threshold.

The selection of the threshold is not an optimal problem. For the currently implemented query, the larger the threshold, the faster the query speed. The only cost is the space occupied by the storage index and the cost of index establishment. Although for local information query, such as general contact query of the new crown epidemic, fast query can be guaranteed as long as the limit threshold is a specified number of hops, but in practical applications, it is often impossible to easily predict the required query scale, so the threshold should be chosen based on acceptable storage overhead.

Algorithm 1 gives the pseudocode of the new index algorithm. First we fill the edge information of the graph into the index (lines 1–7). Second we iterate over the interior vertices of each vertex until the complete subgraph information is stored (lines 8–20). We can build an index as shown in Fig. 2(b).

3.2 Path Query Algorithm

Algorithm 2. Path Query

Input: SgIndex *I*; Start Point *start*; End Point *end*; Empty List *path*;
Output: Result Path List *path*;

```
1:  Get IN_s, the interior vertex set of start, from I;
2:  Get BN_s, the boundary vertex set of start, from I;
3:  if end ∈ IN_s then
4:      return path;
5:  else
6:      i ← 0;
7:      while i < |BN_s| do
8:          path ← FindPath(I, v_i, end, path);
9:          if path is found then
10:             return path;
11:         i++;
```

Algorithm 2 shows how to find a path of two points. First we query the index and get the interior set and boundary set of the starting vertex (lines 1–2). If the interior set contains the end vertex *end*, it will return *path* directly (lines 3–4). If not, it will iteratively search in the boundary set, and each round of iteration will use the vertex in the boundary set as the new starting vertex to perform a new path query, until the path is found (lines 5–11). The function $FindPath(I, v_1, v_2, path)$ indicates that by querying the index I, it is judged whether v_2 is in the interior set of v_1. If it exists, *path* is updated. If it does not exist, the next iteration will be performed in the boundary set of v_2.

Algorithm 3. Subgraph Matching

Input: SgIndex I; Query Graph G_q;
Output: Result Subgraph G_s;

1: Decompose subgraph G_q into set $Path$ and get join condition $Join$;
2: $i \leftarrow 0$;
3: **while** $i < |Path|$ **do**
4: $\quad CandidatePath_i \leftarrow Searchpath(I, path_i.hop, path_i.weight)$;
5: $\quad$ i++;
6: $i \leftarrow 0$;
7: **while** $i < |CandidatePath_0|$ **do**
8: $\quad j \leftarrow 0$;
9: $\quad$ **while** $j < |CandidatePath_1|$ **do**
10: $\quad\quad$ **if** $(candidatepath_{0_i}, candidatepath_{1_j}) \in Join$ **then**
11: $\quad\quad\quad G_s \leftarrow Join(candidatepath_{0_i}, candidatepath_{1_j})$;
12: $\quad\quad$ j++;
13: $\quad$ i++;
14: $iternum \leftarrow 2$;
15: **while** $iternum < |Path|$ **do**
16: $\quad G_s \leftarrow Union(G_s, CandidatePath_{internum})$;
17: $\quad iternum++$;
18: return G_s;

3.3 Subgraph Matching Algorithm

The subgraph matching algorithm based on this index can be divided into three steps: 1. Decompose the query graph into multiple paths (line 1); 2. Generate an n-tuple for each path by index (lines 2–5); 3. Compare n-tuples by index to get the resulting subgraphs (lines 6–18).

The function $Searchpath(I, h, w)$ means to search for paths with $v_i.hop = h$ and $v_i.weight = w$ through index I. The function will output a collection of paths. The function $Join(path_1, path_2)$ means that the two paths and the join condition are combined into a subgraph, then the subgraph will be added to the subgraph set G_s. The function $Union(G_s, Path)$ compares subgraphs in the set G_s with paths in the set $Path$, and unions sunbgraphs and paths that meet the join conditions.

4 Experiments

4.1 Experimental Environment and Dataset

The machine used in this experiment has Intel(R) Core(TM) i5-10300H CPU @ 2.50 GHz processor, 16.0 GB memory, 64-bit operating system. The compiler used is Visual Studio 2019.

The datasets are all from KONECT, respectively: wikipedia_link_mi(Wmi), wikipedia_link_lez(Wlez), wikipedia_link_sah(Wsah), wikipedia_link_cy(Wcy), wikipedia_link_bn(Wbn). The scale of these datasets is shown in Table 1 (Fig. 4).

Table 1. Datasets

Name	Vertices	Edges
Wmi	7,996	116,464
Wlez	5,171	204,133
Wsah	15,531	352,209
Wcy	142,648	2,967,435
Wbn	226,501	2,832,143

4.2 Path Query

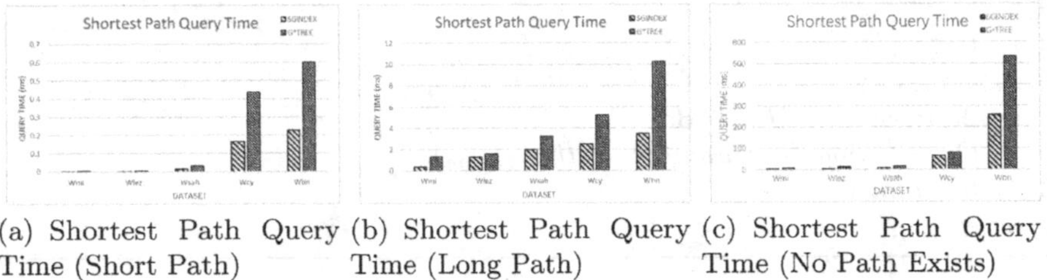

(a) Shortest Path Query Time (Short Path) (b) Shortest Path Query Time (Long Path) (c) Shortest Path Query Time (No Path Exists)

Fig. 3. Shortest path query

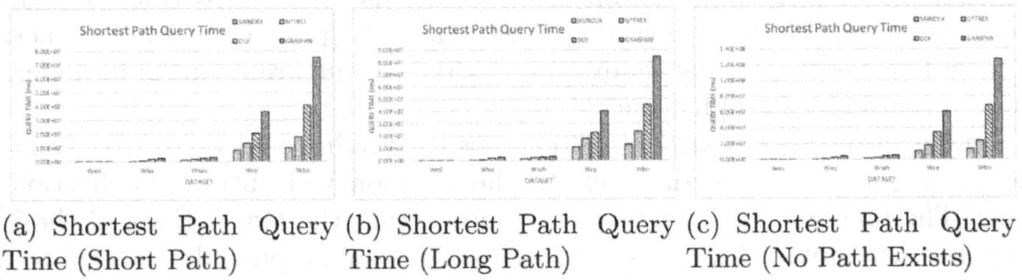

(a) Shortest Path Query Time (Short Path) (b) Shortest Path Query Time (Long Path) (c) Shortest Path Query Time (No Path Exists)

Fig. 4. Shortest path query (compare with system)

Figure 3(a) illustrates that the shortest path query time increases with the size of the dataset. For the smallest dataset Wmi, the query time on SgIndex is 50% of that on G*Tree; while for the largest dataset Wbn, the query time required on SgIndex is 35% of that on G*Tree. SgIndex performs better on larger datasets, which is determined by the structure of the bi-level index. SgIndex summarizes and stores subgraph information so that the shortest path can be obtained without traversing the entire graph. Figure 3(b) shows that SgIndex achieves better results in longer paths. Figure 3(c) is our test for extreme cases. When the required path does not exist, we need to traverse at least all the successor vertices of the starting vertex. Experiments show that SgIndex has withstood this test and achieved an advantage of more than 50%. Figure 3(d) is a comparison between using our index structure and using a mature graph data management system.

4.3 Subgraph Matching

We select the classic SPATH index in the field of subgraph matching and the above G*Tree as a comparison, and use the above real graph data to conduct experiments. Since SPATH takes up too much memory on large graphs, it exceeds the limit supported by our experimental environment. We only conduct experiments on three datasets, Wmi, Wlez and Wsah. We select subgraphs with 5, 6, 8, and 10 vertices for experiments. As can be seen from Fig. 5, SgIndex is more efficient than SPATH in subgraph matching query, especially with the increase of subgraph size. This is because the index established by SPATH still adopts the traditional method of finding candidate vertices and pruning them one by one. However, our method combines the advantages of indexing and invokes indexing in both the search candidate path stage and the pruning stage, thereby improving the overall query efficiency. Figure 6 shows how our index compares to different systems.

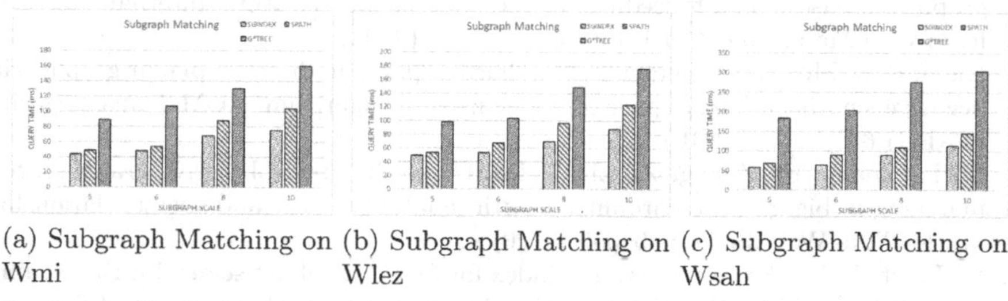

(a) Subgraph Matching on Wmi (b) Subgraph Matching on Wlez (c) Subgraph Matching on Wsah

Fig. 5. Subgraph matching

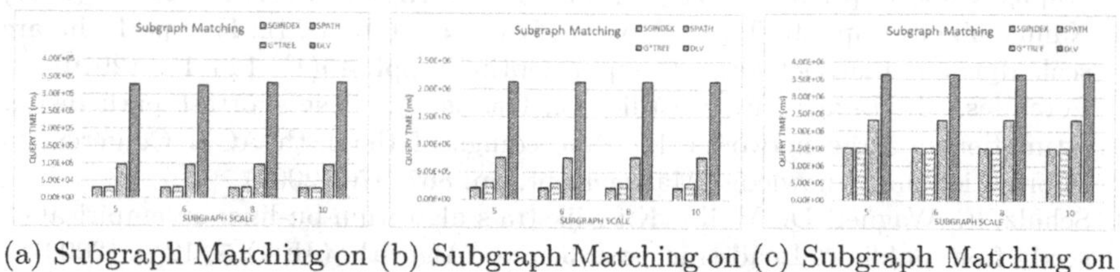

(a) Subgraph Matching on Wmi (b) Subgraph Matching on Wlez (c) Subgraph Matching on Wsah

Fig. 6. Subgraph matching (compare with system)

5 Conclusion

From the perspective of practical application, we propose an index that supports large-scale data sets, and reduces the storage space from the design of the index; at the same time, the index structure we propose is a general index structure that can realize queries on various graph data. Experiments show that our index has advantages in both storage space and query efficiency.

Acknowledgement. This work was supported by the National Natural Science Foundation of China (61972198).

References

1. Arz, J., Luxen, D., Sanders, P.: Transit node routing reconsidered. In: Bonifaci, V., Demetrescu, C., Marchetti-Spaccamela, A. (eds.) SEA 2013. LNCS, vol. 7933, pp. 55–66. Springer, Heidelberg (2013). https://doi.org/10.1007/978-3-642-38527-8_7
2. Cui, W., Xiao, Y., Wang, H., Wang, W.: Local search of communities in large graphs. In: Proceedings of the 2014 ACM SIGMOD International Conference on Management of Data, pp. 991–1002 (2014)
3. He, H., Singh, A.K.: Graphs-at-a-time: query language and access methods for graph databases. In: Proceedings of the 2008 ACM SIGMOD International Conference on Management of Data, pp. 405–418 (2008)
4. Klein, P.N., Mozes, S., Weimann, O.: Shortest paths in directed planar graphs with negative lengths: a linear-space o (n log2 n)-time algorithm. ACM Trans. Algorith. (TALG) **6**(2), 1–18 (2010)
5. Li, L., Zhang, F., Zhang, Z., Li, P., Bu, C.: Multi-fuzzy-objective graph pattern matching in big graph environments with reliability, trust and social relationship. World Wide Web **23**(1), 649–669 (2020)
6. Li, L., et al.: Bindex: a two-layered index for fast and robust scans. In: Proceedings of the 2020 ACM SIGMOD International Conference on Management of Data, pp. 909–923 (2020)
7. Möhring, R.H., Schilling, H., Schütz, B., Wagner, D., Willhalm, T.: Partitioning graphs to speedup Dijkstra's algorithm. J. Exp. Algorith. (JEA) **11**, 2–8 (2007)
8. Nannicini, G., Baptiste, P., Barbier, G., Krob, D., Liberti, L.: Fast paths in large-scale dynamic road networks. Comput. Optim. Appl. **45**(1), 143–158 (2010)
9. Potamias, M., Bonchi, F., Castillo, C., Gionis, A.: Fast shortest path distance estimation in large networks. In: Proceedings of the 18th ACM Conference on Information and Knowledge Management, pp. 867–876 (2009)
10. Schulz, F., Wagner, D., Weihe, K.: Dijkstra's algorithm on-line: an empirical case study from public railroad transport. J. Exp. Algorith. (JEA) **5**, 12-es (2000)
11. Seo, J., Guo, S., Lam, M.S.: Socialite: an efficient graph query language based on datalog. IEEE Trans. Knowl. Data Eng. **27**(7), 1824–1837 (2015)
12. Sommer, C.: Shortest-path queries in static networks. ACM Comput. Surv. (CSUR) **46**(4), 1–31 (2014)
13. Tesfaye, B., Augsten, N., Pawlik, M., Böhlen, M.H., Jensen, C.S.: An efficient index for reachability queries in public transport networks. In: Darmont, J., Novikov, B., Wrembel, R. (eds.) ADBIS 2020. LNCS, vol. 12245, pp. 34–48. Springer, Cham (2020). https://doi.org/10.1007/978-3-030-54832-2_5

14. Yuan, L., Qin, L., Zhang, W., Chang, L., Yang, J.: Index-based densest clique percolation community search in networks. IEEE Trans. Knowl. Data Eng. **30**(5), 922–935 (2017)
15. Zhuge, H., Liu, J., Feng, L., Sun, X., He, C.: Query routing in a peer-to-peer semantic link network. Comput. Intell. **21**(2), 197–216 (2005)

Correction to: Lightweight Model Inference on Resource-Constrained Computing Nodes in Intelligent Surveillance Systems

Zhuohang Wang, Yunfeng Zhao, Yong Wang, Li Yan, Zhicheng Liu, Chao Qiu, Xiaofei Wang, and Qinghua Hu

Correction to: Chapter "Lightweight Model Inference on Resource-Constrained Computing Nodes in Intelligent Surveillance Systems" in: B. Li et al. (Eds.): *Web and Big Data*, LNCS 13421, https://doi.org/10.1007/978-3-031-25158-0_17

The original version of this chapter was inadvertently published with incorrect text in acknowledgement section. This has been corrected.

The updated original version of this chapter can be found at https://doi.org/10.1007/978-3-031-25158-0_17

B. Li et al. (Eds.): APWeb-WAIM 2022, LNCS 13421, p. C1, 2023.
https://doi.org/10.1007/978-3-031-25158-0_46

Author Index

An, Junfeng II-175
Apostolopoulos, George C. I-536

Bian, Xiaofei II-369, II-520
Boots, Robert I-60
Busso, Matteo I-37

Cao, Gang III-446
Cao, Jianjun I-76
Cao, Meng I-91
Chang, Wenjing I-435
Chen, Bing I-150, III-232
Chen, Chunling II-369
Chen, Dongxue III-191
Chen, Fang II-415
Chen, Feng II-356
Chen, Gang II-341
Chen, Guihai I-280
Chen, Haobin III-247
Chen, Hong III-292
Chen, Hongchang I-20
Chen, Hongmei I-417
Chen, Rui III-87, III-191
Chen, Shiping III-318
Chen, Songcan II-459
Chen, Wei III-381
Chen, Weipeng II-520
Chen, Weitong I-60, I-345, II-430
Chen, Xiangyan II-23
Chen, Xiaocong III-117
Chen, Xu III-34
Chen, Yongyong II-175
Chen, Zhaoxin III-42
Chen, Zhenxiang I-443
Chen, Zihao II-385
Chen, Zirui II-323
Cheng, Chufan II-415
Cheng, Guanjie I-313
Cheng, Ning II-144
Cong, Qing II-38
Cui, Lixiao I-180
Cui, Mingxin I-163
Cui, Yuanning II-511
Cui, Yue I-264
Cui, Zhiming III-72

Dai, Haipeng I-280
Dai, Hua I-251, II-301
Dai, Tianlun II-195, II-301
Dai, Yongheng I-461
Delis, Alex I-536
Deng, Hai I-150, II-492, III-301
Deng, ShuiGuang I-313
Deng, Song II-385
Ding, Mingchao I-3
Ding, Panpan III-34
Ding, Xuefeng III-117
Ding, Yingying III-412
Djebali, Sonia II-158
Dong, Xinzhou III-199
Du, Hanwen III-72
Du, Ming II-85
Du, Rong III-309
Du, Xin II-520
Du, Yang III-318
Duan, Lei II-385, III-117

Fan, Qilin I-121
Fang, Yu I-372
Feng, Chong I-296
Feng, Dan I-477
Feng, Liwen II-444
Feng, Shi II-239
Feng, Zhiyong II-38
Fu, Junfeng I-451
Fu, Xiaoming III-57
Fu, Yue III-309

Gabot, Quentin II-158
Gao, Hong II-401
Gao, Jerry III-262
Gao, Kening II-538
Gao, Meng III-451
Gao, Xiaofeng I-280
Gao, Yucen I-280
Gao, Yuemeng III-262

Ge, Chengyue III-334
Ge, Weiyi II-114, III-182
Giunchiglia, Fausto I-37
Gou, Gaopeng I-163
Gu, Chumei I-76
Gu, Jingjing III-149
Gu, Yu II-3
Guan, Donghai I-106, II-401
Guerard, Guillaume II-158
Guo, Bowen III-349
Guo, Hao I-511
Guo, Hongjing III-262
Guo, Ling I-408
Guo, Linliang II-430
Guo, Ye I-296

Han, Aiyang II-459
Han, Xuming II-430
Han, Yifan III-318
Hao, Jie II-492, III-301
Hao, Wenqi I-527
Hao, Xiuzhen II-369
Hao, Yongjing III-72
Hao, Zhiyang III-19
He, Chengxin II-385
He, Shuning II-369, II-520
He, Yanxiao I-372
Hou, Chengshang I-163
Hou, Zhirong II-210
Hu, Feng III-232
Hu, Haibo III-309
Hu, Lingling II-501
Hu, Qinghua I-209
Hu, Wei II-114, III-182
Hu, Yao II-166
Hu, Zheng I-251
Huang, He III-318
Huang, Jiani III-34
Huang, Tao III-292
Huang, Xuqian I-511
Huang, Yuzhou I-553

Ji, Bin II-69
Ji, Ke I-443
Jiang, Dawei II-341
Jiang, Hong I-135
Jiang, Huan III-3
Jiang, Weipeng III-117
Jiang, Xiaoxia II-114, III-182
Jiang, Zhiyuan I-545
Jiang, Zhuolun I-382
Jin, Beihong III-199
Jing, Huiyun III-349

Ke, Weiliang III-334
Kong, Chuiyun I-451
Kong, Dejun I-280
Kong, Lingjun III-232

Lai, Riwei III-191
Leng, Zhaoqi II-12
Li, Aoran III-57
Li, Beibei III-199
Li, Bixin II-286
Li, Changyu I-451
Li, Chunying I-28
Li, Dan III-34
Li, Hanzhe III-149
Li, Hao I-451
Li, Jing II-195, II-511
Li, Kaiwen III-132
Li, Li III-87
Li, Longfei II-474
Li, Longhai II-385
Li, Meng II-474
Li, Min II-210
Li, Qianmu III-349
Li, Rong-Hua I-461
Li, Shasha II-69, II-356
Li, Ting III-397
Li, Xi III-87
Li, Xiaodong I-399, II-225
Li, Xiaoyue I-37
Li, Xinyi II-356
Li, Yang I-121
Li, Yanhui I-461
Li, Yanqiang I-443
Li, Yi I-391
Li, Yin II-286
Li, Ying II-210, III-420
Li, Yingjian II-175
Li, Yusen I-180
Li, Zhao II-323
Li, Zhen I-163
Li, Zhiwei II-38
Liakos, Panagiotis I-536
Lian, Zhichao III-349
Liang, Ji III-247
Liang, Luopu III-349
Liang, Rongjiao II-483

Liao, Jiajun I-296
Lin, Wuhang II-69
Lin, Zhenxi II-255
Liu, Chang III-42, III-451
Liu, Chunyan I-121
Liu, Honghao III-3
Liu, Hu I-224
Liu, Jiaye I-296
Liu, Jing I-199
Liu, Jun II-271
Liu, Kun I-224
Liu, Lu II-101
Liu, Minglong III-42
Liu, PanFei I-190, I-199
Liu, Pengkai I-527
Liu, Shang I-477
Liu, Xiaofan III-277
Liu, Xiaoguang I-180
Liu, Xin II-511
Liu, Yang II-114
Liu, Yanheng II-12
Liu, Yi III-57
Liu, Yu I-495, II-129
Liu, Yuanlong I-251
Liu, Zhe I-426
Liu, Zhicheng I-209
Long, Guodong I-237
Lou, Wenlu III-441
Lu, Guangming II-175
Lu, Guangquan II-483
Lu, Jiayi III-57
Lu, Jie I-20
Lu, Peng II-341
Lu, Shiyin II-166
Lu, Xinjiang III-149
Lu, Yanping I-3
Luo, Qian III-42

Ma, Chenrui III-87
Ma, Jian III-451
MA, Jie I-237
Ma, Jingyuan I-477
Ma, Jun II-69
Ma, Kun I-443
Ma, Zhuo I-106
Ma, Zongmin II-546
Mao, Jiali I-296, III-435
Meng, Qing I-106, III-57
Miao, Hao I-382
Miao, Yuan II-166

Ni, Qian I-190
Ning, Bo II-23
Ning, Zefei I-382
Niu, Yuqing I-426

Pan, Haiwei II-369, II-520
Pan, Jiayi II-315
Pan, Qingfeng I-45
Peng, Huang I-511
Peng, Tao II-101
Peng, Yuwei III-3
Peng, Zhiyong III-3, III-217

Qi, Zhixin III-451
Qian, Luhong II-511
Qian, Weining I-296
Qian, Yudong II-183
Qin, Hongchao I-461
Qin, Xiaolin I-553, III-247
Qin, Yuzhou I-527
Qiu, Chao I-209

Rao, Guozheng II-38
Ren, Wei III-277
Ren, Zebin I-461
Rodas-Britez, Marcelo I-37

Sang, Yu III-102, III-381
Sha, Yuji I-372
Shao, Jie I-451
Shen, Shaofei I-60
Shen, Shifan II-501
Shen, Wenyi III-435
Shen, Ziming III-451
Sheng, Victor S. III-72
Shi, Hui III-72
Shi, Qitao II-474
Shi, Yao I-495
Shi, Yinghui II-255
Shi, Yue III-435
Shi, Zhan I-477
Song, Hongtao III-191
Song, Jiageng III-199
Song, Jingkuan II-129
Song, Shuangyong II-23
Song, Wei III-217, III-334
Song, Yang II-3
Song, Yanyan I-527
Su, Yuhang II-3
Sui, Wenzheng III-182
Sun, Yu-E III-318

Tan, Zhen I-511
Tan, Zongyuan II-85
Tang, Jintao II-356
Tang, Jinyun II-483
Tang, Jiuyang I-511
Tang, Yong I-28
Tang, Zhikang I-28
Tao, Chuanqi III-262
Tran, Vanha I-417, III-441
Tu, Xuezhen II-401

Wang, Baowei I-76
Wang, Chao II-23
Wang, Chenghua III-102
Wang, Chengwei II-101
Wang, Chenxu II-323
Wang, Chongjun I-91
Wang, Daling II-239
Wang, Fang I-477
Wang, Gang I-180
Wang, Gaoxu II-195, III-57
Wang, Guoren I-461
Wang, Hongya II-85
Wang, Hongzhi III-451
Wang, Hui II-492
Wang, Jianzong II-144
Wang, Jie III-166
Wang, Jinghui III-132
Wang, Junchen II-385
Wang, Li I-382
Wang, Limin II-430
Wang, Lisong II-501
Wang, Lizhen I-417, III-441
Wang, Meiling II-210
Wang, Meng II-195, II-255
Wang, Mengda I-76
Wang, Ran II-492
Wang, Shaopeng I-391
Wang, Tiexin I-135, II-195
Wang, Ting II-356
Wang, Tongqing II-538
Wang, Xiaofei I-209
Wang, Xican II-12
Wang, Xin I-372, I-527, II-38, II-323
Wang, Xinmeng II-501
Wang, Xueying II-12
Wang, Yani III-174
Wang, Yichen III-87, III-191
Wang, Yingying I-399
Wang, Yong I-209
Wang, Yu I-163
Wang, Yutong III-451
Wang, Zhengqi I-357
Wang, Zhenyu I-545
Wang, Zhuohang I-209
Wei, Ming III-277
Wei, Rukai II-129
Wen, Hao II-301
Wen, Yingying I-313
Wu, Di I-224
Wu, Fan I-280
Wu, Jiajun II-195, II-301, III-57
Wu, Jun I-443
Wu, Lisha II-239
Wu, Pangjing I-399, II-225
Wu, Tao III-435
Wu, Yutong I-477
Wu, Zhixiang I-545

Xiang, Yuxuan II-195, II-301
Xiao, Edward II-144
Xiao, Jing II-144
Xie, Guicai II-385
Xie, Haoran I-399, II-225
Xie, Min II-53
Xie, Ming I-237
Xie, Yanzhao II-129
Xie, Zhongle I-331
Xiong, Gang I-163
Xu, Bin II-538
Xu, Bingliang II-529
Xu, Chen I-45, III-456
Xu, Fangzhi II-271
Xu, Feng I-190, I-199
Xu, Hui I-3
Xu, Jianqiu III-397, III-412
Xu, Juan I-224
Xu, Kai I-545
Xu, Miao I-60
Xu, Ming I-91
Xu, Shidong II-195
Xu, Shuai I-106, III-57
Xu, Yuanbo III-420
Xu, Yuxin I-76

Yan, Hao III-451
Yan, Hui I-408
Yan, Li I-209
Yan, Xinhua I-135
Yan, Yu III-451

Yang, Chenze I-150
Yang, Geng I-251
Yang, Jingyuan III-149
Yang, Ping II-166
Yang, Qun II-301
Yang, Xinxing II-474
Yang, Yan I-357
Yang, Yi I-45, III-456
Yang, Yongjian III-420
Yang, Zhenhua III-456
Yang, Zilu III-364
Yao, Chang I-331
Yao, Junjie I-28
Yao, Siyu II-271
Ye, Chunyang III-132
Ye, Haizhou II-529
Ye, Qingqing III-309
Ye, Zhiwei I-3
Yi, Zibo II-69
Yin, Jianwei I-313
Yin, Shaowei I-357
Ying, Haochao III-149
Yu, Ge II-3, II-239
Yu, Jianjun I-435
Yu, Jie II-69
Yu, Xiaoguang II-23
Yu, Xinying II-444
Yu, Yisong III-199
Yu, Zhenglin I-408
Yuan, Jiabin I-224
Yuan, Jinliang I-91
Yue, Lin I-60, I-345, II-101

Zang, Yalei III-57
Zeng, Lingze I-331
Zeng, Weixin I-511
Zhai, Shichen II-546
Zhai, Xiangping I-224
Zhang, Baoming I-91
Zhang, Chen II-69
Zhang, Chengqi I-237
Zhang, Daoqiang II-529
Zhang, Guixian II-483
Zhang, Haoran III-451
Zhang, Hongying III-42
Zhang, Huihui I-135
Zhang, Ji III-174
Zhang, Jiale III-364
Zhang, Jie II-85
Zhang, Jijie I-357
Zhang, Jinpeng III-441
Zhang, Kejia II-369, II-520
Zhang, Kejun II-444
Zhang, Li II-38
Zhang, Lijun II-166
Zhang, Meihui I-331
Zhang, Qingpeng I-527
Zhang, Rongtao II-492
Zhang, Tingxuan III-166
Zhang, Wang I-477
Zhang, Wenzhen II-483
Zhang, Xulong II-144
Zhang, Xv III-420
Zhang, Yalin II-474
Zhang, Yanchun I-28
Zhang, Yang III-174
Zhang, Yifei II-239
Zhang, Yihang II-430
Zhang, Yue II-101, II-474
Zhang, Yuxuan I-345
Zhang, Zhaowu II-538
Zhang, Zhe I-345, III-3
Zhang, Zhebin II-341
Zhang, Zhen I-20
Zhang, Zheng II-175, II-529
Zhang, Ziheng II-255
Zhang, Zirui I-553
Zhao, Buze III-301
Zhao, Changsheng III-217
Zhao, Deji II-23
Zhao, Lei III-381
Zhao, Pengpeng III-72
Zhao, Suyun III-292
Zhao, Tianzhe II-271
Zhao, Wenzheng II-114
Zhao, Xiaowei I-345
Zhao, Yan I-264
Zhao, Yanchao III-57, III-364
Zhao, Yunfeng I-209
Zhao, Yunlong I-121
Zheng, Jiaqi I-280
Zheng, Jiping III-19
Zheng, Kai I-264
Zheng, Weiguo II-183, II-315
Zheng, Yefeng II-255
Zhong, Limin I-451
Zhong, Xueyan I-372
Zhou, Hongfu I-45
Zhou, Jin I-443
Zhou, Jun II-474

Zhou, Ke II-129
Zhou, Lu I-426
Zhou, Qian I-251, II-301
Zhou, Shuo III-199
Zhou, Xiangmin III-174
Zhou, Xiaojun I-435
Zhou, Xu II-12
Zhou, Yuqi III-446
Zhu, Chen I-264
Zhu, Feng II-474
Zhu, Guoying III-381
Zhu, Hui I-545
Zhu, KaiXuan III-435
Zhu, Li III-166
Zhu, Qi II-529
Zhu, Shibiao I-553
Zhu, Wei II-511
Zhuo, Wei III-199
Zou, Bo II-23
Zou, Chenxin II-225
Zou, Lei I-495, III-446
Zou, Muquan I-417